Acoustical Signal Processing in the Central Auditory System

Acoustical Signal Processing in the Central Auditory System

Edited by

Josef Syka
Academy of Sciences of the Czech Republic
Prague, Czech Republic

Plenum Press • New York and London

Library of Congress Cataloging-in-Publication Data

International Symposium on Acoustical Signal Processing in the Central
  Auditory System (1996 : Prague, Czech Republic)
    Acoustical signal processing in the central auditory system /
  edited by Josef Syka.
       p.   cm.
    "Proceedings of an International Symposium on Acoustical Signal
  Processing in the Central Auditory System, held September 4-7, 1996,
  in Prague, Czech Republic"--CIP t.p. verso.
    Includes bibliographical references and index.
    ISBN 0-306-45608-7
    1. Auditory pathways--Congresses. 2. Auditory cortex--Congresses.
  I. Syka, Josef. II. Title.
  QP461.I58 1996
  573.8'919--dc21   97-25247
                                                              CIP

Proceedings of an International Symposium on Acoustical Signal Processing in the Central Auditory System, held September 4–7, 1996, in Prague, Czech Republic

ISBN 0-306-45608-7

© 1997 Plenum Press, New York
A Division of Plenum Publishing Corporation
233 Spring Street, New York, N. Y. 10013

http://www.plenum.com

10 9 8 7 6 5 4 3 2 1

Printed in the United States of America

To the memory of R. Bruce Masterson
our friend and colleague

PREFACE

The symposium on Acoustical Signal Processing in the Central Auditory System which was held in Prague on September 4–7, 1996 was the third in a series organized in Prague, after the Neuronal Mechanisms of Hearing symposium in 1980 and Auditory Pathway - Structure and Function symposium in 1987. Approximately 100 scientists registered for the symposium and presented 82 separate papers and posters. The present volume contains 53 of these contributions, mostly presented at the symposium as invited review papers.

Several essential changes occurred since the previous meeting in 1987. In auditory neuroscience, recently developed methods opened new horizons in the investigation of the structure and function of the central auditory pathway. Methods like c-fos tracing techniques and monoclonal antibodies for neurotransmitters and their receptors, like the introduction of electrophysiological recording from brain slices have made possible new insights into the function of individual neurons and their interconnections, particularly in the cochlear nuclei and in the superior olivary complex. Integrative approaches towards understanding the central auditory function started to dominate in the field. It is not easy at the present time to differentiate between purely morphological and neurochemical approaches; similarly electrophysiological approaches are accompanied inevitably by behavioral and psychophysical studies. The understanding of human brain function advanced significantly during the last several years, mainly due to the contribution of magneto-encephalography, positron emission tomography and functional nuclear magnetic resonance imaging. Studies of human auditory function profited significantly from the introduction of these techniques, being inevitably combined with studies of speech perception and speech production. Great attention also has been paid to problems of acoustical space localization. However, the most significant changes appeared in opinions about brain plasticity. Supported by the results of similar studies in the visual and somatosensory systems, auditory physiologists started to realize that the functional organization of the brain is far more plastic and changeable than previously considered. It was the unique experience with auditory neuroprostheses in deaf patients, among others, which gave auditory neuroscientists new ideas about the mechanisms and extent of brain plasticity.

It is not only auditory neuroscience which has changed since the last meeting in 1987. Also, life in the Czech Republic went through a period of essential changes, returning after tens of years of totalitarian regimes into the style of life which is normal in developed democratic countries. The new democratic regime gave us a unique chance to organize the symposium in the middle of Prague's greatest historical monument - Prague Castle. And although both previous symposia happened to be organized in the historical premises of Prague monasteries or galleries, the scientific gathering in the middle of the

Prague castle was a symbolic one, representing a positive relationship of the president and government towards science and their belief in the important role of science in the future of the Czech Republic.

I wish to express my gratitude to all participants of the symposium, including those who were unable to contribute to this volume but whose active participation in the meeting helped significantly to make the symposium a success. I would like to acknowledge support for the symposium provided by the Institute of Experimental Medicine, Academy of Sciences of the Czech Republic as well as the main sponsors of the symposium - International Union of Physiological Sciences, Ministry of Education of the Czech Republic and First Prague Municipal Bank.

In the final stage of the preparation of this volume the sudden death of one of the most outstanding auditory neuroscientists, R. Bruce Masterton, surprised and grieved us. Bruce was always a leader in the field of auditory neuroscience, one of those who paved new pathways. He participated in all three Prague symposia, and he was the excellent co-editor of the book of proceedings of the 1987 symposium. With the consent of the contributors to the present volume, we dedicate this book to his memory.

Josef Syka

CONTENTS

Superior Olivary Complex and Lateral Lemniscus

Auditory Midbrain

Auditory Forebrain

Mechanisms of Sound Localization

Processing of Vocalization and Speech

Plasticity and Pathological Processes

ROLE OF THE MAMMALIAN FOREBRAIN IN HEARING

R. Bruce Masterton

Program in Neuroscience
Florida State University
Tallahassee, Florida 32306-1051

The study of the role of auditory cortex in hearing by ablation-behavior methods is now over a century old (James, 1896). At first, the question of assessing cortical function seemed easily answerable. Animals surgically deprived of auditory cortex acted as if they were deaf — not responding to very loud and unexpected sounds (Ferrier, 1876). But after several repetitions of this experiment yielded variable results, a more careful and systematic ablation-behavior inquiry into auditory cortex began. This inquiry proceeded hesitantly while several sources of acoustic, surgical, and behavioral artifact were eliminated from the experimental procedures one at a time (e.g., Kryter and Ades, 1943). However, by the 1950's, the labs of Neff and of Ades had invented behavioral and acoustical presentation methods, and experimental surgery techniques with histological verification that have had to change little in ensuing years (e.g., Neff, 1968; Ades, 1959). Interesting and often surprising results soon followed.

Most current attempts to summarize what is known about the role of auditory cortex in hearing usually focus on electrophysiological experimentation alone; with few exceptions, avoiding the results of ablation-behavior experimentation (but see Clarey et al., 1991). One of the reasons for this side-step is that the history of the study of auditory cortex through ablation-behavior methods can be confusing to those not directly involved in it. Although the individual experiments themselves are rarely confusing, the results collected across decades frequently seem contradictory and often confounded with variation in non-auditory factors. In general, the results of cortical ablation-behavior experiments are very difficult to gather into a format allowing unarguable conclusions. In the past, it has fallen on Dewey Neff to summarize and integrate this line of research into a useable whole (e.g., see Neff, 1968; Neff et al., 1975). But since his retirement there has been only a rare critical summary. Because my lab, colleagues, students and I have contributed to the confusion, I thought I might take the opportunity provided by this Symposium to try to unscramble it for those interested in making use of it in their own studies.

Toward this end, I would like first to discuss briefly two landmark experiments on auditory cortex and frequency discrimination and then two others on auditory cortex and sound-source localization. From that overview will emerge several non-auditory factors which affect the appearance of behavioral deficits after auditory cortex ablation. As it turns out, these factors include so-called 'task-specific' variables such as the choice of the behavioral responses employed, the choice of the reinforcements used to maintain the response, and the choice of the experimental animals. Finally, I will describe a little known anatomical model of the forebrain auditory system and suggest how its integration seems to mitigate, if not eliminate, many of the past difficulties in the interpretation of the auditory cortex literature.

1. AUDITORY CORTEX AND FREQUENCY DISCRIMINATION

Figure 1A, B are schematic diagrams of the forebrain auditory system as it was known in the 1950's and '60's. The essential feature is that the inferior colliculus projects to the medial geniculate which, in turn, projects to auditory cortex — that is, a strictly serial, two-step projection from midbrain to cerebral cortex. It was with this anatomy in mind that Bob Butler, Irving Diamond and Dewey Neff did their famous experiment on the role of auditory cortex in frequency discrimination (Butler et al., 1957). They ablated auditory cortex bilaterally and tested cats on their ability to detect a change in tone frequency. As it turned out, the bilateral ablation of auditory cortex did *not* result in any serious deficit in frequency discrimination. Although the cats had to re-learn the discrimination postoperatively, they did so just as quickly as they did pre-operatively and

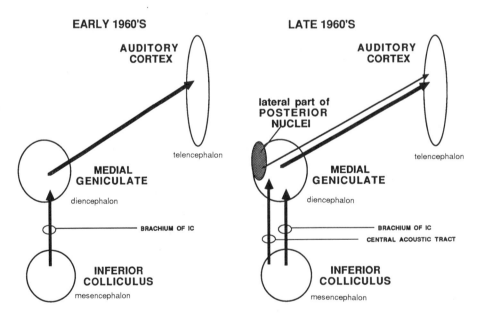

Figure 1. Forebrain auditory system as seen in early (left) and late 1960's (right). Note serial arrangement of Inferior Colliculus, Medial Geniculate, and Auditory Cortex in both. On right, the 'Central Acoustic Tract' and the 'Posterior Nuclei of the Thalamus' have been added to the system (e.g., see Imig and Morel, 1985). Both the 'Central Acoustic Tract' and the 'Brachium of the Inferior Colliculus' have to be sectioned for cats to lose sound-evoked potentials at auditory cortex (Galambos et al., 1961) and to lose the ability to discriminate frequency (Goldberg and Neff, 1961).

the postoperative performance of the frequency discrimination was as good as it had been preoperatively.

This experiment was greeted with great disbelief by the scientific community. To begin with, it contradicted what everybody thought at the time was true; namely, that auditory cortex was necessary for auditory perception itself and therefore certainly, for frequency discrimination. But this critical experiment had another more pervasive effect: it showed that the 'representation' of an auditory dimension in a central auditory structure (in this case, frequency or tonotopic maps in auditory cortex) is *not* sufficient to conclude that the structure is necessary for behavioral discrimination along that dimension. That is, showing electrophysiological 'representation' is *not* the same as showing that the structure is *necessary* for the implied discrimination. In short, the Butler, Diamond and Neff study showed that electrophysiology was no short-cut to conclusions regarding behavioral function.

At first, the electrophysiological community was particularly doubtful of this implication. The field was coming into full blossom under Clinton Woolsey and Jerzy Rose at Wisconsin (Rose and Woolsey, 1958) — electrophysiological equipment and technique were being improved rapidly and frequency maps were being demonstrated in almost every central auditory structure and in a wide variety of animals (e.g., Aitkin et al., 1984). However, decades later and with the experiment now repeated many times in many different animals with ever-increasing size of the cortical lesion, the Butler, Diamond and Neff result holds (e.g., Neff, 1968; Thompson, 1960). That auditory cortex is *not* necessary for frequency discrimination in most experimental animals seems inescapable. That it might be necessary in monkeys and humans is still an open question to which we will return below (Heffner and Heffner, 1990).

A few years later, using the same behavioral, surgical, and histological verification techniques as Butler, Diamond and Neff, Jay Goldberg and Neff (Goldberg and Neff, 1961) showed that transection of the projection from inferior colliculus to medial geniculate is indeed necessary for frequency discrimination. Since such a lesion, depending on its depth, can divorce the forebrain from the inferior colliculus and ultimately from the ears themselves, it seems safe to conclude that the auditory *forebrain* is necessary for frequency discrimination even though auditory *cortex* is not necessary.

One of the effects of this later experiment was to vindicate the behavioral, acoustical, surgical and histological methodology used in Neff's former experiments on auditory cortex. But Goldberg went a step further by showing that section of the superficial brachium of the inferior colliculus (or BIC) was *not* sufficient to result in the frequency deficit. The surgical section had to penetrate deeper — to include the so-called 'central acoustic tract' of Papez (or 'CAT') underlying the brachium. With this deeper lesion, auditory evoked potentials disappeared from the ipsilateral auditory cortex (Galambos et al., 1961) and the animal showed a complete and permanent deficit in frequency discrimination. In contrast, if only the superficial brachium was cut, neither auditory evoked potentials at auditory cortex nor the ability to perform a frequency discrimination was lost. Thus Goldberg and Neff established the necessity of an intact 'central acoustic tract' as well as the necessity of the forebrain (Fig. 1B).

This behavioral result of Goldberg and Neff has also withstood the test of time (Heffner and Heffner, 1984). The central acoustic tract (hodologically verified by Kent Morest, 1965), has proved to be a functional necessity for most auditory discriminations. Though often still called a 'lesion at the level of the brachium of the inferior colliculus', it has been found necessary to transect both the BIC and the CAT medial to it in order to be behaviorally effective (e.g., Thompson and Masterton, 1978; Heffner and Heffner, 1984; Kelly and Judge, 1985). Comparing the deficit resulting from the BIC + CAT lesions with the lack of deficit resulting from cortex lesions, Goldberg reasoned that there must be some projection or target of the auditory system within the forebrain other than the one to medial geniculate and then to auditory cortex. We will return to this conclusion below.

Turning from experiments in frequency discrimination, Neff shifted the focus of his laboratory to sound-source localization. He and his colleagues showed that the one acoustical task truly fragile to auditory cortex ablation was not frequency (nor intensity) discrimination but instead, sound-source localization (Neff et al., 1956). Because of this special fragility and its still current relevance (Masterton, 1992), the remainder of this review focuses on sound localization tasks to illustrate the persistence of difficulties in interpreting the effects of auditory cortex ablations.

In their original studies on the effect of auditory cortex ablation and sound source localization, Neff's group used a very simple apparatus. They placed two speakers with two food boxes attached, one to the left and one to the right of the animal's midline. At the onset of a trial, a sound (usually a buzzer) would be presented from one of the two speakers and if the cat proceeded to that speaker, it would be rewarded with food. On the other hand, if it responded to the silent speaker, it would not be rewarded and would have to be returned to the start point to begin the next trial. Using this task, Neff and his colleagues showed that bilateral ablation of auditory cortex did indeed result in a profound, if not permanent, deficit in the cat's ability to make this discrimination (Neff et al., 1956). Although through the years Neff and his colleagues varied this procedure somewhat, we will see that the important part of his procedure seems to be the requirement for the cats to walk to the sound source in order to receive their reward (see below) (Neff, 1968). And indeed this experiment, too, has been repeated again and again, already in cats, bushbabies, monkeys and ferrets, all with the same result (e.g., Ravizza and Diamond, 1974; Heffner and Masterton, 1975; Jenkins and Masterton, 1982; Cortez et al., 1983; Kavanagh and Kelly, 1987).

This role of auditory cortex in sound-source localization in cats has been re-visited many times, most notably by Jenkins and Merzenich (1984). Because Jenkins (Jenkins and Masterton, 1982) had previously shown that any unilateral lesion of the auditory system above the level of superior olive results in a strictly contralateral deficit in sound localization (the so called 'acoustic chiasm'), it is possible to do two experiments within each cat, one involving the left hemisphere, the other the right hemisphere. Figure 2A, B summarizes Jenkins and Merzenich's (1984) most important result. When a unilateral strip of cortex representing one or another tone was ablated throughout its dorsal-ventral length in AI, that is, within primary auditory cortex, a cat suffered a sound source localization deficit: 1) only in the contralateral hemifield of auditory space and, 2) only for that tone. The ipsilateral hemifield of space and all other tones were unaffected. Adding the inverse to this result (Fig. 2C, D), they also showed that ablation of an area much wider than AI but leaving a small strip of cortex within AI unablated, the cat lost sound-source localization in the hemifield contralateral to the lesion for all tones *except* for the tone represented in the unablated island of cortex. Thus, Jenkins and Merzenich (1984) had shown that auditory cortex was one area of tonotopic representation that was both necessary and sufficient for sound source localization in the contralateral hemifield. Since their results spoke both to sound localization and to tonotopic representation, this experiment has rightly become a bench-mark for further experimentation into the role of auditory cortex in sound-source localization. However, still more recent experiments have shown that even the conclusion based on this provocative result probably has to be modified. To make this point, I will introduce some experimentation that has not been previously published.

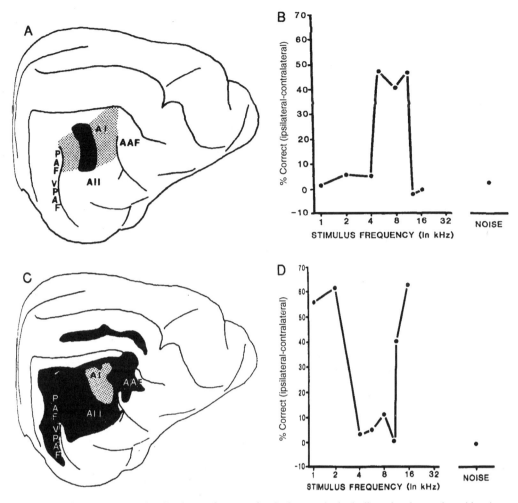

Figure 2. Difference in sound-localization performance for single tones in the ipsilateral and contralateral hemi-fields of auditory space (B & D) after the unilateral lesions shown in (A & C). Note tone-frequency on abscissa of B and D. Cortical lesion shown in 'C' was intended to include most auditory cortex except for an unablated island of tissue in AI to serve as a loose inverse to lesion shown in 'A'. This inverse lesion resulted in inverse behavioral performance (see 'B' vs. 'D'). From Jenkins and Merzenich (1984).

2. SHOCK AVOIDANCE TESTING OF SOUND LOCALIZATION

The sound-source localization method of Neff (1968) or of Jenkins and Masterton (1982) or of Jenkins and Merzenich (1984) requires testing of the animals that approaches heroic on the part of the experimenter. Due to most animals' reluctance to return to the center starting point to begin a new trial, 60 minutes or more are often required to obtain only 10 trials. Because several 100's of trials are often necessary to train and properly test the animal's capacities, testing usually must continue for many months. For this reason alone, we sought to modify the behavioral method in order to automate it and be able to obtain many more trials per daily session and less gruelling requirements on the experi-

FFT - matched wide range speakers

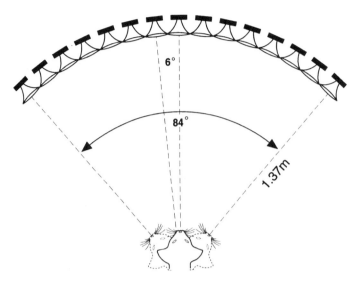

Figure 3. Array of 15 closely matched speakers centered on the midpoint of cat's interaural line. Random-intensity noise pulses at 5/sec emanated from one or another speaker to serve as a 'safe' signal. Alternation of noise pulses between one speaker and any other served as the signal warning of impending shock. Cat avoided the shock by withdrawing momentarily from food trough. This is a very easy task for a normal cat to learn.

menter. To do this we arrayed 15 FFT-matched speakers in front of the animal (Fig. 3). However, we did *not* require them to walk to the source. Instead, they would remain at the center of the circular array of speakers in order to lick a food trough for a slurry of cat food. This response served to stabilize the head at the center of the sound field. The sounds to be localized were presented from the speakers on the perimeter bar almost 2 meters away (Fig. 3). The task for the cat was to avoid a shock by detecting a change in the direction of a train of noise pulses. That is, a train of brief white noise pulses emanated from one or another speaker. As long as the source of the pulses did not change, the animal had nothing to do except collect food by continuing to lick the food trough. However, if the noise pulses alternated between any two of the speakers, the cat would receive an electrical stimulus three seconds later. Although this reinforcing electrical stimulus was quite weak, the cats soon learned to avoid it entirely by withdrawing momentarily from the food trough when they detected a change in direction of the noise pulses and to not return to the trough until the electrical stimulus had terminated and the train of sounds was pulsing from only one speaker once more. This shock-avoidance testing, invented by Henry Heffner in 1976 (Heffner and Whitfield, 1976), has proved to be very useful because instead of 10's of trials per day with the walk-to-the-source test, literally dozens of trials can be obtained each day without apparent wear or tear on the animal or on the experimenter. But the results obtained from this test are both surprising and interesting when they are contrasted to the 'walk-to-the-source' experiments such as those conducted by Neff (1968) and his group or by Jenkins and Masterton (1982) or by Jenkins and Merzenich (1984).

The results of a long series of ablation-behavior experiments ranging from the cochlear nucleus to auditory cortex including tests of sound-source elevation (Masterton and

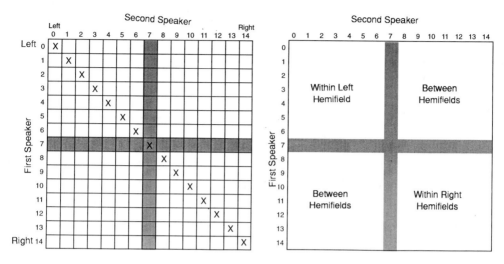

Figure 4. The array of 15 speakers with speaker # 7 on midline, #'s 0–6 in the left hemifield, and #'s 8–14 in the right hemifield allow the results to be unambiguously illustrated with a 15x15 matrix (left). For the present purpose, performance involving the midline speaker (#7) can be ignored (see gray). The 'x''s along the diagonal in the left matrix indicate that these 15 combinations of each speaker alternating with itself constituted the 'safe' signals. Performance along this diagonal yields a highly refined estimate of the probability of a 'false alarm' and serves as a base line for assessing the avoidance responses for any of the 210 other possible combinations. The depiction of the animal's behavioral performance particularly *within* each of the two hemifields can be contrasted by showing its performance in the upper left and lower right quadrants of the matrix (right).

Sutherland, 1994; Sutherland and Masterton, 1994) as well as azimuth (Masterton and Granger, 1988; Masterton et al., 1992) have now been obtained using this variety of shock-avoidance testing. To begin with, it is important to note that some previous conclusions remain firmly intact, viz., if a detectable deficit does result from a lesion above the superior olives, whether superficial or profound in degree, or whether temporary or permanent in duration, it is *always* contralateral to the lesion. This is the so-called 'chiasmatic effect' found by Jenkins and Masterton (1982) (Fig. 4). Secondly, the deficit resulting from a lesion, histologically-verified as complete, at any level from cochlea to medial geniculate is usually sharp, profound, and permanent. Three such results are illustrated in Figures 5, 6, and 7 which show mild contralateral deficits in cats with relatively small unilateral lesions either of the brachium of the inferior colliculus or the medial geniculate. Therefore, the serial connections of the IC via BIC + CAT to MG can be quickly re-established with this type of testing (Fig. 1).

However, if the unilateral lesion is confined to auditory cortex alone, the result is unexpected (Figs. 5, 6, 7, 8). No permanent deficit results. To be sure, there is a temporary deficit (contralateral to the lesion), but with continued testing, the cat seems to recover completely. Therefore, there is a sharp difference between the effect of unilateral auditory cortex lesions and unilateral subcortical lesions up to, and including, the medial geniculate.

Making use of the fact that any sound-source localization deficit is always in the contralateral hemifield, two contrasting experiments can be done in one and the same cat. For example, Figures 6 and 7 show that if a lesion is made first in a cat's auditory cortex, there is no permanent deficit in sound localization. If a second lesion is then made at the level of the brachium of the inferior colliculus or medial geniculate, either ipsilateral or contralateral to the auditory cortex lesion, then the cat suffers a deficit contralateral only

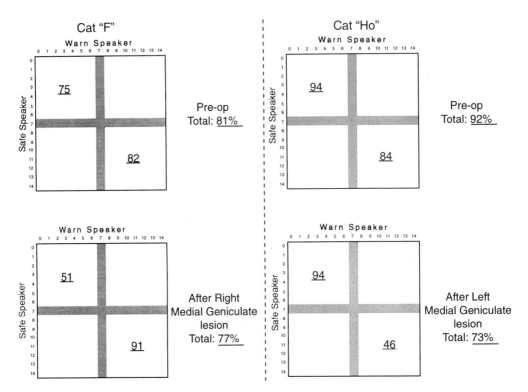

Figure 5. Behavioral performance of 2 cats ('F' and 'Ho') before and after a unilateral lesion of the medial geniculate. Cat 'F' (on left) received a lesion in its *right* MG. As a consequence, its performance fell from 75% preoperatively to 51% postoperatively in the *left* (contra) hemifield while increasing from 82% to 91% in the right (ipsi) hemifield. Cat 'Ho' (on right) received a lesion of its *left* MG and suffered a deficit (84% to 46%) only in its *right* (contra) hemifield.

to the subcortical lesion. No deficit contralateral to the auditory cortex lesion and no deficit ipsilateral to the subcortical lesion appears. Figure 8 shows the same result obtained in an opossum.

Among other things, these results mean that as the auditory system ascends toward the auditory cortex, ablation of the nuclei or tracts at successive levels in the pathway are *not* necessarily the same in effect. The successive structures in the pathway from colliculus to cortex are *not* serial. Furthermore, the system becomes 'not serial' somewhere between the medial geniculate and auditory cortex. The critical result for this conclusion is that a unilateral lesion of the medial geniculate results in a (contralateral) deficit while a lesion confined to auditory cortex results in no permanent deficit whatever in either cats or opossums. What is the difference between the two lesions? This will prove to be the key question in the anatomical discussion to be opened below. But for the present purpose, if auditory cortex is necessary for sound-source localization, as shown by Jenkins and Merzenich (1984), why is there no localization deficit in the cats after cortex ablation as shown in Figures 5, 6 and 7? In seeking the reason why cats in the shock avoidance task have no deficit after auditory cortex lesions while they do show a (contralateral) deficit in the 'walk-to-the-source' test, one can note several differences in the two testing procedures. It is these procedural, non-sensory, differences that have proved instructive for interpreting the past literature on several auditory discriminations.

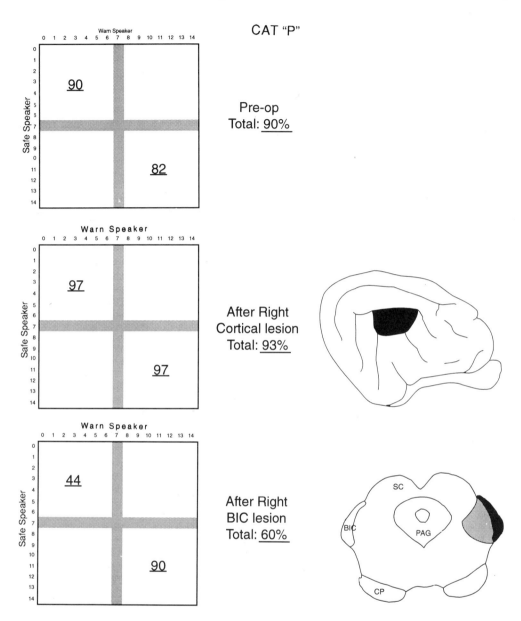

Figure 6. Double-lesion experiment in Cat 'P'. Top: behavioral results before either lesion; Middle: results after a right unilateral cortical lesion; Bottom: results after a second (right) unilateral lesion at the level of the BIC. Note drop in left hemifield performance after BIC + CAT lesion but no change in performance after cortical lesion.

The first experiments to make the point that auditory cortex deficits were not strictly 'auditory' in nature were done by Henry Heffner who trained and tested monkeys on sound-source localization before and after bilateral lesions of their auditory cortex (Heffner, 1973; Heffner and Masterton, 1975). The special feature of the Heffner experiments was that the very same animals with the very same lesions were tested on three different procedural varieties of sound-source localization. The three tasks each had essentially the same sensory requirements. They differed only in their motor requirements and reinforcement consequences.

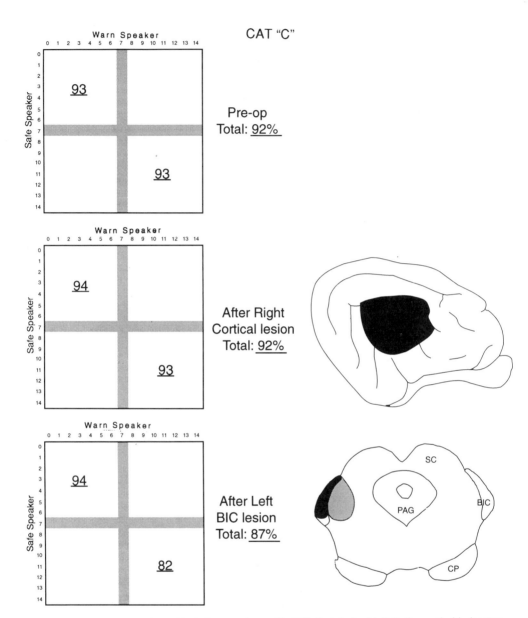

Figure 7. A second example of a double-lesion experiment (Cat 'C'). Note lack of deficit after cortical lesion (top vs. middle) and small but significant deficit in right hemifield after (left) partial BIC + CAT lesion (middle vs. bottom).

One of the tests matched the previous experimentation of Neff: namely, two sound speakers plus a reward mechanism attached. In this case, Heffner's monkeys showed a total loss in their ability to walk to the source of the sound after bilateral auditory cortex lesions just as Neff's cats had shown. Given this result, Heffner began to narrow the interpretation: is there *anything* monkeys can do in this type of test? The answer is yes. For example, if the sound was turned on and left on, the monkey could simply 'home' on the proper speaker and could learn to do so without error. Heffner concluded that the monkeys were *not* suffering a motor

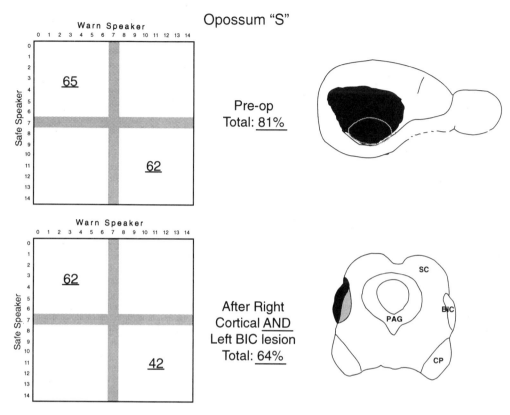

Figure 8. Double-lesion experiment in an Opossum ('S'). Note similar performance scores in two hemifields after (right) unilateral cortical lesion and deficit contralateral to the later (left) BIC + CAT lesion.

deficit that prevented them from responding correctly. To Heffner, this meant, by subtraction, that the deficit was either auditory or audito-motor in nature. To narrow the interpretation still further, Heffner did two other experiments on the same monkeys. He used the shock-avoidance procedure, much like that just described in cats, to see if the monkeys could signal a change in direction from left to right or the difference between left and right sounds. He found the monkeys to be quite similar to the cats previously described. That is, even though the monkeys could not find the source of a brief sound and walk to it to collect a reward, they could detect changes in the sound's direction and they could successfully avoid shock to signify that. Heffner proceeded with yet another task: this time the monkey sat in a primate chair and put its mouth on a water spout. This action, called an 'observing response', opened a gate which resulted in a sound being presented either from a left or a right speaker. The animal was rewarded if it responded to the left lever if the sound was from the left or to the right if the sound was from the right. Importantly, the monkeys could do this task also. Taken together, these three results indicate that the very same monkeys, with the very same lesions, could *not* find the source of a brief sound but they could detect its change from left to right and they could identify whether the sound was from the left or from the right.

In the words of Heffner, "...monkeys with lesions that include primary auditory cortex bilaterally can indicate the direction of the source of a click with a learned (i.e., nonreflexive) response, but they cannot locate (i.e., move to) the source. This dissociation of abilities shows that the deficit in sound localization that results from auditory cortex ablation depends not on the sensory requirements of the task, but on the motor requirements; and therefore, it suggests that the role of auditory cortex in sound localization is probably not sensory nor perceptual but, rather, auditomotor or associative." (Heffner and Masterton, 1975).

Therefore, what the Heffner experiments showed was that an auditory cortex lesion does *not* produce a deficit which is understandable by referring only to the auditory requirements of the behavioral task. Monkeys can perform some sound-source localization tasks and completely fail in others. Therefore, it is *not* the ability to appreciate the auditory requirements of a sound localization task that is lost with an auditory cortex lesion, it is something more subtle. One possibility is that the monkeys suffer neither a strictly sensory nor a strictly motor deficit but instead, have some sort of perceptual deficit, perhaps in the attachment of 'auditory space' to 'motor space' (Heffner and Heffner, 1990). This conclusion is still tenable; that is, there has been no experimentation yet conducted which has contradicted this interpretation. But in more general terms, what the experiments by Heffner had shown was that the deficit suffered by monkeys with auditory cortex lesions was 'task-specific' or 'response-specific' and not 'sound-localization specific'.

One of the implications of this conclusion is relevant to the clinic. Noting that clinical testing of human hearing usually relies on verbal instructions and verbal responses and noting further, that the appreciation and generation of appropriate verbal responses probably requires at least the speech areas in cortex, Heffner's conclusion of 'response-specificity' also suggests one reason why human patients seem to suffer more profound hearing deficits than do experimental animals. We will return to this point below.

In pursuing the difference between the cats in the Jenkins and Merzenich (1984) experiments, which suffer a sharp deficit with an auditory cortex lesion, and those in our shock-avoidance sound localization experiments which do not suffer a permanent deficit, several other factors present themselves. For example, in the Jenkins "walk-to-the-source" experiments (Jenkins and Masterton, 1982; Jenkins and Merzenich, 1984) and in the Neff or Heffner experiments of the same kind, the animals were rewarded for responding correctly. In the Masterton experiments and in one of the Heffner experiments, the animals were punished for responding incorrectly. That is, they were 'reinforced' with shock when they did *not* detect the change in the source of a sound. Therefore, the spectre of 'reinforcement-type' also enters the picture at this point with the possibility that reward versus punishment training tasks are fragile to different lesions, perhaps because they require different pathways of the forebrain. If this is the case, different choices of reinforcement might yield different ablation-behavior results once more. We will also return to this question below.

Up to this point, we have seen that there are at least two factors other than the sensory requirements of a sound localization task that seem to be relevant in explaining deficits or lack of deficits after auditory cortex lesions. These are listed in Figure 9. Other than the site of the forebrain lesion, which is obviously crucial, they include response choice, reinforcement choice, and to be complete, the species of animals used; whether cats, rats, opossums, ferrets, monkeys, etc. This last factor is included because rats, at least, do not seem able to localize sound sources *within* one hemifield even though they can do so *between* hemifields (Kelly, 1980; Kavanagh and Kelly, 1986). With this list of factors which differentiate the shock-avoidance type experiments from the Neff-type reward experiments, perhaps there is less wonder that the outcome of the two experimental procedures might be different. However, a new possibility has arisen which may provide a potential explanation encompassing each of these three, non-sensory factors: That is, there may be only one reason for the marked difference in behavioral outcomes and that reason is to be found in the neuroanatomy of the forebrain auditory system.

SUMMARY OF FACTORS INFLUENCING BEHAVIORAL DEFICITS AFTER AUDITORY FOREBRAIN ABLATIONS:

1) LOCUS OF LESION (e.g. Cortex vs. MG vs. BIC vs. IC)

2) NATURE OF AUDITORY TASK (e.g., Frequency or Intensity Discrimination vs. Sound Source Localization)

3) REQUIRED MOTOR RESPONSE (e.g. Walk to the sound source vs. Indicate the direction of the source)

4) REINFORCEMENT CHOICE (e.g. Reward vs. Punishment)

5) ANIMAL CHOICE (e.g. Cat vs. Rat vs. Monkey)

Figure 9. Summary of factors that seem to affect the presence or absence of behavioral deficits after ablation of auditory cortex. Note that the last 3 factors are *not* auditory in nature. It is variation in these factors that seems to have made the ablation-behavioral literature on auditory cortex appear confusing. A unifying explanation may be found in an updated view of the telencephalic auditory system. See Figure 10.

3. NEUROANATOMY OF THE FOREBRAIN AUDITORY SYSTEM

Whereas Figure 1 shows the serial view of the forebrain auditory system (that is, IC to MG to auditory cortex) as conceived in 1950–1960, evidence provided by Ford Ebner (1967, 1969) has shown that there is missing from that drawing, a second significant tract projecting to a second telencephalic target. It is shown in Figure 10. The important difference between Figure 1A, B and Figure 10 is the presence of the lateral amygdala — a subcortical structure — receiving its own auditory projection from medial geniculate. Although this tract seems to have been first discovered almost thirty years ago (Ebner 1967, 1969), it has not been mentioned in auditory system literature until recently. Expanding on Ebner's (1969) original demonstration of this tract's presence since that time, we have shown that the 'extra' forebrain auditory target is present in most laboratory mammals including two genera of opossums, hedgehogs, rats, armadillos, tree shrews, and bush babies (Frost and Masterton, 1992).

Perhaps more important for the present purpose, this tract is quite substantial. Indeed, in opossums it is very large, originating in fully one-third to one-half of their medial geniculate and projecting independently to the lateral amygdala and perhaps putamen and caudate, but *not* to the auditory cortex (Kudo et al., 1986). As it turns out, there are two separate populations of cells in the medial geniculate of most mammals: one projects to cortex, one projects to lateral amygdala and perhaps to other nearby basal nuclei. For example, retrograde HRP-labeling experiments from auditory cortex label almost all of the cells in the anterior medial geniculate. In contrast, retrograde HRP-labeling experiments with injections in the lateral amygdala label the cells only in the caudal medial geniculate. Although we make the 'rostral' versus 'caudal' distinction here because of our own studies, Golgi studies of MG by Morest (Morest and Winer, 1986) and by Winer (et al., 1988) have shown that the proper distinction may be between 'ventral' versus 'dorsal' divisions of MG. In these terms, it is the 'dorsal division' usually found in the caudal or medial medial geniculate that projects to the lateral amygdala and not to the cortex. Conversely, it is the 'ventral division' found in the rostral MG that projects to auditory cortex and not to lateral amygdala.

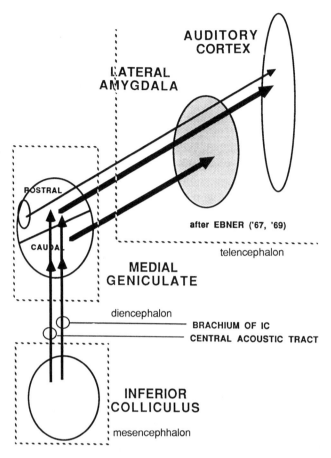

Figure 10. Schematic update of forebrain auditory system. The significant difference from Figure 1 is the presence of the lateral amygdala (gray) with its own independent projection from the medial geniculate. First discovered by Ebner (1967, 1969), this projection has now been shown in a wide variety of mammals, perhaps excepting only higher Primates (Frost and Masterton, 1992). Together with Le Doux's (Le Doux et al., 1984; 1985) demonstration of its involvement in fear-evoked training and testing methods, this 'extra' auditory projection may explain many of the difficulties in interpretation of the ablation-behavior experimentation on auditory cortex.

In addition to being quite large, this geniculo-amygdala tract is probably independent of the one to cortex. Double-labeling experiments show that though the cortical tract passes close by the lateral amygdala, only a trivial number of cells are double-labeled with one retrograde tracer injected in auditory cortex and the other in amygdala. Therefore, the two targets, cortex and lateral amygdala, are *not* innervated by collaterals: apparently, they are quite separate projections into the telencephalon (Frost and Masterton, 1992).

In summary, the presence of Ebner's (1967, 1969) geniculo-lateral amygdala tract in the forebrain of most laboratory mammals demonstrates that there are two telencephalic projections of the auditory system, not just one. The second one, medial geniculate to lateral amygdala, is essentially an independent projection. It follows that Ebner's tract might be a functional alternative as well as a parallel or an adjunct to the one to cortex. Finally, noting that the tract is largest in opossums and hedgehog (Ebner, 1969), we have shown in a further series of experiments on selected mammals that this tract begins large but fades in size along the mammalian lineage leading to anthropoids and eventually to humans (Frost and Masterton, 1992). This apparently evolutionary trend seems to leave only the MG-to-cortex tract remaining in higher Primates and probably in humans. For the present purpose, the presence of Ebner's geniculo-amygdala tract is what the auditory cortex ablation-behavior literature now seems to have suggested for years. That is, there is an 'extra' auditory target in the telencephalon of most mammals. The presence of this target suggests that medial geniculate lesions which denervate it and cortical lesions which neither include it nor denervate it might well result in two different syndromes. This is what Goldberg's experiments into frequency discrimination and the recent sound-source localization experiments have implied.

It should be noted that there is yet another reason why this tract may be one that explains the seemingly contradictory results of cortical ablation-behavior studies. In a series of studies originally pursuing a somewhat different line of inquiry, LeDoux et al. (1984; 1985) have shown that the medial-geniculate to lateral amygdala tract may be necessary for full-blown 'fear responses' evoked by auditory stimuli. Therefore, this tract may be necessary and sufficient for exactly the sort of shock-avoidance responses that have been most widely used in previous auditory cortex ablation-behavior experimentation. That is, even the earliest systematic experimentation into auditory cortex by Neff or by Ades and their colleagues, was invariably shock-avoidance testing. With the lateral amygdala *not* included in a typical 'cortical lesion', it seems possible for animals tested with shock-avoidance methods to perform the simple auditory tasks being investigated at the time. Notably, lower level lesions, either in colliculus, medial geniculate, or still lower auditory centers, this same target is at least partially disconnected from the ears and the attachment of fear-induced avoidance responses to sounds is weakened if not lost entirely. If LeDoux's (LeDoux et al., 1984; 1985) idea proves to be correct, one might expect a survival of sound localization ability when tested by shock avoidance even though there is a severe deficit in sound localization in food reward testing; that is, with no 'fear' involved.

Finally, what makes Ebner's MG-lateral amygdala tract still more attractive as a plausible explanation of the variability in past auditory cortex literature is that this tract seems to have faded over geological time, perhaps disappearing entirely at the level of higher Primates. Therefore, a difference in the ablation behavior results on auditory cortex between opossums, rats, cats and monkeys, and especially in clinical tests of humans, might also have the same anatomical basis (Ravizza and Masterton, 1972; Heffner and Masterton, 1975; Kelly and Kavanagh, 1986; Kavanagh and Kelly, 1988).

Therefore, the presence of Ebner's tract and the fact that it is reduced over the human lineage suggests a means for explaining many of the apparent contradictions in the ablation-behavior literature on auditory cortex. In non-anthropoids, the tract is large and seems to support shock-avoidance testing of any auditory discrimination yet attempted, including sound source localization. In macaques and perhaps humans, where the tract may not be present, auditory cortex may be the only remaining telencephalic target of the auditory system. Hence, auditory cortex ablation can result in deficits of sound source localization whether they are tested with reward or by shock avoidance methods (Heffner et al., 1992).

4. SUMMARY

The many and various contradictions and confounds often leading to confusion in the auditory cortex ablation-behavior literature seem to be at least partly explainable by the presence of a little known and seldom mentioned tract. This tract projects from medial geniculate to lateral amygdala and thus, involves the lateral amygdala as part of the telencephalic auditory system. Present in most mammals, this medial geniculate-lateral amygdala pathway constitutes a significant parallel to the geniculocortical pathway: that is, the lateral amygdala appears to be a second and independent telencephalic target of the auditory system in most animals in addition to the one terminating in auditory cortex. Direct experimentation into what auditory capacities the lateral amygdala can and can not support in behavioral testing is an open question. But if this 'extra' target proves to be capable of supporting some auditory tasks in the absence of auditory cortex, perhaps most of the apparent contradictions of the role of auditory cortex in hearing in humans as well as in experimental animals will eventually disappear.

5. ACKNOWLEDGMENTS

The author thanks Dr. Jack B. Kelly for his comments on an early version of this paper. This research was supported in part by N.I.H. grants DC00197 and NS07726.

6. REFERENCES

Ades, H. W. 1959 Central auditory mechanisms. In J. M. Brookhart and V. B. Mountcastel (Eds.), Handbook of Physiology, Section I: The Nervous System, Vol. III, Sensory Processes, Part 2. Bethesda, Maryland: American Physiological Society, pp. 675–737.

Aitkin, L. M., Irvine, D. R. F., and Webster, W. R. 1984 Central neural mechanisms of hearing. In J. Field, H. W. Magoun, and V. E. Hall (Eds.), Handbook of Physiology, Section I: Neurophysiolgy, Vol. 1. Washington, DC: American Physiological Society, pp. 585–613.

Butler, R. A., Diamond, I. T., and Neff, W. D. 1957 Role of auditory cortex in discrimination of changes in frequency. J. Neurophysiol. 20:108–120.

Clarey, J. D., Barone, P., and Imig, T. J. 1991 Physiology of thalamus and cortex. In A. N. Popper and R. R. Fay (Eds.), The Mammalian Auditory Pathway: Neurophysiology. New York: Springer-Verlag, pp. 232–334.

Cortez, A. M., Thompson, G. C., and Diamond, I. T. 1983 Sound localization in primates following ablation of auditory cortex. Assoc. Res. Otolaryngol. 6:32.

Ebner, F. F. 1967 Medial geniculate nucleus projections to telencephalon in the opossum. Anat. Rec. 157:238.

Ebner, F. F. 1969 A comparison of primitive forebrain organization in metatherian and eutherian mammals. Ann. N.Y. Acad. Sci. 167:241–257.

Ferrier, D. 1876 The Cronnian Lecture—Experiments on the brain of monkeys (Second Series). Phil. Trans. Royal Soc. of London for the year 1875, 165, Part II, 433–488.

Frost, S. B. and Masterton, R. B. 1992 Origin of auditory cortex. In D. B. Webster, R. R. Fay, and A. N. Popper, (Eds.), The Evolutionary Biology of Hearing. New York: Springer-Verlag, pp. 655–671.

Galambos, R., Meyers, R. E., and Sheatz, G. C. 1961 Extralemniscal activation of auditory cortex in cats. Amer. J. Physiol. 200:23–28.

Goldberg, J. M. and Neff, W. D. 1961 Frequency discrimination after bilateral section of the brachium of the inferior colliculus. J. Comp. Neurol. 116:265–290.

Heffner, H. E. 1973 The effect of auditory cortex ablation on sound localization in the monkey (Macaca mulatta). Ph.D. Dissertation, Florida State University.

Heffner, H. and Masterton, B. 1975 Contribution of auditory cortex to sound localization in monkey (Macaca mulatta). J. Neurophysiol. 38:1340–1358.

Heffner, H. and Whitfield, I. C. 1976 Perception of the missing fundamental by cats. J. Acoust. Soc. Am. 59:915–919.

Heffner, H. E., Carroll, B. A., and Kovach, D. R. 1992 Effect of unilateral ablation of auditory cortex of sound localization by macaques. Asso. Res. Otolaryngol. 15:52.

Heffner, R. S. and Heffner, H. E. 1984 Hearing loss in dogs after lesions of the brachium of the inferior colliculus and medial geniculate. J. Comp. Neurol. 230:207–217.

Heffner, R. S. and Heffner, H. E. 1990 Effect of bilateral auditory cortex lesions on sound localization in Japanese macaques. J. Neurophysiol. 64:915–931.

Imig, T. J. and Morel, A. 1985 Tonotopic organization in lateral part of posterior group of thalamic nuclei in the cat. J. Neurophysiol. 53:836–851.

James, W. 1896 The Principles of Psychology. New York: Henry Holt.

Jenkins, W. M. and Masterton, R. B. 1982 Sound localization: Effects of unilateral lesions in central auditory system. J. Neurophysiol. *47*:987–1016.

Jenkins, W. M. and Merzenich, M. M. 1984 Role of cat primary auditory cortex for sound-localization behavior. J. Neurophysiol. *52*:819–847.

Kavanagh, G. L. and Kelly, J. B. 1986 Midline and lateral field sound localization in the albino rat (Rattus norvegicus). Behav. Neurosci. *100*:200–205.

Kavanagh, G. L. and Kelly, J. B. 1987 Contribution of auditory cortex to sound localization by the ferret (Mustela putorius). J. Neurophysiol. *57*:1746–1766.

Kavanagh, G. L. and Kelly, J. B. 1988 Hearing in the ferret (Mustela putorius): Effects of primary auditory cortical lesions on thresholds for pure tone detection. J. Neurophysiol. *60*:879–888.

Kelly, J. B. 1980 Effects of auditory cortical lesions on sound localization by the rat. J. Neurophysiol. *44*:1161–1174.

Kelly, J. B. and Judge, P. W. 1985 The effects of medial geniculate lesions on sound localization by the rat. J. Neurophysiol. *53*:361–372.

Kelly, J. B. and Kavanagh, G. L. 1986 The effects of auditory cortical lesions on pure tone sound localization by the albino rat. Behav. Neurosci. *100*:569–575.

Kryter, K. D. and Ades, H. W. 1943 Studies on the function of the higher acoustical nervous centers in the cat. Am. J. Psychol. *56*:501–536.

Kudo, M., Glendenning, K. K., Frost, S. B., and Masterton, R. B. 1986 Origin of mammalian thalamocortical projections. I. Telencephalic projections of the medial geniculate body in the opossum (Didelphis virginiana). J. Comp. Neurol. *245*:176–197.

LeDoux, J. E., Ruggiero, D. A., and Reis, D. J. 1985 Projections to the subcortical forebrain from anatomically defined regions of the medial geniculate body in the rat. J. Comp. Neurol. *242*:182–213.

LeDoux, J. E., Sakaguchi, A., and Reis, D. J. 1984 Subcortical efferent projections of the medial geniculate nucleus mediate emotional responses conditioned to acoustic stimuli. J. Neurosci. *4*:683–698.

Masterton, R. B. 1992 Role of the central auditory system in hearing: The new direction. Trends Neurosci. *15*:280–285.

Masterton, R. B. and Granger, E. Moreland 1988 Role of the acoustic striae in hearing: Contribution of dorsal and intermediate striae to detection of noises and tones. J. Neurophysiol. *60*:1841–1860.

Masterton, R. B., Granger, E. M., and Glendenning, K. K. 1992 Psychoacoustical contribution of each lateral lemniscus. Hear. Res. *63*:57–70.

Masterton, R. B. and Sutherland, D. P. 1994 Discrimination of sound source elevation in cats: I. Role of dorsal/intermediate and ventral acoustic striae. Assoc. Res. Otolaryngol. 17:84.

Morest, D. K. 1965 The lateral tegmental system of the midbrain and the medial geniculate body: A study with Golgi and Nauta methods in cats. J. Anat. *99*:611–634.

Morest, D. K. and Winer, J. A. 1986 The comparative anatomy of neurons: Homologous neurons in medial geniculate body of the opossum and the cat. Adv. Anat. Embryol. Cell Biol. *97*:1–96.

Neff, W. D. 1968 Localization and lateralization of sound in space. In A. V. S. deReuck and J. Knight (Eds.), Ciba Foundation Symposium on Hearing Mechanisms in Vertebrates. London: J. & A. Churchill Ltd., pp. 207–231.

Neff, W. D., Diamond, I. T., and Casseday, J. H. 1975 Behavioral studies of auditory discrimination: Central nervous system. In W. D. Keidel and W. D. Neff (Eds.), Handbook of Sensory Physiology, Vol. 2. New York: Springer-Verlag, pp. 307–400.

Neff, W. D., Fisher, J. F., Diamond, I. T., and Yela, M. 1956 Role of auditory cortex in discrimination requiring localization of sound in space. J. Neurophysiol. *19*:500–512.

Ravizza, R. and Diamond, I. T. 1974 Role of auditory cortex in sound localization: A comparative ablation study of hedgehog and bushbaby. Fed. Proc. *33*:1917–1919.

Ravizza, R. J. and Masterton, R. B. 1972 The contribution of neocortex to sound localization in the opossum (Didelphis virginiana). J. Neurophysiol. *35*:344–356.

Rose, J. E. and Woolsey, C. N. 1958 Cortical connections and functional organization of thalamic auditory system of cat. In H. F. Harlow and C. N. Woolsey (Eds.), Biological and Biochemical Bases of Behavior. Madison: Univ. Wisconsin Press, pp. 127–150.

Sutherland, D. P. and Masterton, R. B. 1994 Discrimination of sound source elevation in cats: II. Role of inferior colliculus, medial geniculate, and auditory cortex. Assoc. Res. Otolaryngol. *17*:84.

Thompson, G. C. and Masterton, R. B. 1978 Brainstem auditory pathways involved in reflexive head orientation to sound. J. Neurophysiol. *41*:1183–1202.

Thompson, R. F. 1960 Function of auditory cortex of cat in frequency discrimination. J. Neurophysiol. *23*:321–334.

Winer, J. A., Morest, D. K., and Diamond, I. T. 1988 A cytoarchitectonic atlas of the medial geniculate body of the opossum, Didelphys virginiana, with a comment on the posterior intralaminar nuclei of the thalamus. J. Comp. Neurol. *274*:422–448.

2

STRUCTURAL BASIS FOR SIGNAL PROCESSING

Challenge of the Synaptic Nests

D. Kent Morest

Department of Anatomy
Center for Neurological Sciences
The University of Connecticut Health Center
Farmington, Connecticut 06030

1. INTRODUCTION

The synaptic nest represents a structural pattern and a concept whose time have come for serious consideration in the effort to understand the mechanisms responsible for signal coding in the auditory system. Synaptic nests have an organization which makes it difficult to explain mechanisms of signal processing based on current assumptions about the primacy of cell types and their discrete circuits. Synaptic nests occur in many other parts of the central nervous system, including other sensory nuclei and the cerebellum, although this is not very well appreciated at present. Thus their functional significance has general implications for understanding the function of the nervous system.

1.1. What Is a Synaptic Nest?

Synaptic nests contain tight aggregations of synaptic endings. A critical feature of a synaptic nest is its lack of the astrocytic wrappings that ordinarily insulate individual synaptic endings and provide efficient uptake mechanisms for neurotransmitters. For most synapses this feature permits a relatively discrete integration of the output from each axonal ending. Synaptic nests seem to have another purpose. Often there is a heterogeneity of synaptic types in conjunction with terminal dendrites. Another outstanding feature of the synaptic nests is their extent. Not strictly barricaded from the surrounding neuropil by glial sheaths, the nests seemingly extend indefinitely, like so many rivulets in a receding flood, digresssing, yet conforming to a pattern in each region.

Acoustical Signal Processing in the Central Auditory System
edited by Syka, Plenum Press, New York, 1997

1.2. Where Are Synaptic Nests Located in the Auditory System?

Synaptic nests have been identified in several parts of the auditory system, including the medial geniculate body, the medial trapezoid nucleus, and the cochlear nucleus, especially in the small cell shell and the molecular layer of the dorsal nucleus.

1.3. Why Are Synaptic Nests Interesting?

Such a diffuse, non-differential arrangement of synaptic endings could provide the structural basis for an ongoing or slowly fluctuating background activity which could shift the ambient baseline levels in the synaptic nests and bias the input-output transforms of whole groups of cells. Viewing such remarkable synaptic patterns in this light, one cannot help but wonder if we are looking at the synaptic machinery for contextual modifications of processing, e.g., matching and sensory masking functions.

2. THE SYNAPTIC NESTS OF THE MEDIAL GENICULATE BODY

The synaptic nest was discovered and first described in the medial geniculate body of the cat (Morest, 1970, 1971, 1975).

2.1. Ventral Nucleus of the Medial Geniculate Body

In the ventral nucleus of the cat there are two types of neurons (Fig. 1 LEFT). 1) The principal cell type, or thalamo-cortical neuron, receives the main ascending input from the inferior colliculus and projects to the primary auditory cortex. It receives the collicular input via synapses which target the second-order dendritic branches. 2) A local interneuron, the small Golgi type II cell, receives collicular input on its distal dendrites and forms dendro-dendritic synapses which target the same region of the principal cell intermediate dendrites as the inferior colliculus. All of these axonal endings, together, form the synaptic nests.

2.1.1. Synaptic Nests Have an Unusual Organization. The synaptic nest consists of a population of synaptic endings which are packed together without glial processes to separate them (Fig. 1 RIGHT: <u>Top</u>). In the synaptic nests of the medial geniculate body the inhibitory dendrites synapse directly on the principal cell dendrite, while excitatory inputs from the inferior colliculus synapse on both kinds of dendrites. These triadic formations of junctions provide the basis for sequential synapses, so-called because they define a spatial and temporal sequence in which inputs can affect the postsynaptic potentials of the target cell. Thus one feature of the synaptic nest is to ensure the proximity of both types of inputs and to permit interactions between them. This pattern of synaptic organization differs from that observed generally in the CNS.

2.1.2. Digital Processing. In the usual CNS configuration the synaptic endings are individually separated from the neighboring ones by astrocytic processes (Fig. 1 RIGHT: <u>Bottom</u>). Astrocytes often possess high-affinity uptake transporters for chemical neurotransmitters, including the amino acids (Hamberger et al, 1976; Wilkin et al, 1982; Danbolt et al, 1992; Gundersen et al, 1995). Astrocytes probably account for most of the initial uptake of amino acid transmitters released by CNS synapses, including glutamate by the cochlear nerve endings in the cochlear nucleus and by granule cell endings in the small cell shell (Oliver et al, 1983). This arrangement helps to confine the sphere of influence of transmitter molecules released by synaptic activity, both spatially and temporally. The result is that the target neuron can function optimally as an integrator for the singular, more or less punctate, synaptic events that impinge upon it from instant to instant. This continual integration of the instantaneous postsynaptic potential provides the basis for much of our current thinking about, and modeling of, the cellular basis for information processing. It may be called the digital mode of signal processing.

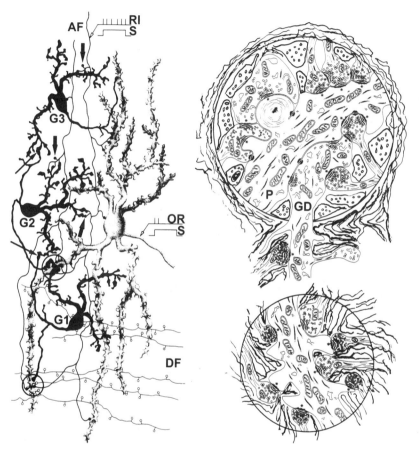

Figure 1. (LEFT)Synaptic organization of the ventral nucleus of the medial geniculate body of the cat. The large principal cell (thalamo-cortical neuron) sends its axon (OR) to the primary auditory cortex and receives synapses of ascending fibers (AF) from the inferior colliculus. Most ascending axons synapse in the synaptic nests on intermediate dendrites (e.g., large circle), fewer on distal dendrites (e.g., small circle), and very few on the soma. Many ascending afferents also synapse on distal dendrites (e.g., thick arrows) of small Golgi II cells (G1, G2, G3), fewer on intermediate dendrites, and relatively few on the soma. Descending fibers (DF) from the cortex end predominantly on distal dendrites of principal cells and on intermediate dendrites and somata of Golgi II cells. Golgi II distal dendrites form dendro-dendritic synapses on principal cell intermediate dendrites in the synaptic nests. Golgi II axons synapse on other Golgi II or principal cell dendrites (thin arrows). Where the AF input is excitatory and the Golgi II input, inhibitory, some duration responses conveyed by AF (RI/S) could be transformed to onset responses by the principal cell (OR/S). [from Morest (1975): Fig. 21] (RIGHT)TOP. Organization of a synaptic nest (see Fig. 1 Left: large circle). P, principal cell dendrite with ball and claw appendages; GD, small Golgi II cell dendrite with pre- and post- synaptic appendages. *, synaptic sites. The nest receives synaptic endings with excitatory (round vesicles) and inhibitory (pleomorphic vesicles) morphologies, representing heterogeneous inputs, including ascending and local circuits, which share the same extracellular compartment with each other, and make sequential synapses as well. Astrocyte processes (thick wavy lines) border the nest but do not penetrate it, leaving the synaptic endings in direct contact with each other. BOTTOM. Distal dendrite of principal cell (see Fig. 1 Left: small circle). Its synaptic complexes, including corticogeniculate inputs, are individually insulated by astrocytic processes. [From Morest (1975): Fig. 22].

2.1.3. Sequential Synapses. If one regards the sequential synaptic arrangement of the ventral nucleus according to the digital mode of processing, then one may propose the following hypothesis. An initial excitatory input ascending from the inferior colliculus to the principal cell could be followed rather quickly by an inhibitory period. This might have the effect of transforming an input coding a duration response pattern into an onset type (Fig. 1 LEFT). In fact, there is evidence for many such units in the ventral nucleus (Morel, 1980; Toros-Morel et al., 1981). However, not all units behave like that. This is not surprising, since other hypotheses, based on other assumptions, can be proposed. For one thing, there are several arrangements of collicular and non-collicular axonal endings in the synaptic nests. While some of these may involve sequential synapses, including both excitatory and inhibitory endings (Fig. 1 RIGHT: Top), not all of the synaptic endings in the nests can form sequential synapses. In fact, in the ventral nucleus of the rat medial geniculate, sequential synapses have not been found, and small Golgi II cells are rare or apparently lacking. Nevertheless, synaptic nests of heterogeneous endings are abundant in the rat, where they gather around the distal dendrites of principal cells (Morest, unpublished).

2.1.4. Analogue Processing. In the synaptic nests the bulk of the synaptic endings are in non-junctional contact with many other endings, where they are assembled "cheek by jowl," as it were. Without the intervention of glial processes, the neurotransmitters released by these endings would be expected to linger longer than usual, since the axonal endings would be expected to utilize less vigorous uptake transporters. This is likely to have important consequences for synaptic function in the nest, with the result that the effects of synaptic activity could persist for relatively prolonged periods, as long as free transmitters had access to receptors. Finally, the presence of presynaptic receptors, such as the metabotropic receptors, would raise the potential for synaptic modifiability in the nests, as discussed below in the small cell shell of the cochlear nucleus. In this scenario the singular events postulated for digital processing would reflect local changes with longer time constants and more of a blending of synaptic activity would result. This may be called the analogue mode of signal processing.

2.1.5. Extent of the Synaptic Nests. The importance of the analogue mode of processing in the medial geniculate body should not be underestimated, because the synaptic nests are extensive. Considering that they are limited only by the glial margins, one could not find an end to them in series of electron micrographs in the cat. Thus it is entirely possible, and likely, that they extend widely in far-reaching continuity. At the light microscopic level, these synaptic nests correspond to the axonal nests observed in rapid Golgi preparations (Fig. 2). Different kinds of axonal endings contribute to the axonal nests. Traced in three dimensions, the axonal nests really seem to weave throughout the neuropil, where they decorate and interconnect the principal cell proximal dendrites. Since the synaptic nests segregate ascending input to one particular region of the principal cell dendrites, the axonal nest geometry is not arrayed strictly parallel to the fibro-dendritic laminae, but crosses over them [see Morest and Winer (1986): Fig. 22]. In other words, although the axonal nests are formed by axons that ascend primarily in the fibro-dendritic laminae, their terminations aggregate in the nests. The synaptic nests extend in continuous fashion with thicker nodes that are continued by thinner strands of endings. Presumably the glial envelope assumes a sleeve-like configuration surrounding the axonal nests. With this arrangement the synaptic nests would be positioned to modify the processing events in the thalamo-cortical pathway in an analogue mode. Hence the signal transformations occurring in the principal cells would be subject to modification depending on the balance of excitatory and inhibitory transmitter concentrations collecting

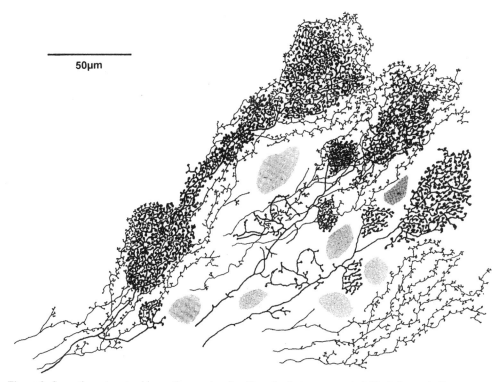

Figure 2. Synaptic nests extend in continuous strands with nodes that are connected. Shaded areas indicate neuronal cell bodies. Pars ovoidea, ventral nucleus of the medial geniculate body. Transverse section, 100-day old opossum pouch young, rapid Golgi method. Scale = 50 µm.

over a period of time. This could provide for a transitory memory, so to speak, which would bias the integrative activity of the principal cells according to prior events, reflecting precedence and sensory context.

2.2. Dorsal Division of the Medial Geniculate Body

The synaptic nest, as a structural basis for synaptic modification, is even more compelling in the case of the dorsal division of the medial geniculate body. Here the neuropil is a veritable sea of synaptic nests conspicuous in both Golgi impregnations (Fig. 3) and electron micrographs (Winer and Morest, 1984: Fig. 15). Although axo-dendritic junctions prevail, sequential synapses are not the major component, but rather there seems to be an enormous aggregation of small synaptic endings of differing vesicle morphology in apposition with each other and without intervening glial processes. Consequently, it appears that the dorsal division would be equipped with the cellular machinery for analogue processing and synaptic modifications par excellence.

2.3. Medial Division of the Medial Geniculate Body

A similar case can be made for the medial division of the medial geniculate body (Winer and Morest, 1983). Interestingly, it is the dorsal and medial divisions, in which a major role of the endogenous neurons in acoustic and multimodal conditioning has been proposed (e.g., Gerren and Weinberger, 1983; LeDoux et al., 1984, 1985). More research on the structure and function of the dorsal and medial divisions and its synaptic nests will be required to elucidate its signal processing modes.

2.4. Summary

The synaptic nests of the medial geniculate body provide the structural basis for the aggregation of a large number of axonal endings, segregating them into particular compartments of the neuropil. This provides the opportunity for sequential synapses, which could contribute ordered events in the digital processing mode. However, since the interior matrix of the synaptic nests generally lacks glial investment, a role for an analogue processing mode is postulated. This is likely to play a major role in thalamic function, since the synaptic nests spread in continuity throughout the neuropil, contacting both the thalamo-cortical neurons and the endogenous interneurons.

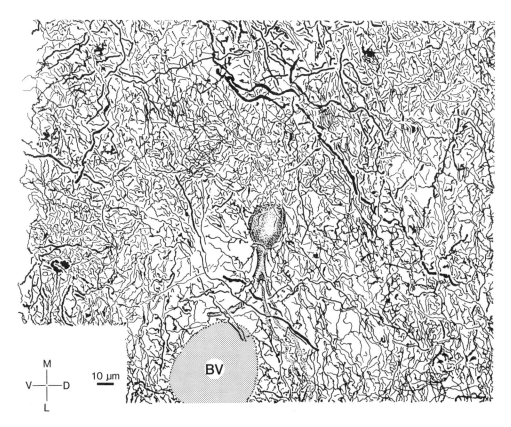

Figure 3. Axonal plexus of the dorsal nucleus of the medial geniculate body. The profile of one bushy principal cell (stippled) is shown next to a blood vessel (BV). The peridendritic plexus is dominated by thin axons and endings in a sea of synaptic nests. Rapid Golgi, 41-day old cat. D, M, V, L, dorsal, medial, ventral, and lateral directions. [from Winer & Morest (1984): Fig. 14.]

3. THE SYNAPTIC NESTS OF THE COCHLEAR NUCLEUS

3.1. Small Cell Shell (Fig. 4)

The mammalian cochlear nucleus contains one of the best developed examples of synaptic nests, namely, the small cell shell. The small cell shell, otherwise known as the small cell cap, consists of a more or less complete encirclement of the ventral cochlear nucleus (Osen, 1969, Brawer et al, 1974; Mugnaini et al, 1980) (Fig. 4). This collection of small neurons forms a conspicuous crust on the free surface of the nucleus (external lamella), extends between the ventral and dorsal cochlear nuclei (internal lamella), and assumes a less conspicuous situation on the medial aspect of the nucleus facing the inferior cerebellar peduncle (medial lamella). The small cell shell is continuous with the molecular layer of the dorsal cochlear nucleus, which contains a number of similar cell types and structures, including synaptic nests. However, the molecular layer is considered as an integral part of the dorsal cochlear nucleus, since it is interconnected with the neurons of the deeper layers. The small cell shell consists of granule cells, small stellate cells, elongate cells, and mitt cells, which are largely intermingled with each other and with the dendritic terminals of large stellate, elongate, antenniform, clavate, and giant cells in the centrally located subdivisions of the cochlear nucleus (Figs. 5–7). While it has a general ground plan throughout its extent, there are known to be regional variations, depending on the part of the cochlear nucleus it borders (Hutson and Morest, 1994).

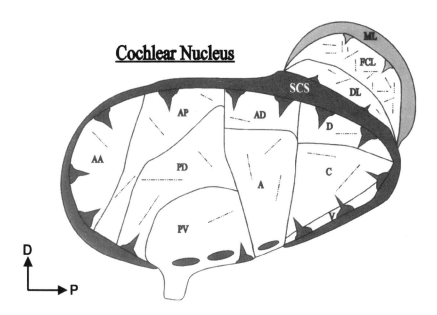

Figure 4. Distribution of synaptic nests in the mammalian cochlear nucleus as seen in a diagrammatic projection on the sagittal plane. (P, posterior; D, dorsal). The heaviest concentrations (dark shading) occur in the small cell shell (SCS), which completely surrounds the ventral cochlear nucleus, except where penetrated by fiber bundles (e.g., the cochlear nerve root at the ventral border of PV). A moderate concentration of synaptic nests (light shading) occurs in the molecular layer of DCN (ML). Extensions of the synaptic nests spill over into the bordering regions of the central nuclear subdivisions, which are further permeated by strands of mininests (broken lines). Abbreviations identify the subdivisions of the cochlear nucleus (see Morest et al, 1990)

3.1.1. Granule cells (Fig. 5, G). Granule cells are the most numerous component of the small cell shell. Among the smallest neurons in the cochlear nucleus, these excitatory interneurons receive excitatory synapses from mossy fiber endings (rosettes) by means of their terminal dendritic claws which extend from several different granule cells into the synaptic glomeruli (Fig. 5, GL). Hence they provide the structural basis for a divergence and diffuse spread of excitatory activity by means of the thin granule cell axons, most of which form synaptic endings within the small cell shell or other parts of the cochlear nucleus. The mossy fiber ending has a characteristic cytology resembling that of the mossy fiber endings of the cerebellum (Palay and Chan-Palay, 1974) and the dorsal cochlear nucleus (Kane, 1974). In the present context the glomerulus can be regarded as a variety of synaptic nest (micronest). Although a few small, presumably inhibitory, axo-dendritic synapses occur on its periphery, the glomerulus is dominated by the mossy fiber endings, which, in a sense, can be called a micronest. This micronest is enveloped by astrocytic processes, which, however, do not invade the interior of the glomerulus.

3.1.2. Mitt cells (Fig. 5, M). Mitt cells, so-named because their single thick dendrite ends in an enormous cup-like expansion resembling a baseball catcher's mitt, were first observed in the cochlear nucleus of the cat (shrub cells of Brawer et al, 1974) and in the small cell shell and dorsal cochlear nucleus of the chinchilla (catcher's mitt cell of Hutson and Morest, 1994). Their fine structure has been described in the small cell shell of the chinchilla (Hutson and Morest, 1996). Somewhat similar cell types have been reported in the cerebellum and dorsal cochlear nucleus of other species (Altman and Bayer, 1977; Mugnaini and Floris, 1994; Wright and Ryugo, 1996). In the chinchilla small cell shell, a single large mossy fiber rosette ends within the dendritic mitt by means of multiple synaptic junctions with excitatory cytological features resembling the mossy fiber endings of the glomerulus (Fig. 5, SM). A few small axons synapse on the periphery of the mitt with either small round vesicles or pleomorphic ones, presumably representing excitatory endings from parallel fibers and small cochlear nerve fibers. Hence the synaptic mitt, like the glomerulus, is a type of micronest. It is surrounded by glial processes, but these do not invade the interior of the mitt. Its synaptic organization would tend to preserve the identity of the afferent mossy fiber input and its signal coding with considerable temporal precision. The exclusion of glial processes from the synaptic mitt might also tend to allow for a certain amount of temporal summation.

3.1.3. Small stellate cells (Fig. 5, SS). The synaptic nests reach their most extensive development in association with the terminal dendritic branches of the small stellate cells, which populate all regions of the small cell shell. Many of these small stellate cells form local circuits in the small cell shell and dorsal and ventral cochlear nuclei. Their transmitters are not completely identified, but many of these cells contain glycine or GABA and could function as inhibitory interneurons (Kolston et al., 1992). The synaptic nests are extensively developed and contain numerous endings of both excitatory and inhibitory cytology (Fig. 5), including mossy fiber endings (arising from multiple sources, including the cochlea and somatic sensory nuclei), medium-sized endings with large spherical vesicles like those of the type I fibers of the cochlear nerve, small endings with large or small spherical vesicles, like those of the thin fiber collaterals of the cochlear nerve and olivo-cochlear bundle, and very small endings with small spherical vesicles, like those of the parallel fibers arising from granule cells and other small fibers, presumably type II cochlear nerve fibers. There is also a large contingent of synaptic endings of all sizes and shapes containing pleomorphic vesicles like those usually associated with GABA and arising in multiple sites, including the dorsal cochlear nucleus and superior olivary complex (including olivo-cochlear collaterals). However, there are relatively few endings containing flat vesicles.

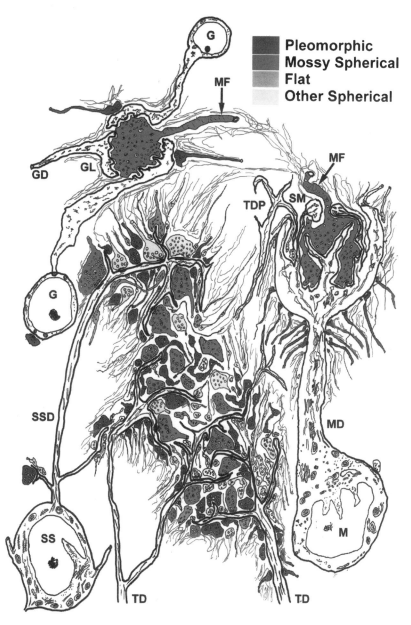

Pleomorphic
Mossy Spherical
Flat
Other Spherical

Figure 5. Drawing to illustrate the characteristic features of the synaptic organization of the small cell shell. The synaptic nest proper (macronest) occupies a relatively large volume associated primarily with the terminal dendritic plumes of small stellate cells (TDP). Other terminal dendrites (TD) belonging to large stellate, elongate, and giant cells, among others, also invade the nest. This heterotypic synaptic nest contains a heterogeneous population of synaptic endings with both excitatory and inhibitory morphology and different origins. Although these endings form synaptic junctions on the distal dendrites (SSD) of small stellate cells (SS) and neighboring granule (G) and mitt cells (M), most of their surface area apposes the synaptolemma of other endings without visible junctions, and they are not separated by astroglia, whose processes (thin wiggly lines) border but do not invade the macronest. The synaptic mitt (SM) formed by the mitt cell dendrite (MD) and the glomerulus (GL) formed by granule cell dendrites (GD) engage mossy fiber synaptic endings (MF), having excitatory morphology. These micronests are contacted in their periphery by a small number of endings from presumed inhibitory and excitatory inputs that are wrapped by astrocytic processes. There is a limited continuity between the micronests and macronests (e.g., between TDP and SM). The gray scale represents endings with inhibitory (Pleomorphic, Flat) and excitatory (Mossy Spherical and Other Spherical) vesicle types.

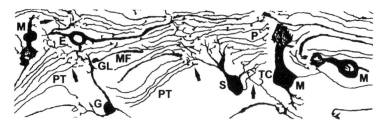

Figure 6. In the small cell shell the axonal plexus forms synaptic nests (<u>arrows</u>), connected by strands that extend throughout the neuropil. <u>E</u>, elongate cell; <u>G</u>, granule cell; <u>M</u>, mitt cells; <u>MF</u>, mossy fibers; <u>P</u>, parallel fibers; <u>PT</u>, primarily thin axons; <u>S</u>, small stellate cell; <u>TC</u>, thin collateral axons.

3.1.4. No Astrocytic Processes in the Synaptic Nests (Fig. 5). Notably lacking from the interior of the synaptic nests are glial processes, which otherwise wrap up individual synaptic endings outside the nests. The astrocytic processes envelop the glomeruli, the synaptic mitts, and form the borders of the synaptic nests associated with stellate cell dendrites (Fig. 5, <u>SSD</u>). These latter could be regarded as heterotypic macronests. The synaptic nests of the cochlear nucleus have a certain similarity to those described in the medial geniculate body, in that they contain aggregations of numerous synaptic endings and the nests continue in a

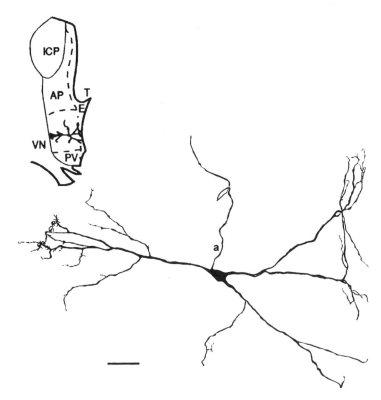

Figure 7. Large elongate cell from the dorsal part of the anteroventral cochlear nucleus of an adult gerbil (*Inset*: location in a transverse section). The dendritic field is flattened in the transverse plane. The lateral dendrites have terminal branches in the external lamella of the small cell shell (<u>E</u>), while the medial dendrites form endings in the medial lamella. The cell was reconstructed after intracellular filling with horseradish peroxidase and visualized by histochemistry. <u>a</u>, axon. Scale = 50 μm. [From Ostapoff et al. (1994): Fig. 13.]

widely ranging pattern of axonal strands permeating the entire small cell shell (Figs. 4,6). They differ from the nests of the cat medial geniculate body in that those in the small cell shell are not characterized by sequential synapses, and there is a much greater heterogeneity of synapses, by their morphologies, their connections, and the endogenous cell types involved. In fact, the resemblance to the dorsal division of the medial geniculate body is more striking than to the ventral nucleus. The unifying principle, however, is readily apparent in both structures. For the absence of intrinsic glia argues for an analogue mode of signal processing. In this respect it is worth noting that the neuropil of the small cell shell is also rich in glutamate receptors which have been implicated in synaptic modifiability, e.g., NMDAR1 (Bilak et al, 1996) and the metabotropic receptor, mGluR1 (Bilak and Morest, 1996) (Fig. 8). In fact mGluR1 is especially abundant in the preterminal portions of axons ending in the synaptic nests. Since glutamate and GABA are presumably the predominant amino acid transmitters released in this region, the absence of the high-affinity transporters usually associated with astrocytes would favor the accumulation of these transmitters in the extracellular space of the nests, until the transporters also associated with axonal endings could remove them (Hamberger et al, 1976; Oliver et al, 1983; Danbolt et al, 1992). For all of these reasons, the synaptic nests of the small cell shell could provide a microenvironment with properties favoring both pre- and post-synaptic modifiability and an analogue mode of processing.

3.1.5. Analogue Processing in the small Cell Shell May Involve the Ventral Cochlear Nucleus. The synaptic nests of the small cell shell are bound to play a role in signal processing by the larger neurons located in the central parts of the ventral cochlear nucleus. In particular, there are many cells, belonging to various non-bushy multipolar types, including the large and small stellate, antenniform, clavate, large and small elongate, and giant cells (Ostapoff et al, 1994). Characteristic of these cells are the distinctive terminal den-

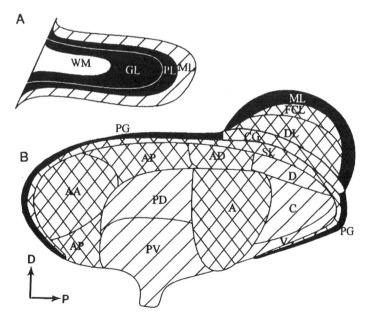

Figure 8. Diagram similar to that in Fig. 4 showing the distribution of NMDA receptor protein (NR1) in the cerebellar cortex (**A**) and cochlear nucleus (**B**) of the mouse. High levels of expression (black) occur in the granular layer of the cerebellum (GL), the external lamella of the small cell shell (PG), and the molecular layer of the dorsal cochlear nucleus (ML), all regions with a high concentration of synaptic nests. [from Bilak et al, (1996): Fig. 11]

dritic branches which form distinct patterns that often extend into the small cell shell. Our findings indicate that these dendrites participate fully in the synaptic nests of the small cell shell (Fig. 7). While some effort has been devoted to deciphering the digital processing modes of these neurons (Morest, 1993), the present findings cry out for more attention to the analogue mode.

3.2. Synaptic Nests in the Dorsal and Ventral Cochlear Nuclei

Other varieties and venues of the synaptic nests exist in the cochlear nucleus. First of all, it appears that the synaptic nests of the small cell shell extend into the molecular layer of the dorsal cochlear nucleus (Fig. 4). In addition, small extensions trickle down from the small cell shell and molecular layer into the outer fringes of the ventral cochlear nucleus and more extensively into the deep layers of the dorsal cochlear nucleus. However, within the central subdivisions of the ventral cochlear nucleus, there appears to be a more limited distribution of synaptic nests, although their extent is not very well documented at present. Nevertheless, numerous examples of relatively small, compact collections of synaptic endings have been reported in contact with the somata of a number of neuron types, including the octopus cell (Kane, 1973), the globular bushy cell (Tolbert and Morest, 1982; Ostapoff and Morest, 1991), and the spherical bushy cell (Cant and Morest, 1979). These aggregations of axosomatic synapses so far have been seen to contain endings of the same cytological type, presumably inhibitory in every case. They might be regarded as homotypic mininests. Mininests have also been described in synaptic association with the terminal dendrites of bushy cells (Cant and Morest, 1979; Ostapoff and Morest, 1991) and octopus cells (Kane, personal communication). These mininests were found to be heterotypic. It is likely that the terminal dendritic plumes of stellate cells and most other non-bushy multipolar neurons in the ventral cochlear nucleus engage synaptic nests of some type (see, e.g., Ostapoff and Morest, 1991).

4. SYNAPTIC NESTS IN OTHER PARTS OF THE BRAIN

Synaptic nests have a widespread distribution in the mammalian brain. For example, homotypic mininests have been observed in association with the perikaryon of the principal cells of the medial nucleus of the trapezoid body, while heterotypic nests decorate their terminal dendrites (Jean-Baptiste and Morest, unpublished). Similar structures have been illustrated in a number of thalamic nuclei, other than the medial geniculate body, e.g., the dorsal nucleus of the lateral geniculate body (Peters and Palay, 1966; Morest and Winer, 1986: Fig. 31), ventrobasal complex (Jones and Powell, 1969), and in the spinal trigeminal nucleus (Gobel, 1974), olfactory bulb (White, 1972), and retina (Kidd, 1962; Dowling and Boycott, 1966; Raviola and Raviola, 1967). The best known example of a micronest is the glomerulus of the cerebellar cortex, where mossy fiber endings, the axonal endings of large Golgi II cells, and granule cell dendrites form synapses and are wrapped up in a little ball by glial processes (Palay and Chan-Palay, 1974). The glomerulus in the cochlear nucleus was described above as a micronest. No doubt more examples could be found, but a systematic analysis of the structure and function of synaptic nests in the brain and their roles in signal processing is not available.

5. CONCLUSIONS

The synaptic nests have a diffuse, relatively non-differentiated arrangement of synaptic endings, which could provide the structural basis for analogue processing of ongoing or slowly fluctuating background activity. Such processing might shift the ambient baseline levels in the neuropil and bias the input-output transforms of whole groups of cells. Taken in this light, the synaptic patterns could provide the cellular machinery for contextual modifications of processing, e.g., matching and sensory masking functions, not to mention the better appreciated reflections of synaptic modifiability underlying attention, conditioning, memory, and learning.

6. ACKNOWLEDGMENTS

Supported by NIH grant DC00127. The author acknowledges the contributions to this research made by Drs. Jeffery A. Winer, Kendall A. Hutson, E.-Michael Ostapoff, Jane J. Feng, Eileen C. Kane, Nell B. Cant, Leslie P. Tolbert, and Michel Jean-Baptiste.

7. REFERENCES

Altman J, Bayer SA (1977) Time of origin and distribution of a new cell type in the rat cerebellar cortex. Exp Brain Res 29: 265–274.

Bilak SR, Morest DK (1996) Differential expression of the metabotropic glutamate receptor mGluR1 in the mouse cochlear nucleus: an in situ/immunohistochemistry study of specific neuron types. (Submitted)

Bilak SR, Bilak MM, Morest DK (1996) Differential expression of N-methyl-D-aspartate receptor in the cochlear nucleus of the mouse. Neuroscience 75: 1075–1097.

Brawer JR, Morest DK, and Kane EC (1974) The neuronal architecture of the cochlear nucleus of the cat. J Comp Neurol 155: 251–299.

Cant NB, Morest DK (1979) The bushy cells in the anteroventral cochlear nucleus of the cat: a study with the electron microscope. Neuroscience 4: 1925–1945.

Danbolt NC, Storm-Mathisen J, Kanner BI (1992) A [Na$^+$ + K$^+$]coupled L-glutamate transporter purified from rat brain is located in glial cell processes. Neuroscience 51: 295–310.

Dowling JE, Boycott BB (1966) Organization of the primate retina: electron microscopy. Proc roy Soc B 166: 80–111.

Gerren RA, Weinberger NM (1983) Long term potentiation in the magnocellular medial geniculate nucleus of anesthetized cat. Brain Res. 265: 138–142.

Gobel S (1974) Synaptic organization of the substantia gelatinosa glomeruli in the spinal trigeminal nucleus of the adult cat. J Neurocytol 3: 219–243.

Gundersen V, Shupliakov O, Brodin L, Ottersen OP, Storm-Mathisen J (1995) Quantification of excitatory amino acid uptake at intact glutamatergic synapses by immunocytochemistry of exogenous D-aspartate. J Neurosci. 15:4417–28.

Hamberger A, Nystrom B, Sellstrom A, Woiler CT (1976) Amino acid transport in isolated neurons and glia. Adv Exp Med Biol. 69:221–36.

Hutson KA, Morest DK (1994) Ultrastructure of small cell regions in the cochlear nucleus (CN) of chinchilla. Assoc. Res. Otolaryngol. Abstr. 17 :27.

Hutson KA, Morest DK (1996) Fine structure of the cell clusters in the cochlear nerve root: stellate, granule, and mitt cells offer insights into the synaptic organization of local circuit neurons. J Comp Neurol 371: 397–414.

Jones EG, Powell TPS (1969) Electron microscopy of synaptic glomeruli in the thalamic relay nuclei of the cat. Proc roy Soc B 172: 153–171.

Kane ESC (1973) Octopus cells int he cochlear nucleus of the cat: heterotypic synapses upon homeotypic neurons. Int. J. Neurosci 5: 251–279.

Kane ESC (1974) Synaptic organization in the dorsal cochlear nucleus of the cat: a light and electron microscopic study. J Comp Neurol 155: 301–330.

Kidd M (1962) Electron microscopy of the inner plexiform layer of the retina in the cat and the pigeon. J Anat (Lond) 96: 179–187.

Kolston J, Osen KK, Hackney CM, Ottersen CP, and Storm-Mathisen J (1992) An atlas of glycine- and GABA-like immunoreactivity and colocalization in the cochlear nuclear complex of the guinea pig. Anat Embryol 186: 443–465.

LeDoux JE, Sakaguchi A, Reis DJ (1984) Subcortical efferent projections of the medial geniculate nucleus mediate emotional responses conditioned to acoustic stimuli. J Neuroscience 4: 683–698.

LeDoux JE, Ruggiero DA, Reis DJ (1985) Projections to the subcortical forebrain from anatomically defined regions of the medial geniculate body in the rat. J Comp Neurol 242: 182–213.

Morel A (1980) Codage des sons dans le corps genouillé médian du chat: évaluation de l'organisation tonotopique de ses différents noyaux. Doctoral dissertation, Université de Lausanne. Zurich: Juris Druck + Verlag.

Morest DK (1970) Electron microscopic study of the synaptic organization in the medial geniculate body of the cat. Anat Rec 166: 351.

Morest DK (1971) Dendrodendritic synapses of cells that have axons: the fine structure of the Golgi type II cell in the medial geniculate body of the cat. Z Anat Entwickl-Gesch 133: 216–246.

Morest DK (1975) Synaptic relationships of Golgi type II cells in the medial geniculate body of the cat. J Comp Neurol 162: 157–194.

Morest DK (1993) the cellular basis for signal processing the mammalian cochlear nuclei. In MD Merchan, JM Juiz, DA Godfrey and E Mugnaini (eds), The Mammalian Cochlear Nuclei: Organization and Function. N.Y.: Plenum Press, pp 195–210.

Morest DK, Hutson KA, and Kwok S (1990) Cytoarchitectonic atlas of the cochlear nucleus of the chinchilla, *Chinchilla laniger*. J Comp Neurol 300: 230–248.

Morest DK, Winer JA (1986) The comparative anatomy of neurons: homologous neurons in the medial geniculate body of the opossum and the cat. Adv Anat Embryol Cell Biol 97: 1–96.

Mugnaini E, Floris A (1994) the unipolar brush cell: a neglected neuron of the mammalian cerebellar cortex. J Comp Neurol 339: 174–180.

Mugnaini E, Osen KK, Dahl A-L, Friedrich VL, Jr, Korte G (1980) Fine structure of granule cells and related interneurons (termed Golgi cells) in the cochlear nuclear complex of cat, rat, and mouse. J Neurocytol 9: 537–570.

Oliver DL, Potashner SJ, Jones DR and Morest DK (1983) Selective labeling of spiral ganglion and granule cells with D-aspartate in the auditory system of cat and guinea pig. J Neuroscience 3: 455–472.

Osen KK (1969) Cytoarchitecture of the cochlear nuclei in the cat. J Comp Neurol 135: 453–484.

Ostapoff EM, Feng J, Morest DK (1994) A physiological and structural study of neuron types in the cochlear nucleus. II. Neuron types and their structural correlation with response properties. J Comp Neurol 346: 19–42.

Ostapoff EM, Morest DK (1991) Synaptic organization of globular bushy cells in the ventral cochlear nucleus of the cat: a quantitative study. J Comp Neurol 314: 598–613.

Palay SL, Chan-Palay V (1974) Cerebellar Cortex, Cytology and Organization. Berlin: Springer-Verlag.

Peters A, Palay SL (1966) The morphology of laminae A and A1 of the dorsal nucleus of the lateral geniculate body of the cat. J Anat (Lond) 100: 451–486.

Raviola G, Raviola E (1967) Light and electron microscopic observations on the inner plexiform layer of the rabbit retina. Amer J Anat 120: 403–426.

Tolbert LP, Morest DK (1982) The neuronal architecture of the anteroventral cochlear nucleus of the cat in the region of the cochlear nerve root: electron microscopy. Neuroscience 7: 3053–3067.

Toros-Morel A, de Ribaupierre F, Rouiller E (1981) Coding properties of the different nuclei of the cat's medial geniculate body. In: Neuronal Mechanisms of Hearing (eds. J. Syka & L. Aitkin). NY: Plenum, pp 239–243.

White EL (1972) Synaptic organization in the olfactory glomerulus of the mouse. Brain Res 37: 69–80.

Wilkin GP, Garthwaite J, Balazs R (1982) Putative acidic amino acid transmitters in the cerebellum. II. Electron microscopic localization of transport sites. Brain Res. 244: 69–80.

Winer JA, Morest DK (1983) The medial division of the medial geniculate body of the cat: implications for thalamic organization. J Neurosci 3: 2629–2651.

Winer JA, Morest DK (1984) Axons of the dorsal division of the medial geniculate body of the cat: a study with the rapid Golgi method. J Comp Neurol 224: 344–370.

Wright DD, Ryugo DK (1996) Mossy fiber projections from the cuneate nucleus to the cochlear nucleus in the rat. J Comp Neurol 365: 159–172.

3

MAPPING ACTIVITY IN THE AUDITORY PATHWAY WITH C-FOS

E. M. Rouiller[*]

Institute of Physiology
University of Fribourg
Rue du Musée 5, CH-1700 Fribourg, Switzerland

ABSTRACT

Induction of the immediate-early-gene c-fos occurs in response to a wide variety of external stimuli, and therefore represents a method to functionally map neuronal activity in the brain with cellular resolution. We investigated systematically in rats the distribution of Fos-like activity in the auditory pathway elicited by various types of stimulation of the cochlea.

In response to pure tones (45 minutes exposure), Fos-like immunoreactivity (FLI) was found to form bands reflecting the tonotopic arrangement of neurons in the dorsal (DCN) and posteroventral (PVCN) cochlear nuclei, the lateral superior olive (LSO), the dorsal nucleus of the lateral lemniscus (DLL) and the central nucleus of the inferior colliculus (CIC). Surprisingly, virtually no FLI was observed in the anteroventral cochlear nucleus (AVCN) and the ventral division of the MGB (vMGB). In the primary auditory cortex (Te1), FLI was present but did not reveal its tonotopic organization.

After electrical stimulation of the whole cochlea, FLI was found to cover the entire extent of DCN, PVCN, LSO, DLL as well as non-tonotopic nuclei such as the external (ECIC) and dorsal (DCIC) nuclei of the inferior colliculus, the medial (mMGB) and dorsal (dMGB) divisions of the medial geniculate body. FLI was also broadly and unspecifically distributed in the 3 auditory cortical areas Te1, Te2 and Te3. In contrast, no FLI was seen in AVCN, CIC and vMGB. Even a short period of stimulation (5 minutes) was sufficient to significantly increase FLI in most of the above auditory nuclei. Changing the intensity or duration of stimulation and the survival time could modify substantially the distribution of FLI. Similarly, using another commercial source of antibody against c-fos could make FLI to be detected in AVCN and the CIC, but not in vMGB.

FLI was established in the auditory pathway of rats during the successive steps of the learning of a complex sensorimotor task. At all steps of learning, FLI was constant in the sub-

* Correspondence: Prof. E. M. Rouiller, Institute of Physiology, University of Fribourg, Rue du Musée 5, CH-1700 Fribourg, Switzerland; Tel: +41 26 300 86 09; fax: +41 26 300 97 34; E-mail: Eric.Rouiller@unifr.ch

cortical auditory nuclei, namely in DCN, the posteroventral cochlear nucleus (PVCN), LSO, the dorsal nucleus of the lateral lemniscus (DLL), CIC, ECIC, DCIC, mMGB and dMGB. As observed above for passive listening, no or little FLI was observed in AVCN and vMGB. In contrast to the subcortical auditory nuclei, a differential FLI distribution reflecting the progessive steps of task learning was found in the auditory cortical areas.

Taken together, these data indicate that c-fos is not a general functional marker of activity. C-fos is characterized by a significant degree of selectivity for certain stimuli; in addition, for the same stimulus, c-fos expression depends on the context. Furthermore, FLI appeared to be induced preferentially in some cell types (small neurons) as opposed to others (large neurons). Parameters of stimulation (duration, intensity, survival time) and the age of the animal play also an important role, as well as the type of antibody used to visualize Fos activity.

1. INTRODUCTION

Since about a decade, the immediate early gene *c-fos* has been used in many systems as a functional marker with cellular resolution in response to a variety of external stimuli (e.g. Hunt et al., 1987; Dragunow and Faull, 1989; Morgan and Curran, 1991; Robertson, 1992; Robertson and Dragunow, 1992). This was the case also for the auditory system: the distribution of immunoreactivity (FLI) was established in the auditory pathway in response either to tonal stimuli (Ehret and Fischer, 1991; Friauf, 1992, 1995; Sato et al., 1992, 1993; Rouiller et al. 1992; Reimer, 1993; Pierson and Snyder-Keller, 1994; Brown and Liu, 1995; Adams, 1995; Gleich et al., 1995; Scheich and Zuschratter, 1995; Zuschratter et al., 1995), to electrical stimulation of the cochlea (Vischer et al., 1994, 1995; Zhang et al., 1996) or following audiogenic seizure (Le Gal La Salle and Naquet, 1990; Snyder-Keller and Pierson, 1992). Confirming previous data derived from electrophysiology and from 2-deoxyglucose experiments, FLI elicited by pure tonal stimulation formed bands of positive neurons consistent with the tonotopic organization of the dorsal cochlear nucleus (DCN) and the central nucleus of the inferior colliculus (CIC). Less dense FLI was observed in the posteroventral cochlear nucleus (PVCN), the superior olivary complex (SOC), the lateral lemniscus (LL), the medial geniculate body (MGB) and the auditory cortex. In tonotopically organized nuclei such as the anteroventral cochlear nucleus (AVCN) and the ventral division of the medial geniculate body (vMGB), there was sparse FLI. In an animal model of cochlear implant, the distribution of FLI in the auditory pathway in response to electrical stimulation of the cochlea was largely consistent with that obtained after tonal stimulation, except for the CIC which was virtually free of FLI, although changes of the stimulation parameters could induce *c-fos* in the CIC (Vischer et al., 1994; Zhang et al., 1996). Thus, the distribution of FLI in the auditory system strongly depends on parameters of stimulation. In our previous c-fos studies (Vischer et al., 1994; Zhang et al., 1996), data were derived from experiments in which the Fos antibodies were supplied by Cambridge Research Biochemicals (CRB) and Oncogene Science (OS). Largely comparable results were obtained after treatment of the tissue with one or the other antibody. In contrast, the use of a third antibody, supplied by Santa Cruz Biotechnology (SCB), lead to significantly different results, in some auditory nuclei. As compared with the previous results based on the CRB and OS antibodies, FLI was much denser in some structures after treatment with the SCB antibody, for instance in the CIC. In other structures, however, a similar pattern of FLI was seen for all three antibodies. In addition, one antibody gave a better FLI increase as compared to control animals in some auditory nuclei than the other antibody while this was the reverse in other auditory nuclei. Therefore, besides the effects on FLI due to the parameters of stimulation, the issue of the commercial source of antibody needs to be considered.

The goal of the present report is to review the Fos data available for the auditory system and to point out the advantages as well as the limitations of the Fos mapping method, in particular in the context of the influence of the parameters of stimulation and the commercial source of the antibody. The data presented below represent an overview of experiments performed in our laboratory on rats subjected to various modes of stimulation of the cochlea: passive listening of pure tones, electrical stimulation of the cochlea (model of cochlear implant) and complex tones with behavioral significance (animals performing a sensorimotor learning task).

2. METHODS

2.1. Tonal Stimulation

Experiments were performed on adult Sprague-Dawley rats (see Rouiller et al., 1992 for details). Briefly, a group of acoustically stimulated animals were placed in a sound proofed room the night preceding the acoustic stimulation. The sound source (free field stimulation) was placed above the cage. Pure tone bursts (50 ms duration presented at a rate of 2 per second) at a fixed frequency were delivered during 45 minutes to one hour, at an intensity close to 80 dB SPL, as measured in the center of the cage. The tone frequencies tested were 1, 4 , 16 or 32 kHz. The animal was free to move in the cage during the acoustic stimulation and did not show external signs of stress and discomfort. A second group of rats were used as control animals: the rats were placed in the same experimental conditions but no acoustic stimulation was delivered to the loudspeaker. This group of rats thus served as control animals for the tonal stimulation experiments.

2.2. Electrical Stimulation of the Cochlea: Surgical Preparation and Mode of Stimulation

Adult Sprague-Dawley or Long Evans rats were anaesthetized with sodium pento-barbital (64.8 mg/kg, i.m.) and two custom made stimulating electrodes (Teflon coated platinum iridium wires) were implanted in the left cochlea (unilaterally), as previously described in details (Vischer et al., 1994, 1995). One electrode was positioned in the basal turn of the cochlea while the other was inserted in the apical turn. The animals recovered at least 8 days in the animal-room. Electrically evoked Auditory Brainstem Responses (EABR) were recorded weekly in experiments aimed to confirm efficacy of the implanted electrodes (Vischer et al., 1994; 1995). Stimuli were monophasic pulses of 20 µs duration, delivered with alternating polarity, at a rate of 50 Hz and 300 µA above EABR threshold.

In a final stimulation/recording session (separated by at least seven days from a previous session), the left cochlea was stimulated with electrical pulses, in order to induce *c-fos*. During the stimulation, the animals were sedated with a mixture of ketamine (20 mg/kg, i.m.) and xylazine (6 mg/kg, i.m.). EABR measurements were periodically collected (every five minutes) to check the stability of the evoked responses which can indicate the reliability of the stimulation. Different groups of animals were tested in order to study the effects of changing several experimental parameters such as the duration of stimulation (ranging from 5 minutes to an hour), the survival time (ranging from 0 to 6 hours) and the intensity of stimulation. Another group of animals (control rats) were also connected to the electrical stimulation/recording setup for forty-five minutes, but no electrical current was delivered to the cochlea.

2.3. Tonal Stimulation with Behavioral Significance

In a recent series of experiments, we mapped the distribution of Fos-like activity in Long-Evans rats performing a sensorimotor learning task based on the discrimination of complex tones. These data are presented in the present volume (see Hervé-Minvielle et al.) and, therefore, they will be mentioned here only briefly, essentially for comparision with the results derived from passive stimulation of the cochlea with tones or electrical pulses.

2.4. Fos Immunocytochemistry

At the end of the stimulating period, the animals were placed in a sound attenuated room for survival time (one hour in general), except for a group of electrically stimulated rats in which the influence of survival time was tested. Then, the rats were sacrificed with a lethal dose of sodium pentobarbital and perfused with 100 ml saline followed by 1000 ml of 4% paraformaldehyde in 0.1M phosphate buffer (pH 7.4). The brains were postfixed for three hours and subsequently cryoprotected by immersion in a 10% sucrose solution for one day and in a 30% sucrose solution for 3 days. Frozen sections (50 μm thick) were cut in the frontal plane. A first series of alternate sections were Nissl-stained, whereas the second series of sections were immunocytochemically processed to visualize FLI, as previously described in detail (Rouiller et al., 1992; Wan et al., 1992; Vischer et al., 1994; Zhang et al., 1996). The immunohistochemical reaction was performed using the primary antibody against Fos from CRB for animals subjected to passive listening of pure tones (see Rouiller et al., 1992), and the primary antibody from OS or from SCB for animals electrically stimulated (see Zhang et al., 1996). Finally, the tissue derived from the rats that performed the sensorimotor learning task (see Hervé-Minvielle et al, present volume) was treated with the antibody supplied by SCB. The immunohistochemical procedure was comparable for the three antibodies, and it was described in details in previous reports (Rouiller et al., 1992; Wan et al., 1992; Zhang et al., 1996).

2.5. Morphometry

The auditory structures studied included DCN, PVCN, AVCN, the lateral superior olive (LSO) and the nucleus of medial trapezoid body (NTB), the lateral lemniscus (LL), CIC, the dorsal (dMGB), ventral (vMGB) and medial (mMGB) divisions of the MGB, the peripeduncular nucleus (PP), and the auditory cortex. In the experiments of electrical stimulation and sensorimotor learning, for each structure, three to five consecutive sections were analyzed and the images were captured into a Power Macintosh (7500) using a video camera. First, the boundaries of the auditory nuclei were determined using the Nissl stained sections. The nomenclature and the criteria were those of the atlas of the rat brain and cortex (Paxinos and Watson, 1986; Zilles, 1985). Second, the area of each auditory nucleus within the defined boundaries was then measured on each section. Third, the number of FLI neurons included in this area was established. Area measurements and FLI neuron counts were performed using the software "Image" (NIH 68k/1.58). The density of FLI neurons in a given auditory nucleus on each section analyzed was then calculated by dividing their number by the corresponding area (expressed in number of neurons/mm^2).

Then, for each auditory nucleus, an average density was computed from the values derived from the sections analyzed in each rat. Finally, a mean density value was derived by cumuling animals within each group of rats subjected to the same protocol of stimulation and histochemical treatment.

3. RESULTS

3.1. PASSIVE TONAL STIMULATION

C-fos expression was mapped in the auditory pathway of rats, stimulated acoustically with pure tones. A significantly different FLI was observed in these stimulated animals as compared to the control (unstimulated) rats which exhibited only a low (basal) FLI in the auditory pathway. The distribution of FLI in the rats stimulated with pure tones was described earlier (Rouiller et al., 1992) and can be summarized as follows. In the cochlear nucleus, two clusters of Fos-like immunoreactive neurons, located respectively in the caudal part of DCN and in the granular cell region, did not show clear systematic shift in their position as a function of the tones frequency. On the other hand, in a more rostral part of DCN, a cluster of Fos-like positive neurons moved progressively from dorsal to ventral for decreasing tone frequencies. In PVCN, another cluster of Fos-like positive neurons was observed, whose position also varied with tone frequencies. Surprisingly, no or very rare Fos-like immunoreactive neurons were present in AVCN and in the SOC. In the inferior colliculus, however, Fos-like immunoreactive neurons formed clear isofrequency contours, shifting from dorsolateral to ventromedial for increasing tone frequencies. In the MGB, FLI was restricted to mMGB and dMGB, while there was no FLI in vMGB. The general distribution of FLI in the auditory pathway in response to passive tonal stimulation is schematized in Fig. 1. These data for subcortical auditory nuclei are largely consistent with the results of studies conducted in other laboratories on animals exposed to passive listening of pure tones (e.g. Ehret and Fischer, 1991; Friauf, 1992, 1995; Sato et al., 1992, 1993; Rouiller et al. 1992; Reimer, 1993; Pierson and Snyder-Keller, 1994; Brown and Liu, 1995; Adams, 1995; Gleich et al., 1995). In our hands, there was no clear distribution of FLI in bands in the primary auditory cortex (Te1), reflecting tonotopic organization. From the study conducted by Zuschratter et al. (1995), it was concluded that there was indeed no Fos-like activity reflecting the tonotopic organization of the primary auditory cortex when the pure tones were delivered for a long time (e.g. 45 minutes). In contrast, bands of Fos-like positive neurons consistent with the tonotopy were observed in the primary auditory cortex only after tonal stimulation lasting a short time (e.g. 5 minutes).

The distribution of FLI in response to pure tones in the subcortical auditory pathway (Fig. 1) is generally consistent with the mapping derived from 2-DG experiments (Fig. 2). A strict correspondence of both functional markers has been demonstrated in the same animal in the CIC (Reimer, 1993). Consistency between FLI and 2-DG is also true for the cochlear nucleus (with a stronger labelling in DCN than in PVCN, itself stronger than in AVCN), the SOC and the LL. In sharp contrast, however, an increase of 2-DG uptake was observed in vMGB, while FLI was virtually absent from vMGB, but present in dMGB and mMGB (compare Figs. 1 and 2 for the thalamus).

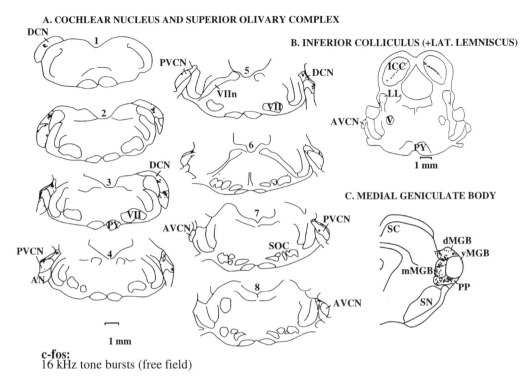

A. COCHLEAR NUCLEUS AND SUPERIOR OLIVARY COMPLEX

B. INFERIOR COLLICULUS (+LAT. LEMNISCUS)

C. MEDIAL GENICULATE BODY

c-fos:
16 kHz tone bursts (free field)

Figure 1. Schematic representation of the distribution of FLI, as seen after passive presentation in free field of pure tones (16 kHz) to rats, in the cochlear nucleus and superior olivary complex (A), in the lateral lemniscus and inferior colliculus (B) and in the medial geniculate body (C). Same scale bar for B and C. In A, sections were arranged from caudal to rostal (1 to 8). Each dot represents a Fos-like positive neuron. AN=auditory nerve; PY=pyramidal tract; SC=superior colliculus; SN=substantia nigra; V=trigeminal motor nucleus; VIIn=facial nerve; VII=facial motor nucleus. See also list of abbreviations.

3.2. Passive Electrical Stimulation of the Cochlea: Fos Distribution and Influence of Experimental Parameters

A highly reproducible FLI was obtained in the auditory pathway of rats subjected to electrical stimulation of the cochlea (Vischer et al., 1994). Therefore, this animal model of cochlear implant was selected to study in details the effects of various experimental parameters on FLI elicited in the auditory pathway. The following parameters were tested: intensity of stimulation, duration of stimulation and survival time separating the offset of stimulation from the sacrifice of the animal (see Zhang et al., 1996 for details). These data can be summarized as follows. As a result of electrical stimulation of the left cochlea, FLI was dense in the ipsilateral DCN and moderate in the contralateral DCN and in PVCN on both sides. The density of FLI in DCN and PVCN was constant after survival times ranging from 0 to 2–3 hours. However, a significant decrease of FLI was observed after longer survival times (5 and 6 hours). In the same nuclei, FLI was increased even by short durations of stimulation (5 and 10 minutes), as compared to control rats, although FLI progressively increased for longer stimulations (20 and 45 minutes). In the auditory thalamus, as observed for pure tones, there was virtually no Fos-like immunoreactive neurons in vMGB, whereas FLI was found mainly in PP, dMGB and mMGB. Such distribution of

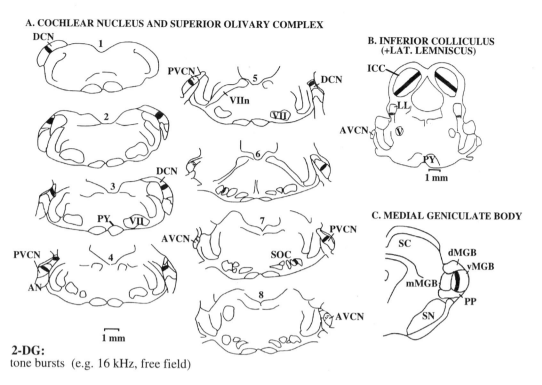

A. COCHLEAR NUCLEUS AND SUPERIOR OLIVARY COMPLEX

B. INFERIOR COLLICULUS
(+LAT. LEMNISCUS)

C. MEDIAL GENICULATE BODY

2-DG:
tone bursts (e.g. 16 kHz, free field)

Figure 2. Schematic representation of 2-DG uptake in the auditory pathway after pure tone stimulation, as derived from data available in the literature (adapted from Webster et al., 1978; Meltzer, 1984; Servière et al., 1984; Reimer, 1993). Dense zone of uptake are represented by bands, while zone of moderate or light uptake are represented by dots (e.g. AVCN). For nomenclature, see list of abbreviations and Figure 1.

FLI in the auditory thalamus was independent of the survival time, ranging from 0 to 6 hours. In both the cochlear nucleus and auditory thalamus, the density of FLI slightly increased in parallel with the intensity of stimulation. In other auditory nuclei, such as the CIC and the DLL, there was no simple relation between the density of FLI and the three tested experimental parameters. From this parametrical study (Zhang et al., 1996), it can be concluded that the distribution and density of FLI did not vary in parallel in the various nuclei of the auditory pathway as a function of the tested experimental parameters; different patterns of FLI changes were instead observed in different auditory nuclei.

3.3 Passive Electrical Stimulation of the Cochlea: Influence of the Fos-Antibody

Using the same animal model of cochlear implant, FLI induced in the auditory pathway by electrical stimulation of the cochlea was established in recent sets of experiments whose goal was to assess the influence of the commercial source of the Fos antibody used to process the brain tissue. The two following antibodies were tested and compared: OS and SCB. The same standard experimental parameters were applied to all animals (45 minutes duration, intensity of 500 µA above EABR threshold and survival time of 1 hour). However, the tissue derived from one subgroup of rats was treated with the OS antibody (using a lot providing optimal results). Later, further lots of the OS antibody were seen to

be less sensitive and, therefore, tissue derived from a second subgroup of rats was processed using the SC antibody (with a lot providing excellent results). Unfortunately, optimal lots of the two antibodies were not available at the same time and therefore it was not
possible to compare directly the two antibodies on the same animal on adjacent sections.

The typical appearance of FLI elicited in the auditory pathway by electrical stimulation
of the cochlea has been previously illustrated in the form of photomicrographs for the DCN,
PVCN, NTB and MGB as well as, for comparison, the FLI present in the same structures in
control animals (see Vischer et al., 1994, 1995; Zhang et al., 1996). Therefore, no photomicrograph of Fos-like positive neurons is shown in the present report. However, illustration of
Fos-like positive neurons are shown in the present volume elsewhere (Hervé-Minvielle et al.)

The general distribution of FLI in the auditory pathway after electrical stimulation
of the cochlea is shown in Figs. 3, 4 and 5, comparing the two antibodies. After treatment
with one or the other antibody, dense FLI was seen in the ipsilateral and contralateral
DCN, with respect to the ear stimulated although, as expected, the number of Fos-like
positive neurons was higher ipsilaterally (Fig. 3, top two panels). In PVCN, a significant
FLI was observed, clearly more prominent on the ipsilateral side than on the contralateral
side (Fig. 3, bottom two panels). In the ipsilateral AVCN, an apparently denser FLI was
observed after treatment with the SCB antibody as compared to the OS antibody (Fig. 4,
top two panels), whereas in the contralateral AVCN FLI was low as seen with both antisera. In the SOC, FLI was observed mainly in the ipsilateral LSO and the contralateral
NTB (Fig. 3, bottom two panels). In the lateral lemniscus and inferior colliculus, a dense
FLI was observed after treatment of the tissue with the SCB antibody while FLI was lower
(light to moderate) after treatment with the OS antibody, in particular in the CIC (Fig. 5).
In contrast, FLI appeared denser in the medial geniculate body (MGB) after treatment
with the OS antibody as compared to the SCB antibody (Fig. 4, bottom two panels).

FLI in DCN, PVCN and SOC

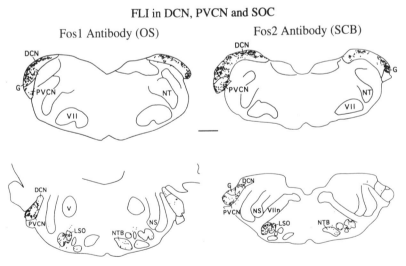

Figure 3. Reconstruction of frontal sections of the brainstem, showing FLI elicited by electrical stimulation of the
left cochlea in DCN, PVCN and the superior olivary complex. Each dot represents a Fos positive neuron. The left
sides of all drawings correspond to the ipsilateral sides with respect to the stimulated cochlea. The two sections on
the left are derived from tissue treated with the OS antibody, while the SCB antibody was used to treat the two
sections shown on the right. Scale bar=1 mm. V=trigeminal motor nucleus; VII=facial motor nucleus; VIIn=facial
nerve; G=granule cell area; NS=sensory nucleus of the trigeminal nerve; NT=nucleus of the spinal tract of the
trigeminal nerve. See also list of abbreviations.

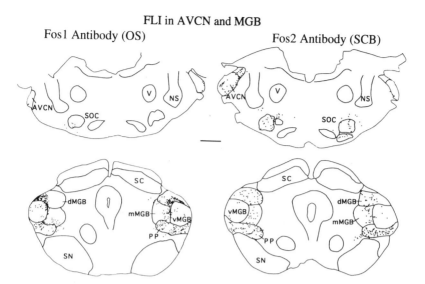

Figure 4. Reconstruction of frontal sections of the brainstem (top two panels) and of the auditory thalamus (botton two panels), showing the distribution of FLI in AVCN and the MGB, respectively. Same conventions as in Figure 3. Each dot represents a Fos-like positive neuron. Scale bar=1 mm. SC= superior colliculus: SN=substantia nigra. See also list of abbreviations.

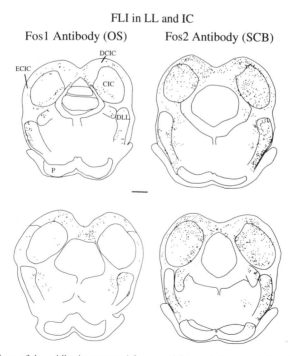

Figure 5. Frontal sections of the midbrain, arranged from caudal to rostral (top to bottom), illustrating the FLI in the LL and IC, as seen after treatment of the tissue with the OS antibody (left column) or with the SCB antibody (right column). Each dot represents a Fos-like positive neuron. Scale bar=1 mm. P=pontine nuclei. See also list of abbreviations.

These data support the idea that the use of one or the other antibody affected the FLI distribution in some auditory nuclei, although the paradigm of stimulation was similar (Figs. 3, 4 and 5). For instance, a stronger FLI was apparently observed in DCN, PVCN, AVCN, LSO and CIC after treatment with SCB antibody as compared to the OS antibody (Figs. 3, 4 and 5). In contrast, FLI observed in the MGB using the OS antibody was denser as compared to that obtained with the SCB antibody (Fig. 4). In other nuclei (e.g. NTB, PP), however, the FLI obtained with the two antibodies appeared generally comparable (Figs. 3 and 4). These observations were confirmed by quantitative measurements of Fos density (Table I).

3.4. Sensorimotor Learning Paradigm

FLI was established in the auditory pathway of rats performing a sensorimotor learning protocol based on the discrimination of complex tones (see present volume, Hervé-Minvielle et al. for details). In contrast to the above paragraph, the stimulation delivered to the cochlea has a behavioral significance to the animal. The essential result of this study is that the distribution of FLI in the subcortical auditory structures was comparable in the group of learning rats and in a group of control rats exposed to the same acoustical stimuli but without significance (the animal had nothing to learn in the latter case). In sharp contrast, a specific FLI activity was observed in the auditory cortex, related to the learning protocol. Another important result of this study was the observation that a minimal duration of exposure to the complex tones (40 burst of 500 ms) was sufficient to elicit Fos-like activity in the auditory pathway that was significantly higher than that observed in unstimulated animals.

Table I. FLI densities (nb. of FLI positive neurons/mm^2) in stimulated rats for the two antibodies tested (SCB and OS)

Auditory nuclei	Location	FLI density stimulated(SCB)		FLI density stimulated(OS)	Statistics	SCB / OS
DCN	IPSI	243.7 ± 35.9	>	189.0 ± 31.6	P < 0.01	1.3
DCN	CONTRA	136.1 ± 35.8	=	117.7 ± 41.2	n.s.	1.2
PVCN	IPSI	126.3 ± 92.3	>	54.2 ± 33.6	P < 0.05	2.3
AVCN	IPSI	130.4 ± 88.2	=	74.7 ± 43.9	n.s.	1.7
AVCN	CONTRA	6.7 ± 2.7	=	13.3 ± 11.2	n.s.	0.5
LSO	IPSI	239.9 ± 67.0	>	103.9 ± 84.4	P < 0.01	2.3
NTB	IPSI	2.7 ± 2.2	=	3.5 ± 2.5	n.s.	1.0
NTB	CONTRA	37.6 ± 32.3	=	62.6 ± 35.8	n.s.	0.6
CIC	IPSI	44.8 ± 29.2	>	8.4 ± 6.8	P < 0.01	5.6
CIC	CONTRA	143.6 ± 75.0	>	28.4 ± 27.5	P < 0.01	5.1
dMGB	IPSI	24.3 ± 11.7	<	95.6 ± 44.1	P < 0.01	0.3
dMGB	CONTRA	21.5 ± 9.6	<	81.5 ± 56.7	P < 0.01	0.3
vMGB	IPSI	12.6 ± 5.3	=	14.8 ± 8.4	n.s.	0.9
vMGB	CONTRA	20.0 ± 4.0	=	15.5 ± 8.1	n.s.	1.3
mMGB	IPSI	34.9 ± 14.1	<	63.6 ± 35.1	P < 0.05	0.5
mMGB	CONTRA	39.7 ± 17.5	=	55.1 ± 30.2	n.s.	0.7
PP	IPSI	119.8 ± 46.5	=	78.8 ± 39.0	n.s.	1.5
PP	CONTRA	98.9 ± 17.5	>	58.0 ± 27.0	P < 0.01	1.7

Note: The contralateral PVCN and LSO were not considered here because the FLI observed in stimulated rats was comparable to that found in control rats, for both antibodies tested.

4. DISCUSSION

A survey of the results presented in the present paper as well as of the studies available in the literature allow to elaborate some characteristics of the FLI elicited in the auditory pathway by stimulation of the cochlea.

4.1. Parameters Influencing Fos Activity

4.1.1. Characteristics of the Stimuli. The effect of the intensity of stimulation (pure tones) was investigated in the CN, SOC and inferior colliculus (IC) of the mouse (Brown and Liu, 1995). These authors reported that, above a threshold ranging from 35–55 dB SPL, the number of Fos-positive neurons increased progressively in the three nuclei for increasing sound pressure levels. A general tendency towards an increase of FLI for increasing intensities of stimulation was also found in our model of electrical stimulation of the cochlea (Zhang et al., 1996), but the dynamic range for such an increase appeared narrower as compared to the dynamic range in response to pure tones. This is consistent with electrophysiological data showing that intensity functions were much steeper in response to electrical stimulation of the auditory nerve than in response to acoustic stimuli (e.g. Kiang and Moxon, 1972; van den Honert and Stypulkowsky, 1987; Horner and Bock, 1984).

The duration of stimulation is also an important parameter affecting FLI elicited in the auditory pathway after stimulation of the cochlea. Empirically, in most studies (see introduction), the duration of stimulation was fixed at 45 to 60 minutes (sometimes 30 or 120 minutes), probably by analogy with 2-DG experiments. There is evidence that such long durations of stimulation are not necessary to elicit FLI in the auditory pathway. In our animal model of cochlear implant, it turned out that a duration of stimulation of 5 minutes was sufficient to elicit a significant FLI, although the density of labelling increased with time, up to 20 minutes (Zhang et al., 1996). In response to tones, it has been demonstrated that a short stimulation of 5 minutes was sufficient (Zuschratter et al., 1995). Such a short stimulation was even a prerequisite to observe tonotopy with Fos in the primary auditory cortex because a longer stimulation would induce a spread of labelling obscuring the tonotopy (Zuschratter et al., 1995). In our paradigm of sensorimotor learning (Hervé-Minvielle et al., present volume), even shorter acoustic stimulations were sufficient to induce a significant FLI, from the cochlear nucleus to the auditory cortex, including the non-primary areas.

There are claims in the literature that Fos is better induced when the stimulus presented is new ("novelty effect"). We have not tested this parameter systematically. In our group of rats stimulated passively with tones, the animals were not exposed before to these stimuli and therefore could be considered as novel. In the electrical stimulation paradigm, this issue was not addressed because the rats were sedated. The question of novelty was addressed indirectly in our sensorimotor learning protocol (Hervé-Minvielle et al., present volume) and it was concluded that significance of the stimulus seems to be a major parameter influencing Fos induction in the auditory pathway, and not novelty.

4.1.2. Survival Time. As for the duration of stimulation, the duration of survival time was fixed empirically by most authors, generally ranging from 0 to 60 minutes. In our animal model of cochlear implant, we have shown that the duration of survival time does not affect dramatically the distribution of FLI, at least for survival times ranging from 0 to 2 hours. For longer survival times, a decrease of FLI was observed in the cochlear nucleus, but not in the auditory thalamus. This means that the time course of Fos induction can vary across auditory nuclei.

4.1.3. Antibody against Fos. Whether the type of antibody used to visualize Fos activity affects the results of the c-fos method is a question that has been addressed previously in the auditory system (Adams, 1995). Two antibodies (OS anti-Fos and UBI anti-Fos) were compared in a semi-quantitative way in the CN, SOC and IC. Our studies confirm quantitatively the conclusions met in the cat (Adams, 1995) and extend them for other nuclei (Table I): "antisera raised against the same nominal antigen can give common results in one nucleus but very different results in another nucleus". The results of both studies (Adams, 1995; present report) may be interpreted in the sense that the Fos antisera do not recognize exclusively the Fos protein, but also related proteins of the same family presenting structural similarities, as discussed by Adams (1995). Then again, it might be that a given antibody can recognize certain antigens in an auditory nucleus and different antigens in another nucleus, while a second antibody can recognize a different combination of antigens in various nuclei.

The strong dependence of FLI on various parameters (stimulation, species) make the comparision of different studies available in the literature difficult. Another important parameter to consider for interpreting c-fos data is the antibody used to visualize the Fos activity, as demonstrated here. If one considers most studies describing Fos activity elicited in the auditory pathway by acoustic or electrical stimulation of the cochlea (Ehret and Fischer, 1991; Friauf, 1992, 1995; Sato et al., 1992, 1993; Rouiller et al., 1992; Reimer, 1993; Pierson and Snyder-Keller, 1994; Vischer et al., 1994, 1995; Brown and Liu, 1995; Adams, 1995; Scheich and Zuschratter, 1995; Zuschrater et al., 1995; Gleich et al., 1995; Zhang et al., 1996), not less than six Fos antisera from different sources have been used and tested.

4.1.4. Anesthesia. FLI can be modified by anesthesia. More precisely, this effect depends on the type of anesthesia (Brown and Liu, 1995). In the cochlear nucleus for instance, Nembutal and Avertin dramatically reduced the number of Fos-like positive neurons while Ketamine had little or no effect. In the present study, no anesthesia was used in the rats subjected to pure tones stimulation as well as in the rats performing the sensorimotor task. In contrast, rats subjected to electrical stimulation of the cochlea were sedated by administration of a low dose of Ketamine, indicating that this light anesthesia did not play an important role, at least for the cochlear nucleus. In the inferior colliculus, it has been observed that different anesthetics gave more variable results than in the cochlear nucleus (Brown and Liu, 1995).

4.1.5 Age of the Animal, Species. There is a report in the literature of an effect of the age of the animal. Friauf (1992) mentioned that a stronger FLI was observed in AVCN in response to tonal stimulation in juvenile rats, as compared to adult rats. Possible differences related to species have not been investigated in detail, since no study was conducted to address specifically this issue. Comparisons across laboratories is difficult because other parameters (e.g. stimulation) may explain discrepancies observed between species, for instance between mouse (Brown and Liu, 1995) and cat (Adams, 1995).

4.2. Can FLI Reflect Inhibition?

Although one cannot be certain that a decrease of electrical activity may lead in some cases to a decrease of Fos-like activity, indeed a significant reduction of FLI associated to the stimulus was found in the ipsilateral NTB and LL after electrical stimulation of the cochlea, as compared to unstimulated (control) animal. Such an inhibition of FLI was observed when using one antibody (SCB), but not the other antibody (OS). An inhibition of Fos activity in the ipsilateral NTB and LL is consistent with the notion that these nuclei receive few inputs or inhibitory inputs from the ipsilateral ear. Tone-evoked inhibition of FLI has been reported recently (Friauf, 1995).

4.3. Identity of Fos-Like Positive Neurons

From several previous studies on the induction of Fos in the auditory system, there is clear evidence for a high degree of specificity of the cell types that respond to stimulation (acoustic or electrical) of the cochlea (e.g. Friauf, 1992; Sato et al., 1992; Rouiller et al., 1992; Vischer et al., 1994; Brown and Liu, 1995; Adams, 1995; Gleich et al., 1995; Zuschratter et al., 1995; Zhang et al., 1996). In other words, only a limited number of cell classes in auditory nuclei do exhibit immunopositivity for Fos in response to cochlear stimulation. In the cochlear nucleus, for instance, there is evidence that Fos positive neurons correspond to small-size neurons, such as the granule cells and small stellate cells (Brown and Liu, 1995; Adams, 1995). A similar trend towards small cells has also been found for FLI in the IC (Adams, 1995). A double-labeling approach allowed to demonstrate that, in the cochlear nucleus of the gerbil, a significant proportion of Fos positive neurons, in response to sound stimulation, were also stained with an antibody directed against GABA (Gleich et al., 1995).

4.4 Comparison with 2-DG

Both c-fos and 2-DG are functional mapping methods. However, they differ on many respects. First, 2-DG provides a regional resolution while c-fos allows to detect individual neurons. Furthermore, 2-DG is believed to reflect, at least in part, an uptake of glucose by astrocytes, in the close vicinity of the neurons activated by the stimulus presented to the animal (Pellerin and Magistretti, 1994). In contrast, Fos corresponds to a direct functional marker for neurons. Another essential difference is that 2-DG is a metabolic marker, indicating which regions of the brain have been activated metabolically. Then 2-DG is a relatively general marker of activity although there are clear differences in the uptake across different activated brain structures. Fos, on the other hand, is a selective marker for specific neuronal populations, with a bias towards small cells (Adams, 1995; Brown and Liu, 1995). This may represent an advantage particularly if one considers that small cells are the ones from which it is most difficult, if not impossible (e.g. granule cells of the cochlear nucleus), to derive single unit electrophysiological recordings. Then, it appears that c-fos and microelectrophysiology are two complementary methods. However, the selectivity of the Fos method for some neuronal subpopulations might be a disadvantage if the goal of the investigation is to establish exhaustively what neuronal populations are activated in relation to a certain stimulus or behaviour. In the auditory system, particulary poor FLI was obtained in the AVCN and vMGB. In the latter nucleus, irrespective of the mode of stimulation, only a very limited number of neurons were Fos-like positive (Rouiller et al., 1992; Vischer et al., 1994; Zhang et al., 1995; Hervé et al., present volume). It appears then that the two functional markers (2-DG and c-fos) are complementary.

4.5 Conclusion

Although the precise role of c-fos is at present not established, besides its general role in a cascade of gene regulations leading to permanent changes of the neuron induced by an external stimulus (suggesting that Fos may reflect plastic changes in the nervous system), this is a valuable method to map activity, in particular in the auditory system. In addition to the advantages mentioned above (cellular resolution, specificity for small cells, etc), one should also consider its low cost as compared to 2-DG. In addition, the c-fos method is easy to apply, even in animals perfoming a complex behavioral protocol (see Hervé-Minvielle et al., present volume), a situation in which chronic electrophysiological recordings are more difficult to obtain. If it will be confirmed that c-fos is a marker of plasticity, it might play an important role in future studies on animals, aimed to understand the mechanisms of functional recovery after trauma (e.g. brain inury, deafness, etc).

ACKNOWLEDGMENTS

The author wishes to thank his collaborators M. Vischer, J. Zhang, A. Hervé-Minvielle, D. Carretta for their significant contribution to the experiments conducted in our laboratory. Thanks are due to V. Moret, A. Tempini and C. Roulin for excellent technical assistance, J. Corpataux and B. Morandi for taking care of the rats in the animal room. The present work was supported by the Swiss National Science Foundation for Scientific Research (grants No 31–25128.88, 31–28572.90, 32–36482.92 and 31–43422.95), the Roche Research Foundation (Basle), the CIBA-GEIGY Jubiläums Stiftung (Basle, Switzerland) and the "Sandoz-Stiftung zur Förderung der medizinisch-biologischen Wissenschaften" (Basel, Switzerland).

ABBREVIATIONS

AVCN, anteroventral cochlear nucleus
CIC, central nucleus of the inferior colliculus
CRB, *c-fos* antibody from Cambridge Research Biochemicals, UK
DCIC, dorsal cortex of the inferior colliculus
DCN, dorsal cochlear nucleus
DLL, dorsal nucleus of the lateral lemniscus
dMGB, dorsal division of the medial geniculate body
EABR, electrically-evoked auditory brainstem responses
ECIC, external cortex of the inferior colliculus
FLI, Fos-like immunoreactivity
IC, inferior colliculus
LL, lateral lemniscus
LSO, lateral superior olivary nucleus
MGB, medial geniculate body
mMGB, medial division of the medial geniculate body
NTB, medial nucleus of trapezoid body
OS, *c-fos* antibody from Oncogene Science, USA.
P, Pontine nuclei
PP, peripeduncular nucleus

PVCN, posteroventral cochlear nucleus
SC, superior colliculus
SCB, *c-fos* antibody from Santa Cruz Biotechnology, USA
SOC, superior olivary complex
VLL, ventral nucleus of the lateral lemniscus
vMGB, ventral division of the medial geniculate body.

REFERENCES

Adams, J.C. Sound stimulation induces Fos-related antigens in cells with common morphological properties throughout the auditory brainstem. J. Comp. Neurol. 361: 645–668; 1995.

Brown, M.C.; Liu, T.S. (1995) Fos-like immunoreactivity in central auditory neurons of the mouse. J. Comp. Neurol. 357: 85–97; 1995.

Dragunow, M.; Faull, R. The use of *c-fos* as a metabolic marker in neuronal pathway tracing. *J. Neurosci. Meth.* 29:261–265; 1989.

Ehret, G.; Fischer, R. Neuronal activity and tonotopy in the auditory system visualized by *c-fos* gene expression. *Brain Res.* 567:350–354; 1991.

Friauf, E. Tonotopic order in the adult and developing auditory system of the rat as shown by *c-fos* immunocytochemistry. *Eur. J. Neurosci.* 4:798–812; 1992.

Friauf, E. C-fos immunocytochemical evidence for acoustic pathway mapping in rats. Behav. Brain Res. *66*: 217–224; 1995.

Gleich, O.; Bielenberg, K.; Strutz, J. Sound induced expression of c-Fos in GABA positive neurones of the gerbil cochlear nucleus. *Neuroreport* 7: 29–32; 1995.

Horner, K. C., Bock, G. R. Inferior colliculus single unit responses to peripheral electrical stimulation in normal and congenitally deaf mice. *Dev. Brain Res.* 15:33–43; 1984.

Hunt, S. P.; Pini, A.; Evan, G. Induction of *c-fos*-like protein in spinal cord neurons following sensory stimulation. *Nature* 328:632–634; 1987.

Kiang, N.Y.S., Moxon, E.C. Physiological considerations in artificial stimulation of the inner ear. *Ann. Otol.* 81: 714–730; 1972.

Le Gal La Salle, G.; Naquet, R. Audiogenic seizures evoked in DBA/2 mice induce *c-fos* oncogene expression into subcortical auditory nuclei. *Brain Res.* 518:308–312; 1990.

Meltzer, P. The central auditory pathway of the gerbil Psammomys Obesus: a deoxyglucose study. Hearing Res. 15: 187–195; 1984.

Morgan, J. I.; Curran, T. Stimulus-transcription coupling in the nervous system: Involvement of the inducible proto-oncogenes *fos* and *jun*. *Annu. Rev. Neurosci.* 14:421–451; 1991.

Paxinos, G.; Watson, C. *The rat brain in stereotaxic coordinates*, New-York:Academic Press, pp. 1–119; 1986

Pellerin, L.; Magistretti, P.M. Glutamate uptake into astrocytes stimulates aeroboc glycolysis: a mechanism coupling neuronal activity to glucose utilization. PNAS 91: 10625–10629; 1994.

Pierson, M.; Snyder-Keller, A. Development of frequency-selective domains in inferior colliculus of normal and neonatally noise-exposed rats. *Brain Res.* 636:55–67; 1994.

Reimer, K. Simultaneous demonstration of Fos-like immunoreactivity and 2-deoxy-glucose uptake in the inferior colliculus of the mouse. *Brain Res.* 616:339–343; 1993.

Robertson, H. A. Immediate-early genes, neuronal plasticity, and memory. *Biochem. Cell Biol.* 70:729–737; 1992.

Robertson, H. A.; Dragunow, M. From synapse to genome: The role of immediate early genes in permanent alterations in the central nervous system. In: Osborne, N.N., ed. Current aspects of the Neurosciences. New-York: Macmillan; 1992:143.

Rouiller, E. M.; Wan, X. S. T.; Moret, V.; Liang, F. Mapping of *c-fos* expression elicited by pure tones stimulation in the auditory pathways of the rat, with emphasis on the cochlear nucleus. *Neurosci. Lett.* 144:19–24; 1992.

Sato, K.; Houtani, T.; Ueyama, T.; Ikeda, M.; Yamashita, T.; Kumazawa, T.; Sugimoto, T. Mapping of the cochlear nucleus subregions in the rat with neuronal Fos protein induced by acoustic stimulation with low tones. *Neurosci. Lett.* 142:48–52; 1992.

Sato, K.; Houtani, T.; Ueyama, T.; Ikeda, M.; Yamashita, T.; Kumazawa, T.; Sugimoto, T. Identification of rat brainstem sites with neuronal Fos protein induced by acoustic stimulation with pure tones. *Acta Otolaryngol. (Stockh.)* 113 Suppl. 500:18–22; 1993.

Scheich, H.; Zuschratter, W. Mapping of stimulus features and meaning in gerbil auditory cortex with 2-deoxyglucose and c-Fos antibodies. Behavioural Brain Res. 66: 195–205; 1995.

Serviere, J., Webster, W.R. and Calford, M.B. Isofrequency labelling revealed by a combined 14C-2-deoxyglucose, electrophysiological and horseradish peroxidase study of the inferior colliculus of the cat. J.Comp.Neurol. 228: 463–477; 1984.

Snyder-Keller, A. M.; Pierson, M. G. Audiogenic seizures induce *c-fos* in a model of developmental epilepsy. *Neurosci. Lett.* 135:108–112; 1992.

Van Den Honert, C., Stypulkowski, P. H. Single fiber mapping of spatial excitation patterns in the electrically stimulated auditory nerve. *Hearing Res.* 29:195–206, 1987.

Vischer, M. W.; Häusler, R.; Rouiller, E. M. Distribution of Fos-like immunoreactivity in the auditory pathway of the Sprague-Dawley rat elicited by cochlear electrical stimulation. *Neurosci. Res.* 19:175–185; 1994.

Vischer, M., Bajo-Lorenzana, V.; Zhang, J.S.; Häusler, R.; Rouiller E.M. (1995) Activity elicited in the auditory pathway of the rat by electrical stimulation of the cochlea. ORL 57, 305–309.

Wan, X. S. T.; Liang, F.; Moret, V.; Wiesendanger, M.; Rouiller, E. M. Mapping of the motor pathways in rats: *c-fos* induction by intracortical microstimulation of the motor cortex correlated with efferent connectivity of the site of cortical stimulation. *Neurosci.* 49:749–761; 1992.

Webster, W.R., Serviere, J., Batini, C. and Laplante, S. Autoradiographic demonstration with 2-14C deoxyglucose of frequency selectivity in the auditory system of cats under conditions of functional activity. Neurosci.Lett. 10: 43–48; 1978.

Zhang, J.S.; Haenggeli, C.A.; Tempini, A.; Vischer, M.W.; Moret , V. and Rouiller, E.M. Electrically induced Fos-like immunoreactivity (FLI) in the auditory pathway of the rats: effects of survival time, duration and intensity of stimulation. *Brain Res. Bull.* 39, 2: 75–82; 1996.

Zilles, K. *The cortex of the rat, a stereotaxic atlas*, Berlin:Springer-Verlag, pp. 1–121; 1985.

Zuschratter, W.; Gass, P.; Herdegen, T.; Scheich, H. Comparison of frequency-specific c-Fos expression and fluoro-2-deoxyglucose uptake in auditory cortex of gerbils (*Meriones unguiculatus*). *Eur.J.Neurosci.* 7:1614–1626; 1995.

4

EVOLUTION OF C-FOS EXPRESSION IN AUDITORY STRUCTURES DURING A SENSORI-MOTOR LEARNING IN RATS

A. Herve-Minvielle,[1,2] D. Carretta,[1,2] V. M. Bajo,[2] A. E. P. Villa,[3] and E. M. Rouiller[1]

[1]Institute of Physiology
University of Fribourg
Rue du Musée 5, CH-1700 Fribourg, Switzerland
[2]Departamento de Biología Celular y Patologia
Facultad de Medicina
Universidad de Salamanca
Avda. del Campo Charro s/n, E-37007 Salamanca, Spain
[3]Institute of Physiology
University of Lausanne
Rue du Bugnon 7, CH-1005 Lausanne, Switzerland

1. INTRODUCTION

The proto-oncogene c-fos is a member of a large class of "immediate early gene" which is rapidly and transiently expressed in neurons in response to a wide variety of external stimuli (Morgan and Curran, 1991). Fos, the protein product of the gene c-fos, acts as a transcription regulator for the expression of other target genes (Morgan and Curran, 1991). In the absence of inputs or strong activations, there is a low basal Fos spontaneous activity (Morgan and Curran, 1991). Thereby, the increase of the c-fos expression can be used as a marker of neuronal activity (Hunt et al., 1987; Sagar et al., 1988; Dragunov and Faull, 1989).

Recently, the Fos mapping technique was applied to explore the auditory pathway and more specifically to identify nuclei involved in the processing of auditory stimuli. FLI was observed in almost all auditory structures, with a variability of density among the structures, following audiogenic seizure (LeGal la Salle and Naquet, 1990; Snyder-Keller and Pierson, 1992) or in response to passive pure tone stimuli (Ehret and Fischer, 1991; Friauf, 1992; 1995; Rouiller et al., 1992; Sato et al., 1992; Adams, 1995; Brown and Liu, 1995; Zuschratter et al., 1995).

Even if c-fos expression has been proposed to be related to plasticity, learning and memory (e.g., Kaczmarek and Nikolajew, 1990; Anokhin and Rose, 1991; Bertaina and

Acoustical Signal Processing in the Central Auditory System
edited by Syka, Plenum Press, New York, 1997

Destrade, 1995; Zhu et al., 1995), few studies have described c-fos expression linked to learning. However, it has been shown that induction of FLI could be obtained following specific learning paradigms; in particular, c-fos was used to study learning processing associated to conditioned taste aversion (Yamamoto, 1993; Schafe et al. 1995; Swank et al. 1995) and the implication of hippocampus in odor discrimination (Hess et al. 1995) and in active avoidance response (Nikolaev et al., 1992). Thus, c-fos expression has been shown to provide an effective index of the sequence of the brain structures which are activated during different learning tasks.

Along the same line, and in contrast to previous auditory Fos studies in which the acoustic stimuli were delivered passively, the aim of the present investigation was to establish in rats how and when distinct auditory structures are activated during the learning of a complex sensori-motor task based on behaviorally relevant sounds. In other words, FLI was used to detect changes in neuronal activity related to learning and significance of acoustic stimuli.

2. MATERIALS AND METHODS

Experiments were performed on 25 young male adult Long Evans rats. The experimental behavioral set-up consisted of a rectangular plastic black box (560mm long x 395mm wide x 395 mm high) placed in a soundproof chamber under dim light. A line of colored tape was affixed to the floor of the box in order to delimit at one extremity of the box a rest area. At the opposite side, the feeder area consisted of a restrained place where the rat had access to food reward. Two loudspeakers were placed each in the lateral walls of the box at the limit of the rest area. A Macintosh computer was used to control the experiments and to store the data. The behavioral paradigm, a "GO/no-GO task" elaborated by Villa et al. (1994) is subdivided into three main phases.

In the phase I, the rats were first trained to make the association between a tone and the food without position limitation. Then, in a second step, the animal had to learn that its position was crucial for the delivery of the stimulus: the tone was presented only when the rat remained behind the colored line. The stimulus used in this phase was a complex tone with two dominant frequencies (4 and 7 kHz).

During the phase II, two complex tones with a dominant frequency of 3 kHz and 10 kHz, respectively, were delivered pseudo-randomly in the right loudspeaker. The rats had to make a discrimination between the two sounds: only the presentation of the higher pitch sound (but not the lower pitch one) was followed by a reward.

During the phase III, the same two sounds used in phase II were delivered simultaneously (one in each loudspeaker) in different combinations. In addition to the spectral discrimination of phase II, the rat had to make a spatial discrimination: only when the higher pitch sound was delivered in the right loudspeaker, the rat was rewarded.

In order to dissociate stimulus-related activity from learning-related activity and therefore to isolate cognitive significance processing, control and learning rats received the same number of each type of stimuli but control animals received the acoustic stimuli pseudo-randomly, without significance (no association between tone and food).

All animals performed daily session of 20–30 minutes except in phase III in which there was until three sessions per day. In all cases, a flashing light was triggered when the rat rested in the feeding area or when the animal made a mistake during the training.

Two learning rats were sacrificed at three different stages of each learning phase (beginning, middle and end). In addition, one rat was used as control for each phase. The day of the sacrifice, immediately at the end of the experimental session, the rat was placed for one hour in a soundproof chamber, in absence of sensory stimulation. Then, the animal was deeply anaesthetised (100mg/kg pentobarbital i.p.) and perfused intracardially with 200 ml of saline solution and 800 ml of 4% paraformaldehyde in phosphate buffer (PB, 0.1M, pH=7.4). Brains were dissected, postfixed for 2 hours in paraformaldehyde and cryoprotected by immersion in a 10% solution of sucrose in PB overnight and later in a 30% solution of sucrose in PB for several days. Then, the brains were cut with a freezing microtome into 50 μm thick frozen sections in frontal plane. One section every 150 μm was stained with cresyl violet and a second every 150 μm section was treated immunocytochemically to reveal FLI, as described elsewhere (Wan et al. 1992). The primary antibody used was supplied by Santa Cruz (diluted 1:3000 in PB-triton X-100 with 1% normal donkey serum). The third series of sections was kept as reserve.

We focused our studies on the auditory structures which presented in our hands a constant and sustained labelling after cochlear stimulation (Rouiller et al., 1992; Zhang et al., 1996) : the dorsal cochlear nucleus (DCN), the lateral nucleus of the superior olivary complex (LSO), the inferior colliculus (IC) subdivided in the central nucleus (CIC), the dorsal cortex (DCIC) and the external cortex (ECIC), the medial geniculate body (MGB) comprising the dorsal (dMGB), ventral (vMGB) and medial (mMGB) divisions, the peripeduncular nucleus (PP), and the fields Te1, Te2, Te3 of the auditory cortex. The immunoreactive neurons were counted in each structures and the data were normalized with respect to the area of the structure examined. Data were analysed at that steps in a semi-quantitative manner.

3. RESULTS

FLI was present in all auditory structures examined except in the ventral division of the MGB where it was virtually absent. In all subcortical structures (from the DCN to the thalamus), learning and control animals presented a comparable FLI distribution. In addition, location and density of FLI in the subcortical structures did not show any clear change related to the different phases of the learning task.

In DCN, FLI neurons were arranged in the form of restricted bands or elongated clusters parallel to isofrequency contours. In LSO, FLI was observed in the lateral and central part of the nucleus, whereas no immunoreactive neurons were seen in the medial part. In CIC, the Fos-positive cells were not uniformly distributed: FLI neurons were clearly more numerous in the dorsal part of the nucleus as compared to the ventral zone. A strong FLI was also detected in DCIC and ECIC. All divisions of the MGB presented a diffuse FLI, except the ventral division where only sparse immunoreactive neurons were present. An intense and homogeneous FLI was systematically observed in PP. This general pattern of FLI in the DCN, LSO, IC, MGB and PP is illustrated in Figure 1.

In contrast to subcortical structures, FLI in the auditory cortical fields (Te1, Te2, Te3) was clearly different in the group of learning animals as compared to the group of control animals. The main differences between learning and control rats were observed in the cortical field Te1, at the middle of phase II and III, and in the cortical area Te2, at the end of phase I and at the middle of phases II and III. In addition, in the group of learning

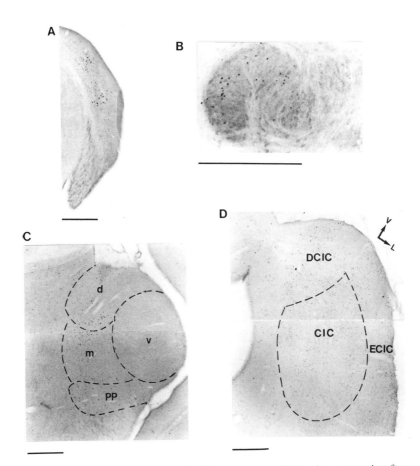

Figure 1. Illustration of typical FLI in subcortical auditory structures: All the pictures are taken from a learning animal sacrificed at the middle of phase II. A similar FLI distribution was observed in all animals : learning and control rats. A. Photomicrograph of FLI in the caudal part of the right DCN. Note the distribution of positive neurons in clusters. B. Photomicrograph of FLI in the left LSO. Note that the medial region of the nucleus is devoid of immunostained cells, as expected from the tonotopy of the LSO. C. Photomicrograph of the right MGB and PP. Note the presence of moderate FLI in the dMGB (d), mMGB (m) and PP, whereas in the vMGB (v) the immunopositive neurons were virtually absent. D. Photomicrograph of FLI taken in the middle of the rostrocaudal extent of the right IC. Note the strong immunostaining in the dorsal and the external cortices. Note also that in the CIC immunoreactive cells are more numerous in the dorsal part of the nucleus as compared to the ventral one. Orientation of IC (V = vertical; L = lateral). Scale Bar = 500 µm.

rats, the distribution and density of FLI varied through the successive phases of learning (Figure 2).

For instance, in the cortical field Te1, FLI showed a sustained high level in phases I and II followed by a decrease in phase III. In contrast, FLI in Te2 and Te3 increased at the end of the phase I and at the beginning of phase II, then decreased during the last part of phase II and in phase III. Clearly, the time course of FLI during the three phases of learning was different for Te1 on one hand and the non-primary areas (Te2 and Te3) on the other hand. The distribution of FLI found in the auditory cortex is summarized in a semi-quantitative way in Table 1.

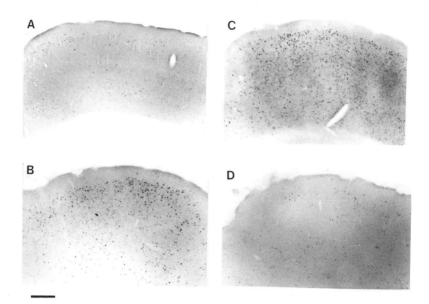

Figure 2. Illustration of FLI changes in the auditory cortical area Te2 between learning and control rats and across the learning phases. Photomicrographs of FLI in the left cortical auditory area Te2 in: A. a control rat sacrificed at phase I. B. a learning rat sacrificed at phase I. C. a learning rat sacrificed at phase II. D. a learning rat sacrificed at phase III. Note the difference of FLI between control and learning rats (Fig 2A and 2B) and the change of FLI as a function of the learning time course (Fig 2B, 2C and 2D). Scale Bar = 500 μm.

4. DISCUSSION

The results of the present experiments showed that it possible to detect c-fos expression in all auditory structures at subcortical and cortical levels in spite of the very short total duration of acoustic stimulation (about 40 bursts of 500 ms). Thus, in a sensori-motor learning paradigm, less than 30 seconds of an acoustic presentation was sufficient to induce a substantial FLI. The location of FLI in such learning conditions was generally comparable in subcortical areas to the FLI observed after passive tonal or electrical stimulations of the cochlea. It is important to mention that such studies were aimed to explore

Table 1. Semi-quantitative analysis of Fos-immunoreactive neurons in the auditory cortical fields Te1, Te2, Te3 of control (c) and learning animals

	Phase 1			Phase 2			Phase 3		
	Beginning	Middle	End	Beginning	Middle	End	Beginning	Middle	End
Te1	++++	++++	++++	++++	++++	+++	++	++	+
	c ++++		c ++++		c +			c +	
Te2	++	+++	++++	++++	++	+	+	+	0
	c ++		c +		c +			c 0	
Te3	+++	+++	++++	++++	+++	++	++	++	+
	c +++		c++		c +			c +	

The density has been normalised with respect to the maximal FLI density observed in each cortical area. Symbols: 0, 0–10%; +, 10–30%; ++, 30–50%; +++, 50–70%; ++++, 70–100%. 10% corresponded to 30 cells/mm² for Te1, 35 cells/mm² for Te2, and 20 cells/mm² for Te3.

the tonotopy of the auditory system and how physical parameters of the stimulus are represented (e.g., Ehret and Fischer, 1991; Rouiller et al., 1992; Sato et al., 1992; Brown and Liu, 1995; Adams, 1995; Friauf, 1995; Zhang et al., 1996). In these studies, the stimuli had no behavioral significance, no cognitive aspects. In contrast, in the present experiments based on a reward paradigm, we focused on the meaning of the stimulus and where it is encoded. We found that cortical areas, but not the subcortical structures, appeared to be involved in the processing of cognitive significance of the stimulus. These data are in agreement with recent results of Scheich and co-workers who concluded that primary auditory cortex in gerbil is a structure implicated in the formation of behaviorally relevant association (Scheich and Zuschratter, 1995; Zuschratter et al., 1995).

It has been claimed that Fos is induced primarily by novel stimuli. Such novelty effect does not play an important role in the present data. For instance, FLI in Te1 was as high in the middle and end of phase I than at the beginning. In Te2 and Te3, even increased at the end of phase I.

In the literature, only an aversive paradigm has been tested using the c-fos technique to address this issue and the correlation between changes in the pattern of expression of FLI and learning (Maleeva et al., 1989; Tichmeyeyer et al., 1990; Campeau et al., 1991; Nikolaev et al. 1992; Yamamoto, 1993; Schafe and Seeley, 1995; Swank et al. 1995).

We found that the time course of Fos induction could be detected in non aversive learning protocol since we observed different FLI in the cortical fields throughout the learning and the information processing. Indeed, FLI changed through the successive phases with a different time course for the three auditory cortical areas. The results indicate that the cognitive aspects of auditory stimuli seem to be processed mainly at the cortical level.

ACKNOWLEDGMENTS

We thank V. Moret and M. Capt for their assistance in immunocytochemistry and histology, J. Eriksson for his contribution to earlier development of the behavioural task. The software for data collection and analysis was designed by C. Eriksson and J. Eriksson. This reseach was supported by EU CHRX-CT93–0269 and Swiss OFES93.0241 grants.

REFERENCES

Adams, J.C. (1995) Sound stimulation induces fos-related antigens in cells with common morphological properties throughout the auditory brainstem. J. Comp. Neurol. 361, 654–668.

Anokhin, K.V. and Rose, S.P.R. (1991) Learning-induced increase of immediate early gene messanger RNA in the chick forebrain. Eur. J. Neurosci. 3, 162–167.

Brown, M.C. and Liu, T.S. (1995) Fos-like immunoreactivity in central auditory neurons of the mouse. J. Comp. Neurol. 357, 85–97.

Bertaina, V. and Destrade, C. (1995) Differential time courses of c-fos mRNA expression in hippocampal subfields following acquisition and recall testing in mice. Cogn. Brain Res. 2, 269–275.

Campeau, S., Hayward, M.D., Hope, B.T., Rosen, J.B., Nestler, E.J. and Davis, M. (1991) Induction of the c-fos proto-oncogene in rat amygdala during unconditioned and conditioned fear. Brain Res. 565, 394–352.

Dragunow, M. and Faull, R. (1989) The use of c-fos as a metabolic marker in neuronal pathway tracing. J. Neurosci. Meth., 29, 261–265.

Ehret, G. and Fischer, R. (1991) Neuronal activity and tonotopy in the auditory system visualized by c-fos gene expression. Brain Res. 567, 350–354.

Friauf, E. (1992) Tonotopic order in the adult and developing auditory system of the rat as shown by c-fos immunocytochemistry. Eur. J. Neurosci. 4, 798–812.

Friauf, E. (1995) C-fos immunocytochemical evidence for acoustic pathway mapping in rats. Behav. Brain Res. 66, 217–224.

Hess, U.S., Lynch, G. and Gall, C.M. (1995) Regional patterns of c-fos mRNA expression in rat hippocampus following exploration of a novel environment versus performance of a well-learned discrimination. J. Neurosci. 15(12), 7796–7809.

Hunt, S.P., Pini, A. and Evan, G. (1987) Induction of c-fos-like protein in spinal cord neurons following sensory stimulation. Nature 328, 632–634.

Kaczmarek, L. and Nikoljew, E. (1990) c-fos protooncogene expression and neuronal plasticity. Acta Neuro. Exp. 50 (4–5), 173–179.

Le Gal La Salle, G. and Naquet, R. (1990) Audiogenic seizures evoked in DBA/2 mice induce c-fos oncogene expression into subcortical auditory nuclei. Brain Res. 518, 308–312.

Maleeva, N.E., Ivolgina, G.L., Anokhin, K.V. and Limborskaja, S.A. (1989) Analysis of the expression of the c-fos protooncogene in the rat cerebral cortex during learning. Genetica 25, 1119–1121.

Morgan, J.I. and Curran, T. (1991) Stimulus-transcription coupling in the nervous system: Involvement of the inducible proto-oncogenes fos and jun. Annu. Rev. Neurosci. 14, 421–451.

Nikolaev, E., Werka, T. and Kaczmarek, L. (1992) c-fos protooncogene expression in rat brain after long term training of two-way active avoidance reaction. Behav. Brain Res. 148, 91–94.

Rouiller, E.M., Wan, X.S.T., Moret, V. and Liang, F. (1992) Mapping of c-fos expression elicited by pure tones stimulation in the auditory pathways of the rat, with emphasis on the cochlear nucleus. Neurosci.Lett. 144, 19–24.

Sagar, S.M., Sharp, F.R. and Curran, T. (1988) Expression of C-Fos protein in brain: metabolic mapping at the cellular level. Science 240, 1328–1331.

Sato, K., Houtani, T., Ueyama, T., Ikeda, M., Yamashita, T., Kumazawa, T. and Sugimoto, T. (1992) Mapping of the cochlear nucleus subregions in the rat with neuronal Fos protein induced by acoustic stimulation with low tones. Neurosci.Lett. 142, 48–52.

Schafe, G.E. and Seeley, R.J. (1995) Forebrain contribution to the induction of a cellular correlate of conditioned taste aversion in the nucleus of the solitary tract. J. Neurosci., 15(10), 6789–6796.

Scheich, H. and Zuschratter, W. (1995) Mapping of stimulus features and meaning in gerbil auditory cortex with 2-desoxyglucose and c-fos antibodies. Behav. Brain Res., 66, 195–205.

Snyder-Keller, A.M. and Pierson, M.G. (1992) Audiogenic seizures induce c-fos in a model of developmental epilepsy. Neurosci. Lett. 135, 108–112.

Swank, M.W. Schafe, G.E. and Bernstein, I.L. (1995) c-Fos induction in response to taste stimuli previously paired with amphetamine or LiCl during taste aversion learning. Brain Res. 673, 251–261.

Tischmeyer, W., Kaczmarek, L., Strauss, M., Jork, R. and Matthies, H. (1990) Accumulation of c-fos mRNA in rat hippocampus during acquisition of a brightness discrimination. Behav. Neural. Biol. 54,165–171.

Villa A.E.P., Eriksson J., Eriksson C. (1994) Analysis of conflict behavior in the rat to detect early communication in the auditory system, Europ. J. Neurosci. Suppl 7: 172.

Wan, X.S., Liang, F., Moret, V., Wiesendanger, M. and Rouiller E.M. (1992) Mapping of the motor pathways in rats: c-fos induction by intracortical microstimulation of motor cortex correlated with efferent connectivity of the site of cortical stimulation. Neuroscience 49, 749–761.

Yamamoto, T. (1993) Neuronal mechanisms of taste learning. Nerosci. Res. 16, 181–185.

Zhang, J.S., Haenggeli, C.A., Tempini A., Visher M.W., Moret V. and Rouiller E.M. (1996) Electrically induced Fos-like immunoreactivity in the auditory pathway of the rat: Effects of survival time, duration and intensity of stimulation . Brain Res. Bull. 39, 75–82.

Zhu, X.O., Brown, M.W., McCabe, B.J. and Aggleton, J.P. (1995) Effects of the novelty or familiarity of visual stimuli on the expression of the immediate early gene c-fos in rat brain. Neuroscience 69, 821–829.

Zuschratter, W., Gass, P., Herdegen, T. and Scheich, H. (1995) Comparison of frequency-specific c-Fos expression and fluoro-2-deoxyglucose uptake in auditory cortex in gerbils (Meriones unguiculatus). Eur. J. Neurosci. 7, 1614–1626.

5

RELATIVE AND ABSOLUTE PITCH PERCEPTION EXPLAINED BY COMMON NEURONAL MECHANISMS

Gerald Langner

Institute of Zoology
THD
Schnittspahnstr. 3, D- 64287 Darmstadt, Germany

1. INTRODUCTION

Pitch is a perceptual attribute of acoustic signals which like the visual attribute colour may be used to distinguish and characterize objects in our environment. However, unlike the fundamentals of colour perception the fundamental neuronal mechanisms of pitch perception are still under debate. Moreover, there is still no consensus about the actual physical parameter corresponding to pitch. Some theories (Terhardt, 1972; Wightman, 1973; Goldstein, 1973) are based on the assumption that frequency components resolved by the auditory analysis are essential for pitch perception. Other theories assume that a temporal analysis of periodicity information has to supplement the restricted frequency analysis.

Nowadays, a variety of arguments are in favor of a neuronal periodicity analysis. It is well known, that many acoustic signal sources generate periodic vibrations. While the waveform of the emitted sounds is crucial for their timbre, the period of their envelope is essential for their pitch. To a first approximation, this 'periodicity pitch' (Schouten, 1940, 1970) is the same for all signals with the same period -including pure tones. Therefore one may say that the relation of pure tone frequency to pitch is unambiguous, because every frequency is related to a certain pitch, but not the other way round: a certain pitch may be elicited by an infinite variety of signals of different frequency composition. While this argument only demonstrates that a periodicity analysis may offer an attractive solution for the pitch problem, results of neurophysiological experiments indicate that periodicity information is indeed coded temporally in the auditory periphery and represented topographically in the central auditory system (Schreiner and Langner, 1988).

It will be discussed below which details of pitch perception and neuronal coding properties may be explained by a neuronal correlation model (Langner, 1981, 1983, 1988, 1992). It is proposed here that this model may even be adequate to explain the relation between absolute and relative pitch perception. While possessors of relative pitch can only judge harmonic relations of sounds, possessors of absolute pitch can tell the absolute value of a tone without use of a particular memory (Barnea et al., 1994, Burns and Campbell, 1994). The immediate perception of absolute pitch without any conscious effort may be compared to that of colour perception, as the pianist Glenn Gould once remarked (Kazdin, 1989). Obviously, the crucial point for any theory of absolute pitch perception is to explain the nature of the internal reference, which is necessary for such effortless pitch 'measurements'. The theory presented provides such a reference and moreover, allows predictions about perceptual differences between absolute and relative pitch which can be verified in psychophysical experiments (Langner and Bleeck, 1996, Bleeck and Langner, this volume).

2. A THEORY OF PERIODICITY CODING

2.1. Temporal Coding of Periodicity in the Auditory Brainstem

The neuronal circuit presented below (Fig. 1) may be thought of as a simplified building block of a correlation network composed of many thousands of similar circuits. The neurons in this network are supposed to represent real neuron types of the auditory brainstem (Langner, 1988) which are known to be much more complicated and to have much more connections than the model suggests. The only purpose of the model is to explain the properties of periodicity coding neurons in the auditory midbrain. For this purpose properties of the peripheral auditory system have also to be included.

For example, the model takes into consideration, that auditory nerve fibers synchronize their activity in response to frequencies up to about 5 kHz and code harmonic signals

A Neuronal Model for Periodicty Analysis

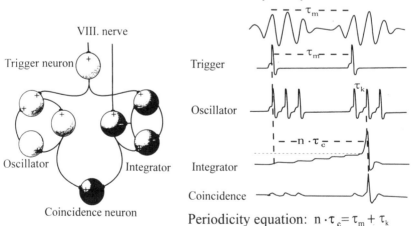

Periodicity equation: $n \cdot \tau_c = \tau_m + \tau_k$

Figure 1. The oscillator and integrator neurons are synchronized due to their common inputs from a trigger neuron to the signal envelopes. The delays resulting for the integrator from temporal integration over the signal fine structure and for the oscillator from intrinsic oscillations have to be compensated by the period of the signal for the coincidence unit to be activated. Therefore the integration time of the integrator neuron is crucial for the best modulation frequency of the circuit (relative mode). The difference for absolute pitch processing lies in the function of the oscillator. When oscillator neurons are used with periods equal to the signal period the oscillator will 'resonate' and thereby may provide the auditory system with an absolute time reference.

or amplitude modulations (AM) by synchronizing their activity to the periodic envelopes of such signals. With amplitude modulations as probe signals, the nerve fibers may be characterized as low pass filters with a cutoff-frequency at 800 Hz (Palmer, 1982)

When judged by their synchronization to the stimulus, the modulation transfer functions of certain types of neurons (onset-, chopper- and pauser-cells) in the ventral cochlear nucleus (VCN) resemble those of low-pass filters at low intensity levels and of bandpass filters at higher intensities (Frisina et al., 1990). Onset-cells are activated in response to pure tones only at the onset. In the model such neurons play an important role for periodicity coding as 'trigger neurons'. After detecting an increase of envelope amplitude (which may include spectral integration over a broad frequency range) these neurons trigger activity in both the oscillator and in the integrator circuit. Whether octopus or certain stellate cells (onset chopper-cells)—which both seem to have the right properties—are really functioning in this sense remains to be shown.

Neuronal recordings in VCN of several species have revealed regular intrinsic oscillations with intervals in the millisecond range (Pfeiffer, 1966). A variety of chopper neurons, corresponding anatomically to stellate cells, with slightly different temporal properties may be distinguished (Rhode et al., 1983; Oertel et al., 1990). Their oscillations may be triggered individually by periodic signals or they may show 'striking interactions' (Frisina et al., 1990) with the stimulus period, addressed here as 'resonances'. In the model, such chopper neurons are assumed to produce intrinsic oscillations with delays at multiples of a synaptic delay of 0.4 ms. For this reason the model includes recurrent connections between chopper neurons with synaptic delays of 0.4 ms. The idea is, that this could be the shortest possible delay for chemical synapses in different species and that such feedback may be suitable to enforce certain preferred intervals in neurons with oscillatory membrane properties. Note, that this preference is essential not for the theory but for the explanation of such preferred intervals found in the auditory midbrain of guinea fowl, mynah bird, cat (Langner, 1981, 1983; Langner and Schreiner, 1988), and also in results of pitch-shift experiments in humans (Fig. 3).

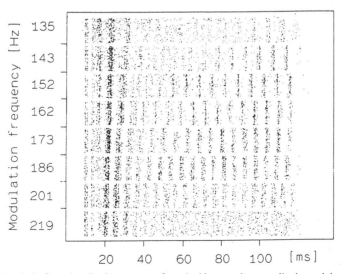

Figure 2. A 'point plot' of synchronized responses of a coincidence unit to amplitude modulated signals (carrier frequency = center frequency = 1.8 kHz) with various modulation frequencies f_m. The onset effect results in temporally increasing synchronization for low f_m (e.g. 143 Hz) and decreasing synchronization for high f_m (219 Hz). This example of a response was obtained from a coincidence neuron in the ICC of Guinea fowl.

The properties of pauser neurons- which correspond to fusiform cells in the dorsal cochlear nucleus- including their inputs from the auditory nerve and from inhibitory interneurons play a role in the 'integrator circuit' of the model. These neurons are largely inhibited by pure tones even at their center frequency unless presented near threshold. In contrast, they synchronize very precisely to amplitude modulations (Zhao and Liang, 1995). In the model, this synchronization to the signal envelope is a result of the trigger input and includes temporal integration of the signal fine-structure (e.g. the carrier frequency of AM). As a consequence of the time constant of their cell membrane, they generate action potentials with delays corresponding to integer multiples of the carrier period. This part of the model makes use of the old, but still recognized, 'volley principle' describing a parallel coding of several nerve fibers which results in an adequate temporal representation of frequencies probably up to 5 kHz (Wever, 1949). Important for the neuronal correlation analysis is that chopper and pauser neurons introduce delays which are necessary for such an analysis of periodic signals, chopper neurons by means of their intrinsic oscillations, and pauser neurons by means of temporal integration.

2.2. Temporal Coding of Periodicity in the Auditory Midbrain

Chopper and pauser neurons are known to project directly to the central nucleus (ICC) of the inferior colliculus (Moore and Osen, 1979) and their projection areas in the so-called isofrequency planes do at least partly overlap (Jähn-Siebert and Langner, 1995). In the model it is assumed that they converge on the same neurons which are functioning as coincidence detectors. Neurons which are tuned to a certain modulation frequency and have response properties appropriate for coincidence detectors were found in the ICC of guinea fowl, cat, chinchilla, and gerbil (Langner, 1992; Heil et al., 1995).

In order to activate coincidence neurons maximally, their input has to be synchronized. Therefore, a high spike rate in coincidence neurons indicates that the periodicity of the acoustic signal is adequate to synchronize their input neurons and to compensate for oscillation and integration delays of their inputs. Only when both synchronization and delay compensation are optimal, a good response of coincidence neurons may be obtained. The modulation period τ_m which is suitable for compensating the delay may be computed from a linear combination of the periods τ_c of the carrier and τ_k of the intrinsic oscillation (Langner, 1983):

$$m*\tau_m = n*\tau_c - \tau_k \text{ (periodicity equation)},$$

where m and n are small integers (m predominantly equal to 1). The integration interval $n*\tau_c$ is the result of integration over activity of n periods of fine structure in the auditory nerve. This interval may be constant over a large intensity range because the responses in many auditory nerve fibers are saturated for sound pressure levels higher than about 30 dB above threshold (Smith, 1979). Only during the on-response is the activation higher and, consequently, should drive the integrator faster to threshold. As a result, during the onset the integration interval and within the required compensating delay are shorter. In other words: the periodicity equation predicts that coincidence neurons prefer high modulation frequencies immediately after stimulus onset and lower ones later. Such onset effects were indeed commonly observed in neuronal recordings of neurons in the ICC (Fig. 2).

The result of coincidence detection is that temporal information is transferred into a rate code. This does not imply that all temporal information is lost, but synchronization is strongly reduced. Obviously, as a result of the temporal analysis, periodicity is represented already in the ICC topographically and orthogonal to the tonotopic organization (Schreiner and Langner, 1988). Finally, using magnetoencephalography, it was found that periodicity pitch and frequency are arranged orthogonally in the human auditory cortex—in accordance with the topographic arrangements in animals (Langner et al., 1997).

3. A THEORY OF PERIODICITY PITCH

3.1. Relative Pitch

The pitch of a harmonic signal corresponds to the fundamental frequency, even when the fundamental frequency is missing ('missing fundamental' effect). Periodicity analysis seems to be the right way to explain this effect. However, only to a first approximation is the pitch of an amplitude modulated signal equal to the pitch of the modulation frequency. To a second approximation pitch is also a function of the carrier frequency (Schouten, 1970). As can be seen by the periodicity equation (see above), this is also true for the optimal response of appropriate coincidence neurons in the auditory midbrain. Consequently, the activation pattern over the auditory midbrain will shift as a function of the carrier frequency with a corresponding effect probably on pitch perception. To a third approximation, periodicity pitch is not a function of the carrier frequency f_c alone. A systematic deviation from a simple relation was found (de Boer, 1956). In the periodicity equation intervals from the intrinsic oscillations contribute to the optimal periodicity which may account for this effect.

The matching lines in Fig. 3 are predictions from the periodicity equation following the assumption that 0.8 ms would be appropriate for the measured pitch range ($f_m = 200$ Hz, $\tau_m = 5$ ms).

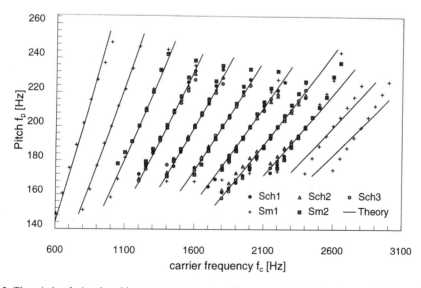

Figure 3. The pitch of signals with a constant envelope frequency of 200 Hz, i.e. amplitude modulations (Schouten et al., 1962) and two-tone complexes (Smoorenburg, 1970). The pitch shifts as a function of carrier frequency (resp., lowest frequency component). The data points were replotted from the original publications. The fitting lines show the approximation computed from the model as explained in the text.

Following the periodicity equation given above, each coincidence neuron which has access to the appropriate integration interval

$$n*\tau_c = \tau_m + 0.8 \text{ ms}$$

would be activated ($m = 1$). From the harmonic case (with $f_c/f_m = \tau_m/\tau_c = n_h$ integer) an average (non-integer) n may be computed: $n = (\tau_m + 0.8 \text{ ms})/\tau_c = (\tau_m + 0.8 \text{ ms}) n_h/\tau_m = 1.16*n_h$.

Assuming that this parameter stays constant for small variations of the carrier frequency, the period corresponding to the pitch may be computed by $\tau_p = 1.16*n_h \tau_c - 0.8$ ms.

As Fig. 3 shows this equation offers a good fit to the data without additional fitting parameters. For large n_h (or large carrier frequencies) the fit could be improved by assuming larger intrinsic oscillation periods.

Finally, the theory predicts another pitch shift which should be observable during signal onset. As described above, the integration delay defines the preferred modulations of coincidence neurons (Fig. 2). The theory suggests that this delay is a function of the activity in the auditory nerve. As a result of the on-response during the first 30 ms there should be a varying integration delay, a change of modulation preferences in coincidence neurons and therefore also a corresponding pitch shift (Langner and Bleeck, 1996, Bleeck and Langner, this volume).

3.2. Absolute Pitch

A theory of absolute pitch perception has to answer the question of the neuronal time reference. The present theory may provide such a reference by use of intrinsic oscillations whose intervals are independent of external stimuli. If possessors of absolute pitch rely exclusively on a temporal analysis based on resonances of intrinsic oscillations, then it might be conceivable how they could tell the absolute value of a tone without making use of a particular tone memory.

Following this hypothesis, the difference between relative and absolute pitch mode is that in relative mode the oscillations are triggered and have smaller periods than the triggering stimulus period (compare Fig. 1), while in absolute mode the resonating oscillations are used with periods equal to that of the triggering waveform of the envelope. Since the period of the waveform is independent of the signal fine structure, in absolute mode the periodicity pitch elicited by an amplitude modulation should stay constant in spite of changes of the carrier frequency. Because the waveform period is also unaffected by the onset response there should also be no onset pitch shift (Langner and Bleeck, 1996, Bleeck and Langner, this volume).

One may argue that the intrinsic oscillations as known from neurons in the cochlear nucleus in various animals are not precise enough to serve as frequency standards. However, it is hard to see where a more precise reference would be available in the brain for this purpose and, in addition, ensemble processing and averaging may improve this situation.

With intrinsic oscillations as a reference for absolute pitch one would expect subjective pitch variations due to physiological conditions. There are indeed reports of effects of the body temperature or hormonal state (Wynn, 1971, Harrer et al., 1979), while corresponding experiments failed (Emde and Klinke, 1977).

4. CONCLUSIONS

- Neurophysiological, anatomical, and psychophysical evidence support a theory of a neuronal correlation or periodicity analysis.
- The correlation theory assumes that signals are represented by synchronized neuronal activity in the auditory nerve and processed by intrinsic oscillations and temporal integration in the cochlear nucleus.
- Neurons in the auditory midbrain act as coincidence detectors and transfer the temporal information into a rate-place code.
- The resulting periodicity or pitch map is arranged orthogonal to the frequency or tonotopic map.
- This correlation theory is able to explain pitch shift effects and may provide an explanation for the difference between relative and absolute pitch perception.

5. REFERENCES

Barnea, A., Granot, R., and Pratt, H. (1994) Absolute pitch-electrophysiological evidence, Int. J. Psychophysiol. 16, 29–38.

Boer, E.de (1956) Pitch of inharmonic signals, Nature 178, 535–536.

Burns, E.M. and Campbell, S.L. (1994) Frequency and frequency-ratio resolution by possessors of absolute and relative pitch: Examples of categorical perception, J. Acoust. Soc. Am., 96 2704–2719.

Emde, C. and Klinke, R. (1977) Does absolute pitch depend on an internal clock? In: M. Portman and J.M. Aran (eds.) INSERM, vol. 68, 145 - 146.

Frisina, R.D., Smith, R.L., and Chamberlain, S.C. (1990) Encoding of amplitude modulation of the gerbil cochlear nucleus: I. A hierarchy of enhancement, Hearing Res. 44, 99–122.

Goldstein, J.L. (1973). An optimum processor theory for the central formation of the pitch of complex tones. J. Acoust. Soc. Am. 54, 1496–1516.

Harrer, G., Harrer, H., Mayr, A. (1979) Die Beeinflussbarkeit des absoluten Gehörs und des rhythmisch-musika-lischen Zeitgefühls durch besondere Einwirkungen (Alkohol bzw. Überwärmung). In: C. Simon (Ed.), Mensch und Musik, Müller, Salzburg.

Heil, P., Schulze, H., and Langner, G. (1995) Ontogenetic development of periodicity coding in the IC of the mongolian gerbil, Aud. Neurosc. 1, 363–383.

Kazdin, A. (1989) Glenn Gould at work, creative lying. E.P. Dutton, New York.

Jähn-Siebert, T.K. and Langner, G. (1995) Afferent innervation and intrinsic connections of isofrequency sheets in the central nucleus of the inferior colliculus (ICC) in the Chinchilla - A double retrograde tracer study. In: N. Elsner and R. Menzel (Eds.), Learning and Memory, Thieme Stuttgart, p. 318.

Langner, G. (1981) Neuronal mechanisms for pitch analysis in the time domain, Exp. Brain Res. 44, 450–454.

Langner, G. (1983) Evidence for neuronal periodicity detection in the auditory system of the guinea fowl: implications for pitch analysis in the time domain, Exp. Brain Res. 52 333–355.

Langner, G. (1988) Physiological properties of units in the cochlear nucleus are adequate for a model of periodicity analysis in the auditory midbrain. In: J. Syka and R.B. Masterton (Eds.) Auditory Pathway - Structure and Function, Plenum Press, New York, London, , pp. 207–212.

Langner, G. (1992) Periodicity coding in the auditory system, Hearing Res. 60, 115–142.

Langner, G. and Schreiner, C.E. (1988) Periodicity coding in the inferior colliculus of the cat. I. Neuronal mechanisms, J. Neurophysiol. 60, 1799–1822.

Langner, G. and Bleeck, S. (1996). Neuronal processing of periodicity pitch may explain the difference between 'relative' and 'absolute' pitch perception. Soc. Neurosci. Abstr. 22, p. 1624.

Langner, G., Sams, M., Heil, P., and Schulze, H. (1997) Frequency and periodicity are represented in orthogonal maps in the human auditory cortex: Evidence from magnetoencephalography. J. Comp. Phys., (submitted).

Moore, J.K. and Osen, K.K. (1979) The human cochlear nuclei. In: O. Creutzfeldt, H. Scheich, and C. Schreiner (Eds.) Hearing Mechanisms and Speech, Springer, Berlin, pp. 36–44.

Oertel, D., Wu, S.H., Garb, M.W., and Dizack, C (1990). Morphology and physiology of cells in slice preparations of the posteroventral cochlear nucleus of mice. J.Comp.Neurol. 295, 136–154.

Palmer, A.R. (1982) Encoding of rapid amplitude fluctuations by cochlear nerve fibres in the guinea pig, Arch. Otorhinolaryngol. 236, 197–202.

Pfeiffer, R.R. (1966) Classification of response patterns of spike discharges for units in the cochlear nucleus: Tone-burst stimulation, Exp. Brain Res. 1, 220–235.

Rhode, W.S., Oertel, D., and Smith, P.H. (1983) Physiological response properties of cells labeled intracellularly with horseradish peroxidase in cat ventral cochlear nucleus. J.Comp.Neurol. 213, 448–463.

Schouten, J.F. (1940) The perception of pitch, Philips Techn. Rev. 5, 286–294.

Schouten, J.F., Ritsma, R.J., and Cardozo, B.L. (1962) Pitch of the residue. J. Acoust. Soc. Am. 34, 1418–1424.

Schouten, J.F. (1970) The residue revisited. In; R. Plomp and G.F. Smoorenburg (Eds.) Frequency Analysis and Periodicity Detection in Hearing, Sijthoff, Leiden, pp. 41–54.

Schreiner, C. and Langner, G. (1988) Periodicity coding in the inferior colliculus of the cat. II.Topographical organization, J. Neurophys. 60, 1823–1840.

Smith, R.L. (1979) Adaptation, saturation, and physiological masking in single auditory-nerve fibers. J. Acoust. Soc. Am. 650, 1660–1780.

Smoorenburg, G.F. (1970) Pitch perception of two-frequency stimuli. J. Acoust. Soc. Am. 48, 924–942.

van Stokkum, I.H.M. (1987) Sensitivity of neurons in the dorsal medullary nucleus of the grassfrog to spectral and temporal characteristics of sound, Hearing Res. 29, 223–235.

Terhardt, E. (1972) Zur Tonhöhenwahrnehmung von Klängen. II. Ein Funktionsschema. Acustica 26, 187–199.

Wightman, F.L. (1973). The pattern-transformation model of pitch. J. Acoust. Soc. Am. 54, 407–416.

Wever, E.G. (1949) Theory of Hearing, Wiley, New York.

Wynn, V.T. (1971) "Absolute" pitch - a bimensual rhythm. Nature 230, 337.

Zhao, H.-B. and Liang, Z.-A. (1995) Processing of modulation frequency in the dorsal cochlear nucleus of the guinea pig: Amplitude modulated tones. Hearing Res. 82, 244–256.

6

PERIODICITY PITCH SHIFT IS ABSENT IN ABSOLUTE PITCH POSSESSORS

Stefan Bleeck[*] and Gerald Langner

Zool. Inst
TH Darmstadt
Schnittspahnstr.3 64289 Darmstadt

1. INTRODUCTION

People possessing absolute pitch perception (AP) are able to label by ear the pitch of harmonic signals according to musical notation. This ability is relatively rare: the occurrence among western musicians is only a few percent, and AP is seldom seen in non-musicians. It is still not clear, how the phenomenon of AP emerges: although it was previously thought that it is inherited, evidence is accumulating that AP can be learned during a sensitive period in childhood (Miyazaki, 1988, Langendorf,1992).

While AP is quiet rare, normal subjects have the capability to judge relative pitch more or less precisely. We therefore denote in the following all people who are not possessors of AP as relative pitch possessors (RP).

In this paper we analyze two different types of pitch shifts: the first is the well known and described first effect of pitch shift ("periodicity pitch shift effect"; Schouten, 1969) observed with non-harmonic amplitude modulated signals and the second one is a theoretically predicted pitch shift elicited by short tones ("onset pitch shift effect"). This effect was already described by Metters and Williams (1973). Both of these pitch shift effects are examined with subjects possessing either AP and RP.

Although the phenomenon of absolute pitch has been known for centuries and was the subject of common interest since, little is known about neuronal processing of AP.

On the basis of a neuronal correlation model of periodicity analysis (Langner, this volume), we suggest a new explanation for the differences in neuronal processing between AP and RP perception. This model explains the percept of the missing fundamental and the pitch shifts elicited by nonharmonic amplitude modulations as well as the onset pitch shift effect.

The periodicity model can function in two modes to which we refer to as "absolute" and "relative" mode. These two modes are characterized mainly by a different functioning of the oscillator unit in the correlation model. With these two modes the theory allows

_____

* Stefan.Bleeck@gl15.bio.th-darmstadt.de http://www.th-darmstadt.de/fb/bio/agl/Welcome.htm.

predictions for pitch shift experiments with possessors of AP and RP: Because of the different role of the oscillator, only possessors of RP should hear the periodicity pitch shift and the onset pitch shift. In contrast, possessors of AP should hear no pitch shifts in both cases.

2. METHODS

8 subjects with AP and 3 subjects with RP participated in our experiments. All subjects possessing AP were musically trained. All subjects possessing RP were experienced in psychophysical experiments and had some musical experience.

Signals were generated digitally by computer and presented monaurally via a headphone. To avoid combination tones the volume was set to 40 dB SPL. The sound transmission characteristic was corrected by a computer. All tones were presented in random order.

2.1 Periodicity Pitch Shift Experiments

In all experiments subjects were given as much time as they wanted to adjust the pitch of a comparison pure tone to that of a probe tone. The probe tone was an amplitude modulated signal with a fixed modulation frequency of 250 Hz. The carrier frequency varied in the experiments with AP over a wide range from 800 Hz to 5000 Hz. The task for the AP was to name the pitch of the complex tone in musical notation. Because all AP subjects were able to discriminate multiple signal components over nearly the whole range, they were also asked to write down all components they could hear. The pitch shift experiments with RP subjects were performed in a smaller range of carrier frequency from 600 to 3100 Hz. Since such experiments are standard and the results are well known, only few of them were performed.

2.2 Onset Pitch Experiments

In the onset pitch experiments the task was to adjust the pitch of a long signal (duration: 250 ms) to the pitch of a short signal (duration: 5 - 20 ms). Probe signals were AM with a modulation frequency of 250 Hz and a carrier of 1000 Hz. The comparison tone was a harmonic complex with the 3rd, 4th and 5th harmonic of a fundamental. The subjects had control over the fundamental frequency of the AM. The short tones were ramped by a squared cosine with a ramp time of 2 ms.

3. RESULTS

3.1 Pitch Shift Experiments

When asked to judge the components of an AM-signal, possessors of AP are able to track the modulation frequency of the harmonic or non-harmonic sounds. Fig. 1 shows in its upper part the results of pitch shift experiments with AP (right) and RP (left) possessors.

Not surprisingly, in our experiments all possessors of RP heard the well known pitch shift, i.e. they heard a change in periodicity pitch although the modulation frequency was kept constant. Possessors of AP did not hear this pitch shift. In contrast to possessors of RP, they perceived the constant modulation frequency for all carrier frequencies tested and made only small errors in estimating the pitch.

The left figure shows one typical experiment with a possessor of RP. As possessors of AP show in the same experiment very similar results, in the right figure only the medium values of all experiments are shown and the data points appear with an error bar.

3.2 Onset Pitch Experiments

In its lower part fig. 1 shows the results from experiments in which subjects matched the pitch of a long amplitude modulation to that of a short one. The duration of the probe signals varied in all experiments between 5 and 20 ms.

Subjects with RP perceive a pitch shift towards lower pitch for short tone duration (left part of figure 1). They perceive a significantly lower pitch when the duration of the signal is between 5 and 10 ms. In fact, the pitch reaches a minimum at 9 ms and increases at shorter durations to the pitch of signals with a long duration. When the duration of the tone is long enough, the perceived pitch is equal to the modulation frequency.

AP possessors do not perceive this shift (right part of the figure). Instead they hear a constant pitch independent of the duration of the AM signal.

As expected, for both possessors of AP and of RP the standard deviation of the perceived pitch becomes smaller when the durations of the signals are longer.

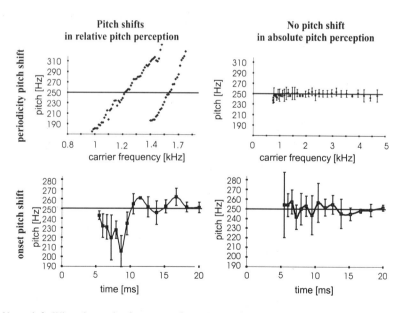

Figure 1. Upper left: When the carrier frequency of an AM tone is changed, periodicity pitch changes for a possessor of RP in spite of a constant modulation frequency (250 Hz), carrier frequency = 950 - 1700 Hz. Upper right: Subjects with AP also perceive a periodicity pitch. However, they do not perceive a pitch shift. Modulation frequency = 250 Hz, carrier frequency = 800 -5000 Hz. Lower left: During signal onset (0 - 20ms) pitch differs for RP from the steady state (duration > 20 ms). Modulation frequency = 250 Hz, carrier frequency = 1250 Hz, 50 dB SPL.Lower right: Subjects with AP do not perceive the onset pitch shift (same signals as for RP).

4. DISCUSSION

According to the neuronal correlation model (Langner, this volume) a coincidence neuron is tuned to the periodicity τ_m of amplitude modulated signals. This periodicity is given by the equation: $m \cdot \tau_m = n \cdot \tau_c + \tau_k$, were τ_C is the carrier period, τ_K is the reference period and m and n are small positive integers. The two neuronal circuits of the model generate intervals which are either multiples of the carrier period (integration interval) or of the reference period. These circuits are triggered by envelope modulations of periodic signals (e.g. AM).

The periodicity model can work in two modes, to which we refer to as "absolute" and "relative" mode. These two modes are characterized by the different functioning of the oscillator units. The described pitch shift effects for possessors of AP and RP can be explained as the result of these different modes.

4.1 Relative Mode

In the coincidence model, the integrator is the key element for the determination of relative pitch. The output of the integrator contains information about the fine structure (carrier frequency) of the input signal. This output coincides at midbrain units with the output from the oscillator. Coincidence units correlate these two inputs and are therefore tuned to signals with a specific periodicity. In relative mode, the oscillator is triggered by each cycle of the envelope since the oscillator intervals are shorter than the envelope period (Fig. 2).

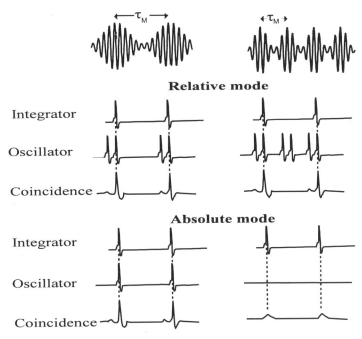

Figure 2. The figure shows that the difference between absolute and relative mode is the function of the oscillator. In relative mode the oscillator is triggered by each envelope cycle. Therefore coinciding activity occurs not only for the fundamental frequency (left side) but also for harmonics (like the octave, right side). In absolute mode the resonance of the oscillator circuit is used which may provide an internal pitch reference. In addition, the tuning of the oscillator prevents responses to harmonics (right side, bottom).

As a consequence of the temporal correlation analysis, the perceived pitch varies as a function of τ_c and τ_k in spite of a constant τ_m (upper left part of fig.1).

4.2 Absolute Mode

The absolute mode is characterized by a dominance of the oscillator input to the co-incidence neurons. The dominance occurs when the envelope period matches the intrinsic period of an oscillator neuron. As a result, a resonance will occur and synchronized activity of this neuron will provide an absolute temporal reference. Subjects, who can make use of this reference have AP perception, because they are able to compare the period of the resonating oscillators with the fundamental period ($\tau_m = \tau_k$, $n = 0$).

While possessors of RP perceive a pitch shift effect, possessors of AP do not perceive such a shift. The results for AP (upper right part of fig.1) can be explained by the fact, that the envelope periodicity of a signal is not affected by variations of the carrier frequency. Provided, the assumption holds that in absolute mode resonating oscillators define the pitch, then pitch does not change with a change of the carrier frequency.

4.3 Fine Structure Analysis

The theory allows for a third possibility in which the period of the resonating oscillators are compared to multiples of the carrier period which are coded by integrator neurons. In this mode the correlation process provides information about high frequency components of a complex signal.

This mode of processing may explain why possessors of AP were able to discriminate the single components of AM frequencies up to the 20th harmonic component (carrier 5000 Hz, modulation 250 Hz) and above. Because of the limited resolution of the cochlea this cannot be explained by spectral analysis.

4.4 Onset Pitch Experiments

The onset properties of coincidence neurons (see Langner in this volume) lead to the prediction of the onset pitch shift for the relative mode of pitch perception. In the theory the integrator is responsible for this shift. Because of higher activity of auditory nerve fibers during onset, the integration interval is shorter for the beginning of the signal. In the relative mode the integration interval is essential for determining pitch. Because possessors of AP do not have to make use of the integration interval and rely mainly on oscillator responses, they do not perceive an onset pitch shift.

4.5 Conclusion

In contrast to subjects with RP, possessors of AP do not perceive a pitch shift:

1. when the carrier frequency of an AM signal is changed,
2. when presented with tones shorter than 10 ms.

Additionally, possessors of AP are able to resolve the components of AM signals even for large ratios of carrier and modulation frequency.

These findings were predicted by a neuronal correlation model. According to the model, AP perception is based on resonances in neuronal oscillatory circuits while the same oscillators are only triggered in the case of RP.

5. REFERENCES

Langner, G. (1992) Periodicity coding in the auditory system, *Hearing Res*, **60**:115–142.

Langendorf, F.G. (1992) Absolute Pitch - Review and speculations". *Medic. Probl. of Perf. Artist* 7:6–13.

Metters, P.J.M. and Williams, R.P. (1973) Experiments on tonal residue of short duration. J. Sound and Vibration **26**:432–436.

Miyazaki, K. Musical pitch identification by absolute pitch possessors. *Perc. Psych.* **44**:501–512, 1988.

Schouten, J.F. (1970) The residue revisited. In *Frequency Analysis and Periodicity Detection in Hearing*, edited by Plomp, R. and Smoorenburg, G.F. (Eds.) Leiden: Sijthoff, pp. 41–54.

7

5-HT INNERVATION OF THE AUDITORY PATHWAY IN BIRDS AND BATS

Alexander Kaiser[1] and Ellen Covey[2]

[1]Institut für Zoologie
Technische Universität München
85747 Garching, F.R.G.
[2]Dept. of Neurobiology
Duke University Medical Center
Durham, North Carolina 27710

1. INTRODUCTION

Serotonin (5-hydroxytryptamine, 5-HT) is a neuroactive substance that is widely distributed in both invertebrates and vertebrates. In vertebrates, 5-HT neurons are located mainly in the raphe nuclei along the midline of the brainstem. Neurons of the raphe nuclei project bilaterally to form a fine network of small terminals that innervate many structures throughout the brain including cell groups within the auditory system. 5-HT-immunoreactive terminals are known to be present in the cochlear nucleus and the inferior colliculus of the rat (Klepper and Herbert, 1991), and in the cochlear nucleus, superior olivary complex, nuclei of the lateral lemniscus and inferior colliculus of the guinea pig, cat and bush baby (Thompson, G.C. et al., 1994; Thompson, A.M. et al., 1995). To the authors knowledge, however, there is no study in birds investigating the distribution of 5-HT fibers in the auditory pathway.

The 5-HT system is associated with a number of vegetative functions such as feeding behavior, thermoregulation, sexual behavior, cardiovascular regulation, sleep, aggression and attention (Glennon, 1990; Wilkinson and Dourish, 1991). Malfunction of the 5-HT system can have dramatic consequences for the overall state and behavior of the organism, and may contribute to diseases such as schizophrenia, depression and migraine (Meltzer and Lowy, 1987; Glennon, 1990). Defects in the 5-HT system can also affect hearing. Some symptoms of migraine, including hyperacusis or tinnitus, may result from depressed function of the 5-HT system (Wang et al., 1996). Thus, although the serotonergic system primarily exerts its effects outside the auditory pathway, it is also capable of modulating auditory function.

The role of 5-HT in modulating auditory function is poorly understood. Only a few studies have examined the effects of 5-HT on responses of auditory neurons. In rat co-

Acoustical Signal Processing in the Central Auditory System
edited by Syka, Plenum Press, New York, 1997

chlear nucleus, microiontophoretic application of 5-HT was shown to produce a decrease in spontaneous activity, a short-latency inhibitory modulation of sound-evoked responses, and a long latency excitatory modulation of sound-evoked responses (Ebert and Ostwald, 1992). These effects might be due to activation of different 5-HT receptor types including those with directly gated ion channels and those with second messenger-activated conductances. Effects similar to those seen in the rat cochlear nucleus have also been reported after 5-HT application in the vestibular nuclei of the rat (Licata et al., 1993; inhibition and excitation) and cat (Kishimoto et al., 1994; inhibition), and in the inferior colliculus of the chicken (Calogero et al., 1977; inhibition).

Comparative studies across vertebrate orders can potentially provide information on the role of 5-HT in audition. Species in which the auditory system has been extensively studied include echolocating bats and birds. In echolocating bats, the auditory brainstem nuclei have the same general organization as in other vertebrate species, but are larger and more highly differentiated (Covey and Casseday, 1995). Furthermore, the function of certain brainstem nuclei is better understood in bats than in any other species. For example, the nuclei of the lateral lemniscus are thought to provide parallel pathways for transmitting information about the timing and the intensity of sound (e.g., Covey and Casseday, 1991; Covey, 1993). The inferior colliculus of bats shows adaptations for echolocation with pronounced expansion of certain frequency laminae. Thus, the bat provides an excellent model for studying 5-HT innervation of the auditory system of a mammal with known specializations. The auditory systems of birds are organized somewhat differently from those of mammals; nevertheless, some auditory structures appear to be homologous in the two vertebrate orders. In birds, the organization and function of brainstem pathways for comparing interaural time and intensity differences is especially well understood (e.g., Konishi, 1993; Carr, 1992). The avian auditory system thus provides an excellent model for comparison with the bat, particularly with regard to auditory pathways specialized for precise timing and for intensity coding.

In this study, we present preliminary data on the distribution of 5-HT-immunoreactive fibers in auditory nuclei of the echolocating bat, *Eptesicus fuscus* and the chicken, *Gallus domesticus*. The nuclei examined include all of the major auditory cell groups of the brainstem and auditory thalamus.

2. METHODS

Bats and young chickens were deeply anesthetized by a lethal dose of sodium pentobarbital (Nembutal, Narcoren) and transcardially perfused with phosphate buffered saline (PBS, pH 7.4) or Ringer's solution followed by cold 2.5 % glutaraldehyde or 4 % paraformaldehyde in 0.1 M PBS. Immediately after fixation brains were removed from the skull and allowed to sink overnight in a cold solution containing 30% sucrose in 0.1 M PBS. Sections were cut at 40 μm on a freezing microtome. Every 4th section was mounted on slides and counterstained with neutral red. Before incubating overnight with 5-HT antiserum (SFRI; Incstar; dilution 1:2,000- 1:16,000), the remaining free floating sections were preincubated with 0.5% hydrogen peroxide in PBST (0.2 % Triton-X-100 in 10 mM PBS, pH 7.2), rinsed in PBST and blocked in 3% normal goat serum in PBST. After several rinses in PBST the sections were incubated with biotinylated goat anti-rabbit IgG (Vector), rinsed and incubated with avidin-biotin complex labeled with peroxidase (Vector Elite). Finally, the sections were reacted with diaminobenzidine, coverslipped, and examined using brightfield microscopy. Every fourth section served as a control by omitting either the primary or the secondary antibody. No labeling was present in control sections.

3. RESULTS

Although the overall intensity of staining was correlated with the concentration of primary antibody used, the relative densities of labeled fibers within the brainstem nuclei that we examined was the same at all antibody concentrations.

3.1. Eptesicus

Figure 1 shows typical photomicrographs of frontal sections immunostained for 5-HT. In the cochlear nuclear complex, the density of 5-HT immunoreactive fibers varied from one subdivision to another (Fig. 1a). Both the anteroventral cochlear nucleus (AVCN not shown) and the dorsal cochlear nucleus (DCN) contained high levels of staining; the posteroventral cochlear nucleus (PVCN) contained only a moderate amount. In Eptesicus, the superior olivary complex is made up of several subdivisions, including the lateral superior olive (LSO), the medial superior olive (MSO), the medial nucleus of the trapezoid body (MNTB), the superior paraolivary nucleus (SPN), and the lateral and ventral nuclei of the trapezoid body (LNTB and VNTB). In all of these cell groups 5-HT fiber density was lower than in adjacent non-auditory structures (Fig. 1b). Toward the more rostral part of the SOC, labeled cell bodies were located very close to the SPN and MSO. In the nuclei of the lateral lemniscus, the density of 5-HT fibers ranged from low to moderate (Fig. 1c). Density was lowest in the columnar division of the ventral nucleus (VNLL), a structure that is thought to provide precise timing information to the inferior colliculus. In the dorsal nucleus of the lateral lemniscus (DNLL), the 5-HT- immunoreactive fibers followed the column-like arrangement of the cell bodies (arrows). Throughout most of the inferior colliculus (IC) immunostaining for 5-HT was moderately dense, except for a clear crescent-shaped band of low density in the ventral and lateral regions (arrows; Fig. 1d). Comparison of this pattern of 5-HT distribution with the tonotopic organization of the inferior colliculus (Casseday and Covey, 1992), shows that the regions of low 5-HT immunostaining correspond to the 20–30 kHz and 50+ kHz frequency laminae (Figs. 1d and e).

Surprisingly, the only auditory structure that contained virtually no 5-HT- immunoreactive fibers was the medial geniculate body (MGB, Fig. 1f), suggesting that the auditory thalamus receives little or no modulatory influence from the serotonergic system.

3.2. Gallus

In birds, the cochlear nuclear complex consists of two primary nuclei (nucleus magnocellularis cochlearis, MCC; nucleus angularis, An) and one secondary nucleus (nucleus laminaris, La). Several lines of evidence suggest that the brainstem auditory system of the bird contains two distinct and parallel binaural pathways, one specialized for processing interaural time differences and the other specialized for analyzing interaural intensity differences (for review see Konishi, 1993; Carr, 1992). The "time pathway" begins in the MCC, which projects bilaterally to the La, which acts as an coincidence detector. The "intensity pathway" begins in the An, which in turn projects to the nuclei of the lateral lemniscus and the inferior colliculus. The nuclei of the lateral lemniscus in the bird are probably not homologous to the nuclei of the same name in mammals.

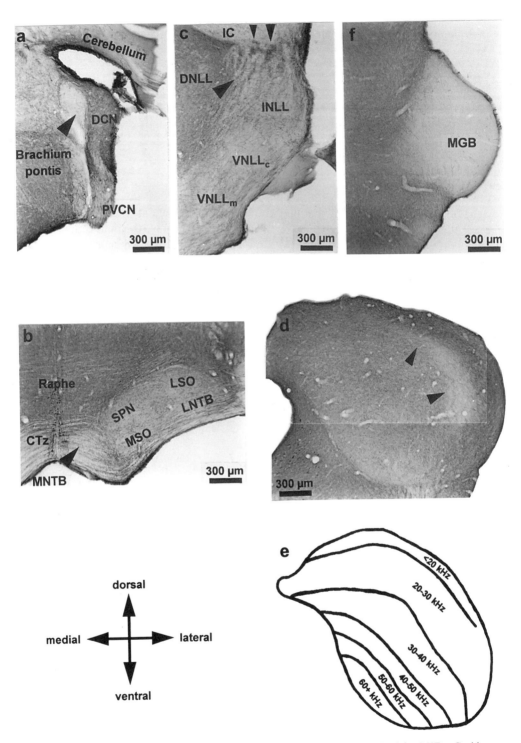

Figure 1. Frontal brain sections of the big brown bat, Eptesicus fuscus, immunostained for 5-HT. **a** Cochlear nuclear complex. **b** Superior olivary complex. **c** Nuclei of the lateral lemniscus. **d** Inferior colliculus. **e** Schematic drawing of the inferior colliculus indicating the isofrequency laminae. **f** Medial geniculate nucleus. See text for further explanation.

The nuclei of the "time pathway", the MCC and the La, did not contain any 5-HT immunoreactive fibers (Fig. 2a). However, the first nucleus of the "intensity pathway", the An, contained a fine network of 5-HT immunoreactive fibers and bouton-like endings (Fig. 2b, arrows). The nuclei of the lateral lemniscus contained many 5-HT immunoreactive fibers (Fig. 2d, arrows). The avian superior olivary nucleus (SO) is not as highly differentiated as in mammals, and is probably not homologous to the mammalian SOC. Its function in audition is unknown. In sections stained for 5-HT immunoreactivity, the SO stands out as a more or less unstained circle in the ventrolateral brainstem (Fig. 2c) with a sparse distribution of 5-HT immunoreactive fibers (arrow). Within the avian inferior colliculus, the nucleus mesencephalicus lateralis, pars dorsalis (MLd), many 5-HT positive fibers were observed, distributed uniformly throughout (Fig. 2e). The high-magnification micrograph in Figure 2f shows a single fiber probably contacting a cell body with bouton-like terminals (arrows). The thalamic auditory structure, the nucleus ovoidalis (OV) contained very few fibers that were immunoreactive for 5-HT (Fig. 2g).

4. DISCUSSION

In both bats and chickens, 5-HT-immunoreactive cell bodies were located in the raphe and other nuclei as has been previously described (Yamada et al., 1984; Klepper and Herbert, 1991; Thompson, G.C. et al., 1994). Within the auditory nuclei of both bats and chickens, the overall 5-HT fiber density was lower than in surrounding brain structures. The relatively low density of 5-HT innervation throughout the auditory system of both vertebrate species suggests that the role of 5-HT in modulating auditory function may be less than its role in modulating other, non-sensory functions. Although the density of 5-HT fibers was low, it was still great enough for there to be clear differences in the amount of 5-HT innervation of different nuclei of the ascending auditory system, and in different regions of the bat inferior colliculus. This observation suggests that different auditory nuclei in both species and different regions of the bat inferior colliculus receive different amounts of serotonergic modulation. This may correspond to a greater or lesser role of attention in the processing performed at these different levels.

In the chicken, 5-HT staining was essentially absent in the nuclei making up the "time pathway", but was moderately dense in the nuclei making up the "intensity pathway". Thus, in birds we would expect more 5-HT-modulation of auditory processing in the nuclei of the "intensity pathway, than in the "time pathway". In the bat, there was no comparable distinction between the binaural pathways via the LSO and MSO, both of which had low levels of 5-HT immunoreactivity. However, differences in 5-HT density were observed in the cochlear nucleus and nuclei of the lateral lemniscus. As in the chicken it seems likely that this differential distribution is correlated with functional properties. 5-HT density was lowest in the PVCN, a region that is thought to be an initial stage in analyzing timing information and that provides input to the columnar region of the VNLL (Covey and Casseday, 1995). The columnar region of the VNLL has been identified as the source of high-precision timing information that could signal the time delay between the emission of a biosonar pulse and the return of an echo (Covey and Casseday, 1991). Taken together with the finding of low 5-HT density in the "time pathway" of the chicken, it appears that there is a general tendency for auditory structures that preserve precise timing information to have a low level of 5-HT innervation.

Within the inferior colliculus of the bat, the distribution of 5-HT terminals was clearly correlated with the expanded 20–30 kHz frequency lamina and and the 50+ kHz

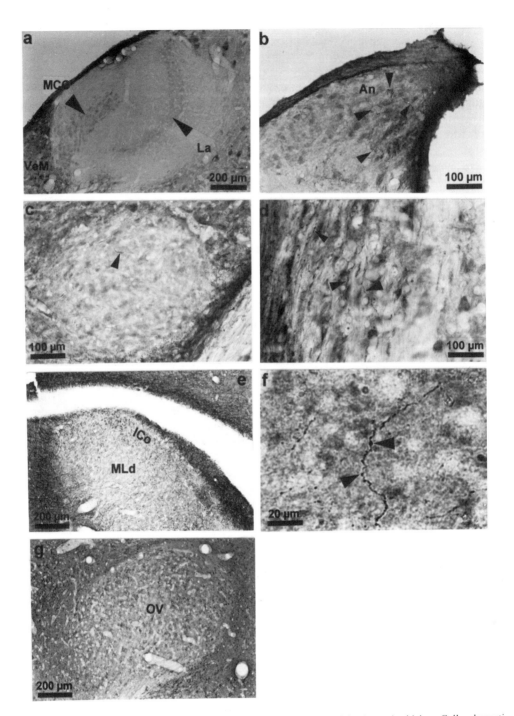

Figure 2. Photomicrographs of 5-HT immunostaining in frontal sections of the domestic chicken, Gallus domesticus. Cochlear nuclei: **a** Nucleus magnocellularis and Nucleus laminaris; **b** Nucleus angularis. **c** Superior olive. **d** Nuclei of the lateral lemniscus. **e** Nucleus mesencephalicus lateralis pars dorsalis (inferior colliculus). **f** higher magnification of the MLd showing a single 5-HT-immunoreactive fiber. **g** Nucleus ovoidalis. See text for further explanation.

frequency lamina, both of which are important for echolocation. Comparison of the distribution of 5-HT terminals with the distribution of GABAergic and glycinergic fibers and terminals (Johnson, 1993) shows that the distribution of GABA immunoreactive fibers overlaps with the regions of highest 5-HT fiber density, while the reverse is true for glycinergic fibers. The glycinergic fibers probably originate, at least in part, from the columnar region of the VNLL (Vater et al., 1996), suggesting that this particular part of the inferior colliculus represents a continuation of the timing pathway from the VNLL. In both bats and chickens, the auditory thalamus was virtually devoid of 5-HT immunoreactivity, suggesting that 5-HT does not play a modulatory role in the function of these nuclei.

In summary, the auditory brainstem and thalamus of both bats and birds contain low levels of serotonin relative to surrounding brain structures. This finding suggests that there has been considerable evolutionary pressure to maintain a baseline level of excitability within the auditory system even during sleep or other states in which 5-HT release would suppress activity. In both bats and birds, the brainstem auditory pathways with the lowest levels of serotonin innervation are those in which precise timing information is conserved. Therefore, we can speculate that these pathways are not only important for analyzing temporal parameters of sound but may also perform some type of alerting function.

5. ACKNOWLEDGMENT

We thank B. Fubara, E. Siegl and G. Schwabedissen for technical assisstance. Supported by the Boehringer Ingelheim Fonds, the DFG/SFB 204 and NIH/DC-00287 and DC-00607.

6. REFERENCES

Calogero, B., Nistico, G., Rotiroti, D., Giannini, P., Olivieri, G. and Cargiulo, G. (1977) Involvement of serotonergic mechanism in the control of chick auditory evoked response. Revue de Laryngologie 98, 233–241.

Carr, C.E. (1992) Evolution of the central auditory system in reptiles and birds. In: The Evolutionary Biology of Hearing. Eds. DB Webster, RF Fay & AN Popper, Springer- Verlag, New York, pp. 511–543.

Casseday, J.H. and Covey, E. (1992) Frequency tuning properties of neurons in the inferior colliculus of an FM bat. J. Comp. Neurol. 319, 34–50.

Covey, E. (1993) The monaural nuclei of the lateral lemniscus: Parallel pathways from cochlear nucleus to midbrain. In: The Mammalian Cochlear Nuclei: Organization and Function. Eds. MA Merchan, JM Juiz, DA Godfrey & E Mugnaini, Plenum Press, New York, pp. 321–334.

Covey, E. and Casseday, J.H. (1991) The monaural nuclei of the lateral lemniscus in an echolocating bat: Parallel pathways for analyzing temporal features of sound. J. Neurosci. 11, 3456–3470.

Covey, E. and Casseday, J.H. (1995) The lower brainstem auditory pathways. In: Hearing by bats. Eds. AN Popper & RF Fay, Springer-Verlag, New York, pp. 235- 295.

Ebert, U. and Ostwald, J. (1992) Serotonin modulates auditory information processing in the cochlear nucleus of the rat. Neurosci. Lett. 145, 51–54.

Glennon, R.A. (1990) Serotonin receptors: Clinical implications. Neurosci. Biobehav. Rev. 14, 35–47.

Johnson, B.R. (1993) GABAergic and glycinergic innervation of the central nucleus of the inferior colliculus of the big brown bat. Ph.D. dissertation, Duke University.

Kishimoto, T., Yamanaka, T., Amano, T., Todo, N. and Sasa, M. (1994) 5-HT1A receptor-mediated inhibition of lateral vestibular nucleus neurons projecting to the abducens nucleus. Brain Res. 644, 47–51.

Klepper, A. and Herbert, H. (1991) Distribution and origin of noradrenergic and serotonergic fibers in the cochlear nucleus and inferior colliculus of the rat. Brain Res. 557, 190–201.

Konishi, M. (1993) Listening with two ears. Sci. Am. 268, 66–73.

Licata, F., Li, V.G., Maugeri, G. and Santangelo, F. (1993) Excitatory and inhibitory effects of 5-hydroxytryptamine on the firing rate of medial vestibular nucleus neurons in the rat. Neurosci. Lett. 154, 195–198.

Meltzer, H.Y. and Lowy, M.T. (1987) The serotonin hypothesis of depression. In: Psychopharmacology: The Third Generation of Progress. Ed. HY Meltzer, Raven Press, New York, pp. 513–526.

Thompson, A.M., Moore, K.R. and Thompson, G.C. (1995) Distribution and origin of serotoninergic afferents to guinea pig cochlear nucleus. J. Comp. Neurol. 351, 104- 116.

Thompson, G.C., Thompson, A.M., Garrett, K.M. and Britton, B.H. (1994) Serotonin and serotonin receptors in the central auditory system. Otolaryngol. Head Neck Surg. 110, 93–102.

Vater , M., Covey, E., and Casseday, J.H. (1996) The columnar region of the ventral nucleus of the lateral lemniscus in the big brown bat (Eptesicus fuscus): Synaptic arrangements and structural correlates of feedforward inhibitory function. Cell and Tissue Res. (in press).

Wang, W., Timsit-Berthier, M. and Schoenen, J. (1996) Intensity dependence of auditory evoked potentials is pronounced in migraine: An indication of cortical potentiation and low serotonergic neurotransmission? Neurology. 46, 1404–1409.

Wilkinson, L.O. and Dourish, C.T. (1991) Serotonin and animal behavior. In: Serotonin Receptor Subtypes: Basic and Clinical Aspects. Ed. SJ Peroutka, Wiley-Liss, Inc., New York, pp. 147–210.

Yamada, H., Takeuchi, Y. and Sano, Y. (1984) Immunohistochemical studies on the serotonin neuron system in the brain of the chicken (Gallus domesticus). I. The distribution of the neuronal somata. Biogenic Amines 1, 83–94.

8

STRUCTURAL ORGANISATION OF NUCLEUS ANGULARIS IN THE PIGEON (Columba livia)

Udo Häusler[*]

German Primate Center
Department of Neurobiology
Kellnerweg 4, 37077 Göttingen

INTRODUCTION

Ramon y Cayal (1908) first described the avian cochlear nuclei to consist of the principal parts nucleus angularis (NA) and nucleus magnocellularis (NM). He considered NM homologue to the ventral cochlear nucleus (VCN) of mammals and NA homologue to the dorsal cochlear nucleus (DCN). Boord and Rasmussen (1963) confirmed this view and further subdivided NA into medial, lateral and ventral parts. Boord (1969) compares NM with the anteroventral cochlear nucleus (AVCN), the medial NA with the posteroventral cochlear nucleus (PVCN) and lateral NA with the DCN.

A functional segregation between NM and NA has first been shown in the red winged blackbird (Sachs and Sinnot, 1978) and more detailed in tyto alba (Sullivan, 1985). Sachs and Sinnot (1978) observed tuning curves with multiple excitatory frequencies, which are typical for the mammalian DCN only in the most lateral part of NA. Sullivan (1985) clearly demonstrated that in the owl NA neurones almost exclusively consist of chopper type neurons well adopted to sound intensity processing. In contrast NM showed very precise temporal coding. Therefore Sullivan (1985) suggested to compare magnocellular neurons with the bushy cell type of the AVCN, and the owl's NA neurones with the stellate cells in the mammalian VCN, which have been shown to respond with a chopper type pattern (Rhode et al., 1983). Analysis of NM neuronal morphology in the chicken (Jhaveri and Morest, 1982; Parks and Rubel, 1978) does support this view. NM almost exclusively consists of a single celltype with a structure comparable to the bushy cells of the mammalian VCN. Therefore our aim was to establish a detailed anatomical description of celltypes in NA, to see wether the same celltypes are found in NA as in the mammalian DCN, and to determine if neuronal morphology is compatible with the physiological response pattern observed in the avian NA.

---

* Tel.: 049-551-3851-256; E-mail: uhausl @www.dpz.gwdg.de.

Acoustical Signal Processing in the Central Auditory System
edited by Syka, Plenum Press, New York, 1997

METHODS

Eight pigeons were perfused with 10% formalin, the brains were removed and embedded in paraffin. Sections of 20 μm were cut in the frontal or horizontal plane and processed for nissl staining. 14 animals were prepared according to the method of Stensaas (1967). After processing of the tissue frontal sections of 120 μm were cut. Camera lucida drawings of stained neurons were made with a Nikon Labophot or Zeiss Universal microscope using a 40X objective. A 100X oil immersion objective was used to check the continuity of the structures. If dendritic trees were not completely included in one section, they were reconstructed from the neighbouring ones. Morphometric analysis was made with NIH Image and statistics with Statview on a Macintosh Quadra 660 AV.

RESULTS

On the basis of nissl stains nucleus angularis could be divided into two major divisions, a medial part (NAm) and a dorso-lateral part that may be further split into two subdivisions, an intermediate (NAi) and lateral (NAl) part

NAm showed a homogenious population of neurons with unique appearance. They have a medium sized, spherical somata with poorly defined cellborders and evenly distributed fine granular nissl substance. Only a few fusiform and multipolar neurons were present in NAm.

Lateral and intermediate parts of NA are heterogenious in nature. They are build of fusiform and multipolar neurons of heterogenious size. The size of the neurons were generally larger in NAi than in NAl. In rostral sections a relatively abrupt change from medial to lateral was observed while in caudal sections the change was more gradual. Medium to large sized neurons with heavily staining nissl substance were scattered throughout both NAi and NAl. All three subdivisions showed significant differences with regard to the distribution of cell sizes (NAi-NAl $p<0.0001$; NAm-NAl $p<0.0001$; NAM-NAi $p<0.007$, Kolmogorow Smirnow-test). For the form parameter (ratio of length to width of the soma), NAm was different from NAi and NAl ($p<0.001$, t-test) while NAi and NAl were not significantly different from each other.

Neuronal Structure in NAm

The main cell type in NAm is the **stubby** cell (Fig.1f). The cells have a relatively large soma size, with a few number of very short unbranched dendrites. These are no longer than 1.5 times the soma diameter. The surface of soma and dendrites is covered with a large number of stubby shaped appendages of 1–2 μm length resembling somatic spines. This neuronal structure is consistent with the appearance the neurons show in the nissl stained material. There they showed a badly defined somatic border which may be due to the large number of stubby appendages. Besides these stubby cells also some neurons with small dendritic trees were found. The relative proportions of the different cell types in NAm are 79% stubby cells, 12% stubby cells with a short dendrite branched once or twice, 4% neurons with a multiply branched dendritic tree inside NAm, 5% neurons outside NAm but with a tree reaching into NAm. Typical synaptic terminals observed in NAm consisted of relatively thick axons with a single large (2–3 μm diameter) varicosity. No en passant varicisities were observed in these axons. A second axon type was much thinner, with small en passant swellings.

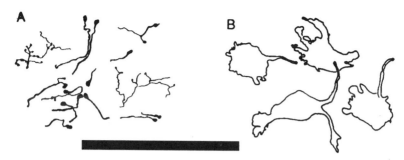

Figure 1. Synaptic terminals observed in NAm (a), neurons observed in NAm (b). Scale bar: 100µm.

Neuronal Structure in NAi and NAl

The remaining parts of NA contained very different cell types than NAm. Their neurons were multipolar and exhibited branched dendritic trees with various degrees of branching and tree size. Soma size and the size of dendritic arbours were roughly correlated (correlation coefficient: 0.495, p < 0,005). Size and form of dendritic arbours were also correlated (correlation coefficient: 0.617, p<0.0001). Small sized dendritic trees were preferentially elongated and large sized ones were of spherical appearance.

Large multipolar neurons (Fig. 1a) were preferentially located in the ventromedial portion of NAi. The thicker main dendrites branch close to the cellbody into longer secondary ones, one or two of which ended in a tufted branching pattern that covers an area of about 100 µm diameter. The form of the dendritic tree was ovoid to spherical. The main orientation of dendrites was in mediolateral direction.

Medium sized multipolar neurons are more heavily branched than large neurons (Fig. 1b). Typical for these neurons is a moderate or heavily tufted branching pattern. The surface of primary and secondary dendrites is smooth. Dendritic shape varies considerably, and depends on the location of the neurons within the nucleus. The overall shape of the dendritic tree could be spherical, but in most cases was polarized. Dendrites were oriented in columns perpendicular to the dorsal surface of the nucleus. The extension of the dendrites along these columns is quite large, while the extension across the colums is small, with an aproximate relation of 1:2.5.

Medium sized bipolar neurons (Fig. 1c) were mainly located at the dorsal border in the rostral half of the nucleus. They were characterized by primary dendrites located at opposing sides of the soma and extending into different directions. Thick main dendrites that extend into heavily branched tufted endings are typical for this type of neuron. The cells are oriented perpendicular to the border of the nucleus. One dendritic tree extends down to the center of the nucleus where the auditory nerve fibers enter and the other one into a layer parallel to the dorsal surface of NA. The surface of the soma and main dendrites are free of spines or other specialized appendages, while at the end of the dendrites irregular shaped dendritic specialisations appear.

Small sized neurons were preferentially found in the lateral aspect of NA. Their typical feature were few dendrites with a low degree of branching. Their dendrites were long compared to the soma size (Fig.1d) and had a smooth surface. Small neurons in the vicinity of the nucleus border could have an elongated tree with the main axis parallel to the border. A second type of small neuron was found in NAi and has short and heavily

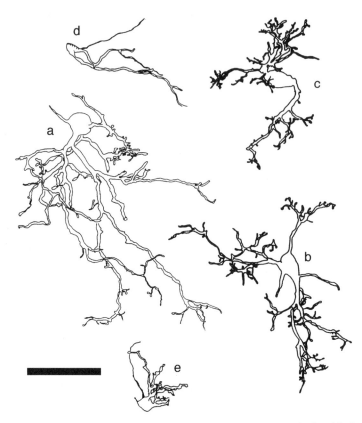

Figure 2. Neurons observed in NAi, NAl. a: Large multipolar type, b: medium sized multipolar type, c: bipolar type, d, e: small neurons.

branched trees, resembling the tufted endings of the medium sized neurons (Fig.1e). This type was not observed in NAl.

Axonal elements in NAi and NAl (Fig. 2b) typically consisted of thin collaterals that branched off thicker axons running parallel to the dendritic columns. Other thin axonal elements showed en passant contacts, and multiple terminations along single dendrites or somas. All synaptic terminals were small with no calyces or endings comparable to NAm.

DISCUSSION

While morphological descriptions of the auditory brainstem have been given for the pigeon (Cayal, 1908; Boord, 1968) and of nucleus magnocellularis in the chicken and owl (Jhaveri and Morest, 1982; Carr and Boudreau, 1993), this study is the first to describe in detail the neuronal architecture of the avian nucleus angularis.

Nucleus angularis has been subdivided into three parts on the basis of nissl stains. Measurements in nissl stains appear to favour the separation into the subdivisions NAm, NAi and NAl as they are significantly different with respect to several parameters. Nevertheless, the morphological changes from medioventral to dorsolateral throughout NAi and NAl may be regarded as gradual ones, possibly associated with tonotopic organisation.

Our observations in golgi stains are largely consistent with the the nissl material. Clear differences between NAm and the remaining parts of NA were observed in both the golgi and nissl preparations. The characteristic cell of NAm in nissl stains is likely to be identical with the stubby cells of NAm in the golgi sections. Therefore it is clear that on neuronal structure it is justified to seperate NAm as an independent subdivision.

All cell types in NAi and NAl are dendritic and can be interpreted as general multipolar neurons. Even though we could show different subtypes of multipolar neurons that are probably distributed differentially throughout NA it appears impossible to assign one of these types to a specific region only. This is similar to the nissl stains. Differences between the regions NAi and NAl could also be due to a general gradient of dendritic tree and cell size throughout the nucleus that is associated with tonotopy. Further electrophysiological mapping may therefore be important to decide if NAi and NAl constitute seperate subdivisions.

One aim of this study was to relate functional properties observed in NA neurons to the cellular morphology of the neurons seen in NA. The initial hypothesis was that NA neurons should exhibit a stellate type morphology as this could account for a lot of the physiological properties observed. This hypothesis has been clearly confirmed for NAi and NAl. The dendritic trees of the multipolar NA neurons on which numerous small synapses terminate, allow them to spatially and temporally integrate the activity of large numbers of input from the auditory nerve. This will result in a loss of temporal precision and phase locking, as is observed for NA in the Owl (Sullivan and Konishi, 1984, 1986; Sullivan, 1985). Furthermore summation of large numbers of auditory nerve fiber inputs could lead to a dynamic range larger than that of single nerve fibers alone. Complex tuning with multiple excitatory frequencies (Sachs and Sinnot, 1978) may also be related to neuronal morphology of the bi- or multipolar neurons. Spatially seperated tufts may pick up input from different regions throughout the frequency representation.

Physiological properties of the stubby cells in NAm, are much more difficult to predict. Due to the lacking dendrites it is clear that a linear integration of synaptic input is unlikely. These neurons should rather have a small dynamic region. The higher numbers of smaller synaptic terminals on the neurons together with the higher membrane capacity compared to NM on the other side should not allow the exact preservation of precise phase information like it is possible in the NM area (Sullivan and Konishi, 1984; Takahashi et al., 1984). If one takes into account that the medial parts of NA receive very low frequency input, NAm may probably be an adaptation to the processing of low frequency or infra sound, that the pigeon uses for orientation (Kreithen and Quine, 1979). Responses to infra sound have been recorded in the pigeon auditory nerve (Schermuly and Klinke, 1990) and cochlear nuclei of the chicken (Warchol and Dallos, 1989). A complementary low frequency specialisation has also been observed in the barn owl NM (Köppl and Carr, 1997).

Comparison of Cell Types in the Avian and Mammalian Cochlear Nuclei

As has been shown by Jhaveri and Morest (1982) neurons of nucleus magnocellularis can be compared to the bushy cell of the mammalian ventral cochlear nucleus (VCN). Our data shows that the majority of NA neurons can be related to the stellate type of the mammalian VCN. The medium sized multipolar neurons observed in NAi and NAl exhibit almost identical morphology as intracellularly stained neurons shown to respond with a chopper pattern (Rhode et al.,1983; Wu and Oertel, 1984; Oertel, 1985). These neurons did not only have a general stellate or multipolar appearance, but show the same moderately tufted dendritic endings as the neurons in the avian NA. Furthermore giant cells of the VCN (Brawer et al,. 1974) could correspond to the large multipolar cells of the medial NAi.

So far only neuron-types from the mammalian VCN could be related to avian counterparts. With regard to the DCN there is no clear evidence for the presence of avian counterparts. Only the bipolar subtype shows some resemblance to the fusiform cell of the DCN, due to their polarized dendritic trees. But as the characteristic features like the dendritic spines of the fusiform cells and the layered organization of DCN are missing in the pigeon, we cannot take this as a corresponding type. It is therefore concluded that an avian counterpart of the DCN or its neurons are lacking.

REFERENCES

Boord, R.L.(1969) The anatomy of the avian auditory system. Ann. N.Y. Acad. Sci. 167,186–198.

Boord, R.L. and Rasmussen, G.L (1963) Projection of the cochlear and lagenar nerves on the cochlear nuclei of the pigeon. J. Comp. Neurol. 120, 463–475.

Brawer, J.R., Morest, K.D. and Kane, E.C. (1974) The neuronal architecture of the cochlear nucleus of the cat. J. Comp. Neurol. 155, 251–300.

Carr, C.E. and Boudreau, R.E. (1993) Organization of the nucleus magnocellularis and the nucleus laminaris in the barn owl: Encoding and measuring interaural time differences. J. Comp. Neurol. 334, 337–355.

Cayal, S.R. (1908) Les ganglions terminaux du nerf acoustique des oiseaux. Travaux Du Laboratorio De Investgationes Biologica (Madrid) 6, 195–224.

Jhaveri, S. and Morest, D.K. (1982) Neuronal architecture in nucleus magnocellularis of the chicken auditory system with observations on nucleus laminaris: a light and electron microscope study. Neuroscience. 7, 809–836.

Köppl, C. and Carr, C. E. (1997) A low-frequency pathway in the barn owl's auditory brainstem. J. Comp. Neurol., in press.

Kreithen, M.L. and Quine, D.B. (1979) Infrasound detection by the homing pigeon: A behavioral audiogram. J. Comp. Physiol. 129, 1–4.

Oertel, D. (1983) Synaptic responses and electrical properties of cells in brain slices of the mouse anteroventral cochlear nucleus. J. Neurosci. 3, 2043–2053.

Oertel, D. (1985) Use of brain slices in the study of the auditory system: spatial and temporal summation of synaptic inputs in cells in the anteroventral cochlear nucleus of the mouse. J. Acoust. Soc. Am. 78, 328–333.

Parks, T.N. and Rubel, E.W. (1978) Organisation and development of the brain stem auditory nuclei of the chicken: primary afferent projections. J. Comp. Neurol. 180, 439–448.

Rhode, W.S., Oertel, D and Smith, P.H. (1983) Physiological response properties of cells labeled intracellularly in cat ventral cochlear nucleus. J. Comp. Neurol. 213, 448–463.

Sachs, M.B. and Sinnot, J.M. (1978) Responses to tones of single cells in nucleus magnocellularis and nucleus angularis of the redwinged blackbird (Agelaius phoeniceus). J. Comp. Physiol. 126, 347–361.

Schermuly L. and Klinke, R. (1990) Infrasound sensitive neurons in the pigeon cochlear ganglion. J. Comp. Physiol. 166, 355–363.

Stensaas, L. J. (1967) The development of hippocampal and dorsolateral pallial regions of the cerebral hemisphere in fetal rabbits. I. Fifteen millimeter stage, spongioblast morphology. J. Comp. Neurol. 129, 59–70.

Sullivan, W.E. (1985) Classification of response patterns in cochlear nucleus of the barn owl: correlation with functional response properties. J. Neurophysiol. 53, 201–216.

Sullivan, W.E. and Konishi, M. (1984) Segregation of stimulus phase and intensity coding in the cochlear nucleus of the barn owl. J. Neuroscience 4, 1787–1799.

Takahashi, T., Moiseff, A. and Konishi, M. (1984) Time and intensity cues are processed independently in the auditory system of the owl. J. Neuroscience 4, 1781–1786.

Warchol, M.E. and Dallos, P. (1989) Neural response to very low-frequency sound in the avian cochlear nucleus. J. Comp. Physiol. 166, 83–95.

Wu, S.H. and Oertel, D. (1984) Intracellular injection with horseradish peroxidase of physiologically characterized stellate and bushy cells in slices of mouse anteroventral cochlear nucleus. J. Neurosci. 4, 1577–1588.

9

MEASUREMENT OF SHORT-TIME SPATIAL ACTIVITY PATTERNS DURING AUDITORY STIMULATION IN THE STARLING

Udo Häusler[*]

German Primate Center
Department of Neurobiology
Kellnerweg 4, 37077 Göttingen, Germany

INTRODUCTION

The field L complex in birds is a highly structured telencephalic area dedicated to auditory processing. It is composed of an thalomo-recipient layer (L2a) surrounded by different types of layered secondary areas arranged in parallel to the forebrain ventricle and the lamina hyperstriatica. It has been described in the pigeon, and the guinea fowl and reaches its highest differentiation in songbirds (Häusler, 1989; Häusler and Leppelsack, in prep; Fortune and Margoliash, 1992, Vates et al.,1996). The field L complex in songbirds has been shown to be tonotopically organized with a tonotopic area that has an increasing bestfrequency from dorsal to ventral (Leppelsack, 1981; Müller and Leppelsack, 1983; Rübsamen and Dörrscheid, 1986). Our attempt was to link the different anatomical areas to physiological response properties Furthermore it is important to see if the different areas do have their own seperated tonotopic organisation or if they form a continuous area only differing in other properties like temporal response patterns.

To achieve this goal a very detailed mapping of auditory responses had to be made. Not only obviously responsive areas had to be mapped but also weakly or nonresponsive areas that anatomically belong to the field L complex. Thefore an automated electrophysiological micromapping method was developed, that allows spatially homogenious sampling with a high spatial and temporal resolution. The spatio-temporal course of the neural activity during auditory stimulation was then directly visualized for a comparison of electrophysiological and anatomical data.

The starling was used as an experimental animal because it is known for its ability to copy a large variety of extraspecific and artificial acoustic signals including human speech. Therefore it is a very good model for generalized auditory analysis, comparable in

* tel.: 049-551-3851-256, E-mail: uhausl@www.dpz.gwdg.de

Acoustical Signal Processing in the Central Auditory System
edited by Syka, Plenum Press, New York, 1997

efficiency with the mammalian system. It is also tempting to find differences and parallels between the mammalian and avian systems.

METHODS

A chronic method (Kirsch, Coles, Leppelsack 1980) was used to record from 7 adult starlings. During recording the animals were awake and located inside a sound-proof unechoic chamber. Auditory stimuli were presented under free field conditions. Stimuly were generated through a computer controlled B&K Generator (Typ 1027). Different series of pure tone pulses (75 dB Spl, 250 ms duration, 4 ms rise fall time) were used with frequencies between 0.1 and 10.0 kHz.

Neuronal multiunit activity was recorded through glass isolated platinum - iridium electrodes with a free tip of 10–20 μm. Action potentials were detected by a window discriminator (Frederic Hair), and fed into a PDP 11/73 microprocessor. Movement artifacts were continously monitored by the computer.

Electrode penetrations were completely computer controlled. Recordings were made every 80 μm along the electrode track. For every recording locus the responses to at least five repetitions of a stimulus series were recorded. If movement artifacts were detected during one repetition this repetition was discarded and repeated until six or eight artifact free repetitions were recorded. If this was not possible within 5 minutes the recording was done anyway but the artifact prone responses were labeled. This procedure was used to assure a minimum amount of data even under bad recording conditions. Data was also recorded irrespective of the actual neuronal response, so that complete response properties were available for all recording locations. This was a necessary requirement for the intended analysis of spatio-temporal activity patterns. Electrophysiological mapping was done in planes of frontal orientation. A row of 10 to 15 dorsoventral penetrations with a medio-lateral distance of 160 or 200 μm were done. This results in up to 800 recordings per plane. In single animals up to 3 different anatomical planes were mapped, and in total 10 complete planes were mapped in 7 animals (2 males, 5 females).

Calculation of Spatio-Temporal Activity Maps

Spatiotemporal activitymaps were calculated for single frontal anatomical planes, in order to have a good correlation with the corresponding histology. Post stimulus time histograms (PSTH) of the responses to all stimuly at all recording sites within one anatomical level were used to calculate maps. PSTH's with 50 ms or 4.5 ms bin width were used. The bins were chosen in a way that one of the bins exactly started at the beginning of the stimulus. Significance of the response during stimulation was tested with a binomial test, comparing pre - stimulus activity with the neural response during and after stimulation.

For a comparison of neural activity at different recording positions the absolute spikerate of a multiunit recording is an inadequate measure. This is mainly due to the fact that the electrode does not always pick up the same number of neural units. Depending on the actual location, the size of the electrode tip and the density of neuronal elements, the recordings may show quite different absolute spike rates. This makes a comparison of neural activity between recordings in seperate penetrations a difficult task. Therefore a relative measure of activation was chosen that may be comparable even between rather different recordings. The measure is the relation of the spontaneous activity, measured during a pre-stimulus interval of 200 ms, to the spikerate within the test bin of the PSTH during stimulation. The measure is expressed in percent activation, with the spontaneous activity taken as 100%. As this measure favours activation over inhibition, the result of a binomial test with a p< 0.05 (two tailed) was used to indicate wether significant inhibition or excitation was present at a certain location. The data of relative activity within single time slices was spatially smoothed and resampled to a regular grid, which then could be plotted as contour maps of relative iso-activity lines. Resampling was necessary as the recording positions could not always be placed on a regular grid.

In addition to the spatio-temporal activity maps a best frequency was determined for every recording site.

RESULTS

Activity maps immediately before the start of the auditory stimulus (Fig. 1 column 1) showed relative activity centered around 100% with little deviation. A significant difference of activity from the total spontaneous activity was observed at less than 2 % of the total number of recorded locations, and was scattered throughout the map. At the onset of stimulation with pure tones (Fig.1 column 2) activity maps generally show a strong activation of large areas throughout the field L complex. Within the activated area a couple of highly activated locations were observed, that showed activation up to 2300 % compared to spontaneous ativity. With ongoing stimulation, the size of the activated area was reduced to several focal regions that in most cases coincided with the highly activated areas during the onset response. Depending on the stimulus and rostrocaudal level within the field L complex, up to 5 spatially seperated focal regions of neuronal activation were observed. The most prominent region was located medial and mediodorsal and was activated with frequencies from 0.1 to about 6 kHz. The location of this region strongly depended on stimulus frequency and therefore is clearly tonotopically organized. A seperated medioventral region was observed during stimulation in the frequency range between 1 and 3 kHz. A dorsolateral as well as a ventrolateral region was present during the stimulus onset throuhout 0.1 to 6 kHz. Responses in regions adjacond to the field L complex as described by Fortune and Margoliash (1992) were observed most prominently in the ventral hyperstriatum, in the paleostriatum as well as the neostriatal regions directly underneath HVc and close to the forebrain ventricle.

With regard to their temporal characteristics the regions differed largely. Phasic tonic responses were observed in the medial and late tonic activation in the medioventral region. Onset activation was present in the dorsolateral and ventrolateral regions. Strongest changes in the extent of the activated areas were observed during the onset of stimulation. The most common case is that neurons at the "outer border" of the activated areas have a shorter time constant than those at the center, therefore activated areas tend to get reduced in size during stimulation. Another case is that different time constants are asymmetrically distributed, resulting in a spatially changing center of the activated area. Such cases are mainly observed in the dorsomedial area during low frequency stimulation.

Frequency Representation within Field L Complex

The largest part of the field L complex showed a best frequency increasing from dorsal to ventral (Fig.2). The low frequency representation formed a dorso-medial triangular shaped zone. A prominent high frequency representation formed a closed medioventral area below which bestfrequency decreases again. Isofrequency lines are running from dorso-medial to ventro-lateral in the low and mid frequency region (Fig.2, frequency bands 0–3), while in the high frequency region they run in mediolateral direction (frequency bands 4–6). Isofrequency lines therefore are oriented parallel to the anatomical layering. In the dorsal aspect of the field L complex best frequency increases from medial to lateral forming a high frequency representation at about 1.8 mm lateral. Lateral to this best frequency again decreases. In the dorsal part the isofrequency lines appear to run from dorsal to ventral, again perpendicular to the anatomical layering. Frequency representation in the ventro-lateral part of the field L complex is less clearly organized but appears as a continuation of the isofrequency-lines observed in the prominent medial region. Therefore four seperated frequency representations were proposed with their iso-frequency lines perpendicular to the anatomical layering.

DISCUSSION

The described automated micromapping method successfully demonstrates spatial activation patterns and the extent and outline of different processing areas in single animals with high spatial resolution. Compared to the deoxyglucose method, which visualizes spatial activation patterns by the amount of glucose consumption in a specific region, the temporal features of the different areas are easily seen by the sequence of the actvity maps. Furthermore inhibition may be seperated from activation. The method allows a much more detailed description of physiological organisation in the field L complex compared to previous investigations. While only one prominent tonotopic area has been recognized by previous investigations (Rübsamen and Dörrscheid, 1986; Müller and Scheich,1985; Müller and Leppelsack, 1985), at least four different auditory processing areas in the starlings field L complex were demonstrated. The different areas show different temporal properties. The largest area shows phasic-tonic responses and may be regarded as a primary processing area. The other areas have either pronounced phasic responses only, or show rather long latencies. They may therefore be regarded as secondary regions. All of the areas are organized tonotopocally.

It is important that the main focal regions are located in different anatomical subdivisions. While the phasic-tonic activation pattern is observed in the subdivision L of Fortune and Margoliash (1992), which represents zone 8 in Häusler (1989), the phasic area is mainly located inside L2b. On the other hand isofrequency layers are running perpendicular to the layering described for L , L2b and their surrounding regions (Häusler, 1989). It is therefore concluded that L and L2b form a continuous anatomical structure in which frequency is represented in columns that cross the layering. The auditory processing areas (L1, L2, L3) that have been demonstrated in the guinea fowl (Bonke et al., 1979; Heil and

→

Figure 1. Examples of spatio-temporal activity maps in response to three different pure tones within a frontal plane through the field L complex of the starling. Demonstrated is the relative neuronal activity throughout the recording positions. The lines indicate locations with similar neuronal activity within field L complex of the starling. Numbers at the isoactivity-lines indicate the percentage of activation or inhibition with respect to spontaneous acitvity (100% = spontaneus activity). Upper row: stimulation with 0.54 kHz, middle row: 1.87 kHz, lower row: 6.43 kHz. Temporal resolution is 50 ms for each map. The first column represents activity from 50 ms before stimulus onset, the second column represents the first and the thind column the second 50 ms of the stimulus. The open squares indicate significant inhibition, filled squares excitation. Anatomical position is indicated on the left side of Fig.2.

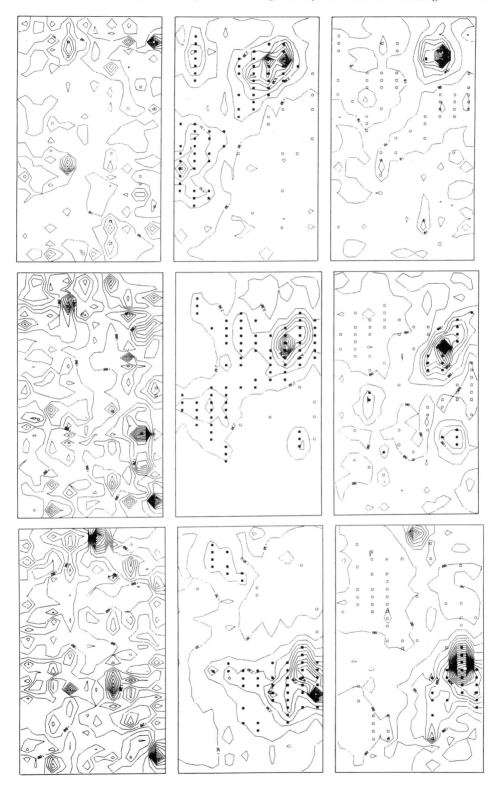

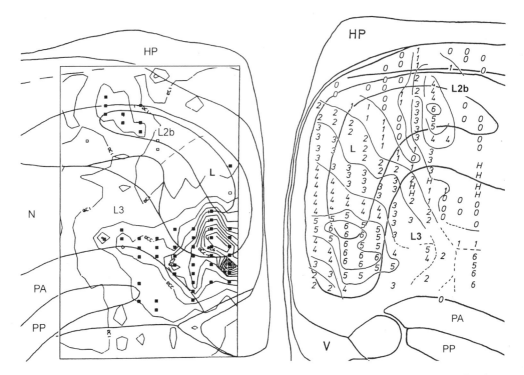

Figure 2. Left side: Anatomical positions of the activity maps in Fig.1 mediolateral distance of recording locations is 200 μm. Right side: Frequency representation in Field L complex demonstrated in a frontal section. Numbers indicate bestfrequency ranges: 0: 0.23 - 0.62 kHz; 1: 0.63 - 1.15 kHz; 2: 1.16 - 1.73 kHz; 3: 1.74 - 2.73 kHz; 4: 2.74 - 3.90 kHz; 5: 3.91 - 4.77 kHz; 6: 4.78 - 5.86 kHz; H-Inhibition only. Lines indicate iso-frequency lines. N -neostriatum, PA- paleostriatum augmentatum, PP-palaeostriatum primitivum, HP- hippocampal area, HV- hyperstriatum ventrale, V-forebrain ventricle. L, L2b, L3- Subdivisions of field L complex.

Scheich, 1985) and various other birds (Müller and Scheich, 1985) fit well into this organisation. The different areas demonstrated in this study are clearly spatially seperated and show separated tonotopic organisation. As the most prominent medial area shows primary phasic tonic responses, one may compare this area with the primary auditory cortex (AI) in mammals, while the other areas may be compared with the mammalian secondary cortical areas.

REFERENCES

Capsius, B. and Leppelsack, H. J.(1996) Influence of urethane anestesia on neural processing in the auditory cortex analouge of a songbird. Hear. Res. 96, 59–70.

Bonke, D., Scheich, H., Langner, G. (1979). Responsiveness of units in the auditory neostriatum of the guinea fowl (Numida meleagris) to species-specific calls and synthetic stimuli. I. Tonotopy and functional zones. J. Comp. Physiol. 132, 243–255.

Fortune, E. and Margoliash, D.(1992). Cytoarchitectonic organisation and morphology of cells of the fiel L complex in male zebra finches (Taenopygia guttata). J. Comp. Neurol. 325, 388–404.

Häusler, U. (1988). Topografy of the thalomotelencephalic Projections in the auditory system of a Songbird. Auditory pathway structure and function. New York and London, Plenum Press. 197 - 202.

Häusler, U. (1989). Die strukturelle und funktionelle Organisation der Höhrbahn im caudalen Vorderhirn des Staren (Sturnus vulgaris). Ph. D. Thesis. Technische Universität München, Institut für Zoologie.

Heil, P. and Scheich, H. (1985). Quantitative analysis and two-dimensional reconstruction of the tonotopic organization of the auditory field L in the chick from 2-deoxyglucose data. Exp. Brain Res. 58, 532–543.

Kirsch, M., Coles, R., Leppelsack, H.J. (1980). Unit recordings from a new auditory area in the frontal neostriatum of the awake starling(Sturnus vulgaris). Exp. Brain Res. 38, 3750–380.

Leppelsack, H. J. (1981). Antwortverhalten auditorischer Vorderhirnneurone eines Singvogels unter besonderer Berücksichtigung des Gesanglernens. Abteilung für Biologie - Ruhr-Universität Bochum.

Müller, C. M. and Leppelsack, H.J. (1985). Feature extraction and tonotopic organization in the avian auditory forebrain. Exp. Brain Res. 59, 587–599.

Müller, S. C. and Scheich, H. (1985). Functional organization of the avian auditory field L. A comparative 2DG study. J. Comp. Physiol. 156, 1–12.

Rübsamen, R. and Dörrscheidt, G.J. (1986). Tonotopic organization of the auditory forebrain in a songbird, the European starling. J. Comp. Physiol. 158, 639–646.

Scheich, H., Bonke,D.and Langner, G. (1979). Tonotopy and analysis of wide-band calls in Field L of the guinea fowl" Exp. Brain Res. Suppl. II, 94–109.

Vates, E.,Broome, M., Mello, V., Nottebohm, F. (1996). Auditory Pathways of caudal Telencephalon and their relation to the song system of adult male zebra finches (Taenopygia guttata). J. Comp. Neurol. 366, 613–642.

10

DISTRIBUTION AND TARGETING OF GLUTAMATE RECEPTORS IN THE COCHLEAR NUCLEUS

R. J. Wenthold,[*] Y. -X. Wang, R. S. Petralia, and M. E. Rubio

Laboratory of Neurochemistry
National Institute on Deafness and Other Communication Disorders, NIH
Bethesda, Maryland

1. INTRODUCTION

Glutamate (or a related excitatory amino acid) is a major excitatory neurotransmitter in the mammalian cochlear nucleus (for reviews, see Wenthold, 1991; Hunter et al., 1992; Wenthold et al., 1993). Two main excitatory pathways which terminate in the cochlear nucleus, the auditory nerve and the parallel fiber pathway, both use glutamate as a neurotransmitter, and, although the evidence is less compelling, most other excitatory synapses in the cochlear nucleus probably also use glutamate as a neurotransmitter. The dominant role of glutamate in the cochlear nucleus is consistent with its role elsewhere in the CNS where it is thought to be a neurotransmitter at most excitatory synapses (Monaghan et al., 1989). In addition to this major role in neurotransmission, two other properties have also contributed to the intense interest in glutamate: 1) Glutamate and glutamate analogs are potent neurotoxins, and this neurotoxicity is mediated through the same receptors which are involved in neurotransmission (for review, see Choi and Rothman, 1990). Human disorders, such as stroke, may involve a release of glutamate which leads to further neuronal loss through excitotoxicity. Therefore, blocking glutamate receptors would be expected to lessen the damage, and animal studies have indicated this may be a promising approach to treatment. Furthermore, neuronal loss that occurs with some neurodegenerative disorders may involve excitotoxicity suggesting that a similar treatment may be promising. 2) Glutamate receptors are involved in synaptic plasticity such as in learning and memory or synapse formation. In one model it is proposed that long-term potentiation (LTP) results from an increase in functional AMPA receptors at the postsynaptic membrane (Liao et al.,

* Correspondence: Dr. Robert J. Wenthold, NIH, Building 36, Room 5D08, Bethesda, MD 20892. Telephone: 301-496-6530; fax: 301-480-2324; E-mail: wenthold@pop.nidcd.nih.gov.

1995). Therefore, knowledge of the mechanisms which regulate the expression of synaptic glutamate receptors will give insights into the general area of synaptic plasticity.

As discussed in more detail below, the glutamate receptor family is complex, consisting of multiple subtypes and subunits which form receptors with different physiological and pharmacological properties. Determining the roles of these different receptors is essential to understanding the function of glutamate receptors in normal neurotransmission, synaptic plasticity, and excitotoxicity. Our research on glutamate receptors in the cochlear nucleus is directed at characterizing the roles of glutamate receptors in neurotransmission in various neuronal populations and at determining the roles of glutamate receptors in synaptic plasticity and excitotoxicity.

2. GLUTAMATE RECEPTORS

Although glutamate receptors had been studied for some time with physiology and ligand binding assays, it was the cloning of the first three members of the glutamate receptor family in 1989 (Gregor et al., 1989; Hollmann et al., 1989; Wada et al., 1989) that dramatically changed this field of study. Efforts to clone additional glutamate receptors over the past six years (for review, see Sommer and Seeburg, 1992; Hollmann and Heinemann, 1994) have identified at least 16 ionotropic receptor subunits and 8 metabotropic receptors (Table 1). Ionotropic receptors contain an ion channel as a part of the receptor molecule and are made up of multiple subunits. Metabotropic receptors do not contain ion channels but exert their actions through intracellular messengers. They function as single subunits and couple to G-proteins. Three different subtypes of glutamate ionotropic receptors have been identified: AMPA receptors, kainate receptors, and NMDA receptors. A fourth type, delta receptors, has also been identified, but these receptors are not functional when expressed in oocytes or transfected cells.

Table 1. Properties of ionotropic and metabotropic glutamate receptors. Only the most common and potent agonists and antagonists are listed. ^{3}H-Glu has been shown to bind to recombinant mGluR1, but may bind to all mGluR's

Receptor subtypes	Subunits	Agonists	Antagonists	Radioactive ligands
AMPA	GluR1-4	AMPA, KA, QA, Glu	Quinoxalinediones	^{3}H-AMPA
Kainate	GluR5-7	KA, QA, DM, Glu	Quinoxalinediones	^{3}H-KA
	KA1&2			
NMDA	NR1	NMDA, Glu	D-APV, CPP	^{3}H-CPP, ^{3}H-Glu
	NR2A-D	Gly	MK801	^{3}H-MK801
Delta	Delta 1&2			
mGluR1 & 5		QA, IBO, Glu		^{3}H-Glu
mGluR2 & 3		Glu, t-ACPD, IBO		
mGluR4, 6, 7 & 8		L-AP4, Glu, L-SOP		

Abbreviations: AMPA, α-amino-3-hydroxy-5-methyl-4-isoxazolepropionate; KA, kainic acid; QA, quisqualic acid; Glu, glutamic acid; DM, domoic acid; NMDA, *N*-methyl-D-aspartic acid; Gly, glycine; APV,2-amino-5-phosphovaleric acid; CPP, (±)-3-(2-carboxypiperazin-4-yl)propyl-1-phosphonic acid; MK801, (+)-5-methyl-10, 11-dihydro-5*H*-dibenzo[*a,d*]cyclohepten-5, 10-imine hydrogen maleate; IBO, ibotenic acid; t-ACPD, trans-1-aminocyclopentane-1,3-dicarboxylate; L-AP4, L-2-amino-4-phosphonobutanoic acid; L-SOP, L-serine-O-phosphate. From Petralia and Wenthold, 1996

2.1. AMPA Receptors

The AMPA receptor is the major non-NMDA glutamate receptor and is responsible for mediating most fast excitatory neurotransmission in the mammalian CNS. AMPA receptors are made up of four subunits, GluR1–4 (or GluR A-D) (for review, see Sommer and Seeburg, 1992; Hollmann and Heinemann, 1994). These subunits have molecular masses of about 100,000 Da and are similar to one another with about 70% amino acid identity. Each subunit can also exist in two major splice variants, flip and flop, with different functional properties (Sommer et al., 1990). The subunits are integral membrane proteins with three transmembrane domains giving a membrane topology such that the N-terminus is extracellular and the C-terminus is intracellular, unlike other ionotropic receptors, such as the $GABA_A$ receptor and the nicotinic acetylcholine receptor, which have four transmembrane domains (for review, see Wo and Oswald, 1995). The ligand binding site is made up of parts of the N-terminus domain and the loop formed between the second and third transmembrane domains (Stern-Bach et al., 1994). Each AMPA receptor subunit can form a functional receptor by itself, or in various combinations of subunits, when expressed in oocytes or transfected cells (Keinänen et al., 1990; Boulter et al., 1990). In brain it is believed that most receptor complexes are made up of more than one subunit, although studies on the distribution of mRNA and protein show some neurons express predominately only one subunit, and would, therefore, be expected to form homomeric receptors (for review, see Petralia and Wenthold, 1996). Ion channels of most native AMPA receptors are permeable to monovalent cations and not to the divalent cations, magnesium and calcium. However, recombinant AMPA receptors GluR1, 3 and 4, expressed in oocytes and transfected cells pass calcium through their ion channels. Calcium does not pass through homomeric GluR2 channels, and GluR2 has a dominant effect when co-expressed with other subunits in forming a channel which is calcium impermeable (Hollmann et al., 1991; Verdoorn et al., 1991). The lack of calcium permeability of GluR2 containing receptors is due to a single amino acid difference in the part of the molecule that lines the ion channel; in GluR2, the amino acid is arginine while in the other AMPA receptors it is glutamine. Therefore, native receptors will vary in their calcium permeability properties based on their subunit compositions, with receptors lacking GluR2 being calcium permeable. The significance of calcium permeable ion channels is that it is believed that excess calcium influx through AMPA receptors may play a major role in excitotoxicity, and neurons that express calcium permeable channels may be more vulnerable to elevated levels of glutamate (Choi and Rothman, 1990). Even low levels of calcium occurring through normal neurotransmission may trigger postsynaptic changes that do not occur with calcium impermeable channels.

2.2. Kainate Receptors

The kainate receptor is made up of five subunits, GluR5–7 and KA1 and KA2 (for review, see Sommer and Seeburg, 1992; Hollmann and Heinemann, 1994). Like the AMPA receptor subunits, the kainate receptor subunits are large, with molecular masses of 100,000 Da or larger. GluR5–7 are similar with 70–80% amino acid sequence identity, and KA1 and KA2 have a 68% sequence identity, but are only about 40% identical to GluR5–7. When expressed in oocytes or transfected cells, GluR5 and 6 respond to kainate and glutamate with rapidly desensitizing currents, while GluR7 and KA1 and 2 are inactive as homomeric receptors. However, these subunits do combine with GluR5 and 6 to form functional receptors. In brain, it is believed that most kainate receptors are heteromeric (Wenthold et al., 1994), but considerably less is known about native kainate receptors than AMPA receptors. Kainate receptors have been difficult to identify physiologically in the CNS. This is due to several factors. They are difficult to distinguish from AMPA receptors since they respond to the same agonists and are blocked by the same antagonists. They can be distinguished by the differential effects of cyclothiazide and ConA on their desensitizations, with cyclothiazide blocking desensitization of AMPA receptors and ConA blocking desensitization of kainate receptors (Partin et al., 1993). While AMPA and kainate receptors cannot be easily differentiated physiologically, their ligand binding properties are very different; kainate receptors bind ^{3}H-kainate while AMPA receptors bind ^{3}H-AMPA (Hampson et al., 1987; Hunter et al., 1990). A second factor contributing to the difficulty in detecting kainate receptors is their relatively low abundance; while distribution studies show their expression is widespread in the CNS, they are about 50 fold less abundant than AMPA receptors, based on total ligand binding (Hampson et al., 1987; Hunter et al., 1990).

2.3 NMDA Receptors

Five NMDA receptor subunits, NR1 and NR2A-D, have been identified (for review, see Sommer and Seeburg, 1992; Hollmann and Heinemann, 1994). The NR1 subunit has a molecular mass of 105,000 Da, while the NR2 subunits are somewhat larger, ranging from 133,000 to 163,000 Da. The NR2 subunits have about a 50% sequence identity with one another and are about 25% identical to NR1. An additional subunit, χ-1 or NMDAR-L, has been reported to affect NMDA receptor activity in oocytes, but there is insufficient information as to whether or not this subunit is normally associated with NMDA receptors (Ciabarra et al., 1995; Sucher et al., 1995). The NR1 subunit forms a functional ion channel, which responds weakly to agonists, when expressed alone. The NR2 subunits are not functional alone, but combine with NR1 to form functional receptors with properties determined by the particular NR2 subunit. As is the case of the other ionotropic glutamate receptors, native NMDA receptors are thought to be heteromers comprised of an NR1 subunit and one or more NR2 subunits. The NMDA receptors are physiologically and pharmacologically distinct from AMPA and kainate receptors. NMDA receptors are voltage dependent and their channels open only when the neuron is depolarized, such as following AMPA receptor activation. A key property of NMDA receptors is that their ion channels are permeable to calcium, and this characteristic has been linked to many of the long-term effects of NMDA receptors. While the NR2 subunits are primarily responsible for the physiological properties of the receptor complex, the NR1 subunit can exist in eight different splice variants, each with different physiological properties and different distributions (for review, see Zukin and Bennett, 1995).

2.4 Delta Receptors

Two delta receptor subunits, delta 1 and delta 2, have been identified (for review, see Sommer and Seeburg, 1992; Hollmann and Heinemann, 1994). The two subunits are 56% identical to one another and about 18–25% identical to other ionotropic receptor subunits. The delta receptors are similar to other ionotropic receptors in size and membrane topology, but do not form functional receptors when expressed either alone, together, or with other ionotropic subunits. Based on sequence similarities, it is assumed that delta 1 and delta 2 are members of the same receptor subfamily and combine to produce a receptor complex. But this has not been established; in fact, the very different distributions of the two subunits rule out a delta1/delta 2 complex in most cases. The most likely explanation for the lack of function of the delta receptors is that additional subunits are required to form functional receptors, but extensive searches have not identified these missing subunits.

2.5 Metabotropic Receptors

The metabotropic receptors differ from ionotropic glutamate receptors in structure, pharmacology, and function (for reviews, see Schoepp et al., 1990; Pin and Duvoisin, 1995). They are members of the G-protein coupled receptor family with a characteristic seven transmembrane topology. mGluR1 and 5 are coupled to the phosphoinositol signal transduction pathway while mGluR2, 3, 4, 6, 7 and 8 are linked to the inhibition of the cAMP cascade. Therefore, the effects of stimulation of metabotropic receptors are slower than those of ionotropic receptors indicating a function as a synaptic modulator. Unlike ionotropic receptors which are predominantly postsynaptic, metabotropic receptors are both pre- and postsynaptic (Petralia and Wenthold, 1996).

2.6 Complexity of the Glutamate Receptor Family

The large variety of receptor subtypes, subunits, and splice variants endows the CNS with the capability of generating a wide range of different receptor complexes and receptor combinations that will give different physiological responses to the same neurotransmitter. The fact that these mechanisms for creating diversity have survived, and likely were enhanced by, the evolutionary process reflects the need for the nervous system to have available a wide range of receptors with distinct functional properties. While all glutamate receptors respond to glutamate, the functional impact of this interaction will vary from synapse to synapse depending on the types and combinations of receptors expressed. The effect of glutamate on AMPA receptors will differ from that on NMDA receptors or mGluR1 receptors, but even within a subtype of ionotropic receptor, very different receptors can be generated based on the subunit combination. The most dramatic example of this is the AMPA receptor that forms a calcium permeable or impermeable channel based on the absence or presence of a single subunit, GluR2 (Hollmann et al., 1991; Verdoorn et al., 1991). But other properties, such as desensitization rate, are also related to subunit composition (Mosbacher et al., 1994). In addition, most subunits (as well as metabotropic receptors) exist in multiple splice variants, each with different properties (Hollmann and Heinemann, 1994). The NR1 subunit has the largest number (eight) of known splice variants among the glutamate receptors; these variants could generate a large number of functionally distinct receptor complexes by combining with the four NR2 subunits in different proportions to form a pentameric structure (Zukin and Bennett, 1995). In analyzing the distribution of subunits of glutamate receptors throughout the brain using either in situ hybridization or immunocytochemistry, one finds that the subunits for a particular subtype have different, but overlapping, patterns of expression, showing that different neurons will have functionally different glutamate receptors (Petralia and Wenthold, 1996).

Most individual neurons express multiple subunits of any particular ionotropic receptor (Petralia and Wenthold, 1996). Also, most neurons receive multiple different excitatory inputs. A fundamental question, therefore, is whether or not each excitatory postsynaptic site has the same subunit composition for any particular ionotropic receptor. An immunoprecipitation study on the CA1 pyramidal neurons suggests that neurons may be able to form multiple receptor complexes with different subunit combinations (Wenthold et al., 1996). In CA1 neurons, which express predominantly GluR1, 2 and 3, few receptor complexes containing both

GluR1 and 3 were found, suggesting most complexes were GluR2/3 and GluR1/2. Some homomeric GluR1 complexes were also present. Therefore, one cannot conclude that a particular subunit combination exists at all synapses based solely on subunits expressed. Analysis of subunits present at specific synaptic populations is required.

Most neurons express multiple glutamate receptor subtypes. For example, Purkinje neurons of the cerebellum express subunits of AMPA, kainate, NMDA and delta ionotropic receptors as well as the metabotropic receptor, mGluR1 (Petralia and Wenthold, 1996). With immunocytochemistry, we have shown that a synaptic population can contain multiple different receptor subtypes. Therefore, in addition to the diversity generated by having multiple subunits and splice variants and different subunit combinations in receptor complexes, synaptic populations can differ based on the combination of subtypes of glutamate receptors that are expressed. If multiple subtypes of receptor are expressed in a neuron, can the neuron selectively sort them to different synaptic populations? To address this question, glutamate receptors expressed postsynaptic of the two excitatory synaptic populations on Purkinje cells, parallel fiber synapses and climbing fiber synapses, were analyzed (Landsend et al., 1997; Zhao et al., 1997). It was found that delta receptors are present only at parallel fiber synapses, while other receptors are present at both populations. Therefore, different synaptic populations on the same neuron may have different combinations of glutamate receptors and, therefore, respond differently to glutamate. Furthermore, the composition of receptors at a particular synapse may vary under different conditions. In the cerebellum, while the delta receptor is expressed only at the parallel fiber synapse in the adult, it is found at both parallel fiber and climbing fiber synapses in the 10 day old rat (Zhao et al., 1997), indicating that the receptor composition at the synapse is developmentally regulated.

3. GLUTAMATE RECEPTOR EXPRESSION IN THE COCHLEAR NUCLEUS

3.1. In Situ Hybridization Studies

While in situ hybridization studies on the cochlear nucleus are still limited, several ionotropic receptor subunits have been studied with in situ hybridization histochemistry, and these results show AMPA and NMDA receptors are the predominant subtypes expressed. AMPA receptor subunits are abundantly expressed throughout the cochlear nucleus, although the subunits expressed vary with cell types (Table 2) (Hunter et al., 1993). Each of the three general cell types, principal neurons, cartwheel cells, and granule cells, expresses a characteristic combination of subunits. Most notable is the virtual absence of GluR1 in principal neurons and the widespread and abundant expression of GluR4. Given its importance in determining ion flux through the AMPA receptor channel, expression of the GluR2 subunit was quantified in several principal cell types and compared with expression in Purkinje cells (Table 3). These results show that large cells of the ventral cochlear nucleus (VCN) express relatively low levels, raising the possibility that the AMPA receptors of these neurons may have calcium permeable ion channels. GluR2 expression in fusiform cells is similar to that of Purkinje cells.

In both the rat and mouse, most neurons abundantly express the NR1 subunit (in mouse it is called the $\zeta1$ subunit) (Watanabe et al., 1994; Sato et al., submitted for publication), although the splice variants of NR1 have a more restricted distribution (Hunter et al., 1995). The NR2 subunits (in mouse, $\varepsilon1$–4) have a more restricted distribution, how-

Table 2. Expression of AMPA receptor subunit mRNA in identified cells of the rat cochlear nucleus

	GluR1	GluR2	GluR3	GluR4
Ventral Cochlear Nucleus				
Large spherical cells	−	+	+	+
Globular cells	−	+	+	+
Octopus cells	−	+	+	+
Granule cells	−	+	+	+
Dorsal Cochlear Nucleus				
Fusiform cells	−	+	+/−	+
Cartwheel cells/				
small stellate cells	+	+	+	+
Granule cells	−	+	−	+

+, selective accumulation of silver grains over cell body. -, silver grains absent over cell bodies. +/-, silver grains present over subpopulations of neurons. Data from Hunter et al., 1993

ever (Watanabe et al., 1994). In mouse, NR2A is the only subunit expressed in the large cells of the ventral cochlear nucleus (VCN) while NR2C is rich in the granule cell cap. In the dorsal cochlear nucleus (DCN), NR2B is expressed in fusiform cells, while NR2A and NR2C are expressed in small and medium sized cells, probably granule cells and cartwheel cells.

Kainate receptor expression is low for all subunits in the rat cochlear nucleus, based on in situ hybridization (Hunter and Wenthold, 1994). GluR7 is the most abundantly expressed subunit and it is found in cartwheel cells, stellate cells, granule cells and large neurons of the VCN. Delta subunit expression in the cochlear nucleus has not been reported.

Metabotropic receptor expression in the cochlear nucleus has been noted as parts of general surveys, but expression in individual cell types has not been reported. From these studies, some receptors, such as mGluR1 and mGluR7, appear to be fairly highly expressed in the cochlear nucleus (Shigemoto et al., 1992; Okamoto et al., 1994).

3.2. Immunocytochemistry

Extensive immunocytochemical localization studies using a pre-embedding immunoperoxidase technique have been carried out on the dorsal cochlear nucleus (DCN; Petralia et al., 1996; Wright et al., 1996); in contrast, only preliminary data on immunocytochemical localization in the ventral cochlear nucleus have been published (VCN; e.g., Wenthold et al., 1993; Wang et al., 1996, and a series of earlier abstracts by this laboratory including Petralia and Wenthold, 1992) (Figs. 1–3). Most of these studies have focused on localization in rats, although a few preliminary studies have been performed on other animals (e.g., gerbil - Schwartz, 1994; mouse - Bilak et al., 1995). In general, immunocytochemical localization of proteins in the cochlear nuclei is similar to distribution of mRNAs, as shown with in situ hybridization (above).

AMPA receptor antibodies (GluR1, GluR2/3, GluR4) are differentially distributed in the DCN. GluR1 antibody stains chiefly the medium-sized cells (cartwheel/stellate cells) and the associated neuropil of the outer DCN. Staining with GluR1 antibody also is found in a small population of small/medium cells of the dorsal acoustic stria of the deep DCN, as well as similar cells in the intermediate acoustic stria. GluR2/3 antibody stains many

Table 3. GluR2 and GluR3 expression in cochlear nucleus neurons of the rat. Data are presented as mean grain density ± SEM. Data from Hunter et al., 1993

Cell type	Grains/100 μm^2	
	GluR2	GluR3
Spherical cells, AVCN	1.8 ± 0.1	4.2 ± 0.2
Globular cells, PVCN	2.7 ± 0.3	3.8 ± 0.6
Octopus cells, PVCN	1.7 ± 0.4	1.9 ± 0.6
Fusiform cells, DCN	8.2 ± 1.3	2.2 ± 0.4
Purkinje cells, cerebellum	9.9 ± 1.1	6.0 ± 0.4

cell types throughout the DCN, including prominent staining in medium neurons (cartwheel/stellate cells) and fusiform cells. Staining with GluR4 antibody is prevalent in the outer neuropil; neuron cell body staining generally is obscured with distinctive staining seen best in large neurons of the deep DCN. With electron microscopy (EM), all three antibodies label postsynaptic membranes and densities of parallel fiber/cartwheel cell synapses of the DCN; GluR1 and GluR4 immunostaining also is found in some glial wrappings of synapses. In the VCN, little staining is seen with GluR1 antibody, while most neuron types are labeled moderately to densely with GluR2/3 and GluR4 antibodies.

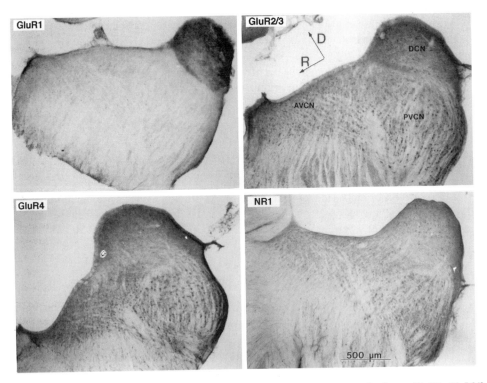

Figure 1. Parasagittal sections of the rat cochlear nucleus immunostained with antibodies to GluR1, GluR2/3, GluR4, and NR1. AVCN, PVCN, DCN = anteroventral, posteroventral, and dorsal cochlear nuclei, respectively. D, dorsal; R, rostral. (GluR2/3 micrograph modified from Petralia et al., 1996).

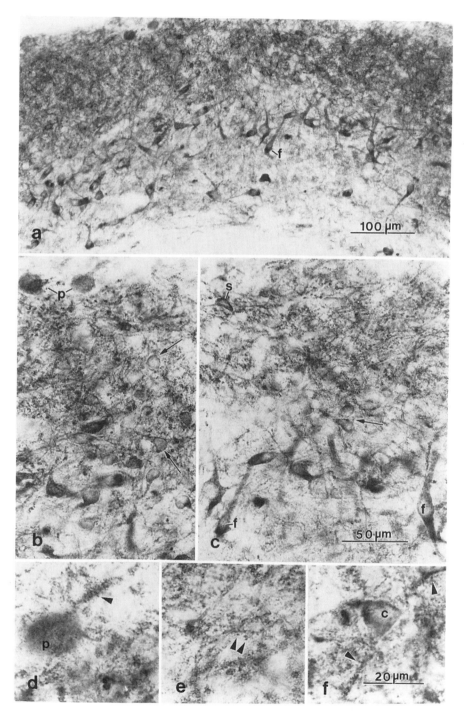

Figure 2. Single coronal section of DCN immunostained with antibody to mGluR1a, shown at medium (**a**), high (**b,c**), and very high (**d-f**) magnifications. c, cartwheel cell; f, fusiform cell; p, possible displaced Purkinje cell; s, stellate cell; arrows, medium-sized neurons (cartwheel/stellate cells). At very high magnification, possible stained dendrite spines (arrowheads) are evident. Dorsal is at top, right; lateral is at top, left. (from Petralia et al., 1996).

With EM, GluR2/3 and GluR4 antibodies produce significant staining in postsynaptic membranes and densities of auditory nerve endbulb synapses on spherical cells of the anteroventral cochlear nucleus (AVCN).

The kainate receptor antibodies, KA2 and GluR6/7, produce only light to moderate staining throughout the DCN and VCN (except for moderately dense staining with KA2 antibody in the auditory nerve nucleus/interstitial nucleus). In the DCN, GluR6/7 antibody produces moderate staining in some cartwheel and fusiform cells and in some medium/large neurons of the deep layers of the DCN. In contrast to these antibodies, a monoclonal IgM antibody, GluR5–7, produces dense staining in many neurons throughout the cochlear nucleus.

The NMDA receptor antibody, NR1, produces moderate staining in many cell types throughout the cochlear nucleus. The monoclonal antibody to NR1 produces an overall similar pattern of staining. In contrast, immunostaining with NR2A/B antibody is more limited; in the DCN, best staining is seen in fusiform cells.

Staining with delta 1/2 antibody is highest in the outer DCN (one of the highest levels in the brain), mainly in the neuropil and in medium-sized neurons (cartwheel/stellate cells). With EM, dense staining is common in postsynaptic membranes and densities of parallel fiber/cartwheel cell spine synapses (figure 3). Recently, in work done in collaboration with the laboratory of Dr. O.P. Ottersen, delta 1/2 immunolabeling has been demonstrated with post-embedding immunogold; with this method, a high-density of gold

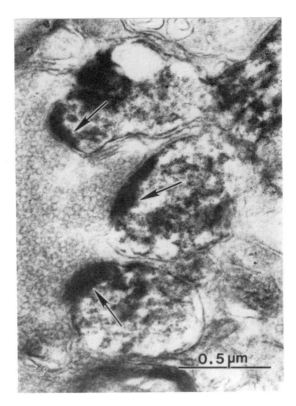

Figure 3. Electron micrograph of the DCN (Sprague-Dawley rat) immunostained (immunoperoxidase method) with delta 1/2 antibody. Dense staining is common in cartwheel cell dendrites, in postsynaptic membranes and densities (arrows) apposed to parallel fiber terminals. (from Petralia et al., 1996).

particles can be seen along the postsynaptic membranes and densities (figure 4). Delta 1/2 antibody produces only light to moderate staining in the VCN.

Each of three metabotropic glutamate receptor antibodies produces very different patterns of staining in the DCN (none produces more than moderate staining in the VCN). Staining with mGluR1a antibody is most prevalent in the outer DCN, including the outer neuropil and both fusiform cells and medium neurons (cartwheel/stellate cells) (figure 2). Staining of puncta in the neuropil of the outer DCN, including those found along the surface of cartwheel cell dendrites, is very distinctive (figure 2d-2f) and corresponds to dense staining seen with electron microscopy in postsynaptic membranes and densities of parallel fiber/cartwheel cell spine synapses. Interestingly, this pattern of dense and common staining seen in parallel fiber/cartwheel cell spine synapses is similar to the pattern seen with delta 1/2 antibody. In addition, both antibodies show a nearly identical pattern of staining in parallel fiber/Purkinje cell spine synapses of the cerebellum, indicative of the structural and functional similarities in these two circuits. mGluR2/3 antibody produces a very different pattern of staining in the DCN, with staining prevalent only in Golgi cells and unipolar brush cells, again nearly identical to the pattern of staining seen in the cerebellum. With mGluR2/3 antibody, stained Golgi and unipolar brush cells also are common throughout the granule cell domain of the cochlear nucleus. Using a different staining method, unipolar brush cells are reported to be labeled with mGluR1a antibody (Wright et al., 1996). With electron microscopy, mGluR2/3 antibody staining is common in unipolar brush cell brush-dendrites and in several kinds of presynaptic terminals, including possi-

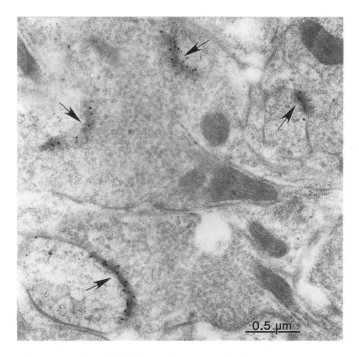

Figure 4. Electron micrograph of the DCN (Wistar rat) immunostained (immunogold method) with delta 1/2 antibody. Fifteen nanometer immunogold particles are highly localized to postsynaptic membranes and densities (arrows) apposed to parallel fiber synapses. Note the identical pattern of localization of immunolabeling seen with the two methods illustrated in Figs. 3 and 4. (R.S. Petralia, Y.-X. Wang, O.P. Ottersen, and R.J. Wenthold, unpublished data).

ble Golgi axon terminals associated with granule cell dendrites in mossy terminal glomeruli; thus mGluR2 and/or mGluR3 receptors may affect mossy terminal functions via various presynaptic and postsynaptic associations with Golgi and unipolar brush cell synapses. In contrast to mGluR1a and mGluR2/3 antibody staining, staining with mGluR5 antibody in the DCN is only light to moderate overall and is uncommon in cell bodies.

3.3. Functional Aspects of Cochlear Nucleus Receptors

The immunocytochemical studies suggest that AMPA receptors are the major receptors mediating transmission at the eighth nerve synapse in the cochlear nucleus and that the major AMPA receptor subunits expressed at this synapse are GluR3 and GluR4. Physiological studies show that AMPA receptors in the chick cochlear nucleus rapidly desensitize (Raman and Trussell, 1992), and in a study comparing desensitization rates of AMPA receptors throughout the nervous system, it was shown that those in the cochlear ganglion and cochlear nucleus desensitized the most rapidly (Raman et al., 1994). The characteristic of rapid desensitization is not limited to the chick or to the cochlear nucleus, since similar desensitization times were seen in the rat medial nucleus of the trapezoid body (Geiger et al., 1995). A rapidly responding and desensitizing receptor would be consistent with the temporal demands of the auditory system, but the question arises as to how the receptors achieve their rapid desensitization properties while AMPA receptors elsewhere in the CNS, which presumably are made up of the same subunits, are more slowly desensitizing. A subunit, or splice variant, expressed only in these neurons could account for this phenomenon. While this cannot be ruled out, extensive screening for cDNAs with homology to known glutamate receptors has not identified such a subunit. It is also possible that posttranslational modification, such as glycosylation or phosphorylation, or the presence of a protein that associates with the receptor complex, contributes to the desensitization properties of these receptors, although this has not been demonstrated. One factor that surely plays a role is the subunit composition of the auditory AMPA receptors. Compared to the rest of the CNS, auditory principal neurons express high levels of GluR4 and low levels of GluR1 and GluR2, and therefore, the predominant receptors are made up of GluR3 and 4. A study on recombinant AMPA receptors showed that the flop splice variants of both GluR3 and GluR4 desensitized much faster that the flip variants of the same subunit and either form of GluR1 and 2 (Mosbacher et al., 1994). Therefore, the subunit composition is likely to be a major factor, and perhaps the only factor, responsible for the uniquely fast desensitizing receptors of the cochlear nucleus.

Another characteristic of principal neurons in the cochlear nucleus is their low expression of GluR2, suggesting that the AMPA receptors on these neurons are calcium permeable. This is consistent with physiological studies showing a high calcium permeability of AMPA receptors in the chick cochlear nucleus (Otis et al., 1995). While the principal neurons of the cochlear nucleus are susceptible to kainate-induced excitotoxicity (Bird et al., 1978), as are most neurons in the CNS, there is no evidence that they are more sensitive to excitotoxins than other neurons or that excitotoxicity plays a role in disorders of the CNS auditory pathways. However, this is an area that should be explored in more detail.

While principal neurons of the cochlear nucleus in general have low levels of GluR2, one cell type that is an exception is the fusiform neuron of the DCN which expresses GluR2 at levels similar to those found in Purkinje neurons. Fusiform neurons also express greater amounts of other glutamate receptors, such as delta1/2 and mGluR1. A key difference between fusiform neurons and principal neurons of the VCN is that fusiform neurons receive two large glutamatergic inputs, the auditory nerve and the parallel fibers, while the other neurons receive only the auditory nerve input. It may be expected, then, that the receptors expressed at these two synaptic populations will be different (Rubio et al., 1996), as seen in the Purkinje neurons.

4. CONCLUSIONS

The in situ hybridization and immunocytochemistry studies clearly demonstrate the complexity of the glutamate receptor family in the cochlear nucleus; most, if not all, receptor types and subunits are expressed and show overlapping, but distinct, cellular distribution profiles. Most neurons express NMDA and AMPA receptors, but the subunit combinations vary and will, therefore, create receptors with different functional properties. Most neurons also express kainate and delta receptors as well as one or more metabotropic receptors. From the information available on the multiple receptors expressed in the cochlear nucleus, no clear pattern of receptor combinations emerges, but as discussed above, the significance of the receptor combinations may lie at the level of the synapse rather than the neuron. The availability of subunit- and subtype-selective antibodies to glutamate receptors and the development of high-resolution colloidal gold detection techniques will allow the identification of glutamate receptors expressed at specific cochlear nucleus synapses.

5. REFERENCES

Bilak, S., Bilak, M., and Morest, D.K. (1995) N-methyl-D-aspartate receptor (NMDAR1) expression in mouse cochlear nucleus (CN) neuronal cell types compared to cerebellar cortex. Assoc. Res. Otolaryngol. Abs. 18:31.

Bird, S.J., Gulley, R.L., Wenthold, R.J. and Fex, J. (1978) Kainic acid injections result in degeneration of cochlear nucleus cells innervated by the auditory nerve. Science 202, 1087-1089.

Boulter, J., Hollmann, M., O'Shea-Greenfield, A., Hartley, M., Deneris, E., Maron, C., and Heinemann, S. (1990) Molecular cloning and functional expression of glutamate receptor subunit genes. Science 249, 1033–1037.

Ciabarra, A.M., Sullivan, J.M., Gahn, L.G., Pecht, G., Heinemann, S., and Sevarino, A. (1995) Cloning and characterization of χ-1: A developmentally regulated member of a novel class of the ionotropic glutamate receptor family. J. Neuroscience 15, 6498–6508.

Choi, D.W. & Rothman, S.M. (1990) The role of glutamate neurotoxicity in hypoxic-ischemic neuronal cell death. Annual Review of Neurology 13, 171–182.

Geiger, J.R.P., Melcher, T., Koh, D.-S., Sakmann, B., Seeburg, P.H., Jonas, P., & Monyer, H. (1995) Relative abundance of subunit mRNAs determines gating and Ca^{2+} permeability of AMPA receptors in principal neurons and interneurons in rat CNS. Neuron 15, 193–204.

Gregor, P., Mano, I., Maoz, I., McKeown, M. and Teichberg, V.I. (1989) Molecular structure of the chick cerebellar kainate-binding subunits of a putative glutamate receptor. Nature 342, 689–692.

Hampson, D.R., Huie, D., and Wenthold, R.J. (1987) Solubilization of kainic acid binding sites from rat brain. J. Neurochem. 49, 1209–1215.

Hollmann, M., Hartley, M. and Heinemann, S. (1991) Ca^{2+} channels of KA-AMPA-gated glutamate receptor channels depends on subunit composition. Science 252, 851–853.

Hollmann, M. and Heinemann, S. (1994) Cloned glutamate receptors. Ann. Rev. Neurosci. 17, 31–108.

Hollmann, M., O'Shea-Greenfield, A., Rogers, S.W. and Heinemann, S. (1989) Cloning by functional expression of a member of the glutamate receptor family. Nature 342, 643–648.

Hunter, C., Doi, K., and Wenthold, R.J. (1992) Neurotransmission in the auditory system. The Otolaryngologic Clinics of North America, 25, 1027–1052.

Hunter, C. and Wenthold, R.J. (1994) Overlapping expression patterns for kainate-, NMDA- and AMPA selective glutamate receptor subunit mRNAs in the rat cochlear nucleus. Assoc. Res. Otolaryngol. Abs. 17, 14.

Hunter, C., Mathura, J. and Wenthold, R.J. (1995) Expression of NMDA-selective glutamate receptor NR1 subunit isoforms in the rat cochlear nucleus. Assoc. Res. Otolaryngol. Abs. 18, 31.

Hunter, C., Petralia, R.S., Vu, T., and Wenthold, R.J. (1993) Expression of AMPA-selective glutamate receptor subunits in morphologically defined neurons of the mammalian cochlear nucleus. J. Neuroscience 13, 1932–1946

Hunter, C., Wheaton, K.D. and Wenthold, R.J.(1990) Solubilization and partial purification of an AMPA binding protein from rat brain. J. Neurochem. 54, 118–125.

Keinänen, K., Wisden, W., Sommer, B., Werner, P., Herb, A., Verdoorn, T.A., Sakmann, B., and Seeburg, P.H. (1990) A family of AMPA-selective glutamate receptors. Science 249, 556–560.

Landsend, A.S., Amiry-Moghaddam, M., Matsubara, A., Bergersen, L., Usami, S., Wenthold, R.J., and Ottersen, O.P. (1997) Differential localization of delta glutamate to receptors in the rat cerebellum: coexpression with AMPA receptors in parallel fiber-spine synapses and absence from climbing fiber-spine synapses. J. Neuroscience 17, 834–842.

Liao, D., Hessler, N.A., and Malinow, R. (1995) Activation of postsynaptically silent synapses during pairing-induced LTP in CA1 region of hippocampal slice. Nature 375, 400–404.

Monaghan, D.T., Bridges, R.J., and Cotman, C.W. (1989) The excitatory amino acid receptors: their classes, pharmacology, and distinct properties in the function of the central nervous system. Annu. Rev. Pharmacol. Toxicol. 29, 365–402.

Mosbacher, J., Schoepfer, R., Monyer, H., Burnashev, N., Seeburg, P.H. and Ruppersberg, J.P. (1994) The molecular determinant for submillisecond desensitization in glutamate receptors. Science, 266, 1059–1062.

Okamoto, N., Hori, S., Akazawa, C., Hayashi, Y., Shigemoto, R., Mizuno, N., and Nakanishi, S. (1994) Molecular characterization of a new metabotropic glutamate receptor mGluR7 coupled to inhibitory cyclic AMP signal transduction. J. Biol. Chem. 269, 1231–1236.

Otis, T.M., Raman, I.M., and Trussell, L.O. (1995) AMPA receptors with high Ca^{2+} permeability mediate synaptic transmission in the avian auditory pathway. J. Physiol. (Lond.) 482, 309–315.

Partin, K.M., Patneau, D.K., Winters, C.A., Mayer, M.L., and Buonanno, A. (1993) Selective modulation of desensitization at AMPA versus kainate receptors by cyclothiazide and concanavalin A. Neuron 11, 1069–1082.

Petralia, R.S., Wang, Y.-X., Zhao, H.-M., and Wenthold, R.J. (1996) Ionotropic and metabotropic glutamate receptors show unique postsynaptic, presynaptic, and glial localizations in the dorsal cochlear nucleus. J. Comp. Neurol. 372, 356–383.

Petralia, R.S. and Wenthold, R.J. (1992) Localization of AMPA-selective glutamate receptor subunits in the cochlear nuclei of the rat using characterized antipeptide antibodies. Assoc. Res. Otolaryngol. Abs. 15:20.

Petralia, R.S. and Wenthold, R.J. (1996) Types of excitatory amino acid receptors and their localization in the nervous system and hypothalamus. In D.W. Brann and V.B. Mahesh (Eds.). Excitatory amino acids: Their role in neuroendocrine function. CRC Press, pp 55–101.

Pin, J.-P. and Duvoisin, R. (1995) The metabotropic glutamate receptors: structure and function. Neuropharmacology 34, 1–26

Raman, I.M. and Trussell, L.O. (1992) The kinetics of the response to glutamate and kainate in neurons in the avian cochlear nucleus. Neuron 9, 173–186.

Raman, I.M., Zhang, S., and Trussell, L.O. (1994) Pathway-specific variants of AMPA receptors and their contribution to neuronal signaling. J. Neuroscience 14, 4998–5010.

Rubio, M.E., Manis, P.B., and Wenthold, R.J. (1996) Types and subunits of glutamate receptors can differ in their distributions in neurons. Neurosci. Abst. 22, 594.

Sato, K., Kuriyama, H., and Altschuler, R.A. Differential distribution of NMDA receptor subunit mRNA in the rat cochlear nucleus. Submitted, Hearing Res.

Schoepp, D.D., Bochaert, J., and Sladeczek, F. (1990) Pharmacological and functional characteristics of metabotropic excitatory amino acid receptors. Trends Pharmacol. Sci. 11, 508–515.

Schwartz, I.R. (1994) Immunocytochemistry of glutamate receptor subunits in the gerbil cochear nucleus and superior olivary complex. Assoc. Res. Otolaryngol. Abs. 17:14.

Shigemoto, R., Nakanishi, S., and Mizuno, N. (1992) Distribution of the mRNA for a metabotropic glutamate receptor (mGluR1) in the central nervous system: An in situ hybridization study in adult and developing rat. J. Comp. Neurol. 322, 212–135.

Sommer, B., Keinänen, K., Verdoorn, T.A., Wisden, W., Burnashev, N., Herb, A., Köhler, M., Takagi, T., Sakmann, B., and Seeburg, P.H. (1990) Flip and flop: A cell-specific functional switch in glutamate-operated channels of the CNS. Science 249, 1580–1585.

Sommer, B. and Seeburg, P.H. (1992) Glutamate receptor channels: novel properties and new clones. Trends Pharmacol. Sci. 13, 291–296.

Stern-Bach, Y., Bettler, B., Hartley, M., Sheppard, P.O., O'Hara, P.J. and Heinemann, S.F. (1994) Agonist selectivity of glutamate receptors is specified by two domains structurally related to bacterial amino acid-binding proteins. Neuron 13, 1345–1357.

Sucher, N.J., Akbarian, S., Chi, C.L., Leclerc, C.L., Awobuluyi, M., Deitcher, D.L., Wu, M.K., Yuan, J.P., Jones, E.G., and Lipton, S.A. (1995) Developmental and regional expression pattern of a novel NMDA receptor-like subunit (NMDAR-L) in the rodent brain. J. Neuroscience 15, 6509–6520.

Verdoorn, T.A., Burnashev, N., Monyer, H., Seeburg, P.H. and Sakmann, B. (1991) Structural determinants of ion flow through recombinant glutamate receptor channels. Science, 252, 1715–1718.

Wada, K., Dechesne, C.J., Shimasaki, S., King, R.G., Kusano, K., Buonanno, A., Hampson, D.R., Banner, C., Wenthold, R.J. and Nakatani, Y. (1989) Sequence and expression of a frog complementary DNA encoding a kainate-binding protein. Nature 342, 684–689.

Wang, Y.-X., Wenthold, R.J., and Petralia, R.S. (1996) Distribution of glutamate receptor subunits associated with auditory nerve terminals in the anteroventral cochlear nucleus. Assoc. Res. Otolaryngol. Abs. 19:168.

Watanabe, M., Mishina, M., and Inoue, Y. (1994) Distinct distributions of five NMDA receptor channel subunit mRNAs in the brainstem. J. Comp Neurol. 343, 520–531.

Wenthold, R.J. (1991) Neurotransmitters of brain stem auditory nuclei. In R.A. Altschuler et al., (Eds) Neurobiology of Hearing: The Central Auditory System (121–139). New York: Raven Press.

Wenthold, R.J., Hunter, C., and Petralia, R.S. (1993) Excitatory amino acid receptors in the rat cochlear nucleus. In M.A. Merchan, J.M. Juiz. D.A. Godfrey, and E. Mugnaini (Eds.), The Mammalian Cochlear Nuclei: Organization and Function (179–194). New York: Plenum Press.

Wenthold, R.J., Petralia, R.S., Blahos, J., and Niedzielski, A.S. (1996) Evidence for multiple AMPA receptor complexes in hippocampal CA1/CA2 neurons. J. Neurosci. 16, 1982–1989.

Wenthold, R.J., Trumpy, V.A., Zhu, W.S., and Petralia, R.S. (1994) Biochemical and assembly properties of GluR6 and KA2, two members of the kainate receptor family, determined with subunit-specific antibodies. J. Biol. Chem. 269, 1332–1339.

Wo, Z.G. and Oswald, R.E. (1995) Unraveling the modular design of glutamate-gated ion channels. TINS 18, 161–168.

Wright, D.D., Blackstone, C.D., Huganir, R.L., and Ryugo, D.K. (1996) Immunocytochemical localization of the mGluR1a metabotropic glutamate receptor in the dorsal cochlear nucleus. J. Comp. Neurol. 364:729–745.

Zhao, H-M., Wenthold, R.J., Wang, Y.X., and Petralia, R.S. (1997) Glutamate receptor subtypes are differentially distributed at parallel and climbing fiber synapses on Purkinje cells. J. Neurochem., 68, 1041–1052.

Zukin, R.S. and Bennett, M.V.L. (1995) Alternatively spliced isoforms of the NMDAR1 receptor subunit. TINS 18, 306–313.

11

ENHANCED PROCESSING OF TEMPORAL FEATURES OF SOUNDS IN BACKGROUND NOISE BY COCHLEAR NUCLEUS SINGLE NEURONS

Robert D. Frisina,[1] Jian Wang,[2] Jonathan D. Byrd,[1] Kenneth J. Karcich,[1] and Richard J. Salvi[2]

[1]Otolaryngology Division
Dept. of Surgery
University of Rochester School of Medicine & Dentistry
Rochester, New York 14642-8629
[2]Hearing Research Laboratory
University of Buffalo
Buffalo, New York 14214

1. INTRODUCTION

Sound envelope temporal fluctuations are important for effective processing of bio-logically-relevant acoustic signals including speech sounds, animal vocalizations, pitch cues, music and sound-source locations. Insights can be gained from previous studies into the nature of how the auditory system processes certain complex sound features. Virtually all of these prior investigations have been carried out utilizing acoustic signals presented in quiet. Representative, key studies will be described that have set the stage for the present investigation. Some of this background has been put forth previously (Frisina et al., 1993).

1.1 Processing of AM by the Auditory Nerve

At low and moderate intensity levels, auditory nerve fibers give significant synchronous responses to amplitude modulated tones. If AM frequency is varied, auditory nerve fibers act as lowpass filters for the encoding of tones and wideband noise whose envelopes are temporally modulated (Møller, 1976a; Javel, 1980; Frisina et al., 1985). The frequency response starts to fall off at AM frequencies above 500 Hz, with some positive relationship observed between best frequency (BF) and the high cutoff for AM.

Acoustical Signal Processing in the Central Auditory System
edited by Syka, Plenum Press, New York, 1997

Auditory nerve fiber phase locking to AM rapidly declines as stimulus intensity is raised (Møller, 1976a; Smith & Brachman, 1980; Brachman, 1980; Evans & Palmer, 1980; Yates, 1987). Usually, an auditory nerve fiber can only give a significant synchronous response to AM over a 30–40 dB range of average intensities. These findings apply to the typical, low threshold, high spontaneous activity rate (SR) auditory nerve fiber. High threshold, low SR nerve fibers are similar except that they have extended dynamic ranges for steady state AM encoding (Joris & Yin, 1992; Cooper et al., 1993; Frisina et al., 1996).

1.2. Effects of Background Noise on Auditory Nerve Encoding of Simple Sounds

Several investigators have measured the effects of background noise on the operating ranges of auditory nerve fibers in regard to their abilities to encode tones and other temporally and spectrally simple sounds (Geisler & Sinex, 1980; Costalupes et al., 1984; Gibson et al., 1985). The majority of evidence indicates that the operating range of most auditory nerve fibers, as measured in steady state rate-intensity functions for tones, shifts to higher average intensities in the presence of wideband noise. Since the operating range **per se** (the sloping portion of the rate-intensity function) is shifted to higher levels, sound intensity coding diminishes at low levels, and can improve at certain higher levels.

1.3. Cochlear Nucleus Processing of AM

Pioneering studies of AM responses in the cochlear nucleus of the rat were done by Møller (1972; 1973; 1974a,b; 1975a,b; 1976a,b,c). His main goal was to perform a linear systems analysis of single-unit responses to determine the fidelity with which AM is encoded. He found that, except at very high stimulus modulation depths, most cochlear nucleus units encode AM with high fidelity. Some units were tuned to different AM frequencies (80–500 Hz) and their tuning properties were relatively insensitive to variations in stimulus parameters such as the depth of modulation, the duration of the stimulus, and whether continuous tones or repetitive bursts were used as stimuli. AM tuning properties remained stable for hours and were found to be independent of the several methodologies used to measure them.

Møller also found that cochlear nucleus units can increase the depth of modulation in the response relative to that of the stimulus at intensities up to 60 dB above threshold. He found, in contrast, that auditory nerve fibers show this amplification over only a 30 dB range. Lastly, he pointed out that the dynamic range for AM encoding can exceed that of the mean rate-intensity function for some units.

As is true of noteworthy pioneering work, Møller's investigations raised many important questions. For instance: Do all cochlear nucleus units encode AM equally? If not, is there a relation between a unit's responses to simple sounds and those to dynamic features of complex sounds such as AM? Where are the units located that preferentially encode AM? What are the possible cellular or neural network mechanisms by which enhanced responses to AM are generated in the cochlear nucleus? What features of AM signals do cochlear nucleus units select for?

Answers to some of these questions come from reports by Kim and coworkers (Kim et al., 1990; Arle & Kim, 1991). They have shown in decerebrate cat that pauser/buildup, Type IV units in the dorsal cochlear nucleus (DCN) show amplified responses to AM. The AM frequency selectivity of this AM encoding is predictable from the unit's autocorrelation function. Also, regularity analyses of pauser/buildup units in the DCN, and units of the ventral cochlear nucleus (VCN), indicate that a strong correlation exists between a unit's regularity of discharge to simple sounds and its ability to effectively encode AM. Modeling efforts, and those of Banks and Sachs (1991) and Hewitt et al. (1992), have shown that part of a cochlear nucleus unit's ability to encode sinusoidally modulated tones in quiet comes from the dendritic filtering properties of stellate and fusiform cell extensive dendritic arbors.

1.3.1. Neural Processing of AM in the VCN. Parts of our results to date help merge two lines of classical research of the cochlear nucleus: 1) Cochlear nucleus coding of simple sounds, and 2) Cochlear nucleus responses to complex sounds such as AM (Frisina, 1983; Frisina et al., 1985; 1990a). To achieve this, single-unit responses were recorded in intermediate regions of the VCN for pure tone and sinusoidal AM stimuli. It was found that all unit types, as well as auditory nerve fibers, show strong synchronous responses or phase locking to 150 Hz AM at low intensities. However, units differ greatly in their abilities to encode AM at high intensities. Onset units tend to show the strongest AM phase locking, followed by chopper, primarylike-with-notch and primarylike, respectively. Qualitatively, auditory nerve fibers respond similarly to primarylike units.

1.3.2. A Hierarchy of Enhancement in VCN for AM Coding. To get a comprehensive understanding of each unit types ability to encode AM, we varied the AM frequency and intensity of stimulation. To quantitatively measure phase locking, Fourier analyses were performed on the spikes generated in response to AM, which yields measures of the first-harmonic and average responses. The percent modulation of the response is defined as the ratio of the first-harmonic response to the average response. The synchronous response was reported in terms of a response gain scale:

$$\text{response gain in dB} = 20 \times \log_{10} \left(\frac{\text{percent modulation of response}}{\text{percent modulation of stimulus}} \right) \tag{1}$$

The results were plotted as 3-dimensional surfaces displaying response gain vs AM frequency and sound level. It was found that throughout their AM response areas, all four VCN unit types show stronger phase locking than auditory nerve fibers. The distances between the response surfaces of VCN neurons and auditory nerve fibers are measures of the amount of enhancement or amplification performed by VCN neurons relative to their ascending input (Frisina et al., 1990a). The distances are greatest for onset units, followed in order by chopper, primarylike-with-notch and primarylike units. The greatest amount of enhancement occurred at high intensities. Rhode and Greenberg (1994) confirmed the existence of a hierarchy of enhancement in the cochlear nucleus of cat. The basic hierarchy was the same as gerbil, with a few details differing. For example, in cat the primarylike-with-notch and on-C units tended to exceed the AM coding capabilities of choppers and on-L units.

1.3.3. AM Response Areas Become Bandpass at High Intensities in VCN. It was also found (Frisina et al., 1985; 1990a) that at low intensities the shapes of the response surfaces are similar for each of the VCN unit types and auditory nerve fibers. The surfaces are lowpass in nature. At high intensities, in contrast, the response surfaces tend to be tuned. Rhode (1994) reported similar findings in cat VCN using AM stimuli with 200% depth of modulation.

1.3.4. Summary of VCN AM Encoding in Quiet. These findings demonstrate that the further a neuron's pure tone responses deviate from a primarylike pattern, the greater its ability to produce strong, synchronous responses to AM. These results, aside from advancing knowledge of the functional organization of the VCN, also provide new clues about the neural correlates of the wide dynamic range of the auditory system. For instance, previous investigations of the auditory nerve reviewed above, which we have built upon and extended, show that most single fibers encode AM over only about a 40 dB intensity range. We found that some onset and chopper units show significant temporal processing of AM over a 90 dB range of average intensities. This is a remarkable operating range for a 2nd order sensory neuron. If a 25 dB spread for absolute thresholds exists within an animal for these unit types in a particular BF region (Bourk, 1976; Frisina, 1983), a population of these units could encode rapid sound-amplitude changes over the auditory system's entire operating range.

1.3.5. Neural Mechanisms for AM Processing in VCN. For auditory nerve fibers, the onset rate-intensity function is a good qualitative and quantitative predictor of a fiber's ability to encode AM. This implies that similar dynamic response characteristics of a neuron are responsible for onset responses to pure tones and encoding of AM. Logically, since onset and chopper units in the VCN have relatively precisely timed onset responses, it seemed worthwhile to examine the relationship between their onset rate-intensity functions and their AM encoding abilities as a function of average sound intensity. We investigated this situation for VCN units (Frisina et al., 1990b). We found that the onset and steady state rate-intensity functions of all VCN units had limited dynamic ranges of 20–30 dB (the exception was On-C units which had extended operating ranges).

As presented above (Frisina et al., 1985; 1990a; Rhode, 1994), AM gain surfaces of VCN units consist of 2 main features: a lowpass characteristic which dominates at low intensities and a bandpass characteristic that is preeminent at high sound levels. The lowpass characteristic is similar to that seen in the responses of auditory nerve fibers. However, the bandpass characteristic asserts itself in varying degrees in the 4 main VCN unit types. The lowpass characteristic dominant at low intensities, and the bandpass characteristic differentially dominant at high intensities, probably result from two or more different neural processing mechanisms in the auditory nerve and cochlear nucleus.

1.3.6. Post-Synaptic Membrane Properties and Biophysical Considerations. As demonstrated previously (Frisina et al., 1990b; 1993), a reduction in average firing rate by VCN units, resulting in an increased response gain to AM in quiet, may result from on-BF or off-BF inhibition from auditory nerve fibers via interneuronal networks or VCN unit inhibitory feedback collaterals. AM frequency selectivity, as manifested in bandpass AM tuning at higher intensities, may result from dendritic filtering properties of VCN cells, or from differential input from low or medium SR auditory-nerve fibers. The abilities of VCN units to abstract temporal envelope information is also probably due to post-synaptic membrane properties and spike generation mechanisms. Pioneering studies of membrane properties and ion channels have been conducted in the cochlear nucleus by Oertel (1983) and Manis (1989; 1990). Membrane properties of Type II cells (bushy), which display non-linear current-voltage relationships (I-V), are well suited for preserving temporal information from the auditory nerve. Primarylike and primarylike-with-notch units are correlated with the bushy/stellate cell anatomical class. In contrast, under appropriate stimulus conditions, Type I cells (stellate), which have linear I-V, have the ability to enhance certain temporal features of complex sounds. Chopper and on-C units correlate with the stellate/multipolar cell anatomical class.

Bioengineering computer modeling simulations of cochlear nucleus units also shed light on how excitatory or inhibitory input channels interact with differential post-synaptic membrane characteristics. For example, Banks & Sachs (1991) and Hewitt et al. (1992) demonstrated how post-synaptic, lowpass dendritic filtering properties can interact with other membrane properties of VCN stellate cells to produce responses of chopper units to pure tones and AM. Arle & Kim (1991) successfully modeled the intrinsic oscillatory behavior of certain cochlear nucleus unit types which is involved with their temporal processing capabilities (e.g. Kim et al., 1990, correlated AM processing with a unit's pure-tone autocorrelation/power spectrum). Arle & Kim accomplished this by modeling various dendritic configurations with the voltage-dependent nonlinear I-V cell membrane conductances of fusiform and bushy cells, or with the voltage-independent characteristics of stellate cells. Building on these studies, Colombo and Frisina (1991; 1992) and Colombo (1994) have found that by using naturalistic auditory nerve inputs, and taking into account differential membrane capacitance properties in the spike generation region (integration time constant of the membrane), that amplified encoding of AM in quiet and background noise can be accurately predicted with biophysical modeling techniques.

Biophysical manifestations of the excitatory and inhibitory channels involved in cochlear nucleus AM coding postulated above are now starting to be known. For instance, in mouse cochlear nucleus slice preparations, intracellular recordings from stellate cells (correlated with the chopper physiological unit type) have shown that electrical stimulation of the auditory nerve generates not only excitatory post-synaptic potentials, but also longer latency inhibitory synaptic events (Oertel, 1983; Oertel et al., 1990). These are mediated through interneurons or principal cell lateral inhibitory collaterals. This inhibitory channel input involves a chloride channel conductance alteration apparently triggered by glycine (Wu & Oertel, 1986). *In vivo* investigations of inhibitory inputs to cochlear nucleus units provide further substantiation of the differences in cellular membrane properties of stellate and bushy cells (Shofner & Young, 1985; Rhode & Smith, 1986). DCN fusiform and giant cells, not only receive off-BF inhibitory channel inputs, but in addition, on-BF inhibitory inputs via Type II interneurons (Voigt & Young, 1980; 1990; Young, 1984). These multiple inhibitory inputs could be a basis for encoding of complex sound features such as spectral notches for sound localization (Young et al., 1992), or abstraction of AM in noise. Neuropharmacological evidence suggests that some VCN neurons also receive on-BF inhibitory channel inputs (Caspary et al., 1994).

1.4. Purpose of the Present Investigation

A major goal of the present study was to pursue an interrelated series of experiments geared towards multiple hypothesis testing concerning the nature of the two independent input mechanisms and post-synaptic cell membrane properties we have proposed for cochlear nucleus encoding of temporal fluctuations of complex sound envelopes. Specifically, extracellular neurophysiological experiments were conducted in the auditory nerve and DCN of the urethane-anesthetized chinchilla using the same stimulus paradigms to determine the abilities of single units to encode AM in background noise. A second goal of the current report was to present some initial data on effects of blocking inhibitory inputs on AM processing in the cochlear nucleus.

2. METHODS

To gain new insights, extracellular single-unit recordings were made in urethane-anesthetized chinchillas using traditional micropipettes and newer multi-barrel electrodes for recording responses after application of bicuculline (GABA antagonist) or strychnine (glycine antagonist). Units were classified and located in the cochlear nucleus according to post-stimulus-time histogram (PSTH) response patterns, 1st spike latencies, shape of BF and noise rate-intensity functions, regularity analyses and in some cases, topographic mapping with HRP injections. BF tone bursts, wideband noise bursts, and AM (20 to 500 Hz) BF tone bursts were employed as stimuli at several sound levels, both in quiet and in the presence of a continuous wideband background noise.

The methods and results of the background noise experiments are described in greater detail in two previous articles. The auditory nerve data come from a report by Frisina et al. (1996), and the DCN data from an article by Frisina et al. (1994). We investigated the effects of a wideband background noise on AM encoding in pauser/buildup units utilizing a stimulus paradigm similar to Gibson et al. (1985). Differences were that we used 125 msec test tones and utilized AM test tones at BF as well as pure tones (Frisina et al., 1991; 1992; 1993; 1994). The novel contribution of the present report is to make some detailed comparisons between the auditory nerve and cochlear nucleus responses, which could not be done before since the DCN studies preceded those of the auditory nerve. The pilot data utilizing multi-barreled electrodes containing inhibitory blockers have not been presented or published elsewhere. For these experiments, the surgical procedures on chinchillas were the same as Frisina et al. (1994; 1996), with the following minor modifications. An initial dose of ketamine preceded the urethane anesthetic, and instead of using a lateral approach to the cochlear nucleus (Frisina et al., 1982), a middle fossa approach was used to allow placement of multi-barreled electrodes directly on a visually-identified division of the cochlear nucleus. The multi-barreled electrode procedures were the same as those described in detail by Caspary and coworkers and Park and Pollak. "Piggy-back" multi-barreled electrodes similar to those fabricated by Caspary (1990; Havey & Caspary, 1980) were made. Recording and microiontophoresis procedures were the same as previous studies (Park & Pollak, 1993a; 1993b; 1994). All of the stimulus conditions for multi-barreled experiments were presented pre- and post-drug (when units were held long enough), including post-drug current and recovery control conditions. Usually, one barrel was filled with bicuculline methiodide (10 mM, pH=3.0), one barrel with strychnine hydrochloride (10 mM, pH=3.5–4.0), one barrel with the drug delivery vehicle alone (0.16 M NaCl, pH=3.5), and one barrel was the balancing or sum channel employed to balance current in the drug barrels and minimize current effects. The drug and balancing barrels were connected to the Neurophore BH-2 drug delivery systems via Ag/AgCl wires.

3. RESULTS AND DISCUSSION

3.1 Effects of Background Noise on Auditory Nerve AM Processing

Surprisingly, there has been only one previous study of auditory nerve AM processing in noise (Frisina et al., 1996). A main finding was that, on the average, auditory nerve fibers preserve their AM coding in the presence of noise due to upward shifts in AM operating ranges and cochlear nonlinearities. However, these shifts cannot fully explain the enhancement of AM coding in DCN cells described below (Frisina et al., 1993; 1994). Also, because auditory nerve synchronous and average AM responses tended to decline in noise, the auditory nerve on-BF activity could not account for the increases in AM synchronous response for DCN units. Some representative findings of just one of many analyses, was in the form of interval histograms (IHs). For 120 Hz AM at moderate intensities (55 dB SPL) background noise at 0 and +6 dB S/N had very little effect on the shape of the IHs to AM relative to those in quiet. Likewise, at high intensities (75 dB SPL), the shape of IHs are relatively unaffected by the presence of the background noise. Current analyses are examining the effects of noise on the regularity (CV analysis) of the responses of auditory nerve fibers to AM in quiet and noise.

3.2 Effects of Background Noise on DCN AM Processing

Some representative findings are given in Figs. 1, 2, 3 & 4 (intensity level: 75 dB SPL, steady state measure of AM encoding). For Figs. 2–4, solid data points show the steady state AM responses in quiet. Open points show the modulation transfer functions (MTFs), under identical stimulus conditions except that a continuous wideband background noise at 75 dB SPL (0 dB signal/noise for Figs. 2–3; +5 dB S/N for Fig. 4) was turned on 1 min before the MTF data were collected. As one might predict, for a fairly intense noise as shown in Fig. 2, at many AM frequencies (in this case 160 - 300 Hz) there is a decline in AM responsivity in the vicinity of the peak of the MTF in quiet. However, it is also noteworthy that at certain AM frequencies (10 - 100 Hz, 400 - 500 Hz), the AM processing of this pauser unit is relatively unaffected by the intense background noise. Thus, at some AM frequencies, preservation of AM processing is achieved despite the presence of acoustic clutter from a fairly intense wideband noise.

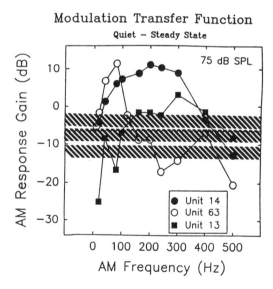

Figure 1. DCN units have MTFs that exhibit relatively high AM response gains and are bandpass in shape at high sound levels. The hatched areas represent the mean and +/- 1 SD for high SR (lowest hatched area), medium SR (middle hatched area), and low SR (highest hatched area) auditory nerve fibers recorded in the same species under identical anesthetic and stimulus conditions. Auditory nerve mean data are from Frisina et al. (1996, Table 1) and the cochlear nucleus data are adapted from Frisina et al. (1994, Fig. 2). Unit 13, threshold = 41 dB SPL, BF = 13.2 kHz; Unit 14, thr. = 47 dB SPL, BF = 8.6 kHz; Unit 63, thr. = 24 dB SPL, BF = 0.36 kHz.

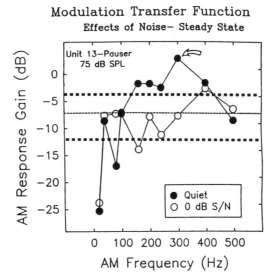

Figure 2. In the presence of a wideband background noise, AM responses in the vicinity of the peak of the response in quiet (arrow) tend to decline, but at very low and high AM frequencies, the AM response is preserved in background noise. The dashed lines show the mean AM response gains for high SR (lowest line), medium SR (middle line), and low SR (highest line) auditory nerve fibers, based on the data presented in Fig. 1. Auditory nerve mean data are from Frisina et al. (1996, Table 1) and the cochlear nucleus data are adapted from Frisina et al. (1994, Fig. 2). Unit 13, threshold = 41 dB SPL, BF = 13.2 kHz.

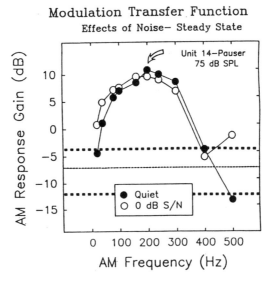

Figure 3. In the presence of a wideband background noise, for some units, AM responses in the vicinity of the peak of the response in quiet (arrow) tend to be preserved in background noise. Dashed lines are the same as Fig. 2. Auditory nerve mean data are from Frisina et al. (1996, Table 1) and the cochlear nucleus data are adapted from Frisina et al. (1994, Fig. 2). Unit 14, thr. = 47 dB SPL, BF = 8.6 kHz.

To gain more insights into this interesting behavior we varied the level of the background stimulation, therefore altering the S/N. The results of lowering the background noise level by 5 dB relative to that of Fig. 2 and otherwise keeping all other stimulus parameters the same, are shown in Fig. 4. The solid data points are the solid points of Fig. 2. The open points show the MTF with the noise at 70 dB SPL (+5 dB S/N). As was discovered for the 0 dB S/N condition, near the peak of the MTF in quiet (200 - 300 Hz), the noise has a deleterious effect on AM processing. However, for low and high AM (10 - 160 Hz, 400 - 500 Hz), not only is AM coding preserved, in noise, but is actually enhanced.

In contrast to the findings of beneficial effects of background noise on AM processing in DCN pauser/buildup units, under some conditions, background noise can be quite devastating to encoding of sound envelope fluctuations (Frisina et al., 1993, 1994). In some cases, the evoked activity in response to an AM stimulus is virtually obliterated. To start to understand whether the obliteration of the evoked response was somehow unique to the AM, we compared the effects of noise on the AM stimulus with effects on a pure tone at the same level and carrier frequency. Comparisons like this show that for some pauser units the background noise seems to inhibit the unit's evoked response regardless of whether the evoking tone is modulated or unmodulated.

Some additional recent findings are that generally in DCN and VCN, MTFs for transient encoding (coding during the initial several msec of the cochlear nucleus neuron response) tend to be lowpass in nature, whereas MTFs for steady-state encoding are lowpass at low intensities and bandpass at higher sound levels; simultaneous background noise interferes more with transient encoding of AM than steady-state encoding; and for 1 kHz carrier frequencies that fall in the "tail" of tuning curves for high-BF units, AM encoding can be quite significant, and robust in the presence of noise (Frisina et al., 1993; 1994).

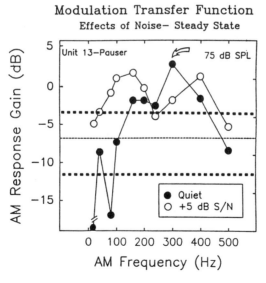

Figure 4. In the presence of a wideband background noise, AM responses in the vicinity of the peak of the response in quiet (arrow) tend to decline, but at very low and high AM frequencies, the AM response is preserved in background noise. Dashed lines are the same as Fig. 2. Auditory nerve mean data are from Frisina et al. (1996, Table 1) and the cochlear nucleus data are adapted from Frisina et al. (1994, Fig. 2). Unit 13, threshold = 41 dB SPL, BF = 13.2 kHz.

3.2.1. Implications of AM Responses in Background Noise for Understanding Neural Mechanisms. Many cochlear nucleus principal cells receive ascending, excitatory channel inputs from auditory nerve fibers with the same BF. Their outputs also depend upon on-BF or off-BF inhibitory inputs via interneurons or output collaterals. This circuitry framework, has spurned some new hypotheses of the present study. For instance, if we are trying to explain the AM encoding capabilities of the cell in the cochlear nucleus principal cells, we hypothesize that the BF input channels increase the synchronous response to AM, and the inhibitory input channels decrease the average response to AM. As tests of the independence of these two AM enhancement mechanisms, we have started analyzing the synchronous and average responses to AM, both in quiet and in the presence of a wideband background noise. Some initial findings, have been obtained for pauser/buildup units. It has been found that at frequencies where the modulation gain is enhanced by the noise the average response is reduced by the noise. At AM frequencies where the modulation gain is reduced by the noise the average response is unaffected by the noise.

A similar analysis was conducted for the effects of noise on the synchronous response. At low AM frequencies, where the response gain increases in the noise the synchronous response also increases in the presence of noise. In the mid frequencies, where the response gain decreases in the noise the synchronous response decreases. And at the high frequencies, the response gain increases in the noise while the synchronous response declines.

From studies such as these, we conclude to date that under most conditions:

1. Background noise decreases the average response to AM.
2. Background noise can increase, preserve or decrease the synchronous response, making it more important for determining whether the AM response gain will increase or decrease in the presence of noise.
3. These findings support the hypothesis put forth above that the synchronous response to AM for cochlear nucleus units (derived from BF excitatory input channels) is relatively independent of the average response (derived from inhibitory input channels).

3.3. Specific Comparisons between Auditory Nerve and DCN AM Coding in Background Noise

Further knowledge concerning cochlear nucleus temporal processing mechanisms can be gleaned by examining AM responses in quiet and noise more closely as obtained in the same species (chinchilla) under identical anesthetic and stimulus conditions. Group responses of auditory nerve fibers (Frisina et al., 1996, Table I) and individual responses from DCN units (Frisina et al., 1994, Fig. 2) are given in Figs. 1–4. The mean responses of a group of high SR (lower hatched area represents mean and +/- 1 standard deviation [SD]), medium SR (middle hatched area represents mean and +/- 1 SD), and low SR (upper hatched area represents mean and +/- 1 SD) auditory nerve fibers are given in Fig. 1. Here they are compared to the MTF responses of three individual DCN units in quiet. It is noteworthy to observe that most of the DCN AM response gains are at the same level or above those of the three groups of auditory nerve fibers. At the intermediate AM frequencies, some of the DCN responses in quiet are as much as 10 dB greater than those of even the low SR fibers, who are the best encoders of AM in the auditory nerve at high intensities such as 75 dB SPL.

On the average for high intensity stimulation, the group coding of AM by auditory nerve fibers do not change significantly in the presence of background noises 0 dB to + 6 dB

S/N (Frisina et al., 1996). The auditory nerve group data from Fig. 1, are replotted in Figs. 2–4 as just dashed lines representing the group means for the three SR groups (SD hatchings are omitted to allow better visualization of DCN MTF individual data points). Fig. 2 displays a case where the AM coding declines at intermediate AM frequencies for the DCN pauser unit. However, even where declines occurred, responses were still in the range of all three groups of auditory nerve fibers (except at two of the three lowest AM frequencies).

Fig. 3 displays a case where the background noise, although greatly decreasing the onset response to the AM tone (not shown), had relatively little impact on the AM processing at all but the highest AM frequency tested (500 Hz). Note that virtually all of the AM response gains are significantly above the responses of all three types of auditory nerve fibers for BF carrier stimulation.

Fig. 4 highlights a case that was unexpected. Here, for AM frequencies of 200 Hz and below and 400 Hz and above, the AM encoding in the presence of the background noise is facilitated relative to quiet. Again, notice that most of the cochlear nucleus responses are at or above the AM response gains of even the best encoders of AM in the auditory nerve (upper dashed line). It is difficult to explain how AM responses of DCN neurons, that are already enhanced relative to their ascending inputs, could be further amplified in the presence of noise due entirely to changes in auditory nerve AM coding in noise. However, some of this improvement could be due to the fact that in the presence of a wideband background noise, auditory nerve fibers with BFs above or below the BF of the cochlear nucleus unit under study could improve their AM coding in noise. Those auditory nerve fibers with BFs very far away from the BF of the cochlear nucleus unit to which they provide inputs, would for the most part have their firing patterns captured by the noise, and thus have very little AM coding relative to quiet to pass on to the cochlear nucleus.

3.4 Cochlear Nucleus Coding with Inhibitory Blocking Agents

Several previous reports have investigated the effects of microiontophoretically-applied neurotransmitters, their agonists or antagonists on cochlear nucleus responses to simple sounds in quiet. The first study revealed that spontaneous activity and driven activity to pure tones can be reduced in the majority of samples of units from the VCN and DCN after application of GABA or glycine (Caspary et al., 1979). Application of baclofen, a lipophilic GABA-mimetic, which reduces the release of excitatory amino acids like glutamate, has been found to inhibit responses of DCN and VCN single units, as did application of GABA (Caspary et al., 1984). The responses of DCN Type IV units, presumably fusiform cells, showed less non-monotonicity of their rate-intensity functions when strychnine was applied to interfere with glycine input channels. Additionally, application of strychnine, glycine antagonist, had the opposite effect on many cells (Caspary et al., 1987). A subsequent study revealed that blockage of glycine neurotransmission with strychnine, and reduction of GABA inputs with bicuculline, can increase single-unit responses in AVCN primarylike and chopper units for both on-BF and off-BF input channels (Caspary et al., 1994).

3.4.1. Inhibitory Blocking and AM Processing. We have initiated multi-barreled electrode, microiontophoretic studies of cochlear nucleus single-unit processing capabilities for complex sound features in quiet and in background noise. These studies started with application of strychnine and bicuculline for units stimulated with tones and AM stimuli at a unit's BF, utilizing stimuli and analysis methods of our recent publications described above. A portion of these findings are displayed in Figs. 5–7.

An example of the effects of strychnine on VCN AM processing is given in Fig. 5 for a chopper unit to an 80 Hz, 100% sinusoidal AM BF tone, prior to application of strychnine. Fig. 5a shows the strong AM coding of this unit in the pre-drug state. Fig. 5b displays the effects of a small amount of strychnine (weak current) on this unit's AM processing capability. A small decline occurs in the AM processing as reflected in slight decreases in the AM response gain, r value and synch spikes (listed in the upper right portion of each PSTH). The to-

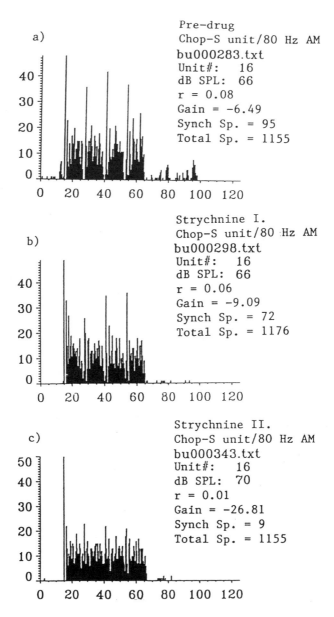

Figure 5. PSTH responses of a VCN chopper unit to AM in quiet for pre- and post-drug conditions. Stimulus was a 50 ms BF-carrier AM tone burst presented for 50 repetitions. Abscissae represent time in ms (stimulus onset at 10 ms), and ordinates represent total # spikes. Gain = AM response gain as defined in Eq. 1; r = vector strength or synchronization index; Synch Sp. = # spikes phase-locked to the AM frequency; and Total Sp. = total spikes in response to the AM stimulus.

tal spikes showed basically no change. However, increasing the application of strychnine (higher current, Fig. 5c) drastically reduced this chopper's AM response as reflected in the major reductions in the AM gain, r value and synch spikes. The total number of spikes remained the same as the pre-drug condition. This units' responses, unlike the next example, supports the notion that the synchronous and average mechanisms for AM processing postulated above are independent.

Fig. 6a displays the control responses of a VCN onset unit for the same stimulus conditions given in the previous figure. Notice the 4 prominent peaks in response to each cycle of the 80 Hz AM stimulus. Fig. 6b displays the responses of the same unit under identical AM stimulus conditions, after the application of strychnine. Note that the synchronous response to the AM significantly declines, as reflected in noticeable decreases in the AM response gain, r value and synch spikes. Notice also that the total number of spikes in the response increases from 438 (pre-drug) to 1165 (post-drug), consistent with blockage of inhibitory input channels. Thus, interfering with glycinergic inputs to this unit significantly disrupts its AM coding abilities in terms of its synchronous and average responses.

The last example shown is for a DCN pauser/buildup (Fig. 7). Here the AM frequency is 120 Hz with 35% modulation depth. As indicated by the AM gain, r value and synch spikes, this unit shows a modest amount of AM coding in the pre-drug control condition (Fig. 7a). The effects of bicuculline are given in Fig. 7b. Here the AM coding is disrupted as indicated by diminishment of the AM gain and r value. This GABA blocker had no effect on the synch spikes, but an increase in the average firing rate occurred. Interest-

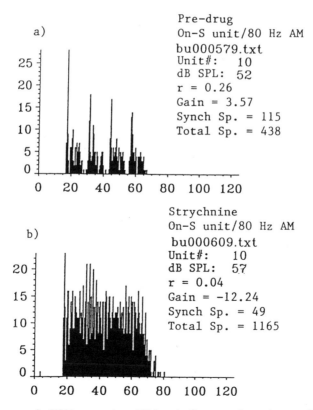

Figure 6. PSTH responses of a VCN onset unit to AM in quiet for pre- and post-drug conditions. Stimulus and PSTH formats the same as Fig. 5.

ingly, strychnine had the same increasing effect on the average firing rate as bicuculline (Fig. 7c). However, strychnine increases the overall AM processing due to significant increases in the synch spikes over the pre-drug control.

4. SUMMARY

1) Cochlear nucleus units can enhance their AM coding relative to the AN even in the presence of relatively loud (0 or +6 dB S/N) background noise; 2) In DCN, this en-

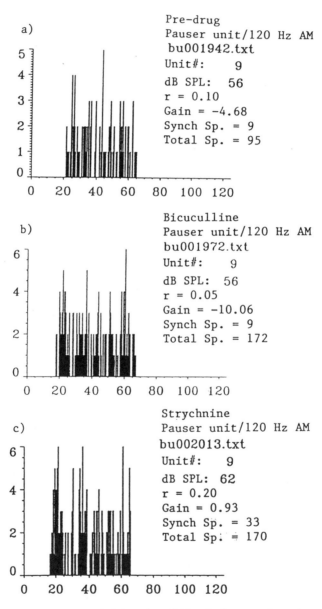

Figure 7. PSTH responses of a DCN pauser/buildup unit to AM in quiet for pre- and post-drug conditions. Stimulus and PSTH formats the same as Fig. 5.

hancement is achieved by lowering the average firing rate, but unlike AN fibers, increasing the AM synchronous response; 3) Multi-barrel data have revealed for example that blocking inhibitory inputs can disrupt AM coding in ventral CN onset units by decreasing synchronous responses to AM. 4) In DCN, blocking inhibitory inputs can in some cases increase AM coding by increasing the synchronous response in greater proportion than the average response. These findings suggest that part of a cochlear nucleus neuron's abilities to enhance AM coding in background noise results from shifts in auditory nerve operating ranges whereas part comes from intrinsic neural circuitry involving inhibitory inputs or post-synaptic cellular membrane properties within the DCN.

5. ACKNOWLEDGMENTS

Supported by NIH-NIDCD Grant #: DC00408 and the International Center for Hearing & Speech Research, Rochester NY.

6. REFERENCES

Arle, J.E. & D.O. Kim (1991) Neural modeling of intrinsic and spike-discharge properties of cochlear nucleus neurons. Biol. Cybernetics, 64, 273–283.

Banks, M.I. & M.B. Sachs (1991) Regularity analysis in a compartmental model of chopper units in the anteroventral cochlear nucleus. J. Neurophysiol. 65, 606–629.

Bourk, T.R. (1976) Electrical responses of neural units in the anteroventral cochlear nucleus of the cat. PhD Dissertation. Cambridge, MA: MIT.

Brachman, M.L. (1980) Dynamic response characteristics of single auditory nerve fibers. In (ed): Dissertation & Special Report ISR-S-19. Syracuse, NY: Institute for Sensory Research.

Caspary, D.M. (1990) Electrophysiological studies of glycinergic mechanisms in auditory brain stem structures. In: Glycine Neurotransmission. O.P. Otterson & J. Storm-Mathisen (Eds). NY: J. Wiley & Sons, Ltd., pp. 453–483.

Caspary, D.M., P.M. Backoff, P.G. Finlayson & P.S. Palombi (1994) Inhibitory inputs modulate discharge rate within frequency receptive fields of anteroventral cochlear nucleus neurons. J. Neurophysiol. 72, 2124–2133.

Caspary, D.M., D.C. Havey & C.L. Faingold (1979) Effects of microiontophoretically applied glycine & GABA on neuronal response patterns in cochlear nuclei. Brain Res. 172, 179–85.

Caspary, D.M., K.E. Pazara, M. Kossl & C.L. Faingold (1987) Strychnine alters the fusiform cell output from the dorsal cochlear nucleus. Brain Res. 417, 273–282.

Caspary, D.M., L.P. Rybak & C.L. Faingold (1984) Baclofen reduces tone-evoked activity of cochlear nucleus neurons. Hearing Res. 13, 113–122.

Celio, M.R. (1990) Calbindin D-28k and parvalbumin in the rat nervous system. Neurosci. 35, 375- 475.

Colombo, J. (1994) Physiological Modeling of Responses to Amplitude Modulated Tones in Background Noise by Cochlear Nucleus Neurons using Naturalistic Inputs. Ph.D. Dissertation, U. Rochester Press.

Colombo, J., Frisina, R.D. & Karcich, K.J. (1992) Quantitative models of ventral cochlear nucleus neurons: Pure tone response predictions. Assoc. Res. Otolaryngology Abstr. 15, 27.

Colombo, J., Frisina, R.D., Karcich, K.J. & Swartz, K.P. (1991) Computational models of ventral cochlear nucleus neurons. International Brain Res. Org. - World Congress Abstr. 3, 250.

Cooper, N.P., D. Robertson & G.K. Yates (1993) Cochlear nerve fiber responses to amplitude-modulated stimuli: Variations with spontaneous rate and other response characteristics. J. Neurophysiol. 70, 370–386.

Costalupes, J.A., E.D. Young & D.J. Gibson (1984) Effects of continuous noise backgrounds on rate response of auditory-nerve fibers in cat. J. Neurophysiol. 51, 1326–1344.

Evans, E.F. & A.R. Palmer (1980) Dynamic range of cochlear nerve fibers to amplitude modulated tones. J. Physiol. 298, 33–34P.

Evans, E.J. & P.G. Nelson (1973) The responses of single neurones in the cochlear nucleus of the at as a function of their location and the anesthetic state. Exp. Brain Res. 17, 402–427.

Frisina, R.D. (1983) Enhancement of responses to amplitude modulation in the gerbil cochlear nucleus: Single-unit recordings using an improved surgical approach. In: Dissertation & Special Report ISR-S-23. Syracuse, NY: Institute for Sensory Research, p. 203.

Frisina, R.D., S.C. Chamberlain, M.L. Brachman, & R.L. Smith (1982) Anatomy and physiology of the gerbil cochlear nucleus: an improved surgical approach for microelectrode studies. Hearing Res. 6, 259–275.

Frisina, R.D., Karcich, K.J., Sullivan, D., Tracy, T., Colombo, J. & Walton, J.P. (1996) Preservation of amplitude modulation coding in background noise by auditory-nerve fibers. J. Acoustical Society Am. 99, 457–490.

Frisina, R.D., O'Neill, W.E. & Zettel, M.L. (1989) Functional organization of mustached bat inferior colliculus. II. Connections of the FM_2 region. J. Comparative Neurology 284, 85–107.

Frisina, R.D., R.L. Smith, and S.C. Chamberlain (1985) Differential encoding of rapid changes in sound amplitude by second-order auditory neurons. Exp. Brain Res. 60, 417–422.

Frisina, R.D., Smith, R.L. and Chamberlain, S.C. (1990a) Encoding of amplitude modulation in the gerbil cochlear nucleus: I. A hierarchy of enhancement. Hearing Research. 44, 99–122.

Frisina, R.D., Smith, R.L. and Chamberlain, S.C. (1990b) Encoding of amplitude modulation in the gerbil cochlear nucleus: II. Possible neural mechanisms. Hearing Research. 44, 123–141.

Frisina, R.D. & Walton, J.P. (1991) Processing of rapid changes in sound amplitude in the cochlear nucleus in quiet and in the presence of noise. International Brain Res. Organization - World Congress Abstr. 3, 250.

Frisina, R.D., Walton, J.P., Karcich, K.J. & Colombo, J. (1992) Effects of background noise on the processing of AM in the cochlear nucleus. Assoc. Res. Otolaryngol. Abstr. 15, 78.

Frisina, R.D., Walton, J.P. & Karcich, K.J. (1993) Differential abilities to extract sound-envelope information by auditory nerve and cochlear nucleus neurons. In: Sensory Research: Multimodal Perspectives. R.T. Verrillo (Ed.) (Hillsdale, NJ: L. Erlbaum Assoc., Inc.) pp. 151–175.

Frisina, R.D., Walton, J.P. & Karcich, K.J. (1994) Dorsal cochlear nucleus single neurons can enhance temporal processing capabilities in background noise. Exp. Brain Res. 102, 160–164.

Geisler, C.D. & D.G. Sinex (1980) Responses of primary auditory fibers to combined noise and tonal stimuli. Hearing Res. 3, 317–334.

Gibson, D.J., E.D. Young & J.A. Costalupes (1985) Similarity of dynamic range adjustment in auditory nerve and cochlear nuclei. J. Neurophysiol. 53, 940–958.

Havey, D.C. & D.M. Caspary (1980) A simple technique for constructing 'piggy-back' multi-barrel microelectrodes. Electroenceph. Clin. Neurophysiol. 48, 249–251.

Hewitt, M.J., Meddis, R. & Shackelton, T.M. (1992) A computer model of a cochlear-nucleus stellate cell: Responses to amplitude modulated and pure-tone stimuli. J. Acoust. Soc. Am. 91, 2096–2109.

Javel, E. (1980) Coding of AM tones in the chinchilla auditory nerve: implications for the pitch of complex tones. J. Acoust. Soc. Am. 68, 133–146.

Joris, P.X. & Yin, T.C.T. (1992) Responses to amplitude-modulated tones in the auditory nerve of the cat. J. Acoust. Soc. Am. 91, 215–232.

Kane, E.S. (1978) Primary afferents and the cochlear nucleus. In R.F. Naunton and C. fernandez (eds): Evoked electrical Activity in the Auditory Nervous system. New York: Academic Press, pp. 337–352.

Kim, D.O., J.G. Sirianni & S.O. Chang (1990) Responses of DCN-PVCN neurons and auditory nerve fibers in unanesthetized decerebrate cats to AM and pure tones: Analysis with autocorrelation/power-spectrum. Hearing Res. 45, 95–113.

Manis, P.B. (1989) Response to parallel fiber stimulation in the guinea pig dorsal cochlear nucleus. J. Neurophysiol. 61, 149–161.

Manis, P.B. (1990) Membrane properties and discharge characteristics of guinea pig dorsal cochlear nucleus neurons studied *in vitro*. J. Neurosci. 10, 2338–2351.

Møller, A.R. (1972) Coding of amplitude and frequency modulated sounds in the cochlear nucleus of the rat. Acta. Physiol. Scand. 86, 223–238.

Møller, A.R. (1973) Statistical evaluation of the dynamic properties of cochlear nucleus units using stimuli modulated with pseudorandom noise. Brain Res. 57, 443–456.

Møller, A.R. (1974a) Coding of amplitude and frequency modulated sounds in the cochlear nucleus. Acustica 31, 202–299.

Møller, A.R. (1974b) Responses of units in the cochlear nucleus to sinusoidally amplitude-modulated tones. Exp. Neurol. 45, 104–117.

Møller, A.R. (1975a) Dynamic properties of excitation and inhibition in the cochlear nucleus. Acta. Physiol. Scand. 93, 442–454.

Møller, A.R. (1975b) Latency of unit responses in cochlear nucleus determined in two different ways. J. Neurophysiol. 38, 812–821.

Møller, A.R. (1976a) Dynamic properties of primary auditory fibers compared with cells in the cochlear nucleus. Acta. Physiol. Scan. 98, 157–167.

Møller, A.R. (1976b) Dynamic properties of excitation and 2-tone inhibition in the cochlear nucleus studied using amplitude modulated tones. Exp. Brain Res. 25, 307–321.

Møller, A.R. (1976c) Dynamic properties of the responses of single neurones in the cochlear nucleus of the rat. J. Physiol. 259, 63–82.

Oertel, D. (1983) Synaptic responses and electrical properties of cells in brain slices of the mouse anteroventral cochlear nucleus. J. Neurosci. 3, 2043–2053.

Oertel, D., S.H. Wu, M.W. Garb & C. Dizack (1990) Morphology and physiology of cells in slice preparations of the posteroventral cochlear nucleus of mice. J. Comp. Neurol. 295, 136–154.

Park, T.J. & G.D. Pollak (1993a) GABA shapes sensitivity to interaural intensity disparities in the mustache bat's inferior colliculus: implications for encoding sound location. J. Neurosci. 13, 2050–2067.

Park, T.J. & G.D. Pollak (1993b) GABA shapes a topographic organization of response latency in the mustache bat's inferior colliculus. J. Neurosci. 13, 5172–5187.

Park, T.J. & G.D. Pollak (1994) Azimuthal receptive fields are shaped by GABAergic inhibition in the inferior colliculus of the mustache bat. J. Neurophysiol. 72, 1080–1102.

Rhode, W.S. (1994) Temporal coding of 200% amplitude modulated signals in the ventral cochlear nucleus of cat. Hearing Res. 77, 43–68.

Rhode, W.S. & Greenberg, S. (1986) Encoding of amplitude modulation in the cochlear nucleus of the cat. J. Neurophysiol. 71, 1797–1825.

Rhode, W.S. & Smith, P.H. (1986) Encoding timing and intensity in the ventral cochlear nucleus of the cat. J. Neurophysiol. 56, 261–286.

Shofner, W. & Young, E.D. (1985) Excitatory/inhibitory response types in the cochlear nucleus: Discharge patterns and responses to electrical stimulation in the auditory nerve. J. Neurophysiol. 54, 917–939.

Smith, R.L., & M.L. Brachman (1980) Response modulation of auditory-nerve fibers by AM stimuli: effects of average intensity. Hearing Res. 2, 123–144.

Voigt, H.F. & Young, E.D. (1980) Evidence of inhibitory interactions between neurons in the dorsal cochlear nucleus. J. Neurophysiol. 44, 76–96.

Voigt, H.F. & Young, E.D. (1990) Cross-correlation analysis of inhibitory interactions in dorsal cochlear nucleus. J. Neurophysiol. 64, 1590–1610.

Wu, S.H. & Oertel, D. (1986) Inhibitory circuitry in the ventral cochlear nucleus is probably mediated by glycine. J. Neurosci. 6, 2691–2706.

Yates, G.K. (1987) Dynamic effects in the input/output relationship of auditory nerve fibers. Hear. Res. 27, 221–230.

Young, E.D. (1984) Response characteristics of neurons of the cochlear nuclei. In C.I. Berlin (ed.): Hearing Science, Recent Advances. San Diego: College Hill Press, pp. 423–460.

Young, E.D., Spirou, G.A., Rice, J.J. & Voigt, H.F. (1992) Neural organization and responses to complex stimuli in the dorsal cochlear nucleus. Phil. Trans. Roy. Soc. Lond. B, 336, 407–413.

12

CIRCUITS OF THE DORSAL COCHLEAR NUCLEUS

Donata Oertel and Nace L. Golding[*]

Department of Neurophysiology
University of Wisconsin Medical School
1300 University Ave., Madison, Wisconsin 53706

1. INTRODUCTION

The dorsal cochlear nucleus (DCN) has become more intriguing as more is learned about it because, although its anatomical structure has been examined from many perspectives and the responses of individual cells have been studied for more than 25 years, it is not yet possible to identify a physiological function that is carried out by the DCN. In this paper we will examine what is known about the cells of the DCN and their roles in the circuit in a search for clues concerning their function.

Neurons in the DCN may contribute to the localization of sound. While neurons of the DCN encode timing with less precision than neurons in the ventral cochlear nucleus (VCN) and are therefore unlikely to mediate localization of sound in the horizontal plane (Young and Brownell, 1976; Rhode et al., 1983), neurons in the deep layer of the DCN are known to inhibit cells in the VCN that encode timing with precision. Neurons in the deep layer, the tuberculoventral cells, project topographically to the VCN where they inhibit bushy and stellate cells (Wickesberg and Oertel, 1988, 1990; Oertel and Wickesberg, 1993). Bushy cells project to the superior olivary complex where interaural timing is encoded (Cant and Casseday, 1986). On the basis of the functional characteristics of the projection in vitro, it has been suggested that tuberculoventral cells could contribute to the suppression of echoes in cells that encode interaural timing (Wickesberg and Oertel, 1990). This hypothesis has recently received support from experiments in vivo (Wickesberg, 1996).

To localize sound in the vertical plane, mammals make use of spectral distortions that arise from the interaction of sounds with the pinnae, the head and neck (Musicant et al., 1990; Rice et al., 1992; Wenzel et al., 1993). While distortions in the sound spectra are essential cues for localization in elevation, they could also contribute to localization in the azimuth (Rice et al., 1992). The possibility that the DCN is involved in the analysis of

---

[*] Present address: Department of Physiology, Northwestern University, Evanston, IL.

Acoustical Signal Processing in the Central Auditory System
edited by Syka, Plenum Press, New York, 1997

sound spectra, in particular to notches in the sound spectra, has some experimental support (Spirou and Young, 1991; Nelken and Young, 1994). To make use of spectral cues, an animal must take into account the position of the head and pinna, indicating that in a pathway involved in this task, auditory and somatosensory input would be expected to be combined, as it is known to be in the DCN. Granule cells in the DCN receive somatosensory input from the dorsal column nuclei through mossy terminals (Weinberg and Rustioni, 1987; Itoh et al., 1987; Wright and Ryugo, 1996). Young and his colleagues have shown that the somatosensory input has a powerful influence upon responses to sound (Young et al., 1995). However, behavioral experiments indicate that lesions of the dorsal acoustic stria affect the ability of cats to orient to sounds only subtly (Sutherland, 1991). To what extent the DCN contributes to the localization of sound is therefore unknown.

In addition to being localized, the biological significance of sounds must be interpreted. Very little is known about how interpretation is accomplished; some is certainly accomplished at higher levels. To what extent the integrative processes in the cochlear nuclei contribute to interpretation is not known.

Intracellular recordings from neurons in the DCN provide information about the functional nature of connections. Below we will review some of the results of recordings from the superficial layers of the DCN in vitro. These recordings give an indication of how neuronal circuits in the superficial layers act on the principal cells of the DCN. We will review what is known about the connections among cells with the goal of understanding the circuitry of the DCN: (1) function of individual cells and their inputs, (2) the integration of signals by the principal cells of the DCN.

2. METHODS

Slices of the cochlear nuclei were prepared from CBA mice from 18 to 26 days old as described previously (Golding and Oertel, 1996). Animals were decapitated and the brain was removed from the skull under oxygenated (95% O_2, 5% CO_2) normal saline which contained: 130 mM NaCl, 3 mM KCl, 1.3 mM $MgSO_4$, 2.4 mM $CaCl_2$, 20 mM Na-HCO$_3$, 3 mM HEPES, 10 mM glucose, 1.2 mM KH_2PO_4, pH 7.4. Parasagittal slices 200–350 µm thick were cut with an oscillating tissue slicer (Frederick Haer). Slices were maintained in a chamber where they were submerged in rapidly flowing, oxygenated saline solution, at a temperature of approximately 34°C. Flow rates through the 0.3 ml-chamber ranged from 9 to 12 ml/min (Oertel, 1985). Strychnine, picrotoxin, DL-2-amino-5-phosphonovaleric acid (APV), 6,7-dinitroquinoxaline-2,3-dione (DNQX), and bicuculline methiodide were obtained from Sigma and were dissolved in physiological saline.

Intracellular recordings were made with electrodes of between 120 and 200 MΩ that were filled with 1% biocytin (Sigma) in 2 M potassium acetate. Shocks (0.1 to 100 V amplitude, 100 ms duration) were delivered through a pair of tungsten wires, insulated except at their 50 µm-diameter-tips, to the nerve root, to the surface of the VCN or to the dorsal tip of the DCN. Synaptic responses were digitally sampled at 40 kHz, low pass filtered at 10 kHz, and stored on computer using pClamp software (Axon Instruments). Responses to exogenously applied agonists were sampled between 0.67 and 1.3 kHz.

Cells were routinely labeled with biocytin. Slices were fixed in 4% paraformaldehyde in 0.1 M phosphate buffer (pH 7.4), embedded in a mixture of gelatin and albumin and sectioned at 60 µm. The sections were reacted with avidin conjugated to horseradish peroxidase (Vector ABC kit), and processed for horseradish peroxidase (Wickesberg and Oertel, 1988) and counterstained with cresyl violet.

3. RESULTS

To study the functional connections in the DCN we have used a combination of anatomy, pharmacology and electrophysiology to isolate and examine connections between classes of cells electrophysiologically. This approach has been an interative process. As anatomical studies contributed to defining the classes of cells and their connections, it became possible to eliminate identified classes of inputs with pharmacological blocking agents and to study other classes of inputs in isolation. Each additional anatomical and pharmacological insight has led to refinement in the interpretation of electrophysiological findings (Hirsch and Oertel, 1988a,b; Oertel and Wu, 1989; Zhang and Oertel, 1993a,b,c, 1994; Golding and Oertel, 1996, 1997). Here we will summarize what we know about classes of connections between specific classes of cells from these combined studies. These combined studies reflect recordings from 66 cartwheel, 50 fusiform, 8 giant, and 15 tuberculoventral cells. Two cells are either superficial stellate or Golgi cells; we will refer to them as superficial stellate cells. These classes of cells were initially identified on the basis of electron microscopical studies, making correspondence with features at the light microscopic level somewhat uncertain (Mugnaini et al., 1980; Wouterlood et al., 1984; Mugnaini et al., 1994). While our studies include most classes of neurons of the DCN of mice, they do not include all. We have not yet recorded from a unipolar brush cell (Mugnaini and Floris, 1994; Mugnaini et al., 1994).

While they may be incomplete, the results from these recordings nevertheless form a picture of the circuitry of the DCN. Figure 1 shows a summary of the connections we have studied.

3.1 Parallel Fibers, the Axons of Granule Cells, Are a Major Source of Glutamatergic Excitation to Superficial Stellate, Cartwheel and Fusiform Cells

Parallel fibers, the axons of granule cells, are the major source of excitatory input to the superficial layers of the DCN. The thickenings of parallel fibers with their clusters of small, clear round vesicles form asymmetric contacts, generally associated with excitation, with their targets (Kane, 1974; Mugnaini et al., 1980; Wouterlood and Mugnaini, 1984;

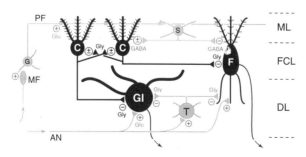

Figure 1. Summary of the connections among cells of the DCN. Mossy fiber (MF) terminals excite granule cells (G). The axons of granule cells, the parallel fibers (PF) excite cartwheel (C), superficial stellate cells (S) and fusiform (F) cells through glutamatergic synapses. Auditory nerve fibers (AN) excite tuberculoventral (T), fusiform (F) and giant (Gl) cells through glutamatergic synapses. Cartwheel cells either excite or suppress firing in other cartwheel cells and they inhibit fusiform and giant cells through glycinergic connections. Superficial stellate cells may be the source of GABAergic inputs to fusiform and cartwheel cells.

Wouterlood et al., 1984). There is no doubt that the spines of cartwheel and fusiform cells are contacted by parallel fibers; no other source of excitatory input to spines on these cells has been identified. Stellate cells receive input from varicosities of parallel fibers on their dendritic shafts and perikarya (Wouterlood et al., 1984).

Mossy fiber inputs from the dorsal column nuclei and elsewhere are another major source of excitatory input in the DCN (Weinberg and Rustioni, 1987; Itoh et al., 1987; Wright and Ryugo, 1996). Mossy fibers generally contact granule and Golgi cells and have not been observed to contact other types of cells.

Activity in granule cells has been examined indirectly because their small size makes sharp microelectrode recordings impossible. Their activity, spontaneous and evoked by shocks, is reflected in their targets: stellate, cartwheel, and fusiform cells. As expected, excitatory input with common features has been observed in each of these targets. In each of these groups of cells, shocks which activate granule cells (to the nerve root, to the lateral surface of the VCN where superficial granule cells are located, and to the parallel fibers in the dorsal DCN) evoke glutamatergic excitation after a delay that varies with the location of stimulation (Zhang and Oertel, 1993a, c, 1994; Golding and Oertel, 1996, 1997). Excitation lasts between 50 and 100 msec. The excitation is eliminated by a combination of AMPA and NMDA receptor blockers and is therefore glutamatergic (Manis 1989; Golding and Oertel, 1996).

Granule cells often fire spontaneously. Many cartwheel and fusiform cells received spontaneous EPSPs that were eliminated by DNQX. As the only known source of glutamatergic input to cartwheel cells is the granule cells, identification of their source in cartwheel cells is particularly clear. The conclusion is strengthened by the finding that fusiform cells also receive spontaneous excitation that just precedes glycinergic IPSPs which presumably arise from cartwheel cells (Zhang and Oertel, 1994).

3.2 Cartwheel Cells Are Glycinergic and Contact Cartwheel, Fusiform and Giant Cells

Cartwheel cells are numerous and prominent in the DCN of most mammals. Anatomical studies were the first to show that cartwheel cells are homologues of Purkinje cells in that they are interposed between parallel fibers and their targets and that they share molecular markers (Berrebi et al., 1990; Berrebi and Mugnaini, 1991). Electrophysiological recordings indicate that cartwheel cells also share electrophysiological characteristics with Purkinje cells in the cerebellum (Manis et al., 1993, 1994; Zhang and Oertel, 1993a; Golding and Oertel, 1996, 1997).

The electrophysiological characteristics of cartwheel cells are unlike those of any other cells of the cochlear nuclei. Like Purkinje cells, cartwheel cells have prominent voltage-sensitive calcium conductances which generate calcium action potentials which give rise to complex action potentials (Llinás and Sugimori, 1980; Hirsch and Oertel, 1988a; Manis et al., 1993, 1994; Zhang and Oertel, 1993a; Golding and Oertel, 1996, 1997). The intrinsic electrical properties of cartwheel cells shape and dominate the responses to synaptic input. In contrast to neurons elsewhere in the cochlear nuclei whose firing reflects precisely the timing of its inputs, cartwheel cells fire relatively independently of the timing of the inputs. Cartwheel cells fire long-lasting complex spikes that can outlast the synaptic input that triggered them by hundreds of milliseconds (Golding and Oertel, 1996; 1997).

One of the molecular markers that cartwheel cells share with Purkinje cells is GAD (Adams and Mugnaini, 1987). We therefore expected cartwheel cells, like Purkinje cells, to be GABAergic. We expected to find GABAergic PSPs in the targets of cartwheel cells whose temporal patterns matched the temporal pattern of the firing of cartwheel cells. Surprisingly, no such GABAergic PSPs have ever been recorded. Instead, all known targets of cartwheel cells receive glycinergic PSPs whose occurrence, spontaneous and evoked, matches the firing patterns of cartwheel cells (Zhang and Oertel, 1994; Golding and Oertel, 1996, 1997). Consistent with the conclusion that cartwheel cells are glycinergic is the result that, like other known glycinergic neurons, cartwheel cells are heavily labeled by antibodies to glycine conjugates (Wenthold et al., 1987; Osen et al., 1990; Wickesberg et al., 1994).

Cartwheel cells are unusual in that they are depolarized by synaptic input that is in other cells inhibitory. Glycinergic and GABA-Aergic inputs that activate chloride conductances can excite cartwheel cells (Golding and Oertel, 1996). While the glycine and GABA-A receptors are conventional, the reversal potential for chloride in cartwheel cells is about -53 mV, near the threshold of firing, compared to -67 mV in other cells of the cochlear nuclei (Golding and Oertel, 1996, 1997). Depolarizing glycinergic and GABAergic PSPs can either trigger complex action potentials or suppress firing depending on the electrophysiological context in which they occur (Golding and Oertel, 1996). The difference between causing firing and suppressing firing reflects a difference in the way synaptic inputs interact with the intrinsic voltage-sensitive conductances under the two conditions and implies that cartwheel cells are not isopotential. While the glycine and GABA-A receptors are conventional, the chloride equilibirum potential in cartwheel cells sits at an unusually depolarized level (Golding and Oertel, 1996).

The characteristic bursts produced by the complex spikes spontaneously and in responses to shocks serve as an electrophysiological tag for identifying the targets of cartwheel cells. Bursts of glycinergic PSPs were found in cartwheel, and fusiform cells, as would have been predicted by the anatomical results of Berrebi and Mugnaini (1991). Giant cells also receive bursts of glycinergic IPSPs, indicating that they, too, are targets of cartwheel cells. Giant cells have cell bodies in the deep layer and dendrites that extend into the fusiform and molecular layers. As the terminals of cartwheel cells in mice lie in the molecular and fusiform cell layers, their input is likely to impinge distally on the dendrites of giant cells. It is not surprising, therefore, that some of the bursts of IPSPs show signs of dendritic filtering (Golding and Oertel, 1997).

Cartwheel cells play a prominent functional role in the circuitry of the DCN. They mediate much of the long-lasting inhibition in responses to shocks and they are the source of most, and perhaps all, spontaneous glycinergic inhibition. Figure 2 shows intracellular recordings of responses to shocks to the dorsal DCN, a condition which favors the activation of parallel fibers and avoids the activation of auditory nerve fibers. Such shocks evoke a long-lasting train of action potentials in cartwheel cells. When the cartwheel cell was hyperpolarized to prevent firing, the underlying PSPs were revealed also to be long-lasting. Long-lasting IPSPs were also detected in fusiform and giant cells but not in the superficial stellate or tuberculoventral cells. Spontaneous PSPs which occurred in bursts and which were blocked by strychnine were observed in the same classes of neurons that had long-lasting responses to shocks.

In addition to receiving glycinergic input from other cartwheel cells, cartwheel cells receive GABAergic input. Shocks to the DCN evoke depolarizing PSPs that are blocked by picrotoxin and bicuculline but not strychnine. What is the source of this input? Immunocytochemical labeling for GAD and GABA reveals two types of neurons that could be GABAergic, superficial stellate and Golgi cells (Mugnaini, 1985; Adams and Mugnaini, 1987; Osen et al., 1990; Kolston et al., 1992).

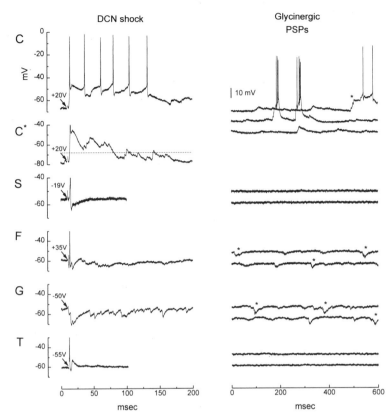

Figure 2. Sample traces show that the timing of firing of cartwheel cells matches the timing of glycinergic PSPs in their targets spontaneously (right) as well as in responses to shocks to the DCN (left). Responses to shocks are shown from anatomically identified cartwheel (C), superficial stellate (S), fusiform (F), giant (G) and tuberculoventral (T) cells. Two responses are shown from the cartwheel cell; the first (C) was recorded with the cell at the resting potential and the second while the cell was hyperpolarized with -0.2 nA current. Hyperpolarization prevents firing and makes it possible to resolve the underlying synaptic events. The cell received bursty inputs that reversed at -52 mV (dotted line). Spontaneous PSPs from the same cells are shown on the right. All cells were selected for having all spontaneous PSPs blocked with strychnine, a blocker of glycinergic inhibition. The glycinergic activity in the cartwheel cell was depolarizing and evoked bursts of action potentials whereas bursts of glycinergic PSPs were conventionally hyperpolarizing in the fusiform and giant cell.

3.3 Superficial Stellate (or Golgi) Cells Are Excited by Parallel Fibers and Terminate in the Molecular Layer

While only two recordings have been made from cells identified as either superficial stellate (or Golgi) cells, they show consistent features (Zhang and Oertel, 1993a; Golding and Oertel, 1997). In our slices from mice no signs of coupling have been observed either electrically or anatomically although it has been reported that superficial stellate cells are connected by gap junctions (Wouterlood et al., 1984). Their dendrites and profuse terminal arbors lie spread in the plane of the molecular layer. While spontaneous depolarizing PSPs were observed, neither labeled cell fired spontaneously. Depolarization causes them to fire with large, regular action potentials. When activated with shocks, these cells fire once or twice within 50 msec of the shock.

The activity in these neurons is consistent with the pattern of GABAergic PSPs in cartwheel and fusiform cells. GABAergic interneurons are activated by shocks to the DCN. Spontaneous GABAergic IPSPs have rarely been observed.

3.4 Tuberculoventral Cells Are Excited by Auditory Nerve Fibers and Inhibited by D Stellate Cells

Tuberculoventral cells, also known as vertical or corn cells, are identified by their projection to the VCN (Wickesberg and Oertel, 1988; Oertel and Wu, 1989; Wickesberg et al., 1991; Zhang and Oertel, 1993c). Their cell bodies and dendrites are confined to the deep layer of the DCN. Their dendrites are aligned along the isofrequency laminae defined by the path of auditory nerve fibers. This anatomical arrangment indicates that tuberculoventral cells receive input from a small population of auditory nerve fibers and that they are expected to be narrowly tuned. In addition to their terminating in the VCN, tuberculoventral cells terminate locally in the deep and fusiform cell layers of the DCN. Their terminals are intermingled with the dendrites of fusiform and giant cells indicating that these cells are potential targets. All criteria have been met to identify the neurotransmitter of tuberculoventral cells to be glycine (Oertel and Wickesberg, 1993).

Electrically, tuberculoventral cells fire regular, overshooting action potentials when depolarized with current (Zhang and Oertel, 1993c). They are activated with monosynaptic EPSPs by shocks to the nerve root, consistent with their being activated directly by the auditory nerve fibers.

Polysynaptic EPSPs and IPSPs were also detected in responses to shocks of the auditory nerve but they were not prominent. Two types of neurons likely to be glycinergic have terminals in the vicinity of tuberculoventral cell dendrites, D stellate cells and tuberculoventral cells (Oertel and Wickesberg, 1993). It is not possible to distinguish between these possible sources electrophysiologically as the firing pattern in responses to shocks in these cells is similar.

3.5 Fusiform and Giant Cells Integrate Inputs from the Deep and Molecular Layers with Contrasting Spatial Patterns and in Different Proportions

Fusiform and giant cells form the two output pathways from the DCN to the inferior colliculi (Osen, 1972; Adams, 1979; Ryugo et al., 1981; Oliver, 1984). They share the same classes of inputs yet their responses to shocks differ from one another. Fusiform and giant cells have dendrites that extend into the superficial as well as into the deep layers. Both are excited by auditory nerve fibers; both are inhibited by cartwheel cells; both are probably also inhibited by tuberculoventral cells. The differences in responses of fusiform and giant cells must arise from differences in the patterns of convergence of inputs.

The convergence of inputs from the auditory nerve is different in fusiform and giant cells. In the deep layer, the basal dendrites of fusiform cells lie in narrow bands. By lying parallel to the tonotopically arranged auditory nerve fibers, fusiform cell dendrites can receive input only from few fibers. In contrast, giant cell dendrites lie spread across the path of auditory nerve fibers, making convergence of input from auditory nerve fibers with a wider range of tuning possible. Electrophysiologically these inputs are reflected as glutamatergic excitation. The difference in convergence in fusiform and giant cells has not been explored electrophysiologically in slices, because it is not possible to separate glutamatergic excitation from parallel and auditory nerve fibers either pharmacologically or physically by the placement of stimulating electrodes.

Similarly, it seems likely that both fusiform and giant cells receive input from tuberculoventral cells. Fusiform cell dendrites are within reach of terminals of fewer tuberculoventral cells than giant cell dendrites. Both respond to shocks with short-latency glycinergic inhibition whose source is unlikely to be the cartwheel cells.

Excitation by parallel fibers is also different in fusiform and giant cells. While fusiform cells have a prominent cluster of apical dendrites which serve as a substrate for input from the molecular layer, giant cells have only the tips of some dendrites in the molecular layer. The tips of the dendrites of giant cells have occasional spines whereas the apical dendrites of fusiform cells characteristically are studded with spines. The fusiform cells therefore receive more input from parallel fibers than giant cells. This conclusion is confirmed by electrophysiological measurements; long-lasting excitation is a common feature of responses to shocks in fusiform but not giant cells. Furthermore, spontaneous glutamatergic EPSPs, likely to arise from granule cells, are common in fusiform (Zhang and Oertel, 1994) but not giant cells (Golding and Oertel, 1997).

Cartwheel cells inhibit both fusiform and giant cells. Both types of cells are in a position to receive input from at least three known sources of glycinergic input, tuberculoventral and cartwheel cells from the DCN and D stellate cells from the VCN. In our experiments these three sources can be only imperfectly differentiated. Pharmacologically they are identically sensitive to strychnine. The position of stimulating electrodes can be used to stimulate parallel fibers selectively. Shocks to the auditory nerve root activate auditory nerve fibers as well as parallel fibers whereas shocks to the dorsal DCN seem to activate parallel fibers preferentially over auditory nerve fibers. (Cartwheel cells serve as assays for the activation of parallel fibers; tuberculoventral cells serve as an assay for the activation of auditory nerve fibers.) To a limited extent, therefore, the prominence of glycinergic inhibition from cartwheel cells can be assessed in responses to shocks of the dorsal DCN. Glycinergic inhibition is prominent in the synaptic responses of both fusiform and giant cells under these conditions indicating that cartwheel cells are an important source of inhibition in both types of principal cells (Figure 2).

Tuberculoventral cells probably also inhibit both giant and fusiform cells. Cartwheel cells are slow to fire in response to shocks. Fusiform and giant cells, however, receive prominent early inhibition whose source could be tuberculoventral cells. Early inhibition is so strong in giant cells that the excitation is masked.

4. DISCUSSION

A striking feature of the anatomical arrangement of the DCN is the orthogonal nature of its two major systems of excitatory and inhibitory inputs. Parallel fibers form an array of excitatory inputs that innervate the DCN in the dorso-ventral direction; auditory nerve fibers form an array of inputs that innervate the DCN in the rostro-caudal direction. It is not known whether the array of parallel fibers is organized rostrocaudally; the array of auditory nerve fibers is organized tonotopically along the dorso-ventral axis. Each of the excitatory arrays is associated with a glycinergic system of inhibition. Cartwheel cells convert the excitation of parallel fibers to glycinergic inhibition in its targets, and tuberculoventral cells convert the excitation of auditory nerve fibers to inhibition.

Fusiform and giant cells, the principal cells which carry the outcome of integration in the DCN to the inferior colliculi, assay the two systems of inputs differently. With their apical dendrites, fusiform cells receive more excitatory input from the molecular layer than the giant cells although inhibition from cartwheel cells is prominent in their responses (Zhang and Oertel, 1993b, 1994; Golding and Oertel, 1997). In the responses of fusiform cells to shocks of the auditory nerve or to the dorsal DCN in slices, excitation and inhibition both are evident. In contrast, inhibition dominates the responses of giant cells under both conditions.

Both classes of principal cells are strongly influenced by cartwheel cells whose role is particularly intriguing. Cartwheel cells convert a short burst of firing in parallel fibers to long-lasting inhibition in fusiform and giant cells (Young et al, 1995; Golding and Oertel, 1996, 1997). Cartwheel cells form an array in the plane of the DCN, beams of cartwheel cells presumably being driven by the activation of a beam of parallel fibers (Manis, 1989). Recordings in slices have shown that cartwheel cells are interconnected through synapses which can be either excitatory or suppressive (Golding and Oertel, 1996). These interactions presumably affect the spatial pattern of activation of the array of cartwheel cells. In being weakly excitatory when cartwheel cells are not firing and weakly inhibitory when they are firing strongly, their interconnections serve to maintain firing at moderate rates. The possibility that cartwheel cells mediate plasticity is raised by their similarities with Purkinje cells of the cerebellum and with homologous structures in electric fishes. Purkinje cells show long term depression when the activation of parallel fiber inputs is coupled with their depolarization (Ito et al., 1982; Linden et al., 1991). In mormyrid electric fishes, neurons which are homologous to Purkinje cells are involved in eliminating the component of the sensory signal that is generated by the animal's own body by an anti-Hebbian plasticity (Bell et al., 1993). We have been unable to demonstrate plasticity in cartwheel cells in slices.

The deep layer carries acoustic information. Auditory nerve fibers terminate topographically in the deep DCN, providing a systematic representation of frequency (Osen, 1970). The acoustic information is transmitted to the fusiform cells (type IV of Young and Brownell, 1976) and reflected in their corresponding tonotopic arrangement (Spirou and Young, 1991). The "central inhibitory areas" of type IV units presumably reflect sharply tuned, tonotopically organized inhibition that arises from type II units, probably tuberculoventral cells (Spirou and Young, 1991; Nelken and Young, 1994). Fusiform cells (type IV units) have a second source of inhibition that is broadly tuned, which could arise through the VCN (Nelken and Young, 1994). Less is known about how giant cells respond to sound as no responses to sound have been reported from anatomically identified giant cells. It seems likely that giant cells are inhibited by sound and correspond to type V units of Evans and Nelson (1973). In such units, the input from the auditory nerve is largely masked by inhibition whose source has not been determined.

What information is carried in the molecular layer is less well understood. The molecular layer of the DCN receives input from many sources, including the inferior colliculi (Caicedo and Herbert, 1993), the cerebral cortex (Feliciano et al., 1993), the vestibular afferents (Burian and Gestoettner, 1988), and the dorsal column nuclei (Itoh et al., 1987, Weinberg and Rustioni, 1987; Wright and Ryugo, 1996) as well as from type II auditory nerve fibers (Berglund and Brown, 1994; Brown et al., 1988; Brown and Ledwith, 1990). It has been suggested that the inputs to granule cells are involved with signaling head and pinna position (Young et al., 1995).

The contribution of the molecular layer upon fusiform and giant cells may until recently not have been detected. Cartwheel cells respond weakly to sound (Parham and Kim, 1995). In contrast, they are strongly inhibited by the somatosensory activation of the dorsal column nuclei (Young et al., 1995). Since almost all single unit recordings must be done under conditions when animals are immobilized, they are done under conditions in which the somatosensory inputs are unnaturally quiet. Recordings of single units in the DCN therefore probably reflect abnormally small contributions from the molecular layer.

The responses of principal cells to sound have been recorded in vivo. The responses of fusiform cells reflect the acoustic input through the deep layers. The combination of excitatory and inhibitory acoustic input is evident in the response areas and persistimulus time histograms. The delicate balance between excitation and inhibition obscures the timing information in auditory nerve inputs (Rhode et al., 1983) and produces response areas in which excitation dominates in responses to weak tones near the characteristic frequency and inhibition dominates in responses to stronger tones at the characteristic frequency (Spirou and Young, 1991). Responses to sound have probably also been recorded from giant cells but a direct and positive correlation of responses with cell type has not yet been published.

In summary, the role of the DCN remains enigmatic. It is clear that the principal cells integrate information from two major systems of inputs which are delivered through the deep and molecular layers respectively. Experiments from slices together with findings from recordings in vivo indicate that because the DCN integrates auditory with non-auditory sensory inputs, this structure is exceptionally difficult to study. The stability necessary for making measurements from single units affects the somatosensory input through the molecular layer. Input through the molecular layer clearly has a profound influence on the principal cells but this influence not understood.

5. ACKNOWLEDGMENTS

We thank Michael Ferragamo for reading and commenting on the manuscript and who was always available for discussion. We are indebted to colleagues upon whose results the present thoughts and conclusions are built, especially Judith Hirsch, Shu Hui Wu, Robert Wickesberg and Sam Zhang, and thank them. The work was supported by a grant from NIH, DC00176.

6. REFERENCES

Adams, J. C. (1979) Ascending projections to the inferior colliculus. J. Comp. Neurol. 183, 519–538.
Adams, J. C. and Mugnaini, E. (1987) Patterns of glutamate decarboxylase immunostaining in the feline cochlear nuclear complex studied with silver enhancement and electron microscopy. J. Comp. Neurol. 262, 375–401.
Bell, C.C., Caputi, A., Grant, K., and Serrier, J. (1993) Storage of a sensory pattern by anti-Hebbian synaptic plasticity in electric fish. Proc. Natl. Acad. Sci. 90, 4650–4654.
Berglund, A. M. and Brown, M. C. (1994) Central trajectories of type II spiral ganglion cells from various cochlear regions in mice. Hear. Res. 75, 121–130.
Berrebi, A. S., Morgan, J. I., and Mugnaini, E. (1990) The Purkinje cell class may extend beyond the cerebellum. J. Neurocytol. 19, 643–654.
Berrebi, A. S. and Mugnaini, E. (1991) Distribution and targets of the cartwheel cell axon in the dorsal cochlear nucleus of the guinea pig. Anatomy & Embryology 183, 427–454.
Brown, M. C. and Ledwith, J. V. (1990) Projections of thin (type-II) and thick (type-I) auditory-nerve fibers into the cochlear nucleus of the mouse. Hear. Res. 49, 105–118.
Brown, M. C., Berglund, A. M., Kiang, N. Y. S., and Ryugo, D. K. (1988) Central trajectories of type II spiral ganglion neurons. J. Comp. Neurol. 278, 581–590.
Burian, M. and Gestoettner, W. (1988) Projection of primary vestibular afferent fibres to the cochlear nucleus in the guinea pig. Neurosci. Lett. 84, 13–17.

Caicedo, A. and Herbert, H. (1993) Topography of descending projections from the inferior colliculus to auditory brainstem nuclei in the rat. J. Comp. Neurol. 328, 377–392.

Cant, N.B. and Casseday, J.H. (1986) Projections from the anteroventral cochlear nucleus to the lateral and medial superior olivary nuclei. J. Comp. Neurol. 247, 457–476.

Evans, E. F. and Nelson, P. G. (1973) The responses of single neurones in the cochlear nucleus of the cat as a function of their location and the anaesthetic state. Exp. Brain Res. 17, 402–427.

Feliciano, M., Saldaña, E., and Mugnaini, E. (1993) Direct projection from the primary auditory cortex to the nucleus sagulum, superior olivary complex and cochlear nucleus of the albino rat. Soc. Neurosci. Abstr. 19, 1427.

Golding, N. L., and Oertel, D. (1996) Context-dependent action of glycinergic and GABAergic inputs in the dorsal cochlear nucleus. J. Neurosci. 16, 2208–2219.

Golding, N.L. and D. Oertel (1997) Physiological identification of the targets of cartwheel cells in the dorsal cochlear nucleus. J. Neurophysiol. (In press).

Hirsch, J.A. and D. Oertel. (1988a) Intrinsic properties of neurones in the dorsal cochlear nuclei of mice, *in vitro*. J. Physiol. (London) 396, 535–548.

Hirsch, J.A. and D. Oertel. (1988b) Synaptic connextions in the dorsal cochlear nucleus of mice, *in vitro*. J. Physiol. (London) 396, 549–562.

Ito, M. Sakurai, M. Tongroach, P. (1982) Long-lasting depression of parallel fiber-Purkinje cell transmission induced by conjunctive stimulation of parallel fibers and climbing fibers in the cerebellar cortex. J. Physiol. (London) 324, 113–134.

Itoh, K., Kamiya, H., Mitani, A., Yasui, Y., Takada, M., and Mizuno, N. (1987) Direct projections from the dorsal column nuclei and the spinal trigeminal nuclei to the cochlear nuclei in the cat. Brain Res. 400, 145–150.

Kane, E. C. (1974) Synaptic organization in the dorsal cochlear nucleus of the cat: A light and electron microscopic study. J. Comp. Neurol. 155, 301–330.

Kolston, J., Osen, K.K., Hackney, C.M., Ottersen, O.P., Storm-Mathisen, J. (1992) An atlas of glycine- and GABA-like immunoreactivity and colocalization in the cochlear nuclear complex of the guinea pig. Anat. Embryol. 186, 443–465.

Linden, D.J., Dickinson, M.H., Smeyne, M., Connor, J.A. (1991) A long-term depression of AMPA currents in cultured cerebellar Purkinje neurons. Neuron 7, 81–89.

Llinás, R. and Sugimori, M. (1980) Electrophysiological properties of in vitro Purkinje cell somata in mammalian cerebellar slices. J. Physiol. Lond. 305, 171–195.

Manis, P. B. (1989) Responses to parallel fiber stimulation in the guinea pig dorsal cochlear nucleus in vitro. J. Neurophysiol. 61, 149–161.

Manis, P.B., Scott, J.C., and Spirou, G.A. (1993) Physiology of the dorsal cochlear nucleus molecular layer. In: The Mammalian Cochlear Nuclei: Organization and Function, edited by M. A. Merchan, J. M. Juiz and D. A. Godfrey. New York: Plenum Publishing Corp., p. 361–371.

Manis, P. B., Spirou, G. A., Wright, D. D., Paydar, S., and Ryugo, D. K. (1994) Physiology and morphology of complex spiking neurons in the guinea pig dorsal cochlear nucleus. J. Comp. Neurol. 348, 261–276.

Mugnaini, E. (1985) GABA neurons in the superficial layers of the rat dorsal cochlear nucleus: light and electron microscopic immunocytochemistry. J. Comp. Neurol. 235, 61–81.

Mugnaini, E. and Floris, A. (1994) The unipolar brush cell: a neglected neurons of the mammalian cerebellar cortex. J. Comp. Neurol. 339, 174–180.

Mugnaini, E., Floris, A., and Wright-Goss, M. (1994) Extraordinary synapses of the unipolar brush cell: An electron microscopic study in the rat cerebellum. Synapse 16, 284–311.

Mugnaini, E., Osen, K.K., Dahl, A.-L., Friedrich, V.L., and Korte, G. (1980) Fine structure of granule cells and related interneurons (termed Golgi cells) in the cochlear nuclear complex of cat, rat, and mouse. J. Neurocytol. 9, 537–570.

Musicant, A. D., Chan, J. C. K., and Hind, J. E. (1990) Direction-dependent spectral properties of cat external ear: New data and cross-species comparisons. J. Acoust. Soc. Am. 87, 757–781.

Nelken, I. and Young, E. D. (1994) Two separate inhibitory mechanisms shape the responses of dorsal cochlear nucleus type IV units to narrowband and wideband stimuli. J. Neurophysiol. 71, 2446–2462.

Oertel, D. (1985) Use of brain slices in the study of the auditory system; spatial and temporal summation of synaptic inputs in cells in the anteroventral cochlear nucleus of the mouse. J. Acous. Soc. Am. 78, 328–333.

Oertel, D. and Wickesberg, R. E. (1993) Glycinergic inhibition in the cochlear nuclei: evidence for tuberculoventral neurons being glycinergic. In: The Mammalian Cochlear Nuclei: Organization and Function, edited by M. A. Merchan, J. M. Juiz and D. A. Godfrey. New York: Plenum Publishing Corp., p. 225–237.

Oertel, D. and Wu, S. H. (1989) Morphology and physiology of cells in slice preparations of the dorsal cochlear nucleus of mice. J. Comp. Neurol. 283, 228–247.

Oertel, D., Wu, S. H., Garb, M.W. and Dizack, C. (1990) Morphology and physiology of cells in slice preparations of the posteroventral cochlear nucleus of mice. J. Comp. Neurol. 295, 136–154.

Oliver, D. L. (1984) Dorsal cochlear nucleus projections to the inferior colliculus in the cat: a light and electron microscopic study. J. Comp. Neurol. 224, 155–172.

Osen, K.K. (1970) Course and termination of the primary afferents in the cochlear nuclei of the cat. Arch. Ital. Biol. 108, 21–51.

Osen, K.K. (1972) Projection of the cochlear nuclei on the inferior colliculus in the cat. J. Comp. Neurol. 144, 355–372.

Osen, K. K., Ottersen, O. P., and Storm-Mathisen, J. (1990) Colocalization of glycine-like and GABA-like immunoreactivities: A semiquantitative study of individual neurons in the dorsal cochlear nucleus of cat. In: Glycine Neurotransmission, edited by O. P. Ottersen and J. Storm-Mathisen. New York: John Wiley and Sons, p. 417–451.

Parham, K. and Kim, D. O. (1995) Spontaneous and sound-evoked discharge characteristics of complex-spiking neurons in the dorsal cochlear nucleus of the unanesthetized decerebrate cat. J. Neurophysiol. 73, 550–561.

Rhode, W.S., Smith, P.H. and Oertel, D. (1983) Physiological response properties of cells labeled intracellularly with horseradish peroxidase in cat dorsal cochlear nucleus. J. Comp. Neurol. 213, 426–447.

Rice, J. J., May, B. J., Spirou, G. A., and Young, E. D. (1992) Pinna-based spectral cues for sound localization in cat. Hear. Res. 58, 132–152.

Ryugo, D. K., Willard, F. H., and Fekete, D. M. (1981) Differential afferent projections to the inferior colliculus from the cochlear nucleus in the albino mouse. Brain Res. 210, 342–349.

Spirou, G. A. and Young, E. D. (1991) Organization of dorsal cochlear nucleus type IV unit response maps and their relationship to activation by bandlimited noise. J. Neurophysiol. 66, 1750–1768.

Sutherland, D.P. (1991) A role of the dorsal cochlear nucleus in the localization of elevated sound sources. Ass. Res. Otolaryng. Abstr. 14, 33–33.

Weinberg, R. J. and Rustioni, A. (1987) A cuneocochlear pathway in the rat. Neurosci. 20, 209–219.

Wenthold, R. J., Huie, D., Altschuler, R. A., and Reeks, K. A. (1987) Glycine immunoreactivity localized in the cochlear nucleus and superior olivary complex. Neurosci. 22, 897–912.

Wenzel, E.M., Arruda, M., Kistler, D., Wightman, F.L. (1993) Localization using nonindividualized head-related transfer functions. J. Acoust. Soc. Am. 94, 111–123.

Wickesberg, R.E. (1996) Rapid inhibition in the cochlear nuclear complex of the chinchilla. J. Acoust. Soc. Am. 100, 1691–1702.

Wickesberg, R. E. and Oertel, D. (1988) Tonotopic projection from the dorsal to the anteroventral cochlear nucleus of mice. J. Comp. Neurol. 268, 389–399.

Wickesberg, R.E. and Oertel, D. (1990) Delayed, frequency-specific inhibition in the cochlear nuclei of mice: a mechanism for monaural echo suppression. J. Neurosci. 10, 1762–1768.

Wickesberg, R.E., Whitlon, D. and Oertel, D. (1991) Tuberculoventral neurons project to the multipolar cell area but not to the octopus cell area of the posteroventral cochlear nucleus. J. Comp. Neurol. 313, 457–468.

Wickesberg, R. E., Whitlon, D., and Oertel, D. (1994) In vitro modulation of somatic glycine-like immunoreactivity in presumed glycinergic neurons. J. Comp. Neurol. 339, 311–327.

Wouterlood, F. G. and Mugnaini, E. (1984) Cartwheel neurons of the dorsal cochlear nucleus: a Golgi-electron microscopic study in rat. J. Comp. Neurol. 227, 136–157.

Wouterlood, F. G., Mugnaini, E., Osen, K. K., and Dahl, A. L. (1984) Stellate neurons in rat dorsal cochlear nucleus studies with combined Golgi impregnation and electron microscopy: synaptic connections and mutual coupling by gap junctions. J. Neurocytol. 13, 639–664, 1984.

Wright, D.D. and Ryugo, D. K. (1996) Mossy fiber projections from the cuneate nucleus to the cochlear nucleus in the rat. J. Comp. Neurol. 365, 159–172.

Young, E. D. and Brownell, W. E. (1976) Responses to tones and noise of single cells in dorsal cochlear nucleus of unanesthetized cats. J. Neurophysiol. 39, 282–300.

Young, E. D., Nelken, I., and Conley, R. A. (1995) Somatosensory effects on neurons in dorsal cochlear nucleus. J. Neurophysiol. 73, 743–765.

Zhang, S. and Oertel, D. (1993a) Cartwheel and superficial stellate cells of the dorsal cochlear nucleus of mice: intracellular recordings in slices. J. Neurophysiol. 69, 1384–1397.

Zhang, S. and Oertel, D. (1993b) Giant cells of the dorsal cochlear nucleus of mice: intracellular recordings in slices. J. Neurophysiol. 69, 1398–1408.

Zhang, S. and Oertel, D. (1993c) Tuberculoventral cells of the dorsal cochlear nucleus of mice: intracellular recordings in slices. J. Neurophysiol. 69, 1409–1421.

Zhang, S. and Oertel, D. (1994) Neuronal circuits associated with the output of the dorsal cochlear nucleus through fusiform cells. J. Neurophysiol. 71, 914–930.

13

CHEMISTRY OF GRANULAR AND CLOSELY RELATED REGIONS OF THE COCHLEAR NUCLEUS

Donald A. Godfrey,[1] Timothy G. Godfrey,[1] Nikki L. Mikesell,[1] Hardress J. Waller,[2] Weiping Yao,[1] Kejian Chen,[1] and James A. Kaltenbach[3]

[1]Department of Otolaryngology
Medical College of Ohio
Toledo, Ohio
[2]Department of Neurological Surgery
Medical College of Ohio
Toledo, Ohio
[3]Department of Otolaryngology
Wayne State University
Detroit, Michigan

1. COCHLEAR NUCLEUS GRANULAR REGIONS

Granule cells of the cochlear nucleus have been described as having very small somata, about 6–10 μm diameter, containing a thin rim of cytoplasm around the nucleus, and 2–4 dendrites (Brawer et al., 1974; Kane, 1974; Mugnaini et al., 1980a,b; Weedman et al., 1996). They are located in many parts of the cochlear nucleus but are especially concentrated in granular regions, which we will define as collections of densely packed granule cells readily recognized in Nissl-stained sections. Most granular regions are located near the boundaries of the cochlear nucleus or its subdivisions: the anteroventral cochlear nucleus (AVCN), posteroventral cochlear nucleus (PVCN), and dorsal cochlear nucleus (DCN). The fusiform soma layer of the DCN is exceptional in that it lies within a subdivision and contains prominent large fusiform (pyramidal) cells as well as densely packed granule cells (Brawer et al., 1974; Mugnaini et al., 1980b; Wouterlood & Mugnaini, 1984; Osen, 1988). The small cell cap, or marginal zone, borders some granular regions and contains granule cells as well as many non-granule small cells, some of which have dendrites that enter the overlying granular layer (Brawer et al., 1974; Cant, 1993; Benson & Brown, 1990). It is sometimes described together with granular regions as a small cell or marginal shell (Zhao et al., 1995), but we will not include it in our consideration of granular regions because it lacks densely packed granule cells. The cochlear nucleus granular regions resemble those of the cerebellar cortex but with less dense packing. Our unpublished measurements in cat of DNA concentrations, which should be proportional to density of cell nuclei, gave values in the granular region dorsolateral to AVCN about a third of those in the cerebellar flocculus granular layer and about three times those in the AVCN.

The unmyelinated axons of many, if not all, granule cells form parallel fibers in the molecular layer of the DCN, where they innervate cartwheel and fusiform cells, as well as small stellate cells (Brawer et al., 1974; Kane, 1974; Mugnaini et al., 1980b; Lorente de Nó, 1981; Wouterlood & Mugnaini, 1984; Wouterlood et al., 1984; Osen, 1988; Berrebi & Mugnaini, 1991). Granule cells are excitatory to cartwheel and fusiform cells, and cartwheel cells are inhibitory to fusiform cells (Zhang & Oertel, 1994). Fusiform cells project the major output from the DCN, to the contralateral inferior colliculus (Osen, 1988; Berrebi & Mugnaini, 1991).

Extrinsic inputs to granular regions, probably in most cases ending on granule cells, converge from such diverse sources as auditory nerve type II spiral ganglion cells (Kane, 1974; Brown & Ledwith, 1990; Berglund et al., 1996), higher auditory regions including the auditory cortex (Feliciano et al., 1995; Weedman & Ryugo, 1996), inferior colliculus (Caicedo & Herbert, 1993; Saldaña, 1993), and superior olivary complex (Shore et al., 1995) including olivocochlear neurons (Brown et al., 1988; Benson & Brown, 1990; Benson et al., 1996), somatosensory nuclei including the dorsal column and trigeminal nuclei (Itoh et al., 1987; Weinberg & Rustioni, 1987; Wright & Ryugo, 1996), vestibular nerve fibers and small neurons in the vestibular nerve root (Burian & Gstoettner, 1988; Kevetter & Perachio, 1989; Zhao et al., 1995), raphe nuclei (Klepper & Herbert, 1991; Thompson et al., 1995), and locus coeruleus (Kromer & Moore, 1980; Moore, 1988; Klepper & Herbert, 1991).

The mountain beaver cochlear nucleus has remarkably large granular regions (Merzenich et al., 1973). Our measurements (Benjamin et al., 1995) indicate that granular regions occupy about half the volume of the cochlear nucleus in the mountain beaver, as compared to about 10% in other mammals. The DCN occupies another approximately 40% of the volume. It shows little evidence of layering, but has a rather homogeneous appearance with scattered neuron somata and a lipid content comparable to that of the superficial DCN of other mammals, i.e., the molecular and fusiform soma layers, which probably contain most of the neurons to which granule cells project. Thus, the regions containing granule cell bodies, axons, and terminals occupy as much as 90% of the volume of the mountain beaver cochlear nucleus.

2. NEUROTRANSMITTER OF GRANULE CELLS

There is now considerable evidence that granule cells release the amino acid glutamate as an excitatory neurotransmitter (Table 1). If so, then it may be possible to detect and correlate glutamate release with activity of granule cell axons. This has been accomplished in a rat brain slice preparation by collecting effluent downstream from the DCN during electrical stimulation of parallel fibers (Waller et al., 1994) (Fig. 1). We have also been able to detect, in the same experiments, increased glycine release during parallel fiber stimulation. This is consistent with considerable evidence (Wenthold et al., 1987;

Table 1. Evidence for glutamate as a neurotransmitter of granule cells

---

Presence of elevated amounts (compared to baseline values) of glutamate:
- Glutamate (but not aspartate) concentration is high in granular regions and superficial DCN; that in superficial DCN is not decreased by auditory nerve lesions (Godfrey et al., 1977a, 1978; Wenthold, 1978).
- Granule cell bodies show glutamate immunoreactivity (as, however, do those of most other cochlear nucleus neurons) (Ottersen & Storm-Mathisen, 1984).
- Glutamate-like immunoreactivity is localized in terminals that appear to be those of parallel fibers (Juiz et al., 1993; Osen et al., 1995).
- Transection of the connection from AVCN to DCN leads to reduction of glutamate in the superficial part of the DCN molecular layer (Godfrey et al., 1995).

Presence of elevated amounts of synthetic enzymes:
- Intense labeling of granule somata by antibodies against AAT and glutaminase is not affected by auditory nerve lesions (Wenthold & Altschuler, 1983; Altschuler et al., 1984).
- High glutaminase and AAT activities are found in granular regions and superficial DCN; values are higher in DCN than in VCN and do not decrease (unlike VCN) after auditory nerve lesions (Wenthold, 1980; Godfrey et al., 1994).

Increased release of glutamate upon appropriate stimulation:
- Release of glutamate increases during stimulation of parallel fibers (Waller et al., 1994).
- Glutamate-like (but not aspartate-like) immunoreactivity is concentrated in parallel fiber terminals and is decreased by high-potassium depolarization in a partially calcium-dependent manner (Osen et al., 1995).

Preferential uptake of glutamate:
- Preferential uptake and retrograde transport of D-aspartate occur in axons of granule cells (Oliver et al., 1983).
- Uptake and release of D-aspartate in DCN are decreased less than in AVCN and PVCN after cochlear ablation (Potashner, 1983).
- Preferential uptake of ^{3}H-labeled glutamate occurs into terminals in the deep part of the DCN molecular layer (Schwartz, 1981).

Presence of elevated amounts of glutamate receptors:
- Dense immunoreactivity for glutamate receptor subtypes occurs in cartwheel and small stellate cells of superficial DCN (Wenthold et al., 1993; Petralia et al., 1996).
- Dense immunoreactivity for AMPA and metabotropic glutamate receptor subtypes occurs on spines of cartwheel cells postsynaptic to parallel fiber terminals in superficial DCN (Wright et al., 1996; Petralia et al., 1996).
- High AMPA glutamate receptor subunit mRNA expression occurs in cartwheel and stellate neurons in superficial DCN (Hunter et al., 1993).
- Glutamate receptor binding for non-NMDA and, to a lesser extent, NMDA receptors is dense in DCN molecular layer (Greenamyre et al., 1984; Monaghan & Cotman, 1985; Li et al., 1997).

Presence of appropriate glutamate pharmacology:
- Blockers of glutamate receptors, mainly non-NMDA, block activation of neurons by parallel fiber stimulation (Manis, 1989; Manis & Molitor, 1996; Waller et al., 1996).

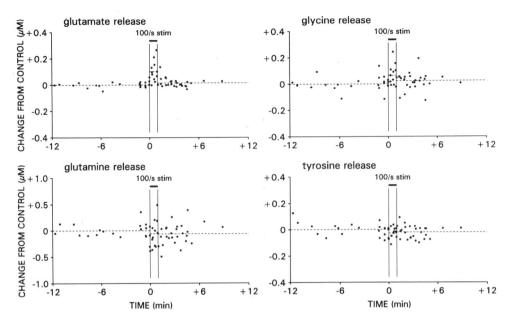

Figure 1. Changes of glutamate, glutamine, glycine, and tyrosine release into effluent during electrical stimulation of parallel fibers in the DCN of brain slices. Stimulation was by 0.1 msec pulses at 100/sec for 1 min through a concentric bipolar electrode with a central conductor diameter of 80 μm. The electrode was placed on the molecular layer, and the polarity was selected for the lower threshold. Dots are differences for individual samples from the control mean for the same slice, which was calculated for all samples before stimulation. Data are from 4 different experiments. Each sample was collected downstream from the DCN via a microsiphon (Godfrey & Waller, 1992) over approximately 20 sec during or 40 sec before and after stimulation. Times are midpoints of sampling periods, corrected for flow delay time in the microsiphon. Dashed lines are means for samples taken before, during, and after stimulation. By t-tests, glutamate release during stimulation was significantly higher than before (P<0.001) or after (P<0.005), and glycine release during stimulation was significantly higher than before (P<0.05). Other differences were not statistically significant.

Osen et al., 1990; Saint Marie et al., 1991; Kolston et al., 1992; Golding & Oertel, 1996) that cartwheel cells release glycine as a neurotransmitter.

3. NEUROTRANSMITTERS OF INPUTS TO GRANULE CELLS

Inputs to granule cells have been described as terminating mainly on their dendrites (Kane, 1974; Weedman et al., 1996), but the possibility also of synapses onto their axon terminals in the DCN molecular layer has not been ruled out. Several neurotransmitters have been implicated in synapses within granular regions, in some cases onto granule cells (Fig. 2).

Glutamate immunoreactivity has been noted in terminals onto granule cell dendrites deriving from the cuneate nucleus (Wright & Ryugo, 1996), and granule cells express mRNA and immunoreactivity for AMPA-type glutamate receptor subunits (Hunter et al., 1993; Petralia et al., 1996).

Puncta immunoreactive for γ-aminobutyrate (GABA) and for the enzyme which catalyzes GABA synthesis, glutamate decarboxylase (GAD), are dense in granular regions, relative to other cochlear nucleus regions (Mugnaini, 1985; Peyret et al., 1986;

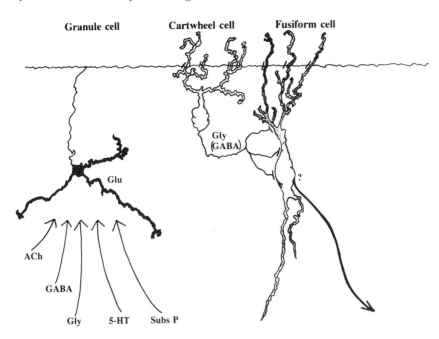

Figure 2. Diagram of the granule cell-cartwheel cell-fusiform cell circuit with some of the neurotransmitters involved. Drawings of cell types have been adapted from those shown in Brawer et al. (1974). The diagram is highly simplified to show only examples of the circuitry. The function of the actual circuit depends on the topography of connections among the entire populations of the cell types. Synaptic connections of granule cell parallel fibers onto cartwheel and fusiform cells and of cartwheel cells onto fusiform cells are represented by short dashes. There is evidence for GABA in cartwheel cells as well as glycine, but the evidence for glycine as a neurotransmitter is stronger (Mugnaini, 1985; Wenthold et al., 1987; Osen et al., 1990; Saint Marie et al., 1991; Kolston et al., 1992; Golding & Oertel, 1996). Convergence upon granular regions of some pathways is represented only crudely because of our limited knowledge about their exact sites of termination. ACh, acetylcholine; 5-HT, serotonin (5-hydroxytryptamine); Subs P, substance P.

Adams & Mugnaini, 1987; Moore & Moore, 1987; Roberts & Ribak, 1987; Osen et al, 1990; Kolston et al, 1992). GABA concentrations (Godfrey et al., 1977, 1978) (Fig. 3) and GAD activities (Wenthold & Morest, 1976) measured in granular regions and superficial DCN are moderate to high compared to other cochlear nucleus regions. Transection of centrifugal pathways leads to about 40% decrease of these GABA concentrations in cats (Godfrey et al., 1988) and rats (Shannon-Hartman et al., 1993). Such lesions in guinea pigs lead to 40–45% decrease of GABA uptake and release in DCN (Potashner, 1985). High levels of GABA receptor binding have been reported in granular regions and superficial DCN compared to elsewhere in the cochlear nucleus (Frostholm & Rotter, 1986; Glendenning & Baker, 1988; Juiz et al, 1994). The binding in the DCN molecular layer included a large proportion of the $GABA_B$ subtype, which is often associated with presynaptic terminals (Juiz et al., 1994).

Numerous glycine-immunoreactive puncta have been reported in granular regions, including the DCN fusiform soma layer (Osen et al, 1990; Kolston et al, 1992), and moderate to high glycine concentrations have been measured in granular regions and superficial DCN (Godfrey et al., 1977a, 1978). Transection of centrifugal pathways leads to 40–50% decrease of glycine uptake and release in DCN of guinea pigs (Staatz-Benson & Potashner, 1988). Glycine receptor immunoreactivity is dense in granular regions and su-

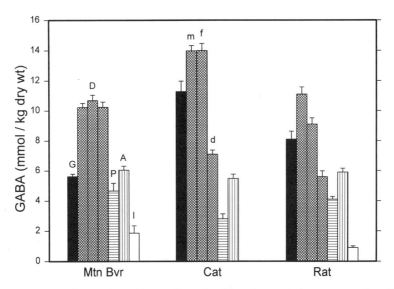

Figure 3. Concentrations of GABA in cochlear nucleus subregions of mountain beaver, cat, and rat. Bar heights and error bars indicate means and SEs of GABA concentrations, based usually on more than 20 samples per region. Although the mountain beaver DCN shows no layering, the data have been divided into 3 bars according to distance from the ependymal surface, to approximate the locations of the layers in the other species. There are no data for interstitial nucleus of cat. Abbreviations are: A, AVCN; D, DCN, with 3 bars representing molecular (m), fusiform soma (f), and deep (d) layers in cat and rat; G, granular region dorsolateral to AVCN; I, interstitial nucleus (auditory nerve root); P, PVCN.

perficial DCN compared to other cochlear nucleus regions (Altschuler et al, 1986). High glycine receptor binding has been reported in granular regions of cat (Glendenning & Baker, 1988), but not of mouse (Frostholm & Rotter, 1986) or gerbil (Sanes et al, 1987).

There have also been reports of fibers and terminals in granular regions which are immunoreactive for serotonin (5-hydroxytryptamine) in rats and guinea pigs (Klepper & Herbert, 1991; Thompson et al, 1995), substance P in cats (Adams, 1993), and dopamine-ß-hydroxylase, the enzyme which catalyzes the synthesis of norepinephrine, in rats and guinea pigs (Kromer & Moore, 1980; Moore, 1988; Klepper & Herbert, 1991).

Acetylcholine is the neurotransmitter most extensively implicated in granular region inputs (Table 2). Results from our slice experiments suggest that granule cells have receptors for acetylcholine. We have found that the excitation of bursting neurons (probably cartwheel cells) by the cholinergic agonist carbachol can be blocked by glutamate receptor blockers, including the general blocker kynurenic acid (1 mM) and the non-NMDA blockers DNQX (4 µM) and CNQX (5 µM) (Chen et al., 1994b). This suggests that carbachol activates granule cells, which then excite cartwheel cells via glutamatergic synapses. If this is true, then it might be possible to detect increased release of glutamate from the DCN during cholinergic excitation. We have recently observed increased release of glutamate from slices during exposure to carbachol, as well as increased release of glycine in the same experiments (Fig. 4).

Muscarinic receptor subtypes 2 and 4 may be mainly involved in cholinergic effects on granule cells (Chen et al., 1995). Muscarinic subtype 2, which is often associated with receptors on presynaptic terminals, is prominent in the DCN molecular layer (Yao & God-

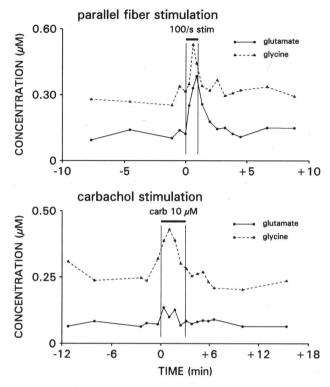

Figure 4. Glutamate and glycine release into effluent during a 3-minute exposure of a brain slice to 10 μM car-bachol, and during parallel fiber stimulation of another slice. Samples were collected downstream from the DCN via a microsiphon and represent approximately 40 sec of release, except those during parallel fiber stimulation, which represent approximately 20 sec. Sampling times in the carbachol experiment were corrected for delay time of carbachol flow to the slice chamber as well as flow delay time in the microsiphon. Stimulation current 0.38 mA; other details as in Fig. 1.

frey, 1996a; Yao et al., 1996), so it is possible that these receptors are present on parallel fiber terminals.

4. VARIATIONS OF CHOLINERGIC INPUT TO GRANULAR REGIONS

Despite the evidence for cholinergic input to granule cells, a more detailed evaluation of the available data for activities of choline acetyltransferase (ChAT), the enzyme of synthesis for acetylcholine and a reliable marker for cholinergic structures (Godfrey et al., 1977b, 1988), suggests considerable variability of this input across species (Fig. 5). The mountain beaver, with its large granular regions and a large DCN resembling the superficial DCN of other mammals, appears to have only sparse cholinergic innervation of those regions. The ChAT activities in cat and rat granular regions and superficial DCN are much higher. However, there are differences in the relative contributions of different

Table 2. Evidence for cholinergic input to granular regions

Presence of elevated amounts of the synthetic enzyme choline acetyltransferase (ChAT):
- Moderate to high ChAT activities occur in granular regions of rats, cats, guinea pigs, and hamsters; these decrease greatly in rats and cats after transection of centrifugal pathways (Wenthold & Morest, 1976; Godfrey et al., 1987a,b, 1990, 1996a).
- Antibodies to ChAT label fibers in granular regions of rat and cat (Tago et al., 1989; Adams, 1993; Godfrey, 1993; Vetter et al., 1993).
- Mossy fiber terminals, which likely terminate on granule cell dendrites in cats, rats, and mice, have round clear vesicles (which are characteristic of excitatory, including cholinergic, terminals). (Mugnaini et al., 1980a; Weedman et al., 1996).
- ChAT immunoreactivity occurs in some mossy fiber terminals which synapse with granule cell dendrites in rats (Vetter et al., 1993).

Release of acetylcholine upon stimulation of inputs:
- There is evidence for acetylcholine release from the cochlear nucleus as a whole during stimulation of the lateral superior olive (Comis & Davies, 1969) and from the DCN surface during stimulation of the crossed olivocochlear bundle (Comis & Guth, 1974). The collection from the DCN surface may preferentially measure granular and DCN release.

Presence of elevated amounts of the degradative enzyme acetylcholinesterase (AChE):
- High AChE activities occur in granular regions of rats, cats, and mountain beavers; these decrease in rats after transection of centrifugal pathways (Godfrey et al., 1977b, 1983, 1987a, 1996b).
- Dark staining for AChE activity occurs in granular regions of chinchillas, rats, cats, and mice (Osen & Roth, 1969; McDonald & Rasmussen, 1971; Osen et al., 1984; Martin, 1981; Godfrey et al., 1977b, 1987a; Yao & Godfrey, 1995; Yao et al., 1996).
- Dark staining for AChE activity occurs in DCN molecular layer of cats and around terminals in DCN molecular layer of rats (Rasmussen, 1967; Osen & Roth, 1969; Godfrey & Matschinsky, 1976; Yao & Godfrey, 1996b).
- Most mossy fiber terminals in granular region next to AVCN of cats and chinchillas stain for AChE activity (McDonald & Rasmussen, 1971).

Presence of acetylcholine receptors:
- Muscarinic cholinergic receptor immunoreactivity occurs in granular regions of rats (Yao & Godfrey, 1995).
- Dense immunoreactivity for muscarinic receptor subtype 2 occurs in and near granular regions of rats (Yao et al., 1996).
- High muscarinic cholinergic receptor binding occurs in granular regions and DCN molecular layer of cats, rats, and mice; binding in rats involves multiple subtypes, including M_2 and M_4 (Wamsley et al., 1981; Frostholm & Rotter, 1986; Glendenning & Baker, 1988; Yao & Godfrey, 1996a).

Presence of appropriate acetylcholine pharmacology:
- Since granule cells have not been recorded from, their activity has been inferred by monitoring responses of cartwheel cells, which in slices correspond closely to bursting neurons, and which are strongly activated by electrical stimulation of granule cell axons (Zhang & Oertel, 1994; Manis & Molitor, 1996; Waller et al., 1996). Bursting neurons can be activated by carbachol, and the effect is blocked in low calcium, high magnesium solutions (Chen et al., 1994a). Muscarinic receptor subtypes particularly involved in activation of bursting neurons by carbachol are M_2 and M_4 (Chen et al., 1995). Blockers of glutamate receptors, particularly non-NMDA, block activation of bursting neurons by carbachol (Chen et al., 1994b).

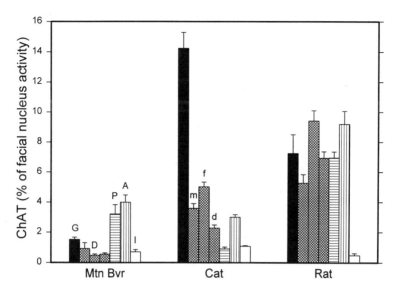

Figure 5. Activities of ChAT in cochlear nucleus subregions of mountain beaver, cat, and rat. To normalize the data for differences in ChAT specific activity of cholinergic neurons in different species, the activities are divided by those of the facial nucleus, a collection of cholinergic motoneurons. Other details as in Fig. 3 except that means and SEs for rat and cat, which are from previously published studies (Godfrey et al., 1983, 1990), are based on numbers of animals (usually 6) rather than numbers of samples.

sources to ChAT activities in cats and rats (Godfrey et al., 1987a,b, 1990). Most of the ChAT activity in cat granular regions appears to derive from branches of the olivocochlear system, whereas that in rats derives primarily from other centrifugal fibers which enter the nucleus from the trapezoid body (Sherriff & Henderson, 1994).

Although the various granular regions within the cochlear nucleus are often assumed to be equivalent, there is evidence that this is not true for their cholinergic innervation. In both rat and cat cochlear nucleus, the granular region adjacent to AVCN has higher ChAT activity than that adjacent to PVCN (Godfrey et al., 1977b, 1987b, 1990). In cat, gradients of ChAT activity occur within granular regions adjacent to AVCN (Godfrey et al., 1977b, 1990), and the presumed olivocochlear contribution to ChAT activity in granular regions has a pronounced decrease from rostral AVCN through caudal AVCN into PVCN and the DCN fusiform soma layer (Godfrey et al, 1990). There is also evidence that muscarinic cholinergic receptor subtypes differ in prevalence among granular regions, including the DCN fusiform soma layer, of rat cochlear nucleus (Yao & Godfrey, 1996a; Yao et al., 1996).

5. COMPARISONS WITH GRANULAR REGIONS OF OTHER BRAIN STRUCTURES

On a larger scale, there is evidence that granular regions in different brain structures within a species differ in their neurotransmitter chemistry. As shown in Figure 6, differences between two cochlear nucleus granular regions are much less than between either and the cerebellar flocculus or olfactory bulb granular layer. The cerebellar flocculus granular layer has much lower glycine and ChAT, and the olfactory bulb granular layer

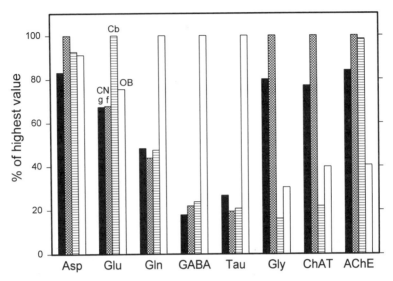

Figure 6. Relative average concentrations of amino acids and activities of enzymes of acetylcholine metabolism across 4 granular regions of rat. The amino acids are aspartate (Asp), glutamate (Glu), glutamine (Gln), γ-aminobutyrate (GABA), taurine (Tau), and glycine (Gly). The enzymes are choline acetyltransferase (ChAT) and acetylcholinesterase (AChE). The granular regions are that dorsolateral to AVCN (g) and the DCN fusiform soma layer (f) of the cochlear nucleus (CN), the cerebellar flocculus granular layer (Cb), and the olfactory bulb granular layer (OB). Data for ChAT and AChE and some for aspartate, glutamate, and GABA are in previous publications (Godfrey et al., 1980, 1983, 1994; Godfrey & Matschinsky, 1981); the rest are not yet published. Highest (100%) values are, for amino acids in mmol/kg dry wt, 14 for Asp, 58 for Glu, 41 for Gln, 41 for GABA, 55 for Tau, and 25 for Gly, and for enzymes in mmol/kg dry wt/min, 0.43 for ChAT and 63 for AChE.

has much higher glutamine, GABA, and taurine and much lower glycine, ChAT, and AChE. The difference for GABA correlates with the evidence that olfactory bulb granule cells use GABA as neurotransmitter, whereas cerebellar and cochlear nucleus granule cells use glutamate (Halász & Shepherd, 1983; Ottersen & Storm-Mathisen, 1984; Table 1). Other differences most likely relate to different neurotransmitters of input pathways.

6. CONCLUSION

The granule cells receive converging input from many sources, which employ a wide variety of neurotransmitters. Changes in their activity then can modulate the activity of major circuits affecting the output of the DCN (Fig. 2). The relative contributions of different pathways to granule cell activity may vary across mammalian species, perhaps correlating with different behavioral characteristics. Such species differences concerning granular regions were mentioned at our previous meeting in Prague (Osen, 1988). Additionally, relative contributions of different pathways to granule cell activity may vary across different parts of the granule cell domain in the same cochlear nucleus, so that generalizations about the entire domain from study of one part should be suggested only tentatively.

7. ACKNOWLEDGMENTS

The authors thank Wei Kong for technical assistance and Barbara McConnell for the final drawing of Figure 2. Supported by NIH grant DC00172.

8. REFERENCES

Adams, J.C. (1993) Non-primary inputs to the cochlear nucleus visualized using immunocytochemistry. In: M.A. Merchán, J.M. Juiz, D.A. Godfrey and E. Mugnaini (Eds.), The Mammalian Cochlear Nuclei: Organization and Function, Plenum, New York, pp. 133–141.

Adams, J.C. and Mugnaini, E. (1987) Patterns of glutamate decarboxylase immunostaining in the feline cochlear nuclear complex studied with silver enhancement and electron microscopy. J. Comp. Neurol. 262, 375–401.

Altschuler, R.A., Wenthold, R.J., Schwartz, A.M., Haser, W.G., Curthoys, N.P., Parakkal, M.H. and Fex, J. (1984) Immunocytochemical localization of glutaminase-like immunoreactivity in the auditory nerve. Brain Res. 291, 173–178.

Altschuler, R.A., Betz, H., Parakkal, M.H., Reeks, K.A. and Wenthold, R.J. (1986) Identification of glycinergic synapses in the cochlear nucleus through immunocytochemical localization of the postsynaptic receptor. Brain. Res. 369, 316–320.

Benjamin, L.C., Godfrey, D.A., Heffner, R.S. and Kaltenbach, J.A. (1995) Volumes of cochlear nucleus regions in mountain beaver compared with other rodents. Otolaryngol. Head Neck Surg. 113, P100.

Benson, T.E. and Brown, M.C. (1990) Synapses formed by olivocochlear axon branches in the mouse cochlear nucleus. J. Comp. Neurol. 295, 52–70.

Benson, T.E., Berglund, A.M. and Brown, M.C. (1996) Synaptic input to cochlear nucleus dendrites that receive medial olivocochlear synapses. J. Comp. Neurol. 365, 27–41.

Berglund, A.M., Benson, T.E. and Brown, M.C. (1996) Synapses from labeled type II axons in the mouse cochlear nucleus. Hearing Res. 94, 31–46.

Berrebi, A.S. and Mugnaini, E. (1991) Distribution and targets of the cartwheel cell axon in the dorsal cochlear nucleus of the guinea pig. Anat. Embryol. 183, 427–454.

Brawer, J.R., Morest, D.K. and Kane, E.C. (1974) The neuronal architecture of the cochlear nucleus of the cat. J. Comp. Neurol. 155, 251–300.

Brown, M.C. and Ledwith III, J.V. (1990) Projections of thin (type-II) and thick (type-I) auditory-nerve fibers in the cochlear nucleus of the mouse. Hearing Res. 49, 105–118.

Brown, M.C., Liberman, M.C., Benson, T.E. and Ryugo, D.K. (1988) Brainstem branches from olivocochlear axons in cats and rodents. J. Comp. Neurol. 278, 591–603.

Burian, M. and Gstoettner, W. (1988) Projection of primary vestibular afferent fibers to the cochlear nucleus in the guinea pig. Neurosci. Lett. 84, 13–17.

Caicedo, A. and Herbert, H. (1993) Topography of descending projections from the inferior colliculus to auditory brainstem nuclei in the rat. J. Comp. Neurol. 328, 377–392.

Cant, N.B. (1993) The synaptic organization of the ventral cochlear nucleus of the cat: the peripheral cap of small cells. In: M.A. Merchán, J.M. Juiz, D.A. Godfrey and E. Mugnaini (Eds.), The Mammalian Cochlear Nuclei: Organization and Function, Plenum, New York, pp. 91–105.

Chen, K., Waller, H.J. and Godfrey, D.A. (1994a) Cholinergic modulation of spontaneous activity in rat dorsal cochlear nucleus. Hearing Res. 77, 168–176.

Chen, K., Waller, H.J. and Godfrey, D.A. (1994b) Glutamate antagonists affect cholinergic activation of bursting neurons in rat dorsal cochlear nucleus. Abstr. Assoc. Res. Otolaryngol., 13.

Chen, K., Waller, H.J. and Godfrey, D.A. (1995) Muscarinic receptor subtypes in rat dorsal cochlear nucleus. Hearing Res. 89, 137–145.

Comis, S.D. and Davies, W.E. (1969) Acetylcholine as a transmitter in the cat auditory system. J. Neurochem. 16, 423–429.

Comis, S.D. and Guth, P.S. (1974) The release of acetylcholine from the cochlear nucleus upon stimulation of the crossed olivo-cochlear bundle. Neuropharmacology 13, 633–641.

Feliciano, M., Saldaña, E. and Mugnaini, E. (1995) Direct projections from the rat primary auditory neocortex to nucleus sagulum, paralemniscal regions, superior olivary complex and cochlear nuclei. Aud. Neurosci. 1, 287–308.

Frostholm, A. and Rotter, A. (1986) Autoradiographic localization of receptors in the cochlear nucleus of the mouse. Brain Res. Bull. 16, 189–203.

Glendenning, K.K. and Baker, B.N. (1988) Neuroanatomical distribution of receptors for three potential inhibitory neurotransmitters in the brainstem auditory nuclei of the cat. J. Comp. Neurol. 275, 288–308.

Godfrey, D.A. (1993) Comparison of quantitative and immunohistochemistry for choline acetyltransferase in the rat cochlear nucleus. In: M.A. Merchán, J.M. Juiz, D.A. Godfrey and E. Mugnaini (Eds.), The Mammalian Cochlear Nuclei: Organization and Function, Plenum, New York, pp. 267–278.

Godfrey, D.A. and Matschinsky, F.M. (1976) Approach to three-dimensional mapping of quantitative histochemical measurements applied to studies of the cochlear nucleus. J. Histochem. Cytochem. 24, 697–712.

Godfrey, D.A. and Matschinsky, F.M. (1981) Quantitative distribution of choline acetyltransferase and acetylcholinesterase activities in the rat cochlear nucleus. J. Histochem. Cytochem. 29, 720–730.

Godfrey, D.A. and Waller, H.J. (1992) Sampling fluid from slice chambers by microsiphoning. J. Neurosci. Meth. 41, 167–173.

Godfrey, D.A., Carter, J.A., Berger, S.J., Lowry, O.H. and Matschinsky, F.M. (1977a) Quantitative histochemical mapping of candidate transmitter amino acids in cat cochlear nucleus. J. Histochem. Cytochem. 25, 417–431.

Godfrey, D.A., Williams, A.D. and Matschinsky, F.M. (1977b) Quantitative histochemical mapping of enzymes of the cholinergic system in cat cochlear nucleus. J. Histochem. Cytochem. 25, 397–416.

Godfrey, D.A., Carter, J.A., Lowry, O.H. and Matschinsky, F.M. (1978) Distribution of γ-aminobutyric acid, glycine, glutamate and aspartate in the cochlear nucleus of the rat. J. Histochem. Cytochem. 26, 118–126.

Godfrey, D.A., Ross, C.D., Herrmann, A.D. and Matschinsky, F.M. (1980) Distribution and derivation of cholinergic elements in the rat olfactory bulb. Neuroscience 5, 273–292.

Godfrey, D.A., Park, J.L., Rabe, J.R., Dunn, J.D. and Ross, C.D. (1983) Effects of large brain stem lesions on the cholinergic system in the rat cochlear nucleus. Hearing Res. 11, 133–156.

Godfrey, D.A., Park-Hellendall, J.L., Dunn, J.D. and Ross, C.D. (1987a) Effect of olivocochlear bundle transection on choline acetyltransferase activity in the rat cochlear nucleus. Hearing Res. 28, 237–251.

Godfrey, D.A., Park-Hellendall, J.L., Dunn, J.D. and Ross, C.D. (1987b) Effects of trapezoid body and superior olive lesions on choline acetyltransferase activity in the rat cochlear nucleus. Hearing Res. 28, 253–270.

Godfrey, D.A., Parli, J.A., Dunn, J.D. and Ross, C.D. (1988) Neurotransmitter microchemistry of the cochlear nucleus and superior olivary complex. In: J. Syka and R.B. Masterton (Eds.), Auditory Pathway, Plenum, New York, pp. 107–121.

Godfrey, D.A., Beranek, K.L., Carlson, L., Parli, J.A., Dunn, J.D. and Ross, C.D. (1990) Contribution of centrifugal innervation to choline acetyltransferase activity in the cat cochlear nucleus. Hearing Res. 49, 259–280.

Godfrey, D.A., Ross, C.D., Parli, J.A. and Carlson, L. (1994) Aspartate aminotransferase and glutaminase activities in rat olfactory bulb and cochlear nucleus; comparisons with retina and with concentrations of substrate and product amino acids. Neurochem. Res. 19, 693–703.

Godfrey, D.A., Farms, W.B., Dunn, J.D. and Godfrey, T.G. (1995) Amino acid concentrations in cat dorsal and anteroventral cochlear nucleus subregions following transection of pathways connecting them. Abstr. Assoc. Res. Otolaryngol., 35.

Godfrey, D.A., Mikesell, N.L., Fulcomer, A.B., Godfrey, T.G. and Kaltenbach, J.A. (1996) Chemistry of the hamster cochlear nucleus. J. Acoust. Soc. Amer. 100, 2787.

Godfrey, D.A., Mikesell, N.L., Godfrey, T.G. and Kaltenbach, J.A. (1996) Neurotransmitter chemistry in the cochlear nucleus of the mountain beaver. Soc. Neurosci. Abstr. 22, 126.

Golding, N.L. and Oertel, D. (1996) Context-dependent synaptic action of glycinergic and GABAergic inputs in the dorsal cochlear nucleus. J. Neurosci. 16, 2208–2219.

Greenamyre, J.T., Young, A.B. and Penney, J.B. (1984) Quantitative autoradiographic distribution of L-[^{3}H]glutamate-binding sites in rat central nervous system. J. Neuroscience 4, 2133–2144.

Halász, N. and Shepherd, G.M. (1983) Neurochemistry of the vertebrate olfactory bulb. Neuroscience 10, 579–619.

Hunter, C., Petralia, R.S., Vu, T. and Wenthold, R.J. (1993) Expression of AMPA-selective glutamate receptor subunits in morphologically defined neurons of the mammalian cochlear nucleus. J. Neurosci. 13, 1932–1946.

Itoh, K., Kamiya, H., Mitani, A., Yasui, Y., Takada, M. and Mizuno, N. (1987) Direct projections from the dorsal column nuclei and the spinal trigeminal nuclei to the cochlear nuclei in the cat. Brain Res. 400, 145–150.

Juiz, J.M., Rubio, M.E., Helfert, R.H. and Altschuler, R.A. (1993) Localizing putative excitatory endings in the cochlear nucleus by quantitative immunocytochemistry. In: M.A. Merchán, J.M. Juiz, D.A. Godfrey and E. Mugnaini (Eds.), The Mammalian Cochlear Nuclei: Organization and Function, Plenum, New York, pp. 167–177.

Juiz, J.M., Albin, R.L., Helfert, R.H. and Altschuler, R.A. (1994) Distribution of $GABA_A$ and $GABA_B$ binding sites in the cochlear nucleus of the guinea pig. Brain Res. 639, 193–201.

Kane, E.C. (1974) Synaptic organization in the dorsal cochlear nucleus of the cat: A light and electron microscopic study. J. Comp. Neurol. 155, 301–330.

Kevetter, G.A. and Perachio, A.A. (1989) Projections from the sacculus to the cochlear nuclei in the mongolian gerbil. Brain Behav. Evol. 34, 193–200.

Klepper, A. and Herbert, H. (1991) Distribution and origin of noradrenergic and serotonergic fibers in the cochlear nucleus and inferior colliculus of the rat. Brain Res. 557, 190–201.

Kolston, J., Osen, K.K., Hackney, C.M., Ottersen, O.P. and Storm-Mathisen, J. (1992) An atlas of glycine- and GABA-like immunoreactivity and colocalization in the cochlear nuclear complex of the guinea pig. Anat. Embryol. 186, 443–465.

Kromer, L.F. and Moore, R.Y. (1980) Norepinephrine innervation of the cochlear nuclei by locus coeruleus neurons in the rat. Anat. Embryol. 158, 227–244.

Li, H., Godfrey, D.A. and Rubin, A.M. (1997) Quantitative autoradiography of 5-[^{3}H]-6-cyano-7-nitro-quinoxaline-2,3-dione (CNQX) and (+)-3-[^{3}H]-dibenzocyclohepteneimine (MK-801) binding in rat vestibular nuclear complex after unilateral deafferentation, with comparison to cochlear nucleus. Neuroscience, 77, 473–484.

Lorente de Nó, R. (1981) The Primary Acoustic Nuclei. Raven, New York.

Manis, P.B. (1989) Responses to parallel fiber stimulation in the guinea pig dorsal cochlear nucleus in vitro. J. Neurophysiol. 61, 149–161.

Manis, P.B. and Molitor, S.C. (1996) N-methyl-D-aspartate receptors at parallel fiber synapses in the dorsal cochlear nucleus. J. Neurophysiol. 76, 1639–1656.

Martin, M.R. (1981) Acetylcholinesterase-positive fibers and cell bodies in the cochlear nuclei of normal and reeler mutant mice. J. Comp. Neurol. 197, 153–167.

McDonald, D.M. and Rasmussen, G.L. (1971) Ultrastructural characteristics of synaptic endings in the cochlear nucleus having acetylcholinesterase activity. Brain Res. 28, 1–18.

Merzenich, M.M., Kitzes, L. and Aitkin, L. (1973) Anatomical and physiological evidence for auditory specialization in the mountain beaver (Aplondontia rufa). Brain Res. 58, 331–344.

Monaghan, D.T. and Cotman, C.W. (1985) Distribution of N-methyl-D-aspartate-sensitive L-[^{3}H]glutamate- binding sites in rat brain. J. Neurosci. 5, 2909–2919.

Moore, J.K. (1988) Cholinergic, GABA-ergic, and noradrenergic input to cochlear granule cells in the guinea pig and monkey. In: J. Syka and R.B. Masterton (Eds.), Auditory Pathway, Plenum, New York, pp. 123–131.

Moore, J.K. and Moore, R.Y. (1987) Glutamic acid decarboxylase-like immunoreactivity in brainstem auditory nuclei of the rat. J. Comp. Neurol. 260, 157–174.

Mugnaini, E. (1985) GABA neurons in the superficial layers of the rat dorsal cochlear nucleus: light and electron microscopic immunocytochemistry. J. Comp. Neurol. 235, 61–81.

Mugnaini, E., Osen, K.K., Dahl, A-L., Friedrich, V.L. and Korte, G. (1980a) Fine structure of granule cells and related interneurons (termed Golgi cells) in the cochlear nuclear complex of cat, rat, and mouse. J. Neurocytol. 9, 537–570.

Mugnaini, E., Warr, W.B. and Osen, K.K. (1980b) Distribution and light microscopic features of granule cells in the cochlear nuclei of cat, rat, and mouse. J. Comp. Neurol. 191, 581–606.

Oliver, D.L., Potashner, S.J., Jones, D.R. and Morest, D.K. (1983) Selective labeling of spiral ganglion and granule cells with D-aspartate in the auditory system of cat and guinea pig. J. Neurosci. 3, 455–472.

Osen, K.K. (1988) Anatomy of the mammalian cochlear nuclei; a review. In: J. Syka and R.B. Masterton (Eds.), Auditory Pathway, Plenum, New York, pp. 65–75.

Osen, K.K. and Roth, K. (1969) Histochemical localization of cholinesterases in the cochlear nuclei of the cat, with notes on the origin of acetylcholinesterase-positive afferents and the superior olive. Brain Res. 16, 165–185.

Osen, K.K., Mugnaini, E., Dahl, A.-L. and Christiansen, A.H. (1984) Histochemical localization of acetylcholinesterase in the cochlear and superior olivary nuclei. A reappraisal with emphasis on the cochlear granule cell system. Arch. Ital. Biol. 122, 169–212.

Osen, K.K., Ottersen, O.P. and Storm-Mathisen, J. (1990) Colocalization of glycine-like and GABA-like immunoreactivities: A semiquantitative study of individual neurons in the dorsal cochlear nucleus of cat. In: O.P. Ottersen and J. Storm-Mathisen (Eds.), Glycine Neurotransmission, Wiley & Sons, New York, pp. 417–451.

Osen, K.K., Storm-Mathisen, J., Ottersen, O.P. and Dihle, B. (1995) Glutamate is concentrated in and released from parallel fiber terminals in the dorsal cochlear nucleus: a quantitative immunocytochemical analysis in guinea pig. J. Comp. Neurol. 357, 482–500.

Ottersen, O.P. and Storm-Mathisen, J. (1984) Neurons containing or accumulating transmitter amino acids. In: A. Björklund, T. Hökfelt and M.J. Kuhar (Eds.), Handbook of Chemical Neuroanatomy. Vol. 3: Classical Transmitters and Transmitter Receptors in the CNS, Part II, Elsevier, North Holland, pp. 141–246.

Petralia, R.S., Wang, Y.-X., Zhao H.-M. and Wenthold, R.J. (1996) Ionotropic and metabotropic glutamate receptors show unique localizations in the dorsal cochlear nucleus. J. Comp. Neurol. 372, 356–383.

Peyret, D., Geffard, M. and Aran, J.-M. (1986) GABA immunoreactivity in the primary nuclei of the auditory central nervous system. Hearing Res. 23, 115–121.

Potashner, S.J. (1983) Uptake and release of D-aspartate in the guinea pig cochlear nucleus. J. Neurochem. 41, 1094–1101.

Potashner, S.J. (1985) Uptake and release of γ-aminobutyric acid in the guinea pig cochlear nucleus after axotomy of cochlear and centrifugal fibers. J. Neurochem. 45, 1558–1566.

Rasmussen, G.L. (1967) Efferent connections of the cochlear nucleus. In: A.B. Graham (Ed.), Sensorineural Hearing Processes and Disorders, Little, Brown & Co, Boston, pp. 61–75.

Roberts, R.C. and Ribak, C. (1987) GABAergic neurons and axon terminals in the brainstem auditory nuclei of the gerbil. J. Comp. Neurol. 258, 267–280.

Saint Marie, R.L., Benson, C.G., Ostapoff, E.-M. and Morest, D.K. (1991) Glycine immunoreactive projections from the dorsal to the anteroventral cochlear nucleus. Hearing Res. 51, 11–28.

Saldaña, E. (1993) Descending projections from the inferior colliculus to the cochlear nuclei in mammals. In: M.A. Merchán, J.M. Juiz, D.A. Godfrey and E. Mugnaini (Eds.), The Mammalian Cochlear Nuclei: Organization and Function, Plenum, New York, pp. 153–165.

Sanes, D.H., Geary, W.A., Wooten, G.F. and Rubel, E.W. (1987) Quantitative distribution of the glycine receptor in the auditory brain stem of the gerbil. J. Neurosci. 7, 3793–3802.

Schwartz, I.R. (1981) The differential distribution of label following uptake of ^{3}H-labeled amino acids in the dorsal cochlear nucleus of the cat. Exp. Neurol. 73, 601–617.

Shannon-Hartman, S., Godfrey, D.A., Dunn, J.D., Godfrey, T.G. and Farms, W.B. (1993) Effects of large brain stem lesions on amino acid chemistry in the rat cochlear nucleus. Soc. Neurosci. Abstr. 19, 533.

Sherriff, F.E. and Henderson, Z. (1994) Cholinergic neurons in the ventral trapezoid nucleus project to the cochlear nuclei in the rat. Neuroscience 58, 627–633.

Shore, S.E., Moore, J.K. and Wu, B.J.-C. (1995) Brainstem input to the cochlear granule cell area, as demonstrated by retrograde transport of fluorogold. Abstr. Assoc. Res. Otolaryngol., 37.

Staatz-Benson, C. and Potashner, S.J. (1988) Uptake and release of glycine in the guinea pig cochlear nucleus after axotomy of afferent or centrifugal fibers. J. Neurochem. 51, 370–379.

Tago, H., McGeer, P.L., McGeer, E.G., Akiyama, H. and Hersh, L.B. (1989) Distribution of choline acetyltransferase immunopositive structures in the rat brainstem. Brain Res. 495, 271–297.

Thompson, A.M., Moore, K.R. and Thompson, G.C. (1995) Distribution and origin of serotoninergic afferents to guinea pig cochlear nucleus. J. Comp. Neurol. 351, 104–116.

Vetter, D.E., Cozzari, C., Hartman, B.K. and Mugnaini, E. (1993) Choline acetyltransferase in the rat cochlear nuclei: immunolocalization with a monoclonal antibody. In: M.A. Merchán, J.M. Juiz, D.A. Godfrey and E. Mugnaini (Eds.), The Mammalian Cochlear Nuclei: Organization and Function, Plenum, New York, pp. 279–290.

Waller, H.J., Godfrey, D.A., Chen, K. and Godfrey, T.G. (1994) Release of glutamate from cochlear nucleus of rat brain stem slices by electrical stimulation of parallel fibers. Soc. Neurosci. Abstr. 20, 137.

Waller, H.J., Godfrey, D.A. and Chen, K. (1996) Effects of parallel fiber stimulation on neurons of rat dorsal cochlear nucleus. Hearing Res. 98, 169–179.

Wamsley, J.K., Lewis, M.S., Young, W.S. III and Kuhar, M.J. (1981) Autoradiographic localization of muscarinic cholinergic receptors in rat brainstem. J. Neurosci. 1, 176–191.

Weedman, D.L. and Ryugo, D.K. (1996) Projections from auditory cortex to the cochlear nucleus in rats: synapses on granule cell dendrites. J. Comp. Neurol. 371, 311–324.

Weedman, D.L., Pongstaporn, T. and Ryugo, D.K. (1996) Ultrastructural study of the granule cell domain of the cochlear nucleus in rats: mossy fiber endings and their targets. J. Comp. Neurol. 369, 345–360.

Weinberg, R.J. and Rustioni, A. (1987) A cuneocochlear pathway in the rat. Neuroscience 20, 209–219.

Wenthold, R.J. (1978) Glutamic acid and aspartic acid in subdivisions of the cochlear nucleus after auditory nerve lesion. Brain Res. 143, 544–548.

Wenthold, R.J. (1980) Glutaminase and aspartate aminotransferase decrease in the cochlear nucleus after lesion of the auditory nerve. Brain Res. 190, 293–297.

Wenthold, R.J. and Altschuler, R.A. (1983) Immunocytochemistry of aspartate aminotransferase and glutaminase. In: L. Hertz, E. Kvamme, E.G. McGeer and A. Schousboe, (Eds.), Glutamine, Glutamate, and GABA in the Central Nervous System, Alan R. Liss, Inc., New York, pp. 33–50.

Wenthold, R.J. and Morest, D.K. (1976) Transmitter related enzymes in the guinea pig cochlear nucleus. Soc. Neurosci. Abstr. 2, 28.

Wenthold, R.J., Huie, D., Altschuler, R.A. and Reeks, K.A. (1987) Glycine immunoreactivity localized in the cochlear nucleus and superior olivary complex. Neuroscience 22, 897–912.

Wenthold, R.J., Hunter, C. and Petralia, R.S. (1993) Excitatory amino acid receptors in the rat cochlear nucleus. In: M.A. Merchán, J.M. Juiz, D.A. Godfrey and E. Mugnaini (Eds.), The Mammalian Cochlear Nuclei: Organization and Function, Plenum, New York, pp. 179–194.

Wouterlood, F.G. and Mugnaini, E. (1984) Cartwheel neurons of the dorsal cochlear nucleus: a Golgi-electron microscopic study in rat. J. Comp. Neurol. 227, 136–157.

Wouterlood, F.G., Mugnaini, E., Osen, K.K. and Dahl, A.-L. (1984) Stellate neurons in rat dorsal cochlear nucleus studied with combined Golgi impregnation and electron microscopy: synaptic connections and mutual coupling by gap junctions. J. Neurocytol. 13, 639–664.

Wright, D.D. and Ryugo, D.K. (1996) Mossy fiber projections from the cuneate nucleus to the cochlear nucleus in the rat. J. Comp. Neurol. 365, 159–172.

Wright, D.D., Blackstone, C.D., Huganir, R.L. and Ryugo, D.K. (1996) Immunocytochemical localization of the mGluR1α metabotropic glutamate receptor in the dorsal cochlear nucleus. J. Comp. Neurol. 364, 729–745.

Yao, W. and Godfrey, D.A. (1995) Immunohistochemistry of muscarinic acetylcholine receptors in rat cochlear nucleus. Hearing Res. 89, 76–85.

Yao, W. and Godfrey, D.A. (1996a) Autoradiographic distribution of muscarinic acetylcholine receptor subtypes in rat cochlear nucleus. Aud. Neurosci. 2, 241–255.

Yao, W. and Godfrey, D.A. (1996b) Electron microscopic histochemistry for acetylcholinesterase in the dorsal cochlear nucleus of rat. J. Acoust. Soc. Amer. 100, 2629.

Yao, W., Godfrey, D.A., and Levey, A.I. (1996) Immunolocalization of m2 muscarinic acetylcholine receptors in rat cochlear nucleus. J. Comp. Neurol. 373, 27–40.

Zhang, S. and Oertel, D. (1994) Neuronal circuits associated with the output of the dorsal cochlear nucleus through fusiform cells. J. Neurophysiol. 71, 914–930.

Zhao, H.B., Parham, K., Ghoshal, S. and Kim, D.O. (1995) Small neurons in the vestibular nerve root project to the marginal shell of the anteroventral cochlear nucleus in the cat. Brain Res. 700, 295–298.

14

SYNAPTIC RELATIONSHIPS IN THE GRANULE-CELL ASSOCIATED SYSTEMS IN DORSAL COCHLEAR NUCLEUS

Eric D. Young,[1] Kevin A. Davis,[1] and Israel Nelken[2]

[1]Department of Biomedical Engineering and Center for Hearing Sciences
Johns Hopkins University
Baltimore, Maryland 21205
[2]Department of Physiology
Hebrew University-Hadassah Medical School
Jerusalem 9120 Israel

1. CIRCUITRY OF THE DORSAL COCHLEAR NUCLEUS

The dorsal cochlear nucleus (DCN) is the most complex part of the cochlear nucleus, from the standpoint of interneuronal circuitry (Lorente de Nó, 1981; Osen et al., 1990). Figure 1 shows a summary of part of the DCN's circuitry. The principal cells of the nucleus are the pyramidal and giant cells, which project their axons to the contralateral inferior colliculus (Adams, 1979; Osen, 1972). The DCN is layered, with three layers running parallel to its free surface; the pyramidal cell somata form the second layer; these cells are bipolar, as indicated schematically in the figure, with apical dendrites in the superficial layer and basal dendrites in the deep layer (Blackstad et al., 1984; Kane, 1974; Smith and Rhode, 1985). The apical and basal dendrites of pyramidal cells encounter two different afferent systems. In the deep layer, there are primarily auditory afferents, including auditory nerve fibers (not shown; Osen, 1970) and collaterals of multipolar cells from PVCN (Oertel et al., 1990; Smith and Rhode, 1989). In addition to pyramidal basal dendrites, these afferent systems contact the other DCN principal cell type, giant cells, as well as vertical cells, which are an inhibitory interneuron (Saint Marie et al., 1991; Wenthold et al., 1987; Wickesberg and Oertel, 1990).

In the superficial layer, by contrast, the inputs are carried by parallel fibers (PFs), which are the axons of granule cells (Mugnaini et al., 1980a). Granule cells receive limited input from auditory nerve fibers (Brown and Ledwith, 1990; Liberman, 1993) but receive inputs from a variety of other auditory and non-auditory sources (reviewed in Weedman et al., 1996). PFs form terminals on three varieties of inhibitory interneurons in superficial DCN (Mugnaini et al., 1980b; Wouterlood and Mugnaini, 1984; Wouterlood et al., 1984). The best understood of these is the cartwheel cell; cartwheel cells stain for both glycine- and GABA-related molecules (reviewed by Osen et al., 1990). Their axons termi-

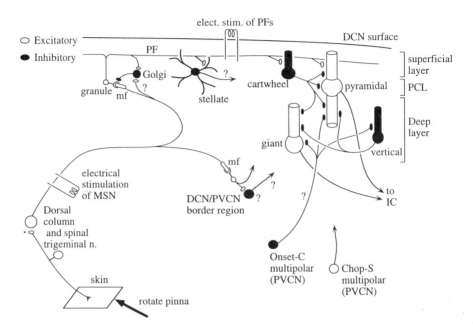

Figure 1. Schematic diagram of the interneuronal circuitry of the DCN, showing the systems associated with the granule cell axons (parallel fibers, *PF*) in the superficial layer. Unfilled cells are excitatory, filled cells are inhibitory. Question marks show elements and connections about which there is uncertainty. Layering of the DCN is indicated at right; *PCL* means pyramidal cell layer. Some authors divide the deep DCN into two layers, which are not shown here. The principal cells are the giant and pyramidal cells, which project to the inferior colliculus (*IC*). The somatosensory projection from the medullary somatosensory nuclei (MSN) is shown terminating as mossy fibers (*mf*) on granule cells in the PCL and in the lamina between the DCN and PVCN (Wright and Ryugo, 1996). The circuitry associated with granule cells in the lamina is not well understood, but neurons responsive to somatosensory stimuli have been recorded in this area (Young et al., 1995). The electrical stimulation sites used to gather the data in this paper are shown. Among the known sources of input to the DCN that are not shown are type I auditory nerve fibers, which terminate in the deep layer.

nate on the somata and dendrites of pyramidal cells (Berrebi and Mugnaini, 1991) and probably also on giant cells. IPSPs are recorded in principal cells which have the properties expected of cartwheel cell inputs (Golding and Oertel, 1995; Zhang and Oertel, 1994). Thus the DCN principal cells receive both direct excitatory and indirect inhibitory inputs from the granule cells.

One source of mossy fiber inputs to granule cells is from the somatosensory dorsal column and spinal trigeminal nuclei (below called MSN for medullary somatosensory nuclei; Itoh et al., 1987; Wright and Ryugo, 1996). This input forms mossy fiber terminals throughout the granule cell domains. Activation of the somatosensory inputs, either by electrical stimulation in MSN or by manually-applied somatosensory stimuli, produces inhibition of most DCN units (Young et al., 1995); in particular, type IV units, which are recorded from DCN principal cells (Young, 1980), are strongly inhibited. The somatosensory effects are largest when the pinna or portions of MSN connected to the pinna are stimulated.

By using a paradigm in which multiple near-threshold shocks are applied in a train, the apparently simple inhibitory response to electrical stimulation of the MSN can be decomposed into three components, two inhibitory and one excitatory (Davis et al., 1996a;

Young et al., 1995). This result suggests that the principal cells receive a balance of excitatory and inhibitory inputs from the granule cell system. The experiments described below were designed to characterize these excitatory and inhibitory effects and to identify the sources of the various components. We show that cartwheel cells are likely to generate one of the inhibitory components and that the properties of the excitatory component are consistent with direct activation of principal cells by PFs.

2. METHODS

Experiments were done in decerebrate cats. To prevent reflex movements when the MSN were stimulated, animals were paralyzed with gallamine triethiodide (10 mg/h iv); animals were not paralyzed until at least 4 hours after the decerebration, to allow signs of incomplete decerebration (voluntary movements) to appear. Single units were recorded in the DCN using Pt-Ir extracellular electrodes. Biphasic electrical stimuli (100 ms per phase) were applied through a bipolar tungsten electrode. The stimuli were applied in trains of four pulses, separated by 50 ms. Stimulation was done at two sites, illustrated schematically in Fig. 1: 1) in the MSN at a position, ≈ 3 mm lateral to the obex, where neurons respond to touching the pinna or to movement of the pinna; and 2) on the surface of the DCN ≈ 1 mm medial to the recording electrode. Stimulating the MSN should activate the granule cells orthodromically through the mossy fibers whereas stimulating the surface of the DCN should activate granule cells antidromically through the PFs. These stimulus sites are different in that PF stimulation may not activate presynaptic components of the granule cell circuit which are activated by MSN stimulation. Comparison of responses in the two cases may allow detection of the effects of synaptic interactions in the complex mossy-fiber/granule cell synapse (Mugnaini et al., 1980b).

3. RESPONSES OF TYPE IV NEURONS TO SOMATOSENSORY AND PF STIMULATION

Responses of four type IV units to the four-pulse stimulus paradigm are shown in Fig. 2. Three components are seen in various combinations in the unit responses (bottom trace in each plot). The components are defined relative to the latency of the evoked potential at the recording site (top trace in each plot). For reference, the vertical dashed lines mark the time of onset of the evoked potential for the third stimulus pulse. Fig. 2A shows all three components, for MSN stimulation: 1) a short-latency inhibition which precedes the evoked-potential onset; 2) an excitatory peak which occurs just after the evoked-potential onset; and 3) a long-latency inhibitory component that occurs after the excitation.

Approximately half (34/75) the type IV neurons studied display all three components in response to MSN stimulation, as in Fig. 2A. Most of the remaining type IV neurons (36/75) showed only the long-latency inhibitory component, as in Fig. 2C. Five units gave no response. In most of the cases with long-latency inhibition (34/36), the inhibition started at or after the onset of the evoked potential. That is, with MSN stimulation the short-latency inhibition and the excitation are either seen together or are not observed and the long-latency inhibition is seen in almost all cells. In response to PF stimulation (Figs. 2B and 2D), the short-latency inhibitory component was not observed, but the excitatory and long-latency inhibitory components were seen. Most type IV units (27/34) showed long-latency inhibition and many of those (11/27) also showed an excitatory peak, as in

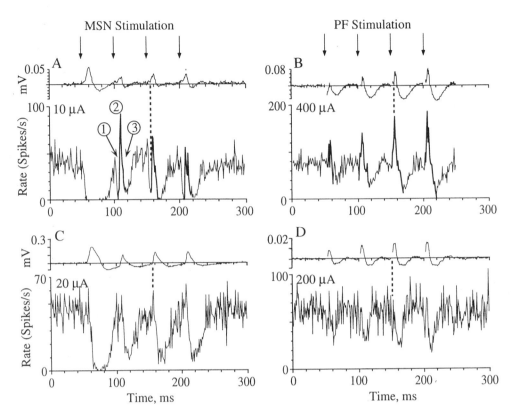

Figure 2. Responses of DCN type IV units to MSN stimulation (A, C) and to PF stimulation (B, D). In each plot, the top trace shows the evoked potential produced by the stimulus at the recording site in DCN; stimulus artifacts have been removed from these traces. The bottom plot shows the average discharge rate in response to 400 repetitions of the stimulus. The effects of the stimulus on spontaneous activity are shown. Arrows at the top mark the times at which electric stimuli were delivered and the current strength is given at left above each PST histogram. Data from four different units are shown. The vertical dashed lines mark the onset of the evoked potential in response to the third stimulus pulse. The circled numbers in A identify the three response components discussed in the text.

Fig. 2B. The properties of excitation and long-latency inhibition were similar for MSN and PF stimulation. In particular, the latency of the excitation and the latency and the adaptation characteristics (for pulses 2–4, discussed below) of the long-latency inhibition formed overlapping distributions for MSN and PF stimulation (Davis et al., 1996a). On this basis, we assume that the excitatory and long-latency inhibitory components reflect the same mechanisms when produced by MSN and PF stimulation.

The response components display characteristic adaptation behavior which is illustrated by the examples in Fig. 2. This behavior is simplest for PF stimulation, where a monotonic increase is seen through pulses 1–4 in the amplitudes of both the excitatory (Fig. 2B) and long-latency inhibitory (Figs. 2B and 2D) components. This behavior resembles the synaptic facilitation that has been demonstrated for PF synapses on pyramidal cells using an *in vitro* preparation (Manis, 1989).

The adaptation behavior is more complex for MSN stimulation (Figs. 2A and 2C). The short-latency inhibitory component changes little through the four pulses. The long-latency inhibitory component is largest for the first pulse, decreases to a small value for the second pulse, and then shows a buildup through pulses 2–4. The buildup of long-latency inhibition for pulses 2–4 is similar to the facilitation seen with PF stimulation; in fact, measures of these two processes are statistically indistinguishable (Davis et al., 1996a). The excitatory component behaves inversely to the long-latency inhibition: excitation is very small for the first pulse, is largest for the second pulse and then decreases. The adaptation behavior of the evoked-potential magnitude is similar to that of the long-latency inhibitory response for both MSN and PF stimulation. Indeed, the adaptation of the evoked potential is statistically indistinguishable from that of long-latency inhibition (Davis et al., 1996a).

The data in Fig. 2 show that the short-latency and long-latency inhibitory components have different properties. The two inhibitions differ in their adaptation behavior; in addition, long-latency inhibition occurs with or without short-latency inhibition across different units or in different stimulus situations. These differences imply that short- and long-latency inhibition are produced by different mechanisms.

4. SOMATOSENSORY EVOKED POTENTIALS IN THE DCN

Interpretation of the response components shown in Fig. 2 requires some comments about the evoked potential, which serves as the reference for the latency measurements. Based on its properties, it has been argued that the evoked potential is produced by the postsynaptic currents in the pyramidal and cartwheel cells (Young et al., 1995). If this is true, then the onset of the evoked potential signals the arrival, at the postsynaptic neuron, of the volley of activity in the PFs. The arguments for this interpretation are summarized using Fig. 3.

In the absence of high-pass filtering of the electrode signal and stimulus artifacts, the evoked potential has a monophasic shape like those in Figs 2A and 2C. Similar evoked potentials are seen with PF stimulation (Fig. 2B and 2D), but their morphology is less clear because of large stimulus artifacts in the records (artifacts have been removed from the figures). The shape of the evoked-potential waveform stays roughly constant as the recording electrode is driven into the DCN. Fig. 3B shows an example of a typical recording track; the track is approximately parallel to this sagittal section and passes perpendicularly through the PCL (shaded). The amplitude of the potential as a function of depth along this track is shown in Fig. 3C. Note that the amplitude peaks near the boundary of the PCL and the deep layer of the DCN.

The evoked potential is interpreted in terms of the model in Fig. 3A; the actual quantitative model is fully described elsewhere (Young et al., 1995). The model is based on standard analyses of evoked potentials in structures that have a sheet of oriented cells wrapped into a sphere or a portion of a sphere (Klee and Rall, 1977). In the case of the DCN, the PCL has the shape of a hemicylinder, rather than a sphere, but the results of the model are qualitatively the same when the calculations are done for a hemicylinder. Fig. 3A shows postsynaptic currents in pyramidal and cartwheel cells only. Pyramidal cells have predominantly radially directed dendrites (Blackstad et al., 1984; Smith and Rhode, 1985) and their dendritic trees are oriented roughly in parallel. The dendrites of cartwheel cells are more complex (Berrebi and Mugnaini, 1991; Manis et al., 1994; Wouterlood and Mugnaini, 1984; Zhang and Oertel, 1993a), but are included in the model because they could contribute radial currents. Assuming that the granule cells are activated uniformly throughout the DCN, which is consistent with the termination pattern of MSN projections (Itoh et al., 1987) and with the fact that virtually all type IV units are affected by MSN

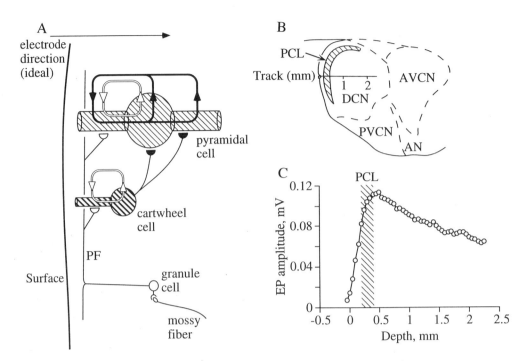

Figure 3. A summary of the model used to explain the properties of the evoked potential. A. Schematic of currents produced by a volley of spikes in the PF. Only two neuron types are considered, the pyramidal cell and cartwheel cell. White currents are produced by excitatory PF synapses on cartwheel and pyramidal cell dendrites; black currents are produced by inhibitory cartwheel-cell synapses on the pyramidal cell. B. Drawing of a histological section showing a reconstruction of a recording electrode track through the DCN. The PCL (shaded) is shown in cross section in this sagittal section. The orientation of the schematic in A is the same as the drawing in B. C. Amplitude of the evoked potential plotted versus depth along the recording track in B. The abscissa scale is the same as the scale on the track in B; 0 depth is the DCN surface.

stimulation, the cartwheel and pyramidal cells should also be uniformly activated, meaning uniform throughout the PCL sheet. With uniform activation, the extracellular currents associated with postsynaptic activity should be predominantly perpendicular to the surface, as drawn. Tangential currents (i.e. parallel to the DCN surface) from the cartwheel and pyramidal cells will be minimized because the uniformity of activation of these cells minimizes voltage gradients along the tangential direction.

Activation of PFs will produce current sinks in the superficial layer (white arrows), due to excitatory PF synapses on apical dendrites; current sources in the deep layer (black arrows) will also occur, due to the inhibitory cartwheel-cell synapses on the somata and basal dendrites of pyramidal cells. Current-source density analysis of the evoked potentials is consistent with this distribution of sources and sinks (Young et al., 1995). The evoked potential will be produced by the radially-flowing currents in the extracellular space. Assuming that the surface of the DCN is near ground potential, the extracellular currents will produce a positive evoked potential as an electrode travels into the nucleus along a track like the one shown by the arrow. The amplitude of the potential is expected to increase through the region occupied by the currents. This is the behavior shown by the data (Fig. 3C). Deeper in the DCN, below the level at which significant membrane currents are flowing, the potential amplitude is determined by the geometry of the sheet of current sources and the spread of current. With the geometry of the DCN, the model predicts a decline in potential similar to the one shown by the data in Fig. 3C (Young et al., 1995).

With the interpretation of Fig. 3, the evoked potential can be used to mark the arrival of the granule cell volley at the postsynaptic cells. Thus the excitatory effects of granule cells on principal cells (in this case, type IV neurons) should occur at or soon after the onset of the evoked potential and the inhibitory effect of cartwheel cells on principal cells should follow after a delay for the synapse and for postsynaptic integration in the cartwheel cell. The data in Figs. 2A and 2B are typical in that the excitatory peak occurs just after the evoked potential onset (latencies of excitation relative to the onset of the evoked potential are 1.4±2.4 ms for MSN and -0.2±1.1 ms for PF stimulation, mean ± SD; (Davis et al., 1996a). Data in Figs. 2C and 2D are also typical in that the long-latency inhibition, seen in isolation from the excitatory peak, occurs with a slightly longer latency (2.4±2.5 ms for MSN and 2.7±1.5 ms for PF stimulation).

Thus, our data are consistent with the hypothesis that the excitatory peak reflects the direct excitatory action of granule cells on pyramidal cells. The fact that a significant number of pyramidal cells do not show an excitatory peak can be explained by three factors: 1) the excitatory effects of parallel fibers on pyramidal cells is reported to be much weaker than on cartwheel cells (Waller et al., 1996; Zhang and Oertel, 1994), so PF inputs to some pyramidal cells may be subthreshold; 2) with MSN stimulation, not all granule cells need be connected to the somatosensory inputs and not all principal cells need be connected to granule cells that do respond to the somatosensory inputs; and 3) with PF stimulation, the stimulus produces activity in a "beam" of PFs (Manis, 1989) and if the type IV neuron from which recordings are being made is not on that beam, it will not receive a direct granule cell input.

5. PRESUMED CARTWHEEL CELLS HAVE PROPERTIES CONSISTENT WITH THE LONG-LATENCY INHIBITOR

The latency data quoted above suggest that the long-latency inhibition could be produced by an inhibitory interneuron, like the cartwheel cell, which is activated by the PFs. In this section it is shown that the response properties of cartwheel cells correspond closely to the characteristics of long-latency inhibition in type IV units. Recordings from presumed cartwheel cells were made in the superficial layer of the DCN. Single units that are presumed to be cartwheel cells have two characteristics: 1) they were recorded with the electrode tip within 500 µm of the surface of the DCN, which corresponds to the range within which cartwheel cells are found (Berrebi and Mugnaini, 1991; Wouterlood and Mugnaini, 1984); and 2) the spike trains show a mixture of single spikes and short bursts of spikes (Fig. 4A). The bursts of spikes resemble the complex action potentials recorded intracellularly in cartwheel cells (Manis et al., 1994; Zhang and Oertel, 1993a). Cartwheel cells are the only neurons in the cochlear nucleus that have been shown to have such complex action potentials.

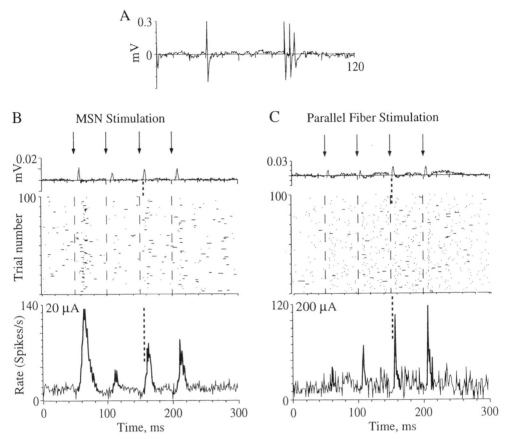

Figure 4. Characteristics of complex-spiking neurons. A. Short portion of the electrode signal from a complex-spiking neuron showing one simple spike and one complex spike. B. Response of a complex-spiking neuron to stimulation in the MSN. C. Response of a different complex-spiking neuron to PF stimulation. For B and C, the stimuli are the same as in Fig. 2, delivered at the times marked by the arrows. Top traces are the evoked potential at the recording site. Stimulus artifacts have been removed from these traces. Dot displays of the units' responses are shown in the middle. Light vertical dashed lines mark the stimulus occurrence times. Dots mark simple spikes and short horizontal lines connect the spikes of bursts corresponding to complex action potentials. The bottom traces show PST histograms of responses to the stimuli, computed from the data in the dot display plus 300 additional repetitions of the stimulus. Heavy vertical dashed lines show the onset of the evoked potential for the third stimulus pulse.

Responses of two complex-spiking neurons to MSN (Fig. 4B) and PF (Fig. 4C) stimulation are shown. The format for these data is the same as in Fig. 2, except that dot displays of 100 repetitions of the stimulus are also shown. These recordings are near the surface (310 and 280 mm), so the amplitudes of the EPs are small. In the dot displays, complex spikes are indicated by connecting the dots marking their component simple spikes with a horizontal line. The two examples shown in Fig. 4 are recorded from two different units, but they are typical in that complex spikes are more likely to be evoked with MSN than with PF stimulation.

The PST histograms show primarily excitatory responses from both stimulus sites. These examples are typical of the whole population; excitatory responses were seen in 24/26 complex-spiking units examined with MSN stimulation and 15/17 units examined with PF stimuli. Inhibitory responses were not seen in complex-spiking units, except as decreases in rate following the excitatory peaks. The excitatory peak occurs immediately after the onset of the evoked potential (heavy vertical dashed lines at the third stimulus of the train) with an average latency of 1.0±2.0 (MSN) and 0.3±1.4 (PF) ms re the onset of the evoked potential. Thus the latency of the excitation in complex-spiking neurons precedes the latency of inhibition in type IV units, as is required if the complex spiking units are the source of the type IV units' long-latency inhibition.

Further evidence that complex-spiking units are the long-latency inhibitor is provided by the adaptation behavior of the units. With PF stimulation, there is a monotonic increase through the four pulses in the size of the long-latency inhibition of type IV units (Figs. 2B and 2D); this increase is qualitatively and quantitatively similar to the monotonic increase in size of the excitatory peak in the complex-spiking units (Fig. 4C; Davis et al., 1996b). Presumably the increase in both cases reflects facilitation at the PF-cartwheel cell synapse. With MSN stimulation, by contrast, the largest response occurs on the first pulse (compare Figs. 2A and 2C with Fig. 4B); the response to the second pulse is smallest and is followed by a monotonic increase, which is the same behavior shown by long-latency inhibition in type IV units. Finally, virtually all type IV units show the long-latency inhibitory response with both MSN and PF stimulation. The fact that almost all type IVs are affected is consistent with cartwheel cells as the inhibitory source because 1) almost all cartwheel cells are excited by MSN stimuli; and 2) cartwheel-cell axons spread widely within an isofrequency sheet, i.e. perpendicular to the direction of travel of the PFs (Berrebi and Mugnaini, 1991). Thus even if only a beam of cartwheel cells are activated by the PF stimulus, their axons could contact principal cells over a wide range.

6. DISCUSSION

The evidence presented above argues for the hypotheses that the excitatory peak in type IV units represents direct activation of principal cells by PFs and that the long-latency inhibition is produced by inhibitory inputs from cartwheel cells. There is no obvious source for the third component of the responses to MSN stimulation, the short-latency inhibition. A detailed discussion of the reasons why known sources of inhibition in DCN seem inadequate is given elsewhere (Davis et al., 1996a).

One feature that is not immediately consistent with the hypotheses discussed above is the adaptation behavior of the excitatory peak with MSN stimulation. In type IVs, this component behaves like the inverse of the long-latency inhibition (Fig. 2A), whereas it is expected that it should show the same adaptation behavior as the other components. However, the excitatory component is weak in type IV units, as discussed above. Given this weakness and given that the long-latency inhibition is a strong effect, it is likely that the amplitude of the excitatory peak is controlled mainly by the amplitudes of the two adjacent inhibitory components; thus as the long-latency inhibition gets stronger, the excitatory peak gets weaker.

A second feature that is not accounted for by the hypotheses above is the adaptation behavior with MSN stimulation; in particular, it is not clear why the response to the first pulse is so strong, for both the evoked potential and the long-latency inhibition. The fact that the strong first-pulse response is seen with MSN but not with PF stimulation suggests that orthodromic activation is necessary for its occurrence. One possible explanation is that an excitatory pathway to the cartwheel cells is activated by orthodromic stimulation which is not activated by antidromic stimulation of the granule cells. This could be a parallel pathway which receives synaptic terminals from the somatosensory inputs. Alternatively, it could be that a very strong adaptive effect occurs in the glomerular synapse, such that the granule cell response to stimuli after the first is greatly attenuated.

It is important to point out that type IV units in both the PCL and the deep DCN are affected in essentially identical ways by both MSN and PF stimulation. Because type IV units in deep DCN are likely to be giant cells, these results raise the question of how granule cell activity is coupled to the giant cells, given that their dendrites make only minimal contact with the superficial DCN (Kane et al., 1981; Ryugo and Willard, 1985; Zhang and Oertel, 1993b). The long-latency inhibitory component is not such a problem since cartwheel cell axons project sufficiently deeply to contact giant cells (Berrebi and Mugnaini, 1991). Granule cell axons could contact giant cells either *en passant* while passing through deep DCN on the way to the superficial layer or through, currently unknown, circuitry associated with the granule cells in the lamina between DCN and PVCN. There are scattered granule cells in the principal cell areas of DCN (Liberman, 1993; Mugnaini et al., 1980a) which could contact the giant cells, but the trajectory of their axons is currently unknown.

An understanding of the role of the granule-cell associated systems discussed here requires a knowledge of the information carried to the DCN principal cells by the mossy fiber inputs. In the case of the somatosensory mossy fibers, the information appears to involve the pinna (Young et al., 1995), and appears to be related to movement or position of the pinna, rather than to touching the skin or hairs on the pinna (Kanold and Young, 1996). Because pinna movements have significant consequences for the shape of the head-related transfer function from different directions in space (Young et al., 1996), moving the pinna can change the spectrum of a fixed sound source as well as spectral sound localization information. Thus, the somatosensory inputs to DCN could be serving a role in coordinating acoustic signal processing with the directional state of the cat's mobile pinna.

7. ACKNOWLEDGMENT

This research was supported by NIH grants DC00115 and DC00979.

8. REFERENCES

Adams, J.C. (1979) Ascending projections to the inferior colliculus. J. Comp. Neurol. 183, 519–538.

Berrebi, A.S. and Mugnaini, E. (1991) Distribution and targets of the cartwheel cell axon in the dorsal cochlear nucleus of the guinea pig. Anat. Embryol. 183, 427–454.

Blackstad, T.W., Osen, K.K., and Mugnaini, E. (1984) Pyramidal neurones of the dorsal cochlear nucleus: A Golgi and computer reconstruction study in cat. Neuroscience 13, 827–854.

Brown, M.C. and Ledwith, J.V. (1990) Projections of thin (type-II) and thick (type-I) auditory-nerve fibers into the cochlear nucleus of the mouse. Hearing Res. 49, 105–118.

Davis, K.A., Miller, R.L., and Young, E.D. (1996a) Effects of somatosensory and parallel-fiber stimulation on neurons in dorsal cochlear nucleus. J. Neurophysiol. 76, 3012–3024.

Davis, K.A., Miller, R.L., and Young, E.D. (1996b) Presumed cartwheel cells in the cat dorsal cochlear nucleus (DCN) are excited by somatosensory and parallel fiber stimulation. Abst. Assoc. for Res. in Otolaryngol. 19, 171.

Golding, N.L. and Oertel, D. (1995) Evidence that cartwheel cells of the dorsal cochlear nucleus excite other cartwheel cells and inhibit principal cells with the same neurotransmitter, glycine. Abst. Soc. for Neurosci. 21, 399.

Itoh, K. et al. (1987) Direct projection from the dorsal column nuclei and the spinal trigeminal nuclei to the cochlear nuclei in the cat. Brain Res. 400, 145–150.

Kane, E.C. (1974) Synaptic organization in the dorsal cochlear nucleus of the cat: a light and electron microscopic study. J. Comp. Neurol. 155, 301–329.

Kane, E.S., Puglisi, S.G., and Gordon, B.S. (1981) Neuronal types in the deep dorsal cochlear nucleus of the cat: I. Giant neurons. J. Comp. Neurol. 198, 483–513.

Kanold, P.O. and Young, E.D. (1996) Deep somatosensory receptors associated with the pinna provide static inhibition of principal neurons in the cat dorsal cochlear nucleus (DCN). Abst. Assoc. for Res. in Otolaryngol. 19, 170.

Klee, M. and Rall, W. (1977) Computed potentials of cortically arranged populations of neurons. J. Neurophysiol. 40, 647–666.

Liberman, M.C. (1993) Central projections of auditory nerve fibers of differing spontaneous rate, II: Posteroventral and dorsal cochlear nuclei. J. Comp. Neurol. 327, 17–36.

Lorente de Nó, R. (1981) The Primary Acoustic Nuclei. Raven Press, New York.

Manis, P.B. (1989) Responses to parallel fiber stimulation in the guinea pig dorsal cochlear nucleus in vitro. J. Neurophysiol. 61, 149–161.

Manis, P.B., Spirou, G.A., Wright, D.D., Paydar, S., and Ryugo, D.K. (1994) Physiology and morphology of complex spiking neurons in the guinea pig dorsal cochlear nucleus. J. Comp. Neurol. 348, 261–276.

Mugnaini, E., Osen, K.K., Dahl, A.L., Friedrich Jr., V.L., and Korte, G. (1980b) Fine structure of granule cells and related interneurons (termed Golgi cells) in the cochlear nuclear complex of cat, rat, and mouse. J. Neurocytol. 9, 537–570.

Mugnaini, E., Warr, W.B., and Osen, K.K. (1980a) Distribution and light microscopic features of granule cells in the cochlear nuclei of cat, rat, and mouse. J. Comp. Neurol. 191, 581–606.

Oertel, D., Wu, S.H., Garb, M.W., and Dizack, C. (1990) Morphology and physiology of cells in slice preparations of the posteroventral cochlear nucleus of mice. J. Comp. Neurol. 295, 136–154.

Osen, K.K. (1970) Course and termination of the primary afferents in the cochlear nuclei of the cat. Arch. Ital. Biol. 108, 21–51.

Osen, K.K. (1972) Projection of the cochlear nuclei on the inferior colliculus in the cat. J. Comp. Neurol. 144, 355–372.

Osen, K.K., Ottersen, O.P., and Storm-Mathisen, J. (1990) Colocalization of glycine-like and GABA-like immunoreactivities. A semiquantitative study of individual neurons in the dorsal cochlear nucleus of cat. In: O.P. Ottersen and J. Storm-Mathisen (Eds.), Glycine Neurotransmission, John Wiley & Sons, New York, pp. 417–451.

Ryugo, D.K. and Willard, F.H. (1985) The dorsal cochlear nucleus of the mouse: A light microscopic analysis of neurons that project to the inferior colliculus. J. Comp. Neurol. 242, 381–396.

Saint Marie, R.L., Benson, C.G., Ostapoff, E.M., and Morest, D.K. (1991) Glycine immunoreactive projections from the dorsal to the anteroventral cochlear nucleus. Hearing Res. 51, 11–28.

Smith, P.H. and Rhode, W.S. (1985) Electron microscopic features of physiologically characterized, HRP-labeled fusiform cells in the cat dorsal cochlear nucleus. J. Comp. Neurol. 237, 127–143.

Smith, P.H. and Rhode, W.S. (1989) Structural and functional properties distinguish two types of multipolar cells in the ventral cochlear nucleus. J. Comp. Neurol. 282, 595–616.

Waller, H.J., Godfrey, D.A., and Chen, K. (1996) Effects of parallel fiber stimulation on neurons of rat dorsal cochlear nucleus. Hearing Res. 98, 169–179.

Weedman, D.L., Pongstaporn, T., and Ryugo, D.K. (1996) Ultrastructural study of the granule cell domain of the cochlear nucleus in rats: Mossy fiber endings and their targets. J. Comp. Neurol. 369, 345–360.

Wenthold, R.J., Huie, D., Altschuler, R.A., and Reeks, K.A. (1987) Glycine immunoreactivity localized in the cochlear nucleus and superior olivary complex. Neuroscience 22, 897–912.

Wickesberg, R.E. and Oertel, D. (1990) Delayed, frequency-specific inhibition in the cochlear nuclei of mice: a mechanism for monaural echo suppression. J. Neurosci. 10, 1762–1768.

Wouterlood, F.G. and Mugnaini, E. (1984) Cartwheel neurons of the dorsal cochlear nucleus. A Golgi-electron microscopic study in the rat. J. Comp. Neurol. 227, 136–157.

Wouterlood, F.G., Mugnaini, E., Osen, K.K., and Dahl, A.-L. (1984) Stellate neurons in rat dorsal cochlear nucleus studied with combined Golgi impregnation and electron microscopy: synaptic connections and mutual coupling by gap junctions. J. Neurocytol. 13, 639–664.

Wright, D.D. and Ryugo, D.K. (1996) Mossy fiber projections form the cuneate nucleus to the dorsal cochlear nucleus of rat. J. Comp. Neurol. 365, 159–172.

Young, E.D. (1980) Identification of response properties of ascending axons from dorsal cochlear nucleus. Brain Res. 200, 23–38.

Young, E.D., Nelken, I., and Conley, R.A. (1995) Somatosensory effects on neurons in dorsal cochlear nucleus. J. Neurophysiol. 73, 743–765.

Young, E.D., Rice, J.J., and Tong, S.C. (1996) Effects of pinna position on head-related transfer functions in the cat. J. Acoust. Soc. Am. 99, 3064–3076.

Zhang, S. and Oertel, D. (1993a) Cartwheel and superficial stellate cells of the dorsal cochlear nucleus of mice: Intracellular recordings in slices. J. Neurophysiol. 69, 1384–1397.

Zhang, S. and Oertel, D. (1993b) Giant cells of the dorsal cochlear nucleus of mice: Intracellular recordings in slices. J. Neurophysiol. 69, 1398–1408.

Zhang, S. and Oertel, D. (1994) Neuronal circuits associated with the output of the dorsal cochlear nucleus through fusiform cells. J. Neurophysiol. 71, 914–930.

15

THE GERBIL COCHLEAR NUCLEUS

Postnatal Development and Deprivation Effects

Otto Gleich,[1] Celia Kadow,[1] Marianne Vater,[2] and Jürgen Strutz[1]

[1]ENT-Department
University of Regensburg
93042 Regensburg, Germany
[2]Department of Zoology
University of Regensburg
93040 Regensburg, Germany

1. INTRODUCTION

Auditory deprivation experiments in several vertebrate species have demonstrated that the auditory pathway shows profound morphological alterations if the activity of the auditory nerve is abolished and that the effects are most pronounced in young animals during a "sensitive period" (e.g. Hashisake and Rubel, 1989; Moore, 1990; 1994; Nordeen et al., 1983; Pasic and Rubel, 1989; 1991; Trune, 1982; Tucci and Rubel, 1985). Although comparative anatomical data are not available in humans, it is well documented that severe deafness in young prelingual children affects speech development and, indeed, that the delayed onset of speech leads occasionally to the detection and diagnosis of a severe hearing problem in children (Ling and Ling, 1978).

A high percentage of children suffers for various periods from less severe conductive hearing losses due to middle ear infections during the first years of their life (Casselbrant et al., 1985; Zeisel et al. 1995). Such a conductive hearing loss due to recurrent episodes of chronic otitis media with effusion may affect speech development and associated cognitive skills (Gravel et al., 1995). This issue, however, still remains controversial due to methodological problems associated with such studies in human subjects (Lous, 1995). For example the socio-economic status is correlated with the rate of occurrence of middle ear infections and also directly affects some of the parameters that were used to document the performance of different groups of children.

Thus animal studies under controlled conditions should help to resolve the question as to whether a conductive hearing loss might also affect the development of the auditory pathway and, subsequently, speech and cognitive skills in children. In order to understand the effects of auditory deprivation during development, it is essential to study and describe the normal postnatal development of the auditory pathway. As a first step, we concentrated our efforts on the cochlear nucleus (CN; the first relay station of the ascending auditory pathway) since it has been shown in previous experiments that a conductive hearing loss induced smaller somata of some cell types and a smaller total volume, especially in the ventral portion of the CN (Blatchley et al., 1983; Coleman et al., 1982; Conlee and Parks, 1981; Webster, 1988). Most of these earlier studies used mice and rats, that have a fairly poor low-frequency hearing. Contrary to these data from mouse and rat, Moore et al (1989) in the ferret and Doyle and Webster (1991) in the rhesus monkey found no significant effects of a conductive type of deprivation. We used gerbils (*Meriones unguiculatus*) as a model for a systematic investigation, since this rodent species shows a good low frequency representation similar to that of humans. Here we report data on the expression of GABA- and Glycine-immunoreactivity and on the increase in volume of the CN subdivisions during normal postnatal development, as well as the effects of a conductive hearing loss on the CN volume as a function of the onset time of the deprivation.

2. METHODS

Age-graded gerbils were obtained from our own breeding colony. For the histological analysis, animals were transcardially perfused by the respective fixative after they had received a lethal dose of anaesthetic (approximately 150mg/kg ketamine and 40mg/kg xylacine).

The expression of GABA and glycine were evaluated in 25μm-thick immunostained sections through the CN of 36 gerbils at various ages between birth and 16 weeks of age. The immunostaining procedure including fixatives and buffers followed the protocols given by the producer of the glutaraldehyde coupled GABA and glycine antibodies (SFRI Laboratoire, Berganton, 33127 Saint Jean Dillac, France). After binding of the primary antibody, a biotinylated secondary antibody followed, that was visualized using a standard avidin-horseradishperoxidase conjugate (ABC, Vectastain) with a subsequent DAB reaction. The immunostaining of somata as well as fibres and puncta in the neuropil was evaluated with the light microscope. The concentrations of the primary antibodies and the incubation time were selected to give a very intense staining in adult animals (GABA 1:8000 and Glycine 1:12000 incubation for 16 hours). The processing conditions were kept constant across all ages in order to allow direct comparison of the stains.

For the analysis of normal CN growth, we used series of 10μm-thick, Nissl-stained frontal paraffin sections and analyzed 4–6 cochlear nuclei from each of 12 age groups between birth and almost three years. Sections at 50μm intervals were selected and digitized through a video camera attached to a microscope. With an image-analysis system, the outlines of the dorsal, postero-ventral and antero-ventral CN were traced on the video image and, after calibration, the corresponding areas were calculated. By multiplying the area with the distance between the sections, we obtained the volume of each CN subdivision contained in a given section. Adding up the values from all sections examined gave the volume of the respective subdivision.

The chain of middle-ear ossicles was interrupted by disarticulating and removing the incus of the right ear in 19 gerbils during surgical anaesthesia at ages between 12 days and three months of age (40–60mg/kg ketamine and 10–15mg/kg xylacine) to induce a conductive hearing loss. The deprived animals survived until they reached an age of six months and were subsequently processed with four age-matched, unoperated controls. For the analysis of the expression pattern of the immediate early gene product c-Fos (not to be reported here), animals were exposed to broad-band noise for 90 minutes. Immediately after the exposure they were fixed with 4% paraformaldehyde and serial frontal kryostat sections processed for the c-Fos demonstration. Here we report volume measurements obtained from these sections. To evaluate the effect of the deprivation, we calculated the percent volume difference between the left and the right side relative to the undeprived left side.

3. RESULTS

The over-all GABA and glycine labelling pattern in the adult gerbil (around three months) conformed to the general mammalian pattern (Wenthold, 1991). A prominent network of GABA-positive fibers was already present at postnatal day (pnd) one throughout the CN and became apparently adult-like (at the level of the light microscope) within the first two weeks of life. Much of this development occurred before the onset of hearing (pnd 12). At pnd one and pnd 6, there were only few and faintly GABA-labelled somata in the CN, except for the granule cell region, where presumed Golgi cells showed an intense expression of GABA even in the youngest animals. The number of weakly-labelled cells increased during the first 2 postnatal weeks and around the onset of hearing, the first intensely GABA-labelled cells occurred. Their number increased over the next two weeks and soma-labelling appeared adult-like in four-week old animals. Throughout the CN, GABA-labelled somata tended to be small. In the AVCN and PVCN, the average cross-sectional area was 70–75µm². Average somatic size in the deep DCN was 61µm² and in the upper layers of DCN 104µm². While the size of GABA-labelled somata increased significantly with age in all subnuclei studied, after the onset of hearing we found a substantial growth only in the molecular and fusiform layers of DCN. Glycine labelling appeared less distinct and more diffuse in young animals as compared to GABA. The number of weakly-labelled somata increased after birth. Intensely glycine-labelled somata appeared in time slightly before intense GABA around the onset of hearing. Soma staining appeared adult-like in four-week old animals. The cross-sectional area of glycine-labelled somata increased significantly with age in all CN subdivisions, but a substantial growth after the onset of hearing was only observed in the outer layers of the DCN and the large glycinergic cells of the ventral CN.

The whole CN showed a volume increase from 0.15±0.04mm³ (n=5) at pnd one to 0.98±0.09mm³ (n=10) in 4–8 months old animals. Figure 1 shows the increase in volume for the DCN and the AVCN. Between birth and pnd 7 the gerbil CN volume increased only moderately. However, between pnd 7 and pnd 12 (onset of hearing), all subdivisions showed a prominent period of growth. In DCN most of the growth occurred before the onset of hearing. Between pnd one and pnd 12, the DCN-volume increased approximately 4-fold and the volume of AVCN increased by a factor of five. However, after the onset of hearing, the DCN showed only a very moderate further growth, reaching a maximum of five times the value found in newborn animals. In contrast, the volume of AVCN almost doubled after the onset of hearing and reached a maximum in four-months old animals, corresponding to a 9-fold increase relative to pnd one. The distribution of data points in Fig. 1, each representing one CN, indicates an increased scatter during the period of the most dynamic growth around the onset of hearing between pnd 9 and pnd 21. Preliminary results indicate that the increase in AVCN volume after the onset of hearing is due to the neuropil, because the somata of principal cells reached an adult size by pnd 12 and packing density decreased in older animals (Gleich et al., 1997).

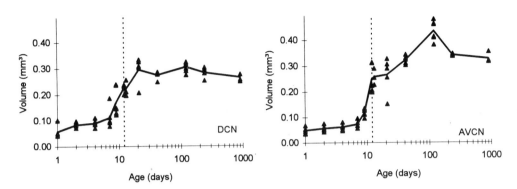

Figure 1. The postnatal increase of the DCN (left) and AVCN (right) volume. The heavy black lines indicate average values. The dotted vertical lines label pnd 12, the onset of hearing.

Unilateral induction of a conductive hearing loss caused no significant difference in the volume of the contralateral CN between the various groups of animals. To determine age-dependent effects of the onset of a conductive hearing loss, we plotted the volume difference as a percentage of the undeprived left side versus the age at deprivation (Fig. 2). Positive values indicate a smaller volume on the deprived side. Because it is known that a deprivation is more effective in young as compared to older animals (e.g. Webster, 1988), we compared the volume difference in animals deprived at 12–14 days with the data from unoperated controls. The AVCN was the only subdivision where deprivation around the onset of hearing induced a significant reduction in volume (Mann Whitney U). In addition, the AVCN was the only subdivision where we found a significant and systematic age-dependent effect of the deprivation (Spearman rank correlation). The volume reduction due to the hearing loss was on average 20% if the deprivation occurred between the second and third week of life. Deprivation in young, adults (three-months) had no significant effect on the volume of AVCN.

4. DISCUSSION

It has been shown in the gerbil that cochlear removal (Hashisake and Rubel, 1989; Nordeen et al., 1983) severely affects AVCN volume and cell size and that these effects are more pronounced in young as compared to older animals. Furthermore, Pasic and Rubel (1989; 1991) demonstrated that the reduction of AVCN cell size occurs within a very short period after blocking neural activity of the auditory nerve by application of tetrodotoxin. They concluded that the loss of auditory nerve activity triggers the cascade of events leading to a reduction of soma size. An unresolved and, in recent years controversially-discussed topic, is whether a less severe than total reduction of auditory nerve activity also affects the development of the CN (Tucci and Rubel, 1985; Moore et al., 1989). This is a relevant issue for evaluating the detrimental potency of a conductive hearing loss during development, since normal spontaneous activity is present and only the auditory input to external stimuli is reduced by a conductive hearing loss (see discussion in Tucci and Rubel, 1985).

We used two strategies to address these questions. One was to identify and characterize maturational processes during normal development that proceed beyond the onset of

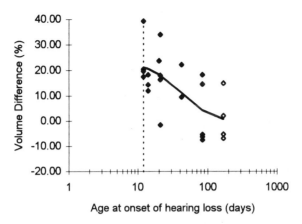

Figure 2. The volume difference between the deprived and undeprived AVCN as a function of the age at the onset of a conductive hearing loss. Open symbols indicate data from 6 months old unoperated control animals. The heavy black line shows a 3rd order polynomial fit.

hearing and might thus depend on the presence of elevated activity by external auditory stimulation. Second, we studied the age-dependent effects of a conductive hearing loss and relate these to the normal development. In addition to measurements of volume and cell size, we investigated the development of the inhibitory system by GABA- and glycine-immunostaining.

In the first two weeks after the onset of hearing, the staining intensity and thus the expression of GABA and glycine increased in most inhibitory somata from weak or moderate to heavy. Large glycinergic cells that probably project to the contralateral CN (Wenthold, 1987) increased in size well beyond pnd 12. Our observation that GABA- and glycinergic somata in the fusiform and molecular layers of the DCN showed a clear growth after the onset of hearing contrasts with the finding that there is only minimal increase in DCN volume over a comparable period. The effects of deprivation on these parameters are currently under investigation.

Our data on the reduction of CN volume after a conductive hearing loss in young gerbils are consistent with previous work in mice and rats (e.g. Webster, 1988; Coleman et al., 1982). The end of the sensitive period for a conductive hearing loss in gerbils occurs between 6 and 12 weeks (Fig. 2) and corresponds with the end of the normal postnatal growth (Fig. 1). The gerbil data support the view that the residual activity of the auditory nerve after a conductive hearing loss in 2–3 week old gerbils is not sufficient to provide for a normal development. The lack of increased activity (induced by external auditory stimulation) results on average in a reduced growth of the volume of the AVCN on the affected side in young animals but does not induce a decrease in volume in older animals.

In contrast, three studies performed in the chicken (Tucci and Rubel, 1985), ferret (Moore et al., 1989) and rhesus monkey (Doyle and Webster, 1991) did not reveal effects of a conductive type of hearing loss on cell size or on VCN volume. One suggestion was that a purely conductive hearing loss may not be effective and thus the observed reductions in the rodent studies would be due to inadvertant induction of a sensorineural hearing loss. We suggest alternative explanations for these discrepancies. Cell size, the only measure used by Tucci and Rubel (1985), appears less sensitive than measuring volumes of the CN subdivisions (Moore et al., 1989). Webster (1988) found a reduction of VCN volume but no change in soma size if a conductive hearing loss was introduced well after the onset of hearing. A previous study by Conlee and Parks (1981) observed cell-size reduction in chickens that received earplugs on E18–19. The discrepancies between both chicken studies may be related to a small age difference (E18–19 vs E19). In addition Conlee and Parks (1981) veryfied that the earplug filled the whole ear canal (without disrupting the tympanic membrane), while the earplugs in the Tucci and Rubel (1985) study never extended to the tympanic membrane. Earplugs filling the whole meatus are, however, more effective in reducing air-conducted sound (Tucci and Rubel; 1985). The study by Moore et al. (1989) found no effect of an earplug on the volumes of the CN subdivisions or on AVCN soma size in ferrets. However, their plugs induced only moderate attenuation below 3 kHz and the plugs had to be adjusted and replaced during the development of the animals in order to provide a tight seal. Thus the auditory input in these experiments may not have been reduced sufficiently to affect the growth of the CN. In the rhesus monkey, Doyle and Webster (1991) found only a small but not significant reduction of the VCN volume after induction of a long term conductive hearing loss in neonatal monkeys. They argued that the development of the CN in animals with precocial hearing may be fairly independent of the auditory input. In the related pigtailed monkey, however, there is only a very moderate postnatal growth of the CN as compared to humans or rodents (Sutton et al., 1991). If the growth pattern in the rhesus monkey is similar, one might not see a measurable effect of a conductive hearing loss on CN volume due to the small rate of postnatal growth. Finally, two elegant experiments argue that a purely conductive loss is sufficient to affect the development of the CN. Raising mice with a conductive hearing loss in a sound-amplified environment prevented the detrimental effects of the deprivation (Webster, 1988). In addition, raising mice (without manipulation in the auditory apparatus) in a low-noise environment by avocal mothers induced similar effects to those of a conductive hearing loss (Webster and Webster, 1979).

What are the conclusions for the situation in humans? A re-analysis of the data on the pre- and postnatal growth of the human VCN from Nara et al. (1993) and Konigsmark and Murphy (1972) demonstrates that the VCN volume almost tripels from below 3mm^3 at birth to 9mm^3 at an age of 10 years. Based on our data from gerbils and these observations of the postnatal growth of the human VCN, we suggest that in children, a conductive hearing loss may exert its largest adverse effects during the first 5 years of life.

5. ACKNOWLEDGMENTS

We thank G.A. Manley for his comments and improving the English and K. Bielenberg and S. Indlekofer for technical assistance. Supported by the DFG (Str. 275 / 2–1) and SFB 204 (Gehör) TP4.

6. REFERENCES

Blatchley, B.J., Williams, J.E. and Coleman, J.R. (1983) Age-dependent effects of acoustic deprivation on spherical cells of the rat anteroventral cochlear nucleus. Exp. Neurol. 80, 81–93.

Casselbrant, M.L., Brostoff, L.M., Cantekin, E.I., Flaherty, M.R., Doyle, W.J., Bluestone, C.D. and Fria, T.J. (1985) Otitis media with effusion in preschool children. Laryngoscope 95, 428–436.

Coleman, J.R., Blatchley, B.J. and Williams, J.E. (1982) Development of the dorsal and ventral cochlear nuclei in rat and effects of acoustic deprivation. Dev. Brain Res. 4, 119–123.

Conlee, J.W. and Parks, T.N. (1981) Age- and position-dependent effects of monaural acoustic deprivation in nucleus magnocellularis of the chicken. J. Comp. Neurol. 202, 373–384.

Doyle, W.J. and Webster, D.B. (1991) Neonatal conductive hearing loss does not compromise brainstem auditory function and structure in rhesus monkeys. Hear. Res. 54, 145–151.

Gleich, O., Kadow, C. and Strutz, J. (1997) The postnatal growth of principal cell somata in the AVCN of the Mongolian gerbil. Abstr. 20th ARO Midwinter Res. Mtg. p 137.

Gravel, J.S., Wallace, I.F. and Ruben, R.J. (1995) Early otitis media and later educational risk. Acta Otolaryngol. Stockh. 115, 279–281.

Hashisaki, G.T. and Rubel, E.W. (1989) Effects of unilateral cochlea removal on anteroventral cochlear nucleus neurons in developing gerbils. J. Comp. Neurol. 283, 465–473.

Konigsmark, B.W. and Murphy, E.A. (1972) Volume of the ventral cochlear nucleus in man: Its relationship to neuronal population and age. J. Neuropath. Exp. Neurol. 31, 304–311.

Ling, D. and Ling, A.H. (1978) Aural habilitation: The foundation of verbal learning in hearing-impaired children. The Alexander Graham Bell Association for the Deaf Inc., Washington.

Lous, J. (1995) Secretory otitis media in schoolchildren. Dan. Med. Bull. 42, 42–99.

Moore, D.R., Hutchings, M.E., King, A.J. and Kowalchuk, N.E. (1989) Auditory brain stem of the ferret: Some effects of rearing with a unilateral ear plug on the cochlea, cochlear nucleus, and projections to the inferior colliculus. J. Neurosci. 9, 1213–1222.

Moore, D.R. (1990) Auditory brainstem of the ferret: Early cessation of developmental sensitivity of neurons in the cochlear nucleus to removal of the cochlea. J. Comp. Neurol. 302, 810–823.

Moore, D.R. (1994) Auditory Brainstem of the ferret: long survival following cochlear removal progressively changes projections from the cochlear nucleus to the inferior colliculus. J. Comp. Neurol. 339, 301–310.

Nara, T., Goto, N., Nakae, Y. and Okada, A. (1991) Morphometric development of the human auditory system: ventral cochlear nucleus. Early Human Dev. 32, 93–102.

Nordeen, K.W., Killackey, H.P. and Kitzes, L.M. (1983) Ascending projections to the inferior colliculus following unilateral cochlear ablation in the neonatal gerbil, *Meriones unguiculatus*. J. Comp. Neurol. 214, 144–153.

Pasic, T.R. and Rubel, E.W. (1991) Cochlear nucleus cell size is regulated by auditory nerve electrical activity. Otolaryngol. Head Neck Surg. 104, 6–13.

Pasic, T.R. and Rubel, E.W. (1989) Rapid changes in cochlear nucleus cell size following blockade of auditory nerve electrical activity in gerbils. J. Comp. Neurol. 283, 474–480.

Trune, D.R. (1982) Influence of neonatal cochlear removal on the development of mouse cochlear nucleus: I. number, size and density of its neurons. J. Comp. Neurol. 209, 409–424.

Tucci, D.L. and Rubel, E.W. (1985) Afferent influences on brain stem auditory nuclei of the chicken: Effects of conductive and sensorineural hearing loss on N. magnocellularis. J. Comp. Neurol. 238, 371–381.

Webster, D.B. and Webster, M. (1979) Effects of neonatal conductive hearing loss on brain stem auditory nuclei. Ann. Otol. Rhinol. Laryngol. 88, 684–688.

Webster, D.B. (1988) Conductive hearing loss affects the growth of the cochlear nuclei over an extended period of time. Hear. Res. 32, 185–192.

Webster, D.B. (1988) Sound amplification negates the central effects of a neonatal conductive hearing loss. Hear. Res. 32, 193–195.

Wenthold, R.J. (1987) Evidence for a glycinergic pathway connecting the two cochlear nuclei: An immunohisto-chemical and retrograde transport study. Brain Res. 415, 183–187.

Wenthold, R.J. (1991) Neurotransmitters of brainstem auditory nuclei. In: R.A. Altschuler, R.P. Bobbin, B.M. Clopton and D.W. Hoffman (eds) Neurobiology of Hearing: The central auditory system. New York: Raven Press, pp. 121–139.

Zeisel, S.A., Roberts, J.E., Gunn, E.B., Riggins, R. Jr., Evans, G.A., Roush, J., Burchinal, M.R. and Henderson, F.W. (1995) Prospective surveillance for otitis media with effusion among black infants in group child care. J. Pediatr. 127, 875–880.

16

NORADRENERGIC MODULATION OF THE PROCESSING OF AUDITORY INFORMATION INSIDE THE COCHLEAR NUCLEI

Coupled Biochemical and Functional Analysis Using DSP-4 Neurotoxin in Adult Rats

H. Cransac,[1,2] S. Hellström,[2] L. Peyrin,[1] and J. M. Cottet-Emard[1]

[1]UPRESA CNRS 5020 / Laboratoire de Physiologie
Faculté de Médecine
8 avenue Rockefeller, F-69373 Lyon, France
[2]Department of Otorhinolaryngology
University Hospital of Umeå
S-901 85 Umeå, Sweden

1. INTRODUCTION

Cochlear nuclei receive a dense noradrenergic innervation originating mainly from the locus coeruleus (Klepper and Herbert, 1991). We have previously documented the high turnover rate of noradrenaline in the antero-ventral cochlear nucleus (AVCN) and in the posterior part of the nucleus in Sprague-Dawley rat at the basal state level (Cransac et al., 1995).

In agreement with previous histological findings (Klepper and Herbert, 1991), noradrenaline is more concentrated in AVCN than in the dorsal cochlear nucleus and postero-ventral cochlear nucleus (DCN + PVCN); however, the turnover of noradrenaline was faster in the dorsal nuclei. These "basal state" data have suggested that the functional involvement of monoamines may be different in cochlear subnuclei.

Another biochemical study in young and old animals showed that noradrenergic neurons in sensory nuclei seems to be differently affected by aging possibly as a compensatory response to dysfunction of sensory input and processing. The main noradrenergic changes in brainstem auditory structures of the aged animals were a selective noradrenergic increase in AVCN. In contrast there was no significant change in the DCN + PVCN. The increase of noradrenaline stores in the AVCN of aged rats is in line with the elevated auditory brainstem threshold reported in old rats and could improve the signal to noise ratio (Cransac et al., 1996).

The functional implication of this noradrenergic innervation originating from the locus coeruleus on the processing of auditory information remains quite speculative.

Acoustical Signal Processing in the Central Auditory System
edited by Syka, Plenum Press, New York, 1997

The arrangement of monoaminergic terminals within the various parts of cochlear nuclei suggest that noradrenaline exerts a modulatory influence on the processing of auditory information (Klepper and Herbert, 1991). Local application of noradrenaline to the surface of the cochlear nuclei led to an increase in absolute and masked auditory thresholds in a behavioural paradigm (Pickles, 1976). Electrical stimulation of the locus coeruleus inhibited DCN neurons, presumably by noradrenaline release within DCN (Chikamori et al., 1980). The differential distribution of noradrenergic fibres within the cochlear nucleus has been related in bats to the tonotopic frequency representation: areas devoted to high-frequency sounds are less densely innervated than those responsive to lower frequencies. In the cochlear nuclei of rats and mice there were no such morphological heterogeneities (Klepper and Herbert, 1991; Kössl et al., 1988).

This experimentation was conducted in order to evaluate the role of locus coeruleus noradrenergic regulation on the processing of auditory information in the cochlear nuclei. We have explored the auditory brainstem response (ABR) to pure tone stimuli and the biochemical content of noradrenaline in cochlear nuclei after destruction of the noradrenergic innervation emanating from locus coeruleus by the mean of DSP-4 neurotoxin.

2. MATERIAL AND METHODS

2.1. Animals

Male Sprague Dawley rats (245±8 g) were housed in a temperature-controlled room at $22 \pm 1°$ C with a 12:12 h light-dark cycle and were allowed free access to food and water. All animal experiments were carried out according to the recommendations of Declaration of Helsinki and the Animal Welfare Guidelines of the Society for Neuroscience, February 1992. The experiments were approved by the local ethical committee for use of animals in research.

2.2. DSP-4 treatment

To perform the lesion we have used the N-(2-chloroethyl)-N-ethyl-2-bromobenzylamine (DSP-4, Research Biochemical International, Natick, USA), one single 50 mg/kg i.p. injection 15 days before experimentation.

2.3. Tissue Preparation and Biochemical Analysis

The rats were killed by cervical dislocation and then decapitated, 8 animals had received the DSP-4 and 8 animals an equivalent volume of saline solution 9‰. The brains were rapidly dissected out, frozen on dry ice and stored at - 80°C. Serial coronal sections (320 μm) were cut in a cryostat with the decussation of seventh cranial nerve as an anatomical reference. Individual nuclei were dissected using micro punches technique (Palkovits and Brownstein, 1988) using a needle and the atlas of Paxinos and Watson as a general guide (1986). We dissected the antero-ventral cochlear nuclei (AVCN) and the two parts of the posterior part of the cochlear nuclei i.e. postero-ventral cochlear nuclei (PVCN) + dorsal cochlear nuclei (DCN). The punches were stored at - 80°C until assay.

Noradrenaline was assayed by HPLC. The punches were homogenized by ultrasonic disruption over ice and the excess of perchloric acid was removed by KOH / formate. The homogenates were centrifuged and 40 μl of the supernatant without any further purification were injected into the HPLC; for details see Cransac et al., 1995.

2.4. ABR Recordings

The ABR thresholds were studied in the same group of 8 Sprague-Dawley rats before treatment with DSP-4 and 15 days after the injection of the drug. The animals were anesthetized with sodium pentobarbital (40 mg/kg i.p.) given intraperitoneally, supplementary doses of 30 mg/kg i.p. were given every 30 min.

The rat was placed in a prone position in a dark, electrically shielded specially designed soundproof box (Tegner AB, Sweden). Body temperature was kept constant at 37°C with a thermostat-controlled heating pad.

Stimuli were delivered via a Beyer DT 48 telephone through a probe in the right ear canal. The stimulus consisted of tone bursts of alternating polarity with a trapezoidal envelope of 3ms overall duration, with linear rise and fall times of 1 ms presented at a rate of 16/s (Tektronix SG 505 Oscillator and Somedic TMS 110 Gate). The attenuator (Somedic TMS 111) covered a range of 0–90 dB and had an approximatively ± 100 Hz bandwidth gating. The frequencies tested were 2, 4, 6, 8, 10, 12, 16, 20 and 31.5 kHz. The response signals were recorded with subcutaneous needle electrodes. The active electrodes were placed in the vertex and in the right front limb with the ground electrode in the left front limb. Cardiac activity was monitored altrough the experiments.

For each measurement, 1024 epochs were stored in an averager after 20000-fold amplification and 100–3000 Hz bandwidth filtering. Signals exceeding 50 μV were automatically rejected before storage in the memory. Each series of stimuli began at a supra-threshold level and the sound pressure level was decreased in 5 dB steps until wave II of the ABR responses could not be detected. The ABR thresholds were defined as the lowest value in 5 dB attenuator steps giving response.

Statistics are conducted using non-parametric Wilcoxon signed rank test for paired values.

3. RESULTS

3.1. Biochemical Effects of DSP-4

Table: Biochemical variations due to DSP-4 treatment on the endogenous contents of noradrenaline in the locus coeruleus, the antero-ventral cochlear nucleus (AVCN), the dorsal cochlear nuclei and the postero-ventral cochlear nuclei (DCN + PVCN) and the frontal cortex. Results are expressed in % versus saline treated animals.

Fifteen days after DSP-4 treatment the noradrenergic content of the AVCN and DCN + PVCN decreased 55 % and 68 % respectively when compared to the saline group treated animals. The noradrenergic contents of the frontal cortex decreased by 69 %, and this structure was used as a comparison being a well known target structure of locus coeruleus innervation.

3.2 Effect of DSP-4 on Auditory Brainstem Response

The ABR study in anaesthetised animals demonstrated significant changes in the threshold level at the intermediate frequencies (4, 6, 12, 16 kHz) before and after DSP-4 treatment. For the frequencies 2, 20 and 31.5 kHz the thresholds were not significantly changed. The maximum shift in ABR threshold (-36 ± 8 % when compared to the saline treated animals) occurred for 12 kHz pure tone stimulation.

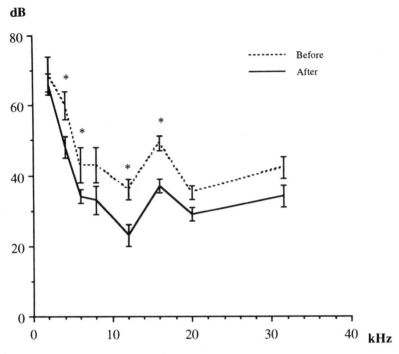

Figure 1. Means ± standard error of the ABR thresholds before and 15 days after DSP-4 treatment. * p<0.05.

4. DISCUSSION

4.1. Biochemistry

Axonal noradrenergic fibers inside the cochlear nuclei represent 93 % of the noradrenergic innervation of cochlear nuclei (Klepper and Herbert, 1991). Biochemical results show that the maximal decrease in terms of content of the cochlear nuclei after chemical destruction of the fibers emanating from the locus coeruleus is 68 % in the PVCN + DCN. Similar decrease is observed in the frontal cortex (-69 %) which receives a well known noradrenergic innervation originating exclusively from the locus coeruleus.

DSP-4 has been demonstrated to be a specific neurotoxin which selectively destroy axonal nerve fibers coming from locus coeruleus in young and old Sprague-Dawley rats (Schuerger and Balaban, 1995; Riekkinen et al., 1992), and causes ablation of nearly all locus coeruleus axons terminals within two weeks after administration (Fritschy and Grzanna, 1992b). Regarding selectivity the drug is a highly selective noradrenergic neurotoxin with little effect on other neurotransmitters (Zhong et al., 1985).

Even if there is an extensive recovery of noradrenergic axons in forebrain areas 2 weeks after DSP-4 treatment it has been shown that there is a lack of noradrenergic axon reinnervation of the cerebellum and brainstem (Fritschy and Grzanna, 1992b), and that there was a nearly complete loss of noradrenaline axon staining in sensory nuclei in the brainstem (Fritschy and Grzanna, 1992a).

In cochlear nuclei, 15 days after DSP-4 treatment at least one third of the level of noradrenaline remains. Since there is no reinnervation of the brainstem by locus coeruleus, the axonal fibers of the locus which has not been destroyed are in part responsible of the remaining

Table 1. Biochemical variations due to DSP-4 treatment on the endogenous contents of noradrenaline in the locus coeruleus, the antero-ventral cochlear nucleus (AVCN), the dorsal cochlear nuclei and the postero-ventral cochlear nuclei (DCN + PVCN) and the frontal cortex. Results are expressed in % versus saline treated animals

	Cochlear nuclei		
	AVCN	DCN + PVCN	Frontal cortex
Noradrenaline	- 55 %	- 68 %	- 69 %

content of noradrenaline, but furthermore, fibers originating from other noradrenergic cell groups than locus coeruleus could be also involved. Two noradrenergic cell groups located in the pons - i.e. A5 and A7 noradrenergic cell groups - give rise to 7 % of noradrenergic fibers found in the cochlear nuclei (Klepper and Herbert, 1991). In the saline treated animals this innervation is not supposed to be strongly involved in the total amount of noradrenaline which could be detected. After the neurotoxic treatment these fibers could also be hyper-active in a compensatory process of the missing noradrenergic innervation of locus coeruleus and be strongly involved in the remaining levels of noradrenaline.

The global content of noradrenaline in one structure correspond to both endogenous stores and extracellular pool which is more likely related to the functionality of the innervation. Then the decrease of global noradrenaline content in one particular nuclei could not be directly related to such a decrease of the extracellular levels since it has been shown that 50 % loss of noradrenergic content in the hippocampus unaffected the extracellular levels of noradrenaline (Abercrombie and Zigmond, 1989). But we assumed that such a perturbation of noradrenergic fibers inside cochlear nuclei could affect the role of these structures on the processing of auditory information, so we test the functionality of auditory pathways after depletion of noradrenergic content of cochlear nuclei.

4.2. Auditory Brainstem Response

The functional study showed a decrease of the threshold to pure tone stimuli after removal of a great part of the noradrenergic innervation of the cochlear nuclei for frequencies 4, 6, 12 and 16 kHz. Moreover, this noradrenergic control of the incoming information in central auditory pathway appears to be connected with the frequencies at which the rat is most sensitive.

But a central compensatory mechanisms could have ocurred between the injection of the drug and the ABR recording period since it was shown, after DSP-4 treatment that a marked alpha-2 adrenoreceptor proliferation occurs in the rat brain within two weeks after noradrenergic lesioning (Heal et al., 1993). This increase of alpha-2 adrenergic receptors is much more likely to derive from receptor proliferation on post-synaptic sites than the regeneration of prejunctional autoreceptors. In this case the functional effect that was observed after destruction of noradrenergic terminals could result either in a lack of noradrenergic regulation, or in an hypersensitivity of post-synaptical target cells and receptors to the resting noradrenergic activity. To confirm one of these hypotheses further experiments should be conducted, acute pharmacological manipulation of alpha-2 adrenergic receptors for instance will give more information and allow to test the thresholds before and after treatment without inbetween compensatory period.

According to the fact that in the AVCN and PVCN noradrenergic fibers are located nearby the fibers of the auditory nerve (Klepper and Herbert, 1991), our results strongly suggests that the locus coeruleus noradrenergic control of the processing of auditory information occurs at the first synaptic level of the central auditory pathways, i.e. in the cochlear nuclei. This regulation could be to extract the auditory signal out of the background information, by producing an enhanced resolution of the neuronal signal-to-noise ratio, this noradrenergic modulation by locus coeruleus of auditory processing could not be involved in the same way for all frequencies.

ACKNOWLEDGMENTS

Supported by European Neuroscience Programme 1995 (Grant N° 201), INSERM and Swedish Medical Research Council.

5 REFERENCES

Abercrombie, E.D. and Zigmond, M.J. (1989) Partial injury to central noradrenergic neurons: reduction of tissue norepinephrine content is greater than reduction of extracellular norepinephrine measured by microdialysis. J. Neurosci. 9, 4062–4067.

Chikamori, Y., Sasa, M., Fujimoto, S., Takaori, S. and Matsuoka, I. (1980) Locus coeruleus induced inhibition of dorsal cochlear nucleus neurons in comparison with lateral vestibular nucleus neurons. Brain Res. 194, 53–63.

Cransac, H., Cottet-Emard, J.M., Pequignot, J.M. and Peyrin, L. (1995) Monoamines (noradrenaline, dopamine, serotonin) in the rat cochlear nuclei: endogenous levels and turnover. Hear. Res. 90, 65–71.

Cransac, H., Peyrin, L., Cottet-Emard, J.M., Farhat, F., Pequignot, J.M. and Reber, A. (1996) Aging effects on monoamines in rat medial vestibular and cochlear nuclei. Hear. Res. 100, 150–156.

Fritschy, J.M. and Grzanna, R. (1992a) Degeneration of rat locus coeruleus neurons is not accompanied by an irreversible loss of ascending projections. Evidence for reestablishment of forebrain innervation by surviving neurons. Ann. N. Y. Acad. Sci. 648, 275–278.

Fritschy, J.M. and Grzanna, R. (1992b) Restoration of ascending noradrenergic projections by residual locus coeruleus neurons: compensatory response to neurotoxin-induced cell death in the adult rat brain. J. Comp. Neurol. 321, 421–441.

Heal, D.J., Butler, S.A., Prow, M.R. and Buckett, W.R. (1993) Quantification of presynaptic alpha 2-adrenoceptors in rat brain after short-term DSP-4 lesioning. Eur. J. Pharmacol. 249, 37–41.

Klepper, A. and Herbert, H. (1991) Distribution and origin of noradrenergic and serotoninergic fibers in the cochlear nucleus and inferior colliculus of the rat. Brain Res. 557, 190–201.

Kössl, M., Vater, M. and Schweizer, H. (1988) Distribution of catecholamine fibers in the cochlear nucleus of horseshoe bats and mustache bats. J. Comp. Neurol. 269, 523–534.

Palkovits, M. and Brownstein, M.J. (1988) Maps and guide to microdissection of the rat brain. Elsevier, New York.

Paxinos, G. and Watson, C. (1986) The rat brain in stereotaxic coordinates. Academic Press, Orlando.

Pickles, J.O. (1976) The noradrenaline containing innervation of the cochlear nucleus and the detection of signals in noise. Brain Res. 105, 591–596.

Riekkinen, P.J., Riekkinen, M., Valjakka, A., Riekkinen, P. and Sirvio, J. (1992) DSP-4, a noradrenergic neurotoxin, produces more severe biochemical and functional deficits in aged than young rats. Brain Res. 570, 293–299.

Schuerger, R.J. and Balaban, C.D. (1995) N-(2-chloroethyl)-N-ethyl-2-bromobenzylamine (DSP-4) has differential efficacy for causing central noradrenergic lesions in two different rat strains: Comparison between Long-Evans and Sprague-Dawley rats. J. Neurosci. Methods 58, 95–101.

Zhong, F.X., Ji, X.Q. and Tsou, K. (1985) Intrathecal DSP4 selectively depletes spinal noradrenaline and attenuates morphine analgesia. Eur. J. Pharmacol. 116, 327–330.

17

INHIBITORY AND EXCITATORY BRAINSTEM CONNECTIONS INVOLVED IN SOUND LOCALIZATION: HOW DO THEY DEVELOP?

Eckhard Friauf,[*] Karl Kandler, Christian Lohmann, and Martin Kungel

Zentrum der Physiologie
Klinikum der Universität Frankfurt
Theodor-Stern-Kai 7, D-60590 Frankfurt, Germany

INTRODUCTION

Interaural time differences and interaural intensity differences are the two major cues that enable vertebrates to localize the direction of a sound source. Small mammalian species with a correspondingly small head width (distance between the two pinna ca. 2–4 cm), such as most rodents, do not experience interaural time differences longer than 60–120 μs and, therefore, they generally rely on interaural intensity differences (IID). High frequency hearing in the ultrasound range, which is common to these small animals, is advantageous to localize sound sources, because the wave lengths of ultrasounds are too short to bend around the head and because sound shadowing by the head becomes better with increasing sound frequency.

The neural substrate for analyzing the information from both ears and converting IID into neuronal activity are parallel fiber tracts originating from the cochlear nuclear complex on both sides of the brainstem plus their target neurons in the superior olivary complex (SOC). The lateral superior olive (LSO) is the postsynaptic target of these convergent inputs. The ipsilateral afferents to the LSO are excitatory and most likely glutamatergic (Boudreau and Tsuchitani, 1968; Goldberg and Brown, 1969; Suneja et al., 1995), whereas the contralateral afferents, which are relayed via the medial nucleus of the trapezoid body (MNTB), are inhibitory and glycinergic (Aoki et al., 1988; Bledsoe et al., 1990; Helfert et al., 1992; Wu and Kelly, 1992). The two types of afferents result in a predominant response type in the LSO of ipsilateral excitatory/contralateral inhibitory (EI; Tsuchitani, 1977). In fact, among all brainstem neurons, LSO neurons are particularly densely decorated with glycine receptors (Friauf et al., 1994, 1997). Importantly, the best frequen-

* Correspondence to the above address. Tel: +49 69 6301-6020; fax: +49 69 6301-6023; E-mail: friauf@em.uni-frankfurt.de.

cies of the excitatory and inhibitory afferents converging on individual LSO neurons are closely matched (Caird and Klinke, 1983; see also Figure 1). Low and high frequencies are represented laterally and medially, respectively, resulting in a tonotopic organization (Guinan et al., 1972; Sanes et al., 1990). As the excitatory and inhibitory pathways are anatomically distinct, the microcircuits are well suited to address principle questions regarding the mechanisms of their development. How these precise topographic connections (excitatory as well as inhibitory) are wired up during development is still an enigma. In this review we want to summarize recent findings from our laboratory and those of others that have contributed to our current understanding on the mechanisms of development and allowed us to draft a working hypothesis for future experiments.

DOES NEURONAL ACTIVITY AFFECT THE DEVELOPMENT OF THE MICROCIRCUITS?

Over the past decade, work on the maturation of the auditory brainstem nuclei has concentrated on ferrets, rats, and gerbils. These animals are well-suited to address physiological and morphological aspects of the developing LSO microcircuit, because they are still deaf at birth and have a long postnatal period of neural maturation. In rats and gerbils, for example, physiological hearing does not begin until the end of the second postnatal week (postnatal days P12-P14; e.g. Uziel et al., 1981; Blatchley et al., 1987; Puel and Uziel, 1987; Kelly, 1992). Morphological data obtained in gerbils by Dan Sanes´ group in New York have demonstrated that the development of precision in the glycinergic input to the LSO requires proper neurotransmission. A functional denervation of this input, either by applying the glycine receptor antagonist strychnine or by surgically removing the contralateral cochlea, results in abnormal dendritic arbors of LSO neurons and also of the terminal fields of the MNTB axons (Sanes and Chokshi, 1992; Sanes et al., 1992; Sanes and

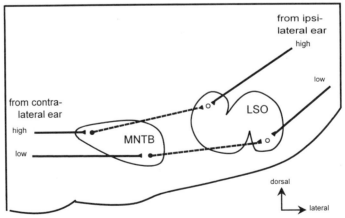

Figure 1. Schematic representation of the rat´s superior olivary complex illustrating the topographically organized input connections to the lateral superior olive (LSO). Glutamatergic input (closed lines) from the ipsilateral ear and glycinergic input (dashed lines) from the contralateral ear, which is relayed by the medial nucleus of the trapezoid body (MNTB), converge in the LSO in a highly topographic fashion.

Takacs, 1993). Similar results were found by David Moore and coworkers in Oxford (e.g. Moore, 1992). Thus it appears that activity-dependent processes play a considerable role in the maturation of the inhibitory input to the LSO. From the activity-dependent maturation of excitatory connections it was concluded that pre- and postsynaptic activity has to occur in conjunction in order to stabilize a synapse (Constantine-Paton et al., 1990; Goodman and Shatz, 1993; Shatz, 1994; see O'Leary et al., 1994, for an essay on the controversial role of neural activity). Such Hebb synapses (Hebb, 1949) are almost certainly present in the visual cortex during the period in which ocular dominance columns form (Fox and Zahs, 1994). In analogy, an inhibitory input neuron, such as an MNTB neuron, would act antagonistically, i.e. its transmitter release would cause a hyperpolarization of the membrane potential in its target neurons through opening chloride channels, thereby decreasing neuronal activity. Thus, the mechanism of an activity-induced depolarization seen at excitatory synapses, which activates a cascade that involves an increase of the intracellular free calcium concentration and ultimately results in synapse augmentation, should be quite different from that present in inhibitory synapses. We will come back to this point later on in the review. At first, however, we will report on the initial steps of SOC development, such as neurogenesis and axon outgrowth.

NEUROGENESIS OF SOC NEURONS

In rats, neurons of the major SOC nuclei, i.e. the LSO, MNTB, medial superior olive (MSO), and superior paraolivary nucleus (SPN), are generated prenatally. We analyzed neurogenesis, defined by the last mitotic division of precursor cells, by means of the 5-bromo-2'-deoxyuridine (BrdU) technique. BrdU is a thymidine analogue which, when applied to the tissue, is incorporated into the DNA of mitotic cells and can be detected immunocytochemically (Miller and Nowakowski, 1988). In order to determine the time course of neurogenesis in the SOC, we injected a single pulse of BrdU intraperitoneally into gravid rats between embryonic days E10-E21 (day of conception equals E0, day of birth equals E22, see Figure 2) or into rat pups between postnatal days P0-P3 and performed immunocytochemical reactions on the brains at juvenile ages (P12-P20). The time frames that we found for the production of SOC neurons differ only slightly between the individual nuclei. For example, neurons in the LSO and the SPN are born between E12 and E14, those in the MSO at E11-E12, and those in the MNTB at E15. These results

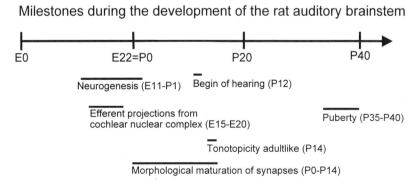

Figure 2. Temporal pattern of major developmental changes in the rat auditory brainstem.

show that rat SOC neurons are generated more than one week before the animals are born, consistent with previous results obtained with tritiated thymidine labeling (Altman and Bayer, 1980). Since the great majority of CN neurons are also generated between E13-E15 (Altman and Bayer, 1980; Weber et al., 1991; Friauf et al., 1993), there is considerable overlap in the production of SOC and CN neurons. Taken together, all essential neuron populations that compose the IID sound localization microcircuit in the rat appear to be generated at about the same time, i.e. within a short period of 5 days in the second week of fetal life.

FORMATION OF AXONAL PROJECTIONS

Once postmitotic, neuroblasts must migrate from their generation site to reach their final location within the nervous system and must grow axonal projections toward their targets. We have not addressed the issue of cell migration in the auditory brainstem, but it appears that this developmental step follows neurogenesis with hardly any delay, because outgrowing axons from the CN can be seen to have entered the ventral acoustic stria and crossed the midline as early as E15, i.e. maximally 2 days after the first generation of CN neurons is born. We investigated axon growth in fetal rats by using crystals of the lipophilic carbocyanine dye DiI which were implanted into the anlage of the CN of aldehyde-fixed brains under visual control. Thereafter, DiI was allowed to diffuse anterogradely within the axonal membranes (Kandler and Friauf, 1993). Remarkably, we found that the process whereby axons initially select the correct neuronal targets is highly precise from the outset, because we detected no aberrant projections into nuclei that are not innervated by CN neurons in the adult. The results of the DiI labeling study also showed that axon collaterals, forming highly branched axon arbors in the ipsilateral SOC, are present as early as E18. As numerous axonal swellings were found on these arbors, this suggested to us that functional synapses might be present at this early stage.

ELECTROPHYSIOLOGICAL RECORDING AND FUNCTIONAL ASPECTS OF SYNAPSE MATURATION

In order to investigate whether functional synapses are indeed present in the rat´s SOC nuclei during early development, we performed electrophysiological recordings from in vitro brainstem slice preparations in fetal and early postnatal rats (Kandler and Friauf, 1995). The recording micropipettes had been filled with potassium acetate in order to circumvent the problem of a perturbed intracellular chloride concentration that may affect the glycine-gated responses which are mediated by Cl⁻ ions (Langosch et al., 1990; Aprison et al., 1996). Intracellular recordings were obtained from LSO neurons which were morphologically identified by the injection of biocytin and the ipsilateral, glutamatergic and contralateral, glycinergic inputs were activated electrically in order to elicit postsynaptic responses (PSPs). We observed several age-dependent changes in the synaptic transmission, most strikingly in the glycinergic pathway. Both types of input are functional as early as E18 in LSO neurons. Regardless of age, ipsilaterally elicited PSPs were always depolarizing. They had a positive reversal potential and could be blocked by the non-NMDA receptor antagonist CNQX. In contrast, contralaterally elicited PSPs were depolarizing from E18 to P4 and able to elicit action potentials before P8. As these contralateral PSPs could be reversibly blocked by strychnine, they were identified as glycine-evoked PSPs. Whereas the resting membrane potential of LSO neurons remained constant at -60 mV throughout the period of investigation, the reversal potential of the glycinergic PSPs changed during development. It was positive to -20 mV before E22, became more negative with increasing age, reached the value of the membrane potential at around P8, and shifted to -73 mV by P10. Application of glycine and its agonist β-alanine to the bath solution confirmed the transitory depolarizing action of glycine (Kandler and Friauf, 1995; Backus and Friauf, 1996). The responses were associated with a considerable increase in conductance and, thus, it was conceivable that the depolarizing effect might result in a shunting inhibition (Nishi et al., 1974; Nicoll and Alger, 1980). However, as the depolarization of the membrane potential was sufficient to elicit action potentials in a significant number of LSO neurons around birth, following both synaptically induced and pharmacologically induced activation, the action of glycine was indeed excitatory and did not result in a shunt. Thus, it can be concluded that glycine acts as an excitatory neurotransmitter during early development at a time when the synaptic connections in the LSO are morphologically refined. By the time of hearing onset, glycinergic neurotransmission has become hyperpolarizing and the microcircuits have reached a stage of relative physiological maturity that allows them to perform their basic function of IID analysis.

A transient period of depolarizing action during early development has also been seen in several other neuronal systems and was described not only for glycine, but also - and even more often - for GABA (Obata et al., 1978; Barker et al., 1982; Mueller et al., 1984; Janigro and Schwartzkroin, 1988; Cherubini et al., 1990, 1991; Luhmann and Prince, 1991; Wu et al., 1992; Ben-Ari et al., 1994; Gao and Ziskind-Conhaim, 1995; Loturco et al., 1995; Obrietan and van den Pol, 1996; Owens et al., 1996). These observations suggest that a transient excitatory effect of the two major inhibitory neurotransmitters in the central nervous system may be a common phenomenon during development. It is thought that a decreased effectiveness of the chloride extrusion machinery in the immature neurons results in a positive shift of the chloride equilibrium potential and, thereby, in depolarizing PSPs. In adult animals, this extrusion machinery actively transports chloride ions out of the cells (most probably via a Cl^-/K^+ co-transport) and maintains the high extracellular/low intracellular chloride gradient across the neuronal membrane which is necessary to obtain hyperpolarizing responses (Lux, 1971; Thompson et al., 1988, 1989). In line with this assumption, a blockade of the outward transport of chloride ions by furosemide or a stimulation by raising the temperature affected the reversal potential of GABA-induced PSPs (Misgeld et al., 1986; Zhang et al., 1991). So far, we have not performed such manipulations of the chloride gradient in LSO neurons to see if this would affect their responses to glycine and, therefore, we cannot say whether developmental changes in the intracellular chloride concentration contribute to the observed switch from depolarizing to hyperpolarizing activity in these cells. Recently, such age-related decreases of the intracellular chloride concentration have been reported for early postnatal rat neocortical neurons (Owens et al., 1996). As optical imaging of intracellular chloride is now possible (Schwartz and Yu, 1995), it would be interesting to use this technique in the developing auditory brainstem nuclei in order to address the question of whether an elevated chloride concentration is present in the cytoplasm of perinatal LSO neurons.

In line with the observed depolarizing effects of GABA and glycine in immature neurons, some authors have reported that these transmitters can also induce a rise of the intracellular free calcium concentration ($[Ca^{2+}]_i$) via activation of voltage-gated calcium channels (Brown and Feldman, 1978; Yuste and Katz, 1991; Segal, 1993; Lin et al., 1994; Reichling et al., 1994; Wang et al., 1994; Leinekugel et al., 1995). Interestingly, the ability to elevate $[Ca^{2+}]_i$ appears to be transient and restricted to early developmental stages, and a developmental reversal from Ca^{2+} elevating to depressing has been reported (Obrietan and van den Pol, 1995). Based on these findings it was concluded that the rise in $[Ca^{2+}]_i$, as it is induced by GABA and glycine, may underly the trophic role of these 'inhibitory' neurotransmitters in the synapse formation. We think that such a role may also be attributable to glycine during the process of synapse refinement in the LSO, which is characterized by a loss of exuberant dendritic branches, predominantly taking place between P4 and P10 (Rietzel et al., 1995).

HYPOTHESIS: EXCITATORY ACTION OF GLYCINE PLAYS A DEVELOPMENTAL ROLE

From the above findings we hypothesize that the inhibitory synapses in the LSO develop through a mechanism which starts with depolarizing activity, similar to that present in classical excitatory synapses. If this assumption is true, the glycinergic and glutamatergic inputs to LSO neurons are likely to act synergistically, not antagonistically, during the

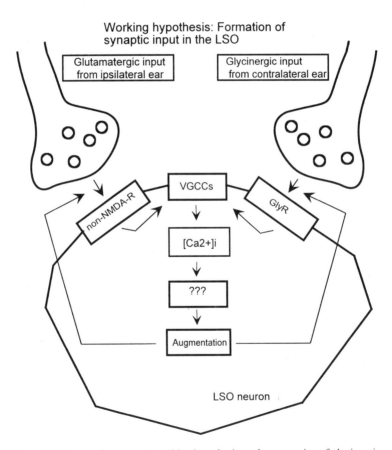

Figure 3. Schematic diagram of our current working hypothesis on the maturation of glycinergic and glutamatergic input to the LSO. Both types of input are depolarizing in perinatal animals and may, therefore, result in the opening of voltage-gated Ca^{2+} channels (VGCCs), inducing an influx of Ca^{2+} into the LSO neurons and finally leading to an activity-dependent augmentation of the synapses.

period of synapse maturation (Figure 3). The depolarizing action of glycine may gate volt-age-sensitive Ca^{2+} channels, thereby triggering signal cascades that require Ca^{2+} as a second messenger. By means of Fura2 measurements following K^+-induced depolarizations (Schmanns and Friauf, 1994) as well as voltage-clamp recordings we could show that SOC neurons of newborn rats indeed have functional Ca^{2+} channels with an activation threshold of -40 mV (Kungel and Friauf, 1996). As the resting potential in these neurons is ca. -60 mV and as the amplitudes of glycine-induced PSPs can reach 25 mV (Kandler and Friauf, 1995), these channels may, therefore, be indirectly gated by glycine. In addition, the opening threshold of -40 mV indicates that they belong to the subtype of high-threshold channels with long-lasting opening times.

INVOLVEMENT OF CA^{2+} IONS IN DEVELOPMENT

As to the present, it is unclear whether glycine receptor activation results in an in-flux of Ca^{2+} ions via gating voltage-sensitive Ca^{2+} channels during the period in which glycine is depolarizing in the SOC (i.e. before P8). We need to do calcium imaging experiments in brain slices with Ca^{2+}-sensitive dyes to address this issue and to analyze the possible role of Ca^{2+} ions during the maturation of the glycinergic input from the MNTB to the LSO. Nevertheless, there is some indirect evidence for an involvement of Ca^{2+} ions in synapse maturation within the SOC. This evidence comes from our findings that LSO neurons are heavily, yet only transiently, immunopositive for the Ca^{2+}-binding proteins calretinin and calbindin (Friauf, 1993; Lohmann and Friauf, 1996). The maximum level of calretinin and calbindin immunoreactivity can be seen at P4 and P7, respectively. As these proteins have been shown to be involved in Ca^{2+} buffering (Batini et al., 1993; Chard et al., 1993), our findings indicate that the regulation of Ca^{2+} homeostasis is relevant to developing LSO neurons and achieved by a consortium of effective buffering molecules. We have recently begun to perform experiments in organotypic slice cultures to further tackle the issue of activity-related mechanisms that are involved in wiring up the topographic connections to the LSO. These experiments have shown that LSO neurons appear to be critically dependent on an optimal internal Ca^{2+} concentration, because their morphological integrity was drastically reduced and the MNTB-LSO projection was severely impaired if we did not tonically depolarize them with an elevated extracellular K^+ concentration (25 mM; Lohmann and Friauf, 1997). When we applied the L-type Ca^{2+} channel antagonist nifedipine in the presence of an elevated extracellular K^+ concentration, the beneficial effect could be abolished. This indicates that the depolarization of the membrane potential results in a Ca^{2+} influx which is mediated through a channel subtype with long-lasting open times. It further suggests that Ca^{2+} ions act as a second messenger in the integrity promoting cascade. We conclude from these results that developing LSO neurons need to be electrically active to stay healthy and that they must be sufficiently depolarized to guarantee an adequate $[Ca^{2+}]_i$. In a slice culture, they are deprived from their peripheral input (i.e. from cochlear afferents) and, therefore, they may be devoid of synaptic activation and need to be activated pharmacologically to substitute for this deficit. Since our data on the transitorily depolarizing effect of glycine show that both types of input to the LSO, the ipsilateral and the contralateral, are able to depolarize the cells and induce excitation, they are in accord with the idea that excitation during development is a prerequisite for the maintenance of integrity and the formation of properly organized inputs.

SUMMARY AND CONCLUSIONS

In summary, our results show that the transiently depolarizing effect of glycine may be a first answer to the question of how the precisely organized inhibitory and excitatory connections develop that form the neuronal microcircuits for the detection of binaural intensity differences. It appears that the two types of transmitters, glycine and glutamate, which act antagonistically in the adult animals and thereby enable the LSO neurons to analyze the information from both ears and to convert IID into neuronal activity, act synergistically during early development. Both may thus activate similar, if not the same, second messenger pathways that result in the activity-dependent stabilization of synaptic contacts. Our findings lead us to suggest that Ca^{2+} ions play an important role in the formation of the IDD microcircuits in the SOC. Especially LSO neurons appear to crucially depend on Ca^{2+} ions whose intracellular concentration has to be kept in an optimal range to ensure their integrity. Taken together, some insights into the site of action of the developing afferents to the LSO have been established. However, the molecular mechanism by which topography is achieved in the glycinergic input, the glutamatergic input, and among both types of input still remains elusive.

ACKNOWLEDGMENTS

The original work presented in this article was supported in part by grants from the Deutsche Forschungsgemeinschaft (Fr 772/1–3 and SFB 269).

REFERENCES

Altman, J. and Bayer, S.A. (1980) Development of the brain stem in the rat. III. Thymidine-radiographic study of the time of origin of neurons of the vestibular and auditory nuclei of the upper medulla. J. Comp. Neurol. 194, 877–904.

Aoki, E., Semba, R., Keino, H., Kato, K. and Kashiwamata, S. (1988) Glycine-like immunoreactivity in the rat auditory pathway. Brain. Res. 442, 63–71.

Aprison, M.H., Galvezruano, E., Robertson, D.H. and Lipkowitz, K.B. (1996) Glycine and GABA receptors: Molecular mechanisms controlling chloride ion flux. J. Neurosci. Res. 43, 372–381.

Backus, K.H. and Friauf, E. (1996) Effects of synchronous depolarization on glycine-induced currents in developing rat auditory brainstem neurons. Soc. Neurosci. Abstr. 256, 6.

Barker, J.L., McBurney, R.N. and MacDonald, R.L. (1982) Fluctuation analysis of neutral amino acid responses in cultured mouse spinal neurones. J. Physiol. (Lond). 322, 365–387.

Batini, C., Palestini, M., Thomasset, M. and Vigot, R. (1993) Cytoplasmic calcium buffer, calbindin-D28k, is regulated by excitatory amino acids. Neuroreport 4, 927–930.

Ben-Ari, Y., Tseeb, V., Raggozzino, D., Khazipov, R. and Gaiarsa, J.L. (1994) γ-Aminobutyric acid (GABA): a fast excitatory transmitter which may regulate the development of hippocampal neurones in early postnatal life. Prog. Brain. Res. 102, 261–273.

Blatchley, B.J., Cooper, W.A. and Coleman, J.R. (1987) Development of auditory brainstem response to tone pip stimuli in the rat. Dev. Brain. Res. 32, 75–84.

Bledsoe, S.C.J., Snead, C.R., Helfert, R.H., Prasad, V., Wenthold, R.J. and Altschuler, R.A. (1990) Immunocytochemical and lesion studies support the hypothesis that the projection from the medial nucleus of the trapezoid body to the lateral superior olive is glycinergic. Brain. Res. 517, 189–194.

Boudreau, J.C. and Tsuchitani, C. (1968) Binaural interaction in the cat superior olive S segment. J. Neurophysiol. 31, 442–454.

Brown, R.D. and Feldman, A.M. (1978) Pharmacology of hearing and ototoxicity. Annu. Rev. Pharmacol. Toxicol. 18, 233–252.

Caird, D. and Klinke, R. (1983) Processing of binaural stimuli by cat superior olivary complex neurons. Exp. Brain. Res. 52, 385–399.

Chard, P.S., Bleakman, D., Christakos, S., Fullmer, C.S. and Miller, R.J. (1993) Calcium buffering properties of calbindin-D28k and parvalbumin in rat sensory neurones. J. Physiol. (Lond). 472, 341–357.

Cherubini, E., Rovira, C., Gaiarsa, J.L., Corradetti, R. and Ben-Ari, Y. (1990) GABA mediated excitation in immature rat CA3 hippocampal neurons. Int. J. Dev. Neurosci. 8, 481–490.

Cherubini, E., Gaiarsa, J.L. and Ben-Ari, Y. (1991) GABA: an excitatory transmitter in early postnatal life. Trends. Neurosci. 14, 515–519.

Constantine-Paton, M., Cline, H.T. and Debski, E. (1990) Patterned activity, synaptic convergence, and the NMDA receptor in developing visual pathways. Annu. Rev. Neurosci. 13, 129–154.

Fox, K. and Zahs, K. (1994) Critical period control in sensory cortex. Curr. Opin. Neurobiol. 4, 112–119.

Friauf, E. (1993) Transient appearance of calbindin-D_{28k}-positive neurons in the superior olivary complex of developing rats. J. Comp. Neurol. 334, 59–74.

Friauf, E., Hammerschmidt, B., Kirsch, J. and Betz, H. (1994) Development of glycine receptor distribution in the rat auditory brainstem: transition from the 'neonatal' to the 'adult' isoform. Assoc. Res. Otolaryngol. Abstr. 17, 10.

Friauf, E., Hammerschmidt, B. and Kirsch, J. (1997) Development of adult-type glycine receptors in the central auditory system of rats. J. Comp. Neurol. (in press).

Friauf, E., Kandler, K. (1993) Cell birth, formation of efferent connections, and establishment of tonotopic order in the rat cochlear nucleus. In: M.A. Merchán, J.M. Juiz, D.A. Godfrey and E. Mugnaini (Eds.), The Mammalian Cochlear Nuclei: Organization and Function, Plenum, New York, pp. 19–28.

Gao, B.X. and Ziskind-Conhaim, L. (1995) Development of glycine- and GABA-gated currents in rat spinal motoneurons. J. Neurophysiol. 74, 113–121.

Goldberg, J.M. and Brown, P.B. (1969) Response of binaural neurons of dog superior olivary complex to dichotic tonal stimuli: Some physiological mechanism of sound localization. J. Neurophysiol. 32, 613–636.

Goodman, C.S. and Shatz, C.J. (1993) Developmental mechanisms that generate precise patterns of neuronal connectivity. Neuron 10, 77–98.

Guinan, J.J.J., Norris, B.E. and Guinan, S.S. (1972) Single auditory units in the superior olivary complex. II: Locations of unit categories and tonotopic organization. Int. J. Neurosci. 4, 147–166.

Hebb, D.O. (1949) The Organization of Behavior. John Wiley and Sons; New York.

Helfert, R.H., Juiz, J.M., Bledsoe, S.C., Bonneau, J.M., Wenthold, R.J. and Altschuler, R.A. (1992) Patterns of glutamate, glycine, and GABA immunolabeling in four synaptic terminal classes in the lateral superior olive of the guinea pig. J. Comp. Neurol. 323, 305–325.

Janigro, D. and Schwartzkroin, P.A. (1988) Effects of GABA and baclofen on pyramidal cells in the developing rabbit hippocampus: an 'in vitro' study. Dev. Brain. Res. 41, 171–184.

Kandler, K. and Friauf, E. (1993) Pre- and postnatal development of efferent connections of the cochlear nucleus in the rat. J. Comp. Neurol. 328, 161–184.

Kandler, K. and Friauf, E. (1995) Development of glycinergic and glutamatergic synaptic transmission in the auditory brainstem of perinatal rats. J. Neurosci. 15, 6890–6904.

Kelly, J.B. (1992) Behavioral development of the auditory orientation response. In: R. Romand (Ed.), Development of Auditory and Vestibular Systems 2, Elsevier, Amsterdam, London, New York, Tokyo, pp. 391–418.

Kungel, M. and Friauf, E. (1996) Patch-clamp studies of auditory brainstem neurons: Developmental changes in physiology and glycinergic pharmacology. 1. Kongress der Neurowiss. Gesellschaft. Berlin.

Langosch, D., Becker, C.-M. and Betz, H. (1990) The inhibitory glycine receptor: A ligand-gated chloride channel of the central nervous system. Eur. J. Biochem. 194, 1–8.

Leinekugel, X., Tseeb, V., Ben-Ari, Y. and Bregestovski, P. (1995) Synaptic GABA$_A$ activation induces Ca^{2+} rise in pyramidal cells and interneurons from rat neonatal hippocampal slices. J. Physiol. (Lond). 487, 319–329.

Lin, M.H., Takahashi, M.P., Takahashi, Y. and Tsumoto, T. (1994) Intracellular calcium increase induced by GABA in visual cortex of fetal and neonatal rats and its disappearance with development. Neurosci. Res. 20, 85–94.

Lohmann, C. and Friauf, E. (1996) Distribution of the calcium-binding proteins parvalbumin and calretinin in the auditory brainstem of adult and developing rats. J. Comp. Neurol. 367, 90–109.

Lohmann, C. and Friauf, E. (1997) An organotypic slice culture from a network of developing inhibitory and excitatory connections. Submitted.

Loturco, J.J., Owens, D.F., Heath, M.J.S., Davis, M.B.E. and Kriegstein, A.R. (1995) GABA and glutamate depolarize cortical progenitor cells and inhibit DNA synthesis. Neuron 15, 1287–1298.

Luhmann, H.J. and Prince, D.A. (1991) Postnatal maturation of the GABAergic system in rat neocortex. J. Neurophysiol. 65, 247–263.

Lux, H.D. (1971) Ammonium and chloride extrusion: hyperpolarizing synaptic inhibition in spinal motoneurons. Science 173, 555–557.

Miller, M.W. and Nowakowski, R.S. (1988) Use of bromodeoxyuridine-immunohistochemistry to examine the proliferation, migration and time of origin of cells in the entral nervous system. Brain. Res. 457, 44–52.

Misgeld, U., Deisz, R.A., Dodt, H.U. and Lux, H.D. (1986) The role of chloride transport in postsynaptic inhibition of hippocampal neurons. Science 232, 1413–1415.

Moore, D.R. (1992) Trophic influences of excitatory and inhibitory synapses on neurones in the auditory brain stem. Neuroreport 3, 269–272.

Mueller, A.L., Taube, J.S. and Schwartzkroin, P.A. (1984) Development of hyperpolarizing inhibitory postsynaptic potentials and hyperpolarizing response to τ-aminobutyric acid in rabbit hippocampus studied in vitro. J. Neurosci. 4, 860–867.

Nicoll, R.A. and Alger, B.E. (1980) Presynaptic inhibition: Transmitter and ionic mechanisms. Annu. Rev. Neurosci. 3, 227–268.

Nishi, S., Minota, S. and Karczmar, A.G. (1974) Primary afferent neurones: The ionic mechanism of GABA-mediated depolarization. Neuropharmacology 13, 215–219.

O'Leary, D.D.M., Ruff, N.L. and Dyck, R.H. (1994) Development, critical period plasticity, and adult reorganizations of mammalian somatosensory systems. Curr. Opin. Neurobiol. 4, 535–544.

Obata, K., Oide, M. and Tanaka, H. (1978) Excitatory and inhibitory actions of GABA and glycine on embryonic chick spinal neurons in culture. Brain. Res. 144, 179–184.

Obrietan, K. and van den Pol, A.N. (1995) GABA neurotransmission in the hypothalamus: developmental reversal from Ca^{2+} elevating to depressing. J. Neurosci. 15, 5065–5077.

Obrietan, K. and van den Pol, A.N. (1996) Growth cone calcium elevation by GABA. J. Comp. Neurol. 372, 167–175.

Owens, D.F., Boyce, L.H., Davis, M.B.E. and Kriegstein, A.R. (1996) Excitatory GABA responses in embryonic and neonatal cortical slices demonstrated by gramicidin perforated- patch recordings and calcium imaging. J. Neurosci. 16, 6414–6423.

Puel, J.L. and Uziel, A. (1987) Correlative development of cochlear action potential sensitivity, latency, and frequency selectivity. Brain. Res. 465, 179–188.

Reichling, D.B., Kyrozis, A., Wang, J. and Mac Dermott, A.B. (1994) Mechanisms of GABA and glycine depolarization-induced calcium transients in rat dorsal horn neurons. J. Physiol. (Lond). 476, 411–421.

Rietzel, H.-J., Friauf, E. (1995) Development of dendritic morphology in the rat auditory brainstem: bipolar and multipolar cells in the lateral superior olive. In: N. Elsner, R. Menzel (Eds.), Proc.Göttingen Neurobiol.Conf. Vol. 23, Thieme Verlag, Stuttgart,

Sanes, D.H., Goldstein, N.A., Ostad, M. and Hillman, D.E. (1990) Dendritic morphology of central auditory neurons correlates with their tonotopic position. J. Comp. Neurol. 294, 443–454.

Sanes, D.H., Markowitz, S., Bernstein, J. and Wardlow, J. (1992) The influence of inhibitory afferents on the development of postsynaptic dendritic arbors. J. Comp. Neurol. 321, 637–644.

Sanes, D.H. and Chokshi, P. (1992) Glycinergic transmission influences the development of dendrite shape. Neuroreport 3, 323–326.

Sanes, D.H. and Takacs, C. (1993) Activity-dependent refinement of inhibitory connections. Eur. J. Neurosci. 5, 570–574.

Schmanns, H. and Friauf, E. (1994) K^+- and transmitter-induced rises in $[Ca^{2+}]_i$ in auditory neurones of developing rats. Neuroreport 5, 2321–2324.

Schwartz, R.D. and Yu, X. (1995) Optical imaging of intracellular chloride in living brain slices. J. Neurosci. Methods 62, 185–192.

Segal, M. (1993) GABA induces a unique rise of [Ca]i in cultured rat hippocampal neurons. Hippocampus 3, 229–238.

Shatz, C.J. (1994) Role for spontaneous neural activity in the patterning of connections between retina and LGN during visual system development. Int. J. Dev. Neurosci. 12, 531–546.

Suneja, S.K., Benson, C.G., Gross, J. and Potashner, S.J. (1995) Evidence for glutamatergic projections from the cochlear nucleus to the superior olive and the ventral nucleus of the lateral lemniscus. J. Neurochem. 64, 161–171.

Thompson, S.M., Deisz, R.A. and Prince, D.A. (1988) Outward chloride/cation co-transport in mammalian cortical neurons. Neurosci. Lett. 89, 49–54.

Thompson, S.M., Deisz, R.A. and Prince, D.A. (1989) Relative contributions of passive equilibrium and active transport to the distribution of chloride in mammalian cortical neurons. J. Neurophysiol. 60, 105–124.

Tsuchitani, C. (1977) Functional organization of lateral cell groups of cat superior olivary complex. J. Neurophysiol. 40, 296–318.

Uziel, A., Romand, R. and Marot, M. (1981) Development of cochlear potentials in rats. Audiology 20, 89–100.

Wang, J., Reichling, D.B., Kyrozis, A. and Mac Dermott, A.B. (1994) Developmental loss of GABA- and glycine-induced depolarization and Ca^{2+} transients in embryonic rat dorsal horn neurons in culture. Eur. J. Neurosci. 6, 1275–1280.

Weber, F., Zillus, H., Friauf, E. (1991) Neuronal birth in the rat auditory brainstem. In: N. Elsner, W. Singer (Eds.), Synapse, Transmission, Modulation. Proceedings of the 19th Göttingen Neurobiology Conference, Georg Thieme Verlag, Stuttgart, New York, pp. 123

Wu, S.H. and Kelly, J.B. (1992) Synaptic pharmacology of the superior olivary complex studied in mouse brain slice. J. Neurosci. 12, 3084–3097.

Wu, W.I., Ziskind-Conhaim, L. and Sweet, M.A. (1992) Early development of glycine- and GABA-mediated synapses in rat spinal cord. J. Neurosci. 12, 3935–3945.

Yuste, R. and Katz, L.C. (1991) Control of postsynaptic Ca^{2+} influx in developing neocortex by excitatory and inhibitory neurotransmitters. Neuron 6, 333–344.

Zhang, L., Spigelman, I. and Carlen, P.L. (1991) Development of GABA-mediated, chloride-dependent inhibition of CA1 pyramidal neurones of immature rat hippocampal slices. J. Physiol. (Lond). 444, 25–49.

18

DIVERSITY IN GLYCINE AND NMDA RECEPTOR SUBUNIT COMPOSITION IN THE RAT COCHLEAR NUCLEUS AND SUPERIOR OLIVARY COMPLEX AND CHANGES WITH DEAFNESS

Richard A. Altschuler,[*] Kazuo Sato, Jerome Dupont, Joann M. Bonneau, and Hironobu Nakagawa

Kresge Hearing Research Institute
Department of Otolaryngology
University of Michigan
Ann Arbor, Michigan 48109-0506

1. INTRODUCTION

Much of the processing of ascending auditory information in the auditory brain stem is mediated by synapses using an excitatory or an inhibitory amino acid as transmitters and acting at an amino acid receptor. When adjacent sections are immunostained for gamma aminobutyric acid (GABA), glycine or glutamate, over 90% of the terminals on ventral cochlear nucleus and lateral superior olivary complex principal cells are immunoreactive for one or more of these amino acids (Altschuler et al, 1993, Juiz et al, 1996, Helfert et al, 1992). While other neurotransmitters such as acetylcholine (e.g. Godfrey et al, 1993) and neuropeptides (e.g. Adams et al, 1993) also have important roles in the auditory brain stem and may often be co-contained in terminals with an amino acid transmitter, certainly amino acid transmitters and receptors have a major role in acoustic signal processing. There is now increasing information that the properties of excitatory and inhibitory amino acid synapses can be influenced by the composition of their receptors. The excitatory amino acid, most likely to be glutamate and the inhibitory amino acids GABA and glycine all act at ionotropic receptors whose properties can vary as a function of their subunit composition. These ionotropic receptors are pentamers with five mem-

* Correspondence to: Richard Altschuler, Ph.D., Kresge Hearing Research Institute, The University of Michigan, 1301 E. Ann Street, Ann Arbor, MI 48109–0506, U.S.A.. Tel: (313) 763–0060 fax: (313) 764–0014. E-mail: shuler@umich.edu.

Acoustical Signal Processing in the Central Auditory System
edited by Syka, Plenum Press, New York, 1997

brane spanning subunits clustered around an ion channel. Depending upon which subunits a neuron uses to compose the complete receptor there can be different binding properties, actions on the ion channel and recovery at the synapses.

The availability of different subunits to compose receptors might be used to "fine-tune" individual synapses and provide different properties as required and/or provide the ability to make on-line changes in receptor properties in response to changes in needs. We are examining receptor subunit compositions in the CN and SOC, first to determine if there is a diverse composition of receptor subunits in different regions and cell types reflecting differences in processing requirements and second to see if subunit expression can react and adapt to differences in processing requirements caused by changes in input, such as unilateral deafness. In the present studies we examined glycine receptors (GlyR) because of the major presence in the CN and SOC and the N-methyl-D-aspartate (NMDA) excitatory amino acid receptor, because it has been shown to be involved in plasticity in several central nervous system regions (e.g. Cotman and Iversen, 1987; Collingrade et al., 1988; Collingrade et al, 1990)

There is considerable evidence for a role of glycine in the auditory brain stem (see Caspary et al, 1986, 1993; Godfrey et al, 1988; Wenthold, 1991; Adams, 1993; Altschuler et al 1993; Evans & Zhao 1993; Oertel & Wickesberg 1993; Potashner et al 1993; Saint Marie et al, 1993; Wickesberg & Oertel, 1993 for reviews). The GlyR has been localized in the auditory brainstem using immunocytochemistry (e.g. Adams, in press; Altschuler et al 1986,1993; Wenthold et al 1988; Wenthold & Hunter, 1990), receptor binding autoradiography (e.g. Zarbin et al 1981; Frostholm & Rotter 1986; Sanes et al 1987; Glendenning & Baker 1988) and pharmacology (e.g. Caspary et al, 1979, 1993; Wu and Oertel, 1986; Oertel & Wickesberg, 1993).

Molecular biological studies indicate that the GlyR is composed of three α and two ß subunits (Kuhse et al 1990a; Betz et al 1991). The α subunit which is ligand binding and strychnine-sensitive has three subtypes: $\alpha1$, $\alpha2$ and $\alpha3$. The $\alpha2$ subunit is the ligand-binding subunit of the immature neonatal Gly R isoform, and is replaced by the $\alpha1$ subunit during postnatal development (Kuhse et al 1990b; Malasio et al 1991). GlyR containing $\alpha3$ subunits have lower strychnine sensitivities than those with $\alpha1$ subunits (Kuhse et al, 1990b) and $\alpha3$ subunit mRNA is expressed in fewer regions of the rat brain than the $\alpha1$ subunit (Malosio et al 1991). While the β subunit is not ligand binding it's addition increases single channel conductances and decreases picrotoxin sensitivity (Meyer et al 1994).

There is also considerable evidence for the role of an excitatory amino acid, most likely to be glutamate, in the auditory brain stem (see Caspary et al, 1986, 1991; Juiz et al, 1993, Godfrey et al, 1988; Wenthold et al, 1993 for reviews). NMDA receptors are one of three classes of excitatory amino acid receptors defined by selective agonists (the other two classes are α-amino-3-hydroxy-5 methyl-4 isoxazolepropinate (AMPA) receptors and kainate receptors). The NMDA receptor also can be modulated by glycine, activated by polyamines, inhibited by Zn^{2+} and has a voltage-dependent Mg^{2+} channel block (Monaghan et al., 1989; Moriyoshi et al., 1991; Kutsuwada et al., 1992; Meguro et al., 1992; Monyer et al., 1992). The ionotropic NMDA receptor is composed of NMDAR1 and NMDAR2A,B,C & D subunits (Moriyoshi et al., 1991; Monyer et al., 1992; Nakanishi et al., 1992; Kutsuwada et al., 1992; Meguro et al., 1992). While NMDAR 1 subunits, by themselves, can form a functional receptor, there is naturally a heteromeric configuration with NMDAR 2 subunits (Monyer et al, 1992; Nakanishi, 1992). Different NMDAR 2 subunits provide functional variability in physiological and pharmacological properties of the NMDA receptor (Monyer et al, 1992; Cik et al, 1993; Durand et al, 1993; Hollmann et al, 1993). High expression of NMDAR2A correlates with antagonist preferring pharmacology and high NMDAR2B with agonist preferring pharmacology (Buller et al 1994; Mishina et al 1993). High NMDAR2C confers high sensitivity to 7-chlorokynurenate (7CK) while high NMDAR2A confers high sensitivity to D-2-amino-5-phosphonovalerate (APV) (Mishina et al, 1993). High expression of NMDAR2D confers sensitivity to glycine and L-glutamate (Mishina et al, 1993).

For both the NMDAR and GlyR, there is therefore a potential for diversity in responses to ligands based on which subunits are used to compose the receptor. This might be used to provide specific synapses with the characterisitics for optimal processing under normal conditions. It might also provide the ability to adapt and react to changes in processing requirements by changing subunit compositions and therefore synaptic properties. *In situ* hybridization was used to examine expression of different receptor subunits in the principal cells of six regions of the superior olivary complex (SOC) and in six cell types in the cochlear nucleus (CN) chosen based on our ability to easily differentiate them based on size, shape and location criteria. Expression was examined in normal hearing rats and in rats which received unilateral deafening by cochlear ablation.

2. METHODS

Studies were carried out in male Sprague-Dawley rats. *In situ* hybridization was performed using 45 mer antisense oligonucleotide probes against published sequences (Malasio et al, 1991 for glycine receptor subunits; Tolle et al, 1993, Moriyoshi et al, 1991 and Monyer et al, 1992 for NMDAR2 receptors subunits). Probes were endlabeled with α-^{35}S dATP New England Nuclear (NEN) using a 3' endlabeling kit (NEN) and applied to 15~18μM thick cryostat sections. Sections were cut from unfixed brains, postfixed in paraformaldehyde and standard *in situ* hybridization methods followed. Slides were dipped in Kodak autoradiography emulsion and exposed for 4 weeks. Control sections were prepared with the labeled probe in the presence of 200-fold excess of unlabeled probe in the hybridization mixture.

Quantitative analysis of the number of silver grains over identified neurons was performed using the MetaMorph Image Acquisition and Analysis System. Neurons had to meet size, shape and location criteria for identification. The number of silver grains over an identified neuron was counted and divided by the cell area to give density of labeling for each neuron tabulated as counts/μm^2. The background labeling, determined by counting silver grains over regions of the nuclei without cells, was subtracted. For NMDAR2 subunits, labeling density of 3–4 x 10^{-2} counts/μm^2 was defined as low (+), 5–6 x 10^{-2} counts/μm^{-2} as medium (++) and 7–8 x 10^{-2} counts/μm^{-2} as high (+++) and above 9 x 10^{-2} counts/μm^{-2} as very high (++++). For the glycine receptor low labeling (+) was defined as 3–4 counts/μm^2, moderate (++) as 5–7 counts/μm^2 and heavy (+++) as 8+ counts/μm^2.

Deafening was accomplished through unilateral cochlear ablation, and groups of rats assessed one, five and 20 days after the deafening. *In situ* hybridization was performed as described above and comparison made between the sides ipsilateral and contralateral to the ablated cochlea as well as to untreated normal hearing rats.

The care and use of animals reported in this study was approved and supervised by The University of Michigan Unit on Laboratory Animal Medicine.

3. RESULTS

3.1 Glycine Receptor

3.1.1 Normal Hearing Rats. The down regulation of the α2 immature subunit of the glycine receptor and upregulation of the mature α1 receptor occurs later in the auditory brain stem than in spinal cord. While this turnover/replacement occurs at the end of the third postnatal week in the rat spinal cord, in the VCN and SOC high expression of the α2 was still seen at postnatal week 3 and it was not until 8 weeks of age that α2 expression was no longer seen in normal hearing animals (Sato et al, 1996). Expression of the α1 subunit increased from 3 weeks to 8 weeks postnatally and in fact continued to increase at 6 months of age.

A differential expression of the mature subunits was seen in the SOC and CN of the mature rat. In the SOC three patterns of expression were seen (Table 1). The lateral and medial superior olive (LSO and MSO) and the lateral nucleus of the trapezoid body (LNTB) all had higher α1 than α3. The superior paraolivary nucleus (SPN) had equivalent levels, while the medial and the ventral nuclei of the trapezoid body (MNTB, VNTB) both had higher α3 than α1. Since α1 has higher strychnine sensitivity than α3, this would suggest that the glycine receptor is less sensitive in the MNTB and VNTB than in the LSO, MSO, LNTB & SPN. This correlates well with the amount of glycinergic input which is much higher in LSO, MSO, LNTB and SPN than MNTB and VNTB. The composition and properties of the glycine receptor therefore accentuate the differences in amount of inputs.

Table 1. Expression of glycine receptor subunits

Cell type	α1	α2	α3	β
In the mature rat cochlear nucleus				
SphBC	+++	–	+++	+++
SmC	+++	–	+++	+++
FusC	+++	–	+++	+++
OctC	++	–	+++	+++
CornC	++	–	+++	+++
GrnC	–	–	++	++
In the mature rat superior olivary complex				
LSO	+++	–	++	++
MSO	+++	–	++	++
LNTB	+++	–	++	++
SPN	++	–	++	++
MNTB	++	–	+++	+++
VNTB	++	–	+++	+++

Charts showing relative expression of glycine receptor subunits in six cell types of the the mature rat cochlear nucleus (CN) and six regions of the mature rat superior olivary complex (SOC). SphBC = spherical bushy cells, SmC = small cells of the small cell cap, FusC = Fusiform cells, OctC = octopus cells, CornC = corn cells, GrnC = granule cells between DCN and VCN. LSO, MSO = lateral and medial superior olives, SPN = superior paraolivary nucleus, LNTB, MNTB, VNTB = lateral, medial and ventral nuclei of the trapezoid body.

In the CN (Table 1) there were two patterns of Gly R subunit expression. Spherical and globular bushy cells of the VCN, fusiform cells of the DCN and small cells of the small cell cap over the VCN all had high $\alpha 1$ expression. Octopus cells of the PVCN and corn (elongate/tuberculoventral) cells of the DCN and granule cells located between the ventral CN and DCN had higher $\alpha 3$ than $\alpha 1$ subunit expression. The relative glycinergic input to all these CN cell types has not been completely determined, however there is clearly greater glycinergic input to spherical and globular bushy cells and fusiform cells than there is to octopus cells (Juiz et al, 1996; in press; Kolston et al, 1992; Wickesberg & Oertel, 1993), and so, at least in this case, the composition of the glycinergic receptor may be accentuating and compounding the effects of differences in amount of input, as in the SOC.

3.1.2. Unilateral Deafness. Our preliminary studies indicate that five days after deafening the $\alpha 2$ subunit which is down regulated in normal hearing animals after 8 weeks of age, becomes upregulated. This, however, is not accompanied by down regulation of the $\alpha 1$ subunit. There are no obvious qualitative changes in $\alpha 1$ and $\alpha 3$ subunits, however additional more quantitative studies are necessary.

3.2 NMDA Receptor

3.2.1. Normal Hearing Rats. Abundant labeling for NMDAR1 subunit mRNA was observed in all major CN and SOC neuronal types. NMDAR2 subunits were detected at lower levels than NMDAR1, with diversity in the expression of NMDR2A-D subunits (Table 2).

Table 2. Expression of NMDR2 subunits

cell type	2A	2B	2C	2D	
In the rat cochlear nucleus					
SphBC	++	+	+	+	Higher A
SmC	+++	+++	++++	+	Higher C
FusC	+	+	++++	+	
CornC	+	++	+++	+	
GranC	+	+	+	+	All Low
OctC	+	+	+	+	
The rat superior olivary complex					
LSO	+++	++	+++	+	High A&C
LNTB	+++	++	+++	+	
MNTB	+	++	+	+	Higher B
SPN	++	++	++	+	Moderate A-C
MSO	++	++	++	+	
VNTB	+++	+++	+++	++	High A-C

Charts showing relative expression of NMDAR2 subunits in six cell types of the the rat cochlear nucleus (CN) and six regions of the rat superior olivary complex (SOC). SphBC = spherical bushy cells, SmC = small cells of the small cell cap, FusC = Fusiform cells, OctC = octopus cells, CornC = corn cells, GrnC = granule cells between DCN and VCN. LSO, MSO = lateral and medial superior olives, SPN = superior paraolivary nucleus, LNTB, MNTB, VNTB = lateral, medial and ventral nuclei of the trapezoid body

In the SOC four distinct patterns of labeling were observed. The LSO and LNTB showed higher expression of NMDAR2A&C than other subunits. The MNTB showed higher NMDAR2B expression than other subunits. The MSO, SPN and VNTB showed relatively equivalent expression of NMDAR2A,B and C subunits, the SPN and MSO at moderate levels the VNTB at high levels. NMDAR2D had low expression, except in VNTB.

In the CN, several patterns of NMDR2 mRNA expression were noted. Spherical bushy cells of the anteroventral cochlear nucleus had moderate 2A and low levels of other NMDAR2 subunits. Fusiform cells of the DCN, small cells of the small cell cap and corn cells all had higher 2C than other subunits. Fusiform cells had very high 2C and low levels of other NMDAR2 subunits. Small cells of the small cell cap had very high 2C and moderate levels of 2A&B, while corn cells had high 2C, moderate 2B and low 2A. Octopus cells and granule cells had low levels of all NMDR2 subunits. NMDAR2D mRNA had low expression in all six cell types assessed.

3.2.2 Unilateral Deafening. No changes in NMDAR2A-D expressions were seen 1, 5 or 20 days after unilateral deafening. NMDAR1, however, showed deafness related changes in some auditory brain stem cells and regions and not in others. Significant ipsilateral decreases in the NMDAR1 expression in the LSO (Figure 1A) and LNTB of the SOC and in spherical bushy (Figure 1B) and octopus cells of the VCN were seen 5 days after unilateral deafening, accompanied by contralateral increases. Interestingly, these changes were transient, returning to normal levels 20 days post-deafening.

4. DISCUSSION

Diversity in expression is seen in the expression of NMDAR2 receptor subunits and glycine receptors subunits in the CN and SOC. Such diversity has also been seen in the expression of AMPA receptor subunits in the rat CN (Hunter et al, 1993), in NMDAR1 in the mouse CN (Bilak et al, 1996) and in the expression of NMDAR1 isoforms in the rat CN (Hunter et al, 1995). The differences we observe in NMDAR2A-D expression in the rat CN and SOC would suggest that there will be differences in the pharmacology of the receptor for different auditory brain stem neurons and regions. Low expression of NMDAR2D in five of the six SOC regions examined and in all CN cell types assessed suggests that glycine modulation may not be a characteristic of NMDA synapses for these neurons.

The patterns of expression observed for glycine receptor subunits largely correlates with differences in the amount of glycinergic input. Low glycinergic input is accompanied by the α3 subunit, with lower sensitivity. The composition of the glycine receptor thus may serve to accentuate differences in amount of input, with less sensitivity for neurons receiving less input.

The changes we observe in NMDA and glycine receptors seen after unilateral deafening add to the accumulating evidence demonstrating deafness induced changes in amino acid transmitters and receptors in the auditory brain stem. These changes include changes in GABA after deafness (Dupont et al, 1994; Bledsoe et al, 1995, chapter this volume) and with age-related hearing loss (Caspary et al, 1995) along with changes in GABA binding (Milbrandt et al, 1994). There are changes in glycine (Dupont et al, 1994,1995) as well as changes in glycine receptor binding with deafness (Benson et al, 1995a,b,) and aging (Milbrandt & Caspary, 1995) There are deafness induced effects on uptake and release of in-

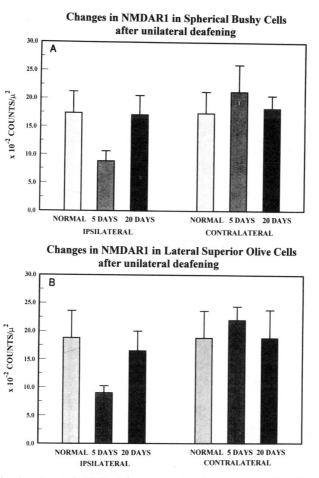

Figure 1. Graphs showing changes in NMDAR1 expression in spherical bushy cells of the rat cochlear nucleus (3A) and principal cells of the rat lateral superior olive (LSO) (3B), 5 and 20 days after unilateral cochlear ablation. At 5 days after deafness there is a unilateral decrease and contralateral increase in expression and a return to normal levels of expression at 20 days.

hibitory and excitatory amino acids (Suneja et al, 1995) and changes in expression of AMPA receptor subunits (Hunter et al, 1995).

Changes in transmitters and receptors may underlie functional central auditory plastic changes shown to be induced by deafness (e.g. Rajan et al, 1993; reviews by Rubel et al, 1990; Moore, 1991; Miller et al, 1992, Rajan & Irvine) or overstimulation (e.g. Boettcher and Salvi, 1993, Salvi et al, 1996).

The upregulation of the immature form of the glycine receptor may reflect a return to the role it has during development of circuitry and it may reflect plasticity. It is interesting that the changes seen in NMDAR1 expression were transient, returning to normal levels 20 days after deafening. Since the expression only reflects production of the receptor subunit, it is possible that there are changes in amount and/or the placement of the receptor protein that are not transient. The return to normal levels of subunit expression could then just reflect the production for the replacement of the cytoplasmic pool of receptor subunits. Immunocytochemical studies of receptor subunit changes will be necessary to address this.

ACKNOWLEDGMENTS

This work was supported by NIDCD grant DC00383.

REFERENCES

Adams, JC: Non-primary inputs to the cochlear nucleus visualized using immunocytochemistry, In: Merchan MA, Juiz JM, Godfrey DA & Mugnaini E (eds): Mammalian Cochlear Nuclei: Organization and Function , (Plenum Press, New York, 1993) 133–42.

Adams, JC (in press) Distribution of some cytochemically distinct cells in the ventral cochlear nucleus of cat and human with emphasis on octopus cells and their projections in "Advances in Speech, Hearing and Human Processing" Vol 3, W.A. Ainsworth (Ed) JAI Press, London.

Altschuler, R. A., Betz, H., Parakkal, M. H., Reeks, K. A. and Wenthold, R. J. (1986) Identification of glycinergic synapses in the cochlear nucleus through immunocytochemical localization of the postsynaptic receptor. Brain Res., 369–320

Altschuler, RA, Juiz, JM, Shore, SE, Bledsoe, SC, Helfert. RH, Wenthold RJ (1993) Inhibitory amino acid synapses and pathways in the ventral cochlear nucleus. In: Merchan MA, Juiz JM, Godfrey DA & Mugnaini E (eds): Mammallian Cochlear Nuclei: Organization and Function , (Plenum Press, New York, pp. 211–224.

Benson CG & Potashner SJ: Modulation of strychnine binding in the lateral lemniscal nucleus after cochlear ablation. Neurosciences Abstract (1995) 21:404

Benson CG, Suneja SK and Potashner SJ (1995) Long-term regulation of glycine receptors in the adult guinea pig cochlear nucleus and superior olivary complex after unilateral cochlear ablation. Neurosciences Abstract 21:403

Betz, H., Langosch, D., Hoch, W., Prior, P., Pribilla, I., Kuhse, J., Schmieden, V., Malosio, M.-L. Matzenbach, B. and Holzinger, F. (1991) Structure and expression of inhibitory glycine receptors. Ads. Exp. Med. Biol., 287, 421–429.

Bilak MM, Bilak SR and Morest DK (1996) Differential expression of N-methyl-D-aspartate receptor in the cochlear nucleus of the mouse. Neuroscience 75:1075–1098.

Boettcher FA, Salvi RJ (1993) Functional changes in the ventral cochlear nucleus following acute acoustic overstimulation. J Acoust Soc Am 94:2123–34

Buller AL, Larson HC, Schneider BE, Beaton JA, Morrisett RA, Monaghan DT (1994) The molecular basis of NMDA receptor subtypes: Native receptor diversity is predicted by subunit composition. J Neurosci 9:5471–5484.

Caspary, D.M., D.C. Havey, and C.L. Faingold (1979) Effects of microiontophoretically applied GLY and GABA on neuronal response patterns in the cochlear nuclei. Brain Res. *172*:179–185.

Caspary, DM, Finlayson PG (1991) Superior Olivary Complex: Functional neuropharmacology of the major cell types. In RA Altschuler, RP Bobbin, BC Clopton and DW Hoffman (Eds), Neurobiology of Hearing: The Central Auditory System, Raven Press NY, pp 141–163

Caspary, D. M., Palombi, P. S., Backoff, P.M., Helfert, R. H. and Finlayson, P. G. (1993) GABA and glycine inputs control discharge rate within the excitatory response area of primary-like and phase-locked AVCN neurons, In M.A. Merchan, J.M. Juiz, D.A. Godfrey & E. Mugniani (Eds) The Mammalian Cochlear Nuclei: Organization and Function, Plenum Press, New York, pp. 239–252.

Caspary, D.M. (1986) Cochlear nuclei: Functional neuropharmacology of the principal cell types. In R.A. Altschuler, D.W. Hoffmann, and R.P. Bobbin (Eds): Neurobiology of Hearing: The Cochlea. New York: Raven Press, pp. 303–332.

Caspary DM, Milbrandt JC and Helfert RH (1995) Central auditory aging: GABA changes in the inferior colliculus. Exp Gerontol. 30:349–60/

Cik M, Chazot PL, Stephenson FA (1994) Expression of NMDAR1–1a (N598Q)/NMDAR2A receptors results in decreased cell mortality. Eur J Pharmacol 226:1–3.

Collingridge GL, Herron CE, Lester RAJ (1988) Frequency-dependent N-methyl-D-aspartate receptor mediated synaptic transmission in the rat hippocampus. J Physiol (Lond) 399:301–312.

Collingridge GL, Singer W (1990) Excitatory amino acid receptor and synaptic plasticity. Trends Pharmacol Sci 11:290–296.

Cotman CW, Iversen LL (1987) Excitatory amino acids in the brain focus on NMDA-receptor. Trends Neurosci 10:263–265.

Durand GM, Gregor P, Zheng X, Bennett MVL, Uhl GR, Zukin RS (1992) Cloning of an apparent splice variant of the rat *N*-methyl-*D*-aspartate receptor NMDAR1 with altered sensitivity to polyamines and activators of protein kinase C. Proc Natl Acad Sci USA 89:9359–9363.

Dupont J, Bonneau JM, Altschuler RA, Aran JM (1994) GABA and glycine changes in the guinea pig brain stem auditory nuclei after total destruction of the inner ear. ARO Abstracts.

Dupont J, Young C, Sapan A, Bonneau JM, Altschuler RA (1995) Plasticity of glycine immunoreactivity and cell size changes in the superior olivary complex after unilateral deafferentation. ARO Abstr.

Evans, E. F. and Zhao, W (1993) Neuropharmacological and neurophysiological dissection of inhibition in the mammalian dosal cochlear nucleus, In M.A. Merchan, J.M. Juiz, D.A. Godfrey & E. Mugniani (Eds) The Mammalian Cochlear Nuclei: Organization and Function, Plenum Press, New York, pp 253–266.

Frostholm, A. and Rotter, A. (1985) Glycine receptor distribution in mouse CNS: autoradiographic localization of [^{3}H] strychnine binding sites. Brain Res. Bull., 15, 475–486.

Glendenning, K. K. and Baker, B. N. (1988) Neuroanatomical distribution of receptors for three potential inhibitory neurotransmitters in the brainstem auditory nuclei of the cat. J. Comp. Neurol., 275, 288–308.

Godfrey, D.A., J.A. Parli, J.D. Dunn, and C.D. Ross (1988) Neurotransmitter microchemistry of the cochlear nucleus and the superior olivary complex. In J. Syka, and R.L. Masterton (eds): Auditory Pathways. Plenum Press, NY

Godfrey DA: Comparison of quantitative and immunohistochemistry for choline acetyltransferase in the rat cochlear nucleus. In: Merchan MA, Juiz JM, Godfrey DA & Mugnaini E (eds): Mammalian Cochlear Nuclei: Organization and Function , (Plenum Press, New York, 1993) 267–278.

Helfert RH, Juiz JM, Bledsoe SC, Bonneau J, Wenthold RJ, Altschuler RA (1992) Patterns of glutamate, glycine and GABA immunolabeling in four synaptic terminal classes in the lateral superior olive of the guinea pig. J Comp Neurol 323:305–325.

Hollmann M, Boulter J, Maron C, Beasley L, Sullivan J, Pecht G, Heinemann S (1993) Zinc potentiates agonist-induced currents at certain splice variants of the NMDA receptor. Neuron 10:943–954.

Hunter C, Mathura J, Wenthold RJ (1995) Expression of NMDA-selective glutamate receptor NR1 subunit isoforms in the rat cochlear nucleus. Abst Assoc Res Otolaryngol pp31.

Hunter C, Petralia RS, Vu T, Wenthold RJ (1993) Expression of AMPA-selective glutamate receptor subunits in morphologically defined neurons of the mammalian cochlear nucleus. J. Neurosci 13:1932–46.

Juiz JM, Helfert RH, Bonneau JM, Wenthold RJ, Altschuler RA (1996) Three classes of inhibitory amino acid terminals in the cochlear nucleus of the guinea pig. J. Comp Neurol., 373:11–26, 1996.

Juiz JM, Rubio ME, Helfert RH, Altschuler TA (1993) Localizing putative excitatory amino acid endings in the cochlear nucleus by quantitative immunocytochemistry. In M.A. Merchan, J.M. Juiz, D.A. Godfrey & E. Mugniani (Eds) The Mammalian Cochlear Nucleus: Organization and Function, Plenum Press, New York, pp 167–178.

Kolston, J., K.K. Osen, C.M. Hackney, O.P. Ottersen, and J. Storm-Mathissen (1992) An atlas of glycine and GABA-like immunoreactivity and colocalization in the cochlear nuclear complex of the guinea pig. Anat. Embryol. 186:443–465.

Kuhse, J., Schmieden, V. and Betz, H. (1990) Identification and functional expression of a novel ligand binding subunit of the inhibitory glycine receptor. J. Biol. Chem., 265, 22317–22320.

Kuhse, J., Schmieden, V. and Betz, H. (1990b) A single amino acid exchange alters the pharmacology of neonatal rat glycine receptor subunit. Neuron, 5, 867–873.

Kutsuwada T, Kashiwabuchi N, Mori H, Sakimura K, Kushiya E, Araki K, Meguro H, Masaki H, Kumanishi T, Arakawa M, Mishina M (1992) Molecular diversity of the NMDA receptor channel. Nature 358:36–41.

Malosio, M.-L., Pouey, B. M., Kuhse, J. and Betz, H. (1991) Widespread expression of glycine receptor subunit mRNAs in the adult and developing rat brain. EMBO J., 10, 2401–2409.

Meguro H, Mori H, Araki K, Kushiya E, Kutsuwada T, Yamazaki M, Kumanishi T, Arakawa M, Sakimura K, Mishina M (1992) Functional characterization of heteromeric NMDA receptor channel expressed from cloned cDNAs. Nature 357:70–74.

Milbrandt JC & Caspary DM (1995) Age-related reduction of strychnine binding sites in the cochlear nucleus of Fischer 344 rat. Neuroscience 67:713–9.

Milbrandt JC, Albin RL and Caspary DM (1994) Age-related decrease in GABAB binding in the Fischer 344 rat inferior colliculus. Neurobiol-Aging 15:699–703.

Miller JM, Altschuler RA, Niparko JK, Hartshorn DO, Helfert RH, Moore JK (1991) Deafness-induced changes in the central auditory system and their reversibility and prevention. In: A Dancer, D Henderson, RJ Salvi, RP Hamernik (Eds) Noise induced hearing loss. St. Louis: Mosby Year Book

Mishina M, Mori H, Araki K, Kushiya E, Meguro H, Kutsuwada T, Kashiwabuchi N, Ikeda K, Nagasawa M, Yamazaki M, Masaki H, Yamakura T, Morita T, Sakimura K (1993) Molecular and functional diversity of the NMDA receptor channel. In H Higashida, T Yoshioka, K Mikoshiba, (Eds) Molecular basis of ion channels and receptors involved in nerve excitation, synaptic transmission and muscle contraction The New York Academy of Sciences, New York, pp136–152.

Monaghan DT, Bridges RJ, Cotman CW (1989) The excitatory amino acid receptor: their classes, pharmacology and distinct properties in the function of the central nervous system. Rev Pharmacol Toxicol 29:365–402.

Monyer H, Sprengel R, Schoepter R, Herb A, Higuch M, Lomel H, Burnashev N, Sakmann B, Seeburg PH (1992) Heteromeric NMDA-Receptor ; Molecular and functional distinction of subtypes. Science 256:1217–1221.

Moore DR (1991) Development and plasticity in the ferret auditory system, in RA Altschuler, DW Hoffman, BC Clopton & RW Bobbin, (Eds), Neurobiology of Hearing: The Central Auditory System, Raven Press, NY,

Moriyoshi K, Masu M, Ishii T, Shigemoto R, Mizuno N, Nakanishi S (1991) Molecular cloning and characterization of the rat NMDA receptor. Nature 354:31–37.

Nakanishi S (1992) Molecular diversity of glutamate receptors and implications for brain function. Science 258:597–603

Oertel, D. and Wickesberg, R. E. (1993) Glycinergic inhibition in the cochlear nuclei: evidence for tuberculoventral neurons being glycinergic, . In M.A. Merchan, J.M. Juiz, D.A. Godfrey & E. Mugniani (Eds) The Mammalian Cochlear Nuclei: Organization and Function, Plenum Press, New York, p 225–237.

Potashner, S. J., Benson, C. G., Ostapoff, E.-M., Lindberg, N. and Morest, D. K. (1993) Glycine and GABA: transmitter candidates of projections descending to the cochlear nucleus, . In M.A. Merchan, J.M. Juiz, D.A. Godfrey & E. Mugniani (Eds) The Mammalian Cochlear Nuclei: Organization and Function, Plenum Press, New York, p 195–210.

Rajan R, Irvine DR Wise LZ Heil P: Effect of unilateral partial cochlear lesions in adult cats on the representation of lesioned and unlesioned cochleas in primary auditory cortex. J Comp Neurol (1993) 338:17–49.

Rajan R, Irvine DRF (1996) Features of and boundary conditions for lesions induced reorganization of adult auditory cortical maps. In RJ Salvi, D Henderson, F Fiorini and V Colletti (Eds), Auditory System Plasticity and Regeneration, Thieme Medical Publishers, Inc, NY pp 224–237.

Rubel EW, Hyson RL, Durham D. Afferent regulation of neurons in brain stem auditory system. J Neurobiol 1990;21:169–96.

Salvi RJ, Wang J, Powers N (1996) Rapid functional reorganization in the inferior colliculus and cochlear nucleus after acute cochlear damage. In RJ Salvi, D Henderson, F Fiorini and V Colletti (Eds), Auditory System Plasticity and Regeneration, Thieme Medical Publishers, Inc, NY pp 275–296.

Sanes, D. H., Geary, W. A., Wooten, G. F. and Rubel, E. W. (1987) Quantitative distribution of the glycine receptor in the auditory brain stem of the gerbil. J. Neurosci., 7, 3793–3802.

Suneja SK, Benson CG, Potashner SJ: Cochlear ablation: long-term effects on uptake and release of D-aspartate, glycine and GABA in brain stem auditory nuclei. Neurosci Abstr. 400:11, 1994.

Tõlle, T. R., Berthele, A., Zieglgänsberger, W., Seeburg, P. H. and Wisden, W. (1993) The differential expression of 16 NMDA and non-NMDA receptor subunits in the rat spinal cord and in periaqueductal gray. J. Neurosci., 13, 5009–5028.

Wenthold, R.J., and C. Hunter (1990) Immunocytochemistry of glycine and GABA receptors in the central auditory system. In O.P. Ottersen, and J. Storm-Mathissen (Eds): Glycine Neurotransmission. Chichester: J. Wiley and Sons, pp.391–416.

Wenthold R. J., Parakkal, M. H., Oberdorfer, M. D. and Altschuler, R. A. (1988) Glycine receptor immunoreactivity in the ventral cochlear nucleus of the guinea pig. J. Comp. Neurol., 276, 423–435.

Wenthold, R.J. (1991) Neurotransmitters of brainstem auditory nuclei. In R.A. Altschuler, R.P. Bobbin, B.M. Clopton, and D.W. Hoffman (eds): Neurobiology of Hearing, The Central Auditory System. New York: Raven Press, pp.121–140.

Wenthold RJ, Hunter C and Petralia RS (1993) Excitatory amino acids in the rat cochlear nucleus. In M.A. Merchan, J.M. Juiz, D.A. Godfrey & E. Mugniani (eds) The Mammalian Cochlear Nucleus: Organization and Function, Plenum Press, New York, pp 179–195.

Wickesberg, R.E., and D. Oertel (1993) Intrinsic connection in the cochlear nuclear complex studied in vitro and in vivo. In M. Merchan, J. Juiz, D. Godfrey, and E. Mugnaini (Eds): The Mammalian Cochlear Nuclei: Organization and Function. New York: Plenum Press. pp.77–90.

Wu, S.H., and D. Oertel (1986) Inhibitory circuitry in the ventral cochlear nucleus is probably mediated by glycine. J. Neurosci. 6:2691–706.

Zarbin, J. M., Wamsley, J. K. and Kuhar, M. J. (1981) Glycine receptor: light microscopic autoradiographic localization with [^{3}H] strychnine. J. Neurosci., 1, 532–547.

19

ACTION OF PUTATIVE NEUROTRANSMITTERS AND NEUROMODULATORS ON NEURONES IN THE VENTRAL NUCLEUS OF THE TRAPEZOID BODY

Donald Robertson and Xueyong Wang

The Auditory Laboratory
Department of Physiology
The University of Western Australia
Nedlands, Western Australia 6907

1. INTRODUCTION

Neurones of the ventral nucleus of the trapezoid body (VNTB) contribute variously to the ascending and descending auditory pathways (Winter et al., 1989; Warr & Beck, 1996; Spangler et al., 1987). By virtue of this connectivity, the VNTB has the potential to influence processing of auditory information in a variety of ways, either by contributing to processing at the brainstem level, or by altering the features of incoming afferent information at lower stages, even at the level of the organ of Corti.

Despite its potentially pivotal role in auditory information processing, so far there is little known of the functional pharmacology of the VNTB. For this reason, we chose to study the effects of a variety of neurotransmitters and neuromodulators on neurones in the VNTB using the *in vitro* slice technique.

2.MATERIALS AND METHODS

Coronal slices of the auditory brainstem were prepared from Wistar and PVG/c rats ranging in age from 3 to 8 weeks of age, using methods described previously (Robertson, 1996). All slices were 300–400 μm in thickness and were maintained in artificial cerebro-spinal fluid at room temperature. Recordings were made from single neurones in the VNTB using high resistance microelectrodes. The responses of cells to standard depolarizing current pulses were monitored before, during and after the application of pharmacological agents dissolved in the fluid bathing the recording chamber. For morphological

Acoustical Signal Processing in the Central Auditory System
edited by Syka, Plenum Press, New York, 1997

reconstruction, cells were injected with biocytin through the recording microelectrode. Subsequent processing and reconstruction of the labeled cells used standard techniques. In a number of experiments we pre-labeled olivocochlear efferent neurones in the VNTB by intracochlear injection of Fast Blue 24–48hrs prior to slice preparation. Microelectrodes filled with Lucifer Yellow were then used to impale single Fast Blue-labeled olivocochlear efferents under direct visual observation using UV epi-illumination fluorescence microscopy.

3.RESULTS

Figure 1 shows a diagramatic representation of the region of auditory brainstem from which slice recordings were made. The structures shown in the diagram could all be visualized in slices and electrodes could be inserted accurately into the VNTB under visual control.

Typical results obtained from a single neurone in the rat VNTB are shown in Figure 2. In this example, the same cell is shown exhibiting excitatory responses to bath application of substance P (1 μM) and noradrenaline (5 μM). The excitatory effects consisted of a small, but measureable depolarization of the resting membrane potential, accompanied by a dramatic increase in the rate of action potentials evoked by a fixed amplitude, suprathreshold depolarizing current pulse injected into the cell through the recording microelectrode.For both substance P and noradrenaline, there was also an accompanying increase in the cell's access resistance, though this was most obvious in the case of substance P.

We have used this basic technique to study the effects on a large number of VNTB cells of a variety of neuroactive substances. We have found that when effective, substance P is always excitatory in its action, whereas enkephalin is always inhibitory. The bioamines noradrenaline and serotonin can be either excitatory or inhibitory. The octapeptide choleycystokinin (CCK_8) was found to be either excitatory or inhibitory at very low con-

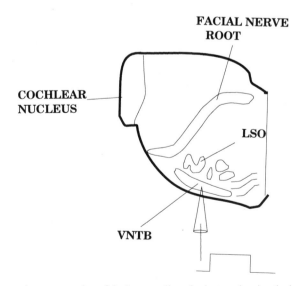

Figure 1. Diagrammatic representation of the lower auditory brainstem showing the location of the VNTB in relation to other structures. LSO, lateral superior olivary nucleus.

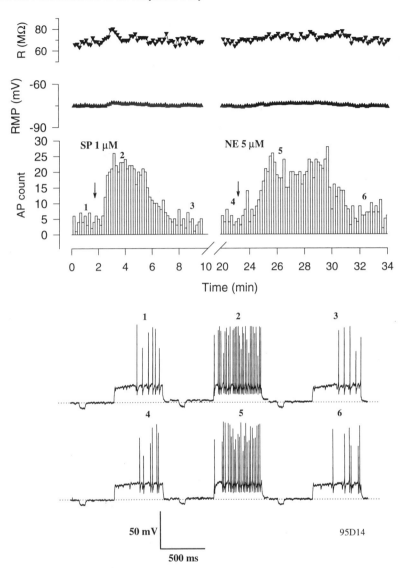

Figure 2. Typical examples of the responses of a single VNTB neurone to bath application of substance P (SP) and noradrenaline (NE) in micromolar concentrations. Top two rows of panels show cell access resistance and resting membrane potential (RMP). Third row of panels from top shows plots of the number of action potentials evoked by a test depolarizing pulse injected through the recording microelectrode. Note dramatic but reversible increase in action potential firing rate caused by drug application in both cases. Lower panels show actual records of cell membrane potential. Numbers correspond to indicated time points on third row of panels. Variations in action potential height are an artifactual consequence of digital sampling limitations.

centrations in a small number of cells (Fig. 3) , whilst another peptide, somatostatin, had little or no effect.

Of particular interest was the very strong and consistent excitatory effect of substance P. This was found in 100% of neurones exhibiting the action potential shape classified previously as AHP2 (Robertson, 1996). Neurones exhibiting the other principal action potential type found in VNTB cells (AHP1) were unresponsive to substance P. This situ-

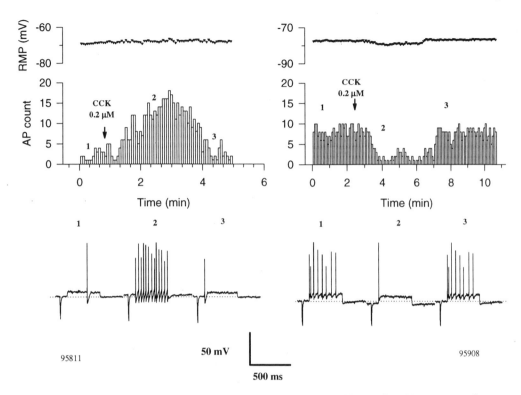

Figure 3. Responses of two different VNTB cells to CCK$_8$ applied in the bathing medium. Note strong excitatory effect in one case and strong inhibition in the other.

ation was in contrast, for example, to noradrenaline, for which about equal numbers of AHP2 and AHP1 neurones showed excitatory effects (Table 1).

It has been suggested that the AHP2 action potential shape may be characteristic of medial olivocochlear neurones (Robertson, 1996). Supportive evidence comes from the axon trajectories of biocytin-injected cells. Figure 4 shows a partial histological reconstruction of a substance P-sensitive neurone (AHP2 action potential type) in the rat lower auditory brainstem. The cell was clearly located in the VNTB and its main axon ascended dorsally out of the superior olivary complex and headed towards the floor of the fourth ventricle. We found that most substance P-responsive cells of the AHP2 type exhibited this kind of axon trajectory and in addition showed morphological features such as thick, slowly tapering, spinous dendrities that had previously been associated with the AHP2 category.

Table 1. Proportions of VNTB cells in different action potential categories responsive to noradrenaline and Substance P

	AHP2 Type	AHP1 Type
Responsive to noradrenaline	77%	82%
Responsive to substance P	100%	0%

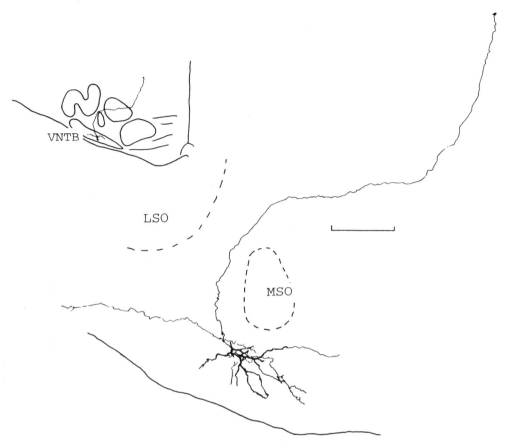

Figure 4. An example of a partial histological reconstruction of a substance P-sensitive AHP2 cell in the VNTB. Scale bar is 0.6 mm in small inset and 0.1 mm in more magnified drawing. Note long, dorsally-ascending axon. LSO, lateral superior olivary nucleus; MSO, medial superior olivary nucleus.

These features of substance P-responsive cells were distinct from those of other cells in the VNTB. Thus, for example, cells with the action potential shape referred to as AHP1, were usually unresponsive to substance P, had thin, rapidly tapering dendrites, and axons that formed terminal arborizations within the VNTB and immediately adjacent nuclei of the superior olivary complex, but did not send their axons dorsally towards the floor of the fourth ventricle.

By making prior Fast Blue injections into the cochlea, we succeeded in obtaining some recordings from pre-labeled olivocochlear neurones. These experiments were difficult to perform and in most cases the physiology of the cells was not stable for more than a few minutes after impalement, for reasons that we have not as yet established. This meant that the responses to substance P could not be reliably tested. However, we found that when action potentials could be recorded from such pre-labeled, identified neurones, they were of the AHP2 type in most cases (Fig. 5).

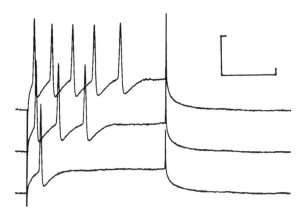

Figure 5. Example of action potentials recorded from a single VNTB neurone that was pre-labeled by intra-co-chlear injection of Fast Blue 24 h before slice preparation. The action potential shape is of the AHP2 category as described by Robertson (1996). The responses to three different levels of current injection are shown, with each trace being artificially displaced for the purposes of clarity. The resting potential of the cell was -70 mV and the vertical and horizontal scale bars represent 40 mV and 20 ms respectively.

4. DISCUSSION

The results obtained in this *in vitro* investigation indicate that cells in the VNTB are potentially subject to a variety of neurochemical influences, some inhibitory and some excitatory. Much, however, remains to be done before a functional role for substances like noradrenaline, serotonin, substance, P, enkephalin and CCK_8 can be firmly established. Thus the presence of nerve terminals containing these substances, and the presence of specific receptors on VNTB neurones both need to be demonstrated. In addition, the source of any innervation by these neuroactive substances remains to be discovered. mRNA for substance P precursors and for CCK have been demonstrated in the inferior colliculus, whereas mRNA for preproenkephalin is prominent in cells of the VNTB itself (Wynne et al., 1995). Thus, the inhibitory action of enkephalin might normally be employed by intrinsic circuitry of the VNTB, whereas other actions might reflect a possible role for descending inputs to this nucleus (Thompson & Thompson, 1993; Vetter et al., 1993). A noradrenergic innervation of the VNTB has also been demonstrated immunocytochemically (Wynne & Robertson, 1996), though its exact origin is unknown.

The data also suggest discrete pharmacological differences between different functional classes of VNTB neurones. Thus the available evidence indicates that medial olivocochlear neurones may possess receptors for substance P, whereas interneurones are much less likely to exhibit responsivenes to this peptide. On the other hand, roughly equal percentages of the different neurone types in VNTB show excitatory responses to noradrenaline. Once again, it remains to be determined how these different profiles of pharmacological sensitivity relate to different roles in auditory signal processing.

5. ACKNOWLEDGMENTS

Supported by grants from the NH&MRC of Australia, The Australian Research Grants Commission and The University of Western Australia.

6. REFERENCES

Robertson, D. (1996). Physiology and morphology of cells in the ventral nucleus of the trapezoid body and rostral periolivary regions of the rat superior olivary complex studied in slices. Audit. Neurosci., 2, 15–31.

Spangler, K.M., Cant, M.B., Henkel, C.K., Farley, G.R. and Warr, W.B. (1987). Descending projections from the superior olivary complex to the cochlear nucleus of the cat. J. Comp. Neurol., 259, 452–465.

Thompson, A. M. and Thompson, G. C. (1993). Relationship of descending inferior colliculus projections to olivo-cochlear neurons. J. Comp. Neurol. 335, 402–412.

Vetter, D.E., Saldana, E. and Mugnaini, E. (1993). Inputs from the inferior colliculus to medial olivocochlear neurons in the rat: A double label study with PHA-L and Cholera Toxin. Hear.Res. 70, 173–186.

Warr, W.B. and Beck, J.E. (1996). Multiple projections from the ventral nucleus of the trapezoid body in the rat. Hear.Res. 83–101.

Winter, I.M., Robertson, D. and Cole, K.S. (1989). Descending projections from auditory brainstem nuclei to the cochlea and cochlear nucleus of the guinea pig. J. Comp. Neurol. 280, 143–157.

Wynne, B. and Robertson, D. (1996) Localization of dopamine-β-hydroxylase-like immunoreactivity in the superior olivary complex of the rat. Audiol. Neuro-otol. 1, 54–64.

Wynne, B. Harvey, A.R., Robertson, D. and Sirinathsinghji, D.J.S. (1995). Neurotransmitter and neuromodulator systems of the rat inferior colliculus and auditory brainstem studied by in situ hybridization. J. Chem. Neuroanat. 9, 289–300.

20

THE NUCLEI OF THE LATERAL LEMNISCUS

Old Views and New Perspectives

Miguel A. Merchán,[1] Manuel S. Malmierca,[1] Victoria M. Bajo,[1] and Jan G. Bjaalie[2]

[1]Laboratory of the Neurobiology of Hearing
Department of Cellular Biology and Pathology
Faculty of Medicine
University of Salamanca
Spain
[2]Department of Anatomy
Institute of Basic Medical Sciences
University of Oslo
Norway

1. PREFACE

A unique feature of the auditory brainstem is the divergent/convergent nature of the pathways from the auditory nerve to the inferior colliculus (IC, reviewed in Irvine, 1992). Some of the projections from the cochlear nucleus complex to the IC are direct while others are indirect via the superior olivary complex and the nuclei of the lateral lemniscus (NLL). In these nuclei, significant monaural and binaural information are extracted from the auditory signal.

In this chapter, we will review the current knowledge and recent new ideas on the anatomical organisation of the NLL and their possible implications for auditory processing. In the past, cat was the most widely used experimental animal in auditory research, including the NLL (Aitkin et al., 1970; Brugge et al., 1970; Guinan et al., 1972; Adams, 1979; Glendenning et al., 1981; Henkel and Spangler 1983; Whitley and Henkel, 1984; Shneiderman et al., 1988). In the last decade, however, the majority of studies on the anatomy and physiology of the NLL have been carried out in bats (Covey and Casseday, 1986, 1991; Covey, 1993a,b; Markovitz and Pollak, 1993, 1994; Yang and Pollak, 1994a,b; Yang et al., 1996) and rats (Bajo et al., 1993; Merchán et al., 1994; Wu and Kelly, 1995a,b, 1996; Fu et al., 1996; Merchán and Berbel, 1996). Because the comparative study of the central nervous system rests on the determination of valid homologies among nuclei and fibre tracts in different species (Ebner, 1963; Bullock, 1984; Morest and Winer, 1986), we will highlight the similarities and differences observed in these three species. We shall also describe studies under way in our laboratory that combine tract-tracing and computer-assisted 3-D reconstructions on the cat NLL, which add a new perspective to the previously published studies.

2. NOMENCLATURE

The parcellation and terminology of the NLL represent a problem. Several cytoarchitectonic maps have been used in different species by different authors (e.g., cat: Adams, 1979; Glendenning et al, 1981; bat: Covey and Casseday, 1986,1991; Covey 1993a,b; Yang et al., 1996; rat: Paxinos and Watson, 1986; Bajo et al., 1993; Caicedo and Herbert, 1993; Merchán et al., 1994; Merchán and Berbel, 1996). The three standard terms: ventral, intermediate and dorsal nuclei of the lateral lemniscus (VNLL, INLL, and DNLL, respectively) have been used ambiguously. Most of the ambiguity refers to the cell groups ventral to the well defined DNLL, i.e. INLL and VNLL. In the present review we shall collectively refer to these ventral groups as the ventral complex of the lateral lemniscus (VCLL, Fig. 1A) . This categorisation conforms also with the presence of two distinct functional systems, a monaural ventral system and a binaural dorsal system (Aitkin et al., 1970; Brugge et al., 1970; Guinan et al., 1972).

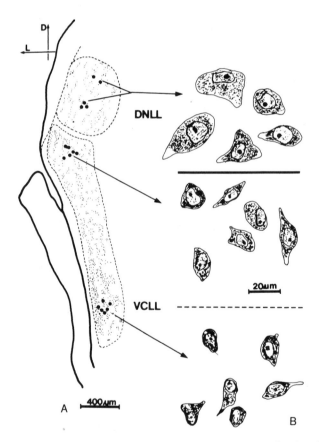

Figure 1. A. Camera lucida drawings from a Nissl stained section in the rat showing the relative density of neurones in the VCLL and DNLL. B. Details at high magnification of Nissl stained somata. Note the variety of shapes of NLL cell bodies, which are relatively larger in the DNLL as compared with those of the VCLL (redrawn from Merchán and Berbel, 1996, J. Comp. Neurol. 372, 245–263).

There are some connectional, neurochemical and physiological properties which are unique to these two systems. In addition, there may be some functional interaction between them.

3. THE MONANURAL SYSTEM: THE VENTRAL COMPLEX OF THE LATERAL LEMNISCUS

3.1. Intrinsic Organisation and Connections

The VCLL represents the groups of neurones embedded within the LL (Fig. 1A) and located between the superior olivary complex and DNLL. It receives inputs mainly from the contralateral ear, as opposed to the DNLL which receives inputs from both ears. In cat, the VCLL has been considered as a single nucleus, the VNLL (Adams, 1979) with three zones (ventral, middle and dorsal), each one having a distinct pattern of Nissl stained cells. Glendenning et al. (1981), on the contrary, consider the dorsal zone a distinct nucleus, the INLL characterised by a projection from the medial nucleus of the trapezoid body (MNTB). This projection has been confirmed after injections of tritiated Leucine in the MNTB (Spangler et al., 1985). In the *Eptesicus fuscus* bat, the VCLL has likewise been subdivided into a VNLL and an INLL (Covey, 1993a). The VNLL is further subdivided into a multipolar (VNLLm) and a columnar (VNLLc) division. Although the same terms VNLL and INLL are used both for cat and bat, the correspondence is not strait forward. The unique projection from the MNTB, which distinguishes the INLL from the VNLL in cat (Glendenning et al., 1981), is not solely confined to the INLL in bat. The MNTB in bats, thus projects both to the INLL and the VNLLc (Covey, 1993a).

Thus far, the most elaborated study of the VCLL in rat is that recently published by Merchán and Berbel (1996). They have shown that the cells exhibit a variety of shapes and sizes in Nissl stained sections (Fig. 1B) in accordance with previous studies in cat (Adams, 1979; Glendenning et al., 1981). Nevertheless, Merchán and Berbel (1996) hypothesise that the variation is due to different orientation of a single type of stellate cell. The same authors, using tract-tracing methods combined with computer-assisted 3-D reconstructions, have shown that these cells are organised in laminae. Whether the dendritic arbors are flat, as those demonstrated for the pyramidal cells of the cat dorsal cochlear nucleus (DCN, Blackstad et al., 1984) and the cells in the rat central nucleas of the inferior colliculus (CNIC, Malmierca, 1991, Malmierca et al., 1993), remains to be demonstrated. After injections in the CNIC of the tracer biotinylated dextran amine (BDA, which is transported both anterogradely and retrogradely), Merchán and Berbel (1996) showed that the labelled cells form clusters along the extent of the VCLL without an apparent orientation (Fig. 2A) as previously described in other species (e.g., Glendenning et al., 1981). But 3-D reconstructions, comprising every section through the VCLL demonstrate that the cells form a partly continuous structure, i.e., a lamina. When the injection of BDA is located at the low frequency region of the CNIC, the labelled cells are situated in the periphery of this complex, while an injection placed at the high frequency region rendered labelled cells in the core of the complex (Fig. 3). Thus, Merchán and Berbel (1996) concluded that the rat VCLL is a single nucleus, made of isofrequency laminae extending the whole dorsoventral and rostrocaudal length of the complex. It is interesting to mention in this context, that injections made in the guinea pig CNIC with biocytin (Malmierca et al., 1996) resulted with a similar pattern of labelled patches in the VCLL (Fig. 2B), suggesting that the complex in guinea pig may be organised as in the rat.

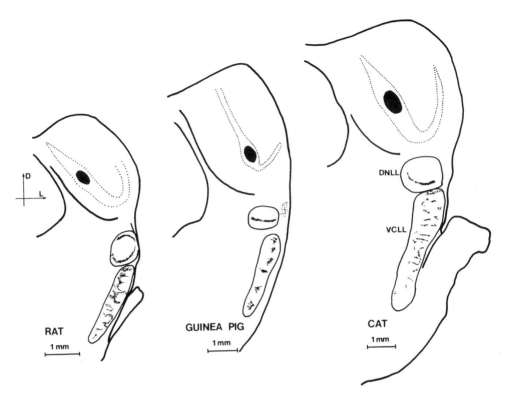

Figure 2. Camera lucida drawings from the labelling observed in the VCLL and DNLL after restricted injections of neuroanatomical tracers into the CNIC of the rat (BDA injection Merchán et al. 1994, Merchán and Berbel 1996), guinea pig (biocytin injection Malmierca et al. 1996), and cat (BDA injection). Patches of labelling in the VCLL with no apparent orientations are observed. The labelling in the DNLL forms an annular band in the rat while the labelling form bands horizontally oriented (slightly concave) in the guinea pig and cat.

From the study of Merchán and Berbel (1996) it is also clear that virtually all cells in the rat VCLL project to the CNIC as it is the case in cat (Whitley and Henkel, 1984). The afferent projections to the VCLL arise mainly from the contralateral ventral cochlear nucleus (VCN, van Noort, 1969; Glendenning et al. 1981). Using degeneration techniques, these authors have shown in cat that the VCN projections produce a patchy pattern of degenerated boutons, that matches those of cells seen after the injections of retrogradely tracers into the CNIC of rat (Merchán and Berbel, 1996) and guinea pig (Malmierca et al., 1996). Careful inspection of drawings by Oliver (1987), following injections of tritiated amino acids in the cat VCN, reveals similar patches of terminal fields in the VCLL.

Different cell types from the VCN project to the VCLL (Adams, 1979; Glendenning et al., 1981, Covey, 1993a). The octopus cells project to the ventral part of the complex only while the stellate and globular cells project to the whole complex (Adams, 1979; Glendenning et al., 1981; Friauf and Ostwald, 1988; Smith et al., 1991). The MNTB projects to the dorsal portion of the complex (Glendenning et al., 1981; Spangler et al., 1985). Others sources of projections to the VCLL are the periolivary nuclei on the ipsilateral side (Glendenning et al., 1981). One of the missing key pieces in this puzzle is the organisation of the local connections arising within the VCLL neurones. It has been shown that many auditory neurones have extensive local connections in addition to their main projection

94040

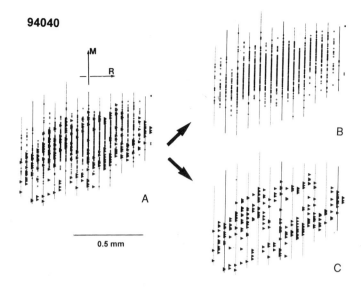

Figure 3. Plots of retrogradely labelled neurons in the rat VCLL from a case where biocytin and horseradish peroxidase (HRP) have been injected in the same animal at two different CNIC locations corresponding to the low- and high-frequency regions, respectively. A. Horizontal view of the VCLL showing together the distribution of HRP-labelled (open circles) and biocytin-labelled neurons (filled triangles). Separate plots of the distribution of HRP-labelled neurons (open circles) projecting to the high-frequency region of the CNIC and that of biocytin-labelled neurons (filled triangles) projecting to the low-frequency region of the CNIC are shown in B and C, respectively (redrawn from Merchán and Berbel, 1996, J. Comp. Neurol. 372, 245–263). The high frequency injection rendered cells in the core of the VCLL, while the low frequency one produced labelled cells in the periphery.

(e.g., Morest and Winer, 1986; Oliver et al., 1991; Malmierca, 1991; Wickesberg and Oertel, 1993; Malmierca et al., 1995), and some preliminary results of intracellularly labelled cells from gerbil brain slices have shown similar findings (Tuck et al., 1993) in all NLL nuclei.

3.2. Neurochemistry and Physiological Significance of the VCLL

Studies of the transmitter chemistry of the VCLL are limited (Saint-Marie, 1993; Winer et al., 1995; González-Hernández, 1996). Studies under way in our laboratory support previous preliminary findings in the cat (Saint-Marie, 1993) that the majority of cells in the ventral part of the complex are glycine and/or GABA immunoreactive, although the incidence of colocalisation of the two transmitters seems considerably higher than previously estimated. Our findings (Riquelme et al., unpublished observations) also indicate that the dorsal part of the complex is composed exclusively of glycine and GABA negative cells. But it remains to be demonstrated the excitatory nature of these cells (Saint-Marie, 1996). The larger amount of glycine-positive pericellular puncta in the dorsal part of the complex agrees with the denser input from the MNTB.

The afferents and intrinsic connections of the VCLL influence the discharge properties of the VCLL neurones. The heavy bias of the VCLL input to the contralateral VNC is reflected in the response properties of the VCLL neurones, i.e., most of them respond only to contralateral stimulation, and the VCLL is primarily devoted to the analysis of monaural properties of the sound (Aitkin et al., 1970; Brugge et al., 1970; Guinan et al., 1972). (A weak projection form the ipsilateral CNC has been described -Glendenning et al., 1981

- and recently Batra and Fitzpatrick, 1997 - have shown preliminary data of binaural responses in some neurones located in the medial margin of the complex.)

Whether there is a tonotopical organisation or not in the VCLL is a matter of discussion. Aitkin et al. (1970) found a dorsoventral gradient of responses with the low frequency ones located dorsally and the high frequency ones are located ventrally. By contrast, Guinan et al., (1972) failed to demonstrated such an organisation. The anatomical studies using anterograde tracers (e.g., Whitley and Henkel, 1984) have suggested that the VCLL lacks a tonotopical organisation.

Neurones in the ventral portion of the complex have shown a peristimulus histogram (PSTH) of the onset type after stimulation with pure tones (Aitkin et al., 1970; Covey, 1993a). This is consistent with the anatomically demonstrated input from the octopus cells which are also onset units (Oertel et al., 1990). Large terminals resembling the calyces of Held of the MNTB have been shown in the ventral portion of the complex and, most probably originate from the octopus cells (Adams, 1979). Thus, it seems that these cells would be suited to convey precise and secure temporal information. In addition, both primary-like and chopper responses have been observed (Covey, 1993a) after stimulation with pure tones, consistent with the idea that also globular and multipolar cells take part in this projection (Smith et al., 1991). The VCLL thus may be acting as a relay station in the projection to the IC but these responses may also be generated *de novo* within the VCLL.

Preliminary reports obtained from intracellular recordings from VCLL cells based on brain slice preparations in young rats (Wu, 1996) indicate that there are at least two functional types of cells in the VCLL. One type has phasic responses with only one action potential; the other shows tonic responses with a series of several action potentials after injection of a current to the cell membrane. These findings agree with the cell activities shown *in vivo* (Covey, 1993a). Interestingly, no reports of regularly firing cells (Wu, 1996) are mentioned in this study despite the observation of chopper units in *in vivo* studies.

In addition to these differences in the response pattern, there are some common physiological features of responses to tonal stimuli, at least in bat (Covey, 1993a). These include little or no spontaneous activity, constant latency, short integration times and broad tuning curves. All this together implies that the VCLL neurones show a clear suitability for encoding temporal events. By this ability to detect variations in the temporal features of the auditory input, the VCLL may be a fundamental component of the neural circuitry involved in vocalizations and speech-like communications.

4. THE BINAURAL SYSTEM: THE DORSAL NUCLEUS OF THE LATERAL LEMNISCUS

4.1. Cytoarchitecture and Intrinsic Organisation

The DNLL is a distinctive group of neurones embedded within the dorsal part of the lateral lemniscus (Fig. 1A). Dorsally, it is separated from the IC by a region devoid of neurones, ventrally, it limits with the VCLL through a narrow region of flat neurones. Laterally, the DNLL is separated from the pial surface by the nucleus sagulum, which is populated by small cells (Henkel and Shneiderman, 1988). Because the DNLL also contains some small cells, the lateral limit can not be established accurately at every rostrocaudal level (Shneiderman et al., 1988; Bajo et al., 1993; Merchán et al., 1994). In contrast to the VCLL, the DNLL receives input from both ears, and it projects to both ICs (Fig. 4) and to the homologue nucleus on the opposite side (Fig. 5) through the commissure of Probst. The DNLL cells are, therefore, influenced binaurally (Aitkin et al., 1970; Brugge et al., 1970).

The DNLL in rat is cube-shaped, and in contrast to the VCLL, the similarity among species is apparent, except for minor variations (Adams, 1979; Kane and Barone, 1980; Glendenning et al., 1981; Covey and Casseday, 1986; Iwahori, 1986; Shneiderman et al., 1988; Hutson et al., 1991; Bajo et al., 1993; Merchán et al., 1994; Wu and Kelly, 1995b; Yang et al., 1996). Although the DNLL borders are fairly well established, the neuronal types populating the DNLL (Fig. 1B) are a matter of discussion. Depending on the species and the criteria used for describing the neurones, several classifications have been proposed. In cat, Kane and Barone (1980) described as many as nine types, while Adams and Mugnaini (1984) described only two types. In rat, Bajo et al. (1993) described four types; similar classes have been described in mouse (Willard and Ryugo, 1982) and opossum (Willard and Martin, 1984). Few studies have dealt with the dendritic arbors of the DNLL cells impregnated with the Golgi method (Ramón y Cajal, 1909–1911; Morest and Oliver, 1984; Iwahori, 1986). More recently, using intracellular injection of biocytin in brain slice preparations of rat DNLL, Wu and Kelly (1995b) have identified five types of cell based on the size and shape of both soma and dendritic arbors. This latter study, in addition, provides with a wealth of data concerning the membrane properties of the DNLL cells in young rats. Despite their morphological diversity, all injected cells had similar membrane

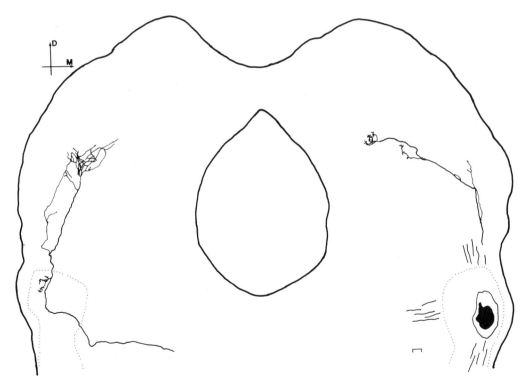

Figure 4. Serial reconstruction from 15 adjacent sections of two individual fibers labelled after biocytin injection in the rat DNLL (redrawn from Bajo et al., 1993, J. Comp. Neurol. 334, 241–262). Note that the DNLL on the right side projects to the contralateral DNLL and bilaterally to the IC. The terminal fields in the CNIC seems to be confined to a restricted area of the IC. Calibration bar = 50 μm.

properties with a sustained series of regular action potentials, indicating that they may form only one functional cell type.

In a recent study, Merchán et al. (1994) have revealed a novel principle of organisation of the DNLL after injections of BDA into the rat IC. They showed that a single, small injection of tracer into the IC produced a distinct pattern of neuronal and fibrilar labelling in the ipsi- and contralateral DNLL. They found more neurones in the contralateral DNLL and more fibres in the ipsilateral DNLL. The labelled structures formed an annular band both in transverse (Figs. 2A and 6) and sagittal sections. This band extends over serial sections and so they conclude that the DNLL in rat has a concentric organisation of layers like that of an onion. Furthermore, their findings strongly suggested that the DNLL is primarily composed of neurones with flattened dendritic arbors as also depicted from Golgi studies (Morest and Oliver, 1984; Iwahori, 1986), but so far 3-D reconstruction of dendritic arbors are still pending. They suggest that the DNLL neurones belong to a single type. This hypothesis is consistent with the study of Wu and Kelly (1995b) in which all neurones studied possessed similar membrane properties. The reason why different authors have described different cells types could be that in sections the shapes and sizes change according to the orientation of the cells (Merchán et al.,1994), as suggested also for the VCLL.

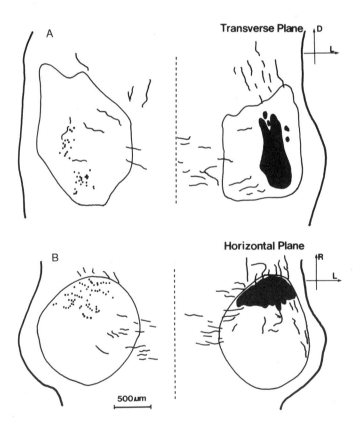

Figure 5. Camera lucida drawings from two injections into the DNLL in two different rats (redrawn from Bajo et al., 1993, J. Comp. Neurol. 334, 241–262). A. Biocytin injection. B. *Phaseolus vulgaris*-leucoagglutinin injection. The labelling arising from the DNLL is symmetric and mirrors the location of the injection site in the contralateral DNLL.

This concentric organisation may be fundamental for the tonotopical representation in the DNLL. A focal injection of BDA in the low frequency region of the IC produces a labelling with the shape of a sphere in the centre of the DNLL (Fig. 6B), while an injection in the high frequency region of the IC produces another sphere of a larger diameter and located at the periphery of the DNLL (Fig. 6A). A similar organisation has been recently described in gerbil by Zook et al. (1996). In cat (Aitkin et al., 1970; Shneiderman et al., 1988) and guinea pig (Malmierca et al., 1996) the tonotopical sequence differs slightly from the rodent model. An injection of biocytin in the low frequency region of the IC gives rise to a sheet of labelling at the dorsal region of the DNLL, while injections at more high frequency regions produce sheets progressively more ventrally placed in the DNLL (Malmierca et al., 1996). Nevertheless, these minor variations may simply reflect evolutionary changes across species reflecting their audible range (Heffner and Masterton, 1990) rather than a genuine change in the intrinsic organisation of the DNLL. Rodents are sensitive to high frequencies (up to 60 kHz), which are outside the hearing range of the guinea pig and cat (Heffner and Masterton, 1990). Therefore, it is logic to think that rodents may need or devote more tissue for the coding of high frequencies than the two mentioned species. Hence, the concentric model of the rodent's DNLL (Fig. 6, Merchán et al., 1994) may be the result of an expanded high frequency lamina that encircles the dorsal lower frequency laminae. A typical example of hypertrophied frequency region is found in the so called "auditory fovea" representing the echolocation signal frequency of certain bats (Rübsamer, 1992).

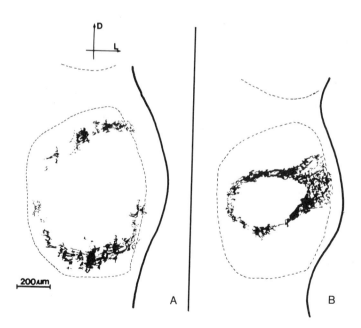

Figure 6. Camera lucida drawings illustrating a single section in two different cases. In A, BDA was injected in the ventromedial, high-frequency region of the rat IC and in B, BDA was injected in the dorsolateral, low-frequency region of the rat IC (redrawn from Merchán et al., 1994, J. Comp. Neurol. 342, 259–278). The labelling is made of fibres, terminal boutons and neurones, and forms a annular band in each section.

4.2. Connections and Neurochemistry of the DNLL

The DNLL receives input from both ears, contralaterally from the VCN and DNLL, ipsilaterally from the medial superior olive (MSO) and VCLL, and bilaterally from the lateral superior olive (LSO, van Noort, 1969; Glendenning and Masterton, 1983; Glendenning et al., 1981). Generally speaking, the DNLL receives a copy of the same afferences that innervate the IC. As was the case for the VCLL, careful inspection of the drawings from studies that are concerned with connections of the VCN (Oliver, 1987), MSO and LSO (Henkel and Spangler, 1983; Shneiderman et al., 1988; Vater et al., 1995) also shows that the projection from these nuclei to the DNLL is tonotopic. Furthermore, the elegant study by Vater et al. (1995) in bats, following injections of two different tracers at the same frequency regions of MSO and LSO in the same animal, shows that the output of these olivary nuclei overlaps extensively in the DNLL as well as in the IC. This opportunity for convergence at the DNLL plus the fact that the DNLL projects to the contralateral DNLL and bilaterally to the IC, suggest that binaural processing takes place in multiple iterative steps. The known chemical nature of the involved nuclei projecting to the DNLL is in good agreement with the GABA and glycine immunoreactive perisomatic boutons observed around the DNLL neurones (Saint Marie, 1993, Winer et al., 1995; Riquelme et al., 1996, unpublished observations). The excitatory nature of the afferences, however, remains to be shown.

The DNLL projection to the IC is laminar (Shneiderman et al., 1988) and bilateral to the two ICs, with a predominant projection to the contralateral one (Shneiderman et al., 1988; Bajo et al., 1993; Merchán et al., 1994).

The neurochemistry of the DNLL cells is relatively clear. Adams and Mugnaini (1984) first described that most, if not all DNLL cells use GABA as transmitter. Since their pioneering study, several reports have confirmed the GABAergic nature of the DNLL cells and their inhibitory influence on the IC responses in different animal species (Hutson et al., 1991; Saint Marie, 1993; Shneiderman et al., 1993; Yang et al., 1992; Faingold et al., 1993; Yang and Pollak, 1994a,b; Winer et al., 1995). More recently, Shneiderman et al. (1996) have demonstrated that the cat DNLL has two distinct populations of neurones according to their location in the DNLL and their projection to the ipsi- or contralateral IC. They have shown preliminary data where a crossed projection arise from the cells located laterally in the DNLL. Up to 60% of the cells have their somata surrounded by a large number of GABAergic perisomatic puncta. In contrast, the uncrossed projection arises from the medial part of the DNLL, and the neuropil surrounding these cells is less dense in GABAergic puncta than in the lateral part. These findings are in agreement with our tract-tracing studies in cat (further details are described below in section 5).

4.3. Physiological Significance of the DNLL

An essential issue in the central auditory system is how the origin of a sound is localised. Whereas functions such as pitch or intensity coding can be performed by one ear alone, the location of a sound needs two ears and depends on the comparison between the inputs arriving to both of them (for review see Kuwada and Yin, 1987, and Yin and Chan, 1988). The afferent and efferent connections of the DNLL shows that this nucleus must play an important role in binaural processing, i.e., sound localisation. The early studies of Aitkin et al., (1970) and Brugge et al., (1970) demonstrated that the DNLL possesses a tonotopical organisation and most of its neurones were driven when stimulated binaurally.

The studies of Aitkin, Brugge and co-workers showed that many DNLL neurones were nonmonotonic, indicating that inhibitory processes were present in the DNLL. More recently, the bat studies (Covey, 1993b; Markovitz and Pollak, 1993, 1994; Yang et al., 1996) have confirmed these earlier findings and have extended the knowledge on the response properties of single neurones in the DNLL. Perhaps, the most interesting finding is the large proportion of choppers responses registered after stimulation with pure tones, both monaurally and binaurally (Markovitz and Pollak, 1993, 1994; Yang et al., 1996). Chopper responses are characterised by having a trend of regularly spaced action potentials (Bourk, 1976). This is in agreement with the *in vitro* studies in the rat brain slices preparation by Wu and Kelly (1995a,b) who have shown that most intracellularly recorded cells in the DNLL exhibit a series of regularly spaced action potencials. The degree of regularity has also been quantified (Covey, 1993b). The role of regularity in the IC neurones has been discussed in depth by Le Beau et al. (1995, 1996) and Rees et al. (1997). A possible role of this type of neurones is to enhance the signalling of intensity. In this regard it is of interest to mention that one of the cues used for sound localisation is the interaural intensity differences (Kuwada and Yin, 1987, and Yin and Chan, 1988). Chopper neurones in the DNLL, therefore, may have an important role in sound localisation as demonstrated in the barn owl studies (Sullivan and Konishi, 1984; Takahashi and Konishi, 1988), where a clear dichotomy between the time and intensity pathways exits. The nucleus angularis in this species is populated by stellate cells having chopper responses and they carry the interaural intensity differences, as compared with the nucleus magnocellularis which carries the interaural time differences.

The cat and bat studies (Aitkin, et al. 1970; Brugge et al. 1970, Covey, 1993b, Markovitz and Pollak, 1993,1994) show that the majority of cells in the DNLL, in addition to being choppers, are of EI type. Hence, they are excited by the contralateral ear and inhibited by the ipsilateral ear. A minor proportion of the DNLL cells are of the EE type, i.e., they are excited by both ears. EI cells are first constructed at the LSO, while EE cells at the MSO (Yin and Chan 1988). Recently, Vater et al., 1995 have demonstrated that the input from LSO and MSO overlap extensively at the DNLL and IC, indicating that binaural processing takes place at several steps. Therefore the DNLL may sharpen some of the binaural features transmitted from the superior olivary complex SOC, but it may also create new features which are transmitted to the IC. This is discussed in detail in Yin and Chan (1988) and Markovitz and Pollak (1994) when comparing the binaural responses of the IC with those of SOC and DNLL.

5. NOVEL FINDINGS IN THE CAT NLL

In this section of the review is mainly concerned with our preliminary results achieved with a multidisciplinary approach combining tract-tracing methods and computer-assisted 3-D reconstructions in cat. The present findings, which still are under development in our laboratory, basically confirm and further extend those described in rat (Merchán et al., 1994; Merchán and Berbel, 1996).

In this account, we present the pattern of labelling in both VCLL and DNLL following stereotaxically guided iontophoretic injections of the tracer BDA made into the cat CNIC (van Noort, 1969; Rockel and Jones, 1973; Oliver and Morest, 1984). The injections were placed at different *loci* of the CNIC corresponding to different frequency-band laminae (Merzenich and Reid, 1974; Servière et al., 1984). The resulting labelling in the VCLL and DNLL was made up of fibres, axonal terminal plexuses, and neurones (Fig. 2C). The labelling was ipsilateral for the VCLL and bilateral for the DNLL. Viewed in single sections, the labelling in the VCLL seems to form patches with no preferential orientation whereas labelling in the DNLL forms horizontal bands (Fig. 2C). They are made of retrogradely labelled neurones whose dendritic trees seems to be flat and oriented parallel to each other. Moreover, the ipsilateral lamina contains labelled terminal fibres parallel to the labelled dendrites.

The distribution of the labelled terminal axonal plexuses and neuronal somata were reconstructed in 3-D (Fig. 7), dynamically rotated and analysed with custom software developed at the Department of Anatomy, University of Oslo (Leergaard et al., 1995; Bjaalie et al., 1997). The software runs on Silicon Graphics workstations.

The computer reconstructions of the labelled patches in the VCLL reveal their mutual organisation. A number of observations are common to all cases: 1) the patches form a single continuous plexus, 2) the retrogradely labelled somata in the VCLL are confined to the domain of the axonal plexus, and 3) there is a gradient in the amount of axonal and neuronal labelling, being denser ventrally (Fig. 7).

The shape and location of the plexuses differ according to the location of the injections into topographically low- and high frequency regions in the CNIC, respectively. In the low frequency case, the plexus resembles a tube located at the periphery of the VCLL.

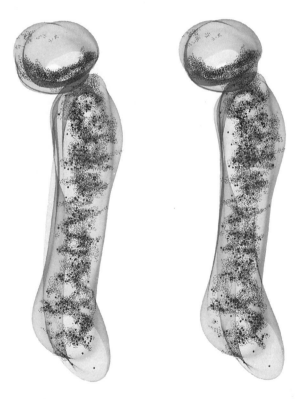

Figure 7. Computer reconstruction showing a stereo pair from the VCLL and DNLL in the cat (left image is tilted 7 degrees from the right). Labelling within the nuclei resulted from an injection into the central region of the CNIC. The cloud of small points represent the axonal labelling in the VCLL. It forms a continuos plexus that is laminar in the rostrocaudal and ventrodorsal length of the VCLL. The large dots represents retrogradely labelled cells. Note that these cells are confined to the domain of the axonal plexus. The labelling in the DNLL forms a distinct lamina that extends rostrocaudally.

In the high frequency case, by contrast, it is like a sheet located in the core of the nucleus. Thus, the low and high frequency plexuses occupy complementary locations in the VCLL, and we conclude that the tube and sheet are extreme forms of the same principle structure, i.e., a lamina. It is tempting to speculate that these VCLL laminae are the structural basis for the VCLL tonotopy. Future electrophysiological studies are needed for outlining this tonotopical arrangement. Our reconstructions, however, show that the recording electrodes should be advanced in a lateromedial direction, rather than the classical ventromedial direction. Indeed, Aitkin et al. (1970) already suggested that another orientation of the electrodes in their cat experiments might have lead to a better map of the tonotopical organisation of the VCLL.

With 3-D reconstructions of the bands of labelling seen in each DNLL it becomes evident that they represent rostrocaudally oriented laminae (Fig. 7). These laminae occupy symmetrical positions in the ipsi- and contralateral DNLLs. The location and shape of these laminae change with a shift of injection site along the tonotopical axis of the CNIC, as in the VCLL. After injections in the ventromedial, high frequency region of the CNIC, the laminae are located at the ventromedial surface of the DNLL. After injections in the dorsolateral, low frequency region of the CNIC, the laminae are located in the dorsolateral corner of the DNLL. Furthermore, the shape of the laminae in these two extreme cases is also different. The high frequency DNLL laminae is an elongated sheet with the largest axis oriented rostrocaudally. The low frequency DNLL lamina resembles a rostrocaudal tube.

6. CONCLUDING REMARKS

We have reviewed the current knowledge of the nuclei of the lateral lemniscus and have presented new evidence that these nuclei exhibit an essentially laminar organisation. This is particularly interesting in the case of the VCLL where a tonotopical organisation has been argued in the past. The present review strongly suggests that the NLL follow the general principle of organisation of the auditory system, i.e., they process information on a frequency-frame basis, and they preserve the spectral decomposition of the signal performed by the cochlea. Nevertheless, there are unanswered questions that would apply to both the VCLL and DNLL. What is the intrinsic organisation of the laminae? Are there spatially segregated laminae? How do the cells and the afferent and efferent projections interact within a single laminae? How are response properties of single neurones generated and/or sharpened by the intrinsic and extrinsic connections? These and other questions need to be explored in future studies.

7. ACKNOWLEDGMENTS

We thank Kirsten Osen for critical reading of the manuscript and I. Plaza, F.R. Nodal and E.O. Andersen for technical assistance. Financial support was provided by the Spanish DGES (PB95–1129) to M.M. and the Research Council of Norway to J.G.B. M.S.M. was supported by the Spanish MEC and the Commission of the E.U., V.M.B, by the MEC.

8. REFERENCES

Adams, J. C. (1979) Ascending projections to the inferior colliculus. J. Comp. Neurol. 183, 519–538.

Adams J.C. and Mugnaini, E. (1984) Dorsal nucleus of the lateral lemniscus: a nucleus of GABAergic projection neurons. Brain Res. Bull. 13, 585–590.

Aitkin, L.M., Anderson, D.J. and Brugge, J.F. (1970) Tonotopic organization and discharge characteristics of single neurons in nuclei of the lateral lemniscus in cat. J. Neurophysiol. 33, 421–440.

Bajo, V.M., Merchán, M.A., López, D.E., and Rouiller, E.M. (1993) Neuronal morphology and efferent projections of the dorsal nucleus of the lateral lemniscus in the rat. J. Comp. Neurol. 334, 241–262.

Batra, R. and Fitzpatrick, D. C. (1997) The ventral nucleus of the lateral lemniscus contains neurons sensitive to interaural temporal disparities. ARO Abstr. 20, 164.

Bjaalie, J. G., Dælen, M. and Stensby, T. V. (1997) Surface modelling of biomedical data. In: Numerical methods and software tools in industrial mathematics. A. Tveito and M. Dælen Eds. Birkhauser Publishing Cooperation, Boston. pp. 9–26 (in press).

Blackstad, T.W., Osen, K.K. and Mugnaini, E. (1984) Pyramidal neurons of the dorsal cochlear nucleus: a Golgi and computer reconstruction study in cat. Neurosci. 13, 827–854.

Brugge, J. F., Anderson, D. J. and Aitkin, L. M. (1970) Responses of neurons in the dorsal nucleus of the lateral lemniscus of cat to binaural tonal stimulation. J. Neurophysiol. 33, 441–458.

Bullock, T.H. (1984) Comparative neuroscience holds promise for quiet revolutions. Science 225, 473–478.

Bourk, T. R. (1976) Electrical responses of neural units in the anteroventral cochlear nucleus of the cat. Ph.D thesis, MIT, Cambridge.

Caicedo, A. and Herbert, H. (1993) Topography of descending projections from the inferior colliculus to auditory brainstem nuclei in the rat. J. Comp. Neurol. 328, 377–392.

Covey, E. (1993a) The monaural nuclei of the lateral lemniscus: parallel pathways from cochlear nucleus to midbrain. In: The Mammalian Cochlear Nuclei: Organization and Function. M. A. Merchán, J. M. Juiz,D. A. Godfrey and E. Mugnaini NATO Series. New York: Plenum Publishing Co. pp. 321–334.

Covey, E. (1993b) Response properties of single units in the dorsal nucleus of the lateral lemniscus and paralemniscal zone of an echolocating bat. J. Neurophysiol. 69, 842–859.

Covey, E. and Casseday, J. H. (1986) Connectional basis for frequency representation in the nuclei of the lateral lemniscus of the bat *eptesicus fuscus*. J. Neurosci. 6, 2926–2940.

Covey, E. and Casseday, J.H. (1991) The monaural nuclei of the lateral lemniscus in an echolocating bat: parallel pathways for analyzing temporal features of sound. J. Neurosci. 11, 3456–3470.

Ebner, F. F. (1969) A comparison of primitive forebrain organization in metatherian and eutherian mammals. Ann. NY Acad. Sci. 167, 241–257.

Faingold, C.L., Boersma-Anderson, C.A. and Randall, M.E. (1993) Stimulation or blockade of the dorsal nucleus of the lateral lemniscus alters binaural and tonic inhibition in contralateral inferior colliculus neurons. Hearing Res. 69, 98–106.

Friauf, E. and Ostwald, J. (1988) Divergent projections of physiologically characterized rat ventral cochlear nucleus neurons as shown by intra-axonal injection of horseradish peroxidase. Exp. Brain Res. 73, 263–284.

Fu, X.W., Wu, S.H., Brezden, B.L. and Kelly, J.B. (1996) Potassium currents and membrane excitability of neurons in the rat's dorsal nucleus of the lateral lemniscus. Journal of Neurophysiology 2, 1121–1132.

Glendenning, K. K., Brunso-Bechtold, J. K., Thompson, G. C. and Masterton, R. B. (1981) Ascending auditory afferents to the nuclei of the lateral lemniscus. J. Comp. Neurol. 197, 673–703.

Glendenning, K. K. and Masterton, R. B. (1983) Acoustic chiasm: efferent projections of the lateral superior olive. J. Neurosci. 3, 1521–1537.

González-Hernández, T., Mantolan, B., González, B., and Pérez, H. (1996) Sources of GABAergic input to the inferior colliculus of the rat. J. Comp. Neurol. 372, 309–326.

Guinan Jr, J.J., Norris, B.E. and Guinan, S.S. (1972) Single auditory units in the superior olivary complex II: locations of unit categories and tonotopic organization. Int. J. Neurosci. 4,147–166.

Heffner R. S. and Masterton R. B. (1990) Sound localization in Mammals: brain-stem mechanisms. In: Comparative perception. Vol. I Basic mechanisms. M. A. Berkley and W. C. Stebbins (eds.) John Wiley and sons. Inc. 9, pp. 285–314.

Henkel, C.K. and Shneiderman, A. (1988) Nucleus sagulum: projections of a lateral tegmental area to the inferior colliculus in cat. J. Comp. Neurol. 271, 577–588.

Henkel, C. K. and Spangler, K. M. (1983) Organization of the efferent projections of the medial superior olivary nucleus in the cat as revealed by HRP and autoradiographic tracing methods. J. Comp. Neurol. 221, 416–428.

Irvine D. R. F. (1992) Physiology of the auditory brainstem. In: The mammalian auditory pathway: neurophysiology. A. N. Popper and R. R. Fay (eds.). Springer-Verlag, New York. pp. 153–231.

Iwahori, N. (1986) A Golgi study on the dorsal nucleus of the lateral lemniscus in the mouse. Neurosci. Res. 3, 196–212

Kane, E. S. and Barone, L. M. (1980) The dorsal nucleus of the lateral lemniscus in the cat: neuronal types and their distributions. J. Comp. Neurol. 192, 797–826

Kuwada, S. and Yin, T. C. T. (1987) Physiological mechanisms of directional hearing. In: Directional Hearing. W. A. Yost and G. Gourevitch (eds.). Springer-Verlag, Berlin. pp. 146–176.

Le Beau, F.E.N., Malmierca, M.S. and Rees, A. (1995) The role of inhibition in determining neuronal response properties in the inferior colliculus. In: Advances in Hearing Research. G. A. Manley, G. Klump, C. Köppl, H. Fastl and H. Oeckinghaus (eds) , World Scientific Press, Singapore, pp. 300–313.

Le Beau, F.E.N., Rees, A. and M. S. Malmierca (1996) Contribution of GABA- and glycine-mediated inhibition to the monaural temporal response properties of neurons in the inferior colliculus. J. Neurophysiol. 75, 902–919.

Leergaard, T.B. and Bjaalie, J.G. (1995) Semi-automatic data acquisition for quantitative neuroanatomy. Micro-Trace-computer programme for recording of the spatial distribution of neuronal populations. NeuroReport 22, 231–243.

Malmierca, M.S. (1991) Computer-assisted three-dimensional reconstructions of Golgi impregnated cells in the rat inferior colliculus. Doctoral thesis, Universities of Oslo and Salamanca.

Malmierca, M.S., Blackstad, T. W., Osen, K. K. Karagulle, T. and Molowny, R. L. (1993) The central nucleus of the inferior colliculus in rat: A golgi and computer reconstruction study of neuronal and laminar structure. J. Comp. Neurol. 333, 1–27.

Malmierca, M. S., Rees, A., Le Beau, F. E. N. and J. G. Bjaalie (1995) Laminar organization of frequency-defined local axons within and between the inferior colliculi of the guinea pig. J. Comp. Neurol. 357, 124–144.

Malmierca, M. S., Le Beau, F. E. N. and Rees, A. (1996) The topographical organization of descending projections from the central nucleus of the inferior colliculus in guinea pig. Hearing Res. 93, 167–180.

Markovitz, N. S. and Pollak, G. D. (1993) The dorsal nucleus of the lateral lemniscus in the mustache bat: Monaural properties. Hearing Res. 71, 51–63.

Markovitz, N. S. and Pollak, G. D. (1994) Binaural processing in the dorsal nucleus of the lateral lemniscus. Hearing Res. 73, 121–140.

Merchán M. A. and Berbel, P. (1996) Anatomy of the ventral nucleus of the lateral lemniscus in rats: A nucleus with a concentric laminar organization. J. Comp. Neurol. 372, 245–263.

Merchán, M. A., Saldaña, E. and Plaza, I. (1994) Dorsal nucleus of the lateral lemniscus in the rat: Concentric organization and tonotopic projection to the inferior colliculus. J. Comp. Neurol. 342, 259–278.

Merzenich, M. M. and Reid, M. D. (1983) Representation of the cochlea within the inferior colliculus of the cat. Brain Res. 77, 397–415.

Morest, D. K. and Oliver, D. L. (1984) The neuronal architecture of the inferior colliculus in the cat: Defining the functional anatomy of the auditory midbrain. J. Comp. Neurol. 222, 209–236.

Morest, D. K. and Winer, J. A. (1986) The comparative anatomy of neurons: Homologous neurons in the medial geniculate body of the opossum and the cat. In: Advances in Anatomy, Embriology and Cell Biology, Vol 97. Beck, F. et al. Eds., Springer-Verlag Berlin Heidelberg.

Oertel. D., Wu, S.H., Micah, W. G. and Dizack, C (1990) Morphology and Physiology of Cells in Slice Preparations of the Posteroventral Cochlear Nucleus of Mice. J. Comp. Neurol. 295, 136–154.

Oliver, D. L. (1987) Projections to the inferior colliculus from the anteroventral cochlear nucleus in the cat: Possible substrate for binaural interaction. J. Comp. Neurol. 264, 24–46.

Oliver, D.L., Kuwada, S., Yin, T.C.T., Haberly, L.B. and Henkel, C.K. (1991) Dendritic and axonal morphology of HRP-injected neurons in the inferior colliculus of the cat. J. Comp. Neurol. 303, 75–100.

Oliver, D. L. and Morest, D. K. (1984) The central nucleus of the inferior colliculus in the cat. J. Comp. Neurol. 222, 237–264.

Paxinos, G. and Watson, C. (1986) The rat brain in stereotaxic coordinates. 2nd ed. Academic Press. Sydney, New York, London.

Ramón y Cajal, S. (1909–1911) Textura del sistema nervioso del hombre y de los vertebrados. 3 vols., facsimile. Alicante: E.G. Obra Ramón y Cajal S. L: and Instituto de Neurociencias.

Rees, A. Sarbaz, A., Malmierca, M. S. and Le Beau, F. E. N. (1997) Regularity of firing of neurons in the inferior colliculus. J. Neurophysiol. (in press).

Rockel, A. J. and Jones, E. G. (1973) The neuronal organization of the inferior colliculus of the adult cat. I: the central nucleus. J. Comp. Neurol. 147, 11–60.

Rübsamer, R. (1992) Postnatal development of central auditory maps. J. Comp. Physiol.-A 170, 129–143.

Saint Marie, R. L. (1993) GABA and glycine immunoreactivities define principal subdivisions of the cat lateral lemniscal nuclei. Soc. Neurosci. Abstr. 19, 1204.

Saint Marie, R.L. (1996) Glutamatergic connections of the auditory midbrain: Selective uptake and axonal transport of D-[H-3] Aspartate. J. Comp. Neurol. 373, 255–270.

Servière, J., Webster, W.R. and Calford, M.C. (1984) Isofrequency labeling revealed by a combined 14C-2-deoxiglucose, electrophysiological, and horseradish peroxidase study of the inferior colliculus of the cat. J. Comp. Neurol. 228, 463–477.

Shneiderman, A., Chase, M. B., Rockwood, J. M., Benson, C. G. and Potashner, S. J. (1993) Evidence for a GABAergic projection from the dorsal nucleus of the lateral lemniscus to the inferior colliculus. J. Neurochem. 60, 72–82.

Shneiderman, A. and Henkel, C. K. (1987) Banding of lateral superior olivary nucleus afferents in the inferior colliculus: A possible substrate for sensory integration. J. Comp. Neurol. 266, 519–534.

Shneiderman, A., Oliver, D. L. and Henkel, C. K. (1988) Connections of the dorsal nucleus of the lateral lemniscus: An inhibitory parallel pathway in the ascending auditory system? J. Comp. Neurol. 276, 188–208.

Shneiderman, A., Stanforth, D. A. and Saint Marie, R. L. (1996) A comparison of the crossed and uncrossed, GABA-immunoreactive projections of the cat dorsal nucleus of the lateral lemniscus. Soc. Neurosci. Abstr. 22, 124.

Smith, P. H., Joris, P. X., Carney, L. H. and Yin, T. C. T. (1991) Projections of physiologically characterized globular bushy cell axons from the cochlear nucleus of the cat. J. Comp. Neurol. 304, 387–407.

Spangler, K., Warr, W. B. and Henkel, C. K. (1985) The projections of principal cells of the medial nucleus of the trapezoid body in the cat. J. Comp. Neurol. 238, 249–262.

Sullivan, W. E. and Konishi, M. (1984) Segregation of stimulus phase and intensity coding in the cochlear nucleus of the barn owl. J. Neurosci. 4, 1787–1799.

Takahashi, T. T. and Konishi, M. (1988) Projections of nucleus angularis and nucleus laminaris to the lateral lemniscal nuclear complex of the barn owl. J. Comp. Neurol. 274, 212–238.

Tuck, G. P., Zook, J. M. and Kuwabara, N. (1993) Neurons with intrinsic axonal terminations in the gerbil's lateral lemniscus. Soc. Neurosci. Abstr. 19, 1204.

van Noort, J. (1969) The structure and connections of the inferior colliculus. An investigation of the lower auditory system. van Gorcum & Comp, Leiden.

Vater, M., Casseday, J. H., Covey, E. (1995) Convergence and divergence of ascending binaural and monaural pathways from the superior olives of the mustached bat. J. Comp. Neurol. 351, 632–646.

Wickesberg, R. E. and Oertel, D. (1993) Delayed, frequency-specific inhibition in the cochlear nuclei of mice: a mechanism for monaural echo suppression. J. Neurosci. 10(6), 1762–1768.

Whitley, J.M. and Henkel, C.K. (1984) Topographical organization of the inferior collicular projection and other connections of the ventral nucleus of the lateral lemniscus in the cat. J. Comp. Neurol. 229, 257–270.

Willard, F. H. Ruygo, D. K. (1982) Anatomy of the central auditory system. In: The Auditory Psychobiology of the Mouse. J. F. Willot (ed.). C. Thomas Inc. Springfield. pp. 201–304.

Willard, F. H. and Martin, G. F. (1984) Collateral innervation of the inferior-colliculus in the north american opossum: a study using fluorescent markers in a double-labeling paradigm. Brain Res. 303, 171–182.

Winer, J. A., Larue, D. T. and Pollak, G. D. (1995) GABA and glycine in the central auditory system of the mustache bat, structural substrates for inhibitory neuronal organization. J. Comp. Neurol. 355, 317–353.

Wu, S. H. (1996) Intrinsic membrane properties and firing characteristics of neurons in the ventral nucleus of the lateral lemniscus studied in rat brain slices. Soc. Neurosci. Abstr. 22, 647.

Wu, S. H. and Kelly, J. B. (1995a) In vitro brain slide studies of the rat's dorsal nucleus of the lateral lemniscus. I. membrane and synaptic response properties. J Neurophysiol. 73, 780–793.

Wu, S. H. and Kelly, J. B. (1995b) In vitro brain slide studies of the rat's dorsal nucleus of the lateral lemniscus. II. Physiological properties of biocytin-labeled neurons. J Neurophysiol. 73, 794–809.

Wu, S. H. and Kelly, J. B. (1996) In vitro brain slices studies of the rat's dorsal nucleus of the lateral lemniscus. III. Synaptic Pharmacology. J Neurophysiol. 75(3),1271–1282.

Yang, L., Liu, Q. and Pollak, G. D. (1996) Afferent connections to the dorsal nucleus of the lateral lemniscus of the mustache bat: evidence for two functional subdivisions. J. Comp. Neurol. 373, 575–592.

Yang, L. and Pollak, G. D. (1994a) The roles of GABAergic and Glycinergic inhibition on binaural processing in the dorsal nucleus of the lateral lemniscus of the mustache bat. J. Neurophysiol. 71, 1999–2013.

Yang, L. and Pollak, G. D. (1994b) Gaba and Glycine have different effects on monaural response properties in the dorsal nucleus of the lateral lemniscus of mustache bat. J. Neurophysiol. 71, 2014–2024.

Yang, Y., Pollak, G. D. and Resler, C. (1992) GABAergic circuits sharpen tuning curves and modify response properties in the mustache bat inferior colliculus. J. Neurophysiol. 68, 1760–1774.

Yin, T. C. T. and Chan, J. C. K. (1988) Neural mechanisms underlying interaural time sensitivity to tones and noise. In: Auditory Function. G. M. Edelman, W. E. Gall y Cowan, W. M: (eds.). Wiley, New York. pp. 385–430.

Yin, T. C.T. Chan, J. C. K., and Carney, L. H. (1987) Effects of interaural time delays of noise stimuli on low-frequency cells in the cat's inferior colliculus. III: evidence for cross-correlation. J. Neurophysiol. 58, 562–583.

Zook, J. M., Schanbacher, B. and Kuwabara, N. (1996) Afferent and intrinsic projections to the dorsal nucleus of the lateral lemniscus. Soc. Neurosci. Abstr. 22, 128.

21

THE ROLE OF GABA IN SHAPING FREQUENCY RESPONSE PROPERTIES IN THE CHINCHILLA INFERIOR COLLICULUS

Donald M. Caspary, Robert H. Helfert, and Peggy Shadduck Palombi

Departments of Pharmacology and Surgery
School of Medicine
Southern Illinois University
801 N. Rutledge St., Springfield, Illinois 62702

In this chapter, we summarize data from several recent single unit iontophoretic studies on the role of $GABA_A$ receptor-mediated inhibition in the central nucleus of the inferior colliculus (CIC). Specifically, we will address the role of GABA circuits in shaping acoustically evoked monaural responses in the frequency domain.

Understanding the relative strength and functional arrangement of excitatory and inhibitory inputs impinging upon CIC neurons is an important step in efforts to successfully model acoustic information processing. It is our contention that models of acoustic processing in central auditory structures which simply incorporate pure lateral inhibition, without a near characteristic frequency (CF) inhibitory component, cannot accurately describe the coding of acoustic signals by brainstem auditory structures. Some recent models of cochlear nucleus have incorporated inhibitory influences focused within the excitatory response area (Rothman and Young, 1996; Rothman et al., 1993).

The CIC is an important auditory processing center, receiving inputs from lower brainstem nuclei and contralateral inferior colliculus (IC) neurons as well as indirect inputs from higher auditory and nonauditory structures (for review, see Irvine, 1986, Oliver and Huerta, 1992, Oliver and Shneiderman, 1991). Several studies have shown that most of the major synaptic inputs to the CIC are arranged into tonotopically organized, frequency band laminae (Malmierca et al., 1995; Saldaña and Merchán, 1992). Bands of excitatory and inhibitory inputs from a variety of sources in the lower auditory brainstem and midbrain form overlapping synaptic domains within each lamina (for review, see Oliver and Huerta, 1992). The tonotopic organization is such that the lowest frequency laminae are located dorsolaterally in the CIC, and from this area the frequency response of the laminae increases progressively the further ventromedially they are located. The CIC is surrounded by the external and dorsal cortices which are also tonotopically organized (Faye-Lund and Osen, 1985; Morest and Oliver, 1984; Rockel and Jones, 1973; Saldaña and Merchán, 1992).

Acoustical Signal Processing in the Central Auditory System
edited by Syka, Plenum Press, New York, 1997

GABA AS A CIC NEUROTRANSMITTER

GABA is a major inhibitory neurotransmitter in the IC. Its levels are among the highest found in the brain (Banay-Schwartz et al., 1989; Mugnaini and Oertel, 1985). Several studies suggest that a high percentage of neurons and synaptic endings in the CIC immunolabel for GABA (Carr et al., 1989; Caspary et al., 1990; Moore and Moore, 1987; Mugnaini and Oertel, 1985; Oliver and Huerta, 1992; Oliver et al., 1994; Roberts and Ribak, 1987; Tachibana and Kuriyama, 1974; Thompson et al., 1985), and that virtually every IC neuron receives significant GABAergic inputs (Oliver et al., 1994; Roberts and Ribak, 1987). Physiologic and functional pharmacologic studies support an important role for GABA acting through $GABA_A$ receptors in coding acoustic signals in the CIC (Faingold et al., 1989; Goldsmith et al., 1995; LeBeau et al., 1996; Pollak and Park, 1993; see Caspary et al., 1995 for review). Recent binding studies show a large number of presumptive $GABA_A$ receptors in the IC (Glendenning et al., 1992; Milbrandt et al., 1996).

ORIGIN AND ALIGNMENT OF IC GABA PATHWAYS

In addition to the ascending excitatory lemniscal inputs, numerous GABA-containing extrinsic, intrinsic, and commissural projections appear to terminate as either focused or diffuse inputs onto CIC neurons (Adams and Mugnaini, 1984; Helfert et al., 1989; Huffman and Henson, 1990; Saint Marie et al., 1989; Shneiderman and Oliver,1989; Shneiderman et al., 1988, 1993). The major GABAergic inputs are shown in Figure 1. GABA-containing neurons within the IC project intrinsically and to the contralateral CIC (Caspary and Helfert, 1993; Oliver et al., 1994; Roberts and Ribak, 1987; Thompson et al., 1985). The predominant extrinsic sources of GABAergic input to the CIC are from the contralateral, and to a lesser extent the ipsilateral, dorsal nucleus of the lateral lemniscus (DNLL) (Figure 1) (Adams and Mugnaini, 1984; Carr et al., 1989; Glendenning et al., 1992; Saint Marie et al., 1989; Shneiderman et al., 1993; Vater et al., 1992b). Studies by Li and Kelly (1992) and Faingold et al. (1993) have shown that electrically or chemically manipulating DNLL output alters binaural and monaural inhibitory processing by IC neu-

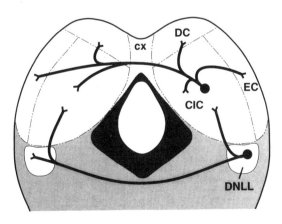

Figure 1. Schematic of the major GABAergic projections of the CIC. CX=cortex, DC=dorsal cortex of the IC, EC=external cortex of the IC, CIC=central nucleus of the IC, DNLL=dorsal nucleus of the lateral lemniscus.

rons. Neurochemical studies have demonstrated the GABAergic nature of the DNLL projection and of a subset of the IC commissural projection (Caspary and Helfert, 1993; Schneiderman et al., 1993; Smith, 1992). Most germane to the thesis put forth in this chapter are the findings of recent tracer injection studies which suggest that GABAergic projections to CIC neurons are tonotopically aligned with their CIC neuronal targets (Caspary and Helfert, 1993; Malmierca et al., 1996; Merchán et al., 1994, and this volume; Saldaña and Merchán, 1992).

PHYSIOLOGICAL ATTRIBUTES OF CIC INHIBITION

There is some evidence for lateral components of CIC inhibition. Early work by Katsuki and coworkers (1958, 1959) suggested that IC neurons were more sharply tuned than neurons in lower parts of the auditory brainstem (for further discussion, see Suga, 1995). This concept was consistent with observations of center-surround neurons (Hubel and Weisel, 1962) and GABA enhancement of orientation selectivity in the visual system (Sillito et al., 1980; Tsumoto et al., 1979). It was thus logical that auditory physiologists would attempt to incorporate lateral inhibition into their thinking about the auditory system. More recent visual system studies suggest a broader role for inhibition (see Henry et al., 1994 for a recent review and Somers et al., 1995 for a recent model). A reevaluation is indicated in the auditory system as well.

There is considerable evidence for potent near CF inhibition onto CIC neurons. A high percentage of CIC units display nonmonotonic rate/intensity functions and onset temporal response patterns to contralateral CF tones, as well as inhibitory responses to ipsilateral CF tones (for review, see Aitkin, 1985; Caird, 1991; Irvine, 1992). The neuron shown in Figure 2 is one of 55% of chinchilla CIC units which display strongly nonmonotonic rate/intensity functions in response to CF stimuli (Palombi and Caspary, 1996). Closed tuning curves, as observed in some bat CIC units, provide additional evidence of near CF inhibition (Fuzessery and Hall, 1996; Suga and Tsuzuki, 1985; Vater et al., 1992a; Yang et al., 1992).

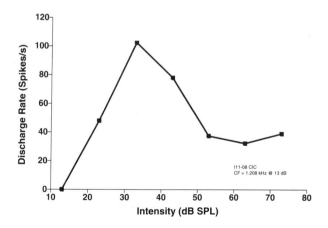

Figure 2. This chinchilla CIC unit, with a characteristic frequency of 1208 Hz and a threshold of 13 dB SPL displays a nonmonotonic rate/intensity function.

IONTOPHORETIC METHODS AND DATA ANALYSIS

The role of GABA in shaping contralateral monaural response areas of CIC neurons has been investigated in several iontophoretic studies. Although presentation and analysis techniques differed, the results were similar. A recent study by Palombi and Caspary (1996) used six-barreled piggy-back electrodes (Havey and Caspary, 1980) to study CIC neurons in ketamine/xylazine anesthetized chinchillas. The recording barrel was filled with 2 M potassium acetate and balancing barrel was filled with horseradish peroxidase (HRP).

GABA inhibition was examined by recording isointensity contours at several intensity levels prior to, during, and after $GABA_A$ receptor blockade. Iontophoretic application of the receptor antagonist bicuculline methiodide (BMI) was used to block $GABA_A$ receptors. Almost all neurons tested in the CIC were sensitive to GABA and BMI (Palombi and Caspary, 1996). Effects of $GABA_A$ receptor blockade on a family of isointensity contours gathered from a typical CIC neuron can be seen in Figure 3. The greatest effects of BMI application onto this neuron are seen around the CF, within the excitatory response area (Figure 3).

Since discharge rate varies considerably across frequency, with just a few spikes occuring near the edges of the response area and relatively high discharge rates near the center of the response area, one must carefully consider methods of data analysis used in iontophoretic studies. At present, it is not possible to determine whether percent change in discharge rate or absolute change in discharge rate is the key factor in the processing of neural codes by higher centers; it may depend on the particular circuitry, pathways involved and/or information being processed. A clear statement about assumptions made on this fundamental issue is vital to interpretation of iontophoretic data. Because of anatomic data suggesting that GABAergic circuits are in tonotopic register with their CIC targets, we have focused on the near CF inhibitory effects of GABA and have, therefore, made the

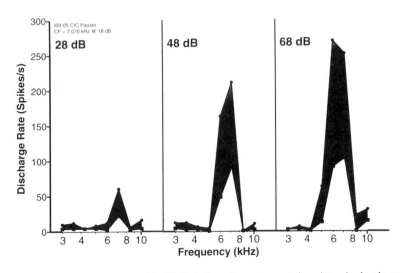

Figure 3. Control and bicuculline methiodide (BMI) isointensity contours at three intensity levels are shown for this chinchilla CIC neuron. Shaded area represent difference between control and BMI condition. CF = 7070 Hz. Threshold = 18 dB SPL.

assumption that a change in absolute discharge rate has a greater impact on coding of response areas in the CIC than does percent change.

To that end, changes in mid-level isointensity contours with BMI application were quantified in two ways. First, the breadth, or bandwidth (in octaves), of each isointensity contour was calculated between the two frequency points of zero discharge. If either limb (low or high side) did not reach zero, points were extrapolated. These bandwidths were compared before and during drug application. Second, Z-score analysis (based on Edeline and Weinberger's analysis across frequency, Edeline and Weinberger, 1991a, b) was used to evaluate the effects of $GABA_A$ blockade as a function of frequency. Z-score analysis expresses the difference between the control and drug responses at any one frequency as a number standard deviations from the mean difference for that unit. We did not analyze percentage changes in discharge rate with GABA blockade because at the edges of the response area, where spike counts are low, extremely large percentage changes can arise from very small absolute changes. We felt that absolute changes, as reflected by Z-score analysis, better indicated where in the frequency receptive field GABA was exerting its greatest effects on firing rate. A positive Z-score indicates a change from control greater than the mean change across all frequencies tested; a negative Z-score indicates a smaller change from control than the mean change across all frequencies tested. Units were categorized based on the Z-score analysis; units with positive Z-scores restricted to the frequency region surrounding CF were considered to display near CF drug effects while units displaying negative Z-scores for frequencies near CF and positive scores at the edges of the frequency response area would be considered as displaying off CF drug effects.

TWO DIFFERENT INHIBITORY INFLUENCES ON CIC NEURONAL RESPONSE AREAS

As previous studies have described (LeBeau et al., 1995; Vater et al., 1992a; Yang et al., 1992), inhibitory influences affecting frequency response areas of chinchilla CIC neurons can be separated into two categories: 1) inhibition with the greatest effect near the unit's CF and completely contained within the excitatory response area (Group 1) and 2) inhibition focused near CF, but extending beyond the excitatory response area, usually on the low frequency side (Group 2). In the chinchilla, approximately 41% of the neurons fall into Group 1 while 49% are classified as Group 2 (Palombi and Caspary, 1996). A few neurons showed complex BMI effects for which frequency bandwidth could not be clearly quantified (Palombi and Caspary, 1996).

The Z-score plot at 30 dB above threshold from the same unit displayed in Figure 3 is shown in Figure 4. Typical of Group 1 units, Z-score analysis clearly shows that application of 40 nA of BMI increases near CF discharge rate with little or no expansion of the excitatory response area (Figure 5); this increase is greatest at high intensities. Of the 40 chinchilla CIC neurons tested using BMI to block $GABA_A$ receptors, 16 (41%) showed BMI effects similar to the neuron in Figures 3 and 4.

In a study of chinchilla CIC neurons, Palombi and Caspary (1996) found that 19 (49%) CIC neurons displayed some broadening of the midlevel isointensity contours upon BMI application. An example of a chinchilla CIC neuron with a near CF and low side effect is seen in Figure 5. Panel A shows the midlevel isointensity contour while panel B shows the Z-score plot. These chinchilla results are consistent with similar studies by Vater et al. (1992a) in horseshoe bats, Yang et al. (1992) in mustache bats, and LeBeau et

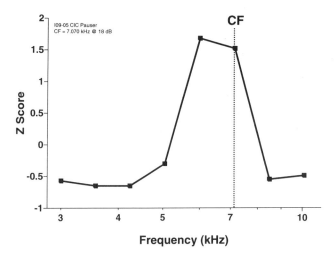

Figure 4. The Z-score is plotted for the same unit shown in Figure 3. CF is indicated by the dotted line.

al. (1995) in guinea pigs. These iontophoretic studies in other species all find response area maps suggestive of inhibition at higher intensities and with greatest inhibition centered near CF. Vater and colleagues (1992a) showed in horseshoe bats that BMI application altered the shape of tuning curves obtained from CIC. Their findings suggested that both lateral and near CF inhibition was affected with $GABA_A$ receptor blockade. Figure 7 of Vater et al. (1992a) showed an example of a closed tuning curve opened by GABA receptor blockade. Yang et al. (1992) found that 42% of mustache bat neurons showed broadening of sharply tuned neurons in the 60 kHz region. We assume these authors found that 58% of the neurons in their sample did not show broadening but rather showed changes confined to the excitatory response area near CF as shown in the examples in figures 7 and 8 of Yang et al. (1992). Fuzessery and Hall (1996) in the pallid bat CIC also show similar findings to those in other bat species.

LeBeau et al. (1995) formalized separation of responses to $GABA_A$ receptor blockade in guinea pig CIC neurons into two distinct groups. In their study, LeBeau et al. (1995) visually quantified the effects of BMI on response areas using response area plots and subtraction techniques similar to those developed by Evans and Zhao (1993). These plots reveal changes from the pre-BMI control in the frequencies which make up the excitatory response area. These authors also describe two populations of cells revealed upon $GABA_A$ receptor blockade: Group 1 neurons which show increases confined within the excitatory response area and Group 2 neurons which display a broadening of the response area. Examples of their Group 1 CIC neurons, neurons whose response areas did not change greatly with GABA blockade can be found in figures 1 and 2 of LeBeau et al. (1995) while an example of a Group 2 neuron with low side expansion can be seen in their figure 3. Examination of plots from LeBeau et al. (1995) suggests that both Group 1 and 2 neurons' maximum GABA inhibition is centered within the excitatory response area.

The above-mentioned studies all present similar data in different ways. To illustrate the importance of these differences, we plotted the Z-score from one of our chinchilla CIC units classified as having an on/low effect (Figure 6) and plotted the same unit as a response area map as was done by Yang et al. (1992). The resulting figure (Figure 7) shows similarity to figure 10 of Yang et al. (1992) as well as to figure 3 of LeBeau et al. (1995). Changes induced by BMI blockade of GABA were large both near CF and below.

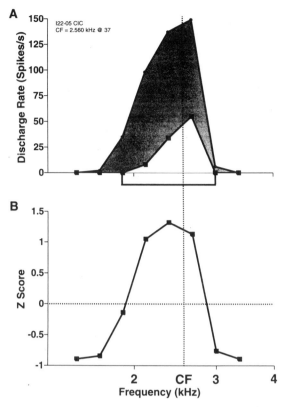

Figure 5. Panel A shows the control and BMI isointensity contours from a CIC unit with a CF of 2560 Hz and a threshold of 37 dB SPL. Panel B shows the Z-score plot for the same isointensity contour. CF and zero level Z-score are indicated by dotted lines.

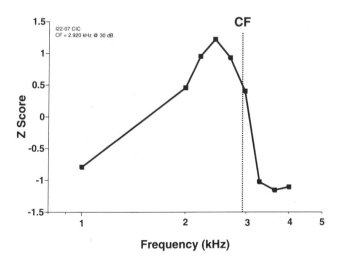

Figure 6. The Z-score for the 60 dB isointensity contour is plotted for a unit with low side expansion upon BMI application. CF is indicated by the dotted line. CF = 2920 Hz. Threshold = 30 dB SPL.

As noted above, anatomic studies suggest that inhibitory circuits are tonotopically aligned with their CIC targets (Caspary and Helfert, 1993; Malmierca et al., 1996; Merchán et al., 1994, this volume; Saldaña and Merchán, 1992). In this chapter we reviewed results from several iontophoretic studies with data suggesting that the strongest GABAergic inhibition is focused near the center of the excitatory response area, which appears consistent with the anatomic studies noted above. In about half the neurons, inhibition appears to extend beyond the excitatory response area as well, usually on the low side. Other physiologic data supports the existence of strong inhibition centered near CF as well. A large percentage of CIC neurons display strongly nonmonotonic input/output functions in response to CF stimuli (for review, see Aitkin, 1985, Caird, 1991, Irvine, 1992).

TWO GROUPS OF CIC NEURONS: IS THERE AN ANATOMIC SUBSTRATE TO THE DIFFERENT GABAERGIC EFFECTS ON RESPONSE AREAS?

Several possibilities exist. Perhaps the most conservative explanation is the possibility that Group 1 neurons are the discoid cells of the CIC whose dendrites are completely contained within a given lamina and which probably receive inhibition from GABAergic neurons in tonotopic register with their CIC targets. CIC neurons showing BMI effects

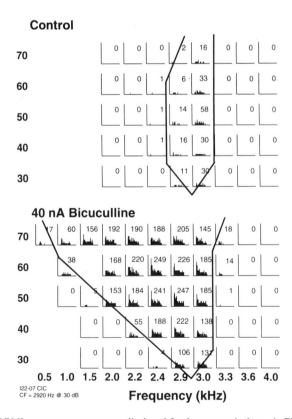

Figure 7. Control and BMI response area maps are displayed for the same unit shown in Figure 6. Dark lines indicate the 10% of maximum firing levels as in Yang et al. (1992).

without any lateral spread of the inhibition appear not to receive inhibitory inputs from GABAergic projections from other laminae. Group 2 CIC neurons may be stellate cells whose dendrites cross the laminar boundaries and thus receive GABAergic inputs both from neurons in tonotopic register and from neurons in neighboring laminae. These neurons would show effects both near CF and at frequencies extending beyond the target neuron's excitatory response area. Alternatively, Group 2 neurons could receive major near CF inhibitory inputs from the DNLL and contralateral CIC, but also receive interlaminar inhibitory inputs from neighboring isofrequency CIC lamina. These interlaminar inputs must arise primarily from the more dorsal (low frequency) lamina since 13/40 (32.5%) of the neurons showed low side broadening in our study, whereas only 3 units showed high side broadening.

Figure 8 shows models of four possible relationships between the excitatory and inhibitory frequency inputs to a CIC neuron. Theoretical tuning curves (top) are shown for the excitatory (black) and inhibitory (grey) input neurons. The horizontal line indicates a mid-level intensity. Theoretical isointensity contours (bottom) are shown before and during BMI application at this intensity. In panel A, the inhibitory inputs are aligned tonotopically with the excitatory inputs and do not extend lateral to them; blockade of GABA$_A$ receptors would increase rate within the excitatory response area, greatest near CF, without increasing frequency extent. As noted, this result was observed for 45% of the neurons tested (Group 1). In panel B, the inhibitory inputs flank the excitatory inputs without overlapping with the CF; blockade by BMI would increase the width of the response area with

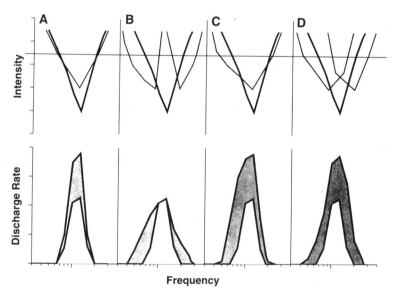

Figure 8. Models of four possible relationships between the excitatory and inhibitory frequency inputs to a CIC neuron are shown. Theoretical tuning curves (top) are shown for the excitatory (black) and inhibitory (grey) input neurons. The horizontal line indicates a mid-level intensity. Theoretical isointensity contours (bottom) are shown before and during BMI application at this intensity. In panel A, the inhibitory inputs are aligned tonotopically with the excitatory inputs and do not extend lateral to them; blockade of GABA$_A$ receptors would increase rate within the excitatory response area, greatest near CF, without increasing frequency extent. In panel B, the inhibitory inputs flank the excitatory inputs without overlapping with the CF; blockade by BMI would increase the width of the response area with no discharge rate increase near CF. In panel C, the inhibitory inputs are aligned with, but extend lateral to, the excitatory inputs and, in panel D, the inhibitory inputs are both lateral to and overlapping with the excitatory inputs, summing to have their largest effect near CF.

no discharge rate increase near CF. None of the CIC units tested displayed this pattern of response. In panel C, the inhibitory inputs are aligned with, but extend lateral to, the excitatory inputs and, in panel D, the inhibitory inputs are both lateral to and overlapping with the excitatory inputs, summing to have their largest effect near CF. Either model C or D could be used to explain the Group 2 responses (40% of the neurons tested). Given the anatomical data available about GABAergic projections to CIC neurons, model C would appear to be a logical explanation of the results.

In conclusion, for all CIC neurons, substantial inhibition appears to occur near the center of the excitatory response area. This physiological observation is supported by anatomical evidence suggesting that the auditory brainstem is organized into tonotopically aligned circuits. Therefore, inhibitory inputs projecting onto CIC neurons appear to exert their greatest effects on near CF dynamic range adjustment, a function important in the coding of complex signals and signal detection in noise. Understanding this arrangement should help us better model the coding of complex responses in the IC.

REFERENCES

Adams, J.C. and Mugnaini, E. (1984) Dorsal nucleus of the lateral lemniscus: A nucleus of GABAergic projection neurons. Brain Res. Bull. 13, 585–590.

Aitkin, L. (1985) The Auditory Midbrain: Structure and Function in the Central Auditory Pathway. Clifton, NJ: Humana.

Banay-Schwartz, M., Lajtha, A. and Palkovits, M. (1989) Changes with aging in the levels of amino acids in rat CNS structural elements I: Glutamate and related amino acids. Neurochem. Res. 14, 555–562.

Caird, D. (1991) The central auditory system. In: R.A. Altschuler, R.P., Bobbin, B.M. Clopton, and D.W. Hoffman (Eds.), Neurobiology of Hearing, New York, Raven Press, pp. 253–292.

Carr, C.E., Fujita, I. and Konishi, M. (1989) Distribution of GABAergic neurons and terminals in the auditory system of the barn owl. J. Comp. Neurol. 286, 190–207.

Caspary, D.M. and Helfert, R.H. (1993) Two populations of commissural projections connecting like regions of the inferior colliculus immunostain for different amino acid transmitters. Soc. Neurosci. Abstr. 19, 1425.

Caspary, D.M., Milbrandt, J.C. and Helfert, R.H. (1995) Central auditory aging: GABA changes in the inferior colliculus. Exp. Gerontol. 30, 349–360.

Caspary, D.M., Raza, A., Armour, B.A.L., Pippen, J. and Arnerić, S.P. (1990) Immunocytochemical and neurochemical evidence for age related loss of GABA in the inferior colliculus: Implications for neural presbycusis. J. Neurosci. 10, 2363–2372.

Edeline, J.-M., and Weinberger, N.M. (1991a) Subcortical adaptive filtering in the auditory system: associative receptive field plasticity in the dorsal medial geniculate body. Behav. Neurosci. 105, 154–175.

Edeline, J.-M. and Weinberger, N.M. (1991b) Thalamic short-term plasticity in the auditory system: associative retuning of receptive fields in the ventral medial geniculate body. Behav. Neurosci. 105, 618–639.

Evans, E.F. and Zhao, W. (1993) Neuropharmacological and neurophysiological dissection of inhibition in the mammalian dorsal cochlear nucleus. In: M.A. Merchán, J. Juiz, D.A. Godfrey and E. Mugnaini, (Eds.), The Mammalian Cochlear Nuclei: Organization and Function, Plenum Press, New York, pp 373–384.

Faingold, C;L., Boersma Anderson, C.A. and Randall, M.E. (1993) Stimulation or blockade of the dorsal nucleus of the lateral lemniscus alters binaural and tonic inhibition in the contralateral inferior colliculus neurons. Hearing Res. 69, 98–106.

Faingold, C.L., Gehlbach, G. and Caspary, D.M. (1989) On the role of GABA as an inhibitory neurotransmitter in inferior colliculus neurons: Iontophoretic studies. Brain Res. 500, 302–312.

Faye-Lund, H. and Osen, K.K. (1985) Anatomy of the inferior colliculus in rat. Anat. Embryol. 171, 1–20.

Fuzessery, Z.M. and Hall, J.C. (1996) Role of GABA in shaping frequency tuning and creating FM sweep selectivity in the inferior colliculus. J. Neurophsyiol. 76(2), 1059–1073.

Glendenning, K.K., Baker, B.N., Hutson, K.A. and Masterton, R.B. (1992) Acoustic chiasm. V: inhibition and excitation in the ipsilateral and contralateral projections of LSO. J. Comp. Neurol. 319, 100–122.

Goldsmith, J.D., Kujawa, S.G., McLaren, J.D., Bledsoe, Jr., S.C. (1995) In vivo release of neuroactive amino acids from the inferior colliculus of the guinea pig using brain microdialysis. Hearing Res. 83, 80–88.

Havey, D.C. and Caspary, D.M. (1980) A simple technique for constructing "piggy-back" multibarrel electrodes. Electroencephalograph. Clin. Neurophysiol. 48, 249–251.

Helfert, R.H., Bonneau, J.M., Wenthold, R.J. and Altschuler, R.A. (1989) GABA and glycine immunoreactivity in the guinea pig superior olivary complex. Brain Res. 501, 269–286.

Henry, G.H., Michalski, A., Wimborne, B.M. and McCart, R.J. (1994) The nature and origin of orientation specificity in neurons of the visual pathways. Progress Neurobiol. 43, 381–437.

Hubel, D.H. and Wiesel, T.N. (1962) Receptive fields, binocular interaction and functional architecture in the cat's visual cortex. J. Physiol., Lond. 160, 106–154.

Huffman, R.F. and Henson, O.W., Jr. (1990) The descending auditory pathway and acousticomotor systems: Connections with the inferior colliculus. Brain Res. Rev. 15, 295–323.

Irvine, D.R.F. (1992) Physiology of the auditory brainstem. In: A.N. Popper, and R.R. Fay (Eds.), The mammalian auditory pathway: Neurophysiology, New York: Springer-Verlag, pp.153–231.

Irvine, D.R.F. (1986) The Auditory Brainstem: A Review of the Structure and Function of Auditory Processing Mechanisms. Berlin: Springer-Verlag.

Katsuki, Y., Sumi, T., Uchiyama, H. and Watanabe, T. (1958) Electric responses of auditory neurons in cat to sound stimulation. J. Neurophysiol. 21, 569–588.

Katsuki, Y., Watanabe, T. and Maruyama, N. (1959) Activity of auditory neurons in upper levels of brain of cat. J. Neurophysiol. 22, 343–359.

Le Beau, F.E.N., Malmierca, M.S. and Rees, A. (1995) The role of inhibition in determining neuronal response properties in the inferior colliculus. In: G.A. Manly, G.M. Klumpke. Chapel, H. Fast and H. Oeckinghaus (Eds.), Advances in Hearing Research, World Scientific Publishing Co. Pte. Ltd., p.300–313.

Le Beau, F.E.N., Rees, A. and Malmierca, M.S. (1996) Contribution of GABA- and glycine-mediated inhibition to the monaural temporal response properties of neurons in the inferior colliculus. J. Neurophysiol. 75, 902–919.

Li, L. and Kelly, J. B. (1992) Inhibitory influence of the dorsal nucleus of the lateral lemniscus on binaural responses in the rat's inferior colliculus. J. Neurosci. 12, 4530–4539.

Malmierca, M.S., Le Beau, F.E.N. and Rees, A. (1996) The topographical organization of descending projections from the central nucleus of the inferior colliculus in guinea pig. Hearing Res. 93, 167–180.

Malmierca, M.S., Rees, A., LeBeau, F.E.N., and Bjaalie, J. (1995) Laminar organisation of frequency-defined local axons within and between the inferior colliculi of the guinea-pig. J. Comp. Neurol. 357, 124–144.

Merchán, M.A., Bajo, V.M., Malmierca, M.S. and Bjaalie, J.G. (this volume) The laminar organization of the nuclei of the lateral lemniscus (NLL): A tracer and computer reconstruction study in cat. In: J. Syka (Ed.), Acoustical Signal Processing in the Central Auditory System, New York, Plenum Publishing.

Merchán, M.A., Saldaña, E. and Plaza, I. (1994) Dorsal nucleus of the lateral lemniscus in the rat: Concentric organization and tonotopic projection to the inferior colliculus. J. Comp. Neurol. 342, 259–278.

Milbrandt, J.C., Albin, R.L., Turgeon, S.M. and Caspary, D.M. (1996) GABA$_A$ receptor binding in the aging rat inferior colliculus. Neurosci. 73, 449–458.

Moore, J.K. and Moore, R.Y. (1987) Glutamic acid decarboxylase-like immunoreactivity in brainstem auditory nuclei of the rat. J. Comp. Neurol. 260, 157–174.

Morest, D.K. and Oliver, D.L. (1984) The neuronal architecture of the inferior colliculus in the cat: Defining the functional anatomy of the auditory midbrain. J. Comp. Neurol. 222, 209–236.

Mugnaini, E. and Oertel, W.H. (1985) An atlas of the distribution of GABAergic neurons and terminals in the rat CNS as revealed by GAD immunohistochemistry. In: A. Björklund and T. Hökfelt (Eds.), Handbook of Chemical Neuroanatomy. Vol. 4: GABA and Neuropeptides in the CNS, Part I, Amsterdam, Elsevier Science Publishers B.V., pp. 436–608.

Oliver, D.L. and Huerta, M.F. (1992) Inferior and superior colliculi. In: D.B. Webster, A.N. Popper, and R.R. Fay (Eds.), The Mammalian Auditory Pathway: Neuroanatomy, New York, Springer-Verlag, pp. 168–221.

Oliver, D.L. and Shneiderman, A. (1991) The anatomy of the inferior colliculus: a cellular basis for integration of monaural and binaural information. In: R.A. Altschuler, R.P. Bobbin, B.M. Clopton, and D.W. Hoffman (Eds.), Neurobiology of Hearing: The Central Auditory System, New York, Raven, pp. 195–222.

Oliver, D.L., Winer, J.A., Beckius, G.E. and Saint Marie, R.L. (1994) Morphology of GABAergic neurons in the inferior colliculus of the cat. J. Comp. Neurol. 340, 27–42.

Palombi, P.S. and Caspary, D.M. (1996) GABA inputs control discharge rate primarily within frequency receptive fields of inferior colliculus neurons. J. Neurophysiol. 75, 1–11, 1996.

Pollak, G.D. and Park, T.J. (1993) The effects of GABAergic inhibition on monaural response properties of neurons in the mustache bat's inferior colliculus. Hearing Res. 65, 99–117.

Roberts, R.C. and Ribak, C. (1987) GABAergic neurons and axon terminals in the brainstem auditory nuclei of the gerbil. J. Comp. Neurol. 258, 267–280.

Rockel, A.J. and Jones, E.G. (1973) The neuronal organization of the inferior colliculus of the adult cat. I: The central nucleus. J. Comp. Neurol. 147, 11–60, 1973.

Rothman, J.S. and Young, E.D. (1996) Enhancement of neural synchronization in computational models of ventral cochlear nucleus bushy cells. Aud. Neurosci. 2, 47–62.

Rothman, J.S., Young, E.D. and Manis, P.B. (1993) Convergence of auditory nerve fibers onto bushy cells in the ventral cochlear nucleus: Implications of a computational model. J. Neurophysiol. 70, 2562–2583.

Saint Marie, R.L., Morest, D.K. and Brandon, C.J. (1989) The form and distribution of GABAergic synapses on the principal cell types of the ventral cochlear nucleus of the cat. Hearing Res. 42, 97–112.

Saldaña, E. and Merchán, M.A. (1992) Intrinsic and commissural connections of the rat inferior colliculus. J. Comp. Neurol. 319, 417–437.

Shneiderman, A., Chase, M.B., Rockwood, J.M., Benson, C.G. and Potashner, S.J. (1993) Evidence for a GABAergic projection from the dorsal nucleus of the lateral lemniscus to the inferior colliculus. J. Neurochem. 60, 72–82.

Shneiderman, A. and Oliver, D.L. (1989) EM autoradiographic study of the projections from the dorsal nucleus of the lateral lemniscus: A possible source of inhibitory inputs to the inferior colliculus. J. Comp. Neurol. 286, 28–47.

Shneiderman, A., Oliver, D.L. and Henkel, C.K. (1988) Connections of the dorsal nucleus of the lateral lemniscus: An inhibitory parallel pathway in the ascending auditory system? J. Comp. Neurol. 276, 188–208.

Sillito, A.M., Kemp, J.A., Milson, J.A. and Berardi, N. (1980) A re-evaluation of the mechanisms underlying simple cell orientation selectivity. Brain Res. 194, 517–520.

Smith, P.H. (1992) Anatomy and physiology of multipolar cells in the rat inferior collicular cortex using the in vitro brain slice technique. J. Neurosci. 12, 3700–3715.

Somers, D.C., Nelson, S.B. and Sur, M. (1995) An emergent model of orientation selectivity in cat visual cortical simple cells. J. Neurosci. 15, 5448–5465.

Suga, N. (1995) Sharpening of frequency tuning by inhibition in the central auditory system: tribute to Yasuji Katsuki. Neurosci. Res. 21, 287–299.

Suga, N. and Tsuzuki, K. (1985) Inhibition and level-tolerant frequency tuning in the auditory cortex of the mustached bat. J. Neurophysiol. 53, 1109–1145.

Tachibana, M. and Kuriyama, K. (1974) Gamma-aminobutyric acid in the lower auditory pathway of the guinea pig. Brain Res. 69, 370–374.

Thompson, G.C., Cortez, A.M., and Lam, D.M.-L. (1985) Localization of GABA immunoreactivity in the auditory brainstem of guinea pigs. Brain Res. 339, 119–122.

Tsumoto, T., Eckart, W. and Creutzfeldt, O.D. (1979) Modification of orientation sensitivity of cat visual cortex neurons classified on the basis of their responses to photic stimuli. Brain Res. 61, 351–363.

Vater, M., Habbicht, H., Koessl, M. and Grothe, B. (1992a) The functional role of GABA and glycine in monaural and binaural processing in the inferior colliculus of horseshoe bats. J. Comp. Physiol. [A] 171, 541–553.

Vater, M., Koessl, M. and Horn, A.K.E. (1992b) GAD- and GABA-immunoreactivity in the ascending auditory pathway of horseshoe and mustached bats. J. Comp. Neurol. 325, 183–206.

Yang, L., Pollak, G.D. and Resler, C. (1992) GABAergic circuits sharpen tuning curves and modify response properties in the mustache bat inferior colliculus. J. Neurophysiol. 68, 1760–1774.

22

THE INFLUENCE OF INTRINSIC OSCILLATIONS ON THE ENCODING OF AMPLITUDE MODULATION BY NEURONS IN THE INFERIOR COLLICULUS

A. Rees and A. Sarbaz

Department of Physiological Sciences
The Medical School
University of Newcastle upon Tyne
Newcastle upon Tyne, NE2 4HH, United Kingdom

1. INTRODUCTION

Amplitude modulations are a characteristic feature of many biologically significant sounds including animal vocalisations and human speech. Plomp (1983) showed the modulation-frequency spectrum for a sample of connected speech extends up to 12 Hz with a maximum around 4 Hz, the frequency of word production. These modulations play an important role in auditory communication. In noisy environments listeners must use some means to identify which of many frequencies reaching the ear originate from the same source and should therefore be perceived as a single auditory object. This process, variously termed auditory grouping, streaming or scene analysis (Bregman 1993) depends on several cues. One of these is the similar pattern of amplitude change shared by frequency components with a common origin. Evidence that common modulation patterns play a role in grouping is suggested by the phenomenon of co-modulation masking release (Hall et al., 1984; see Moore, 1997 for review).

Modulations may also be important for the temporal analysis of pitch. Speech sounds consist of harmonically related frequencies. The peripheral auditory system that processes these signals behaves like a bank of overlapping bandpass filters. At low frequencies the filter bandwidth is sufficiently narrow to resolve individual harmonics. At higher frequencies the bandwidth widens and two or more components will pass through the same filter. These components interact and the amplitude of the output signal is modulated at the fundamental frequency of the harmonic series. The analysis of such modulations may be important for extracting the pitch of complex sounds (Langner 1992, and this volume) thus providing an additional cue for grouping.

Acoustical Signal Processing in the Central Auditory System
edited by Syka, Plenum Press, New York, 1997

Given the importance of AM in auditory processing, one might expect to find specific neural mechanisms for its extraction and analysis. Modulation transfer functions (MTFs) determined for single units in the cochlear nerve and nuclei show that they respond to amplitude modulated sounds with a discharge synchronized to the modulation envelope of the stimulus. Even at these most peripheral levels of the auditory pathway, there is enhancement of the response to amplitude modulation relative to the input stimulus (Møller 1976, Frisina et al., 1990 and this volume, Rhode and Greenberg 1994).

In the main midbrain nucleus, the inferior colliculus (IC), the gain of the synchronized response increases relative to that at other brainstem sites, and there is a downward shift in the range of modulation rates to which the units respond (Rees and Møller 1983, Rees and Møller 1987). Responses to AM in the IC also reveal pronounced variations in mean firing rate as a function of modulation frequency (Langner and Schreiner 1988, Rees and Palmer 1989). About 40% of IC units have rate modulation transfer functions that peak at the same frequency as the MTF of the synchronized response (Rees and Palmer 1989). This may represent the re-encoding of information from a temporal to a rate based code.

A neuronal property that influences the response of units to AM is the regularity of neural discharge. Regularity has been studied extensively for units in the cochlear nuclei (Young et al., 1988; Blackburn and Sachs, 1989; Parham and Kim, 1992). Regularity is readily apparent in units that have a chopper PSTH pattern, but regular firing also occurs in units with other PSTHs, e.g. pause-build (Parham and Kim, 1992). The action potentials of some highly regular units occur within a narrow band of frequencies that are independent of the frequency and temporal properties of the stimulus. Kim et al. (1992) reported such intrinsic oscillations (IOs) in the discharge patterns of pause-build and sustained chopper units, where the frequency of the IO correlated closely with the unit's best modulation frequency (BMF). Rhode and Greenberg (1994) noted that highly regular units were among the most effective in retaining a synchronized response to AM as the mean level of the stimulus was increased.

In a recent study (Rees et al., 1997), we measured the discharge regularity of neurons in the central nucleus of the IC (CNIC) and found that regular firing occurs in several PSTH types. Under chloralose anaesthesia about 25% of units in the CNIC of guinea pig were highly regular (CV ≤ 0.35). Such units often have an intrinsic oscillation in their discharge.

In the present study we sought to determine whether in the IC regular firing and the intrinsic oscillations present in highly regular units correlate with their responses to amplitude-modulated sounds.

2. METHODS

2.1 Anaesthesia and Surgical Preparation

The experiments were performed on mature pigmented guinea pigs of either sex. Anaesthesia was induced with HYPNORM (Janssen, fentanyl citrate 0.315 mg/ml and fluanisone 10 mg/ ml; dose 1.5 ml/kg s.c.) and midazalem (HYPNOVAL, Roche, 1.5 ml/kg s.c., Flecknell 1988) and maintained with α-chloralose (Sigma, 75 mg/kg i.p). Atropine sulphate (0.05 mg/kg s.c.) was given to reduce bronchial secretions. A tracheotomy was performed and the trachea cannulated. The animals were artificially respired and end-tidal CO_2 was monitored and maintained at 5%. The animal's temperature was monitored with a rectal thermometer and maintained at 37°C.

The animal was fixed in a stereotaxic frame in which the earbars were replaced by perspex speculi that fitted tightly into the external auditory meatuses. The skull overlying the colliculus was exposed and the bone was removed to make an opening approximately 5mm in diameter. The cortex was exposed and covered with 2% agar to prevent desiccation.

2.2 Stimulus Generation and Presentation

Pure tones were generated by a Hewlett Packard 3325A Waveform Synthesiser under computer control. The signals were trapezoidally-gated with rise/fall times of 5 ms. The phase of the tone was not locked to the tone-burst envelope. The signal was attenuated independently to the two ears by a pair of digitally controlled attenuators.

Amplitude-modulated tones were generated using a Hewlett Packard 3314A Function Generator to amplitude modulate the output of the Waveform Synthesiser. The modulation frequency, percent depth and mean level of the modulated stimulus could be controlled independently.

The animal was placed inside a sound attenuating booth and stimuli were delivered through closed acoustic systems based on Sony MDR 464 earphones (Rees 1990) Stimuli were presented monaurally to the ear contralateral to the recorded colliculus. The maximum output of the system was flat from 0.1–9 kHz (100 ± 5 dB SPL) and then fell with a slope of approximately 15 dB/octave.

2.3 Single Unit Recording

Single units were recorded in the IC with glass coated tungsten electrodes. Vertical electrode penetrations were made stereotaxically through the cerebral cortex. The electrode was advanced through the IC by a remotely controlled stepping motor.

Extracellularly recorded action potentials were amplified (x10,000) and filtered (0.3–3 kHz) by a preamplifer (World Precision Instruments, DAM 80). After window discrimination and further amplification, the spikes were converted to logic pulses timed to an accuracy of 10 µs by a CED-1401 Laboratory Interface (Cambridge Electronic Design). This device also generated and stored timing and trigger pulses. The spike and synchronization times were passed to the computer for storage and analysis.

The positions of recorded neurons and electrode tracks were marked with electrolytic lesions for subsequent histological verification.

2.4 Data Collection and Analysis

Single neurons were isolated using 50-ms noise- or tone-bursts as search stimuli, and BF and threshold were identified audiovisually. A rate intensity function was generated for the unit based on 20 presentations of a 75-ms tone at several sound levels.

A unit's response to AM was determined by recording the response to a five second amplitude-modulated tone, at several modulation frequencies. The carrier frequency for the modulation was always the neuron's best frequency (BF) and, unless specified otherwise, the modulation depth was 25%. For analyses at only one sound level, the mean level of the modulated stimulus was placed about two-thirds up the slope of the unit's rate-level

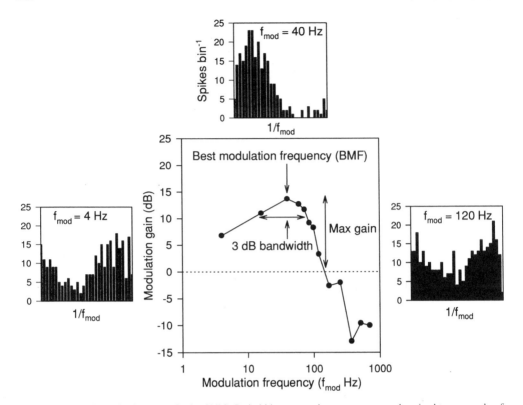

Figure 1. sMTF for a single neuron in the CNIC. Period histograms show response synchronized to one cycle of the modulation at three modulation frequencies. Main figure: sMTF derived from modulation response. Arrows define measures used in the analysis of responses to AM.

function, clear of the plateau and any nonmonotonicity. A period histogram synchronized to a single period of the modulation and comprised of 32 bins was constructed from the response to each modulation frequency (see Fig. 1). The synchronized response modulation in each histogram was obtained by calculating the synchronization coefficient, from which the percent response modulation and the modulation gain with respect to the stimulus could be determined (Rees and Palmer 1989). The modulation transfer function for the synchronized response (sMTF, Fig. 1) was obtained by plotting gain or synchronization coefficient versus the modulation frequency. Several descriptive measures could be derived from the sMTF as indicated in Fig. 1. The maximum modulation response represents the largest gain in the MTF, and the BMF is defined as the modulation frequency at which this maximum occurs. The bandwidth of the MTF is measured 3 dB (or its linear equivalent) down from the maximum. Fourier analysis was also applied to the spike times to determine the first Fourier component in the responses. Rate modulation transfer functions (rMTFs) were plotted from the mean firing rate obtained in response to each modulation frequency.

To determine the regularity of a unit's discharge, spike times were collected in response to up to 2000 presentations of a 75-ms tone at BF, at a repetition rate of 4 s^{-1}. A PSTH (e.g. Fig. 2A) was generated from the spike times and the beginning of the tone-

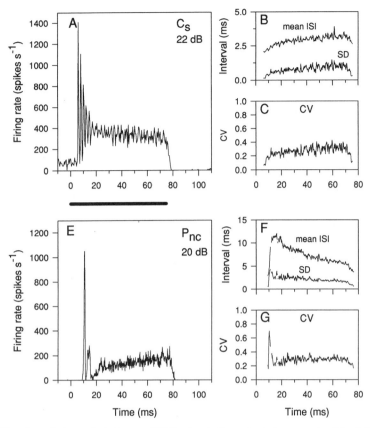

Figure 2. PSTH and regularity analysis for two CNIC units: A-C sustained chopper unit , BF =9.5kHz and E-G pauser-no-chop unit , BF = 18 kHz. Bar between PSTHs marks onset and duration of 75-ms tone stimulus.

driven response was selected by eye using a cursor. Regularity of spike discharge was calculated using the method described by Young et al. (1988). The time interval between each successive spike pair in the train was calculated and stored in a time bin corresponding to the time at which the first spike in the interval occurred. The bin width for the regularity analysis was 100 or 200 µs. The mean and standard deviation (SD) of the interspike intervals (ISIs) in each bin were calculated and the CV was derived from the ratio of the SD:mean ISI. Bins that contained fewer than three intervals were not plotted. To reduce end effects (see Young et al., 1988) only the first 65 ms of the 75-ms response was included in the regularity analysis. The units were classified according to their sustained CV, measured in a 10-ms window beginning 30 ms after response onset. A regular response is defined by a CV <0.5, while a CV ≤0.35 defines a highly regular response. Intrinsic oscillations mainly occurred in highly regular units (CV ≤0.35), so for the purposes of this study the units were divided into two groups, highly regular (CV≤0.35) and less regular/ irregular (CV >0.35).

The frequency of any IO in a unit's response was determined by performing a discrete Fourier transform (DFT) on the spike times obtained in response to an 850-ms pure tone at the unit's BF. Alternatively, a DFT was performed on the autocorrelation of the spike times to obtain the power spectrum.

2.5 Histological Processing and Analysis

At the end of each experiment the animal was given an overdose of anaesthetic and perfused transcardially with rinse solution followed by fixative. The brain was removed fixed for a further period and cryoprotected (see Rees et al. 1997 for details). Tranverse or saggital frozen sections were cut at 70 μm, mounted on slides, stained with cresyl violet and coverslipped.

The subdivisions of the IC [central nucleus (CNIC), dorsal cortex (DCIC) and external cortex (ECIC)] were recognised based on cytoarchitectural criteria we have previously defined for the guinea pig (Malmierca et al., 1995). All of the recordings described here were made in the CNIC.

3. RESULTS

3.1 Regular Firing in IC Neurons

The results are based on recordings from 145 tonically firing single units. We classified the neurons on the basis of their PSTHs at 15–25 dB above threshold. In this level range and above, the characteristic features (e.g. the presence of a pause) in the neurons' firing patterns were well defined. The units were first classified as chopper, pauser, on-sustained or sustained units according to the scheme used in a previous iontophoretic study (Le Beau et al., 1996). Under chloralose anaesthesia pauser units constitute the largest group (44%), followed by on-sustained (42%) chopper (8%) and sustained (6%) units. These categories could be further subdivided on the basis of PSTH characteristics indicating regular firing, such as chopping, that were subsequently confirmed by regularity analysis (see Rees et al., 1997 for further details). Consistent with previous studies (Popelář and Syka, 1982), chopping was seen in relatively few units and these were divided between those that chopped at the onset of the response (5%) and those that showed chopping throughout the response (3%). Regularity analysis confirmed the presence of regular firing during all or part of the responses in these units, as illustrated in Fig. 2 A-C. Most pauser units (39% of the sample) showed no chopping in their PSTHs. Despite the absence of chopping many pauser units fired regularly. The data in Fig. 2 E-G is from a pauser-no-chop unit. Apart from the transiently high CV at onset which is a consequence of the pause, the unit had a low CV (≈ 0.2) indicating highly regular firing in the sustained part of the response.

Across all units in the sample, 68% fired regularly (CV< 0.5) during their sustained response, and 23% were highly regular (CV≤ 0.35). Thus a substantial proportion of units in the CNIC fire regularly and although the presence of chopping is a reliable predictor of regularity, units in other PSTH categories were highly regular despite the absence of chopping.

3.2 Intrinsic Oscillations in the Firing of CNIC Neurons

Regular firing as evidenced by the CV does not imply that all intervals in a neuron's response are of equal duration. A low CV simply indicates that intervals occurring at a particular time in the discharge tend to be constant from sweep to sweep. But most highly regular units (based on CV) do fire with a relatively constant interspike interval, through-

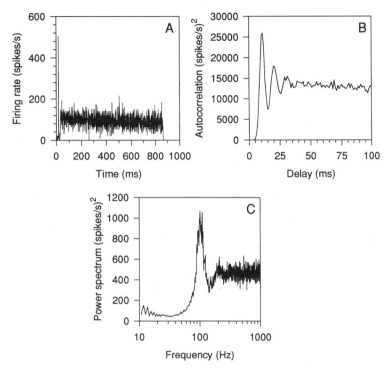

Figure 3. Intrinsic oscillation (IO) in the discharge of a pauser-no-chop unit to a pure tone. BF = 2.95 kHz, level 35 dB re threshold. A. PSTH, B. Autocorrelation of spike discharge, C. Power spectrum, note IO centred on 100 Hz.

out each spike train as well as from one spike train to another. This IO can be revealed by Fourier analysis as described in the Methods. In the example shown in Fig. 3, the autocorrelogram (Fig. 3B) shows a marked periodicity with a delay of about 10 ms which corresponds to the peak centred on 100 Hz in the power spectrum (Fig. 3C). The oscillation frequency was not related to the frequency of the stimulus, but did vary to some degree with tone level. In our sample, 66% of highly regular units had a measurable intrinsic oscillation although the height and breadth of the peak in the power spectrum varied greatly between units. A much weaker intrinsic oscillation was observed in 18% of less regular/irregular units.

3.3 Correlation between IO Frequency and BMF

For 32 units that demonstrated a prominent IO in their response to a pure tone we plotted the MTF from the first Fourier component in the response to AM and determined the unit's BMF. Fig. 4 shows the relationship between IO frequency and BMF. The BMF for most units was below 150 Hz, and only three units had a BMF at a higher frequency. There is a highly significant correlation between the BMF and the frequency of the IO. This relationship demonstrates that the most precise synchronization to the envelope of the amplitude modulation occurs when the modulation frequency coincides with and entrains the unit's IO frequencies.

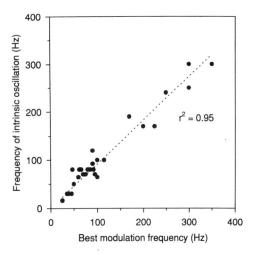

Figure 4. Correlation between BMF for amplitude-modulated tones and IO frequency in the response to pure tone for CNIC units. Line indicates linear regression.

3.4 Correlation between MTF Bandwidth and Regularity

If a unit's IO limits the range of modulation frequencies to which its discharge synchronizes, this should be reflected in the bandwidth of the sMTF. To compare the sMTF bandwidths of units that have IOs with those in which IO was weak or absent, we plotted sMTF bandwidth versus the unit's sustained CV (Fig. 5). Despite the scatter in the data, it is evident that the widest sMTF bandwidths occur in less regular/ irregular units that have little or no IO, whereas the bandwidths for highly regular units with a prominent IO are

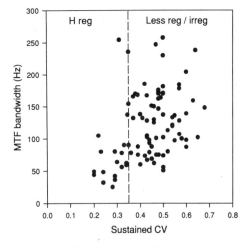

Figure 5. Bandwidth of sMTF (measured 3dB down from peak) as a function of sustained regularity. Modulation depth 25%. H reg = highly regular, less reg/irreg = less regular/irregular.

amongst the narrowest in the sample. Bandwidth may depend on the centre frequency at which it is measured, in this case the BMF. Nevertheless, the difference in sMTF bandwidth between highly regular and less regular/ irregular units persisted when the MTF bandwidth was plotted versus modulation frequency for the two groups, and when a quality factor for each MTF was derived by calculating BMF/ sMTF bandwidth.

3.5 Effects of Modulation Depth

For most analyses, data were collected at 25% modulation depth. For a smaller sample of units we measured responses at additional modulation depths. The synchronization coefficient at the best modulation frequency was plotted as a function of the sustained regularity. Data obtained for four different modulation depths are shown in Fig. 6. The maximum response modulation in the sample increased as stimulus depth was raised from 10 to 40%. At 40% and above, the response of some units saturates as synchronization coefficient approaches 1, the maximum possible synchronization. At the two lowest modulation depths, the biggest responses to modulation occur for units in the less regular/ irregular group. This is particularly evident at 25% depth. At higher depths the difference between the two groups is less apparent as the response of an increasing proportion of the sample approaches saturation. These data show the most precise synchronization to the modulation occurs for units in the less regular/ irregular group.

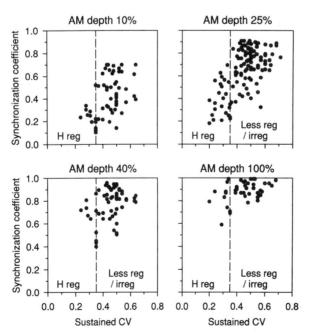

Figure 6. Synchronization coefficient of response to AM as a function of sustained regularity for CNIC units at four modulation depths.

3.6 Effects of Mean Level

The third parameter of the AM stimulus we have investigated is the mean level around which the amplitude fluctuates. As stated in the Methods, the preceding data were gathered at a level on the steeply sloping part of a unit's rate-level function. To investigate the influence of mean level, the maximum modulation gain at BMF was measured for each unit at three different levels above threshold. The data for highly regular units are shown in Fig 7A and those for the less regular/irregular group are shown in Fig 7B. In both cases modulation gain is plotted versus mean level above threshold. For both groups of units, there is some reduction in gain as mean level increases, but the reduction is most apparent for the highly regular group. Analysis of these data shows that the reduction in response modulation with increasing level is significantly smaller for the less regular/ irregular group than for the highly regular units. A similar difference was also apparent in a smaller group of units tested at 100% modulation depth.

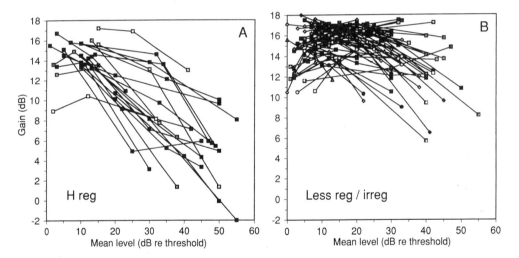

Figure 7. Peak modulation gain as a function of mean stimulus level for A. Highly regular units, B. Less regular/ irregular units

3.7 rMTF Peaks and Regularity

Several different types of rMTF are seen for CNIC units (Sarbaz and Rees unpublished data), but about 40% show a peak in their rMTF that coincides with the peak in the sMTF. The presence and magnitude of these rate responses are not invariant and in most cases depend on

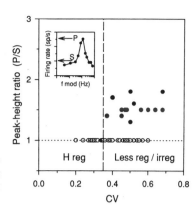

Figure 8. Peak of rMTF (measured as peak-height ratio, peak (P): shoulder (S), see inset) as a function of sustained CV. Open circles: units with no discernible peak (nominally set to 1). Filled circles: units with peaked rMTF.

mean level. To determine whether the presence of a peak in the rMTF correlates with discharge regularity, we compared the sustained regularities of units with peaked rMTFs with those of units in which peaks were absent. The height of the peak was measured by calculating the ratio of the firing rate at the peak to the rate at the shoulder of the rMTF on the low frequency side. Units without a distinct peak in their rMTF occured in both the highly regular and the less regular/ irregular groups. Units with a prominent peak in their rMTF, were restricted to the less regular and irregular group. The data obtained with 25% modulation depth, at a sound level on the upper third of the sloping segment of the rate level function. At higher or lower levels and depths a few peaked rMTFs were observed for highly regular units, but the proportion was always lower than for the less regular and irregular group.

4. DISCUSSION

These data show that several aspects of the responses to AM in CNIC units correlate with discharge regularity. The best modulation frequency of units that display IO correlates with IO frequency. These highly regular units also have narrower sMTF bandwidths than less regular/irregular units. The highest degree of synchronization to the modulation waveform occurs for less regular/ irregular units and these units also maintain their response to modulation at higher sound levels better than highly regular units. Finally a peak in the rMTF is mainly restricted to less regular/ irregular units.

A strong correlation between BMF and IO frequency was reported by Kim et al. (1990) for PVCN and DCN neurons. Evidence for intrinsic oscillations in the IC has been previously reported in the responses of neurons recorded in cat (Langner and Schreiner 1988), but these were only analysed over the first few intervals at the onset of the response. Intracellular studies in the IC and in the cochlear nuclei suggest that the tendency of neurons to fire in this highly regular manner depends on the neuron's intrinsic membrane properties and the spatial distribution of their inputs (e.g. Oertel et al., 1988; Smith and Rhode, 1989)

The relationship between BMF and the frequency of IO, and the narrower sMTF bandwidths of highly regular units are explained by the tendency of highly regular units to fire over a restricted range of frequencies. When the modulation frequency coincides with the frequency range of the IO the unit's firing is entrained and driven by the amplitude fluctuations in the modulation envelope. The unit's intrinsic properties resist attempts to drive it at other frequencies. Consequently the response to modulation frequencies on either side of the BMF decreases more steeply compared with the responses of less regular and irregular units. The weaker regularity in the latter group allows their firing to follow the envelope of the sound more faithfully and their discharge is entrained by a wider range of modulation frequencies.

Given the above, it is surprising that the least regular units show the highest degree of synchronization for measurements made at BMF. One might expect a unit showing intrinsic oscillations to show an almost perfect entrainment to the modulation waveform. It is important to remember, however, that intrinsic oscillation is represented by a broad peak in the spectrum, not a line. Thus, although some of the spike intervals correlate closely with the periodicity of the modulation waveform, others do not. Spikes will therefore occur at times in the period histogram that do not correspond to the peak of the modulation envelope and the response falls short of perfect entrainment. Nevertheless the response modulation is still substantial, some IO units display a 100% response modulation to a stimulus of 25% depth. In contrast, the firing of less regular/ irregular units is more precisely driven by the stimulus and spikes seldom occur at times other than at the peaks of the envelope resulting in an synchronization coefficient ~1 in some cases. This tight synchronization suggests that the firing of less regular/ irregular units is governed by a mechanism, perhaps inhibitory, that prevents the cell from firing except when the excitatory input to the cell is at its maximum.

The less restricted firing frequencies of units in the less regular/ irregular group may also help to explain the differences between the responses of the two regularity groups to mean level. For many CNIC units, regularity increases as a function of mean level (Rees et al., 1997). The tendency of highly regular units to fire with their own intrinsic frequency pattern thus becomes more pronounced as the mean level increases, and the degree of synchronization to the modulation envelope should decrease. In addition, the frequency of IO often changes with level, and this may explain the dependence of BMF on level observed in some IC units (Rees and Møller, 1987). For the less regular/ irregular group, in contrast, the absence of such an intrinsic oscillation enables the unit to be driven at the modulation frequency over a wider range of mean levels. One would also expect response modulation to decrease as level approaches the region where the rate level function saturates, and the difference between the groups might reflect differences in their rate-level slopes (c.f. Cooper et al., 1994).

The effects of level in the CNIC contrast with similar measurements made by Rhode and Greenberg in the cochlear nucleus. With 100% modulation depth they found the responses of units known to fire regularly (e.g. sustained choppers) better retain their AM responses as a function of mean level than irregular units (e.g. primary-like and primary-like-with-notch, compare their Fig. 16E with 16B &C). However, when one compares the absolute synchronization values for units in the two nuclei, highly regular units in the IC have only a slightly higher degree of synchronization compared to sustained choppers in the cochlear nucleus. The biggest difference between the two nuclei is that less regular/ irregular units in the CNIC respond with a much higher degree of synchronization than irregular units in the CN, and they maintain a better response to modulation at higher levels. This finding again suggest that some mechanism, operates in the CNIC to enhance the capacity of irregular neurons to synchronize to modulation.

The difference between the mean-rate responses to modulation in the two groups of units may be, in part, because highly regular units respond with a more invariable firing rate and are less able to change their firing rate as a function of modulation frequency. This cannot completely explain these rate responses since some less regular/ irregular units do not show any evidence of a peak in their mean firing rMTFs. It has been suggested from modelling studies that such responses in the IC reflect the convergence of information from regularly firing units at lower levels in the pathway (e.g. Langner, 1983; Hewitt and Meddis, 1994).

In conclusion, our results suggest that the regularity of neurons in the IC does influence their response to AM stimuli. Units that respond to modulation with the greatest degree of synchronization, and maintain the AM response best at high levels, are irregular. Inhibitory inputs may contribute to some of these responses. In contrast, highly regular units show their largest responses to modulation when the stimulus parameters, in particular the modulation frequency, coincide with the intrinsic firing frequencies of the unit. The precise role of these different response types in the coding of modulation awaits further investigation.

5. ACKNOWLEDGMENTS

AS was supported by the Iranian Ministry of Health and Medical Education.

6. REFERENCES

Blackburn, C.C. and Sachs, M.B. (1989) Classification of unit types in the anteroventral cochlear nucleus: PST histograms and regularity analysis. J. Neurophysiol. 62 1303–1329.

Bregman, A.S. (1993) Auditory scene analysis: hearing in complex environments. In: S. McAdams and E. Gigand (Eds.), Thinking in Sound, The Cognitive Psychology of Human Audition. Clarendon Press, Oxford, pp. 10–36.

Cooper, N.P., Robertson, D. and Yates G.K. (1993) Cochlear nerve fiber responses to amplitude-modulated stimuli variations with spontaneous rate and other response characteristics. J. Neurophysiol. 70, 370–386.

Flecknell, P.A. Laboratory Animal Anaesthesia. London, Academic Press, 1988.

Frisina, R.D., Smith, R.L. and Chamberlain, S.C. (1990) Encoding of amplitude modulation in the gerbil cochlear nucleus. I. A hierarchy of enhancement. Hear. Res. 44, 99–122.

Hall, J.W., Haggard, M.P., and Fernandes, M.A. (1984) Detection in noise by spectro -temporal pattern analysis. J. Acoust. Soc. Am. 76 50–56.

Hewitt, M.J. and Meddis, R. (1994) A computer model of amplitude modulation sensitivity of single units in the inferior colliculus. J. Acoust. Soc. Am. 95, 2145–2159.

Kim, D.O., Sirianni, J.G. and Chang, S.O. (1990) Responses of DCN-PVCN neurons and auditory nerve fibers in unanesthetized decerebrate cats to AM and pure tones: Analysis with autocorrelation/power spectrum. Hear Res. 45 95–113.

Langner, G. (1983) Evidence for neuronal periodicity detection in the auditory system of the guinea fowl: implications for pitch analysis in the time domain. Exp. Brain Res. 52, 333–355.

Langner, G. (1992) Periodicity coding in the auditory system. Hear. Res. 60,115–142.

Langner, G. and Schreiner, C.E. (1988) Periodicity coding in the inferior colliculus of the cat. I. Neuronal mechanisms. J. Neurophysiol. 60,1799–1821.

Le Beau, F.E.N., Rees, A. and Malmierca, M.S. (1996) Contribution of GABA- and glycine-mediated inhibition to the monaural temporal response properties of neurons in the inferior colliculus. J. Neurophysiol.75, 902–919.

Malmierca, M.S., Rees, A., Le Beau, F.E.N., and Bjaalie, J.G. (1995) The laminar organisation of frequency-specific local axons within and between the inferior colliculi of the guinea pig. J. Comp. Neurol. 357, 1–21.

Møller, A.R. (1976) Dynamic properties of primary auditory fibers compared with cells in the cochlear nucleus. Acta. Physiol. Scand. 98, 157–167.

Moore, B.C.J. (1997) An Introduction to the Psychology of Hearing. 4th Edition Academic Press, London.

Oertel, D. Wu, S.H. and Hirsch, J.A. (1988) Electrical characteristics of cell and neuronal circuitry in the cochlear nuclei studied with intracellular recordings from brain slices. In: G.M. Edelman, W.E. Gall, and W.M. Cowan (Eds.), Auditory Function, Wiley, NewYork, pp. 277–312.

Parham, K. and Kim, D.O. (1992) Analysis of temporal discharge characteristics of dorsal cochlear nucleus neurons of unanesthetized decerebrate cats. J. Neurophysiol. 67, 1247–1263.

Plomp, R. (1983) The role of modulation in hearing. In: R. Klinke and R. Hartman (Eds.), Hearing-Physiological Bases and Psychophysics, Springer-Verlag, Berlin, pp. 270–275.

Popelář, J. and Syka, J. (1982) Response properties of neurons in the inferior colliculus of the guinea pig. Acta. Neurobiol. Exp. 42, 299–310.

Rees, A. (1990) A. closed-field sound system for auditory neurophysiology. (Abstract) J. Physiol. 430, 6P.

Rees, A. and Møller, A.R. (1983) Responses of neurons in the inferior colliculus of the rat to AM and FM sounds. Hear. Res. 10, 301–330.

Rees, A. and Møller, A.R. (1987) Stimulus properties influencing the responses of inferior colliculus neurons to amplitude-modulated sounds. Hear. Res. 27, 129–143.

Rees, A. and Palmer, A.R. (1989) Neuronal response properties to amplitude-modulated and pure-tone stimuli in the guinea pig inferior colliculus and their modification by broadband noise. J. Acoust. Soc. Am. 85, 1978–1994.

Rees, A. Sarbaz, A. Malmierca, M.S. and Le Beau, F.E.N. (1997) Regularity of firing of neurons in the inferior colliculus. J. Neurophysiol. 77, In Press.

Rhode, W.S. and Greenberg, S. (1994) Encoding amplitude modulation in the cochlear nucleus of the cat. J. Neurophysiol. 71, 797–1825.

Rhode, W.S, Oertel, D. and Smith, P.H. (1983) Physiological response properties of cells labeled intracellularly with horseradish peroxidase in cat ventral cochlear nucleus. J. Comp Neurol. 213, 448- 463.

Smith, P.H. and Rhode, W.S. (1989 Structural and functional properties distinguish two types of multipolar cells in the cat ventral cochlear nucleus. J. Comp. Neurol. 282, 95–616.

Young, E.D., Robert, J-M., and Shofner, W.P. (1988) Regularity and latency of units in ventral cochlear nucleus. Implications for unit classification and generation of response properties. J Neurophysiol. 60, 1–29.

23

SINGLE CELL RESPONSES TO AM TONES OF DIFFERENT ENVELOPS AT THE AUDITORY MIDBRAIN

Paul W. F. Poon and T. W. Chiu

Department of Physiology
Medical College
National Cheng Kung University
Tainan Taiwan, Republic of China

ABSTRACT

Amplitude modulation (AM) and frequency modulation (FM) are important components of complex acoustic signals including speech. Response to AM tone is found at many places of the central auditory system particularly the inferior colliculus (IC) of the midbrain. With a repetitive sinusoidal AM tone, IC cells could respond best to a preferred modulation rate, indicating the rate of amplitude change is an important response determinant. Notably, most spike responses are found during the rising phase of the AM slope, and rarely during its falling phase. However, cells at the auditory cortex are known to respond to a sound source moving away from the animal, and presumably the sound would contain an AM component corresponding to a drop in amplitude. We speculated that the falling phase of an AM tone could evoke cell response at levels below the cortex, and that the response may depend on certain feature of the AM stimulus. In this study we systematically varied the envelop of an AM tone and examined its effects on single cell response at the IC of urethane-anesthetized rats.

Results from 161 AM cells showed that the response pattern of the same unit could change markedly depending on the modulation envelop. Over one-third of the units showed a second response peak during the falling phase of the envelop. This finding strongly suggested that the shape of the AM envelop is an important determinant of the response of AM cells, and this could account partly for the cortical sensitivity to a sound source traveling away from the ear.

Acoustical Signal Processing in the Central Auditory System
edited by Syka, Plenum Press, New York, 1997

1. INTRODUCTION

Amplitude modulation (AM) and frequency modulation (FM) represent important building blocks of communication signals (Winter et al 1966; Liberman et al., 1967). Modulation of single unit response by AM stimuli of different frequencies is observed in spike count and synchronization functions. Modulation transfer function shows progressive changes towards higher centers of the auditory pathways, e.g., it tends to vary from low-pass to band-pass (Gaese and Ostwald, 1995), with systematic decrease in best modulation rate. AM sensitivity can be detected as early as the primary auditory nerve (Ghoshal et al, 1992; Joris and Yin, 1992; Wang and Sachs, 1992; Cooper et al., 1993; Feng et al, 1994), subsequently at the cochlear nucleus (Wang and Sachs, 1992; Rhode and Greenberg, 1994; Zhio and Liang, 1995), medial superior olive (Grothe 1994), inferior colliculus (Brugge et al, 1993; Condon et al, 1996) and auditory cortex (Schreiner and Urbas, 1986; Phillips and Hall, 1987; Knipschild et al, 1992; Stumpf et al, 1992; Eggermont, 1994; Gaese and Ostwald, 1995). Similar responses to FM sounds were also found (Rees and Moller, 1983; Poon et al, 1991, 1992).

Due to the periodic nature of AM signals in speech, many AM studies adopted repetitive rather than single tone bursts. Furthermore a sinusoidal envelop rather than other modulation waveforms were most commonly used. There are at least two inherent problems with a periodic stimulus: (a) the response to each cycle of stimulus could be contaminated by its response to the preceding cycles; and (b) the sinusoidal envelop, though convenient to implement, may not necessarily be the most effective for the cell. The knowledge on how neurons of the auditory midbrain respond to single AM tones of different envelops will be necessary for understanding responses to repetitive AM stimuli, and the way AM response develops at the cortex.

In this experiment, we recorded single unit responses at the inferior colliculus (IC) of the anesthetized rat to single AM tone bursts. We compared their response pattern to AM tones of different envelops, and attempted to correlate their AM sensitivity to the response area of the cell.

2. METHODS

Sprague Dawley rats (150–250 gm) were anesthetized with urethane (1.5 g/kg b.w., i.p.) before skull was opened on one side to expose the occipital cortex overlying IC. The skull was then fixed to a special head holder with super-glue and placed inside a sound-shielded room. The normal posture of the pinnae were reconstructed by carefully suturing the incised skin. Acoustic stimulus was delivered from a free field speaker (frequency response +/- 10 dB from 500 to 40k Hz) positioned at 30° azimuthal from midline, 70 cm away in the contralateral sound field. Rectal temperature was maintained at 38°C with a heating pad throughout experiment.

Single unit activities from the IC were picked up with glass micropipette (10–30 MΩ) inserted through the cortex with a stepping microdrive (Narishige). Auditory neurons were identified by their response first to a free field click signal delivered at 2/sec. Then their response to sounds such as AM tones were studied systematically under the control of a computer (HP Pentium). The period of the single AM tones was varied from 4 to 40 cps, overlapping much with the range of the modulation rates used in the literature on IC cells.

Linear Scale Logarithmic Scale

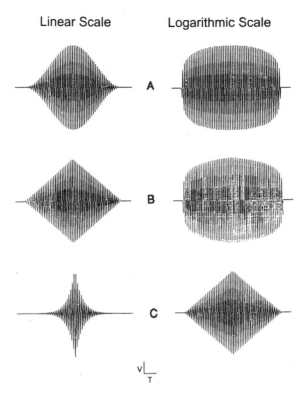

Figure 1. Three time waveforms of AM tone bursts with voltage displayed in different scales: linear (left column) versus logarithmic (right column). A. sinusoidal envelop, B. triangular envelop, and C. exponential envelop. On the logarithmic scale, note the similarity between A and B. In C, note the rate of amplitude change remains constant in the log scale.

Acoustic signal was generated by programming a 16 bit ADDA (Tucker-Davis Technology) that produced synthesized waveforms at 100 kHz. The AM tones were digitally synthesized with either a linear envelop or an exponential envelop (Fig. 1). Spike times of neuronal spikes were captured at 0.1 msec resolution. To characterize the response area of IC cells, a slow FM signal (1 up-and-down sweep / 2 sec) was adopted. This was accomplished by using the ADDA output to modulate the instantaneous frequency of a function generator (Tektronix FG 503) via its VCF control. Unit responses were displayed in dot raster and analyzed in peristimulus time histogram (PSTH) to repeated presentations of the stimuli.

3. RESULTS

Unit response was first collected using single cycles of tone bursts modulated with a triangular envelop. The period of the envelop changed systematically across the series of tone bursts. Auditory units were identified as AM cells when they showed phase locking to the AM modulation (i.e. either low-, high- or band-pass) over a range of modulation frequency (MF) estimated to be equivalent to about 4 to 40 cycle/sec. Fig. 2 shows an ex-

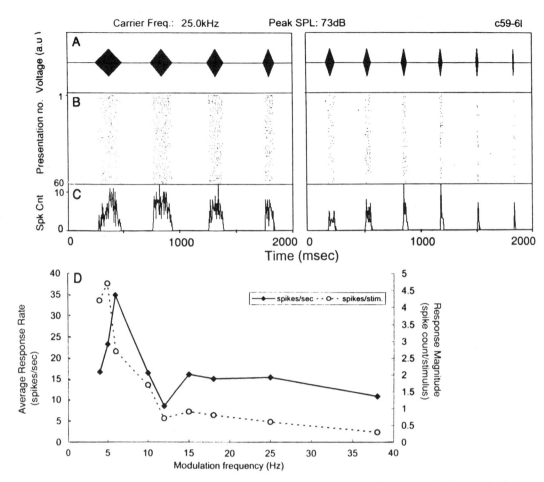

Figure 2. Typical response of an AM unit responding to tone bursts of various periods (A), with spike activity displayed in dot raster (B) and in the corresponding PSTH (C). Tones were delivered 30 dB suprathreshold at the cell's CF. Bin width is 4 msec. D. The results from C replotted to show tuning to the AM tones expressed in equivalent modulation frequencies. Note the spike response showed a band-pass tuning regardless of expressing the response function in terms of average response rate, or spike count/stimulus.

ample of an AM cell displaying a band-pass tuning in rate count function at its characteristic frequency (CF) over a range of stimulus level (i.e. 60 dB). This tuning to stimulus period could be considered as a tuning to the modulation rate similar to the tuning to a repetitive AM tone. Expressing the MF tuning in terms of spike count per unit time of the stimulus rather than on per stimulus basis would give rise to a different best modulation rate. However, the band-pass nature of the response function was not altered. Similar situation applied to cells showing low- or high-pass tuning to modulation frequency.

From a population of 371 click-sensitive units we have studied in the IC, 225 of them (60.6%) were AM cells. For the majority of AM cells (87.5%, n=197), spikes fell within both the rising and falling phases of the modulation envelop. Only 12.7%% (n=28) generated spikes during the rising phase, while none responded exclusively during the falling phase.

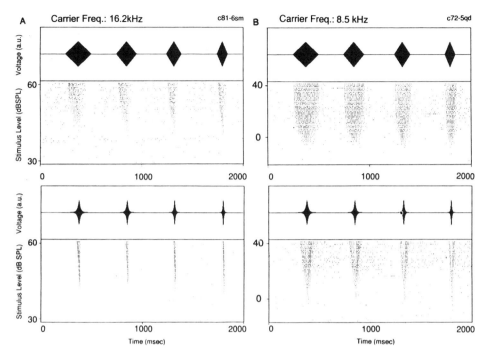

Figure 3. A. An AM unit responding during the rising phase of the modulation envelop. The response pattern remained more or less the same regardless of triangular or exponential envelop, even with stimulus intensity varied over a range of 30 dB. B. Another AM unit showing difference in response pattern when changed from triangular to exponential envelops.

Of the 225 AM units, 161 were fully characterized in terms of their response to both the triangular and exponential AM envelops. A majority of units (65.7%, n=109), the response pattern was apparently not markedly sensitive to the change in modulation envelop. Fig. 3A shows an example of this kind, even when the stimulus level was varied systematically over a range of 30 dB. However, over one-third of the units (34.3%, n=52) showed response difference to the modulation envelop. Specifically, many of them (88.5%%, n=46) would change its response from a single-peaked to a double-peaked pattern when the AM envelop was switched from triangular to exponential. Fig. 3B shows such an example with the stimulus level varied over a range of 60 dB. The double-peaked pattern could be revealed more clearly when the stimulus level was fixed (e.g. at 30 dB suprathreshold) as shown in Fig. 4 for another unit.

The difference in response pattern to the two AM envelops could be partly related to the response area of the cell. To investigate this, the response areas of these cells were characterized by using a slowly varying FM tone over an intensity range of 60 dB. The response areas of those cells showed in Fig. 3A displayed a simple tuning characteristics quite similar to those found at the auditory nerve. However, the response areas for the cells showed in Figs. 3B and 4 looked quite different (Fig. 5).

In Fig. 5A, the unit responded best at the intermediate stimulus level. A non-monotonic rate-level function to pure tone was obvious. In the example shown in Fig. 5B, response areas were skewed (tilted towards the left at high intensity levels) over the frequency-intensity plane.

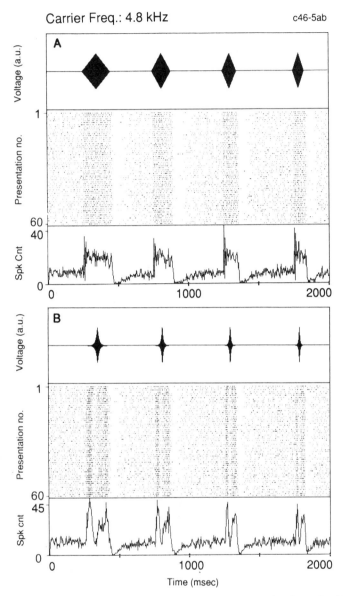

Figure 4. Response pattern of another AM cell similar to Fig. 3B. The stimulus level was fixed (30 dB suprathreshold) at the cell's CF to allow computation of PSTH. Note the clear change from a single-peaked (or plateau) response pattern (A), to a double-peaked response pattern (B).

4. DISCUSSION

The response to exponential envelop is of particular interest. In terms of the total stimulus energy, the exponential AM tone, when compared with the triangular AM tone, would have a lower average energy level. However, the peak of response to the exponential stimulus was often larger (e.g. Fig. 4). This could be related first to the different degree of adaptation to a time-varying stimulus. For example, the exponential AM is likely

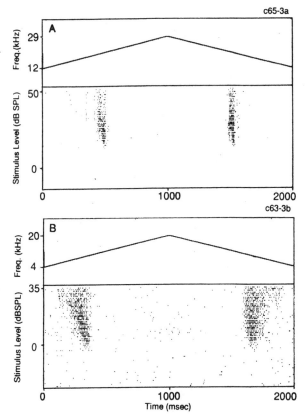

Figure 5. Response areas of two AM cells obtained with a slow FM sweep. The frequency changes across the cell's response area twice during a period of 2 sec (top panels). The stimulus level varied across successive presentations over a range of 60 dB. Note the indication of non-monotonic rate-level function in A, and a skewed response area in B.

harder to adapt to as the slope of modulation is changing more rapidly with time (Fig. 1). And second, this could be further related to the response area of the cell. For example, a non-monotonic rate-level function, as suggested by the response area in Fig. 5A, could easily explain for the double-peaked PSTH. Since it becomes harder for adaptation to take place in the case of an exponential AM stimulus, the effective stimulus level that could be reached would be higher. Hence, the stimulus level could have varied through the whole non-monotonic portion of the rate-level function. The result would be a drop of response during the mid-portion of the tone burst, where energy was also maximal. Similar perception of a sharper change in intensity for exponential AM tone occurred when the sound was played to our own ears.

In the case of monotonic rate-level functions the double-peaked PSTH could also be explained by a skewed response area (e.g. Fig. 5B). Although part of the skewness towards left could be accounted for by an increase in response latency when stimulus is attenuated close to threshold level. However, on the rising and falling phases of the slow FM sweep, this low intensity effect ought to act equally on both response areas. The fact that the skewness was not identical implied that the response area, disregard of low intensity effects, must already be skewed itself. Such skewed response area is probably not surprising as it has been first reported in the cochlear nucleus (Young et al, 1992). A simple convergence of two response areas sharing neighboring CFs but with different Q values could give rise to response areas like this. The AM stimulus is a tone varying in intensity but fixed in frequency. This stimulus would be equivalent to slicing through the cell's response area in a vertical direction on the frequency-intensity plane. Given the skewness, it is obvious that the behvior of rate-level function would resemble that of a non-monotonic function. Consequently the PSTH would have two peaks corresponding to the rise and fall phases of the envelop.

In fact, our own experience of listening to the two AM envelops suggested that the pattern of perceived change in sound level follow the time profile depicted in the logarithmic plot rather than the linear plot (Fig. 1).

Our principal finding is that a substantial portion of the AM cells at IC responded to AM tones quite dependent on the modulation envelop. Specifically the change from a single-peaked to a double-peaked response pattern implied that cells can now detect the falling phases in addition to the rising phase of amplitude modulation. This finding could be related to the animal's ability to discriminate AM signals in general (Moody, 1994) and the detection of sound motions at the auditory cortex (Stumpf et al, 1992; Toronchuk et al, 1992). The response to the falling phase of AM indicated that decreasing sound levels can be easily decoded. That auditory neurons at the cortical level were found to be sensitive to sound sources moving away from the animal can now be explained in part by neurons sensitive to falling phase of an AM sweep. Furthermore the AM rates at which these cells are sensitive to are also consistent with the movement speed of pinnae in these animals. That suggests that as the pinnae move, the pinna transfer functions also change, with the result of creating fast fall and rise in the intensity of a tone within the cell's frequency range. Such AM sensitivity would in theory provide extra sensitivity to the detection of a sound source with relative movement with respect to the pinnae.

That the midbrain is involved in head and pinna orientation to sudden loud sound has been known (Poon 1979). It is also known that FM and AM components are also rich in sounds produced by a moving object across a surface. We would therefore like to speculate that the emergence of these FM and AM cells at the midbrain could well serve to provide the necessary neural circuitry for the detection of an approaching enemy that generates complex sounds. Yet, it remains unclear as to what extent the inferior collicular processing is related to the cortical responses.

Results suggested that the AM sensitivity is more complex than a simple tuning in modulation rate. Specifically, IC cells not only responded to the rising phase of an AM envelop, but also responded to the falling phase as well. In fact, comparing sensitivities between AM and FM, a good analogy can be found. Namely, a receptive space containing (a) the extent of modulation change, (b) the rate of modulation change (both rise- and fall-phases), and (c) the frequency or intensity of the carrier tone. The analogy could provide a unified concept of FM and AM sensitivities at the auditory midbrain. In fact AM and FM are time derivatives of the two most basic stimulus parameters of a monaural pure tone, viz., its intensity (i.e. AM) and its frequency (i.e. FM). It is tempting to speculate that the emergence of FM and AM sensitivities at this level of the brain allows drastic data reduction, namely, a temporal sequence of stimulus events can now be encoded into the firing pattern of single IC neurons.[*]

REFERENCES

Brugge, J.F., Blatchley, B. and Kudoh, M. (1993) Encoding of amplitude-modulated tones by neurons of the inferior colliculus of the kitten. Brain Res. 615, 199–217.

* Supported by National Science Council grants NSC85-2332-B006-064, 86-2314-B006-087.

Condon, C.J., White, K.R. and Feng, A.S. (1994) Processing of amplitude-modulated signals that mimic echoes from fluttering targets in the inferior colliculus of the little brown bat, Myotis lucifugus. J. Neurophysiol. 71, 768–784.

Condon, C.J., White, K.R. and Feng, A.S. (1996) Neurons with different temporal firing patterns in the inferior colliculus of the little brown bat differentially process sinusoidal amplitude-modulated signals. J. Comp. Physiol. A. Sens. Neural Behav. Physiol. 178, 147–157.

Cooper, N.P., Robertson, D. and Yates, G.K. (1993) Cochlear nerve fiber responses to amplitude-modulated stimuli: variations with spontaneous rate and other response characteristics. J. Neurophysiol. 70,: 370–386.

Eggermont, J.J. (1994) temporal modulation functions of AM and FM stimuli in cat auditory cortex. Effects of carrier type, modulating waveform and intensity. Hear. Res. 74, 51–66.

Feng, A.S., Lin, W.Y. and Sun L. (1994) Detection of gaps in sinusoids by frog auditory nerve fibers: importance in AM coding. J. Comp. Physiol. A. Sen. Neural Behav. Physiol. 175, 531–546.

Gaese, B.H. and Ostwald, J. (1995) Temporal coding of amplitude and frequency modulation in rat auditory cortex. Europ. J. Neurosci. 7, 438–450.

Ghoshal, S., Kim, D.O. and Northrop, R.B. (1992) Amplitude-modulated tone encoding behavior of cochlear nucleus neurons: modeling study. Hear. Res. 58, 153–165.

Grothe, B. (1994) Interaction of excitation and inhibition in processing of pure tone and amplitude-modulated stimuli in the medial superior olive of the mustached bat. J. Neurophysiol. 71, 706–721.

Joris, P.X. and Yin, T.C. (1992) Responses to amplitude-modulated tones in the auditory nerve of the cat. J. Acoust. Soc. Am. 91, 215–232.

Knipschild, M., Dorrscheidt, G.J. and Rubsamen, R. (1992) Setting complex tasks to single units in the avian auditory forebrain. I. Processing of complex artificial stimuli. Hear. Res. 57, 216–230.

Liberman, A.M., Cooper, F.S. and Shanweiler, D.P. and Studdert-Kennedy, M. (1967) Perception of the speech code. Psychol. Rev. 74, 431–461.

Moody, D.B. (1994) Detection and discrimination of amplitude-modulated signals by macaque monkeys. J. Acoust. Soc. Am. 95, 3499–3510.

Phillips, D.P. and Hall, S.E. (1987) Responses of single neurons in cat auditory cortex to time-varying stimuli: linear amplitude modulations. Exp. Brain Res. 67, 479–492.

Poon, P.W.F. (1979) Cortical centers and midbrain pathways involved in orientation to sound in space. In: 'Festschrift for W.D. Neff', Univ. Chicago, USA, pp. 154–156.

Poon, P.W.F., Chen, X. and Hwang, J.C. (1991) basic determinants for FM responses in the inferior colliculus of rats. Exp. Brain Res. 83, 598–606.

Poon, P.W.F., Chen, X. and Cheung, Y.M. (1992) Differences in FM responses correlate with morphology of neurons in the inferior colliculus of rats. Exp. Brain Res. 91, 94–104.

Rees, A. and Moller, A.R. (1983) Responses of neurons in the inferior colliculus of rat to AM and FM tones. Hear. Res. 10, 301–331.

Rhode, W.S. and Greenberg, S. (1994) Encoding of amplitude modulation in the cochlear nucleus of the cat. J. Neurophysiol. 71, 1797–1825.

Schreiner, C.E. and Urbas, J.V. (1986) Representation of amplitude modulation in the auditory cortex of the cat. I. The anterior auditory field (AAF). Hear. Res. 21, 227–241.

Stumpf, E., Toronchuk, J.M. and Cynader, M.S. (1992) Neurons in cat primary auditory cortex sensitive to correlates of auditory motion in three-dimensional space. Exp. Brain Res. 88, 158–168.

Toronchuk, J.M., Stumpf, E. and Cynader, M.S. (1992) Auditory cortex neuron sensitive to correlates of auditory motion: underlying mechanisms. Exp. Brain Res. 88, 169–180.

Wang, X. and Sachs, M.B. (1992) Coding of envelope modulation in the auditory nerve and anteroventral cochlear nucleus. Phil. Trans. Royal Soc. London Ser. B. Biol. Sci. 336, 399–402.

Winter, P., Ploog, D. and Latta, J. (1966) Vocal repertoire of the squirrel monkey (Saimiri sciureus) its analysis and significance. Exp. Brain Res. 1: 359–384.

Young, E.D., Spirou, G.A., Rice, J.J. and Voigt, H.F. (1992) Neural organization and responses to complex stimuli in the dorsal cochlear nucleus. Phil. Trans. Royal Soc. London Ser. B. Biol. Sci. 336, 407–413.

Zhao, H.B. and Liang, Z.A. (1995) Processing of modulation frequency in the dorsal nucleus of the guinea pig: amplitude modulated tones. Hear. Res. 82, 244–256.

24

EVIDENCE FOR "PITCH NEURONS" IN THE AUDITORY MIDBRAIN OF CHINCHILLAS

U. W. Biebel and G. Langner

Institute of Zoology
THD
Schnittspahnstr. 3, D-64287 Darmstadt, Germany

1. INTRODUCTION

An important feature of our auditory system is its ability to detect voiced signals even under extremly noisy conditions ("cocktail party effect"). By spectral filtering, the cochlea improves signal-to-noise relations. However, this gives rise to the problem that information about a broadband signal gets scattered over various frequency channels in the auditory system. Psychophysic experiments show that for voiced or harmonic sounds the auditory system seems to make use of periodicity information to recombine this distributed information (Assman and Summerfield, 1990). In the auditory periphery, the envelope of complex tones is coded by phase locking (Shofner et al., 1996; Zhao & Liang, 1995; Ruggero, 1991). In the auditory midbrain this kind of temporal information is degraded and periodicity information is transformed into a rate-place code. In the inferior colliculus (ICC) best modulation frequencies (BMF = maximum of a modulation transfer function) of neurons are represented topographically, roughly orthogonal to the tonotopic organization (Langner, 1992). Modulation frequencies relevant for communication sounds (especially human speech) are in general below 1000 Hz. Therefore neurons in the inferior colliculus, that are tuned to low frequencies (characteristic frequency = CF < 1000 Hz) are likely candidates for spectral integrators of distributed activity representing a broadband signal. The aim of the present investigation was to look for such neurons with low CFs that may integrate particular periodicity information over a broad frequency range.

2. METHODS

112 single and multi units in the inferior colliculus (ICC) were recorded from 4 male awake chinchillas (Chinchilla laniger). Tungsten microelectrodes (about 1 MΩ resistance) were inserted through a chronically implanted chamber above the IC. The recordings were made at about 4 to 8 mm depth in vertical tracks. The wave shapes of the recorded spikes

were stored digitally for separation of single cells from multi unit recordings ("Discovery" program, DataWave™). Amplitude modulated sinusoids (SAM) were produced by summation of three sinusoids by three programmable synthesizers. The periodicity of these signals can be manipulated easily by simply varying the modulation period while holding the carrier frequency constant (and within the spectral center). On the other hand the spectrum of the signal can be varied by changing the carrier frequency, without influencing the envelope periodicity of the sound. Sounds were delivered by way of an electrostatic earphone (STAX, Lambda pro) in a sound-attenuated chamber. The earphone was placed close to the ear (about 4 cm) which was contralateral to the recorded IC. Stimuli were monitored acoustically by an electrostatic microphone and loudspeaker and visually by a spectrum analyzer (Medav, Mosip 3000). Distortion products were absent at the applied loudness levels. Unit response thresholds and characteristic frequencies (CF) were determined audio-visually. Iso-intensity plots and/ or tuning curves were measured as well as modulation transfer functions (MTF). Different combinations of carrier and modulation frequencies were applied for each unit at a level of about 30 dB above its threshold. An electrolytic lesion of the recording site in one animal, the response types of the neurons and reproducible neuronal characteristics in certain recording depths indicated that the recordings were mainly in the central part of the IC (ICC, figure 1). Only neurons having a clear tuning to one characteristic frequency (CF) arranged along a tonotopic gradient from dorsal to ventral were used in this investigation.

3. RESULTS

In addition to conventional response properties most of the neurons with characteristic frequencies up to 3 kHz showed reactions to complex stimuli with frequencies lying spectrally far outside their pure tone response area (87 % of 112 tested). This reaction appeared far beneath loudness levels that are known to evoke combination tones. Several neurons responded at levels as low as 20 dB SPL. 30 % of the responding neurons synchronized their discharge to the modulation frequency. About half of the low frequency neurons were sharply bandpass tuned to a certain amplitude modulation of a carrier frequency, although they did not respond when the carrier was presented without modulation. 45 % of the MTFs of neurons were multipeaked with one or two side maxima in addition to the main maximum (BMF). The BMF was strongly correlated with the CF of the neuron.

Figure 2 shows the response of a neuron with an extremely low CF to tone bursts with different frequencies. Pure tones of 170 Hz elicited the highest spike rate. In figure 2B the reaction of the same neuron is shown to SAM-stimuli with different modulation frequencies applied at the same intensity as in A. The response resembles strongly that to pure tones although the frequency spectrum is totally different in the two cases: the SAM signals lie totally outside the pure tone response area. In contrast to modulated tones, the unmodulated 4 kHz pure tone elicited no response (Figure 2 B, bottom, modulation frequency 0 Hz). Note, that the responses to pure tones continue for 20 - 30 ms after the end of the signal (vertical line), while the response to SAM already gets weaker during the stimulation. In figure 2 C the average spike rates of plots A and B are plotted as a function of frequency, respectively modulation frequency. The average rate of the same neuron to SAM with a carrier of 7 kHz is added. The response curves are very similar in all three cases.

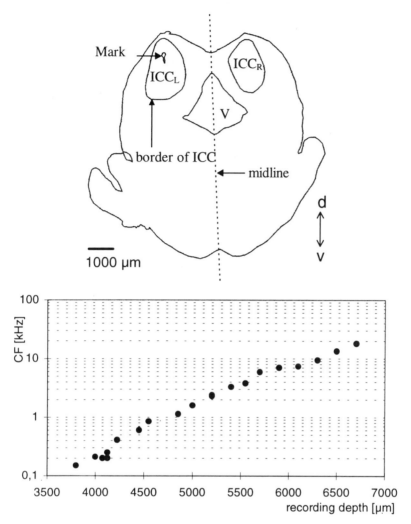

Figure 1. Verification of recording sites. A: Drawing of a transversal Nissl-stained section through the midbrain of a chinchilla including the borders of ICC and the marked recording site (Mark). At this location in the dorsal part of the ICC a recording was made from a "pitch-neuron" with a CF of about 400 Hz. d: dorsal, v: ventral, V: ventricle, ICC_L: left ICC, ICC_R: right ICC. B: Typical logarithmic relationship between recording depth and CF indicating that recordings were located inside the ICC.

In general, neurons responded to amplitude modulations when their characteristic frequency was used as *carrier* frequency. In addition, to respond to SAM stimulation with high frequency carriers the *modulation* frequency had to be in the range of their CF. Figure 3 shows the correlation of best modulation frequency (BMF) with characteristic frequency (CF) for 74 neurons, for which both measures are available. This correlation indicates, that low frequency neurons in the ICC should be characterized not only by their

CF. In addition they may respond to a variety of periodic signals (like SAMs with different carrier frequencies), that is they have a characteristic period.

While the BMF is quite independent of the chosen carrier frequency in these neurons, the strength of the response may change with carrier frequency. Most neurons respond to a quite wide range of carrier frequencies, as long as the modulation frequency is near their BMF. In addition, some neurons seem to be tuned to certain best carrier fre-

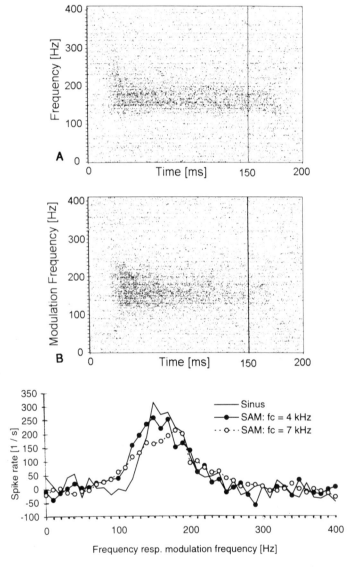

Figure 2. Illustration of a low frequency ICC neuron that responded to modulated sinusoids although the carrier frequency was outside its pure tone response. Point plots of the discharges are shown for stimuli of 150 ms duration and 50 ms of silence. The vertical line indicates the end of the stimulus. A: Response to 10 presentations of pure tones of different frequencies. B: Response to SAM stimuli with a carrier frequency of 4 kHz and modulation frequencies from 10 to 400 Hz. C: Comparison of discharge rates of unit G7–0 in response to stimulation with pure tones and SAM (spontaneous rate substracted). All stimuli were about 30 dB above threshold (40 dB SPL), unit G7–0.

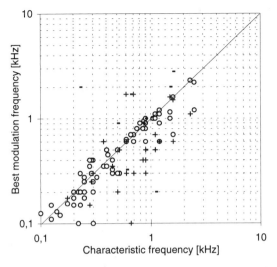

Figure 3. CF and BMF of 74 neurons with CFs between 0,1 and 3 kHz were correlated. Open circles symbolize the main maxima, while + and - stand for second and third order peaks in the modulation transfer functions.

quencies. In Figure 4 the response of a neuron is illustrated, where two modulation frequencies were tested with various carrier frequencies. Average spike rate changes drastically with different carrier frequencies and constant modulation frequency. With a modulation frequency different from BMF the response is still strong and complex, but much weaker than at BMF. At the same time the maximum is shifted towards a lower carrier frequency (3 instead of 5 kHz). In this as well as in most other neurons the carrier frequency of the SAM must not exceed 10 kHz in order to evoke a response.

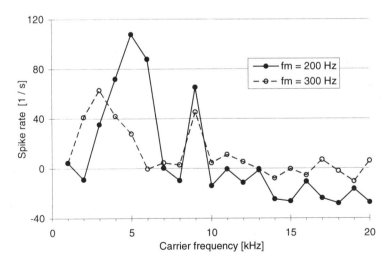

Figure 4. Illustration of the broadband modulation sensitivity of an ICC neuron tuned to 200 Hz. Stimuli were SAM signals with combinations of many different carrier frequencies (between 1 and 20 kHz, abscissa) and two modulation frequencies (see panel). (CF = 200 Hz, threshold: 10 dB SPL, loudness: 40 dB SPL, 8 repetitions, unit J8–3.)

4. DISCUSSION

Since neurons in the ICC respond to a certain envelope frequency (which may elicit a certain pitch) while spectral variations of the stimulus are less important, they may be called "pitch neurons". One can say that such neurons do not respond to a certain characteristic frequency, but rather to a certain characteristic period, either pure tone period or SAM modulation period. In other words: a neuron which may code a certain fundamental frequency when it appears in the spectrum of a sound will still be activated when the fundamental frequency is missing and only some overtones display the corresponding periodicity. In this way pitch neurons actually code the so called "missing fundamental", that is the absent fundamental frequency in SAM or other synthetic stimuli (Ritsma, 1962; McFadden, 1988).

The signals that evoked responses in a pitch neuron were spectrally totally different and therefore activated very different regions in the cochlea. We assume, that in a central part of the auditory pathway many neurons, that are tuned to the same periodicity project to a certain neuronal subpopulation which integrate information across different spectral

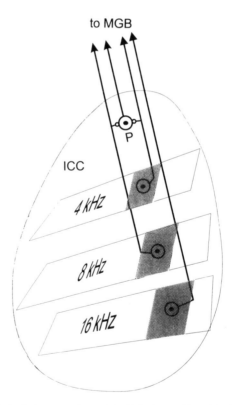

Figure 5. A scheme of possible intrinsic connections of the ICC, that could explain the response characteristics of pitch neurons. Three frequency planes are drawn schematically. As an example the neurons in these planes have CFs of 4, 8 and 16 kHz. Because of the periodotopy neurons with similar BMF (stippled areas) are piled up. Fibers projecting to the thalamus (MGB) give collaterals to neurons lying more dorsally in the ICC and thus having lower CFs. These fibers connect neurons with the same BMF but different CF to pitch neurons (P) with a CF, that corresponds to that BMF (for example 500 Hz). Fibers from neurons with CFs above 10 kHz were not found to innervate these pitch neurons. Therefore no connection is drawn from the 16 kHz neuron to the pitch neuron.

channels. Such neurons may then respond to various complex signals with different spectra but the same periodicity. The pitch neurons might either represent this population or reflect its response.

The periodotopic organization of the isofrequency sheets in the ICC (Schreiner and Langner, 1988) might work as a system that provides topographically organized input to the low frequency neurons. Output fibers of the ICC traverse the laminae roughly orthogonally and make collaterals inside the ICC before they leave to the medial geniculate body (Oliver et al., 1990; Malmierca et al., 1995). We assume that many neurons with different CFs, but tuned to the same modulation frequency, are connected to one pitch neuron lying in the dorsomedial part of the ICC. In figure 5 the proposed intrinsic network of the ICC is drawn as a schematic model.

In the auditory cortex of gerbils (field AI) neurons were found that react quite similar to SAM signals (Schulze, 1995). The main difference is that in the cortex no clear correspondence of CF and BMF was found. Cortical neurons may respond to certain combinations of carrier and modulation frequencies that are apparently independent of one another.

Chinchillas utter individually different harmonic communication calls with fundamental frequencies between 500 and 1000 Hz. The spectral range of these sounds is very broad but seldom reaches 10 kHz (Braun, S., pers. comm.). It is likely that pitch neurons are used for the detection of such calls. Vowels in human speech have much the same structure and probably pitch neurons in the human auditory system may serve as "pitch extractors" for speech sounds. Especially, they could play an important role in detecting a voice in a background of noise or against a competing voice.

This work was supported by the DFG: "Sonderforschungsbereich 269"

5. REFERENCES

Assmann, P.F. and Summerfield, Q. (1990) Modeling the perception of concurrent vowels - vowels with different fundamental frequencies. J.Acoust.Soc.Am. 88:680–697.

Langner, G. (1992) Periodicity coding in the auditory system. Hear. Res. 60, 115–142.

Malmierca, M.S., Rees, A., Le Beau, F.E.N., & Bjaalie, J.G. (1995) Laminar organization of frequency-defined local axons within and between the inferior colliculi of the guinea pig. J. Comp. Neurol. 357, 124–144.

McFadden, D. (1988) Failure of a missing-fundamental complex to interact with masked and unmasked pure-tones at its fundamental-frequency. Hear. Res. 32, 23–40.

Oliver, D.L., Kuwada, S., Yin, T.C.T., Haberly, L.B., & Henkel, C.K. (1990) Dendritic and axonal morphology of HRP-Injected neurons in the inferior colliculus of the cat. J. Comp. Neurol. 303, 75–100.

Ritsma, R.J. (1962) Existence region of the tonal residue I. J. Acoust. Soc. Am. 34, 1224–1229.

Ruggero, M.A. (1991) Physiology and Coding of Sound in the Auditory Nerve. In A.N. Popper & R.R. Fay (Eds.), The Mammalian Auditory Pathway: Neurophysiology, Springer, New York, Berlin, pp. 34–93.

Schreiner, C.E. and Langner, G. (1988) Periodicity coding in the inferior colliculus of the cat. II. Topographical organization. J.Neurophysiol. 60, 1823–1840.

Shofner, W.P., Sheft, S., & Guzman, S.J. (1996) Responses of ventral cochlear nucleus units in the chinchilla to amplitude modulation by low-frequency, two-tone complexes. J. Acoust. Soc. Am. 99, 3592–3605.

Zhao, H.-B., & Liang, Z.-A. (1995) Processing of modulation frequency in the dorsal cochlear nucleus of the guinea pig: Amplitude modulated tones. Hear. Res. 82, 244–256.

25

RESPONSES TO INTENSITY INCREMENTS AND DECREMENTS IN DIFFERENT TYPES OF MIDBRAIN AUDITORY UNITS OF THE FROG

N. G. Bibikov and Oxana N. Grubnik

N. N. Andreev Acoustics Institute
Moscow, Russia

INTRODUCTION

Adaptation seems to be a very important feature of auditory processing. It provides the conservation and the enhancement of small changes of the sound envelope in the enormous dynamic range. In our previous studies of the midbrain auditory units of the frog we used sinusoidally modulated tones to analyse the influence of the long-term adaptation upon the encoding of the small intensity changes. Using the 10% amplitude-modulated (AM) high-intensity (30–60 dB above threshold) tones we calculated the modulation depth of the firing rate in successive periods after the tone onset. The results obtained in many units with a sustained response were the following. During the first 100–200 ms of the 10% AM tone duration the synchronisation of firing with the modulation waveform was not evident. Then, continuously the synchronisation appears. To the end of the first second the modulation depth of the firing rate was rather prominent. It continued to increase till to the 8–10th sec of the tone presentation (Bibikov, 1990; Bibikov and Grubnik, 1991; Bibikov and Nizamov 1996).

The long-term adaptation manifested itself even more dramatically when a amplitude-modulated segment was embedded in a long pure tone. As it was noticed earlier, in many cells a discharge pattern did not change after a substitution of a 200 ms pure tone burst by a 10% AM tone burst. A 10% modulation fragment of the same duration embedded in a long continuous pure tone could, however, evoke a strong response that was synchronised perfectly with the sound envelope (Bibikov, 1989).

In the present study we study the response of the midbrain auditory units of the grass frog (Rana t. temporaria) to modulated fragments of long pure tones more carefully. We inserted in a long pure tone either AM fragments that consist of 12 periods of 20 Hz modulation (Fig. 1 b) or only one period of modulation (Fig. 1 c). In the first case the overall duration of modulation was 600 ms, in the second case it was 50 ms. Furthermore,

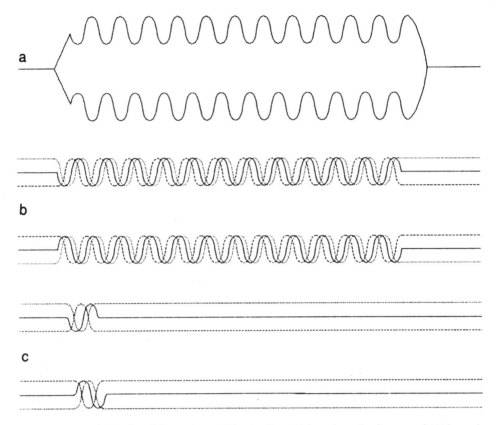

Figure 1. A schematic drawing of the envelopes of the stimuli. a - AM tone burst, b - fragment of AM in continuous pure tone, c - single modulation period in a continuous pure tone. In fig. 1 b and 1 c the AM fragments centred on the steady pure tone level are shown by solid line, the increments are shown by dashed line and decrements are shown by dotted line. For clarity the signals at 33% modulation depth are shown.

the modulation could be presented as intensity increments (dashed line in Fig. 1) or as intensity decrements (dotted line in Fig. 1). Modulation could be also centred around the average tone level (solid lines in Fig. 1 b, c).

This report provides the evidence that many auditory units of the frog's midbrain demonstrate a high sensitivity to small amplitude changes in the adapted state. In some units even decrements of the intensity could evoke well synchronised firing. Many units which display only onset discharge to a 80% modulated tone burst responded vigorously to small amplitude changes in a continuous pure tone.

METHODS

The full description of the method has been done previously (Bibikov, 1990; Bibikov and Grubnik, 1991; Bibikov and Nizamov 1996). Therefore only the main aspects will be repeated here.

The surgery (left side midbrain exposition) was provided under hypothermia. Afterwards, the frog was immobilised with an intramuscular injection of d-tubocurarine chloride (12 mg per kg body weight). For recording the animal was placed in a sound isolating booth with temperatures of 18–20°C. The recordings began when the animal was warmed up to the room temperature.

Electrodes were glass pipettes (tip diameter 0.5–3.0 mkm) filled with either 3 M NaCl or horseradish peroxidase. The micropipette was positioned above the surface of tectum opticum and was moved vertically through the brain by a remote-controlled step-motor microdrive.

Neuronal activity was amplified by high-impedance amplifier. Extracellular spikes of single units were selected by a window discriminator and input to a personal computer (IBM AT/386) for real data acquisition. As a rule, peristimulus time histograms (PSTH) with a bin width of 0.5 ms were obtained on-line after 50 presentation. For the displays presented in the figures of this report, we used a bin width of 4 ms (eight original bins). All stimuli were delivered to the right tympanum via a dynamic earphone through a plastic tube with a tapered fitting. Characteristic frequency (CF) and threshold at the CF was determined audio-visually for each unit using tone burst with a 200 ms duration.

In experimental sessions two formats of stimulus presentation were used. First, tone bursts of 612.5 ms of the whole duration with 5 ms rise-fall time were presented once every 2.2 seconds. Carrier CF signals at an average intensity of 30 dB re the unit's threshold were either unmodulated or amplitude modulated at a frequency of 20 Hz (Fig. 1 a). We classified units under study according to the PSTHs of their responses to bursts at 0%, 10% and 80% modulation depths (Bibikov and Nizamov, 1996). Hereupon the long CF pure tones having approximately the same intensity were presented. In these tones 600 ms or 50 ms AM fragments were inserted once every 2.2 seconds (Fig. 1 b, c). The modulation was sinusoidal with a frequency of 20 Hz. The calculation of the PSTH synchronised with the beginning of modulated fragments was started 5–10 s after the onset of the tone. The overall duration of each tone presentation was in the range 120–125 s and the interval between successive presentations usually exceeded 2 min.

RESULTS

The data reported here are based on recording of 65 TS units in 10 animals. In 11 of 27 units with tonic and tonic-phasic responses to tone burst we observed the effect of enhancement of synchronisation ability from the initial to the terminal modulation periods of a tone burst with a 10% modulation depth (see Bibikov and Nizamov, 1996). The examples of the PSTHs of the responses obtained for one of these units are shown in Fig. 2. The response to pure tone burst was classified as tonic-phasic (Fig. 2 a). For the burst with a 10% modulation depth the synchronisation increased gradually toward the terminal periods of modulation (Fig. 2 b).

All such units demonstrated also a phase-locking response to 10% AM fragments embedded in pure tone. The synchronisation of the response to each modulation period of AM fragments was considerably higher than to the best (usually the last) modulation period of the burst (Fig. 2 b, c). Moreover, the response of these units to a sequence of intensity decrements also demonstrated a prominent phase-locking. For such stimuli the firing probability usually increased from the first decrement toward the following modulation periods (Fig. 2 d). In response to a single modulation period these units usually displayed the same discharge pattern as they did to the first period of a 600 ms AM fragment. There-

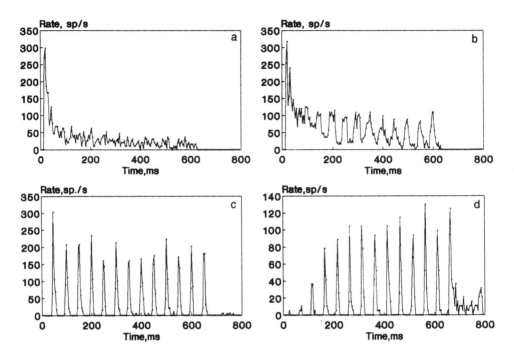

Figure 2. The PSTHs of the responses of the midbrain auditory unit 6–0306. Stimuli: a - pure CF tone burst, b - AM tone burst, modulation frequency (MF) - 20 Hz, modulation depth (MD)-10%, c - sinusoidal 600 ms AM fragment in continuous pure CF tone, MF-20 Hz, MD-10%, d - sequence of 12 sinusoidal intensity decrements, MF - 20 Hz, MD - 10%.

fore, the response to a single intensity decrement was comparatively week. The similar pattern of activity to AM fragments was observed in one unit showing the build-up discharge pattern in response to a pure tone burst.

In 19 of 37 phasic and phasic-tonic units the onset discharge pattern to a tone burst did not change even for stimulus envelopes with a 80% modulation depth (Fig. 3 a,b). Therefore, these cells did not respond to large variations of the amplitude throughout 600 ms tone burst. Nevertheless, many of them responded efficiently to small amplitude variations of a pure tone. Usually they did not reproduce a sequence of modulation periods but revealed a good sensitivity to the first period of the sequence or to a single amplitude change. In the majority of phasic units, which could not reproduce the 80% modulated envelope of a tone burst, a strong reaction to a single 10% intensity increment could be observed (Fig. 3 c). In some of these units even a single decrement in intensity evoked a synchronised response (Fig. 3 d).

Among other types of units (both phasic and tonic ones, demonstrated different phase-locking ability in the gated condition) we observed a great variability of the discharge pattern in response to AM fragments in pure tone. The general tendency of the enhancement of small amplitude changes in adapted state was, however, typical for the majority of units. The typical example is shown in Fig. 4. The cell was classified as a tonic-phasic unit with a week reproduction of the 10% modulated burst (Fig. 4 a, b). In the

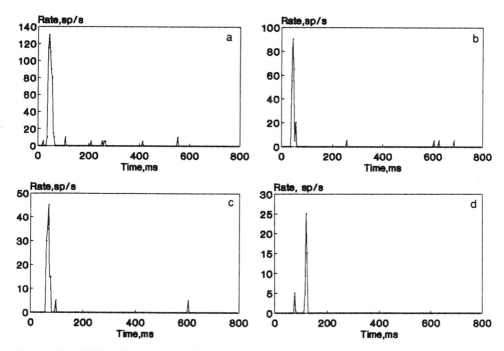

Figure 3. The PSTHs of the responses of the phasic midbrain unit 6–0407. Stimuli: a - AM tone burst, MF - 20 Hz, MD - 80%, b - a single 50 ms intensity increment in continuous pure CF tone, MD - 10%, c - a single 50 ms AM fragment in continuous pure CF tone, MD-10%, d - a single 50 ms intensity decrement in continuous pure CF tone, MD-10%.

adapted state this unit responded both to one period of 10% modulation around the steady state (Fig. 4 c) and to a 5% intensity increment (Fig. 4 d).

DISCUSSION

The present study was aimed to answer two questions. Does the phenomenon of the enhancement of the response to the small amplitude changes depend upon "whether the adapting signal varies in amplitude or not"? Can units with pure phasic response to AM tone burst respond to small increments in continuous pure tone? We compared the unit's response to AM tone burst and to the same AM presented as a fragment in a continuous pure tone. In many units with sustained response to tone burst the response to AM fragments embedded in a pure tone was highly synchronised with a sinusoidal envelope. The synchronisation to the stimulus envelope was considerably better than in a gated paradigm. The strong response was observed also when a small amplitude increment was presented in a continuous pure tone. Therefore, the best reproduction of small amplitude changes in the activity of tonic units was observed when an adapting signal was a pure tone.

The majority of phasic units generated only non-response to AM tone bursts. Nevertheless, they could respond very effectively to small amplitude changes (preferably incre-

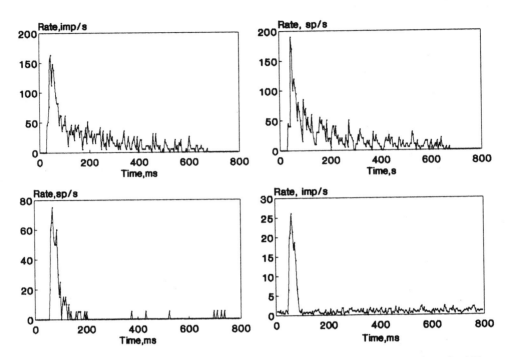

Figure 4. The PSTHs of the responses of the midbrain unit 6–0402. Stimuli: a - pure CF tone burst, b - AM tone burst, MF- 20 Hz, MD-10%, c - a single 50 ms intensity increment in continuous pure CF tone, MD - 10%, d - a single 50 ms intensity increment in continuous pure CF tone, MD - 5% (500 presentations).

ments, but also decrements) in a long pure tone. Therefore, in the majority of phasic units a long pure tone was also an effective adapting signal for a detection of small amplitude changes.

These physiological data has interesting correlates in psychoacoustics. In several studies a considerable decrease of the detection threshold for AM fragments (Bibikov and Makeeva, 1989) and for intensity increments (Viemeister and Bacon, 1984; 1989, Canevet at al., 1991) was observed in the adapted paradigm. Moreover the temporal "sharpening" of response during the adaptation could be important for the effect of sound localisation (Canevet, 1996).

Our results show that many central auditory units of the frog display a very high sensitivity to small increments and even decrements in the intensity of continuous pure tone sound. The enhanced response to AM fragments may be a result of accurate balance between synaptic excitation and inhibition in the adapted state (Bibikov, 1992). In such state incoming synaptic stream could be adjusted just below the level of spike triggering. Thus, even a small disturbance of the resulting input balance would evoke a discharge. From the other hand, it is possible to suggest some intracellular mechanisms of the observed effects. Recently it was shown that both intracellular protein synthesis and concentration of free cytoplasmatic Ca2+ depend critically upon forthcoming orthodromic stimulation (Rubel, 1996). As a result, cells being in adapted state can be more sensitive to synaptic input. Intracellular recordings and manipulations with ionic content can be used for evaluation of cellular mechanisms of the observed effects.

REFERENCES

Bibikov N.G. Extraction of amplitude-modulated segments in a continuous tone by auditory neurones of the frog. Soviet Physics. Acoustics. 1988, 34, 400–401.

Bibikov N.G. Response of frog midbrain neurones to tones amplitude-modulated by pseudorandom noise. Neurophysiology 1990, 22, 180 - 186.

Bibikov N.G., Grubnik O.N. Detection of a periodic component of amplitude modulation against a background of noise by neurones of the torus semicircularis of the lake frog. Sensory systems, 1991, 4, 28–34.

Bibikov N.G. Responses of single auditory units to random stimulation and lateral inhibition. Journal de physique, 1992, IV, 2, 233–236.

Bibikov N.G. and Makeeva I.P. Auditory adaptation and AM detection thresholds. Sov. Phys. Acoust. 1989, 35, 585–588.

Bibikov N.G. Nizamov S.V. Temporal coding of low-frequency amplitude modulation in the torus semicircularis of the grassfrog. Hear. Res. 1996, 101, 23–46.

Canevet G. Auditory adaptation: some new results. Acustica. 1996, 82, S30.

Canevet G., Scharf B., Ward L.M. Detection d'un increment d'intensite et adaptation de sonie. Acustica. 1991, 74, P.69–76.

Rubel E.W. Auditory experience and calcium in the avian brainstem. 1996. Abstracts, symposium acoustical signal processing in the central auditory system, Pragne, p. 5.

Viemeister N.F. and Bacon S.P. Intensity discrimination, increment detection and magnitude estimation for 1 kHz tones. J. Acoust. Soc. Amer. 1984, 84 , 172–178.

26

IS EMERGENT MOTION SENSITIVITY IN THE INFERIOR COLLICULUS A RESULT OF CONVERGENT INPUTS?

David McAlpine, Alan R. Palmer, and Dan Jiang

MRC Institute of Hearing Research
Science Road, University Park
University of Nottingham
Nottingham, NG7 2RD, United Kingdom

1. INTRODUCTION

A model of low-frequency sound localization consists of an array of brainstem neurones, acting as coincidence detectors, receiving input from the two ears via axons of different lengths (Jeffress, 1948). When the external delay of the signal is compensated by an appropriate internal delay (the characteristic delay, CD, see Rose *et al.* 1966; Yin and Kuwada, 1983), inputs from each ear arrive at the same time. The CD can be measured as the (linear) slope of the plot of the best interaural phase (BP) as a function of the stimulus frequency (the phase plot). CDs that correspond with a peak in the delay function indicate that the delay sensitivity is generated by excitatory inputs from both sides, whilst CDs that correspond with a TROUGH indicate the interaction of excitation from one ear with inhibition from the other. Recordings of single neurones in the *medial* superior olive (MSO: Goldberg and Brown, 1959; Yin and Chan, 1990; Spitzer and Semple, 1995) have confirmed the existence of binaurally-sensitive, coincidence detector neurones that fire maximally at a particular interaural delay of the stimulus (peak-type). The low-frequency region of the *lateral* superior olive (LSO) also contains neurones sensitive to interaural-time and -phase (e.g. Goldberg and Brown, 1969; Finlayson and Caspary, 1991; Batra *et al.*, 1995), mediated by excitation from the ipsilateral ear, and a timed inhibition from the contralateral ear (trough-type). The responses of many low-frequency neurones in the inferior colliculus (IC) to interaural time delays are also consistent with a single, fixed delay, or CD. However, for a significant proportion of neurones in the IC, the phase plot is not a single straight line, but is often better fitted by two or more straight-line segments. We hypothesize that such neurones receive convergent input from brainstem. Consistent with this hypothesis, it is possible to suppress the contribution of one input by presenting a tone at its least favourable delay (Palmer *et al.* this volume), leading to domination of the activity by the remaining input(s). It has also been reported that sensitivity to the direction of sound-source motion is an emergent property of the IC neurones (Spitzer and Semple, 1993), as it is not found in at the level of the MSO (Spitzer and Semple, 1992). In this report, we investigate the possibility that the emergent motion sensitivity results from convergent inputs from lower brainstem neurones onto single neurones in the IC.

2. METHODS

Recordings were made from the left IC of 300–450 g guinea pigs anaesthetised with Urethane (1.3 g/kg in 20% solution), with further anaesthesia provided by phenoperidine (1 mg/kg). Animals were placed in a stereotaxic frame with hollow earbars into which fitted 12.7 mm Brüel and Kjær condenser earphones, and 1-mm probe tubes fitted to 12.7 mm Brüel and Kjær microphones. Single-neurone action potentials were measured using tungsten-in-glass microelectrodes (Merrill and Ainsworth, 1972; Bullock *et al.*, 1988).

Responses to binaural beats (dichotic tones differing in frequency, producing continually-varying interaural phase disparities, IPDs) were obtained from low-frequency IC neurones. Responses were obtained when the signal delivered to the contralateral ear was 1 Hz greater than that delivered to the ipsilateral ear, resulting in the IPD at the contralateral ear initially 'leading' that at the ipsilateral, 'lagging' ear. From the responses at each frequency, the neurone's CD, and characteristic phase (CP) values were calculated. The CP is the phase intercept of the regression line fitted to the data points in the phase plot, and gives an indication of the type of binaural interaction. CPs close to 0.0 or 1.0 indicate a peak-type responses, whilst CPs close to +/- 0.5 indicate a trough-type response.

Responses to interaural phase modulation (IPM) stimuli were also obtained, where the phase at one ear was sinusoidally modulated, whilst the phase at the other ear was fixed. The range of IPDs traversed was controlled by adjusting the depth and/or centre of the phase modulation. The same modulation frequency (1 Hz at the right ear) was used throughout, and a change in the depth of IPM produces a concomitant change in the *velocity* of the stimulus.

3. RESULTS

3.1 Neurones with Linear Phase Plots

An example of an IC neurone with a linear phase plot is shown in Fig. 1. This neurone's phase plot (Fig. 1A) indicated a peak-type response (CP = -0.02), with a CD of +499 μs. When a second tone with a fixed delay was presented in order to inactivate one input (see chapter by Palmer *et al.*, this volume), the effect was to reduce substantially the discharge rate evoked by the binaural beats, but leave the BP unchanged. This suggests that the effect of adding a second tone at worst delay to a neurone with a linear phase plot suppresses its input, and that IC neurones with linear phase plots receive input from single brainstem coincidence detectors.

Figures 1B shows the same neurone's response to IPM stimuli at its best frequency (BF = 242 Hz), where the interaural phase was 100% modulated around a centre of 0.0 cycles of IPD (indicated by the vertical bar overlying the direction arrows). Thus, sweeping the IPD from -0.5 through zero and out to +0.5 (solid arrow) produced the solid curve, whilst sweeping the IPD in the opposite direction (dotted arrow) produced the dotted curve. The direction of motion had negligible effect on the response; the IPD func-

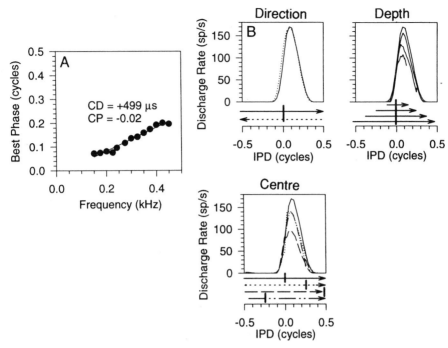

Figure 1. (A) Phase plot of an IC neurone (BF=242 Hz) which was best fitted by a single straight line, and indicated a peak-type (CP = -0.02) response with a CD of +499 μs. (B) Response to 100% IPM centred around 0.0 cycles of IPD (indicated by solid vertical line on arrows). The solid curve indicates the response to excursions of IPD from -0.50 to +0.50 cycles of IPD, and the dotted curve indicates the response to excursions of IPD in the opposite direction. (C) Response to different depths of IPM for one direction of motion (see arrows) centred at 0.0 cycles of IPD (solid vertical line). (D) Response to one direction (see arrows) of 100% IPM, centred at four different IPDs (arrows from top to bottom - 0.0, 0.25, 0.50 and 0.75 cycles).

tion was virtually identical for both directions of motion. Figure 1C shows the response of the same neurone to different depths of IPM for the same direction of motion, centred around an IPD of 0.0 cycles. Because the IPM was sinusoidal, and the modulation frequency was fixed at 1 Hz, the reduction in depth of IPM resulted in the IPD changing more slowly for shallower than for deeper depths. As the depth was reduced from 100% down to 75%, 50% and 25%, the maximum discharge rate evoked was reduced. Although not conclusive, this is consistent with this neurone being sensitive not only to instantaneous IPD, but also to the velocity of sound source motion. Figure 1D shows the response to 100% IPM centred on 0.0, 0.25, 0.50 and 0.75 (-0.25) cycles of IPD. The highest disharge rate was evoked when the interaural phase was modulated around 0.0 cycles of IPD. Lower, and equal, discharge rates were evoked when the interaural phase was modulated around 0.25 and 0.75 cycles of IPD, with the lowest discharge rates being evoked when the IPM was centred at 0.50 cycles of IPD. Thus, this neurone's response was dependent on the centre IPD around which the interaural phase was modulated, which is also consistent with a component of the neurone's response being sensitive to the rate of change of IPD.

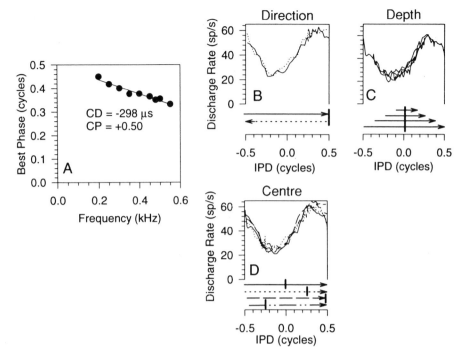

Figure 2. (A) Phase plot of an IC neurone (BF=478 Hz) which was best fitted by a single straight line, and indicated a trough-type (CP = +0.50) response with a CD of -298 µs. (B) Response to 100% IPM centred around 0.0 cycles of IPD (indicated by solid vertical line on arrows). The solid curve indicates the response to excursions of IPD from -0.50 to +0.50 cycles of IPD, and the dotted curve indicates the response to excursions of IPD in the opposite direction. (C) Response to different depths of IPM for one direction of motion (see arrows) centred at 0.0 cycles of IPD (solid vertical line). (D) Response to one direction (see arrows) of 100% IPM, centred at four different IPDs (arrows from top to bottom - 0.0, 0.25, 0.50 and 0.75 cycles).

A second example of a neurone with a linear phase plot is shown in Fig. 2. The phase plot (Fig. 2A) indicated that this neurone was a trough-type (CP = 0.50), with a CD of -298 µs. Again, in response to 100% IPM (Fig. 2B, with a centre IPD of 0.50 cycles) at BF (478 Hz), this neurone showed similar responses to both directions of interaural phase change. In addition, unlike the previous example, this neurone was insensitive to changes in the depth of IPM (modulation centre of 0.0 cycles in Fig. 2C). Given the variation in the rate at which the IPD changed with the change in depth of IPM, this suggests that this neurone was insensitive to sound-source velocity. Finally, this neurone was also insensitive to the centre IPD around which interaural phase was modulated (Fig. 2D), also consistent with it being insensitive to sound-source velocity.

3.2 Neurones with Non-Linear Phase Plots

Figure 3 shows an example of an IC neurone with a complex phase plot (Fig. 3A). In response to binaural beats, the low-frequency range (125–400 Hz) of the phase plot in Fig. 3A provided evidence of a peak-type input (phase intercept = 0.11), with a slope of +1314 µs. Above 400 Hz, however, the phase plot plateaued and started to roll over. Using the second-tone method to inactivate selectively the low-frequency, peak-type input (see chapter by Palmer *et al.*, this volume), we concluded that the complexity of the phase plot

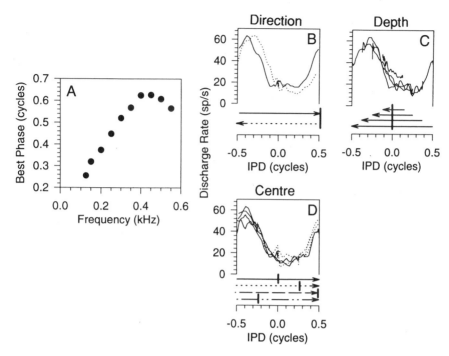

Figure 3. (A) Complex, non-linear phase plot of an IC neurone (BF=582 Hz). (B) Response of neurone in part A to 100% IPM centred around 0.0 cycles of IPD (indicated by solid vertical line on arrows) for both directions of motion (see arrows). (C) Response of neurone in part A to different depths of IPM for one direction of motion (see arrows) centred at 0.0 cycles of IPD (solid vertical line). (D) Response of neurone in part A to one direction (see arrows) of 100% IPM, centred at four different IPDs (arrows from top to bottom - 0.0, 0.25, 0.50 and 0.75 cycles).

was consistent with it receiving a second peak-type input with a CP of 1.02 and a CD of -1023 μs. Thus, the parsimoneous explanation of this complex phase plot is that it is the result of convergence onto a single IC neurone of two peak-type inputs, each showing a linear phase plot.

Figure 3B shows the response to both directions of 100% IPM for a centre IPD of 0.0 cycles at 450 Hz, and indicates that this neurone was sensitive to the direction of IPM, showing non-overlapping IPD functions for the two directions. Changing the depth of IPM (Fig. 3C) and the centre IPD around which the interaural phase was modulated (Fig. 3D) had little or no effect on the neurone's response.

A second example is shown in Figure 4. The phase plot for this neurone (Fig. 4A) was complex, with a clear transition between 400 Hz and 450 Hz. Neither of the local regions of the phase plot (150–400 Hz, and 450–550 Hz) corresponded to a simple peak-type or trough-type input, but had phase intercepts somewhere in between. Using the second tone method, we were able to alter some of the BP values, but were unable to reveal definitively the nature of any one of the binaural inputs to this neurone.

4. DISCUSSION

In general, IC neurones with linear phase plots showed little or no sensitivity to the direction of apparent motion provided by IPM stimuli. The phase plots of such neurones

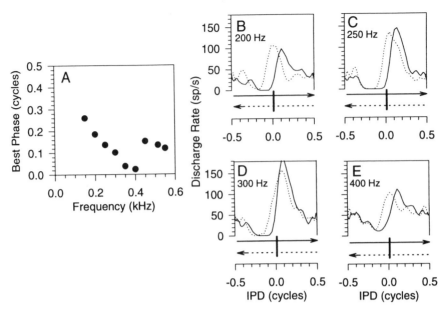

Figure 4. (A) Complex phase plot of an IC neurone (BF=288 Hz) with non-linear slope. (B-E) Response to 100% IPM centred around 0.0 cycles of IPD (indicated by solid vertical line on arrows) for both direction of motion, and for four different frequencies (B = 200 Hz, C = 250 Hz, D = 300 Hz, and E = 400 Hz). Solid curves indicates the response to excursions of IPD from -0.50 to +0.50 cycles of IPD, and the dotted curve indicates the response to excursions of IPD in the opposite direction.This neurone showed particularly strong sensitivity to the direction of IPM, and this sensitivity appeared to be frequency dependent. Figures 4B-E show the responses to 100% IPM for four different frequencies 200 Hz (Fig. 4B), 250 Hz (Fig. 4C), 300 Hz (Fig. 4D) and 450 Hz (Fig. 4E) for both directions of motion, where the centre IPD around which the interaural phase was modulated was 0.0 cycles. The extent to which the IPD functions for the two directions were non-overlapping differed for the different frequencies, being greatest at 200 Hz, and least at 300 Hz.

were not changed by the addition of a second, suppressive tone designed to unmask the presence of secondary inputs. In contrast, neurones with complex, non-linear phase plots often showed marked sensitivity to the direction of motion of IPM stimuli. Often for these neurones it could also be demonstrated that they received convergent input from brainstem coincidence detectors with simple, linear peak-type or trough-type responses (see chapter by Palmer *et al.*, this volume). This suggests that that sensitivity to motion direction observed by us and others (Spitzer and Semple, 1993) in the IC may be dependent upon convergent input from brainstem coincidence detectors onto single IC neurones.

In contrast, sensitivity to motion *depth* (or velocity) was unrelated to whether or not a neurone's phase plot was linear; some IC neurones with linear phase plots were sensitive to the depth of IPM (Fig. 1C), whilst others were not (Fig. 2C). So too, neurones with complex phase plots that were sensitive to motion direction were not necessarily sensitive to depth of IPM (velocity) or the centre IPD around which interaural phase was modulated.

In our paradigm, because the rate of IPM was always 1 Hz, the 'depth of motion' cue is confounded with the 'velocity of motion' cue. However, although not conclusive, another line of evidence that we have presented (the effect of altering the centre IPD around which interaural phase was modulated) suggests that in some instances a component of the response was due to the rate of change of IPD (velocity). For sinusoidal IPM, the velocity of the stimulus at any one IPD is not constant, but is dependent on the IPD around which the phase is modulated. Thus, for example, with 100% IPM and a centre IPD of 0.0 cycles, the rate of change of IPD ('velocity') is fastest around 0.0 cycles, and slowest at the extremities of a sinusoidal IPM excursion (-0.50 and +0.50 cycles in this case) where the direction of motion reverses. However, when the centre IPD is 0.50 cycles of IPD, the velocity is fastest at 0.50 cycles, and slowest around 0.0 cycles. Thus, for a neurone that responds well to IPDs close to 0.0 cycles (the neurone in Fig. 1, for example), interaural phase will change rapidly through the responsive region when the centre IPD is 0.0 cycles, but relatively slowly when the centre IPD is 0.50 cycles. Consistent with the data obtained by changing the depth of IPM, discharge rates were lowest when the IPD changed slowest (0.50 cycles centre IPD) through the neurone's best IPD region, than when the it changed fastest (0.0 cycles centre IPD). 100%-modulated IPM stimuli with intermediate centres of 0.25 and 0.75 (-0.25) cycles of IPD evoked discharge rates in between those evoked by IPM with 0.0 and 0.50 centres. Thus, when the apparent velocity was reduced by changing the depth of IPM, or when the sinusoidal IPM velocity was close to the point at which the direction of motion changed (and thus was slowest), the peak discharge rate was reduced.

Other neurones with linear phase plots that were sensitive to IPM depth, were also sensitive to the centre IPD around which interaural phase was modulated. Neurones that were insensitive to depth of IPM (neurones in Figs. 1 and 3) were also insensitive to the IPD around which interaural phase was modulated. These associated sensitivities are consistent with the same neurones' responses containing a velocity-sensitive component. If, as we have implied, IC neurones with linear phase plots receive a single binaural input from the brainstem, this suggests that neurones in the MSO and LSO may also show sensitivity to velocity of sound-source motion.

Finally, the responses of the neurone in Fig. 4 indicate that the effects of motion direction observed in neurones with complex phase plots may be dependent on frequency. Frequencies where there is relatively little interaction between two binaural inputs would be expected to show little sensitivity to motion direction, whilst frequencies at which there was maximum interaction between two inputs might would be expected to show maximum sensitivity to motion direction.

We conclude that the sensitivity to motion *direction* appears to be most apparent in those neurones that show complex phase plots and, thus, evidence of more than one binaural input. Neurones with linear phase plots show little or no sensitivity to motion direction. However, sensitivity to the depth of sinusoidal IPM, which also changes velocity, and to different centres of sinusoidal IPM, which also produces different velocities at some IPDs depending on the centre of IPM, was not restricted to, nor was it necessarily evident in, those neurones with complex phase plots. Taken together, these findings suggest the following. (1) The sensitivity to the direction of motion of sound sources moving slowly (1 Hz), which is an emergent IC response property, appears to correlate with evidence of convergent input from lower brainstem nuclei. (2) The responses of many IC neurones seem to contain a component that is sensitive to the rate of change of IPD. (3) This velocity sensitivity does not appear to be restricted to those neurones receiving convergent input and, thus, may well be a basic property of lower level coincidence detectors.

REFERENCES

Batra, R., Fitzpatrick, D.C. and Kuwada, S. (1995) Relationship of synchrony to ipsilateral and contralateral tones to ITD-sensitivity in the superior olivary complex. Eighteenth Meeting of the A.R.O.18, 62.

Bullock, D., Palmer, A. R. and Rees, A. (1988) A compact and easy to use tungsten-in-glass microelectrode manu-facturing workstation. Med. and Biol. Eng. and Computing. 26, 669–672.

Finlayson, P. G. and Caspary, D. M. (1991) Low-frequency neurons in the lateral superior olive exhibit phase-sen-sitive binaural inhibition. J. Neurophysiol. 65, 598–605.

Goldberg, J. M. and Brown, P. B. (1969) Response of binaural neurons of dog superior olivary complex to dichotic tonal stimuli: Some physiological mechanisms of sound localization. J. Neurophysiol. 32, 613–636.

Jeffress, L. A. (1948) A place theory of sound localization. J. Comp. Physiol. Psychol. 61, 468–486.

Merrill, E. G. and Ainsworth, A. (1972) Glass-coated platinum-coated tungsten microelectrodes. Med. and Biol. Eng. 10, 662–627.

Rose, J. E., Gross, N. B., Geisler, C. D. and Hind, J. E. (1966) Some neural mechanisms in the inferior colliculus of the cat which may be relevant to localization of a sound source. J. Neurophysiol. 29, 288–314.

Spitzer, M. W. and Semple, M. N. (1992) Responses to time-varying Interaural phase disparity in gerbil superior olive: Evidence for hierarchical processing. Society for Neuroscience Abstracts. 18, 149.

Spitzer, M. W. and Semple, M. N. (1993) Responses of inferior colliculus neurones to time-varying interaural phase disparity: Effects of shifting the locus of virtual motion. J. Neurophysiol. 69, 1245–1263.

Spitzer, M. W. and Semple, M. N. (1995) Neurons sensitive to interaural phase disparity in gerbil superior olive: Diverse monaural and temporal response properties. J. Neurophysiol. 73, 1668–1690.

Yin, T. C. T. and Chan, J. C. K. (1990) Interaural time sensitivity in medial superior olive of cat. J. Neurophysiol. 64, 465–488.

Yin, T. C. T. and Kuwada, S. (1983) Binaural interaction in low-frequency neurons in inferior colliculus of the cat. III. Effects of changing frequency. J. Neurophysiol. 50, 1020–1042.

27

DEVELOPMENT OF AUDITORY SENSITIVITY IN THE INFERIOR COLLICULUS OF THE TAMMAR WALLABY *Macropus eugenii*

Guang Bin Liu, K. G. Hill, and R. F. Mark

Developmental Neurobiology Group
Research School of Biological Sciences
Australian National University
Canberra 2601, Australia

1. INTRODUCTION

The Australian tammar wallaby, a macropod marsupial, is now the subject of established investigations into the development of the visual and somatosensory systems (e.g., Waite et al., 1991, 1994; Mark and Marotte, 1992). Recently, we have commenced studies of the development of the auditory system. The particular advantage of the marsupial preparation is that much of its development occurs both slowly and *ex utero*, the young being carried and suckled in a pouch, so that access is available at very early stages of development. We have now established the timetable for the first appearance and the properties of the scalp-recorded, auditory brainstem response (ABR) in developing wallabies, in several cases monitoring the ABR in longitudinal studies (Cone-Wesson, Hill and Liu, in prep.). Following its first appearance as a simple biphasic wave, in a preparation at around 120 days of pouch life, the ABR progressively becomes more complex, with an increased number of distinct peaks, as the animal matures. A likely reason for the elaboration of the ABR with development is the progressive onset of function in auditory brainstem nuclei, which are candidates as generators of discrete evoked potentials that may sum in the form of the ABR. The present study is concerned with the development of function in the prominent auditory centre, the inferior colliculus (IC). In marsupials, this major midbrain structure receives afferent connections from the more-peripheral auditory nuclei (Aitkin, 1986). The development of the IC has been studied in several mammalian species, rat (Altman and Bayer, 1981), cat (Aitkin and Reynolds, 1975; Moore, 1980), bat (Möller *et al.*, 1978) and rhesus monkey (Cooper and Rakic, 1981), however, little information is available on development of the IC in marsupials (Aitkin *et al.*, 1995).

Acoustical Signal Processing in the Central Auditory System
edited by Syka, Plenum Press, New York, 1997

2. METHODS

Evoked potentials were recorded intracranially, in response to acoustic clicks, from the location of the IC in 23 pouch young and juvenile wallabies of varied age between 114 and 293 postnatal days (PND). Recordings were performed in a sound-isolation booth (Tegner T-Room). Animals were anaesthetised with urethane (20%, 1ml/100g, i.p.; a quarter of initial dose subsequently was used when necessary to maintain anaesthesia). Body temperature was maintained at 37°C with a thermistor-controlled electric blanket. Acoustic clicks at a range of intensities from below threshold to approximately 78 dB peSPL measured at the left ear were presented from a distance of 820mm and about 45° left of the animal's mid-line. ABRs, being the average of scalp-recorded responses to 500–1000 stimulus presentations, were obtained prior to craniotomy, which exposed the right side of the brain for recording more focal, evoked potentials from auditory nuclei. With the animal placed on a stage, with its head positioned so that the line between inside canthi was horizontal, focal recordings were made using tungsten-in-glass, low resistance microelectrodes, that were advanced along vertical penetrations from the cortical surface, on which position was specified by antero-posterior and medio-lateral coordinates. Potentials evoked by click stimuli were monitored as the electrode was lowered in controlled steps (Burleigh 6000 micro-positioner). Responses were progressively assessed according to electrode depth, as a guide to the location of neural generators of auditory evoked potentials. When the electrode had been advanced to the maximum permissible depth, it was withdrawn to that depth at which the evoked potential was of maximum amplitude. The evoked potential was then characterized with respect to click stimulus intensity and tone stimulus frequency and intensity. Following systematic recording at specific locations, some of which were marked with electrolytic lesions, preparations were given urethane overdose and then perfused for subsequent histological verification of electrode position. Evoked potential response magnitude in a specified time segment was calculated as the area under the power spectrum of the recorded waveform for that segment. Evoked potential data were plotted as normalised response magnitude, in separate time segments following the stimulus, versus depth and as recorded waveforms in the spatio-temporal plane.

3. RESULTS

With respect to age, an IC response was first recorded in an animal at age 114 PND. Response magnitude versus depth is shown for this preparation in Figure 1. Response amplitude was unchanged over depth in segment 1 (before the stimulus) and in segments 4 and 5 (more than 28 ms following the stimulus). In segment 3 (approximately 10–28 ms after the stimulus), response amplitude peaked between depths 2000 μm to 4000 μm, which includes the position of the IC. In segment 2 (approximately 4–10 ms after the stimulus), response amplitude increased from 3000 μm to 5000 μm depth, implying a more ventral origin for this component of the response, than that occurring in segment 3.

In general, a click-evoked potential was recorded throughout the electrode penetration, from first contact with the brain surface. Typically, along a track that was histologically-confirmed (post-experiment) to have passed through the IC, the evoked potential increased in amplitude over a restricted range of depth (e.g. Figure 1). Such a component of the evoked potential was attributed to the IC only when its maximum amplitude was re-

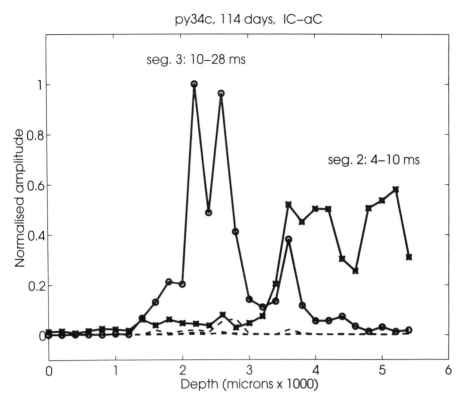

Figure 1. Response amplitude versus depth of recording in separate time segments following the click stimulus. Evoked potentials were recorded in time segments 2 and 3 (solid lines), no responses occurred in other time segments (dashed lines).

stricted to a narrow depth range, which corresponded with the histological depth of the IC and when the marker lesion occurred within the IC. Figure 2 shows the spatio-temporal plot of the evoked potential recorded from the preparation aged 114 PND. Throughout the track, a relatively-short latency (3ms) evoked potential is evident, that becomes of greatest amplitude at the ventral extent of the track. This early potential is attributed to the auditory nerve (N1) and cochlear nucleus responses, so the recording is remote at the brain surface. The evoked potential that corresponds with the IC with respect to depth appears in the track between depths 1500 to 3000 μm. At this age, the IC response was of low amplitude (maximum peak-peak 4.5 μv) and of extended duration (approximately 17ms). The latency of response onset was about 8 ms from the click arrival time at the left ear. Note that the response attributed to the IC is not recorded by the electrode at the start of the track, when in contact with the dorsal brain surface.

The spatio-temporal distribution of the evoked potential recorded in an animal of age 140 PND is shown in Figure 3. The focal IC response was recorded at depths between 4000 and 6000 μm and was confirmed histologically.

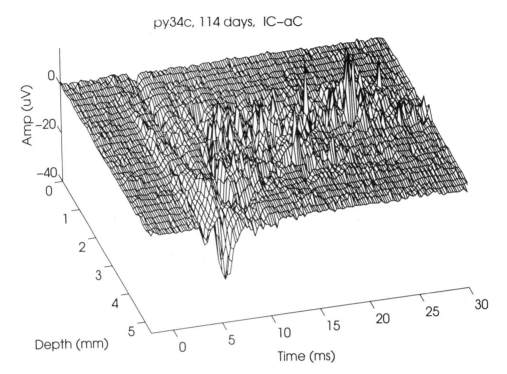

Figure 2. Saptio-temporal distribution of the click-evoked potential recorded from one penetration which passed through the IC of an animal aged 114 PND.

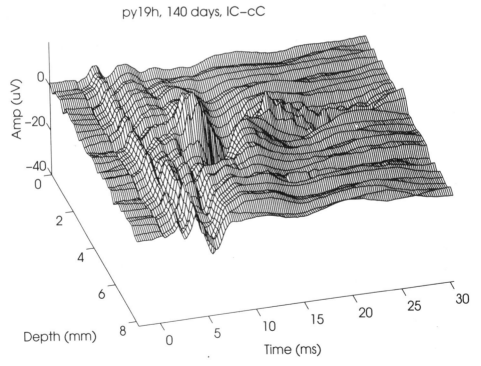

Figure 3. Saptio-temporal distribution of the click-evoked potential recorded from one penetration which passed through the IC of an animal aged 140 PND.

The spatio-temporal distribution of the evoked potential is shown for an animal aged 206 PND in Figure 4. In this case, the focal IC response was recorded at depths between 4500 and 7500 μm.

Obvious trends in Figures 2, 3 and 4 are that the IC evoked potential becomes of greater amplitude, shorter duration and shorter latency with increased developmental age, furthermore, the IC response occurs at progressively greater depth from the dorsal surface. In none of the examples, is there a clear recording of the IC response, when the electrode is first in contact with the brain surface, whereas, the shorter latency potentials (attributed to auditory nerve and cochlear nucleus) clearly appear throughout each penetration.

In Figure 5, the records containing the maximum IC response peak amplitude for each age (middle record) are compared with corresponding records from separate tracks in each animal containing the maximum (confirmed) response from the auditory nerve root (upper record) and the scalp-recorded ABR (lower record). Also shown at top are three such records from a younger (101 PND) animal in which no IC response was detected. In this comparison, the maturation of the IC response can be related to responses recorded at other locations at the same age.

4. DISCUSSION

The reduction in delay between the auditory nerve root response and the IC response in preparations of increasing age indicates reduced conduction time in the central nervous system, implying faster axonal conduction and/or synaptic transmission (cf. Brugge, 1992; Ehret and Romand, 1992). The reduction in duration of the IC response indicates a greater

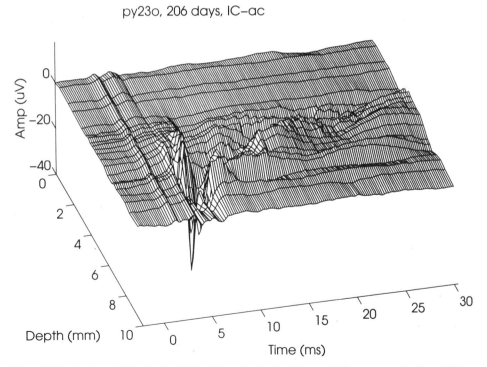

Figure 4. Saptio-temporal distribution of the click-evoked potential recorded from one penetration which passed through the IC of an animal aged 206 PND.

degree of synchrony in the initial excitation of the IC, which could also follow from increased security in drive from the more peripheral auditory centres (Eggermont, 1985; Konishi, 1973). The increased amplitude of the IC evoked potential in part would be expected to result from greater synchrony of onset in the IC response. The increased peak amplitude may also reflect a greater number of responding units in the IC (Jiang et al., 1993). Whereas the IC response in the youngest animals occurs delayed with respect to

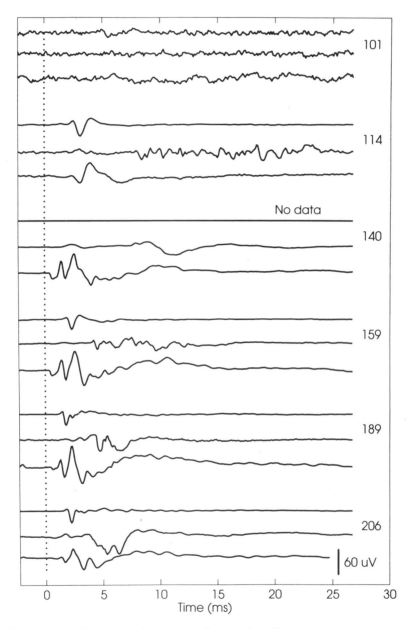

Figure 5. Temporal comparison of evoked potentials attributed to the auditory nerve root (upper traces), IC (middle traces) and the ABR during development. The focal ANR and IC responses are plotted to the same amplitude scale, while the records of the ABR have been magnified ×80. Age of each preparation (PND) is given to the right of each set of traces. Dotted vertical line shows the click arrival time at the left ear.

the ABR, when the IC response latency is substantially reduced, the IC response overlaps the latter part of the scalp-recorded evoked potential. With increased age of the preparation, the ABR becomes more complex and includes a larger number of peaks. There is no clear indication in Figures 3 and 4, however, that the overlap of the IC response with the ABR is the cause of the additional peaks in the ABR. The IC evoked potential appears pronounced at source (comparable with that of the auditory nerve), yet it is not recorded by the microelectrode from the brain surface. The low amplitude of the IC response in the far field suggests that it does not contribute to the ABR waveform.

5. REFERENCES

Aitkin L.M. (1986) The auditory midbrain. Structure and function in the central auditory pathway. Humana Press, Clifton, NJ, pp. 1–246

Aitkin L. (1995) The auditory neurobiology of marsupials: A review. Hear. Res., 82: 257–266

Aitkin L.M. and Reynolds A. (1975) Development of binaural responses in kitten inferior colliculus. Neurosci. Lett., 1: 315–319

Altman J. and Bayer S.A. (1981) Time of origin of neurons of the rat inferior colliculus and the relations between cytogenesis and tonotopic order in the auditory pathway. Exp. Brain Res., 42: 411–423

Brugge J.F. (1992) Development of lower auditory brainstem of the cat. In R. Romand (Ed.), Development of auditory and vestibular systems II., Elsevier, Amsterdam, pp. 273–296

Cone-Wesson B., Hill K.G. and Liu G.B. Development of the auditory brainstem response (ABR) in tammar wallaby (*Macropus eugenii*). Hear. Res., (MS in preparation)

Cooper M.L. and Rakic P. (1981) Neurogenetic gradients in the superior and inferior colliculi of the Rhesus monkey. J. Comp. Neurol., 202: 309–334

Eggermont J.J. (1985) Evoked potentials as indicators of auditory maturation. Acta Otolaryngol., (Suppl), 421: 41–47

Ehret G, Romand R. (1992) Development of tone response thresholds, latencies and tuning in the mouse inferior colliculus. Develop Brain Res, 67: 317–326

Jiang Z.D., Zhang L., Wu Y.Y. and Liu X.Y. (1993) Brainstem auditory evoked responses from birth to adulthood: development of wave amplitude. Hear. Res., 68: 35–41

Konishi M. (1973) Development of auditory neuronal responses in avian embryos. Proc. Nat. Acad. Sci., USA, 70: 1795–1798

Mark R.F. and Marotte, L.R. (1992) Australian marsupials as models for the developing mammalian visual system. TINS, 15: 51–57

Möller J., Neuweiler G. and Zöller H (1978) Response characteristics of inferior colliculus neurons of the awake CF-FM bat *Rhinolophus ferrumequinum*. J. Comp. Physiol., 125: 217–225

Moore D.R. and Irvine D.R.F. (1981) Development of responses to acoustic interaural intensity differences in the cat inferior colliculus. Exp. Brain Res., 38: 103–108

Waite P.M.E., Marotte L.R. and Mark R.F. (1991) Development of whisker representation in the cortex of the tammar wallaby *Macropus eugenii*. Dev. Brain Res., 58: 35–41

Waite P.M.E., Marotte L.R. and Leamey, C.A. (1994) Timecourse of Development of the wallaby trigeminal pathway. 1. Periphery to brainstem. J. Comp. Neurol., 350: 75–95

28

ASCENDING PROJECTIONS TO THE MEDIAL GENICULATE BODY FROM PHYSIOLOGICALLY IDENTIFIED LOCI IN THE INFERIOR COLLICULUS

M. S. Malmierca, A. Rees, and F. E. N. Le Beau

Department of Physiological Sciences
The Medical School
University of Newcastle upon Tyne
United Kingdom

1. INTRODUCTION

The auditory thalamus, the medial geniculate body (MGB), possesses three main divisions: ventral (MGBv), medial (MGBm), and dorsal (Morest, 1964). The MGBv has a tonotopic organization (Aitkin and Webster, 1972; Imig and Morel, 1988, Morel et al., 1987, Wenstrup et al., 1994). In the guinea pig, low frequency regions are located in the caudal, dorsal and medial portions of the MGBv while high frequencies are represented in the rostral and lateral portions of the MGBv (Redies and Brandner, 1991). The main source of ascending input to the MGB is from the central nucleus of the inferior colliculus (CNIC; tree shrew: Oliver and Hall, 1972; cat: Andersen et al., 1980; Kudo and Niimi, 1980; Oliver, 1984; Rouiller and Ribaupierre, 1985; rat: LeDoux et al., 1987). These findings are based on large injections of HRP into the MGB, and injections of tritiated amino acids or WGA-HRP into the inferior colliculus (IC). They demonstrate unequivocally that the CNIC projects to the ventral and medial divisions of the MGB and that the projecting fibres terminate in a topographically ordered manner in the MGB. Dorsolateral regions of the CNIC project to the lateral parts of the MGBv while ventromedial regions of the CNIC project to the medial parts of the MGBv. This tectothalamic pathway originates from both disc-shaped and stellate cells (Oliver, 1984; Malmierca, 1991) and, at least four categories of axons ascending from the IC have been described in Golgi impregnated material of the cat MGBv (Morest, 1975). In the MGBv of the bat (Wenstrup et al., 1994) and ferret (Pallas and Sur, 1994) two types of terminal boutons associated with fibres from the CNIC have been observed following injections of biocytin into the IC or its brachium.

Despite the availability in guinea pig of information concerning the intrinsic, commissural and descending projections of the CNIC (Malmierca et al., 1995, 1996) as well as the tonotopy in the MGBv and its projections to the auditory cortex (Redies and Bradner, 1991, Redies et al., 1989), the organization of the auditory tectothalamic pathway in this species has yet to be described. In this report we studied the ascending projections of the CNIC in guinea pig after making minute injections of the tracer biocytin at physiologically defined frequency regions of the CNIC. The distribution of the labelled terminal fields in the MGB was studied as a function of the frequency recorded at the injection site. We also describe the types of terminal boutons associated with fibres that project from the CNIC to the MGB.

2. MATERIALS AND METHODS

This study used material acquired in the course of two previous studies (Malmierca et al., 1995, 1996). The reader is referred to these earlier papers for a detailed description of the experimental procedures. Here it suffices to mention the essential points.

Adult pigmented guinea pigs were anaesthetised with a Hypnorm/midazolam cocktail or urethane (Malmierca et al., 1995; Flecknell, 1988) The animal was placed inside a sound-attenuating booth and pure-tone stimuli were delivered through a sealed acoustic system. The activity of single neurones or multineurone clusters in the IC was recorded with a glass micropipette (tip diameter 5–10 μm). The micropipettes were filled with biocytin (3.5% solution in 0.05M TRIS buffer, pH 7.6 and 2M NaCl, King et al., 1989). Best frequencies (BF, the frequency requiring the least intensity to drive the neurone) were determined audiovisually, and iontophoretic injections of biocytin were made by passing pulses of positive current (2–4 μA DC, duty cycle 7s) for 10 to 30 minutes.

After 18 to 36 hours survival the animal was deeply anaesthetised with an overdose of anaesthetic and transcardially perfused with rinse solution (0.05 $NaNO_2$ in 0.1M phosphate buffer, pH 7.4) followed by fixative (1.25% glutaraldehyde and 1% paraformaldehyde freshly depolymerized in 0.1M phosphate buffer, pH 7.4). The thalamus, midbrain and brainstem were cut into 50–70 μm sections on a freezing microtome in the transverse or horizontal plane. The sections were collected in 0.1M phosphate buffer and immediately processed to visualise the biocytin with the Avidin D-HRP procedure (King et al., 1989). Every other section was counterstained with cresyl violet.

In this study we subdivided the IC and MGB according to maps available for guinea pig. The IC is parcelled as in Malmierca et al., (1995) and the MGB according to Redies et al. (1989). The MGBv as described in the map of Redies et al. (1989) closely corresponds to that seen in other species (reviewed in Winer, 1992), although in cat the MGBv has been further subdivide into *pars ovoidea* and *lateralis* (Morest, 1964). The medial division of the MGB identified in cat does not correspond directly to the region defined by (Redies et al., 1989) as the caudomedial and rostromedial nuclei in guinea pig. But, for the sake of simplicity, we will use the term medial division of the MGB (MGBm) here when referring to the most medial and caudal parts of the MGB since a comprehensive treatment of the cytoarchitecture of the MGB is out of the scope of the present study.

Light microscopic analysis was performed with an Leitz Medilux microscope. Detailed camera lucida drawings of labelled structures in the MGB were made with the aid of a drawing tube attached to the microscope.

3. RESULTS

In all 7 cases selected for this study, a consistent pattern of labelling was observed in the ventral and medial divisions of the MGB after an injection of biocytin (about 300 μm in diameter) into the CNIC. A band of labelling was observed in the ipsilateral MGBv when single sections were inspected, and a cluster of labelled fibres and terminals in MGBm. The bands extend rostrocaudally along the MGBv, to form a plexus of fibres, *en passant* and terminal boutons. Each band extends over several sections to form a lamina (Figs.1 and 2). At least two distinct types of presumed terminal boutons were observed: small and large (Fig. 3), the latter arranged in grape-like bunches. In the MGBm, clusters of labelling made up of fibres and terminal boutons with no preferential orientation were also observed.

In the following, we shall describe the distribution of the labelled plexus in two cases where the injection sites were located in the dorsolateral corner and the ventromedial border of the CNIC. The BFs at these injection sites in the CNIC were 0.5 and 21 kHz, respectively. The same cases were used previously to study the intrinsic, commissural (Malmierca et al., 1995) and descending (Malmierca et al., 1996) projections of the CNIC.

3.1 Case GP-36 # 0.5 kHz (Fig. 1)

In this case an injection was made into the dorsolateral region of the CNIC (cf. Fig. 2A in Malmierca et al., 1995) adjacent to the external cortex of the inferior colliculus. Fibres arising from the injection site join the brachium of the IC and ascend to the MGB giving rise to terminals of collateral fibres in the nucleus of the brachium of the IC. Rostrally, terminal labelling is present mainly in MGBv but terminal labelling is also located caudally and medially, most probably in the MGBm. The labelling spans 700 μm rostrocaudally. The labelling in MGBm and MGBv overlaps, making it difficult to establish the border between the two subdivisions in our material. It was therefore not possible to estimate precisely the size of the lamina in the MGBv.

3.2 Case GP-21 # 21 kHz (Fig. 2)

This case had two injections (separated by approximately 150 μm) in a single frequency-band lamina located at the ventromedial border of the CNIC. (cf. Fig. 2C in Malmierca et al., 1995). Fibres from both injections run orthogonal to the laminae in the CNIC, coursing in a lateral direction to join the brachium of the IC. These fibres ascend in the brachium where they give collaterals in its nucleus as in the previous case. Further rostrally, they reach the MGB where they enter the ventral portion of the nucleus and terminate in bands of labelling oriented slightly obliquely in the lateral portions of the MGBv. The labelling in the MGBv is about 75–100 μm thick (fIG. 3) and spans 500 μm rostrocaudally. In the most rostral sections with labelling in the MGBv, no labelling is observed in the MGBm. Caudally, there are patches of weak terminal labelling in MGBm.

The injection sites in the five remaining cases had BFs at 2.7, 6, 6.1, 8 and 15 kHz. The labelling in the MGBv seen in these cases forms laminae located between the limits described in the two cases above. The labelling shifts progressively more lateral and rostral as frequency increases. In summary, the position of the laminar plexus in the MGBv varied as a function of the frequency recorded at the injection site in the CNIC. The case injected in the 0.5 kHz region produced a lamina in the caudal and medial part of the MGBv, while the injection in the 21 kHz region produced a lamina located rostrally and laterally with respect to the low frequency case. Labelling in the MGBm was also present in all cases.

4. DISCUSSION

This study shows that in guinea pig, a lamina in the CNIC produces a laminar plexus of labelling in the MGBv. This plexus contains at least two types of terminal boutons. The

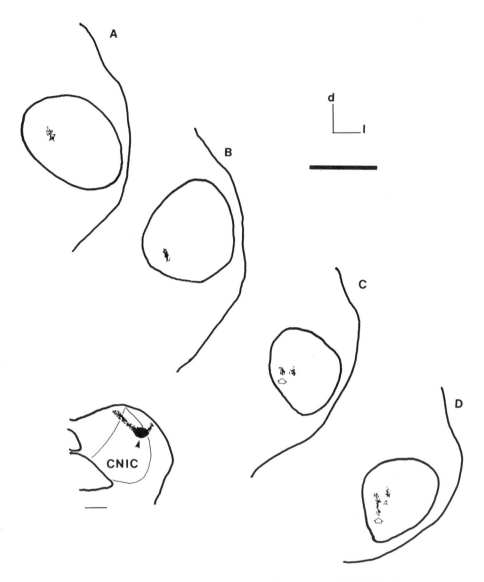

Figure 1. Camera lucida drawings from a case injected in the dorsolateral CNIC (BF, 0.5 kHz). A, rostral, D , caudal. Arrows indicate labelling in the MGBm. d, dorsal; l, lateral. Scale bars 1 mm. Distance between sections, 150–200 μm. Inset redrawn from Fig. 2A in Malmierca et al.1995. Arrowhead indicates the injection site.

change in the position of the lamina with frequency at the injection site shows that the auditory tectothalamic pathway is tonotopically organised. These findings are in good agreement with previous studies in other species (Anderson et al., 1980; Rouiller and Ribaupierre, 1985).

Some methodological limitations may however have influenced our results. The labelling we obtained in the MGB is relatively sparse, and therefore our Results may not show the entire extent of the projection of the CNIC to the MGB. Two reasons may account for the weak labelling. First, because we wanted to restrict our injections to physiologically defined sites, the injections in the CNIC were small compared with other studies (Anderson et al., 1980; Kudo and Niimi, 1980; Wenstrup et al., 1994). A second reason may be the survival

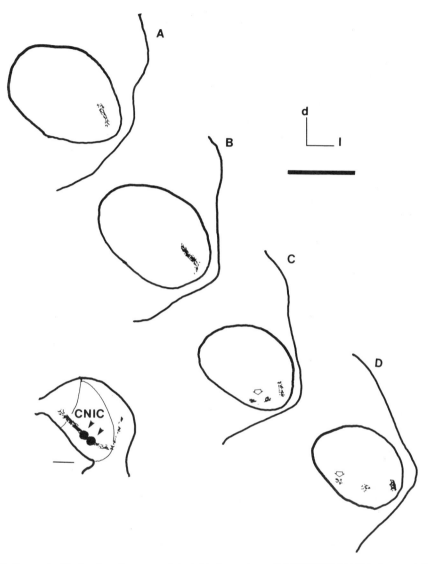

Figure 2. Camera lucida drawings from a case injected in the ventromedial CNIC (BF, 21 kHz). A, rostral, D, caudal. Arrows indicate labelling in the MGBm. d, dorsal; l, lateral. Inset redrawn from Fig 2C in Malmierca et al.(1995). Scale bars are 1 mm. Distance between sections, 150–200 μm. Arrowheads indicate the injection site. Detail from B showing two fibres with different types terminal boutons are shown in Figure 3.

time. It is known that biocytin is metabolised at axon terminals when long survival times are used (McDonald, 1992). It has been shown that the labelling from the auditory cortex to the cat CNIC is weaker when biocytin is used as opposed to *Phaseolus vulgaris*-Leucoagglutinin (Ojima, 1994). Labelling may be absent in the contralateral MBG for the same reasons.

Our results show that the low frequency region of the CNIC projects to the caudal and medial portion of the MGBv (possibly corresponding to pars ovoidea in cat, where the tonotopy mirrors that of pars lateralis) while higher frequency regions of the CNIC project

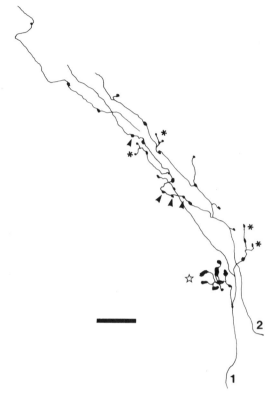

Figure 3. Camera lucida drawings made at high magnification from the labelled lamina in the 21 kHz case (Fig. 2) showing presumed small (asterisks) and large (star) terminals. Arrow heads indicate presumed *en passant* boutons. Scale bar, 50 µm.No retrograde labelled cells were observed either in the ventral or the medial divisions of the MGB. Likewise, no labelling was observed in the MGB contralateral to the injections site in the CNIC.

more rostrally and laterally in the MGBv. The rostrocaudal gradient in frequency representation is consistent with the tonotopical organisation of the guinea pig MGBv described in Redies and Bradner (1991). Redies et al. (1989) demonstrated that neurones populating the caudal and medial portions of the MGBv in the guinea pig project to the low frequency regions of the auditory cortex, while those located more rostrally and laterally project to the high frequency regions of the auditory cortex. Thus, our findings show that the tonotopically organised input to these regions from the CNIC of the guinea pig overlies the source of these cortical projections.

We found both small and large terminal boutons in the projections to the MGB. Similar types were first shown in cat (Morest, 1964) and more recently in bat (Wenstrup et al., 1994) and ferret (Pallas and Sur, 1994). It seems likely that they correlate with different neurotransmitters in these fibres. Hu et al. (1994) demonstrated that IC projections provide the MGB with a fast excitatory pathway mediated by glutamate that acts on both NMDA and non-NMDA receptors. This is supported by the recent study of Saint-Marie (1996) where D-[^{3}H]aspartate was injected into the MGB and retrogradely labelled neurones were observed in the IC. The large terminals that we describe in the MGB are similar to those that have been shown to have an ultrastructure compatible with excitatory synapses (Majorossy and Réthelyi, 1968; Jones and Rockel, 1971; Majorossy and Kiss, 1974). Thus, the large terminals might be glutamatergic.

Some of the small terminals we describe might be GABAergic. In several species, the MGBv possesses many small GABAergic puncta (Winer and Larue, 1996). A significant GABAergic projection from the CNIC to the MGB has also been demonstrated (Winer et al., 1996). This projection constitutes only 20% of the ascending projection in the cat (Winer et al., 1996), but in the rat it is significantly larger, up to around 50% (Peruzzi and Oliver, 1996). Conversely the rat MGB has a smaller population of GAD positive cells compared with the cat (Winer and Larue, 1988). Recent physiological studies have suggested that the GABAergic projection to the MGB may improve spike timing (Barnett and Smith, 1997).

The individual endings described here may contribute to a more complex level of synaptic organisation. Several auditory nuclei, including the MGB, contain synaptic nests, a pattern of organisation first described by Morest (1971, 1975) where terminal boutons of different origins synapse with the dendrites of principal and Golgi type II cells (reviewed in Morest, 1997, Chapter 2 this volume). The large and small terminals of collicular origin shown here may constitute part of this intricate arrangement.

The present results taken together with our previous studies of the intrinsic, commissural (Malmierca et al., 1995) and descending projections of the CNIC (Malmierca et al., 1996) demonstrate that the projections of the CNIC are well adapted to re-distribute auditory information converging from lower auditory nuclei and facilitate the interaction of information ascending from the CNIC with descending information from the auditory cortex.

5. ACKNOWLEDGMENTS

V.M. Bajo, D.L. Oliver and E. Rouiller, kindly commented in a previous version of the manuscript. Financial support was provided by the Royal Society, the University of Newcastle upon Tyne research Committee and the Commission of the EU. MSM was supported by the Spanish DGES (PB95–1129), the MEC and the Commission of the EU. We thank S. McHanwell for the use of his histological facilities. MSM is currently at the Laboratory for the Neurobiology of Hearing. Faculty of Medicine. University of Salamanca. Spain, and FENLB at the Department of Physiology. University of Maryland. Baltimore. USA.

6. REFERENCES

Aitkin, L.M. Webster, W.R. (1972) Medial geniculate body of the cat: Organization and responses to tonal stimuli of neurons in ventral division J. Neurophysiol. 35, 365–380.

Andersen, R.A., Roth, G., Aitkin, L.M., and Merzenich, M.M. (1980) The efferent projections of the central nucleus and the pericentral nucleus of the inferior colliculus in cat J. Comp. Neurol. 191, 479–497.

Bajo, V.M., Rouiller, E.M., Welker E., Clarke S., Villa A.E.P., de Ribaupierre Y., and de Ribaupierre F. (1995) Morphology and spatial distribution of corticothalamic terminals originating from the cat auditory cortex. Hearing Res. 83, 161–174.

Barlett, E.L. and Smith, P.H. (1997) A direct GABAergic input from the inferior colliculus may serve to improve spike timing in the rat medial geniculate body. ARO Abstract (In press).

Flecknell, P.A. (1988) Laboratory Animal Anaesthesia. Academic Press, London.

Hu, B., Senatorov, V., and Mooney, D. (1994) Lemniscal and non-lemniscal synaptic transmission in rat auditory thalamus. J. of Physiol. 479, 217–231.

Imig, T.J. and Morel, A. (1988) Organisation of the cat's auditory thalamus. In Auditory Function, Ed. by G.M. Edelman, W.E. Gall, and W.M. Cowan. John Wiley and Sons.

Jones, E.G. and Rockel, A.J. (1971) The synaptic organization in the medial geniculate body of afferents fibres ascending from the inferior colliculus. Zeit für Zellforsch. und Mikroskop. Anat. 113, 44–66.

King, M.A., Louis, P.M., Hunter, B.E., and Walker, D.W. (1989) Biocytin: a versatile anterograde neuroanatomical tract-tracing alternative. Brain Res. 49, 361–367.

Kudo, M. and Niimi, K. (1980) Ascending projections of the inferior colliculus in the cat: An autoradiographic study. J. Comp. Neurol. 191, 545–556.

LeDoux, J.E., Ruggiero, D.A., Forest, R., Stornetta, R. and Reis, R.J. (1987) Topographic organization of convergent projections to the thalamus from the inferior colliculus and spinal cord in the rat. J. Comp. Neurol. 264, 123–146.

Majorossy, K. and Réthelyi, M. (1968) Synaptic architecture in the medial geniculate body (ventral division) Exp. Brain Res. 6, 306–323.

Majorossy, K. and Kiss, A. (1976) Specific pattern of neuron arrangement and of synaptic articulation in the medial geniculate body. Exp. Brain Res. 26, 1–17.

Malmierca, M.S. (1991) Computer-assisted three-dimensional reconstructions of Golgi impregnated cells in the rat inferior colliculus. Doctoral thesis, Universities of Oslo and Salamanca.

Malmierca, M.S., Le Beau, F.E.N., and Rees, A. (1996) The topographical organization of descending projections from the central nucleus of the inferior colliculus in guinea pig. Hearing Res. 93, 167–180.

Malmierca, M.S., Rees, A., Le Beau, F.E.N., and Bjaalie, J.G. (1995) Laminar organization of frequency-defined local axons within and between the inferior colliculi of the guinea pig. J. Comp. Neurol. 357, 124–144.

McDonald, A.J. (1992) Neuroanatomical labelling with biocytin: a review. NeuroReport 3, 821–827.

Morel, A., Rouiller, E., Ribaupierre, Y, Ribaupierre, F. (1987) Tonotopic organization in the medial geniculate body (MGB) of lightly anesthetized cats. Exp. Brain Res. 69, 24–42.

Morest, D.K. (1964) The laminar structure of the medial geniculate body of the cat. J. Anat. 99, 143–160.

Morest, D.K. (1971) Dendrodendritic synapses of cells that have axons: the fine structure of the Golgi type II cell in the medial geniculate body of the cat. Z. Anat. Entwickl-Gesch, 133:216–246.

Morest, D.K. (1975) Synaptic relationships of Golgi type II cells in the medial geniculate body of the cat. J. Comp. Neurol., 162:157–187.

Morest, D.K. (1997) Structural basis for signal processing: Challenge for synaptic nests. Chapter in this book.

Ojima, H. (1994) Terminal morphology and distribution of corticothalamic fibers originating from layers 5 and 6 of cat primary auditory cortex. Cerebral Cortex 6, 616–663.

Oliver, D.L. (1984) Neuron types in the central nucleus of the inferior colliculus that project to the medial geniculate body. Neurosci. 11, 409–424.

Oliver, D.L. and Hall, W.C. (1978) The medial geniculate body of the tree shrew, *Tupaia glis.* Y. cytoarchitecture and midbrain connections. J. Comp. Neurol. 182, 423–458.

Pallas, S.L., and Sur, M. (1994) Morphology of retinal axon arbors induced to arborize in a novel target, the medial geniculate nucleus. II. comparison with axons from the inferior colliculus. J. Comp. Neurol. 349, 363–376.

Peruzzi, D., Oliver, D.L. (1996) Neurons of the rat inferior colliculus with GABA-like immunoreactivity can project to the medial geniculate body. Soc. Neurosci. Abstr. 22, 1069.

Redies, H., Brandner, S., and Creutzfeldt, O.D. (1989) anatomy of the auditory thalamocortical system of the guinea pig. J. Comp. Neurol. 282, 489–511.

Redies H. and Brandner, S. (1991) Functional organization of the auditory thalamus in the guinea pig. Exp. Brain Res. 86, 384–392.

Rouiller, E.M., and Ribaupierre, F. (1985) Origin of afferents to physiologically defined regions of the medial geniculate body of the cat: ventral and dorsal divisions. Hearing Res. 19, 97–114.

Saint Marie, R.L. (1996) Glutamatergic connections of the auditory midbrain: Selective uptake and axonal transport of D-[^{3}H]Aspartate. J. Comp. Neurol. 373, 255–270.

Wenstrup, J.J., Larue, D.T., and Winer, J.A. (1994) Projections of physiologically defined subdivisions of the inferior colliculus in the mustached bat: targets in the medial geniculate body and extrathalamic nuclei. J. Comp. Neurol. 346, 207–236.

Winer, J.A. (1992) The functional architecture of the medial geniculate body and the primary auditory cortex. In The mammalian auditory pathway: neuroanatomy. Ed. by D.B. Webster, A.N. Popper, and R. Fay. Springer-Verlag.

Winer, J.A. and Larue, D.T. (1988) Anatomy of glutamic acid decarboxylase immunoreactive neurons and axons in the rat medial geniculate body. J. Comp. Neurol. 278:47–68.

Winer, J.A. and Larue, D.T. (1996) Evolution of GABAergic circuitry in the mammalian medial geniculate body. Proc. Natl. Acad. Sci. USA 93, 3083–3087.

Winer, J.A., Saint Marie, R.L., Larue, D.T., and Oliver, D.L. (1996) GABAergic feed-forward projections from the inferior colliculus to the medial geniculate body. Proc. Natl. Acad. Sci. USA 93, 8005–8010.

29

REPRESENTATION OF AMPLITUDE MODULATED SOUNDS IN TWO FIELDS IN AUDITORY CORTEX OF THE CAT

Jos J. Eggermont

Department of Psychology
The University of Calgary
Calgary, Alberta, Canada

1. INTRODUCTION

Temporally structured sounds such as sine or square-wave modulated carriers and periodic click trains are represented somewhat differently in separate cortical fields in cat (Schreiner and Urbas, 1988) but less so in squirrel monkey (Bieser and Müller-Preuss, 1996). Representation of these complex sounds in cortex has been observed as a modulated-rate (AM) wide-band noise (Eggermont, 1993). The average higher BMF in AAF appeared largely due to values obtained in units with CFs above 10 kHz, for lower CFs all BMFs were below 20 Hz. In awake squirrel monkeys Bieser and Müller-Preuss (1996) found a broad range of BMFs from 2 - 128 Hz with a sharp reduction in the number of units with BMFs above 16 Hz. Low modulation frequencies (2 - 64 Hz) were mostly encoded by phase-locked neural responses and higher AM (128 - 512 Hz) sounds showed a distinction in overall-spike-rate variations. A similar view is offered by the recordings of Steinschneider et al. (1980, 1982) in auditory cortex of awake macaques: they found phase-locked activity in the depth-recorded local field potentials up to 250 Hz and for multi-unit activity up to 100 Hz. Because AI and AAF receive similar anatomical projections from the medial geniculate body (MGB; Rouiller et al., 1991) one expects the reason for these apparent differences in the two primary fields of the cat to reside in intrinsic cell properties or in intra-cortical network properties.

For periodic click trains Eggermont (1993) found for ketamine anesthetized cats a best repetition rate of 7.94 ± 2.09 Hz and in barbiturate anesthetized cats Schreiner and Raggio (1996) obtained a similar value of 6.47 ± 3.81 Hz. A comparison with the results from awake squirrel monkeys (Bieser and Müller-Preuss, 1996) and awake macaques (Steinschneider et al., 1980, 1982) suggests that anesthesia affects the BMF and thus likely the best click-repetition rate. This conclusion was also drawn by Goldstein et al. (1959) and Eggermont and Smith (1995a) on basis of evoked potential studies. Under Nembutal anesthesia the limiting rate (50% of maximum EP amplitude) in cats was 6–8

Hz down from 15–20 Hz in awake cats (Goldstein et al., 1959) whereas for light ketamine anesthesia the limiting rate was only slightly less than 15 Hz (Eggermont and Smith, 1995a).

A compilation of typical temporal modulation transfer functions (tMTF) from auditory nerve fibers to auditory cortex is shown in Figure 1, together with two representative results from psychoacoustical studies in humans. The tMTFs are expressed as a modulation gain, defined as G = 20 log (2S/m), where S is the vectorstrength or synchronization index and m is the modulation index of the sound. Note that a modulation gain of 0 dB corresponds for m = 1 (100% modulation of the sound) to S = 0.5. In this case, the envelope of the firings in the period histogram mimics the modulation waveform. One observes that the MTF for auditory nerve fibers (Joris and Yin, 1992) shows a low-pass function with a limiting rate (6 dB down from the gain at the BMF) of about 1 kHz. Neurons in the VCN generally show an appreciable modulation gain with a BMF around 200 Hz (Møller, 1972; Frisina et al., 1985). In the ICC (Rees and Møller, 1983) the gain is somewhat lower and so is the most frequently found BMF at around 50 Hz. For MGB only limited data are available (Rouiller et al., 1981) that suggest that the most frequent BMFs for periodic click trains are observed between 25 and 50 Hz. For auditory cortex, as we have seen the BMF is around 10 Hz and limiting rates tend to be less than 50 Hz. The

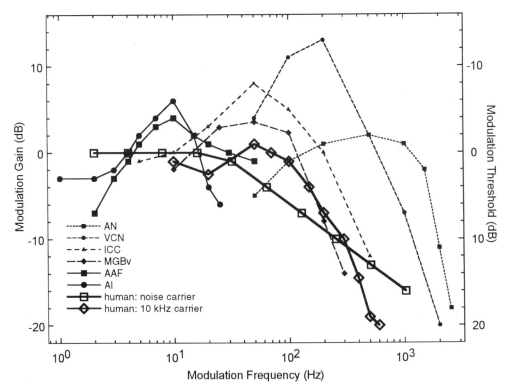

Figure 1. Comparison of psychophysical and electrophysiological Temporal Modulation Transfer Functions. Selected representative tMTFs are shown for stations along the neuraxis together with behavioral modulation threshold curves in humans. The human curves have been arbitrarily aligned to make their low-frequency asymptote equal to zero dB.

psychophysical data, expressed as a modulation threshold, suggest limiting rates up to 150 Hz for pure tone carriers (Fassel and Kohlrausch, 1995) and somewhat lower for noise carriers (Viemeister, 1979). Clearly, psychophysical performance resembles neural responses in ICC and MGB rather than auditory cortex.

Comparison of the effect of various periodic stimuli such as sinusoidal AM and rectangular AM (Schreiner and Urbas, 1986; 1988) with periodic click trains (Eggermont, 1991, 1993; Schreiner and Raggio, 1996) in cat AI suggests that the BMFs are highest for sinusoidal modulation, then for rectangular modulation and lowest for periodic click trains. However, the differences were not dramatic and suggested that repetition rate was the dominant variable.

Phillips et al. (1989) used continuous repetitive tonepips and analyzed them in terms of the response per tonepip, i.e., in terms of entrainment. Eggermont (1991) showed a similar analysis for periodic click stimulation and Schreiner and Raggio (1996) also presented their data in this alternative format in addition to using the modulation transfer function. These entrainment functions are low-pass functions of click (tonepip) repetition rate and they appear to be similar for barbiturate anesthesia and ketamine anesthesia.

Comparing the results in awake and anesthetized animals suggests that level and type of anesthesia are potentially having an effect. It is thus desirable to eliminate the effects of differences in anesthesia between recordings when comparing temporal coding properties in various cortical fields. For this purpose simultaneous recordings were made from AI and AAF in lightly ketamine-anesthetized cats. In two animals pentobarbital was administered later in the experiment to study potential effects on the tMTFs.

2. METHODS

The animal preparation procedures have been described in detail in previous publications (Eggermont, 1991, 1996). Acoustic stimuli were presented in an anechoic room from a speaker placed 55 cm in front of the cat's head. After the frequency tuning properties of the cells at each electrode were determined, periodic click trains and amplitude-modulated noise-bursts were presented once per three seconds. The modulation frequencies were randomly selected between 1–32 or 2–64 Hz at logarithmically equal distance with 4 values per octave. The modulating waveform was an exponentially transformed sinewave (Epping and Eggermont, 1986b) with a maximum modulation depth of 17.4 dB. The sequences of 21 click trains or 21 AM noise-bursts were repeated 10 times resulting in a total stimulus ensemble duration of 630 seconds. The procedures for spike recording were completely similar to those described previously (Eggermont, 1991; 1996).

The click- and AM-following capacity of the neurons was evaluated from the rate- and synchronization modulation transfer functions obtained by Fourier transformation of the period histograms (Eggermont, 1991). The rate-MTF is equal to the DC component and the synchronization-MTF is equal to the amplitude of the first harmonic as a function of the click-repetition or amplitude-modulation rate. The best modulating frequency (BMF) was defined as that click rate or AM rate for which the synchronized firing rate was maximal. The limiting rate was defined as the rate at which the response was 50% of that at the BMF.

3. RESULTS

Results presented are from 62 simultaneous recordings with two electrodes in AI and one in AAF resulting in 186 multi-unit (MU) records and 186 LFP recordings in 13 cats.

3.1 Dotdisplays

To introduce the click following properties of cortical cells three examples from simultaneous MU recordings, composed of clearly distinguishable SUs only, in AI and AAF are shown in Figure 2. The left hand column shows results for AI, the right hand column for AAF. Click rates are shown vertically ranging between 1 and 32 Hz, the time axis runs from 0–1.1s leaving 100 ms after the presentation of the last click in the trains. One of the characteristics of response behavior in AI is clear; often there is a profound rebound and rhythmic firing after clicks presented at rates below about 4/s or above 25/s. This is never as pronounced in AAF. Because of the simultaneous recordings, anesthesia differences which could affect the potential for spindling can be ruled out. Taking the example on the top, visible click following for AI is clear up to about 16 Hz, then the re-

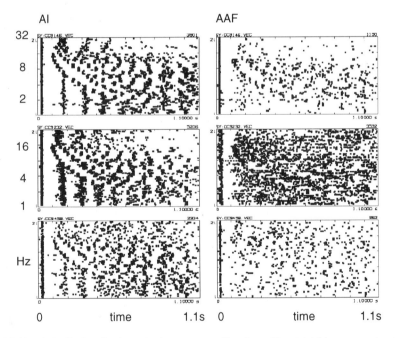

Figure 2. Multi-unit dotdisplays for three simultaneous recordings from AI and AAF in response to periodic click trains. Time axis for each plot is from 0–1.1s and the repetition rate axis is in 21 equal logarithmic increments between 1–32 clicks/s.

sponse skips the second click up to rates of about 25 Hz. For AAF clear following ceases above 11.28 Hz. Furthermore, the response appears to be more brisk for AI. The same can be said for the other examples. Latencies to the first click are shorter by 3–4 ms in AAF than in AI.

3.2 From Period Histograms to Modulation Transfer Functions

A more quantitative representation of periodic click following is shown in Figure 3 for the example in the left top row of the previous Figure. We show the period histogram twice on the left of each part of the Figure, followed by the average spike count per click train (rMTF), the vectorstrength (vMTF) and the temporal modulation transfer function (tMTF). The tMTF is the product of the rMTF and the vectorstrength. For this example from the AI the rMTF is low-pass, the vMTF is high pass and the tMTF is band-pass with a BMF of 9.52 Hz.

In Figure 4A the tMTFs are shown for all six examples from Figure 2, the data for the AI are shown with open symbols connected by full lines and the AAF data with similar filled symbols connected with dashed lines. It is noticed that the synchronized spike rates are higher in AI than in AAF and that the BMFs tend to be slightly higher in AI as well. In Figure 4B we show the mean tMTFs per cat averaged over AI and AAF and at

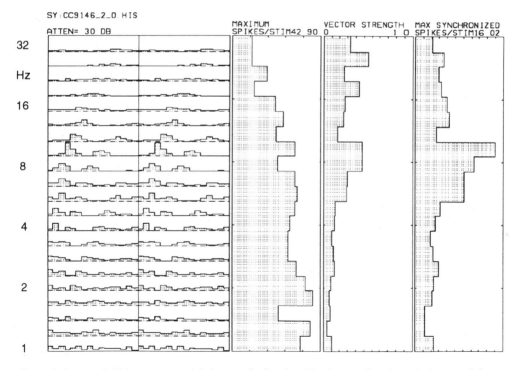

Figure 3. From period histogram to modulation transfer functions. For the recording shown in the upper left corner of Figure 2 the period histograms are shown (twice) for repetition rates between 1 and 32 Hz. The next columns show the rate MTF (total spike count/click train as a function of repetition rate), the vectorstrength and the synchronized-rate MTF.

least 4 individual recordings comprising about 12 units. One observes that the average tMTFs are very similar.

3.3 BMFs and CFs

In order to compare data across stimulus conditions and anesthesia levels we plotted our BMFs (old data from Eggermont, 1993 and those in the present study all for periodic click trains) as a function of CF together with those (for AM-tones) of Schreiner and Urbas (1986, 1988). Despite the differences in stimulation and in anesthesia the data for AI (Figure 5A) largely overlap. For AAF (Figure 5B) Schreiner and Urbas's BMFs are higher than those obtained by us. Taking into account the rather close correspondence in the results for AI, this suggests that the AAF responds differentially to clicks and AM CF-tones.

3.4 Entrainment

In order to facilitate comparison with Phillips et al. (1989) who used continuous repetitive toneburst stimuli and calculated the entrainment we also present our AI data as synchronized rate/click. Figure 6A shows the mean entrainment curves per cat expressed in synchronized spikes/click together with the grand mean curve across 13 cats. The curves are low-pass with a mean limiting entrainment around 10 Hz. Figure 6B shows this grand mean curve together with single unit data (open circles) from Schreiner and Raggio (1996) and from Phillips et al. (1989; filled symbols). The data from Phillips et al are on the lower end of the range found by Schreiner and Raggio (1996) likely because the latter used relatively short duration modulated tonebursts whereas Phillips et al. used semi-continuous repetitive stimuli. Our grand average is somewhere in the middle of this range.

3.5 Similarities between AI and AAF

Figure 7a shows the mean tMTF for all simultaneous multi-unit recordings for click-train stimulation in AI and AAF. This suggests that the AAF has a lower BMF and limiting rate than AI. A pair-wise comparison showed that BMFs in AI (mean 10.18 Hz) were significantly ($p = 0.0008$) higher than in AAF (mean 9.07 Hz) .

Individual cats differed greatly in their average tuning both in AI and AAF. This was not dependent on recording site or depth below the surface, but likely the result of a different impact of the anesthesia in each animal. Figure 7b shows a selection of mean tMTFs intended to show the range of values found. The cat's mean BMF may be as low as 5.64 (cat 156) or as high as 16 Hz (cat 163). One cat (151) showed tMTFs that were very narrowly tuned, another (165) showed rather broad tuning.

3.6 Differences between AI and AAF

The suggestion that AAF may be better responding to sinusoidal CF-tone modulation than to periodic click trains was explored. Figure 8 shows LFP-trigger data for simultaneous recording in AI and AAF for three periodic stimuli: periodic 1s-duration click trains, exponential-sine AM-noise and exponential-sine CF-tone. For all three stimuli the repetition/modulation rates are between 2 and 64 Hz. For AI click following occurs up to 19 Hz, then the response to the second click is skipped but responses to the third and sub-

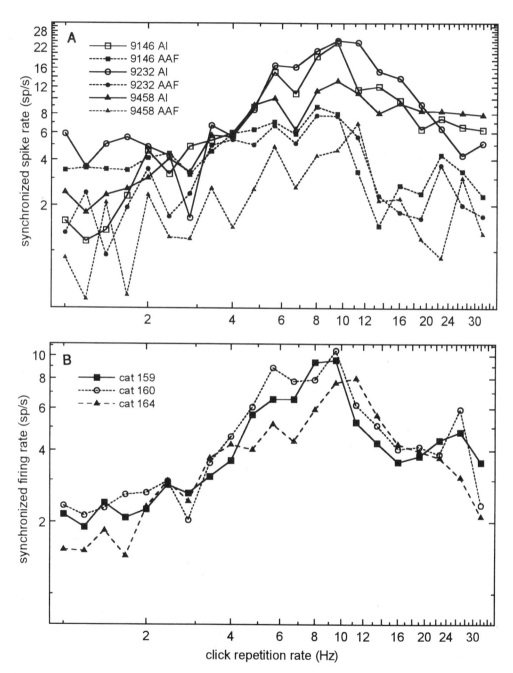

Figure 4. Temporal modulation transfer functions for individual multi-unit recordings for the data shown in Figure 2. A. The full lines connecting open symbols represent the data for AI, the dashed lines connecting filled symbols represent data from AAF. B. Average tMTFs per cat, in which data shown in part A were obtained, across AI and AAF.

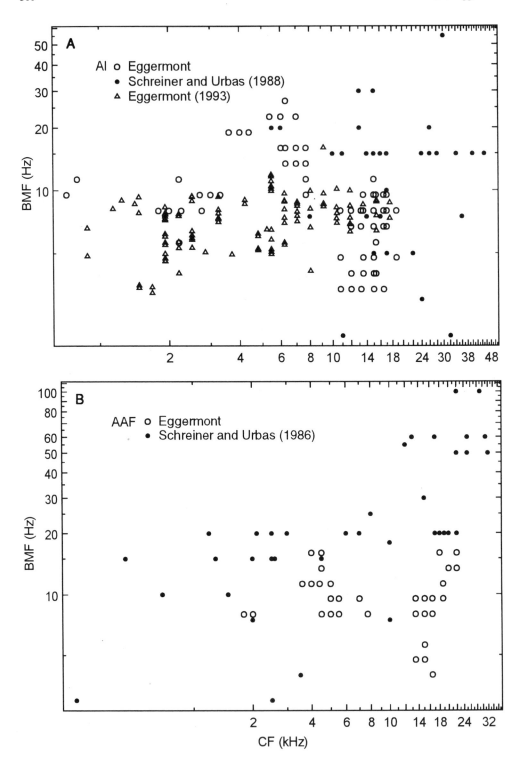

Figure 5. Scatterplots of BMF as a function of CF. A. Results for AI show a close correspondence between the amplitude-modulated tone data (Schreiner and Urbas, 1988) and the two data sets for periodic click trains. B. Results for AAF suggest a discrepancy between AM-tone data and periodic click data.

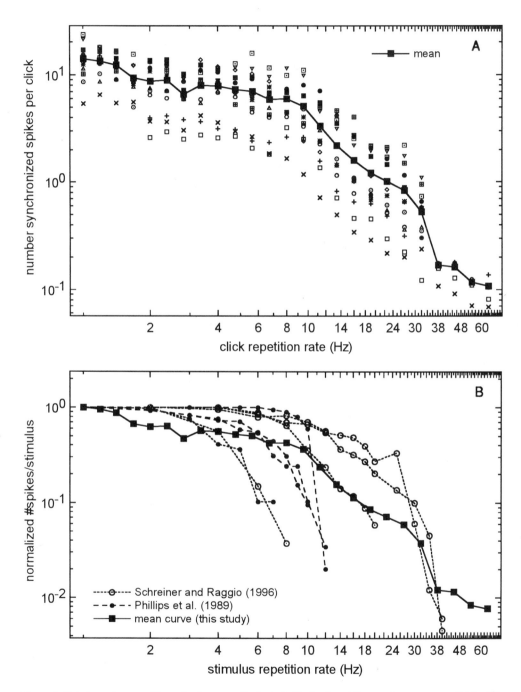

Figure 6. Entrainment curves. Figure 6a shows results for individual cats and the grand mean across all cats. Figure 6b shows a comparison of the grand mean from this study with selected representative individual single neuron examples from the literature.

sequent clicks is visible up to 26 Hz. In AAF click following ceases altogether above 19 Hz. For AM-noise stimulation, reliable locking occurs up to 32 Hz in AI and to 19 Hz in AAF. In contrast, AM-tone stimulation produces much stronger phase locking in AAF up to 32 Hz whereas in AI the locking is poor and limited to frequencies below 16 Hz. The same result was found for the multi units recorded on the same electrode. Similar findings were obtained in three other cats, supporting the notion that AAF is slightly lower tuned

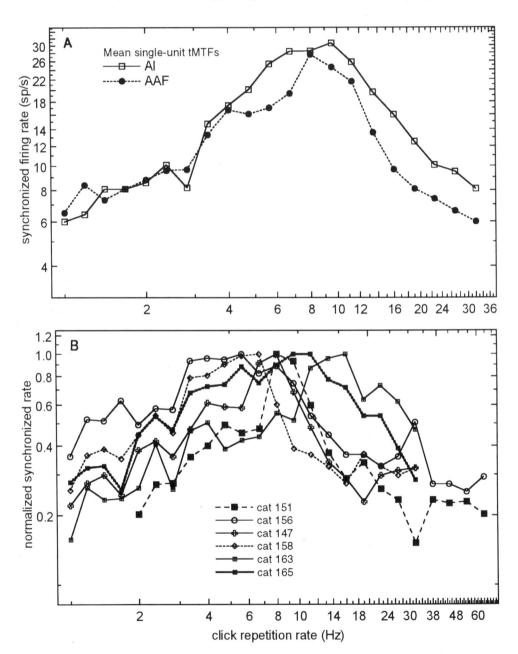

Figure 7. a. Grand mean single-unit tMTFs for AI and AAF for periodic click train stimulation in 13 cats. Figure 7b. Representative normalized average tMTFs for individual cats (AI and AAF combined).

for clicks, responds somewhat less to high AM rates in modulated noise, but is far better responsive to CF carrier tones (up to modulation frequencies of 32 Hz).

3.7 Effect of Pentobarbital

After a prolonged recording session under ketamine anesthesia in cat #166, Sodium pentobarbital was administered i.m. in 6 mg/kg doses every hour. The effect on the tMTF was quite moderate and nearly identical for MU and LFP as well as on AI or AAF. Figure 9A shows the mean across three electrodes under ketamine and pentobarbital as well as their ratio. The effect of pentobarbital was to raise the synchronized spike rate for click-repetition rates below 10 Hz, sharply reduce the synchronized firing rate at 11.28 and 13.44 Hz and hardly affect the results for the higher click rates. As a result the ratio function shows a peak between 8 and 16 Hz. The BMFs remained the same in this cat. Figure 9B shows the mean ratio curves per electrode (this time on a logarithmic amplitude axis) that suggest an interesting differential effect on AAF compared to AI: whereas for both electrodes in AI the response for ketamine is larger than for pentobarbital around 10–12 Hz, the response in AAF is actually better for pentobarbital between 12 and 19 Hz. This was not as apparent in the other cat which received pentobarbital, but should be explored further.

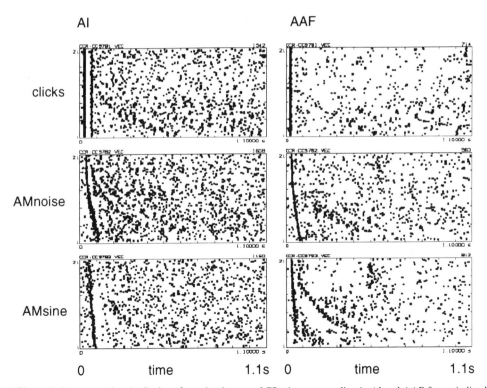

Figure 8. Representative dotdisplays for a simultaneous LFP-triggers recording in AI and AAF for periodic click trains (upper row), exponential-sine AM-noise (middle row) and exponential-sine AM-CFtone (bottom row).

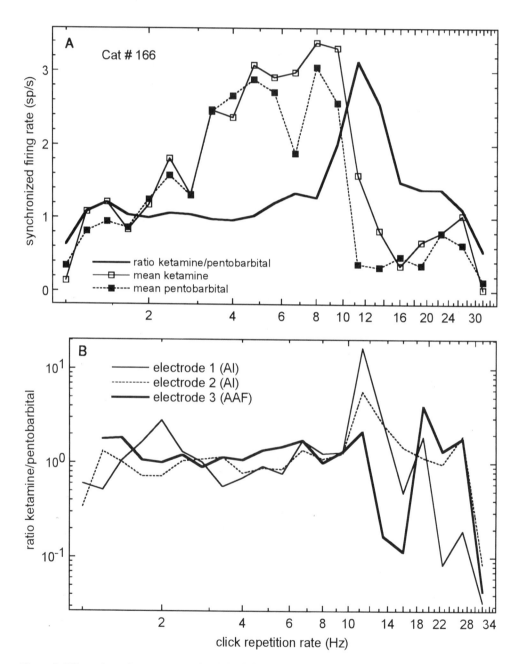

Figure 9. Effect of supplementary pentobarbital administration upon tMTFs. A. mean tMTF under ketamine shows better synchronization between 8 and 16 Hz than under pentobarbital. B. Differential effect of pentobarbital on AI and AAF.

4. DISCUSSION

We have shown that rate MTFs for cortical neurons are low pass functions that are not tuned to modulation frequency whereas synchronized-rate MTFs invariably show tuning expressed in a BMF. Does this mean that synchronized rate is more useful than overall discharge rate in conveying temporal aspects of amplitude modulated sounds? Coding by average discharge rate alone necessitates a 'labeled line' or 'place' coding because there is no other means internal to the spike train itself for conveying what kind of signal or what periodicity in the signal it is representing. Rate codes, both average and synchronized rate, can be interpreted only by neurons or populations of neurons that have long integration times. Neurons can be considered as leaky integrators with time constants that vary throughout the nervous system. In the periphery neurons have short, few ms, time constants and so do most interneurons in the cortex. In contrast pyramidal cells in sensory neocortex have time constants of 10 ms (Mason et al., 1991). Is rate coding therefore limited to the thalamus and cortex? An alternative approach is to consider integration time constants with respect to the average inter-spike intervals (ISI) that have to be processed (König et al., 1996). If the time constant is short relative to the average ISI then rate coding is unlikely (the unit may act as a coincidence detector or more appropriately a co-firing detector) but if the time constant is long with respect to the ISI then rate coding is feasible. For a different and more restricted definition of temporal encoding based on information content of the spike train see Theunissen and Miller (1995). If, as in cortex, firing is characterized largely by bursts with long inter-burst intervals the simple classification based on the relationship between ISI and integration time constants either breaks down or the neuron alternately must act as a rate detector (for the spikes in the bursts) and as a coincidence detector of the bursts. The histograms that show the modulated firing rate for the sounds used have period durations that range from about 16 ms to about 1s and only the relative temporal precision with respect to the period of stimulation remains constant. Thus, it is well possible to describe the results as a modulated rate code.

A system that keeps track of inter-spike intervals could in principle convey far more information than a system relying only on averaged firing rates (e.g., Geisler et al., 1991) but this may simply be the result of the availability of more independent variables (to the experimenter) to solve the stimulus prediction problem. The absolute upper limit to the transmitted information is set by the number of distinguishable spike sequences given some spike-timing precision (jitter). For instance if the spike times in a burst of a certain length matter, then a temporal encoding process may be at work. If, on the other hand, only the number of spikes matters then rate coding transmits all the information. The amount of transmitted information measures the number of stimulus waveforms that can be distinguished from the observation of the spikes. If cortical neurons behave as coincidence detectors and are interconnected in a divergent/convergent fashion, then the timing of spikes can propagate through cortex with great fidelity to convey information and to synchronize other neurons (Abeles, 1982).

In sensory systems that are receptive to stimuli with a periodic structure (the auditory, somatosensory and electric sensory system) action potentials are generally produced synchronously with this stimulus periodicity and thus with each other. Timing and synchrony in neural populations thus contains information about these periodicities. Modern theories for periodicity pitch combine interspike interval distributions from many frequency regions of the auditory nerve to produce pooled interspike interval distributions from which the pitch is than extracted. For the vast majority of periodic complex stimuli it was found that the pitch heard by human listeners corresponded to the most common interspike interval in the auditory nerve and that the salience of the pitch heard corresponded to the peak-to-background ratio in the all-order interval distribution for the population (Cariani, 1994). A problem that remains to be explained is the apparent absence of periodicity following by cortical pyramidal cells for rates above 50/s, with the potential exception of high CF (>10 kHz) fibers in auditory anterior field (Schreiner and Urbas, 1988). These CFs, however, are so much higher than the dominant formant frequencies that they are unlikely to be activated by speech sounds. Thus periodicity pitch detection is either done sub-cortically or the periodicity pitch has been mapped onto the spatial dimension in cortex. A subcortical correlate is suggested by the comparison of psychophysical tMTFs (Viemeister, 1979; Fassel and Kohlrausch, 1995) and those obtained at various nuclei in the auditory system which suggests that the correlation is largest with ICC and MGB.

Largely because the topographical ordering of collicular and cortical maps (likely a consequence of economically wiring a topographically organized sensory periphery to more central parts) it has long been assumed that ICC and cortex are spatial pattern processors and consequently it has been taken for granted that all that is required for sensory coding is a rate-place code. In such a code the distribution pattern of firing rate across the collicular or cortical surface represents the stimulus. From the belief that, e.g., colliculus and cortex are exclusively spatial processors it follows that all information which is not place coded in sensory peripheries (e.g., periodicity pitch, acoustic space) must eventually be converted into the 'common' language of the colliculus and cortex, i.e., in spatial excitation patterns. Because the auditory periphery is characterized to a large extent by precise temporal coding, there must be temporal feature detectors, e.g., in the superior olivary complex and inferior colliculus (Langner, 1992) which perform these time-to-place transformations. In the literature on information processing in the auditory nervous system one finds repeatedly the suggestion that neural synchrony is important for processing in auditory nerve and in the lower brainstem, but generally looses its importance in higher centers to be gradually replaced by a rate coding (Langner, 1992). Or as Capranica and Rose (1983) stated that "time domain processing involves a transformation from a periodicity code in the peripheral auditory system to a temporal filtering assembly in the central auditory system". This may be an interpretation based on the gradual change from an event-type correlation in the auditory periphery to the rate-type correlation one finds in higher centers. For short lag times one may even detect this rate-correlation as not much more than a stimulus induced rate increment.

Rees and Palmer (1989) in the inferior colliculus of the guinea pig and Eggermont (1990) in the torus semicircularis of the leopard frog found that the mean discharge rates were greatest for those modulation frequencies which also elicited maximum synchronization. Thus the rMTF and tMTF were similar in shape as a result of the fact that all firings are time-locked to the stimulus envelope. This is in contrast to more peripheral parts of the auditory system where the degree of synchronization is independent of mean firing rate, and units are not tuned to modulation frequency in terms of average rate. This may arise in cases where the stimulus simply modulated the occurrence of the 'spontaneous' firings. This suggested (Rees and Palmer, 1989) that the auditory midbrain acted as a decoder for temporal information and transformed the degree of synchrony into firing rate. This may ultimately result in a rate-place code for auditory space as occurs in the superior colliculus (Knudsen et al., 1987).

The auditory thalamus and cortex form an apparent superstructure that operates with many synapses on the same cell and uses the rate code from the inferior colliculus as its input. In cortex many inputs affect the pyramidal cell and a single presynaptic spike has potentially little bearing on the exact timing of the postsynaptic spike. As we have shown, single-unit firing rate in cortex is rather independent of click stimulus periodicity, whereas the amount of spike synchronization with the stimulus clearly depends on the repetition rate (see also Eggermont, 1994). This suggests a re-emergence of a synchrony code at the level of the cortex potentially emphasized by the transient character of its responses. One complicating factor, at least in anesthetized animals and potentially also in awake but drowsy animals, is that the optimal synchrony is generally found at the dominant EEG-spindle frequency or tau-rhythm (Hari, this volume) in the preparation, which in itself tends to synchronize spike activity (Puil et al., this volume). Stimulation with periodic clicks with repetition rates around this spindle frequency tends to entrain this frequency (Eggermont, 1992). In the common marmoset, the dominant modulation frequency of its 'twitter' call is in this tau-frequency range. Auditory cortical neurons were found to be very finely tuned to the modulation frequency in the call (Wang et al., 1995). One wonders whether this is an adaptation of the call modulation frequency to a dominant EEG frequency in auditory cortex or that the temporal tuning found is a mere coincidence of the similarity of the tau-frequency and the modulation frequency of the 'twitter' call. Cortical oscillations in the 7–14 Hz range have been suggested to determine the latency difference between express and regular saccadic eye movements (Kirschfeld et al., 1996) through modulation of the neuronal thresholds. Specifically, it was found that stimulus-induced oscillations with a frequency of 10 Hz persisted for a second after the visual stimulus. These oscillations potentially modulate the saccade generating neurons. Related to that the temporal framing, i.e., a sense of apparent simultaneity, was found to be correlated with the alpha-cycle (Varela et al., 1891). Evidence for discrete processing cycles in human perception, correlating with gamma-band frequencies in between 30 and 85 Hz have been found in reaction time studies (DeHaene, 1993). All this suggests that the response of the cortex depends as much on ongoing autonomous activity as it does on stimulus induced activity. We have previously shown that synchrony in the activity of distant neurons is also a reflection of both such global and local components (Eggermont and Smith, 1995a).

Given all this, what is the role of auditory cortex, besides providing a state-dependent context, in complex sound processing? Steady-state vowel discrimination is preserved in case of lesions in primary auditory cortex whereas stop consonant discrimination, depending on precise voice-onset-time (VOT) is impaired (for review: Phillips, 1993). VOT is coded in AI (Eggermont, 1995a) as well as in AAF (Eggermont, unpublished observations). Several other correlates of precise timing, such as gap-detection, can be found in AI whereas others such as auditory fusion for clicks do have a much higher threshold than the perceptual one (Eggermont, 1995b). Is there any specialization for AI and AAF? Our preliminary results presented above suggest that AAF may be specialized in processing time information in narrow-band stimuli whereas AI is showing much better responsiveness to the temporal structure of broad-band stimuli. The latter would agree with the accurate transient timing properties in AI. Temporal resolution is not high enough to permit representation of voice pitch in the firing patterns of individual pyramidal cells or populations thereof. It is possible that the various (inhibitory) interneurons produce time locked firings to modulation frequencies as high as a few hundred Hz (Steinschneider et al., 1980; 1982). How this information is transmitted to speech areas in the brain remains a mystery; it would likely require long axon inhibitory neurons that project outside the auditory cortex. Piesman et al. (1994) in fact suggest the presence of an inhibitory pathway emanating from AI and affecting secondary cortical regions as well as IC. In anatomical studies such long axon inhibitory neurons have been found in ICC and they project well into the MGB (Oliver, this Volume).

ACKNOWLEDGMENTS

This investigation was supported by grants from the Alberta Heritage Foundation for Medical Research and the Natural Sciences and Engineering Research Council of Canada. Denise Bowman and Geoff Smith provided valuable suggestions throughout the experiment. Denise Bowman, Kentaro Ochi and Mutsumi Kenmochi assisted with the data collection

REFERENCES

Abeles, M. (1982) Local cortical circuits: An electrophysiological study. Berlin, Springer Verlag.

Bieser, A. and Müller-Preuss, P. (1996) Auditory responsive cortex in the squirrel monkey: neural responses to amplitude-modulated sounds. Exp. Brain Res. 108, 273–284.

Capranica, R.R. and Rose, G. (1983) Frequency and temporal processing in the auditory system of anurans. In: F. Huber and H. Markl (Eds.), Neuroethology and behavioral physiology, Springer Verlag, Berlin, pp.136–152.

Cariani, P. (1994) As if time really mattered: temporal strategies for neural coding of sensory information. Comm. Cogn. Art. Intell. 12, 157–219.

DeHaene, S. (1993) Temporal oscillations in human perception. Psychological Science 4, 264–270.

Eggermont, J.J. (1990) Temporal modulation transfer functions for single neurons in the auditory midbrain of the leopard frog. Intensity and carrier-frequency dependence. Hearing Research 43, 181–198.

Eggermont, J.J. (1991) Rate and synchronization measures of periodicity coding in cat primary auditory cortex. Hearing Research 56: 153–167.

Eggermont, J.J. (1992) Stimulus induced and spontaneous rhythmic firing of single units in cat primary auditory cortex. Hearing Research 61, 1–11.

Eggermont, J.J. (1993) Differential effects of age on click-rate and amplitude modulation-frequency coding in primary auditory cortex of the cat. Hearing Research 65, 175–192.

Eggermont, J.J. (1994) Temporal modulation transfer functions for AM and FM stimuli in cat auditory cortex. Effects of carrier type, modulating waveform and intensity. Hearing Research 74, 51–66.

Eggermont, J.J. (1995a) Representation of a voice onset time continuum in primary auditory cortex of the cat. J. Acoust. Soc. Amer. 98, 911–920.

Eggermont, J.J. (1995b) Neural correlates of gap detection and auditory fusion in cat auditory cortex. Neuroreport 6, 1645–1648.

Eggermont, J.J. (1996) How homogeneous is cat primary auditory cortex? Evidence from simultaneous single-unit recordings. Auditory Neuroscience 2, 76–96.

Eggermont, J.J. and Smith, G.M. (1995a) Synchrony between single-unit activity and local field potentials in relation to periodicity coding in primary auditory cortex. J. Neurophysiology 73, 227–245.

Eggermont, J.J. and Smith, G.M. (1995b) Separating local from global effects in neural pair correlograms. NeuroReport 6, 2121–2124.

Epping, W.J.M. and Eggermont, J.J. (1986): Sensitivity of neurons in the auditory midbrain of the grassfrog to temporal characteristics of sound. II. Stimulation with amplitude modulated sound. Hearing Research 24: 55–72.

Fassel, R. and Kohlrausch, A. (1995) Modulation detection as a function of carrier frequency and level. IPO Annual Progress Report 30, 21–29.

Frisina, R.D., Smith, R.L. and Chamberlain, S.C. (1985) Differential encoding of rapid changes in sound amplitude by second-order auditory neurons. Exp. Brain Res. 60, 417–422.

Geisler, W.S., Albrecht, D.G., Salvi, R.J. and Saunders, S.S. (1991) Discrimination performance of single neurons: rate and temporal-pattern information. J. Neurophysiology 66, 334–362.

Goldstein, M.H., Kiang, N.Y-S. and Brown, R.M. (1959) Responses of the auditory cortex to repetitive acoustic stimuli. J. Acoust. Soc. Amer. 31, 356–364.

Hari, R. (199·) Temporal aspects of human auditory cortical processing. (this volume)

Joris, P.X. and Yin, T.C.T. (1992) Responses to amplitude-modulated tones in the auditory nerve of the cat. J. Acoust. Soc. Amer. 91, 215–232.

Kirschfeld, K., Feiler, R. and Wolf-Oberhollenzer, F. (1996) Cortical oscillations and the origin of express saccades. Proc. R. Soc. Lond. B 263, 459–468.

Knudsen, E.I., du Lac, G. and Esterly, S.D. (1987) Computational maps in the brain. Ann. Revue of Neuroscience 10, 41–65.

König, P., Engel, A.K. and Singer, W. (1996) Integrator or coincidence detector? The role of the cortical neuron revisited. TINS 19, 130–137.

Langner, G. (1992) Periodicity coding in the auditory system. Hearing Research 60: 115–142.

Mason, A., Nicoll, A. and Stratford, K. (1991) Synaptic transmission between individual pyramidal neurons of the rat visual cortex in vitro. J. Neuroscience 11, 72–84.

Møller, A.R. (1972) Coding of amplitude and frequency modulated sounds in the cochlear nucleus of the rat. Acta Physiol. Scand. 86, 223–238.

Oliver, D.L. (199·) Synaptic domains in the central nucleus of the inferior colliculus. (this volume).

Piesman, M., Chao, E.C., Gruen, E., Woody, C.D. and Zotova, E. (1994) Inhibition of discharge in inferior colliculus, AII cortex and EP cortex after presentations of click stimuli. Brain Research 657, 320–324.

Phillips, D.P. (1993) Representation of acoustic events in the primary auditory cortex. J. Exp. Psychology HEP 19, 203–216.

Phillips, D.P., Hall, S.E. and Hollett, J.L. (1989) Repetition rate and signal level effects on neuronal responses to brief tone pulses in cat auditory cortex. J. Acoust. Soc. Amer. 85, 2537–2549.

Puil, E., Ries, C.R., Schwas, D.W.F., Tennigkeit, F. and Yarom, Y. (199·) Sleep-inducing and anesthetic actions of drugs on neurons of auditory thalamus. (this volume)

Rees, A. and Møller, A.R. (1983) Responses of neurons in the inferior colliculus of the rat to AM and FM tones. Hearing Research 10, 301–330.

Rees, A. and Palmer, A.R. (1989) Neuronal responses to amplitude-modulated and pure-tone stimuli in the guinea pig inferior colliculus, and their modification by broadband noise. J. Acoust. Soc. Amer. 85, 1978–1994.

Rouiller, E.M., de Ribaupierre, Y., Toros-Morel, A. and de Ribaupierre, F.(1981) Neural coding of repetitive clicks in the medial geniculate body of the cat. Hearing Research 5, 81–100.

Rouiller, E.M., Simm, G.M., Villa, A.E.P., de Ribaupierre, Y. and de Ribaupierre, F. (1991) Auditory corticocortical interconnections in the cat: evidence for parallel and hierarchical arrangement of the auditory areas. Exp Brain Res 86, 483–505.

Schreiner, C.E. and Urbas, J.V. (1986) Representation of amplitude modulation in the auditory cortex of the cat. I. The anterior auditory field (AAF). Hearing Research 21, 227–241.

Schreiner, C.E. and Urbas, J.V. (1988) Representation of amplitude modulation in the auditory cortex of the cat. II. Comparison between cortical fields. Hearing Research 32, 49–64.

Schreiner, C.E. and Raggio, M.W. (1996) Neuronal responses in cat primary auditory cortex to electrical cochlear stimulation. II. Repetition rate coding. J. Neurophysiology 75, 1283–1300.

Steinschneider, M., Arezzo, J. and Vaughan, H.G. (1980) Phase-locked cortical responses to a human speech sound and low-frequency tones in the monkey. Brain Research 198, 75–84.

Steinschneider, M., Arezzo, J. and Vaughan, H.G. (1982) Speech evoked activity in the auditory radiations and cortex of the awake monkey. Brain Research 252, 353–365.

Theunissen, F. and Miller, J.P. (1995) Temporal encoding in nervous systems: a rigorous definition. J. Computational Neuroscience 2, 149–162.

Varela, F.J., Toro, A., John, E.R. and Schwarz, E.L. (1981) Perceptual framing and the cortical alpha rhythm. Neuropsychologica 19, 675–686.

Viemeister, N. (1979) Temporal modulation transfer function based upon modulation thresholds, J. Acoust. Soc. Amer. 66, 1364–1380.

Wang, X., Merzenich, M.M., Beitel, R. and Schreiner, C.E. (1995) Representation of a species-specific vocalization in the primary auditory cortex of the common marmoset: temporal and spectral characteristics. J. Neurophysiology 74, 2685–2706.

30

TEMPORAL ASPECTS OF HUMAN AUDITORY CORTICAL PROCESSING

Neuromagnetic and Psychoacoustical Data

Riitta Hari[*]

Brain Research Unit
Low Temperature Laboratory
Helsinki University of Technology
02150 Espoo, Finland

1. BACKGROUND

Time is a quintessential issue in studies of audition. First, the sound stimuli as such are changes in time and, second, audition as a warning sense (or as an attention-triggering mechanism) is very sensitive to any changes in the auditory environment. Both single stimuli and their sequences are variations in time which have to be kept in mind for at least as long as the message is understood. Such temporal buffers are inherently more complicated to be investigated than stationary neural firing patterns.

I will approach temporal aspects of human central auditory processing by first discussing some magnetoencephalographic data used to study the dynamics of cortical signal processing and the duration of the auditory buffer. The psychoacoustical part of the paper describes an illusion of directional hearing which allows to quantify the sluggishness of neural processing preceding conscious auditory percept.

Further details about the MEG method and its application to studies of human cortical functions can be found in previous reviews (Hari and Lounasmaa, 1989; Grandori et al., 1990; Hari, 1990; Sato, 1990; Hämäläinen et al., 1993; Lounasmaa et al., 1996).

2. MAGNETOENCEPHALOGRAPHY

Magnetoencephalography (MEG) provides a noninvasive method for studying macroscopic functional organization of the human auditory cortex. The MEG sensors detect weak

* Fax: +358-9-4512969; E-mail: hari@neuro.hut.fi.

extracranial magnetic fields produced by cerebral electric currents. The method allows differentiation between signals from various cortical regions with excellent temporal and good spatial resolution. Due to its sensitivity to tangential current sources, MEG is well suited for noninvasive investigation of auditory cortical regions embedded within the Sylvian fissures. In humans, these regions include the primary koniocortex and several surrounding belt areas.

Electric currents flowing in the brain generate weak magnetic signals, typically only one part in 10^8 or 10^9 of the Earth's geomagnetic field. The measurements are therefore carried out in a magnetically shielded room. The signals are first detected by an array of superconducting flux transformers, and then sensed by SQUIDs (Superconducting QUantum Interference Devices), which are sensitive to extremely small magnetic fields.

To locate the underlying neural activity, the magnetic field pattern over the scalp must be sampled with a dense spacing. With the new helmet-type magnetometers, covering the whole scalp, it is possible to obtain signals from the whole neocortex at the same time. The data presented in this paper were collected with a 122-channel neuromagnetometer (Neuromag-122™). Planar gradiometers incorporated in this instrument detect the strongest signals just above a locally restricted area of cortical activity.

Whole-head MEG recordings allow comparisons of hemispheric differences in the processing of auditory stimuli, including speech sounds. The ability to monitor the activity of several cortical regions simultaneously facilitates research into the neural basis of human cognitive activity, and it is also possible to investigate subjects displaying special abilities or disorders of auditory processing. In addition to evoked responses, important information may be obtained from recordings of ongoing spontaneous activity.

3. EVOKED AND SPONTANEOUS ACTIVITY FROM THE AUDITORY CORTEX

3.1. Responses to Tones

The early MEG recordings demonstrated that the most prominent deflections of the auditory evoked magnetic fields (AEFs) are generated in the auditory cortex on the superior surface of the temporal lobe (Elberling et al., 1980; Hari et al., 1980). These findings contrasted the existing view about the nonspecificity of all long-latency evoked responses.

Figure 1 shows an example of AEFs recorded with the 122-channel neuromagnetometer. Prominent responses occur over the right and left temporal lobes, reflecting activity of the auditory cortices. The signals are usually interpreted with the help of source models, and models consisting of single or multiple equivalent current dipoles (ECDs) within a sphere are commonly applied. The orientation, strength, and the 3-dimensional location of the ECD best accounting for the measured signal distribution are found by means of a least-squares fit to the measured signals. The physiological interpretation of a current dipole is activation of a cortical area smaller than 2 cm in diameter. In Fig. 1, the sites of two ECDs have been superimposed on a magnetic resonance image (MRI) to combine the functional and structural information. Such an integration implies that the 100-ms response (N100m) is generated in the lower surface of the Sylvian fissure, with major contribution from areas in the vicinity and just posterior to the Heschl's gyrus.

Neuromagnetic activity starts in the supratemporal auditory cortex within 20 ms from a sound onset and continues for a few hundred milliseconds; for a review, see Hari (1990). Successive deflections of the auditory evoked response vary slightly in their source sites and probably reflect time-varying contribution from different cytoarchitec-

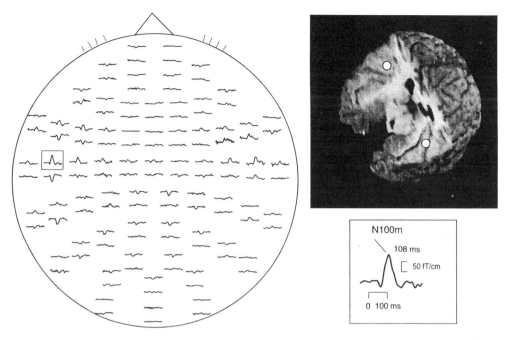

Figure 1. Left: AEFs recorded with the 122-channel neuromagnetometer when 50-ms tones were presented to a healthy subject's right ear once every 4 s. The latitudinal and longitudinal derivatives of the magnetic field are shown over each other. The head is viewed from above and the helmet has been 'flattened' to show responses from the whole head simultaneously; the nose points upwards. The main contralateral response is shown enlarged. The passband is 0.05–40 Hz. Right: ECDs for N100m superimposed on an MRI surface rendering of the same subject, viewed from above and with the frontal lobes removed.

tonic areas. The generator areas agree in general with the sites of intracranially recorded auditory evoked potentials (Liegeois-Chauvel et al., 1990). An infarction extending to deep parts of the temporal lobe abolishes responses in that hemisphere (Mäkelä et al., 1991), thereby supporting the obtained solutions of the neuromagnetic inverse problem.

3.2. Neural Correlates of Sensory Memory

The timing and history of the stimulus sequence influence the evoked responses. The amplitude of N100m increases with increasing interstimulus interval (ISI) and reaches a plateau with ISIs of 4–8 s. This behavior has been suggested to reflect a primitive memory, based on state (refractoriness) changes of the activated neural network, and to have a behavioral correlate in the duration of the psychophysically determined memory for loudness of sound (Lu et al., 1992b).

According to Cowan (1984), auditory stimuli are stored into a 200–300 ms 'short store' and a 10–20 s 'long store'. The duration of the long store agrees with the above neuromagnetic estimates of sensory memory, and correlates of the short store have been found in neuromagnetic studies of responses to paired tones (Loveless et al., 1996).

Another way to study the storage of auditory information is to present infrequent deviant stimuli among a series of monotonously repeated standards. The deviants evoke an additional response, mismatch field (MMF); for reviews, see Näätänen and Picton (1987), and Hari (1990). MMF seems not to be related to stimulus features *per se* but rather to their changes and can therefore reveal memory processes needed for comparing the simi-larity of successive stimuli. Both the MMF and behavioral data suggest that the auditory sensory ('echoic') memory lasts about 10 s (Sams et al., 1993), in good agreement with the estimates based on the ISI dependence of N100m.

MMF is stronger over the right than the left hemisphere for both left-ear and right-ear tones and an extra source area in the right inferior parietal lobe, in addition to bilateral sources in the auditory cortex, is needed for satisfactory explanation of the response distri-bution (Levänen et al., 1996). The right hemisphere thus seems to be more involved in change detection than the left.

According to Levänen et al. (1993), MMFs to simultaneous double deviances (changes in e.g. frequency *and* ISI) resemble closely the arithmetic sum of MMFs to deviances in sin-gle stimulus features (e.g. frequency *or* ISI). Different sound features thus seem to have paral-lel and largely independent representations in the human auditory cortex.

3.3. Responses to Sound Omissions

Figure 2 illustrates responses of 3 subjects to omissions of tones in an otherwise regular stimulus sequence. Interestingly, the attended omissions were associated with tem-porally distinct percepts of 'nothingness' at the time of the omission and also elicited clear responses peaking about 200 ms after the expected sound occurrence (Raij et al., 1997).

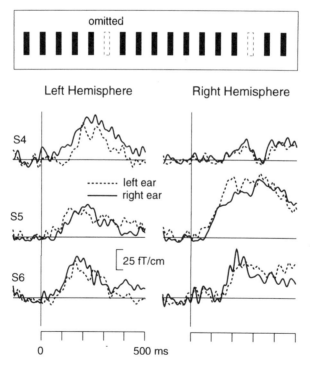

Figure 2. Above: Schematic presentation of the odd-ball paradigm with sound omissions. Tone pips were pre-sented once every 0.5 s, with 7% of them randomly omitted. Below: Responses of 3 subjects to sound omissions over the left and right hemispheres when the stimulus sequence was presented either to the left or the right ear (dashed and solid lines, respectively). The omitted sound was expected to occur at time zero. Adapted from Raij et al. (1997).

Source modelling indicated that the omission responses received major contribution from the supratemporal cortex, thereby indicating that the regularities of the environment are monitored at the level of the auditory cortex. Additional active areas were found in the superior temporal sulcus and in the posterior frontal lobe, again with right-hemisphere dominance.

3.4. Sound-Location-Related Responses

In audition (in contrast to vision and somesthesis), information on stimulus location does not exist on the receptor surface, and sound source locations have to be calculated centrally from the interaural time and intensity differences. McEvoy et al. (1993) studied cortical mechanisms of sound lateralization by recording AEFs to binaural 70-Hz click trains containing interaural time differences (ITDs). ITDs varied in different trains between +0.7 and -0.7 ms and the subjects perceived the stimuli at different left-right locations. In an oddball experiment, the stimuli giving rise to the left-most perception were presented as standards with infrequent deviants perceived at other locations. The MMF amplitude was smallest when the deviants were close to the standard stimulus in perceived location, and increased as the spatial separation between the sounds increased.

These results can be interpreted as reflecting a functional separation between neuronal groups which react to sounds with different ITDs; evidently the neuronal populations activated by spatially close stimuli overlap, and the overlap decreases with increasing spatial separation. However, source locations in the auditory cortex did not show any sign of a systematic large-scale spatial mapping of sound location.

3.5. Spontaneous Activity

The visual and somatosensory cortices are known to display spontaneous rhythmic activity. The auditory cortex also generates a 8–10 Hz spontaneous rhythm, tau, which is strongest when the subject is drowsy (Tiihonen et al., 1991; Lu et al., 1992a; Hari, 1993). Tau is occasionally suppressed by sounds and has a slight right-hemisphere predominance (Lehtelä et al., 1997). The functional significance of the tau rhythm for auditory processing and/or auditory-related orienting remains to be shown.

4. PSYCHOACOUSTICS: ILLUSION OF DIRECTIONAL HEARING

Hari (1995) recently presented a new psychoacoustical illusion as an auditory analogue for the 'cutaneous rabbit' (Geldard and Sherrick, 1972): When binaural clicks are presented in a sequence of 4 left-ear leading clicks, followed by 4 right-ear leading (see Fig. 3A), the subject perceives the clicks as jumping at equidistant steps from left to right at short interstimulus intervals (Fig. 3B).

Figure 3C plots the perceived jump from 4th to 5th click as a function of ISI. In control subjects the saltatory illusion exists up to ISIs of about 150 ms; at longer ISIs the subjects perceive clearly lateralized stimuli. Evidently the later auditory experience (here the right-sided stimuli) can, at ISIs less than 150 ms, affect perception of stimuli presented even hundreds of millisecond earlier; the content of the almost 0.5-s time window seems to be interpreted holistically. It is worth noting that the symmetry about the midline (see Fig. 3B) means that the more recent stimuli do not bias the responses towards their sites, in contrast to what a simplistic memory-trace based model would suggest.

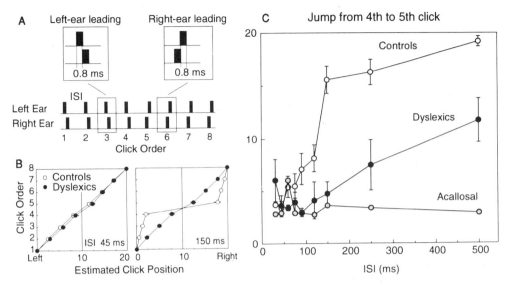

Figure 3. (A) Schematic presentation of stimuli resulting in the directional hearing illusion. Both ears received eight 1-ms clicks: 4 identical left-ear leading clicks were followed by 4 right-ear leading clicks. (B) Mean (± SEM) estimates of the perceived click positions at two interstimulus intervals for 20 control subjects (open symbols) and 10 dyslexics (black symbols). The total left-right distance was coded as 20. (C) Mean (± SEM) sizes of perceived jumps across the midline (from the 4th to the 5th click) as a function of ISI in the control group, the dyslexics (according to Hari and Kiesilä, 1996), and in one subject with congenital callosal agenesis.

The illusion was practically missing for monaural stimuli which, although commonly used in auditory studies, are of course nonphysiological since they do not exist in ecologically valid situations for an organism whose auditory system is intact. For further discussion about the monaural/binaural differences, see Hari (1995).

Interestingly, dyslexic adults were markedly abnormal in their evaluation of the click trains, perceiving the illusory jumping at much longer ISIs than the control subjects do (Fig. 3C; Hari and Kiesilä, 1996), as though they would need a longer time for their conscious percept to arise. This finding supports the existence of a deficit of auditory processing in dyslexic subjects, a topic currently under intensive debate (Tallal et al., 1993; Studdert-Kennedy and Mody, 1995).

Figure 3C also illustrates that a highly-performing subject with congenital callosal agenesis had a very prominent central sluggishness in his auditory perception. Evidently hemispheric interaction is important for this type of performance.

5. CONCLUSIONS

MEG provides a noninvasive method for studying macroscopic functional organization of the human auditory cortex, thereby complementing information obtained from other brain imaging methods and from scalp electroencephalograms. MEG recordings show that the human auditory cortex is extremely sensitive to various acoustic events, such as onsets, offsets, and changes within sounds, including frequency and amplitude modulations which are important constituents of speech. The auditory cortex also seems to be important for sensory memory and for directional hearing.

Top-down approaches, with identical stimuli but with varying tasks, have become feasible with the whole-head instruments. Previous MEG recordings have already shown that identical sounds produce different activation in the auditory cortex depending on the attentional state of the subject (Hari et al., 1989; Rif et al., 1991). MEG recordings, with their excellent temporal resolution, will hopefully open new insights into some key issues of auditory perception, such as extraction of auditory objects from the acoustic inflow. Timing is a key issue also in segmentation of auditory events, especially important for understanding speech.

The 150 ms upper limit for ISIs resulting in the saltation illusion agrees with timing of the main transient cortical responses to sound stimuli and also resembles time windows of forward and backwards masking, apparent auditory motion, loudness enhancement, perceptual grouping, and enhancement of evoked responses with paired stimulus presentation (cf. Altman and Viskov, 1977; Loveless et al., 1989; Zwicker and Fastl, 1990). However, the saltatory illusion also indicates that another time lag, beyond the interval relevant for the above phenomena, precedes the conscious auditory percept. Future studies are expected to clarify which neural processes occupy this time interval.

6. ACKNOWLEDGMENTS

This study has been supported by the Academy of Finland. I thank N. Loveless for discussions, and R. Salmelin and A. Schnitzler for participation in the study of the acallosal subject.

7. REFERENCES

Altman, J. and Viskov, O. (1977) Discrimination of perceived movement velocity for fused auditory image in dichotic stimulation. J. Acoust. Soc. Am. 61, 816–819.

Cowan, N. (1984) On short and long auditory stores. Psychol. Bull. 96, 341–370.

Elberling, C., Bak, C., Kofoed, B., Lebech, J. and Saermark, K. (1980) Magnetic auditory responses from the human brain. A preliminary report. Scand. Audiol. 9, 185–190.

Geldard, F. and Sherrick, C. (1972) The cutaneous "rabbit": a perceptual illusion. Science 178, 178–179.

Grandori, F., Hoke, M. and Romani, G.-L., Eds. (1990) Auditory Evoked Magnetic Fields and Potentials. Adv. Audiol., Vol. 6. Karger, Basel.

Hämäläinen, M., Hari, R., Ilmoniemi, R., Knuutila, J. and Lounasmaa, O.V. (1993) Magnetoencephalography – theory, instrumentation, and applications to noninvasive studies of the working human brain. Rev. Mod. Physics 65, 413–497.

Hari, R. (1990) The neuromagnetic method in the study of the human auditory cortex. In: F. Grandori, M. Hoke, G. Romani (Eds.), Auditory Evoked Magnetic Fields and Potentials. Adv. Audiol., Vol. 6. Karger, Basel, pp. 222–282.

Hari, R. (1993) Magnetoencephalography as a tool of clinical neurophysiology. In: E. Niedermeyer, F. Lopes da Silva (Eds.), Electroencephalography. Basic Principles, Clinical Applications and Related Fields. 3rd edition. Williams & Wilkins, pp. 1035–1061.

Hari, R. (1995) Illusory directional hearing in humans. Neurosci. Lett. 189, 29–30.

Hari, R. and Lounasmaa, O.V. (1989) Recording and interpretation of cerebral magnetic fields. Science 244, 432–436.

Hari, R. and Kiesilä, P. (1996) Deficit of temporal auditory processing in dyslexic adults. Neurosci. Lett. 205, 138–140.

Hari, R., Aittoniemi, K., Järvinen, M.L., Katila, T. and Varpula, T. (1980) Auditory evoked transient and sustained magnetic fields of the human brain. Localization of neural generators. Exp. Brain Res. 40, 237–240.

Hari, R., Hämäläinen, M., Kaukoranta, E., Mäkelä, J., Joutsiniemi, S.L. and Tiihonen, J. (1989) Selective listening modifies activity of the human auditory cortex. Exp. Brain Res. 74, 463–470.

Lehtelä, L., Salmelin, R. and Hari, R. (1997) Evidence for reactive magnetic 10-Hz rhythm in the human auditory cortex. Neurosci. Lett. 222, 111–114.

Levänen, S., Ahonen, A., Hari, R., McEvoy, L. and Sams, M. (1996) Deviant auditory stimuli activate human left and right auditory cortex differently. Cereb. Cortex 6, 288–296.

Levänen, S., Hari, R., McEvoy, L. and Sams, M. (1993) Responses of the human auditory cortex to changes in one vs. two stimulus features. Exp. Brain. Res. 97, 177–183.

Liegeois-Chauvel, C., Musolino, A. and Chauvel, P. (1990) Générateurs des potentiels évoqués auditifs corticaux chez l'homme. Colloque de Physique 51(C2), 135–138.

Lounasmaa, O.V., Hämäläinen, M., Hari, R. and Salmelin, R. (1996) Information processing in the human brain – magnetoencephalographic approach. Proc. Natl. Acad. Sci. USA. 93, 8809–8815.

Loveless, N., Hari, R., Hämäläinen, M. and Tiihonen, J. (1989) Evoked responses of human auditory cortex may be enhanced by preceding stimuli. Electroenceph. Clin. Neurophysiol. 74, 217–227.

Loveless, N., Levänen, S., Jousmäki, V., Sams, M. and Hari, R. (1996) Temporal integration in auditory sensory memory: Neuromagnetic evidence. Electroenceph. Clin. Neurophysiol. 100, 220–228.

Lu, S.-T., Kajola, M., Joutsiniemi, S.-L., Knuutila, J. and Hari, R. (1992a) Generator sites of spontaneous MEG activity during sleep. Electroenceph. Clin. Neurophysiol. 82, 182–196.

Lu, Z.-L., Williamson, S. and Kaufman, L. (1992b) Behavioral lifetime of human auditory sensory memory predicted by physiological measures. Science 258, 1668–1670.

McEvoy, L., Hari, R., Imada, T. and Sams, M. (1993) Human auditory cortical mechanisms of sound lateralization: II. Interaural time differences at sound onset. Hear. Res. 67, 98–109.

Mäkelä, J.P., Hari, R., Valanne, L. and Ahonen, A. (1991) Auditory evoked magnetic fields after ischemic brain lesions. Ann. Neurol. 30, 76–82.

Näätänen, R. and Picton, T. (1987) The N1 wave of the human electric and magnetic response to sound: a review and analysis of the component structure. Psychophysiol. 24, 375–425.

Raij, T., McEvoy, L., Mäkelä, J.P. and Hari, R. (1996) Human auditory cortex is activated by omissions of auditory stimuli. Brain Res., 745, 134–143.

Rif, J., Hari, R., Hämäläinen, M. and Sams, M. (1991) Auditory attention affects two different areas in the human auditory cortex. Electroenceph. Clin. Neurophysiol. 79, 464–472.

Sams, M., Hari, R., Rif, J. and Knuutila, J. (1993) The human auditory sensory memory trace persists about 10 s: neuromagnetic evidence. J. Cogn. Neurosci. 5, 363–370.

Sato, S., Ed. (1990) Magnetoencephalography. Adv. Neurol., Vol. 54. Raven Press, New York.

Studdert-Kennedy, M. and Mody, M. (1995) Auditory temporal perception deficits in the reading-impaired: A critical review of the evidence. Psychon. Bull. & Rev. 2, 508–514.

Tallal, P., Miller, S. and Fitch, R. (1993) Neurobiological basis of speech: a case for the preeminence of temporal processing. Ann. NY Acad. Sci. 682, 27–47.

Tiihonen, J., Hari, R., Kajola, M., Karhu, J., Ahlfors, S. and Tissari, S. (1991) Magnetoencephalographic 10-Hz rhythm from the human auditory cortex. Neurosci. Lett. 129, 303–305.

Zwicker, E. and Fastl, H. (1990) Psychoacoustics. Facts and Models. Springer-Verlag, Berlin.

31

CONTRIBUTION OF THE DORSAL NUCLEUS OF THE LATERAL LEMNISCUS TO BINAURAL PROCESSING IN THE AUDITORY BRAINSTEM

Jack B. Kelly

Laboratory of Sensory Neuroscience
Institute of Neuroscience
Carleton University
Ottawa, Ontario, Canada

1. INTRODUCTION

The focus of this chapter is on the role of the dorsal nucleus of the lateral lemniscus (DNLL) in binaural processing and sound localization. Until recently, relatively little attention has been paid to the DNLL as a significant factor in acoustic signal processing but over the last 10 years it has become increasingly apparent from neuroanatomical, electrophysiological and behavioral studies that the DNLL is actively involved in shaping neural responses and controlling sensory function. This chapter will present electrophysiological and behavioral evidence that the DNLL is an important component of the auditory brainstem circuitry responsible for the neural analysis of binaural input and that it is involved in the representation of the location of sounds in space.

2. ANATOMICAL CONSIDERATIONS

The DNLL is situated just below the inferior colliculus (IC) among the fibers of the lateral lemniscus and receives projections from lower brainstem nuclei which themselves are known to be involved in binaural processing. The bilaterally innervated nuclei of the superior olivary complex (SOC) project heavily to DNLL with parallel projections to the central nucleus of the IC. The lateral superior olive (LSO) projects to DNLL bilaterally (Covey and Casseday, 1985; Glendenning et al., 1981; Labelle and Kelly, 1996, Schneiderman et al., 1988). It is assumed that the ipsilaterally and contralaterally projecting neurons represent largely separate populations in LSO and that very few if any neurons project to DNLL on both sides of the brain. This anatomical pattern has been demonstrated by double labeling studies for the projection from LSO to IC in the cat and a similar projection seems likely from LSO to DNLL via axon collaterals (Glendenning and

Masterton, 1983; Glendenning et al., 1985; 1992). It is also assumed that the ipsilateral and contralateral projections from LSO to DNLL, like those to IC, are histochemically distinct with a large proportion of the ipsilateral connections arising from glycine-positive neurons in LSO (Hutson et al., 1987; Ito et al., 1995; Saint Marie et al., 1989; Saint Marie and Baker, 1990). Contralaterally projecting neurons from LSO are glycine-negative and are thought to serve an excitatory function probably with glutamate as the neurotransmitter. Considering that the vast majority of LSO neurons are excited by ipsilateral acoustic stimulation and are inhibited by contralateral stimulation (Boudreau and Tsuchitani, 1968; Caird and Klinke, 1983; Sanes, 1990; Tsuchitani and Boudreau, 1969; Wu and Kelly, 1991, 1992), one would expect that activation of neurons in LSO would result in a net excitation of the contralateral DNLL with a prominent glycinergic inhibition of the ipsilateral DNLL. Indeed, most neurons in the DNLL of the rat are excited by contralateral stimulation and inhibited by simultaneous ipsilateral stimulation (Buckthought, 1993). Similar binaural response patterns have been reported for the DNLL of the cat, big brown bat and mustache bat (Brugge et al., 1970; Covey, 1993; Covey and Casseday, 1995; Markovitz and Pollak, 1994).

The DNLL also receives projections from the ipsilateral medial superior olive (MSO) in rats, cats and echolocating bats (Covey and Casseday, 1995; Glendenning et al., 1981; Labelle and Kelly, 1996; Vater et al., 1995). In the cat it is well established that the MSO is a binaural nucleus that receives direct projections from both left and right ventral cochlear nuclei (Cant, 1991; Harrison and Feldman, 1970; Stotler, 1953; Warr, 1982). Electrophysiological data indicate that the neurons in MSO are excited by both ipsilateral and contralateral acoustic stimulation and are exquisitely sensitive to the timing and phase relation of sounds in the two ears (Goldberg and Brown, 1969; Irvine, 1992; Yin and Chan, 1990). In effect, the nucleus serves as a delay line with individual neurons maximally excited by specific binaural time differences (Smith et al., 1993). For many MSO cells the most effective stimulus is associated with contralateral lead times produced by sounds located in the contralateral spatial field. Therefore, the ipsilateral projections from MSO, like the contralateral projections from LSO, probably contribute substantially to the excitation of DNLL neurons by sounds from a contralateral source. Relatively little is known about the physiological properties of the rat's MSO, which is proportionally much smaller than the cat's, but the available data indicate that these cells are also binaurally sensitive and probably reflect the same general principles of organization found in the cat (Inbody and Feng, 1981).

The superior paraolivary nucleus (SPN) is a large and distinctive structure in the rat's SOC and has ipsilateral projections to both DNLL and IC (Beyerl, 1978; Coleman and Clerici, 1987; Colombo et al., 1996; Ito et al., 1996; Kelly et al., 1995; Labelle and Kelly, 1996). The physiological properties of the SPN neurons, however, have not been systematically investigated and it is not known whether they are binaural or preferentially sensitive to the location of sounds in space. It seems likely on anatomical grounds, however, that the SPN is a functionally important input to DNLL in the rat's central auditory system.

An additional input to DNLL arises from neurons in the contralateral cochlear nucleus. Injection of retrograde tracers into DNLL results in labeling in the contralateral cochlear nucleus with little or no labeling in the ipsilateral cochlear nucleus (Glendenning et al., 1981; Labelle and Kelly, 1996; Schwartz, 1992, Schneiderman et al., 1988). Thus, acoustic stimulation of the contralateral ear can contribute to excitation of DNLL neurons via direct projections from the cochlear nucleus.

The DNLL sends efferent projections to the contralateral DNLL and to both ipsilateral and contralateral central nuclei of the IC (Adams, 1979; Bajo et al., 1993; Beyerl, 1978; Brunso-Bechtold et al., 1981; Coleman and Clerici, 1987; Merchán et al., 1994; Shneiderman et al., 1988; Zook and Casseday, 1982). The ipsilateral and contralateral projections to IC arise from separate populations of neurons in DNLL. These neurons are partially segregated spatially in the cat's DNLL with more medially located neurons projecting contralaterally and more laterally situated cells projecting ipsilaterally (Shneiderman et al., 1988; 1996). In the rat the two populations appear to be more evenly distributed (Ito et al., 1996; Kelly et al., 1995). The contralaterally projecting neurons in the rat account for approximately 70% of the cells projecting to IC whereas in the cat this population is closer to 60% (Ito et al., 1996; Shneiderman et al., 1988). Immunolabeling studies have shown that the vast majority of neurons in the DNLL are GABAergic (Adams and Mugnaini, 1984; Glendenning and Baker, 1988; Moore and Moore, 1987; Roberts and Ribak, 1987; Thompson et al., 1985; Vater et al., 1992; Winer et al., 1995) and destruction of the DNLL results in a reduction in the evoked release of GABA in the IC with the greatest reduction contralateral to the damaged cells (Shneiderman et al., 1993). EM studies of the terminals of DNLL neurons that give rise to projections to either contralateral DNLL or IC in the cat have revealed flattened or pleomorphic synaptic vesicles characteristic of inhibitory synapses (Oliver and Shneiderman, 1989). Many terminals of neurons projecting to the ipsilateral IC contain flattened or pleomorphic vesicles but some contain rounded synaptic vesicles (Oliver and Shneiderman, 1989; Shneiderman and Oliver, 1989). These anatomical data indicate that two separate populations of cells in DNLL give rise to GABAergic projections: one innervates the contralateral DNLL and IC and the other innervates the ipsilateral IC. As suggested originally by Shneiderman et al. (1988) the crossed inhibitory pathway from DNLL may provide a neural substrate for binaural interactions at the level of the auditory midbrain.

In both rat and bat, and probably also in other species, the rostral DNLL sends projections directly to the deep layers of the superior colliculus (Bajo et al., 1993; Covey and Casseday, 1995; Kudo, 1991; Tanaka et al., 1985). This projection has not been studied in as much detail as the one to IC but it may be functionally significant for auditory spatial perception and sound localization.

The commissure of Probst (CP) is the exclusive pathway for all crossed projections from DNLL to the opposite DNLL and IC (Hutson et al., 1991; Ito et al., 1996). Tract tracing studies show that all contralaterally projecting fibers from DNLL pass through the CP. Surgical transection of the CP in the rat completely eliminates the transport of Fluorogold from the central nucleus of IC to the opposite DNLL (Ito et al., 1996) and blocks the anterograde transport of biocytin to either contralateral DNLL or IC (unpublished observations).

Electrophysiological mapping studies have led to conflicting conclusions about the topographic organization of DNLL in various mammalian species. Recordings from the cat's DNLL suggest a tonotopic gradient with low frequencies represented dorsally and high frequencies ventrally (Aitkin et al., 1970) but no consistent pattern has been found in rat or echolocating bats (Buckthought, 1993; Covey, 1993; Markovitz and Pollak, 1993). In the rat the projections from DNLL to IC are, however, topographically organized as indicated by tract tracing studies. Neurons located along the outer margins of DNLL project to ventral areas in the central nucleus of IC and cells in more centrally located areas of DNLL project to the dorsal part of the central nucleus of IC. The IC itself is tonotopically organized with high frequencies represented ventrally and low frequencies dorsally (Kelly et al., 1991). Therefore, the projection pattern of DNLL in the rat has been characterized

as concentric or "onion-like" with high frequencies represented by relatively large diameter layers of cells and lower frequencies by smaller diameter layers within the nucleus
(Merchán et al., 1994). This concentric organization has been confirmed by Fluorogold injections into electrophysiologically-defined regions of the rat's inferior colliculus (Kelly
et al., 1995). Injections into high-frequency regions of IC result in large diameter rings of
labeled cells in both ipsilateral and contralateral DNLL whereas injections into low frequency regions result in more centrally located rings with smaller diameters. The concentric nature of the organization is particularly apparent in rats with different tracers
(Fluorogold and horseradish peroxidase) injected into high and low frequency regions of
the same inferior colliculus. Rings of cells labeled with the high frequency tracer can be
seen surrounding smaller diameter rings of cells labeled with low frequency tracer.

3. BRAIN SLICE STUDIES OF DNLL

A series of *in vitro* brain slice studies has provided basic information about the intrinsic membrane properties, synaptic responses and pharmacology of neurons in the rat's
DNLL (Wu and Kelly, 1995a, b; 1996). In these studies electrophysiological data were
obtained from 400 μm brain slices taken in the frontal plane through the DNLL. Synaptic
responses were evoked by electrical stimulation of the fibers in the lateral lemniscus
and/or CP and the pharmacology of excitatory and inhibitory postsynaptic potentials was
examined by adding neurotransmitter agonists and antagonists to the saline solution in
which the slice was bathed. Physiologically characterized DNLL neurons were labeled by
intracellular injection of biocytin and somatic and dendritic profiles were identified by microscopic reconstruction.

As shown in Figure 1, three general groups of neurons can be identified in DNLL on
the basis of intracellular biocytin labeling: large multipolar cells, elongate cells and round
cells (Wu and Kelly, 1995b). The elongate neurons have laterally-oriented dendrites,
which in many cases extend across the entire DNLL from its medial to lateral margins,
whereas multipolar and round neurons tend to have radial dendrites without specific orientation. These three broad classes of neurons can also been recognized in Golgi or Nissl
stained tissue (Bajo et al., 1993; Covey, 1983; Kane and Barone, 1980; Shneiderman et
al., 1988; Tanaka et al., 1985). Not represented in the biocytin-labeled sample are much
smaller round cells that can be seen in Nissl stained material (Bajo et al., 1993).

The intrinsic membrane properties of DNLL neurons are quite similar regardless of
their somatic and dendritic morphology. All DNLL neurons sampled by intracellular recording electrodes had linear current-voltage relationships as illustrated in Figure 2. Injection of negative current produced a hyperpolarization and positive current produced a
depolarization of the cell membrane. The relation between the shift in membrane potential
and current strength was essentially linear over a wide range of positive and negative current values (Wu and Kelly, 1995a). All DNLL neurons fired sustained trains of action potentials when sufficiently depolarized by positive current and the number of action
potentials was monotonically related to the degree of depolarization. This response pattern
reflects the ability of DNLL neurons to fire in a graded manner and is consistent with their
role as integrators of converging postsynaptic inputs.

In the brain slice preparation excitatory and inhibitory postsynaptic responses can be
elicited in DNLL neurons by direct electrical stimulation of either the lateral lemniscus or
the CP. Intracellular recordings reveal a high degree of convergence of synaptic input onto
individual neurons (Wu and Kelly, 1995a). The IPSPs elicited by stimulation of the CP are

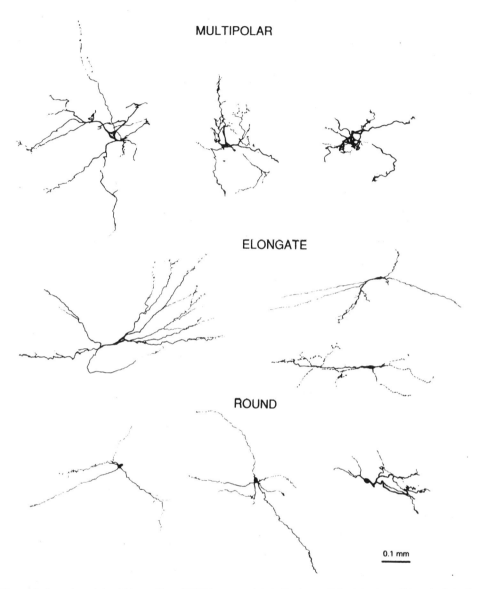

Figure 1. Somatic and dendritic profiles of DNLL neurons labeled by intracellular injection of biocytin. Injections were made after physiological recordings from cells in 400 μm brain slices taken through the rat's DNLL. Adapted from Wu and Kelly (1995b).

mediated by $GABA_A$ receptors and can be blocked by bath application of the $GABA_A$ antagonist, bicuculline, a finding that supports the idea from immunolabeling studies that the crossed projections of DNLL through the CP are inhibitory and GABAergic. In contrast, the IPSPs elicited by stimulation of the lateral lemniscus are mediated by glycine receptors and are blocked by application of the glycine antagonist, strychnine (Wu and Kelly, 1996). One probable source of this lemniscal inhibition is from glycinergic neurons in the ipsilateral LSO.

Short latency EPSPs evoked by electrical stimulation of the lateral lemniscus can be blocked by the non-NMDA receptor antagonist, CNQX. The lemniscal EPSPs are probably derived from various sources including the contralateral LSO, ipsilateral MSO, ipsilateral SPN or contralateral cochlear nucleus. Electrical stimulation of the lateral lemniscus frequently elicits both EPSPs and IPSPs in the same neuron reflecting a convergence of excitatory and inhibitory input from lower brainstem nuclei. It should be pointed out, however, that this does not necessarily mean that the EPSPs and IPSPs are elicited under the same stimulus conditions *in vivo*. For example, in the rat the EPSPs in DNLL are probably evoked by contralateral acoustic stimulation whereas the lemniscal IPSPs, if they originate from glycinergic neurons in LSO, are probably evoked by ipsilateral acoustic stimulation. Thus, the convergence of EPSPs and IPSPs in DNLL reflects the interplay between synaptic events evoked by sounds in different spatial locations in the environment. A single lateralized sound source would likely result in a net excitation of neurons in one DNLL and a simultaneous inhibition of neurons in the other.

Our recent brain slice studies have shown that electrical stimulation of the lateral lemniscus evokes NMDA as well as non-NMDA receptor-mediated excitatory responses in DNLL (Fu et al., 1997; Wu and Kelly, 1996). Both types of excitation can be elicited in

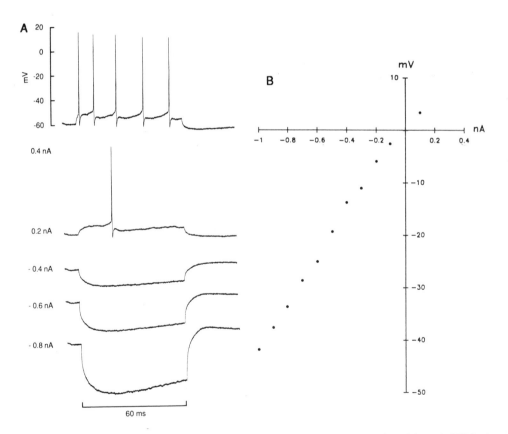

Figure 2. Intracellular recordings made from a typical neuron in a brain slice preparation of the rat's DNLL. A. The shift in membrane potential produced by intracellular injections of positive and negative currents from -0.8 nA to 0.4 nA. B. The current-voltage curve obtained by measuring the peak voltage shift produced by current injections of various strengths. The neurons in DNLL show linear current-voltage relations regardless of morphological class. From Wu and Kelly (1995a).

the same postsynaptic neuron but the time course of the two events is different. Bath application of the non-NMDA receptor antagonist, CNQX, selectively blocks a short latency component of the lemniscally-induced EPSPs whereas application of the NMDA antagonist, APV, blocks a longer lasting component that can persist for many milliseconds. Short latency action potentials associated with the early component of the EPSP can be blocked by CNQX and longer latency action potentials (3 to 24 ms) associated with the later component can be blocked by APV (Wu and Kelly, 1996).

With the brain slice in normal saline the existence of EPSPs can often be obscured by the occurrence of IPSPs that are evoked at the same time by the same level of lemniscal stimulation. The IPSPs can completely cancel the EPSPs and prevent their detection. To investigate the EPSPs in more detail it is necessary to block the glycineric IPSPs by continuously perfusing the brain slice with 0.5 μM strychnine. Under this condition all IPSPs evoked by lemniscal stimulation are blocked and the properties of the EPSPs can be investigated without complication. Patch clamp recordings from DNLL neurons under the influence of strychnine show that virtually every cell expresses both non-NMDA and NMDA receptor-mediated excitatory responses (Fu et al., 1997).

As can be seen in Figure 3, with patch clamp recordings an early non-NMDA response can be blocked by application of CNQX and a longer lasting NMDA response can be blocked by APV. The NMDA response component is voltage dependent and subject to pharmacological block by extracellular magnesium. Thus, the NMDA component is most clearly expressed in brain slices perfused with low magnesium saline solutions or when

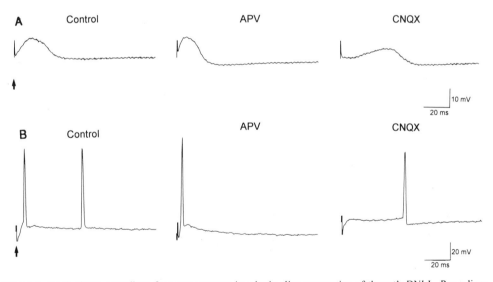

Figure 3. Patch clamp recordings from two neurons in a brain slice preparation of the rat's DNLL. Recordings were obtained with the brain slice submerged in an artificial cerebrospinal fluid containing 0.5 μM strychnine to eliminate the occurrence of glycinergic IPSPs. A. Electrical stimulation of the lateral lemniscus, indicated by the arrow in the first trace, produced an EPSP that contained both NMDA and non-NMDA receptor-mediated components. Addition of 50 μM APV to the solution bathing the slice resulted in expression of an early excitatory response whereas addition of 10 μM CNQX blocked the early component and resulted in the expression of a longer lasting excitatory response. B. Suprathreshold stimulation of the lateral lemniscus evoked EPSPs and both short and long latency action potentials. The short latency action potential was selectively blocked by application of 10 μM CNQX and the long latency action potentials were blocked by application of 50 μM APV. From Fu et al. (1997).

the membrane potential is depolarized by voltage clamp procedures (Fu et al., 1997). Nevertheless, the NMDA component can often be seen when the cell is at or near resting potential and the slice is bathed in normal saline (Wu and Kelly, 1996). These data suggest that the NMDA response of DNLL neurons *in vivo* can be brought into play at membrane potentials near the resting state. Acoustic stimulation would be expected to cause an initial depolarization through activation of non-NMDA receptors. The initial depolarization would then be followed by activation of a longer lasting NMDA receptor-mediated excitation the magnitude of which would depend on the strength of stimulation. The NMDA component would have the effect of extending the time period over which the DNLL neuron remains depolarized. These brain slice data suggest that both NMDA and non-NMDA components normally play a role in synaptic transmission in DNLL and probably contribute to the processing of sensory information.

4. *IN VIVO* PHYSIOLOGICAL STUDIES

In vivo physiological studies show that neurons in DNLL help shape binaural responses recorded from the IC or primary auditory cortex of the rat. Blockade of excitatory synapses in DNLL by local injection of pharmacological agents disrupts binaural responses of neurons in the contralateral IC (Faingold et al., 1993; Kelly and Kidd, 1997; Kidd and Kelly, 1996; Li and Kelly, 1992) and unilateral lesions of DNLL disrupt responses in the contralateral auditory cortex (Glenn and Kelly, 1992).

As shown in Figure 4 binaural responses of IC neurons to interaural level differences are affected by local injection of the non-selective excitatory amino acid antagonist, kynurenic acid, into the contralateral DNLL (Li and Kelly, 1992). Most neurons in the rat's IC are excited by contralateral stimulation and are strongly inhibited by simultaneous ipsilateral stimulation, the so-called EI cells (Kelly et al., 1991). In these cells an increase in the level of ipsilateral stimulation produces a sharp decrease in firing rate that is in some cases preceded by an increase in firing. Interaural intensity difference (IID) curves can be generated for single neurons by plotting their response rate as a function of the relative intensity tone bursts presented to the two ears. After injection of kynurenic acid into the DNLL the IID curves of neurons in the contralateral inferior colliculus are shifted due to a release from the inhibition that is normally produced by acoustic stimulation of the ear ipsilateral to the recording site. The extent of the release is variable from cell to cell but is present to some extent in virtually every neuron sampled. No change is seen in the IID curves obtained from neurons in the IC ipsilateral to the injection site in DNLL. Similar observations have been reported by Faingold et al. (1993) after blockade of activity in DNLL by local injection of lidocaine. These authors also report an enhancement of ipsilateral inhibition in the central nucleus of IC after electrical stimulation of the contralateral DNLL or injection of the excitatory amino acid agonist, kainic acid, into DNLL.

Injection of the excitatory amino acid antagonist, kynurenic acid, into the rat's DNLL has a marked effect on the response of IC neurons to interaural time differences (ITDs) (Kidd and Kelly, 1996). Although most neurons in the rat's inferior colliculus are relatively insensitive to the ongoing phase differences between binaurally presented tones, as might be expected in a species with very limited sensitivity to low frequency sounds (Kelly and Masterton, 1977), they are nevertheless quite sensitive to small time differences between acoustic transients delivered to the two ears. Thus, ITD sensitivity can be determined for a wide range of frequency-specific neurons in IC by manipulating the time difference between binaurally presented clicks. For most EI neurons, the probability of generating an action potential is greatest when a click presented to the contralateral ear leads one presented to the ipsilateral ear by a significant margin regardless of the cell's response to pure tone frequency. As the contralateral click is progressively delayed relative to the ipsilateral, however, the response probability drops from unity to near zero over a range of a few hundred microseconds. Typical ITD curves for IC neurons can be generated by plotting the probability of an action potential as a function of binaural time differences over a range from +/- 1.0 ms.

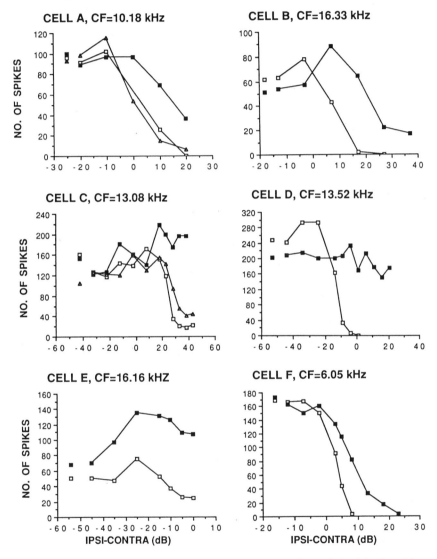

Figure 4. IID curves from neurons in the rat's central nucleus of the IC before and after injection of the excitatory amino acid antagonist, kynurenic acid, into the contralateral DNLL. The characteristic frequency (CF) is shown above the panel for each neuron. The IID curves were generated by stimulating the contralateral ear with a tone burst at a fixed sound pressure level while simultaneously stimulating the ipsilateral ear with tone bursts of increasing sound pressure. The level difference (IID) between the two ears is expressed as ipsilateral-contralateral sound pressure in dB. All tones were 110 ms in duration and were presented at CF at a rate of 1/sec. The ordinate shows the number of spikes over 50 stimulus presentations. Unfilled squares, before injection; filled squares, after injection; unfilled triangles, after recovery. From Li and Kelly (1992).

After injection of kynurenic acid into the DNLL, the ITD curves of neurons in the contralateral inferior colliculus are shifted (see Figure 5). Stimulation of the ipsilateral ear is less effective in lowering the probability of responses evoked by stimulation the contralateral ear. This result is illustrated in Figure 5 for ipsilateral lead times up to 1.0 ms. The resulting ITD curves are less steep and the IC neurons are less sensitive to small binaural time differences than before the injection.

A similar effect has been reported for slow evoked potentials recorded from the rat's primary auditory cortex (Glenn and Kelly, 1992). In this case the amplitude of the responses to binaurally presented clicks is reduced as binaural time differences are gradually shifted in favor of the ipsilateral ear. Unilateral destruction of the DNLL by injection of the excitatory neurotoxin, kainic acid, reduces the degree of binaural suppression and flattens the ITD curves in the hemisphere contralateral to the lesion.

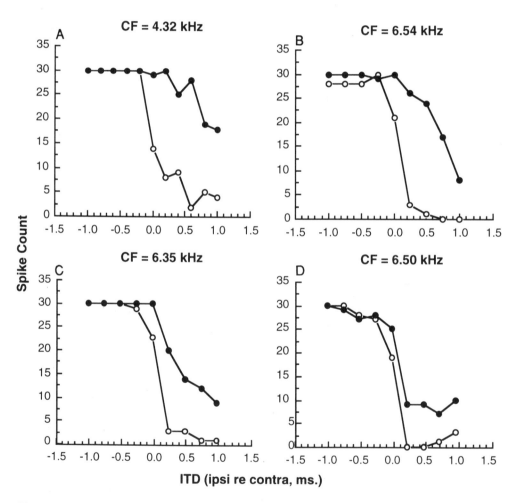

Figure 5. ITD curves from neurons in the rat's IC before and after injection of kynurenic acid into the contralateral DNLL. The characteristic frequency of each neuron is given above its respective panel. To determine ITD sensitivity the cells were stimulated by clicks delivered at equal sound pressure to the two ears. The ordinate shows spike count summed over 30 stimulus presentations. Positive ITDs refer to ipsilateral-leading-contralateral click pairs. Unfilled circles, before injection; filled circles, after injection. Adapted from Kidd and Kelly (1996).

In the rat's inferior colliculus the inhibition of single unit action potentials produced by stimulation of the ipsilateral ear can last for many milliseconds (Kidd and Kelly, 1996). The time course of this effect can be determined with dichotic click pairs by increasing the lead time of the ipsilateral click until the inhibitory effect is no longer evident. With clicks of equal sound pressure in the two ears the inhibition can last from 5 ms to more than 25 ms depending on the cell from which the recordings are made. As shown in Figure 6, injection of kynurenic acid into the DNLL reduces the strength of this long-lasting inhibition for cells located in the inferior colliculus contralateral to the injection site (Kidd and Kelly, 1996). The effect is apparent regardless of the sound frequencies to which the cell is most sensitive or the duration of the inhibitory period. Pharmacological blockade of DNLL neurons results in a release from inhibition at all ITD intervals that normally produce binaural interaction from around 0 ms to more than 25 ms.

Although the DNLL clearly has an anatomical projection to the ipsilateral as well as the contralateral IC, unilateral injection of kynurenic acid into the DNLL has no effect on

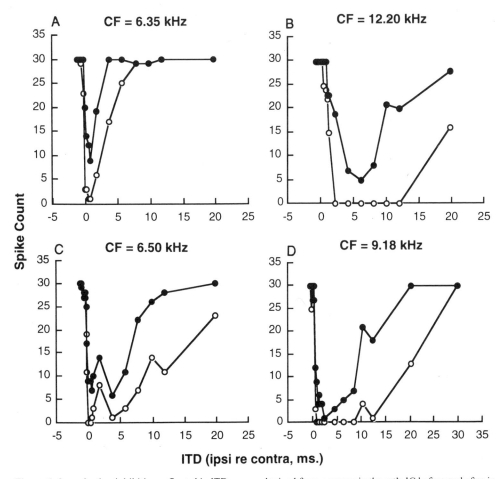

Figure 6. Long lasting inhibition reflected in ITD curves obtained from neurons in the rat's IC before and after injection of kynurenic acid into the contralateral DNLL. Curves were obtained in the same manner as those in Figure 5 but ipsilateral lead times were extended to 20 ms or more to determine the duration of binaural inhibition. Unfilled circles, before injection; filled circles, after injection. Adapted from Kidd and Kelly (1996).

the ITD curves recorded from single neurons in the IC ipsilateral to the injection (Figure 7). There is no release from inhibition at either short (0–1.0 ms) or long (1.0–25 ms) interaural time intervals (Kidd and Kelly, 1996). These results confirm our previous observations showing a lack of effect of kynurenic acid injections into DNLL on the IID curves of neurons in the ipsilateral IC (Li and Kelly, 1993) or unilateral kainic acid lesions of DNLL on ITD curves in the ipsilateral auditory cortex (Glenn and Kelly, 1993).

The exclusively contralateral effect of DNLL blockade is further illustrated by a single case in which responses to binaural time differences were recorded from separate neurons in the left and right IC after injections of kynurenic acid sequentially into the left and right DNLL (Figure 8). After the first (left) DNLL injection there was a release from inhibition recorded in the right IC but there was no effect on the binaural response of a neuron in the left IC. After a second (right) DNLL injection there was a release from inhibition in the left IC but there was no further effect on the binaural response of the neuron in the right IC (Kidd and Kelly, 1996).

Brain slice studies have shown that both NMDA and non-NMDA receptors contribute to excitatory responses in DNLL. Early excitatory responses are mediated by non-NMDA receptors and longer lasting responses by NMDA receptors. As pointed out earlier, the NMDA

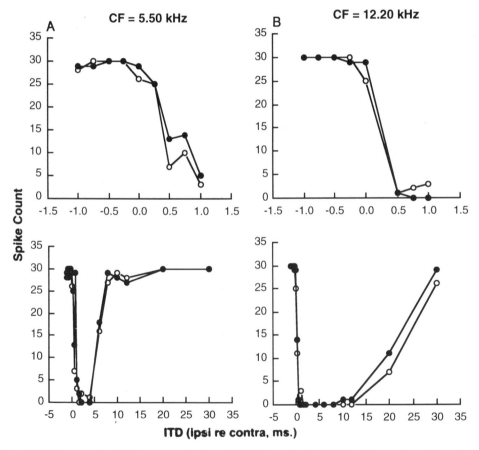

Figure 7. ITD curves obtained from neurons in the rat's IC before and after injection of kynurenic acid into the ipsilateral DNLL. Data are shown for two cells (A and B) with CFs of 5.50 and 12.20 kHz respectively. The IID curves for short intervals (+/- 1.0 ms) are shown in the upper panels and those for longer intervals (up to 30 ms) are shown in the lower panels. Curves were obtained in the same manner as those shown in Figures 5 and 6. Unfilled circles, before injection; filled circles, after injection. Adapted from Kidd and Kelly (1996).

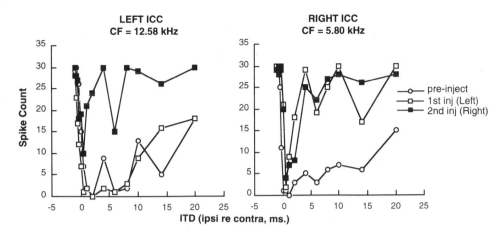

Figure 8. The effects of kynurenic acid injections into the left and right DNLL on long lasting binaural inhibition of responses recorded from two separate neurons in the left and right IC. Injections were made sequentially, first into the left DNLL and then into the right DNLL. Unfilled circles, responses before injection; unfilled squares, responses after the first DNLL injection; filled squares, responses after the second DNLL injection. From Kidd and Kelly (1996).

component of these excitatory responses serves to extend by many milliseconds the period of depolarization produced by activation of the fibers in the lateral lemniscus. Because DNLL neurons are primarily GABAergic, their prolonged excitation would be expected to result in a long lasting inhibition of cells in IC (and elsewhere) that receive their projections (Wu and Kelly, 1996; Fu et al., 1997). Thus, prolonged binaural inhibition in the IC might be due to NMDA receptor-mediated excitation of neurons in DNLL and the selective pharmacological blockade of NMDA receptors in DNLL might result in a release from long lasting inhibition. This prediction is supported by a recent physiological study showing that local injection of the NMDA antagonist, APV, into the DNLL results in a release from long lasting inhibition of binaural neurons in the IC contralateral to the injection site (Kelly and Kidd, 1997). Figure 9 shows an example of the effect produced on one cell by injection of 30 mM APV into the con-

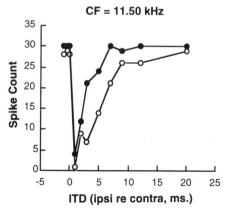

Figure 9. ITD curves obtained from a neuron in the rat's IC before and after injection of 30 mM APV into the contralateral DNLL. Curves were obtained in the same manner as those shown in Figures 5 and 6. Unfilled circles, responses before injection; filled circles, responses after injection.

tralateral DNLL. Recordings from other cells showed that the extent of release is dependent on the concentration of the drug and is not due to the injection per se.

The contralateral influence of DNLL on binaural responses in the rat's IC can be demonstrated in a highly selective fashion by surgical transection of the CP. Single unit studies are problematic because of the technical difficulty of holding neurons while cutting commissural fibers but reliable evoked potentials can be recorded from the IC before and after cutting the CP with a microsurgical knife. As can be seen in Figure 10, evoked response amplitude in normal rats is highly dependent on the time delay between paired clicks delivered to the two ears. Ipsilateral inhibition is reflected in the reduction of evoked response amplitude as the binaural time difference is increased in favor of the ipsilateral ear (i.e., with ipsilateral-leading-contralateral clicks). After transecton of the CP this effect is greatly reduced although it is not completely eliminated (van Adel et al., 1997). These data underscore the importance of the contralateral projection from DNLL for shaping binaural responses in inferior colliculus.

5. BEHAVIORAL STUDIES

Accurate sound localization depends to a large extent on the central processing of binaural cues. As the source of an acoustic stimulus is moved laterally in the horizontal plane, small differences emerge in the intensity and time of arrival of the sound as it impinges on the two ears. These differences contribute to the ability of animals to determine the location of the sound in space. After blockade of neural activity in the DNLL, both binaural time and intensity functions of neurons in the rat's inferior colliculus are compromised. Binaural inhibition is reduced and the neurons become less sensitive to time and

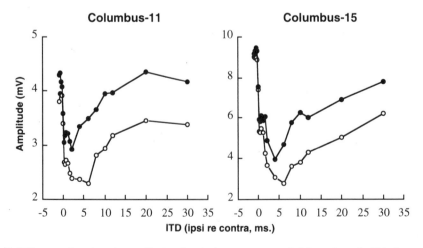

Figure 10. ITD curves based on the amplitude of evoked responses recorded from the rat's IC before and after transection of the CP. Response amplitude was measured from positive to negative peaks of the low-pass filtered evoked response. Ipsilateral-leading-contralateral ITDs are expressed as positive values. Unfilled and filled circles represent recordings obtained before and after surgical transection respectively.

intensity differences at the two ears. Thus, one would expect that destruction of the DNLL or transection of its crossed efferent projections in the CP would result in deficits in sound localization and a degradation in the accuracy of auditory spatial discrimination. These predictions are supported by behavioral studies of sound localization in the rat (Ito et al., 1996; Kelly et al., 1996).

A schematic view of the apparatus used to test the sound localization ability of rats is shown in Figure 11. After the animals had been shaped to initiate trials from a central start area, they were trained to discriminate between brief (45 ms) noise bursts presented from loudspeakers located around the perimeter of the apparatus on the left and right of midline. The rats were then tested for their ability to determine whether a sound came from a loudspeaker on the left or right. Performance was expressed as the percentage of trials in which rats correctly identified the speaker positions (left vs. right) and psychophysical curves were generated from the performance scores at each speaker location.

After unilateral kainic acid lesions of DNLL, sound localization performance was substantially reduced as shown in Figure 12. The lesions, which were made by local injection of the excitatory neurotoxin, kainic acid, destroyed all nerve cell bodies in DNLL without disrupting the fibers of passage that course through and around the nucleus en route to the IC. The integrity of fibers of passage was confirmed after completion of behavioral testing by retrograde transport of Fluorogold from the IC through the lateral lem-

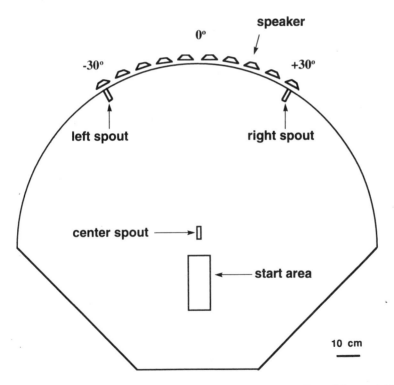

Figure 11. Schematic view of the apparatus used to test sound localization ability. Trials were initiated from a centrally located start area and noise bursts were presented from loudspeakers located around the perimeter of the apparatus. While facing forward rats contacted a center spout to activate stimulus presentation and indicated their choice (left or right) by approaching one of two spouts at +/- 30 degrees azimuth for a water reward. Adapted from Ito et al. (1996).

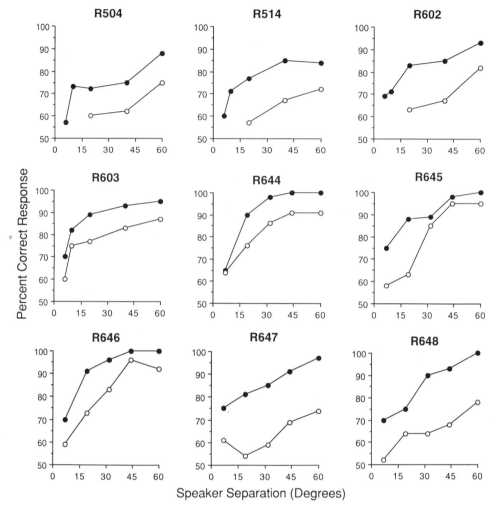

Figure 12. Performance curves for sound localization before and after unilateral kainic acid lesions of the DNLL. Data are shown for animals with complete destruction of DNLL. Unfilled circles, performance before DNLL lesion; filled circles, performance after DNLL lesion. Adapted from Kelly et al. (1996).

niscus to auditory nuclei of the lower brainstem (Kelly et al., 1996). Unilateral lesions of DNLL resulted in a drop in performance on sound localization tests with single noise bursts presented at the beginning of each trial. A similar drop in performance was found in animals with bilateral DNLL lesions (not shown). Neither unilateral nor bilateral lesions resulted in a complete inability to localize sounds but marked deficits in sound localization were apparent in both cases.

Minimum audible angles were determined from the psychophysical curves shown in Figure 12 by estimating the angle of speaker separation that supported a performance of 75% correct response. The average minimum audible angles before and after DNLL lesions are shown in Figure 13. For control purposes minimum audible angles were also ob-

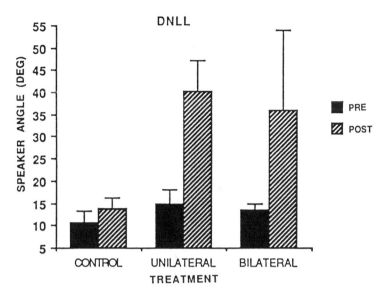

Figure 13. Average minimum audible angles for sound localization before and after kainic acid lesions of the DNLL. Preoperative and postoperative means are shown as solid and shaded columns respectively. Error bars indicate the standard error of the mean. From Kelly et al. (1996).

tained from normal animals on two separate occasions with an intervening time period comparable to that required for recovery from lesions in experimental cases. Unilateral kainic acid lesions of the DNLL resulted in an average elevation in minimum audible angle of approximately 25 degrees (Kelly et al., 1996). Similar deficits were found after bilateral lesions of DNLL. Thus, kainic acid lesions resulted in a degradation in sound localization and an elevation in minimum audible angles. These data support the view that the DNLL is important for maintaining normal auditory spatial acuity.

The functional contribution of DNLL to sound localization has also been examined in animals with surgical transection of the CP. Unlike kainic acid lesions of DNLL, the CP lesions selectively destroy those DNLL neurons that projection to the contralateral IC and other contralateral targets but leave intact the neurons that project to the ipsilateral IC. After transection of the CP retrograde cell loss is apparent in DNLL on both sides of the brain but many neurons remain intact due to their uncrossed connections. Transport of Fluorogold after a large unilateral injection into the central nucleus of the IC reveals normal retrograde labeling of cells in the ipsilateral DNLL but, in contrast to the pattern seen in normal animals, there is no retrograde labeling in the contralateral DNLL. These anatomical data show that midline transection of the CP selectively destroys the crossed projections of DNLL while leaving the ipsilateral projections intact.

Rats with surgical transection of the CP suffer deficits in sound localization comparable in severity to those found after unilateral or bilateral destruction of DNLL. Psychophysical curves obtained before and after CP lesions are shown in Figure 14 together with control data obtained from normal animals. The ability to localize sounds in space is not completely eliminated by transection of the CP, but there is a significant drop in performance that can be attributed to the lesion (Ito et al., 1996).

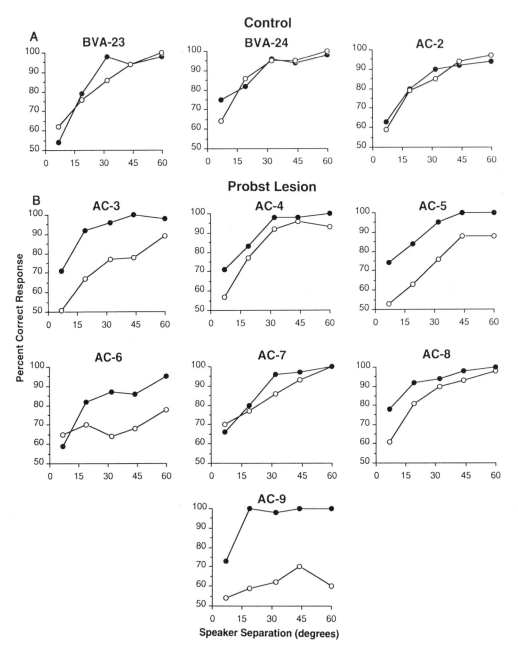

Figure 14. Performance curves for sound localization before and after surgical transection of the CP. Unfilled circles indicate performance before the CP lesion and filled circles indicate performance after the lesion. Control cases were tested before and after a sham surgical procedure. From Ito et al. (1996).

Minimum audible angles for control cases and cases with transection of the CP are shown in Figure 15. Transection of the CP resulted in an average elevation of minimum audible angle of approximately 21 degrees, similar to that produced by kainic acid lesions of the DNLL. These data suggest that the effect of DNLL lesions is due largely to the disconnection of its crossed projections in the CP.

Our behavioral results lead to the conclusion that the DNLL contributes significantly to sound localization and auditory spatial acuity by refining physiological responses to binaural stimulation through the crossed projections of the CP. The GABAergic inhibition from DNLL sharpens binaural responses of neurons in the contralateral IC and enhances the contrast between sounds located in different positions in the horizontal plane. The distinction between sounds located in the left and right spatial fields is strengthened. This sharpening of neural representation may be an important factor in establishing the functional laterality of auditory space at midbrain levels and above (Jenkins and Masterton, 1982; Kavanagh and Kelly, 1987, 1992; Kelly and Kavanagh, 1994).

The question of whether the enhancement of spatial contour is restricted to the boundary between left and right has not been adequately addressed. For the rat, the ability to make auditory spatial discriminations within the lateral fields, i.e., between various positions in the same azimuthal quadrant, is extremely limited. In our earlier studies of the normal rat we were unable to demonstrate significant localization of brief acoustic stimuli presented within the same lateral field (Kavanagh and Kelly, 1986). Therefore, we have tested the effects of DNLL lesions in rats only on discriminations of left vs. right. Further studies should be conducted with species (e.g., cats or ferrets) that are capable of lateral field sound localization.

It has been suggested by several investigators that DNLL might play a role in echo suppression (Carney and Yin, 1989; Fitzpatrick et al., 1995; Yang and Pollak, 1994 a, b; Yin, 1994). Recordings of single unit activity in the bat's DNLL (Yang and Pollak, 1994 a, b) and in the IC of cats (Carney and Yin, 1989; Yin, 1994), rabbits (Fitzpatrick et al., 1995) and rats (Kidd and Kelly, 1996) reveal a long lasting inhibition that affects binaural

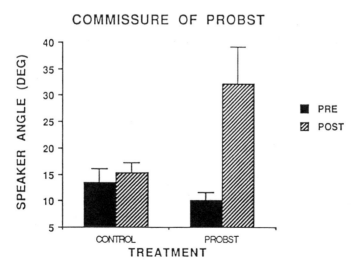

Figure 15. Average minimum audible angles for sound localization before and after CP lesions. Preoperative and postoperative means are shown as solid and shaded columns respectively. Control cases were tested before and after a sham surgical procedure. Error bars indicate the standard error of the mean. From Ito et al. (1996).

responses at time intervals much greater than those associated with the binaural time cues used for simple sound localization. For a single sound source the maximum binaural time difference is determined by the interaural distance and the speed of sound as it travels around the head. For most mammalian species the maximum time difference is substantially less than 1.0 ms. For example, for humans the maximum travel time is approximately 0.6 ms and for rats it is just over 0.1 ms. Therefore binaural interactions that occur at interaural times greater than 1 ms are probably not involved in the localization of single sounds although they may play some role in the analysis of latency shifts associated with interaural intensity differences (Irvine, 1992). A more likely possibility is that the long lasting binaural inhibition of neurons in IC and DNLL contributes to sound localization involving multiple sources.

Our electrophysiological studies have shown that long lasting inhibition in the rat's IC can be attributed at least in part to the contralateral DNLL (Kidd and Kelly, 1996; Kelly and Kidd, 1997). Thus, one would expect that damage to DNLL or transection of the CP would result in a functional impairment in echo suppression. Behavioral tests have shown that rats do indeed suppress echo and that the duration of effective suppression is similar to that found in humans (Kelly, 1974). Also, the period over which the perceptual suppression of echo operates is consistent with the duration of the long lasting inhibition of neurons in the rat's IC. However, appropriate behavioral tests of echo suppression in animals with central nervous system lesions have not yet been conducted.

6. SUMMARY AND CONCLUSIONS

Anatomical, electrophysiological and behavioral studies indicate that the DNLL is an important inhibitory nucleus in the auditory midbrain that contributes to binaural responses in the IC and plays a role in the ability of animals to localize sounds in space. Pharmacological blockade of excitatory synapses in the DNLL reduces the strength of binaural inhibition normally imposed on neurons in IC by acoustic stimulation of the ear ipsilateral to the recording site. Both IID and ITD curves are affected and the sensitivity of neurons to small binaural intensity and time differences is reduced. Blockade of synaptic conduction in DNLL reduces both short-term and long lasting inhibition of IC neurons. The long lasting effect is mediated in part by NMDA receptors in DNLL and is reduced by local injection of the selective NMDA antagonist, APV. The influence of DNLL on binaural responses in the central nucleus of IC is predominantly contralateral at least in the rat. Blockade of DNLL has no apparent effect on binaural responses of EI cells in the ipsilateral IC to either IIDs or ITDs.

Behavioral studies show impaired sound localization and elevated minimum audible angles after either unilateral or bilateral kainic acid lesions of DNLL. Similar deficits in sound localization are produced by surgical transection of the CP. These data lead to the hypothesis that the DNLL contributes to sound localization and auditory spatial acuity by refining binaural responses in the auditory midbrain and enhancing the representation of sound sources in the left and right hemifields.

Disconnection of the DNLL by pharmacological blockade, kainic acid lesions or transection of the CP does not completely eliminate binaural responses or result in a total inability to localize sounds in space. Other structures, no doubt, contribute to the early processing of binaural cues and the neural representation of auditory space. We suggest that the DNLL complements and reinforces the initial binaural analysis provided by the SOC and that the two structures act together to provide accurate auditory spatial responses.

ACKNOWLEDGMENTS

Research was supported by grants from the Natural Sciences and Engineering Research Council of Canada. The author would like to thank Prof. Shu Hui Wu, Brian van Adel, Rosalie Labbelle, and Sean Kidd for their help in preparing this chapter.

ABBREVIATIONS

APV D, L-2-amino-5-phosphonovaleric acid
CNQX 6-cyano-7-nitroquinoxaline-2,3-dione
CF characteristic frequency (sound frequency to which a neuron is most sensitive)
CP commissure of Probst
DNLL dorsal nucleus of the lateral lemniscus
EI excitatory-inhibitory (binaural cell type)
EPSP excitatory postsynaptic potential
GABA gamma-aminobutyric acid
IC inferior colliculus
IID interaural intensity difference
IPSP inhibitory postsynaptic potential
ITD interaural time difference
LSO lateral superior olive
MSO medial superior olive
NMDA N-methyl-D-aspartic acid
SOC superior olivary complex
SPN superior paraolivary nucleus

7. REFERENCES

Adams, J.C. (1979) Ascending projections to the inferior colliculus. J. Comp. Neurol. 183, 519–538.

Adams, J.C. and Mugnaini, E. (1984) Dorsal nucleus of the lateral lemniscus: a nucleus of GABAergic projection neurons. Brain Res. Bull. 13, 585–590.

Aitkin, L.M., Anderson, D.J. and Brugge, J.F. (1970) Tonotopic organization and discharge characteristics of single neurons in nuclei of the lateral lemniscus of the cat. J. Neurophysiol. 33, 421–440.

Bajo, V.M., Merchán, M.A., López, D.E. and Rouiller, E.M. (1993) Neuronal morphology and efferent projections of the dorsal nucleus of the lateral lemniscus in the rat. J. Comp. Neurol. 334, 241–262.

Beyerl, B.D. (1978) Afferent projections to the central nucleus of the inferior colliculus in the rat. Brain Res. 145, 209–223.

Boudreau, J.C. and Tsuchitani, C. (1968) Binaural interaction in the cat superior olive S-segment. J. Neurophysiol. 31, 442–454.

Brugge, J.F., Anderson, D.J. and Aitkin, L.M. (1970) Responses of neurons in the dorsal nucleus of the lateral lemniscus of cat to binaural tonal stimulation. J. Neurophysiol. 33, 441–458.

Brunso-Bechtold, J.K., Thompson, G.C. and Masterton, R.B. (1981) HRP study of the organization of auditory afferents ascending to the central nucleus of the inferior colliculus. J. Comp. Neurol. 197, 705–722.

Buckthought, A.D. (1993) Electrophysiological investigation of the dorsal nucleus of the lateral lemniscus in the auditory system of the albino rat. Master Thesis, Carleton University, Ottawa, Ontario, Canada.

Caird, D. and Klinke, R. (1983) Processing of binaural stimuli by cat superior olivary complex neurons. Exp. Brain Res. 52, 385–399.

Cant, N.B. (1991) Projections to the lateral and medial superior olivary nuclei from the spherical and globular bushy cells of the anteroventral cochlear nucleus. In: R.A. Altschuler, R.P. Bobbin, B.M.Clopton ans D.W.

Hoffman (Eds.) Neurobiology of Hearing: The Central Auditory System, Raven Press, New York, pp. 99–120.

Carney, L.H. and Yin, T.C.T. (1989) Responses of low-frequency cells in the inferior colliculus to interaural time differences of clicks: excitatory and inhibitory components. J. Neurophysiol. 62, 144–161.

Coleman, J.R. and Clerici, W.J. (1987) Sources of projections to subdivisions of the inferior colliculus in the rat. J. Comp. Neurol. 262, 215–226.

Colombo, A., Saldaña, E. and Berrebi, A.S. (1996) Efferent projections of the rat superior paraolivary nucleus. Assoc Res. Otolaryngol. Abstr. 19, 121.

Covey, E. (1993) Response properties of single units in the dorsal nucleus of the lateral lemniscus and paralemniscal zone of an echolocating bat. J. Neurophysiol. 69, 842–859.

Covey, E. and Casseday, J.H. (1995) The lower brainstem auditory pathways. In: A.N. Popper and R.R. Fay (Eds.) Hearing in Bats, Springer Verlag, New York, pp. 235–295.

Faingold, C.L., Boersma-Anderson, C.A. and Randall, M.E. (1993) Stimulation or blockade of the dorsal nucleus of the lateral lemniscus alters binaural and tonic inhibition in contralateral inferior colliculus neurons. Hearing Res. 69, 98–106.

Fitzpatrick, D.C., Kuwada, S., Batra, R. and Trahiotis, C. (1995) Neural responses to simple simulated echoes in the auditory brain stem of the unanesthetized rabbit. J. Neurophysiol. 74, 2469–2486.

Fu, X.W., Brezden, B.L., Kelly, J.B and Wu, S.H. (1997) Synaptic excitation in the dorsal nucleus of the lateral lemniscus: whole-cell patch-clamp recordings from rat brain slice. Neuroscience, 78, 815–827.

Glendenning, K.K. and Baker, B.N. (1988) Neuroanatomical distribution of receptors for three potential inhibitory neurotransmitters in the brainstem auditory nuclei of the cat. J. Comp. Neurol. 275, 288–308.

Glendenning, K.K., Baker, B.N., Hutson, K.A. and Masterton, R.B. (1992) Acoustic chiasm V: Inhibition and excitation in the ipsilateral and contralateral projections of LSO. J. Comp. Neurol. 319, 100–112.

Glendenning, K.K., Brunso-Bechtold, J.K., Thompson, G.C. and Masterton, R.B. (1981) Ascending auditory afferents to the nuclei of the lateral lemniscus. J. Comp. Neurol. 232, 673–703.

Glendenning, K.K., Hutson, K.A., Nudo, R.J. and Masterton, R.B. (1985) Acoustic chiasm. II. Anatomical basis of binaurality in lateral superior olive of cat. J. Comp. Neurol. 232, 261–285.

Glendenning, K.K. and Masterton, R.B. (1983) Acoustic chiasm: efferent projections of the lateral superior olive. J. Neurosci. 3, 1521–1537.

Glenn, S.L. and Kelly, J.B. (1992) Kainic acid lesions of the dorsal nucleus of the lateral lemniscus: effects on binaural evoked responses in rat auditory cortex. J. Neurosci. 12, 3688–3699.

Goldberg, J.M. and Brown, P.B. (1969) Response of binaural neurons of dog superior olivary complex to dichotic tonal stimuli: some physiological mechanisms of sound localization. J. Neurophysiol. 32, 613–636.

Harrison, J.M. and Feldman, M.L. (1970) Anatomical aspects of the cochlear nucleus and superior olivary complex. In: W.D. Neff (Ed.) Contributions to Sensory Physiology, Volume 4, Academic Press, New York, pp 95–142.

Hutson, K.A., Glendenning, K.K. and Masterton, R.B. (1987) Biochemical basis for the acoustic chiasm? Soc. Neurosci. Abstr. 13, 548.

Hutson, K.A., Glendenning, K.K. and Masterton, R.B. (1991) Acoustic chiasm IV: Eight midbrain decussations of the auditory system in the cat. J. Comp. Neurol. 312, 105–131.

Inbody, S.B. and Feng, A.S. (1981) Binaural response of single neurons in the medial superior olivary nucleus of the albino rat. Brain Res. 210, 361–366.

Irvine, D.R.F. (1992) Physiology of the auditory brainstem. In: A.N. Popper and R.R. Fay (Eds.) The Mammalian Auditory Pathway: Neurophysiology, Springer Verlag, New York, 153–231.

Ito, M., Kudo, M. and Kelly, J.B. (1995) Plasticity in the auditory brainstem after unilateral ablation of the inferior colliculus in the neonatal albino rat (Abstract). World Congress of Neuroscience (IBRO), Kyoto, Japan, 4, 197.

Ito, M., van Adel, B. and Kelly, J.B. (1996) Sound localization after transection of the commissure of Probst in the albino rat. J. Neurophysiol. 76, 3493–3502.

Jenkins, W.M. and Masterton, R.B. (1982) Sound localization: effects of unilateral lesions in the central auditory system. J. Neurophysiol. 47, 987–1016.

Kane, E.S. and Barone, L.M. (1980) The dorsal nucleus of the lateral lemniscus in the cat: neuronal types and their distributions. J. Comp. Neuol. 192, 797–826.

Kavanagh, G.L. and Kelly, J.B. (1996) Midline and lateral field sound localization in the albino rat (*Rattus norvegicus*). Behav. Neurosci. 100, 200–205.

Kavanagh, G.L. and Kelly, J.B. (1997) Contribution of auditory cortex to sound localization in the ferret (*Mustela putorius*). J. Neurophysiol. 57, 1746–1766.

Kavanagh, G.L. and Kelly, J.B. (1992) Midline and lateral field sound localization in the ferret (*Mustela putorius*): contribution of the superior olivary complex. J. Neurophysiol. 67, 1643–1658.

Kelly, J.B. (1974) Localization of paired sound sources in the rat: small time differences. J. Acoust. Soc. Amer. 55, 1277–1284.

Kelly, J.B., Glenn, S.L. and Beaver, C.J. (1991) Sound frequency and binaural response properties of single neurons in rat inferior colliculus. Hearing Res. 56, 273–280.

Kelly, J.B. and Kavanagh, G.L. (1994) Sound localization after unilateral lesions of the inferior colliculus in the ferret (Mustela putorius). J. Neurophysiol. 71, 1078–1087.

Kelly, J.B. and Kidd, S.A. (1997) NMDA and non-NMDA excitatory processes in the rat's DNLL shape binaural inhibitory responses in the contralateral inferior colliculus. Assoc. Res. Otolaryngol. Abstr. 20, 176.

Kelly, J.B., Li, L. and van Adel, B. (1996) Sound localization after kainic acid lesions of the dorsal nucleus of the lateral lemniscus in the albino rat. Behav. Neurosci. 110,

Kelly, J.B., Liscum, A., van Adel, B. and Ito, M. (1995) Retrograde labeling in the rat's dorsal nucleus of the lateral lemniscus from frequency specific regions of the central nucleus of the inferior colliculus. Assoc. Res. Otolaryngol. Abstr. 18, 40.

Kelly, J.B. and Masterton, R.B. (1977) Auditory sensitivity of the albino rat. J. Comp. Physiol. Psychol. 91, 930–936.

Kidd, S.A. and Kelly, J.B. (1996) Contribution of the dorsal nucleus of the lateral lemniscus to binaural responses in the inferior colliculus of the rat: interaural time delays. J. Neurosci. 16, 7390–7397.

Kudo, M. (1981) Projections of the nucleus of the lateral lemniscus in the cat: an autoradiographic study. Brain Res. 221, 57–69.

Labelle, R.E. and Kelly, J.B. (1996) A retrograde transport study of the ascending projections to the dorsal nucleus of the lateral lemniscus in the albino rat. Assoc. Res. Otolaryngol. Abstr. 19, 121.

Li, L. and Kelly, J.B. (1992) Inhibitory influence of the dorsal nucleus of the lateral lemniscus on binaural responses in the rat's inferior colliculus. J. Neurosci. 12, 4530–4539.

Markovitz, N.S. and Pollak, G.D. (1993) The dorsal nucleus of the lateral lemniscus in the mustache bat: monaural properties. Hear. Res. 71, 51–63.

Markovitz, N.S. and Pollak, G.D. (1994) Binaural processing in the dorsal nucleus of the lateral lemniscus. Hear. Res. 73, 121–140.

Merchán, M.A.E., Saldaña, E. and Plaza, I. (1994) Dorsal nucleus of the lateral lemniscus in the rat: concentric organization and tonotopic projection to the inferior colliculus. J. Comp. Neurol. 342, 259–354.

Moore, J.K. and Moore, R.Y.(1987) Glutamic acid decarboxylase-like immunoreactivity in brainstem auditory nuclei of the rat. J. Comp. Neurol. 260, 157–174.

Oliver, D.L. and Shneiderman, A. (1989) An EM study of the dorsal nucleus of the lateral lemniscus: inhibitory, commissural, synaptic connections between ascending auditory pathways. J. Neurosci. 9, 967–982.

Roberts, R.C. and Ribak, C.E. (1987) GABAergic neurons and axon terminals in the brainstem auditory nuclei of the gerbil. J. Comp. Neurol. 258, 267–280.

Saint Marie, R.L. and Baker, R.A. (1990) Neurotransmitter-specific uptake and retrograde transport of [^{3}H] glycine from the inferior colliculus by ipsilateral projections of the superior olivary complex and nuclei of the lateral lemniscus. Brain Res. 524, 244–253.

Saint Marie, R.L., Ostapoff, E-M, Morest, D.K. and Wenthold, R.J. (1989) Glycine-immunoreactive projection of the cat lateral superior olive: possible role in midbrain ear dominance. J. Comp. Neurol. 279, 382–396.

Sanes, D.H. (1990) An in vitro analysis of sound localization mechanisms in the gerbil lateral superior olive. J. Neurosci. 10, 3494–3506.

Schwartz, I.R. (1992) The superior olivary complex and lateral lemniscal nuclei. In: D.B. Webster, A.N. Popper and R.R. Fay (Eds.) The Mammalian Auditory Pathway: Neuroanatomy, Springer Verlag, New York, pp. 117–167.

Shneiderman, A., Chase, M.B., Rockwood, J.M., Benson, C.G. and Potashner, S.J. (1993) Evidence for a GABAergic projection from the dorsal nucleus of the lateral lemniscus to the inferior colliculus. J. Neurochem. 60, 72–82.

Shneiderman, A. and Oliver, D.L. (1989) EM autoradiographic study of the projections from the dorsal nucleus of the lateral lemniscus: a possible source of inhibitory inputs to the inferior colliculus. J. Comp. Neurol. 286, 28–47.

Shneiderman, A., Oliver, D.L. and Henkel, C. (1988) Connections of the dorsal nucleus of the lateral lemniscus: an inhibitory parallel pathway in the ascending auditory system? J. Comp. Neurol. 276, 188–208.

Shneiderman, A., Stanforth, D.A. and Saint Marie, R.L. Features of GABA and glycine immunoreactivities in the dorsal nucleus of the lateral lemniscus. Neurosci. Abstr. 21, 403.

Smith, P.H., Joris, P.X. and Yin, T.C.T. (1993) Projections of physiologically characterized spherical bushy cell axons from the cochlear nucleus of the cat: evidence for delay lines to the medial superior olive. J.Comp. Neurol, 331, 245–260.

Stotler, W.A. (1953). An experimental study of the cells and connections of the superior olivary complex of the cat. J. Comp. Neurol. 98, 401–432.

Tanaka, K., Otani, K., Tokunaga, A. and Sugita, S. (1985) The organization of neurons in the nucleus of the lateral lemniscus projecting to the superior and inferior colliculi in the rat. Brain Res. 341, 252–260.

Thompson, G.C., Cortez, A.M. and Lam, D.M.K. (1985) Localization of GABA immunoreactivity in the auditory brainstem of guinea pigs. Brain Res. 339, 119–122.

Tsuchitani, C. and Boudreau, J.C. (1969) Stimulus level of dichotically presented tones and cat superior olive S-segment cell discharge. J. Acoust. Soc. Amer. 46, 978–988.

van Adel, B., Kidd, S.A. and Kelly, J.B. (1997) Transection of the commissure of Probst affects interaural time difference sensitivity in the rat's inferior colliculus. Assoc. Res. Otolaryngol. Abstr. 20, 88.

Vater, M., Casseday, J.H. and Covey, E. (1995) Convergence and divergence of ascending binaural and monaural pathways from the superior olives of the mustached bat. J. Comp. Neurol. 351, 632–646.

Vater, M., Kössl, M. and Horn, A.K.E. (1992) GAD- and GABA-Immunoreactivity in the ascending auditory pathway of horseshoe and mustached bats. J. Comp. Neurol. 325, 183–206.

Warr, W.B. (1982) Parallel ascending pathways from the cochlear nucleus: neuroanatomical evidence of functional specialization. In: W.D. Neff (Ed.) Contributions to Sensory Physiology, Volume 7, Academic Press, New York, pp. 1–38.

Winer, J.A., Larue, D.T. and Pollak, G.D. (1995) GABA and glycine in the central auditory system of the mustache bat: structural substrates for inhibitory neuronal organization. J. Comp. Neurol. 355, 317–353.

Wu, S.H. and Kelly, J.B. (1991) Physiological properties of neurons in the mouse superior olive: membrane characteristics and postsynaptic responses studied in vitro. J. Neurophysiol. 65, 230–246.

Wu, S.H. and Kelly, J.B. (1992) Synaptic pharmacology of the superior olivary complex studied in mouse brain slice. J. Neurosci. 12, 3084–3097.

Wu, S.H. and Kelly, J.B. (1995) In vitro brain slice studies of the rat's dorsal nucleus of the lateral lemniscus. I. Membrane and synaptic response properties. J. Neurophysiol. 73, 780–793. (a)

Wu, S.H. and Kelly, J.B. (1995) In vitro brain slice studies of the rat's dorsal nucleus of the lateral lemniscus. II. Physiological properties of biocytin-labeled neurons. J. Neurophysiol. 73, 794–809. (b)

Wu, S.H. and Kelly, J.B. (1996) In vitro brain slice studies of the rat's dorsal nucleus of the lateral lemniscus. III. Synaptic pharmacology. J. Neurophysiol. 75, 1271–1282.

Yang, L. and Pollak, G.D. (1994) The roles of GABAergic and glycinergic inhibition on binaural processing in the dorsal nucleus of the lateral lemniscus of the mustache bat. J. Neurophysiol. 71, 1999–2013. (a)

Yang, L. and Pollak, G.D. (1994) Binaural inhibition in the dorsal nucleus of the lateral lemniscus of the mustache bat affects responses for multiple sounds. Auditory Neurosci. 1, 1–17. (b)

Yin, T.C.T (1994) Physiological correlates of the precedence effect and summing localization in the inferior colliculus of the cat. J. Neurosci. 14, 5170–5186.

Yin, T.C.T. and Chan, J.C.K. (1990) Interaural time sensitivity in medial superior olive of cat. J. Neurophysiol. 64, 465–488.

Zook, J.M. and Casseday, J.H. (1982) Origin of ascending projections to inferior colliculus in the mustache bat, Pteronotus parnellii. J. Comp. Neurol. 207, 14–28.

32

PROCESSING OF INTERAURAL DELAY IN THE INFERIOR COLLICULUS

Alan R. Palmer, David McAlpine, and Dan Jiang

MRC Institute of Hearing Research
University of Nottingham
University Park, Nottingham, NG7 2RD, United Kingdom

1. INTRODUCTION

The localization of low-frequency (< 1500 Hz) sounds in azimuth is based on the auditory system's sensitivity to small differences in both the arrival time (interaural time difference, or ITD) and the ongoing fine-time structure (interaural phase difference, or IPD) of the sound at each ear (Rayleigh, 1907). The model most commonly used to describe how this is achieved is the coincidence-detection model proposed by Jeffress (1948). In this model, an array of neurones, receiving convergent input from the two ears via axons with different lengths, act as coincidence detectors, discharging when spikes from the two ears arrive at the same time. Electrophysiological recordings from the *medial* superior olive (MSO; Goldberg and Brown, 1969; Yin and Chan, 1990; Spitzer and Semple, 1995) and from the inferior colliculus (IC; e.g. Rose *et al.*1966; Yin and Kuwada, 1983a,b) have confirmed the existence of binaurally-sensitive neurones that behave like coincidence-detectors firing maximally at particular interaural delays of the stimulus.

The coincidence-detection model may also be extended to include delay sensitivity based upon an excitatory input from one ear, and an inhibitory input from the other. In this case, stimulation of the excitatory ear alone is sufficient to cause the neurone to fire maximally, whilst coincident arrival of discharges from the other ear inhibits the neurone in a delay-dependent manner. Neurones that respond in such a manner have been recorded from the low-frequency region of the *lateral* superior olive (LSO) (e.g. Goldberg and Brown, 1969; Finlayson and Caspary, 1991; Batra *et al.*, 1995), as well as from the IC (Rose *et al.*1966; Yin and Kuwada, 1983a,b).

The responses of delay-sensitive units to different frequencies can often be described in terms of a single fixed delay, presumably the result of different axonal lengths to the brainstem coincidence detector. This "characteristic delay" (CD; Rose *et al.*, 1966) can be computed as the rate of change of the most effective or "best" interaural phase difference (BP) with stimulus frequency (the group delay).

Acoustical Signal Processing in the Central Auditory System
edited by Syka, Plenum Press, New York, 1997

Consistent with the coincidence detection model, in neurones that receive excitation from both ears the CD coincides with a peak in the delay function (PEAK-TYPE responses), whilst in neurones that receive excitation from one ear and inhibition from the other the CD coincides with a trough in the delay function (TROUGH-TYPE responses). The point of intersection of the phase plot with the phase axis is termed the characteristic phase (CP) and gives an indication of the type of interaction. The CP is near 0.0 or 1.0 for PEAK-TYPE neurones, and near ± 0.5 for TROUGH-TYPE neurones.

In the MSO and LSO, plots of BP *vs* frequency (phase plots) are generally linear and of the PEAK- and TROUGH-TYPE (Yin and Chan, 1990; Spitzer and Semple, 1995; Batra *et al.*, 1995). However, in the IC, CDs of the majority of delay-sensitive neurones coincide neither with a peak nor a trough in the delay function, but occur on a slope of the delay function (slope-type; Rose *et al.*, 1966; Yin and Kuwada 1983b). Such responses are difficult to reconcile with simple coincidence detection models. Furthermore, Kuwada *et al.* (1987) found that the majority of IC neurones could not be described by a single CD, because the relationship between stimulating frequency and BP was not linear. Thus, delay-sensitivity in the IC appears more complex than in the MSO and LSO.

An obvious explanation for the added complexity of interaural delay processing in the IC is that neurones here receive convergent input from lower brainstem nuclei. The sheer variety of inputs to the IC favours such a suggestion: projections from virtually all brainstem auditory nuclei terminate in the IC (e.g. Brunso-Bechtold *et al.*, 1981). This includes a significant binaural input, chiefly from the Superior Olivary Complex (SOC); bilaterally from the LSO (Adams, 1979; Nordeen *et al.*, 1983; Moore, 1988), and ipsilaterally from the MSO (Beyerl, 1978; Elverland, 1978; Roth *et al.*, 1978; Brunso-Bechtold *et al.*, 1981; Schweitzer, 1981; Nordeen *et al.*, 1983).

In the present paper, we present evidence to suggest that complex, or non-linear, phase plots are the result of convergence from more than one delay-sensitive neurone in the brainstem. By selectively inactivating one of the binaural inputs to an IC neurone, it is possible to reveal the delay characteristics of other inputs to that neurone. The revealed inputs show simpler, linear phase plots, often with PEAK-TYPE or TROUGH-TYPE behaviour.

2. METHODS

Many of the detailed methods have been described previously (Palmer *et al.*, 1990; Caird *et al.*, 1991; McAlpine *et al.*, 1996).

2.1 Preparation

Guinea pigs of 300 - 450 g were premedicated with atropine and anaesthetised with Urethane, with additional analgesia obtained using phenoperidine.

To monitor the condition of the cochlea throughout the experiment, the threshold of the cochlear action potential from 500 - 30000 Hz was measured via a silver wire electrode placed on the round window. Pressure in the sealed bulla of both sides was equalized by a thin tube.

Single-unit recordings were made from the right IC with stereotaxically placed tungsten-in-glass microelectrodes (Bullock *et al.*, 1988).

2.2 Stimuli

Stimuli to each ear were delivered from condenser earphones in closed-field via hollow earbars. A probe-tube microphone was used to calibrated the sound system in dB *re*. 20 μPa a few millimeters from the tympanic membrane. The sound systems for each ear were flat ±5 dB from 100 - 10000 Hz and were matched to within ±2 dB.

When a single unit was isolated, its best frequency (BF) and threshold to binaural tones at zero interaural delay were determined audio-visually. The unit's binaural frequency *vs* level response area was then mapped for frequencies two octaves above, and four octaves below the unit's BF, and over a maximum of 100 dB in 5 dB steps and with different interaural delays.

The interaural-delay sensitivity of IC units to single tones was measured using binaural beats produced when the frequency of the tone at the left ear (contralateral to the recording site) was 1 Hz greater than that at the right (Yin and Kuwada, 1983a). The 1-Hz frequency difference causes the tones to move in and out of phase with each other once every second (the beat frequency) starting with the phase at the contralateral ear leading. Each beat stimulus was 3 seconds in duration, thus presenting three complete cycles of IPD for every repetition of the beat stimulus. The stimulus was presented ten times and the mean best phase (BP) and the vector strength (R) were calculated from the middle two cycles of the beat response (Goldberg and Brown, 1969).

For each neurone, responses to 1-Hz binaural beats were obtained over a range of stimulating frequencies including BF and CD and CP values were obtained from the weighted linear regression of the phase plots (each data point weighted by the vector strength of the response; Kuwada *et al.*, 1987; Spitzer and Semple, 1995). To test for linearity of the phase plots, the data were also fitted with higher order terms and subjected to analysis of variance (ANOVA). Phase plots were considered to be linear when F exceeded the 0.005 level of significance.

Neurones with CPs within ±0.1 cycles of zero were classified as PEAK-TYPE, those with CPs within ±0.1 cycles of ±0.5 were classified as TROUGH-TYPE and other neurones were classified as SLOPE-TYPE or COMPLEX. SLOPE-TYPE were well fitted by a linear regression line, but the CP value was not at 0.0 or ±0.5. COMPLEX responses were not well fitted by a linear regression line.

BP, CD and CP were also measured in the presence of a simultaneous second tone presented at its worst delay (calculated as the best delay + half the period of the second-tone frequency).

3. RESULTS

3.1 Response Types

Responses of 90 delay-sensitive neurones in the guinea pig IC were examined using binaural beats which revealed PEAK, TROUGH, SLOPE and COMPLEX phase plots. Examples of each type of phase plot are shown in Figure 1.

The phase plots shown in Figures 1A-C all met our criterion for linearity at the P < 0.005 level. That is, the simple linear fit was significantly better than that using higher order terms. The CP and CP are shown in each panel in Figure 1. In Figure 1A the phase plot is linear with an intercept at -0.03 cycles and a CD of +854 μs. This unit is therefore a

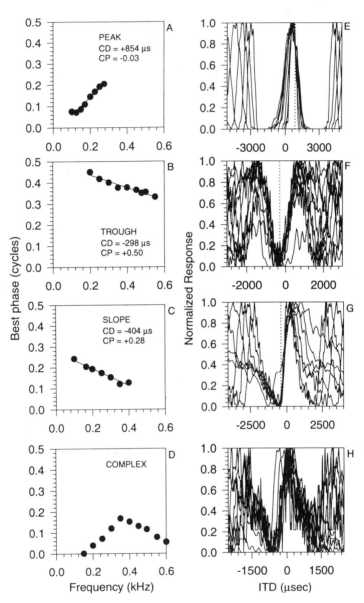

Figure 1. Phase plots of three IC units (A-C) that showed linear behaviour at the 0.005 level of significance and one which was non-linear (D). Weighted regression lines, in which each BP value was weighted by the vector strength of the response at that frequency, were fitted to the data points of phase plots A. phase plot showing positive slope (CD) and CP close to zero, indicating a PEAK-TYPE response BF = 172 Hz. B. phase plot showing negative slope (CD) and CP close to 0.5, indicating a TROUGH-TYPE response BF = 478 Hz. C. phase plot showing a negative CD and a SLOPE-TYPE response BF = 166 Hz. D) non-linear phase plot showing a COMPLEX response BF = 495 Hz. E-H Responses to binaural beats for each of the units in A-D respectively with the curve for each frequency normalized to its maximum. Dotted lines on Figures 1 E-G indicate the delay corresponding to the slope (CD) of the phase plot.

PEAK-TYPE and reflects detection, at the brainstem level of the MSO, of coincident excitatory inputs from the two ears. The value of the CD is consistent with values obtained in previous studies in several species (Stillman, 1971; Yin and Kuwada, 1983b; Kuwada *et al.*, 1987) in that it represents greatest responses when sounds originate in the sound field contralateral to the recording site in the IC. The responses to binaural beats of the unit are normalized and overplotted in Figure 1E. It is clear that all the curves for the different frequencies coincide at a peak which occurs at a position close to that indicated by the dotted line i.e. the CD of +854 μs.

Figures 1B and 1F show the responses of a TROUGH-TYPE unit. The phase plot indicates a CP of 0.5, and the CD of -298 μs indicates that the trough is located at time delays in the sound field ipsilateral to the recordings. This is confirmed in Figure 1F where all of the curves coincide at a trough at the CD shown by the dotted line. Such responses reflect coincidences (probably in the LSO) of an inhibitory input from one ear with an excitatory input from the other.

The responses that we describe as SLOPE-TYPE are illustrated in Figures 1C and 1G. The best fit to the phase plot is obtained by simple linear regression and this provides values of both CD and CP. However, the CP is neither 0.0 nor ±0.5. This fact means that the responses to the binaural beats coincide on the slope of the functions as shown by the dotted line at the CD in Figure 1G. Such responses have been repeatedly reported in the literature (Yin and Kuwada, 1983b; Kuwada *et al.*, 1987) and are not easily explicable in terms of simple coincidence detection at lower brainstem levels.

Finally, in Figure 1D and 1H we show responses for a single unit which did not meet our linearity criterion and which we have termed COMPLEX. Certainly, fitting a single linear regression to the phase plot of Figure 1D is inappropriate, but one possible description of such a plot is that it appears to be constructed two straight-line segments: one from 150 - 350 Hz and one from 350–600 Hz. The adequacy of such an interpretation is provided below. What is quite clear, from the responses to the binaural beats at different frequencies shown in Figure 1H, is that there is no single delay value at which all the curves coincide and hence a single CD value does not describe the behaviour of this type of unit.

Of the 90 units for which we obtained responses to binaural beats at a sufficient number of frequencies 22 (24%) were PEAK- 4 (3.3%) were TROUGH-, 11 (12.2%) were SLOPE- and 54 (60%) were COMPLEX-TYPE responses.

3.2 Complex Responses

Our interpretation of complex phase plots accords with suggestions made by earlier authors who measured many SLOPE-TYPE and COMPLEX-TYPE responses (Yin and Kuwada, 1983b; Kuwada *et al.*, 1987): discrepancies from simple PEAK- or TROUGH-TYPE behaviours reflect convergence of inputs from lower brainstem nuclei onto single IC neurones. In this section, we demonstrate that other response characteristics accord with the irregularities of the phase plots, and support the suggestion that such neurones receive more than one binaural input.

In Figure 2 we present data for an IC unit which had a COMPLEX phase plot that could be well characterized as consisting of two straight-line segments. Figure 2A shows this phase plot and the two linear regression lines we have fitted. Over the frequency range from 150 - 500 Hz the phase plot is linear with a slope of -844 μs and an intercept of 0.56 indicative of a TROUGH-TYPE response. From 500 - 800 Hz the trend of the phase plot reverses and is better fit by a line with a slope of +263 μs and an intercept of

0.01 indicative of a PEAK-TYPE response. In Figures 2B and C we show the binaural beat responses to the low frequencies (B) and the high frequencies (C) separately. The point at which the curves coincide corresponds with the delay value computed from the slope as indicated by the arrows in each case: at a trough in B and at a peak in C. Our interpretation is that the IC unit receives input from brainstem units whose frequency sensitivities are non-overlapping: a PEAK-TYPE (MSO) unit tuned above 500 Hz and a

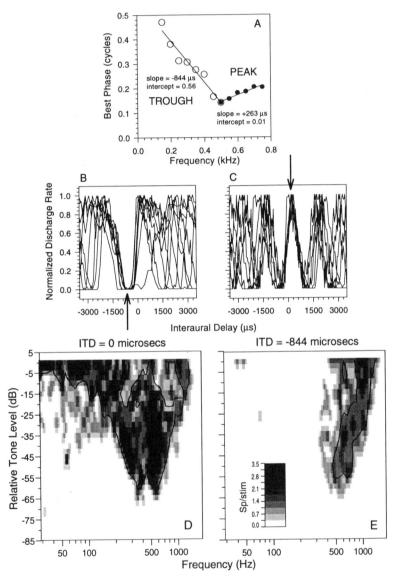

Figure 2. Responses of a single COMPLEX-TYPE IC neurone with BF = 455 Hz. A. phase plot with two non-overlapping frequency regions in which the responses are TROUGH-TYPE at low and PEAK-TYPE at high frequency. B,C. Responses to binaural beats with the curve for each frequency normalized to its maximum. The low and high frequencies are plotted separately in B and C respectively. The arrows indicate the delay computed from the low and high frequency portions of the phase plot. D, E Frequency level response areas plotted as grey scale representations with spike counts indicated by the key. D the response area when the interaural delay of the tones was zero μs and E when the interaural delay of the tones was -844 μs.

TROUGH-TYPE (LSO) unit tuned below 500 Hz. The full frequency/level response areas shown in Figures 2D and E are also in accord with this interpretation. The response area measured at zero interaural delay in Figure 2D shows two regions of high activity separated by a less active region at 500 Hz. The activity at zero delay in both the low- and high-frequency regions is neither at a peak or a trough in the delay function as can be seen by reference to Figures 2B and 2C. However, a radically different response area is obtained if each of the stimulating tones is given an interaural delay of -844 μs, corresponding to the slope of the low-frequency (TROUGH) part of the phase plot. This delay setting virtually abolishes all activity in the low-frequency part of the response area leaving only responses to frequencies above 500 Hz. Reference to Figures 2B and 2C indicates why this should be the case. For the low-frequency region the delay of -844 μs corresponds to the trough and none of these low frequencies will produce any activity, whereas at this delay the high frequencies still generate responses.

In most instances, the separation of the inputs was by no means as clear cut as in Figure 2, as the inputs presumably overlapped in their frequency sensitivities. Nevertheless, even when we inferred that the frequency sensitivities did overlap, there was still a dramatic change in the response area shape depending on the interaural delay used.

The binaural response areas merely confirm what is evident from the phase plots and ITD functions; that different frequency regions exhibit different types of delay sensitivity, and that these influence a neurone's responsiveness to binaural stimuli. In the following section we provide evidence of the nature of these convergent inputs.

3.2 Suppressing One Binaural Input to Reveal the Nature of Another

The delay sensitivity of neurones in the IC appears to be a remarkably linear process and superposition therefore applies. In previous studies, the simple summation of the delay functions obtained with tonal stimulation was shown to provide a good estimate of the delay sensitivity to broadband stimulation (Yin et al., 1986, 1987). We have confirmed this result and have also shown that a Fourier analysis of the response to interaurally delayed noise produces the same non-linear phase plot as that obtained with tonal stimulation. We have used this superposition property to demonstrate the underlying inputs to COMPLEX IC neurones by switching off one input and then measuring the responses to binaural beats of the remaining input(s).

To switch off one input, we first measured the whole phase plot and identified regions in which it was most likely that a single input dominated (i.e. where the phase intercept was close to 0.0 or ±0.5). We then presented a tone at a single frequency in this region at its worst delay, while measuring binaural beat responses to other frequencies. The worst delay was taken as the delay which reduced the response to minimum. For PEAK-TYPE units the worst delay at any frequency occurs at approximately half a period away from the CD, whilst for a TROUGH-TYPE unit the worst delay is equal to the CD. Here we show two examples of this selective suppression of one input: one example where the convergent inputs were both from PEAK-TYPE inputs and the other where the IC neurone received a PEAK and a TROUGH input.

The phase plot of a COMPLEX IC unit is shown in Figure 3A. The phase plot is patently non-linear and appears to consist of at least two regions. In the low-frequency region, below 500 Hz, the points are well fitted by a straight line, and the intercept of this straight line is indicative of a PEAK-TYPE input. The best phases in response to frequencies above 500 Hz are more variable and if fitted with a regression line give a SLOPE-TYPE response. Figure 3B shows the result of remeasuring the best response phases when a fixed tone at 300 Hz and +1886 μs is also present (the best delay at 300 Hz was +219 μs, so the

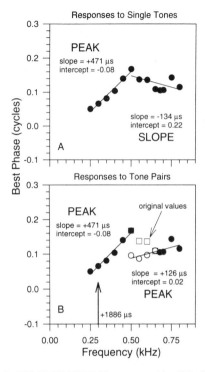

Figure 3. Phase plots of a single COMPLEX-TYPE IC neurone with a BF of 679 Hz. A. shows the phase plot to binaural beats presented alone. The lines are fitted linear regressions with slope and intercept as indicated. B. The filled circles are the original BP values. The unfilled squares show the original BP values for the frequencies re-measured in the presence of a tone at 300 Hz and +1886 μs fixed interaural delay. The open circles show the new BP values measured in the presence of the second tone.

worst delay was 219 + 1667 μs, i.e. best delay + half the period of the 300 Hz). The open circles in Figure 3B show the remeasured best response phases. Binaural beats were not remeasured beyond 650 Hz as no further change in BP was observed. A regression line through the BP values at frequencies above 500 Hz now has a positive slope and an intercept of 0.02 indicative of a PEAK-TYPE response. Our interpretation of these alterations in the units responses to binaural beats is that the unit receives at least two separate inputs from PEAK-TYPE (MSO) inputs. The frequency sensitivities of the two inputs overlap between 500 and 650 Hz and the best response phase in this frequency region results from the interaction of the response phases of the two inputs. When the low-frequency input is switched off, the response phase above 500 Hz represents only the input which remains active, and this demonstrates a PEAK-TYPE response.

Figure 4 shows a second example of selective suppression. In Figure 4A the initial phase plot is certainly non-linear, but in the low-frequency region, below about 350 Hz, it is well fitted by a straight line with a phase intercept of -0.4; just within our criteria for a TROUGH-TYPE response. In Figure 4B we show the effect of remeasuring the BP to binaural beats when a tone at 250 Hz and +1289 μs is also present (Note that because the phase intercept was -0.4, +1289 μs was the worst delay for 250 Hz not the +885 μs ex-pected of a perfectly trough-like input). The BP at 300 Hz was little affected, but the other three remeasured BPs were reduced. A linear regression fitted to the remeasured points now has a phase intercept of 0.04 indicative of a PEAK-TYPE response. Note, however, that the PEAK-TYPE response has a negative slope of -495 μs, which might imply input

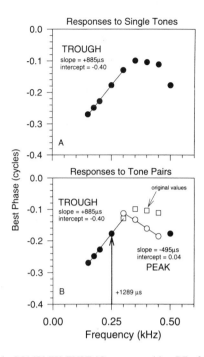

Figure 4. Phase plots of a single COMPLEX-TYPE IC neurone with a BF of 392 Hz. A. shows the phase plot to binaural beats presented alone. The line is a fitted linear regression with slope and intercept as indicated. B. The filled circles are the original BP values. The unfilled squares show the original BP values for the frequencies remeasured in the presence of a tone at 250 Hz and +1289 μs fixed interaural delay. The open circles show the new BP values measured in the presence of the second tone.

from the opposite MSO. The data in Figure 4 therefore indicate that the phase plot of Figure 4A is the result of interaction of a TROUGH-TYPE input which dominates at low frequency and a PEAK-TYPE input with which it overlaps at frequencies above 300 Hz.

We also examined the effect of adding a suppressive second tone on the responses of IC neurones with simple, linear phase plots including some with SLOPE-TYPE responses. Not surprisingly, we were unable to alter the BP of the responses of linear phase plots using a second tone at worst delay. The suppressive tone dramatically reduced or abolished the discharge rate to binaural beats across all frequencies examined. This suggests that IC neurones with linear phase plots receive input from either a single brainstem coincidence detector or, in the case of SLOPE-TYPE neurones, from multiple brainstem coincidence detectors with very similar response properties.

4. DISCUSSION

From the earliest, previous investigations of responses to interaural time delays at the midbrain level have identified activity inconsistent with simple coincidence detection in the superior olivary complex (Rose *et al.*, 1966; Yin and Kuwada, 1983b; Kuwada *et al.*, 1987; Batra *et al.*, 1993). This has led to the suggestion that the more complicated interaural delay sensitivity embodied in SLOPE- and COMPLEX-TYPE phase plots is the result of convergence from lower brainstem neurones onto single IC neurones. Some of the COMPLEX-TYPE phase plots certainly seem more appropriately described by more than one straight line segment and this analysis strongly suggests inputs from neurones with simpler interaural delay sensitivities, in cases where the presumed inputs from the lower level have largely non-overlapping response areas (Figure 2). The shifts in the responsiveness observed on application of a fixed delay to all tones, when measuring the frequency *vs* level response area, also provides support for convergence. The response areas in Figure 2 are striking examples, but we have produced quite radical changes in the frequency *vs* level response areas even in cases in which it appeared that the frequency sensitivities of the putative inputs were partially or completely overlapping: this will inevitably ensue if the delay sensitivities are different.

However, even if IC neurones receive inputs with partially- or totally overlapping frequency sensitivities, their phase plots may still show linear PEAK- or TROUGH-TYPE behaviour, if their delay characteristics are identical. Our methods will not resolve such convergent input, but its functional significance would be the same as a single delay sensitive input.

A novel aspect of the current study is the selective suppression of one of the putative inputs to reveal the characteristics of another input. Often this process allowed an interpretation of COMPLEX-TYPE phase plots in terms of a combination of simpler delay sensitive inputs. There are other methods which have been previously employed to achieve the same aims, although generally applied to unravelling the sensitivity to interaural level differences at high frequencies. Such studies have employed pharmacological techniques to inactivate inputs either locally or by lesioning the site of origin of the inputs (e.g. Faingold *et al.*, 1991; Kelly and Sally, 1993; Sally and Kelly, 1992; Yang *et al.* 1992; Park and Pollak, 1993). The present, simple method of selectively switching off inputs has the disadvantage that it is applicable only in specific circumstances. However, compared to pharmacological and lesion studies it also has a number of advantages. It is technically less demanding than pharmacological methods and is completely and rapidly reversible. It is also specific and highly selective at the site of binaural interaction in the SOC for those neurones projecting to the IC.

IC neurones that receive inputs with identical frequency response areas, but different delay characteristics might account for the SLOPE-TYPE neurones that are observed in the IC. The addition of a second tone to linear SLOPE-TYPE neurones produced a reduction in response amplitude without alteration of the BP. Presumably, the second tone had an equal effect on the convergent inputs preserving their balance.

Finally, the IC is also the target of binaural inputs from other binaural nuclei, such as the dorsal nucleus of the lateral lemniscus (e.g. Oliver, 1984; Shneiderman *et al.*, 1988), which are also sites of convergent binaural input from the brainstem. Such convergence at high frequencies has been demonstrated using pharmacological methods, but as yet we have no indication of whether this type of second order convergence also contributes to interaural delay sensitivity in the IC at low frequencies.

Recent evidence suggests that IC neurones show more complex behaviour than is observed in the MSO. Both Yin and Chan (1990) and Spitzer and Semple (1995) reported that MSO neurones were almost exclusively PEAK-TYPE, with no TROUGH-TYPE responses being recorded in either study. If trough-type neurones are confined to the LSO, as Batra *et al.* (1995) have suggested, then our data provides evidence that single IC neurones receive delay-sensitive input from both MSO and LSO.

We can only speculate as to the functional significance of the increased complexity of delay sensitivity at the IC level that results from convergence of the simpler brainstem inputs. One possibility is suggested by the relatively dramatic changes in the frequency vs level response areas as the interaural delay (a surrogate for azimuthal position) is altered. The alteration in the representation of single or multiple frequency components across the tonotopic axis will depend upon the azimuthal position of the sound source. However, these effects, in real physiological situations, are unlikely to be as marked as those we have shown, since the delay range encountered in the natural world only encompasses the best and worst delays of binaural neurones in the animals with the largest head sizes (see Phillips and Brugge, 1985; Palmer *et al.*, 1990).

A second possible role for IC neurones that receive convergent binaural input is that they are involved in the processing of auditory-motion stimuli. Studies by Spitzer and Semple (1992, 1993) suggest that many low-frequency neurones in the IC are sensitive to the depth and direction of the dynamic IPD cues associated with sound source motion. Neurones in the MSO did not appear to be sensitive to these parameters (Spitzer and Semple, 1992), suggesting that there is a hierarchical processing of these binaural auditory cues. Our present data (McAlpine *et al.*, this volume) provide some support for the suggestion that motion sensitivity at the IC may be a consequence of convergent input from lower levels.

ACKNOWLEDGMENTS

We thank Trevor Shackleton for helpful comments on this manuscript.

REFERENCES

Adams, J. C. (1979) Ascending projections to the inferior colliculus. J. Comp. Neurol. 183, 519–538.

Batra, R., Fitzpatrick, D.C. and Kuwada, S. (1995) Relationship of synchrony to ipsilateral and contralateral tones to ITD-sensitivity in the superior olivary complex. *Eighteenth Meeting of the A.R.O.* 18, 62.

Batra, R., Kuwada, S. and Stanford, T. R. (1993) High-frequency neurons in the inferior colliculus that are sensitive to interaural delays of amplitude-modulated tones: Evidence for dual influences. J. Neurophysiol. 70, 64–80.

Beyerl, B. D. (1978) Afferent projections to the central nucleus of the inferior colliculus in the rat. Brain. Res. 145, 209–223.

Brunso-Bechtold, J. K., Thompson, G. C. and Masterton, R. B. (1981) HRP study of the organization of auditory afferents ascending to central nucleus of inferior colliculus in cat. J. Comp. Neurol. 197, 705–722.

Bullock, D., Palmer, A. R. and Rees, A. (1988) A compact and easy to use tungsten-in-glass microelectrode manufacturing workstation. Med. and Biol. Eng. and Computing. 26, 669–672.

Caird, D. M., Palmer, A. R. and Rees, A. (1991) Binaural masking level difference effects in single units of the guinea pig inferior colliculus. Hear. Res. 57, 91–106.

Elverland, H. H. (1978) Ascending and intrinsic projections of the superior olivary complex in the cat. Exp. Brain Res. 32, 117–134.

Faingold, C. L., Anderson, C. A. B. and Caspary, D. M. (1991) Involvement of GABA in acoustically-evoked inhibition in inferior colliculus neurons. Hear. Res. 52, 201–216.

Finlayson, P. G. and Caspary, D. M. (1991) Low-frequency neurons in the lateral superior olive exhibit phase-sensitive binaural inhibition. J. Neurophysiol. 65, 598–605.

Goldberg, J. M. and Brown, P. B. (1969) Response of binaural neurons of dog superior olivary complex to dichotic tonal stimuli: Some physiological mechanisms of sound localization. J. Neurophysiol. 32, 613–636.

Jeffress, L. A. (1948) A place theory of sound localization. J. Comp. Physiol. Psychol. 61, 468–486.

Kelly, J. B. and Sally, S. L. (1993) Effects of superior olivary complex lesions on binaural responses in rat auditory cortex. Brain. Res. 605, 237–250.

Kuwada, S., Stanford, T. R. and Batra, R. (1987) Interaural phase-sensitive units in the inferior colliculus of the unanesthetized rabbit: Effects of changing frequency. J. Neurophysiol. 57, 1338–1360.

McAlpine, D., Jiang, D. and Palmer, A. R. (1996) Interaural delay sensitivity and the classification of low best-frequency binaural responses in the inferior colliculus of the guinea pig. Hear. Res. 97, 136–152.

Moore, D. R. (1988) Auditory brainstem of the ferret: Sources of projections to the inferior colliculus. J. Comp. Neurol. 269, 342–354.

Nordeen, K. W., Killackey, H. P. and Kitzes, L. M. (1983) Ascending auditory projections to the inferior colliculus in the adult gerbil *Meriones unguiculatus*. J. Comp. Neurol. 214, 131–143.

Oliver, D.L. (1984) Dorsal cochlear nucleus projections to the inferior colliculus in the cat: a light and electron microscopic study. J. Comp. Neurol. 224, 155–172.

Palmer, A. R., Rees, A. and Caird, D. (1990) Interaural delay sensitivity to tone and broad band signals in the guinea-pig inferior colliculus. Hear. Res. 50, 71–86.

Park, T. J. and Pollak, G. D. (1993) GABA shapes sensitivity to interaural intensity disparities in the mustache bat's inferior colliculus: Implications for encoding sound location. J. Neurosci. 13, 2050–2067

Phillips, D.P. and Brugge, J.F. (1985) Progress in neurophysiology of sound localization. Annu. Rev. Psychol. 36, 245–274.

Rayleigh, L. (1907) On our perception of sound direction. Philos. Mag. 13, 214–232.

Rose, J. E., Gross, N. B., Geisler, C. D. and Hind, J. E. (1966) Some neural mechanisms in the inferior colliculus of the cat which may be relevant to localization of a sound source. J. Neurophysiol. 29, 288–314.

Roth, G. L., Aitkin, L. M., Andersen, R. A. and Merzenich, M. M. (1978) Some features of the spatial organization of the central nucleus of the inferior colliculus of the cat. J. Comp. Neurol. 182, 661–680.

Sally, S. L. and Kelly, J. B. (1992) Effects of superior olivary complex lesions on binaural responses in rat inferior colliculus. Brain. Res. 572, 5–18.

Schweitzer, H. (1981) The connections of the inferior colliculus and the organization of the brainstem auditory system in the greater horseshoe bat (*Rhinolophus ferrenequinum*). J. Comp. Neurol. 201, 25–49.

Shneiderman, A., Oliver, D.L. and Henkel, C.K. (1988) The connections of the dorsal nucleus of the lateral lemniscus. An inhibitory parallel pathway in the ascending auditory system. J. Comp. Neurol. 27, 188–208.

Stillman, R.D. (1971) Characteristic delay neurons in the inferior colliculus of the kangaroo rat. Exp. Neurol. 32, 404–412.

Spitzer, M. W. and Semple, M. N. (1992) Responses to time-varying Interaural phase disparity in gerbil superior olive: Evidence for hierarchical processing. Society for Neuroscience Abstracts. 18, 149.

Spitzer, M. W. and Semple, M. N. (1993) Responses of inferior colliculus neurones to time-varying interaural phase disparity: Effects of shifting the locus of virtual motion. J. Neurophysiol. 69, 1245–1263.

Spitzer, M. W. and Semple, M. N. (1995) Neurons sensitive to interaural phase disparity in gerbil superior olive: Diverse monaural and temporal response properties. J. Neurophysiol. 73, 1668–1690.

Yang, L., Pollak, G. D. and Resler, C. (1992) GABAergic circuits sharpen tuning curves and modify response properties in the mustache bat inferior colliculus. J. Neurophysiol. 68, 1760–1774.

Yin, T. C. T. and Chan, J. C. K. (1990) Interaural time sensitivity in medial superior olive of cat. J. Neurophysiol. 64, 465–488.

Yin, T. C. T., Chan, J. C. K. and Carney, L. H. (1987) Effects of interaural time delays of noise stimuli on low-frequency cells in the cat's inferior colliculus. III. Evidence for cross-correlation. J. Neurophysiol. 58, 562–583.

Yin, T. C. T., Chan, J. C. K. and Irvine, D. R. F. (1986) Effects of interaural time delays of noise stimuli on low-frequency cells in the cat's inferior colliculus. I. Responses to wideband noise. J. Neurophysiol. 55, 280–300.

Yin, T. C. T. and Kuwada, S. (1983a) Binaural interaction in low-frequency neurons in inferior colliculus of the cat. II. Effects of changing rate and direction of interaural phase. J. Neurophysiol. 50, 1000–1019.

Yin, T. C. T. and Kuwada, S. (1983b) Binaural interaction in low-frequency neurons in inferior colliculus of the cat. III. Effects of changing frequency. J. Neurophysiol. 50, 1020–1042

33

CONSTRUCTION OF AN AUDITORY SPACE MAP IN THE MIDBRAIN

Andrew J. King,[*] Jan W. H. Schnupp, and Ze D. Jiang

University Laboratory of Physiology
Parks Road, Oxford OX1 3PT, United Kingdom

1. INTRODUCTION

The deeper layers of the mammalian superior colliculus (SC) contain a two-dimensional map of auditory space, which is in register with maps of visual space and of the body surface (King and Carlile, 1995; Stein and Meredith, 1993). This arrangement facilitates the integration of multisensory signals from particular positions in space and the guidance of orientation movements toward both unimodal and multimodal targets. In contrast to the visual and somatosensory maps, which arise from the topographic arrangement of axons leaving their respective receptor surfaces, the map of auditory space has to be computed centrally by neurons that are differentially sensitive to the acoustic localization cues that result from the interaction of sound with the ears and head (King and Carlile, 1995).

The barn owl SC (or optic tectum) receives a topographic projection from a nucleus thought to correspond to the mammalian external nucleus of the inferior colliculus (ICX, Knudsen and Knudsen, 1983) and, in these birds, both structures contain similar maps of auditory space (Knudsen and Konishi, 1978; Knudsen, 1982). In contrast, previous anatomical studies have reported that the mammalian SC is innervated by a number of auditory brainstem nuclei and cortical areas (Oliver and Huerta, 1992). The relative importance of these inputs in determining the auditory spatial tuning of SC neurons is currently unknown. In this study, we have used retrograde tracing and electrophysiological recording techniques to examine the processing steps involved in the emergence of the auditory space map in the SC of the ferret.

* Correspondence to: Dr. A.J. King, University Laboratory of Physiology, Parks Road, Oxford OX1 3PT, UK. Tel: +44-1865-272523; fax: +44-1865-272469; E-mail: ajk@physiol.ox.ac.uk.

Acoustical Signal Processing in the Central Auditory System
edited by Syka, Plenum Press, New York, 1997

2. METHODS

2.1. Anatomical Tract Tracing

Full details are given in Jiang et al. (1996). Five ferrets were anaesthetized with Saffan (2 ml/kg) and the left SC exposed by removal of the overlying cranium and cortex. Rhodamine (red) or green fluorescent latex microspheres were injected into each quadrant of the SC. In 3 animals, one of the tracers was injected into both rostral quadrants, whereas the other tracer was injected into the two caudal quadrants. The other 2 animals received a mediolateral pattern of microsphere injections (Fig. 1). After a survival time of 1–2 days, each ferret was deeply anaesthetized and perfused transcardially with phosphate buffered saline and 4% paraformaldehyde. The brains were sectioned frontally and every third 50 μm section was used to identify retrogradely-labelled neurons using a microscope with an epifluorescence attachment. Neurons containing cytoplasmic granules fluorescing either red or green were distinguished using Zeiss rhodamine and fluorescein filter sets. A second set of sections was Nissl stained to identify the auditory brainstem nuclei.

2.2. Electrophysiological Recording

Full details are given elsewhere (King and Hutchings, 1987; Schnupp et al., 1995). Single-unit recordings were made in the SC and the nucleus of the brachium of the inferior colliculus (BIN) in 5 ferrets. The animals were anaesthetized with Saffan, the trachea was cannulated and a craniotomy performed. They were then paralysed with Flaxedil. Anaesthesia and paralysis were maintained with sodium pentobarbital (1 mg/kg per h) and Flaxedil (20 mg/kg per h) respectively. The animals were artificially ventilated, body temperature was maintained at 39°C, and the electrocardiogram, electroencephalogram and end-tidal CO_2 were monitored.

All recordings were carried out in an anechoic chamber. A tungsten electrode was advanced through the overlying cortex into the superficial layers of the SC. Visual stimuli generated by a flashing LED were used to locate the lateral border of this nucleus. Subsequent electrode penetrations were made at least 500 μm further lateral and at an angle of roughly 10° outward from vertical to allow the electrode to run parallel to the main dorsoventral axis of the BIN. To map auditory spatial receptive fields, 100 ms broadband noise bursts were delivered from a loudspeaker, which was moved around the animal's head using a computer-controlled hoop. The neural signals were filtered, amplified and digitized and single units were discriminated on the basis of spike shape (see Schnupp et al., 1995). Electrolytic lesions were made in each electrode track to allow subsequent histological reconstruction of the recording sites.

3. RESULTS

3.1. Distribution of Retrogradely-Labelled Neurons

Following injections of fluorescent microspheres into the SC, retrogradely-labelled neurons were found in several auditory brainstem nuclei. The great majority were found in some part of the IC, particularly the BIN and the ICX. Some labelling was also found in the nucleus sagulum, the paralemniscal region ventral to the IC, and, to a very limited extent, in the medial nucleus of the trapezoid body and in certain of the periolivary nuclei. In all of these regions, the labelled neurons were found primarily on the side ipsilateral to the injected SC.

3.1.1. Topographic Projection from the Nucleus of the Brachium of the Inferior Colliculus. Injections of red and green microspheres into the rostral and caudal halves of the SC resulted in a clear segregation of labelling along the rostrocaudal axis of the ipsi-

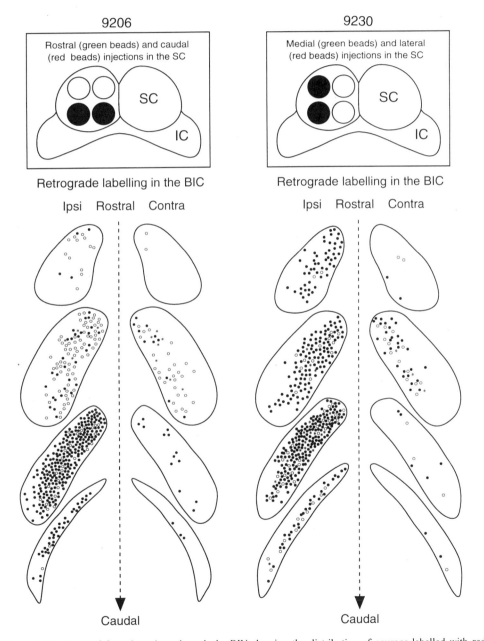

Figure 1. Drawings of frontal sections through the BIN showing the distribution of neurons labelled with red (filled circle) or green (open circle) or both (asterisk) fluorescent microspheres after rostrocaudal (left) or medio-lateral (right) injections in the SC. The top diagrams are dorsal views of the midbrain illustrating the positions of the injections of red (filled circles) and green (open circles) microspheres in the left SC.

lateral BIN (Fig. 1). Most of the neurons labelled by the tracer injected into rostral SC (green beads in 9206) were found in the rostral half of the BIN. More labelled neurons were found in the caudal half of the BIN and almost all of these contained the microspheres (red in 9206) that had been injected into caudal SC. Less than 5% of the labelled neurons in the ipsilateral BIN were double labelled and most of these were concentrated near the central part of the nucleus where the proportion of red and green single-labelled neurons reversed. Considerably less labelling was present on the contralateral side, which was found predominantly in a region just rostral to the central part of the BIN. Most of these cells contained the tracer injected into the rostral half of the SC.

We also examined the distribution of labelling following mediolateral injections of microspheres into the SC (Fig. 1). Retrograde labelling in the ipsilateral BIN was dominated by the tracer injected in the lateral half of the SC (red beads in ferret 9230), but, in contrast to the rostrocaudal pattern of injections, single-labelled neurons containing the two tracers were not obviously segregated in the BIN.

3.2. Auditory Responses of Units in the Nucleus of the Brachium of the Inferior Colliculus

The majority of auditory units (n = 112) were found in the accessory region, while the remaining 6 units were adjudged to be located in the interstitial region of the BIN (see Morest and Oliver, 1984). BIN units had very similar onset latencies and temporal firing patterns to those previously described in the SC (King and Hutchings, 1987).

3.2.1. Representation of Auditory Space. Spatial response profiles were determined at two sound levels, typically 5–10 dB and 25–30 dB above unit threshold. The histograms in Fig. 2 show the distribution of 50% bandwidths of the azimuth profiles for the units recorded in the BIN, together with those measured in a previous study of auditory units in the SC. In both nuclei, the spatial response profiles tended to broaden with increasing sound level. At corresponding sound levels, the SC units were more sharply tuned than those recorded in the BIN (Wilcoxon rank-sum tests, $P < 0.01$).

The majority of BIN units were either tuned to single positions or showed a broad preference for sounds in the contralateral hemifield. In order to examine whether the preferred sound directions vary systematically with recording site, we recorded from most of the rostrocaudal and dorsoventral extent of the region of the BIN that projects topographically to the SC. Linear regression analyses revealed that an axis of +18° relative to the horizontal gave the best correlation between azimuthal best position and recording site. The distribution of best positions along this "azimuth axis" is shown for all the tuned BIN units recorded in the 5 ferrets at supra-threshold and near threshold sound levels in Figs. 3C and D respectively. Figures 3A and B illustrate the equivalent representation of sound azimuth in the SC.

The auditory best positions of SC units change very little with sound level and the topographic representations of sound azimuth at near and supra-threshold sound levels are statistically indistinguishable (Figs. 3A, B). In the BIN, however, a significant correlation between best azimuth and recording site was apparent only for supra-threshold stimuli (Fig. 3C). With near threshold stimulation, most best positions were instead clustered in the anterior quadrant of the contralateral hemifield and particularly just in front of the interaural axis (Fig. 3D).

Many of the auditory units recorded in the BIN were also tuned in elevation. However, in contrast to the SC, where best elevations vary systematically from the medial to the lateral side of the nucleus, we found no indication of any elevation topography in the BIN at either near or supra-threshold sound levels.

3.2.2. Azimuth Coding by Synchronous Activity. Supra-threshold azimuth response profiles were also constructed from "synchronized" spike trains of pairs of BIN units that were recorded simultaneously by the spike-sorting software. Cross-correlograms for these unit pairs revealed that 41% showed clear evidence for coupling of neural activity, featuring a broad, significant peak near zero, which is indicative of a common excitatory input to both neurons of the pair (Fig. 4A). The remaining correlograms suggested no interactions other than the synchronizing effect of a common stimulus (Fig. 4B).

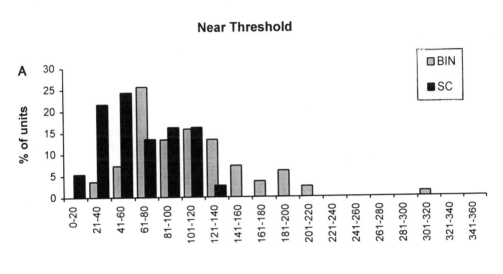

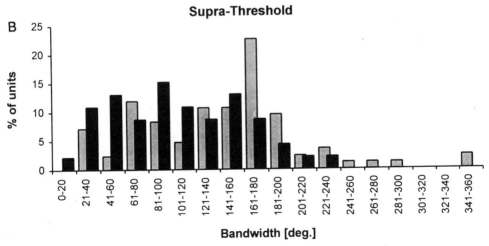

Figure 2. Distribution of 50% bandwidths for the azimuth response profiles of units recorded in the SC (black bars) and BIN (grey bars). (A) Data obtained at 5–10 dB above unit threshold. (B) Data obtained at 25–30 dB above unit threshold. In both nuclei, increasing the stimulus level leads to a broadening of the response profiles, but auditory units in the SC are more sharply tuned than those in the BIN.

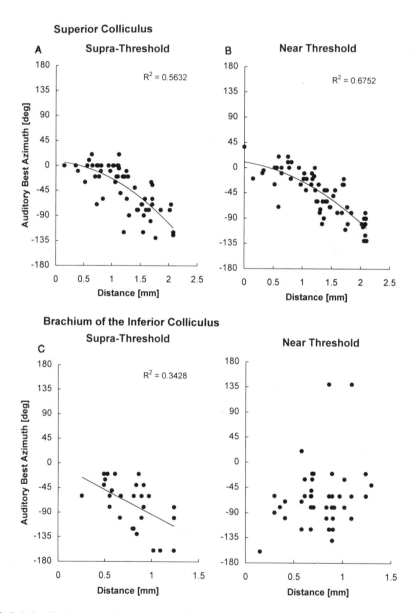

Figure 3. Relationship between auditory best azimuth and unit location within the SC (A, B) and BIN (C, D). The supra-threshold data (A, C) were obtained at 25–30 dB and the near threshold data (B, D) at 5–10 dB above unit threshold. The topographic representation of sound azimuth in the SC is best described by a second-order polynomial and does not change with variations in sound level. The correlation between best azimuth and recording site in BIN is, by contrast, significant only at supra-threshold sound levels.

4. DISCUSSION

Our anatomical data show that several auditory nuclei project to the ferret SC, suggesting that converging inputs from various structures may contribute to the elaboration of its two-dimensional map of auditory space. We found that the largest auditory projection originates in the ipsilateral BIN and is topographically ordered along the rostocaudal axis, implying that the representation of sound azimuth along this axis may be constructed at an earlier stage in the brainstem and then relayed to the SC. We confirmed this by showing that the best azimuths of auditory units recorded in the BIN are also topographically ordered, with anterior directions represented predominantly in rostral BIN and more peripheral sound directions represented in caudal regions of the nucleus. On the other hand, no segregation of inputs was apparent from the BIN to the medial and lateral halves of the SC, which represent superior and inferior sound directions respectively. This is again consistent with our electrophysiological data, which provide no evidence for a systematic variation in elevation tuning within the BIN.

The near threshold tuning of most auditory units in the SC can be attributed to monaural spectral cues, whereas sensitivity to both interaural level differences and spectral cues appears to be responsible for the supra-threshold map of sound azimuth (King and Carlile, 1995). The cues underlying the auditory representation in the BIN are pres-

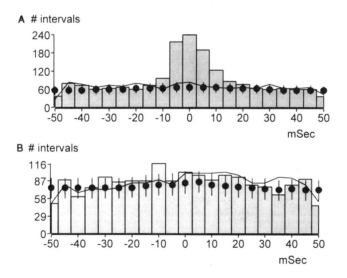

Figure 4. Examples of cross-correlation histograms from pairs of units in the BIN. (A) Correlogram indicative of a "common input" type interaction. (B) Correlogram suggesting no interaction. The correlograms for these unit pairs were calculated from the pooled data for all stimulus positions in the contralateral hemifield. The continuous lines indicate the levels of the "shift predictor" and the filled circles the levels of the "PST-predictor" together with 95% confidence interval error bars for each bin. Shift and PST-predictors represent the level of interaction expected from common stimulation. Shift predictors were computed by correlating spike events from responses that were not simultaneously recorded. PST-predictors correspond to the cross-correlation functions of the PSTHs of the two neurons in the pair. Using a 5 msec integration interval, we found that the bicellular response profiles (see Ghose et al. 1994) were significantly narrower than those of the parent units, suggesting that synchronous activity in the BIN can encode sound direction more precisely than the activity of individual neurons. However, a comparable improvement in tuning was also observed with the response profiles derived from unit pairs whose firing was not significantly correlated and also with "shuffled" response profiles, where any coding based on the precise synchronization of activity in the two neurons is disrupted.

ently unknown, but the lack of both elevation topography and near-threshold azimuth topography, together with a greater incidence of units with spatially ambiguous response profiles, raises the possibility that a full complement of directional pinna cues may not be encoded at this level of the auditory pathway. Presumably afferents from the ICX or other areas of the brainstem or cortex contribute additional information used in the construction of the two-dimensional representation in the SC.

Although the BIN contains a map of sound azimuth at sound levels that provide binaural cues at all stimulus positions and frequencies, this representation is less precise than that found in the SC. Restricting our analysis to spikes occurring within a short time interval in pairs of BIN units gave rise to sharper spatial response profiles with bandwidths much closer to those of SC units. Although this improvement in azimuthal tuning does not appear to require synchronization in the unit pairs other than time locking to a common stimulus, it does suggest that convergence of inputs from several BIN neurons may give rise to the improved spatial selectivity of neurons in the SC. Given the importance of sensory experience in the development of the auditory space map in the SC (King and Carlile, 1995; Schnupp et al. 1995), it is likely that activity-dependent mechanisms are responsible for selecting the most appropriate combination of inputs during early life.

ACKNOWLEDGMENTS

This work was supported by the Wellcome Trust.

REFERENCES

Ghose, G.M., Ohzawa, I. and Freeman, R.D. (1994) Receptive-field maps of correlated discharge between pairs of neurons in the cat's visual cortex. J. Neurophysiol. 71, 330–346.

Jiang, Z.D., King, A.J. and Moore, D.R. (1996) Topographic organization of projection from the parabigeminal nucleus to the superior colliculus in the ferret revealed with fluorescent latex microspheres. Brain Res. 743, 217–232.

King, A.J. and Carlile, S. (1995) Neural coding for auditory space. In: M.S. Gazzaniga, (Ed.) The Cognitive Neurosciences, MIT Press, Cambridge, pp. 279–293.

King, A.J. and Hutchings, M.E. (1987) Spatial response properties of acoustically responsive neurons in the superior colliculus of the ferret: a map of auditory space. J. Neurophysiol. 57, 596–624.

Knudsen, E.I. (1982) Auditory and visual maps of space in the optic tectum of the owl. J. Neurosci. 2, 1177–1194.

Knudsen, E.I. and Knudsen, P.F. (1983) Space-mapped auditory projections from the inferior colliculus to the optic tectum in the barn owl (*Tyto alba*). J. Comp. Neurol. 218, 187–96.

Knudsen, E.I. and Konishi, M. (1978) A neural map of auditory space in the owl. Science 200, 795–797.

Morest, D.K. and Oliver, D.L. (1984) The neuronal architecture of the inferior colliculus in the cat: Defining the functional anatomy of the auditory midbrain. J. Comp. Neurol. 222, 209–236.

Oliver, D.L. and Huerta, M.F. (1992) Inferior and superior colliculi. In: D.B. Webster, A.N. Popper and R.R. Fay (Eds.) The Mammalian Auditory Pathway: Neuroanatomy. Springer-Verlag, New York, pp. 168–221.

Schnupp, J.W.H., King, A.J., Smith, A.L. and Thompson, I.D. (1995) NMDA-receptor antagonists disrupt the formation of the auditory space map in the mammalian superior colliculus. J. Neurosci. 15, 1516–31.

Stein, B.E. and Meredith, M.A. (1993) The Merging of the Senses. MIT Press, Cambridge.

34

SPATIAL RECEPTIVE FIELDS OF SINGLE NEURONS OF PRIMARY AUDITORY CORTEX OF THE CAT

John F. Brugge, Richard A. Reale, and Joseph E. Hind

Department of Neurophysiology and
Waisman Center on Mental Retardation and Human Development
University of Wisconsin
Madison, Wisconsin 53705

1. INTRODUCTION

Neurons in the primary auditory cortical field (AI) have been shown to be sensitive to the direction of a sound when the source is either in an anechoic free field (Middlebrooks et al, 1980; Rajan et al, 1990; Imig et al, 1990) or in anechoic virtual acoustic space (Brugge et al, 1994; 1996a,b). The spatial receptive fields obtained under these stimulus conditions are typically large in size at suprathreshold levels, often exceeding an acoustic hemifield; close to threshold their centers tend to lie on or near the acoustic axis. How large receptive fields centered around the acoustic axis enable AI neurons to encode information about sound direction is not well understood, although it would appear that the time structure of the neuronal discharge within the receptive field plays a role (Middlebrooks et al, 1994; Brugge et al, 1996). In this paper we review and extend our findings on directional sensitivity of isolated AI neurons to transient sound, employing conventional extracellular recording methods (Brugge et al, 1994,1996a) and a technique by which synthesized signals that mimic sounds coming from particular directions in space are delivered at the eardrums of Nembutal-anesthetized cats through a sealed and calibrated sound delivery system (Chan et al, 1993; Reale et al, 1996).

2. METHODS

2.1. Virtual Acoustic Space

The combination of reflections and refraction by the head and associated structures and the directional transmission characteristics of the pinna produce an orderly transformation of free-field sound as a function of sound source direction. Presumably a listener uses the resulting order in complex acoustic cues to localize the source of a sound in space. Although the free-field approach to studying directional hearing takes advantage of the natural sound transformations at the tympanic membranes, it has several limitations which may be overcome by the use of appropriately synthesized dichotic stimuli presented via headphones. Using the headphone approach Wightman and Kistler (1989a,b) found that under anechoic conditions human listeners are as accurate in localizing a virtual sound source as they are in localizing an actual source in the free field. We have modified this headphone delivery method for the cat and have implemented it in studies of cortical mechanisms of directional hearing.

The acoustic transformation created by the head and pinna is formally referred to as the free-field to eardrum transfer function (FETF) or, in human psychophysics, the head-related transfer function (HRTF). The FETF expresses for a given source direction and over a specified range of frequency the transformations of amplitude and phase that occur from a sound measured in the free field in the absence of a subject to the pressure measured when the subject is introduced into the sound field. Musicant et al (1990) varied systematically the direction of a loud speaker in a spherical coordinate system around experimental cats to obtain thousands of FETFs for the animals' left and right ears. From this large empirical database we created what we refer to as a virtual acoustic space (VAS) for the cat. In our earlier work we used this empirical estimate of the FETF obtained by simply calculating the ratio of the discrete Fourier transform of the sound pressure waveform obtained near the eardrum to that obtained in the absence of the animal (Brugge et al, 1994). This approach proved not entirely satisfactory and we thus developed a more objective method that employs finite impulse response (FIR) filters (Chen et al, 1995). Using this approach the FETF is modeled entirely in the time domain as an FIR filter with coefficients determined using least squares error criteria (Chen et al, 1994; Reale et al, 1996). Comparing free-field waveforms empirically recorded to those recovered with FIR filters yields correlation coefficients that typically exceed 0.999.

Originally we synthesized VAS signals at each of the sound directions for which free-field recordings had been made previously (Brugge et al, 1994). This approach limited spatial resolution to that of the original measurement paradigm, which was 4.5 degrees over most of the field. A functional model of VAS was later devised and validated based on the database of free-field measurements on the living cat and on the Kemar model (Chen et al, 1995). By implementing this mathematical model for the cat, an FETF can be synthesized for any direction and, therefore, a VAS can be obtained with any spatial resolution. This is particularly important in simulating sound motion or reverberate environments. The model has been implemented entirely in the time domain. Running on a high-end workstation, it is able to meet the demands of our electrophysiological experiments for on-line quasi-real-time signal generation (Reale et al, 1996).

Figure 1 illustrates the salient features of the VAS stimulus and the neural response to it. The top panel of Fig. 1 shows the time waveforms of signals that appear near the tympanic membrane in each ear as the result of a 10 µsec square pulse applied to a loudspeaker that was located in the right acoustic hemifield in anechoic space. Below are illustrated the corresponding amplitude spectra of these signals. The lower panel shows a single action potential time locked to the onset of an effective VAS stimulus. Under our experimental conditions, an AI neuron usually produces but a single time-locked spike with an average latency between

--→

Figure 1. Top panel: Time waveforms and respective magnitude spectra of signals at the two ears arising from one direction in virtual acoustic space (from Musicant et al, 1990). Bottom panel: Single action potential from a single AI neuron time locked to the VAS stimulus. Inset: Dot raster showing the time locked response to 40 repetitions of a VAS stimulus arriving at the two ears from one direction.

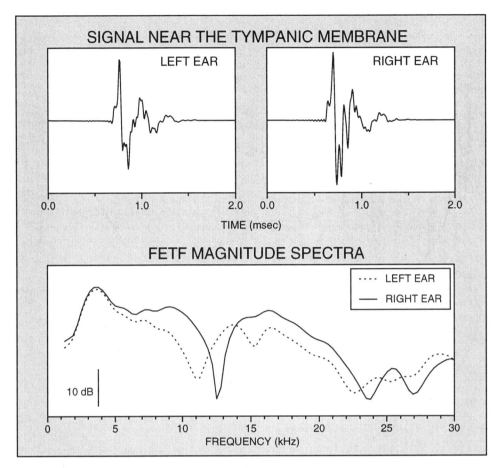

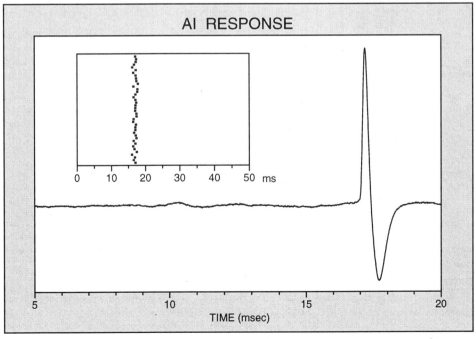

about 10 and 30 ms. For some neurons the response may consist of several spikes the first of which is securely time locked to the stimulus onset. The inset of Fig. 1 is a dot raster illustrating for one AI neuron the degree of time locking to 40 stimuli presented at the same effective direction in VAS.

3. RESULTS

3.1. Virtual Acoustic Space Receptive Fields under Anechoic Conditions

In a typical experiment directional stimuli are presented in random order at a rate of about 2/sec, and we sample all or a chosen segment of the VAS in steps of 4.5 degs or 9 degs. The direction of a VAS signal is referenced to an imaginary sphere, the center of which is occupied by the cat's head. The single-neuron data are typically plotted on a spherical representation of VAS, with the sphere bisected to show the insides of the front and rear acoustic hemifields from the cat's point of view (Reale et al, 1996). Effective directions tend to aggregate to form a virtual space receptive field (VSRF) the location, size and shape of which depend on the neuron's CF, on stimulus level and on binaural interactions (Brugge et al, 1994, 1996a,b). Stimulus intensity level is expressed as dB attenuation (dB Att) from the maximal levels available. For the data presented here signals were attenuated equally at the two ears.

Figure 2 illustrates major features of the VSRF. The data shown were obtained at two stimulus levels from each of four neurons. All directions in VAS were sampled except for those below -36 deg elevation where no measurements were made (see Musicant et al, 1990). The upper VSRF of each pair was obtained some 15–20 dB above threshold and, therefore, shows the most sensitive region in the field; the lower VSRF was obtained at levels 10–20 dB higher than that. Each VSRF is represented by two hemispheres: the left hand hemisphere represents the frontal acoustic hemifield and the right the rear hemifield. We obtained these VSRFs by presenting a single stimulus at each direction. The black squares denote directions for which an evoked time-locked spike was obtained; the dots in the upper receptive field map of Figure 2A indicate the direction of a stimulus for which there was a stimulus but no response (for clarity the dots have been omitted on subsequent plots).

At 15–20 dB above threshold the shape, size, and location of each receptive field differ among AI neurons to the extent that VSRFs fall rather naturally into five classes (Brugge et al, 1994, 1996b). Those VSRFs labeled 'contra hemifield' (Fig. 2B,D) represent neurons responding to sound presented mainly, if not exclusively, in the contralateral acoustic hemifield. Contralateral hemifield VSRFs make up nearly 60% of our sample. About 10% of neurons in our sample have similar VSRFs in the ipsilateral acoustic hemifield (Fig. 2C). About 7% of cells show responses spread across the midline and are termed 'frontal' VSRFs (Fig. 2A). Frontal fields may be formed by neurons that under dichotic conditions would be classified as EE cells, or they may be pure PB neurons whose firing depends on binaural stimulation (see Brugge and Reale, 1985; Phillips et al, 1991 for reviews). The remaining two classes of VSRFs in our sample (not illustrated here) are termed 'omnidirectional' because even at the lowest effective stimulus levels the receptive field covers much if not all of the VAS, or 'complex', meaning that they do not fit into any coherent category. With the possible exception of PB cells, the great majority of neurons recorded had their greatest sensitivity at or near the acoustic axis.

When stimulus level is raised the VSRF usually expands monotonically (Brugge et al, 1996a). For some cells the VSRF eventually comes to fill the entire acoustic space. We refer

UNBOUNDED VSRFS

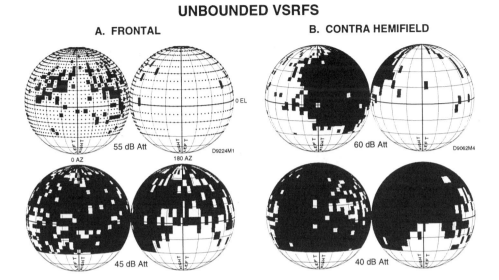

BOUNDED VSRFS

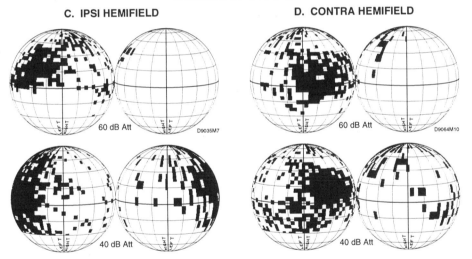

Figure 2. Virtual space receptive fields at two stimulus levels (dB attenuation) from each of four AI neurons. Dots in A (top) represent the loci of VAS stimuli presented to obtain each of the VSRFs shown. Dark squares represent directions at which evoked responses to single VAS stimuli were obtained. A. D9224M1; B. D9062M4; C. D9035M7; D. D9064M10

to such VSRFs as being 'unbounded'. Two unbounded VSRFs are illustrated in Figures 2A and B. As a rule, frontal VSRFs are unbounded whereas hemifield VSRFs need not be. For hemifield VSRFs, raising stimulus level results in increasing size of the VSRF, but for some the increase is confined largely to one or the other hemifield. We refer to the VSRFs with this property as being 'bounded' (Figs 2C and D). The mechanism involved in restricting the VSRF in this way is an inhibition evoked by sounds in one acoustic hemifield (usually the one ipsilateral to the brain hemisphere occupied by the neuron under study). Cells with bounded

VSRFs are thus capable of encoding the laterality of a sound source, although the receptive fields appear too large to provide further information about the direction of the source within that hemifield.

In summary, as a rule at stimulus levels greater than about 20 dB above threshold the VSRF represents a quadrant or more of acoustic space. The size of the field is too large to account for a listener's spatial acuity, which may approach a few degrees of arc in some regions of acoustic space (Mills, 1958; Heffner and Heffner, 1988). Thus, we have turned to studies of the internal structure of the receptive field exploring whether there are information-bearing features to the neuronal discharge within the VSRF that the cell may use to encode sound direction.

3.2. Internal Structure of the VSRF

Our transient VAS signals contain the directional cues of interaural time, interaural intensity and interaural spectrum. In addition, the energy in a VAS stimulus at each eardrum varies, in a frequency-dependent way, as a function of the direction of the sound source. Independently changing interaural time, interaural intensity, or overall intensity of non-directional tones or noise has been shown to alter the onset latency of AI neurons (Brugge et al, 1969; Brugge and Imig, 1978; Phillips, 1989; Phillips and Hall, 1990). Thus, it seemed quite likely that onset latency would be sensitive to sound-source direction where all of the directional cues are represented. If such were the case, one might expect that the latency would vary by several ms across the VSRF. Figure 3 illustrates the *frequency distribution* of first spike latency derived from all responses that made up the VSRF of three neurons; each VSRF was obtained at stimulus levels near the middle of the neuron's dynamic range. The latency distributions are sharply peaked and they have a range of some 3–5 ms.

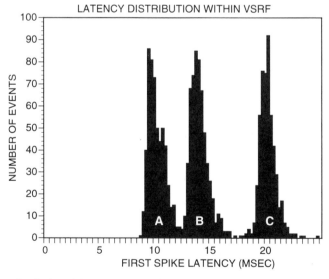

Figure 3. Frequency distribution of first-spike latency from the VSRFs of three AI neurons. A. D92103M10; B. D93011M5; C. D93011M4

The data illustrated in Fig. 3 represent neurons showing some of the shortest (10.5 ms) and longest (20.5 ms) average latencies in our sample. Latency differences of this magnitude and greater may have implications for how directional information is transmitted by an ensemble of AI cells. The probability that a target neuron receiving an AI ensemble volley will in turn fire action potentials depends on the timing of that incoming volley. If all, or most, neurons in the ensemble have the same latency distribution, then in response to a transient acoustic signal the temporal spread of presynaptic input at the target cell would be minimal and the postsynaptic discharge would tend to be precisely time-locked and phasic. If, on the other hand, the ensemble contains AI cells with a wide range of average latency, then under the same stimulus conditions there would exist a more temporally-graded synaptic input to a cell upon which such a population of AI neurons converge. Thus, if there is a spatio-temporal structure to the sound source, as occurs during motion of repeated acoustic transients for example, a spread in onset latency across neurons may provide for smooth integration of directional information by neuronal assemblies.

The question naturally arises as to how a 3–5 ms spread in onset latency is distributed *spatially* across a VSRF. Because our stimulus set consisted of hundreds of signals from closely spaced directions we were able to derive VSRFs with a very fine grain, and by plotting VSRFs based on discharge latency we were able to search for *spatio-temporal gradients* that may be related to stimulus direction.

Data illustrating temporal organization within a VSRF are illustrated in Figs. 4 and 5. For these experiments we presented multiple (5–15) stimuli at each direction in VAS thereby building up sufficient numbers to compute for each effective direction the average latency to the first spike and its standard deviation. On these figures each effective direction is designated by a black square regardless of the number of spikes evoked at that direction, as illustrated previously in Fig. 2; the black dots on each of the plots indicate the directions for which a stimulus was delivered but not effective. Because of time constraints in collecting multiple-trial data at each stimulus direction, only a portion of the receptive field that contained the most effective directions was mapped. In order to determine whether there was an internal structure to the VSRF related to onset latency we plotted the VSRF based only on those stimulus directions that resulted in time locked spikes within selected latency windows. Figures 4 and 5 show the results of that kind of analysis with a series of VSRF maps obtained with different latency windows. The VSRF at the top of each column of receptive field maps illustrates the extent of the map obtained over the full range of discharge latency. Each successive VSRF map in a column represents the spatial distribution of first spike latency within successive time windows after stimulus onset (shown above each plot). For any given neuron the latency window width (in milliseconds) remained constant although it might have been chosen to be slightly different from one neuron to the next.

Figure 4 illustrates for four contralateral hemifield neurons the range of latency structure in our data sample. Only data related to signals in the frontal hemisphere are represented in this figure. Figure 4A illustrates one of the most striking examples we have of an orderly latency representation within an AI spatial receptive field. In this case only the contralateral frontal quadrant of VAS was mapped. All first-spike responses with a latency between 12.00 and 12.80 ms aggregate within a relatively small region at or near the acoustic axis. Directions representing first spike latency between 12.80 and 13.60 ms and between 13.60 and 14.40 ms occupy more distant concentric rings; later spikes occupy the outer fringes of the receptive field. As a whole the spatial distribution of onset latency for this neuron forms a *temporal gradient* in the receptive field with the shortest latency rep-

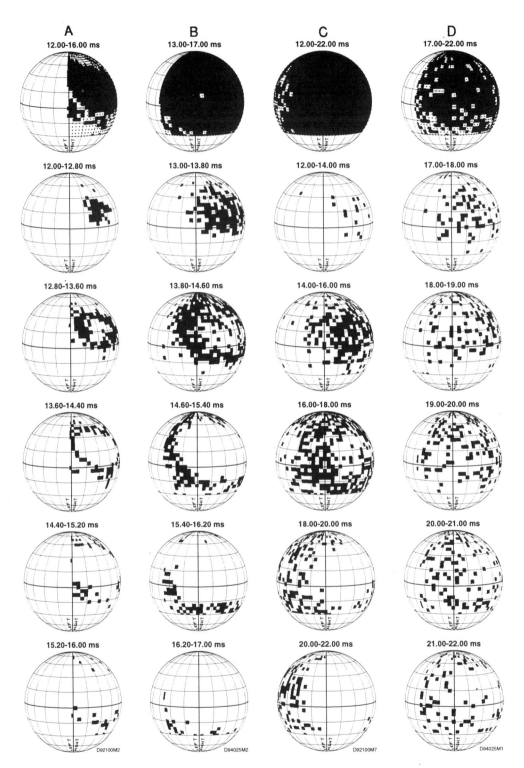

Figure 4. Hemifield VSRFs from four AI neurons showing the spatial distribution of first spike latency. Each VSRF derived from multiple trials at each VAS direction. Only the most sensitive portion of the VSRF in the frontal hemifield was obtained. Top VSRF in each column shows the complete map of responses within the latency range indicated. Dots indicate the loci of stimuli presented. Below each complete VSRF is a set of VSRF representations based on latency criteria given above. A. D92100M2; B. D94025M2; C. D92100M7; D. D94025M1

resented at or near the acoustic axis and successively longer latency radiating out from it. The neuron in Fig. 4B exhibited a VSRF whose latency gradient is slightly different but no less robust than that shown in Fig. 4A. Figure 4C shows a VSRF with a latency gradient that is not as well defined as the previous two, but which is nonetheless clearly in evidence. Within our sample of several hundred neurons, about half had VSRFs that showed sign of an internal time structure similar to that illustrated here; we refer to these as 'ordered' VSRFs. Figure 4D illustrates a VSRF for which such a gradient is barely recognized or is nonexistent. We refer to such VSRFs as being 'disordered' (Brugge et al, 1996a).

Figure 5 shows ordered behavior for onset latency within the VSRFs of two 'frontal' neurons. Both cells responded to CF tone bursts presented to either ear (so-called EE neurons). For the neuron shown in Fig. 5A a latency gradient radiates mainly along lines of elevation from an area of shortest latency that stretches across the midline between 0 degs and 54 degs of elevation. The frontal VSRF in Fig. 5B clearly exhibits two foci of short latency, each on or near the respective left or right acoustic axis. This VSRF also exhibits two nearly mirror-symmetric latency gradients radiating from the two short-latency foci. Neurons that fire only to binaural stimulation, the so-called PB cells, also exhibit ordered VSRFs centered at or near the midline (not shown). As mentioned earlier, frontal VSRFs represent no more than about 7% of our sample.

Obtaining data over a wide expanse of VAS with high spatial resolution and with multiple trials at each direction is usually not achieved simply because the electrode usually does not remain in contact with a neuron for the many hours required to carry out such an experiment. In order to obtain multiple-trial data in a systematic way we adopted a strategy whereby we first derived a single-trial VSRF map and then selected a limited set of directions for presenting multiple stimulus trials. Typically we limited the stimulus set to a single azimuth or elevation. Figure 6 illustrates results from such an experiment. On the left (Fig. 6A) is a column of VSRFs obtained with a single stimulus at each direction. The VSRF is analyzed and arranged as in Figs. 4 & 5. Even with a spatial resolution of 9 degs and with a single stimulus being presented at each direction, a region of short latency evoked activity (10.20–11.20 ms) is evident around the acoustic axis in the right acoustic hemifield. Directions representing onset spikes with successively longer latency are arrayed more or less systematically at increasing distance from the central core area thereby creating a temporal gradient as illustrated previously in Figs. 4 and 5.

Curves in Fig. 6B and C were obtained by presenting 40 consecutive stimuli at a rate of 2/sec at each of 21 directions (9 deg separation) along an azimuth, from -90 deg to +90 deg. At each direction a mean first-spike latency and standard deviation were computed. Fig. 6B shows a plot of mean first-spike latency as a function of stimulus azimuth (latency azimuthal function) for the front acoustic hemifield. Three latency azimuthal functions are shown, one for each of three elevations: 0, 18 and 36 degs (corresponding to arrows on left top VSRF of 6A). As can be seen in Fig. 6B, for stimulus directions between about +27 deg and +72 deg latency was relatively short and relatively unchanging; within this azimuthal range, and at any of the three elevations shown, latency varied by no more than about 0.5 ms, or 0.01 ms/deg. For stimulus directions located closer toward the midline and into the opposite acoustic hemifield the slopes of the curves become steeper, and between +27 and -27 degs, where the curves are steepest, there is a change in latency of nearly 2 ms. This translates into a spatial gradient within this range of about 0.3 ms/deg. Because ordered VSRFs vary somewhat in their spatial configuration from one neuron to the next and even in the same neuron at different stimulus levels (Brugge et al, 1996a), the slope of the gradient measured across a fixed set of azimuths and elevations and the spatial location of the steepest gradient will vary somewhat. For VSRFs with latency gradi-

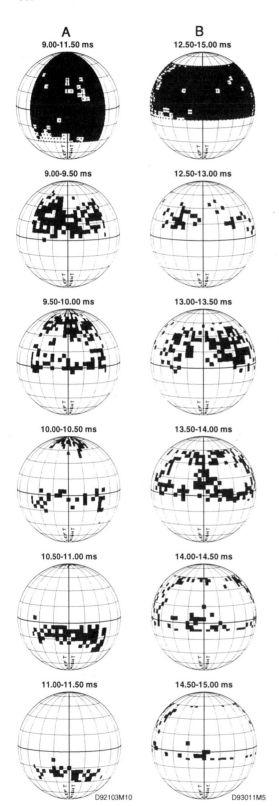

Figure 5. Frontal VSRFs from two AI neurons showing the spatial distribution of first spike latency. Each VSRF derived from multiple trials at each VAS direction. Only the most sensitive portion of the VSRF in the frontal hemifield was obtained. Top VSRF in each column shows the complete map of responses within the latency range indicated. Dots indicate the loci of stimuli presented. Below each complete VSRF is a set of VSRF representations based on latency criteria given above. A. D92103M10; B. D93011M5

ents that radiate in all directions away from the acoustic axis (e.g. Fig. 4A, 6A), the azimuthal function that cuts through the center of the acoustic axis will tend to exhibit a temporal gradient that is steeper than the one seen at higher or lower elevations. On the other hand, for VSRFs such as shown in Fig. 5A one would expect the gradient to take a different orientation, emphasizing the elevational component of the stimulus.

To the right of the azimuthal functions shown in Fig. 6D is shown a family of latency histograms derived from the same data plotted in Fig. 6B,C, at 18 degs elevation. Accompanying each histogram is the corresponding mean latency (+/- one standard deviation). Latency histograms are typically narrow for stimuli originating from directions in the region of the

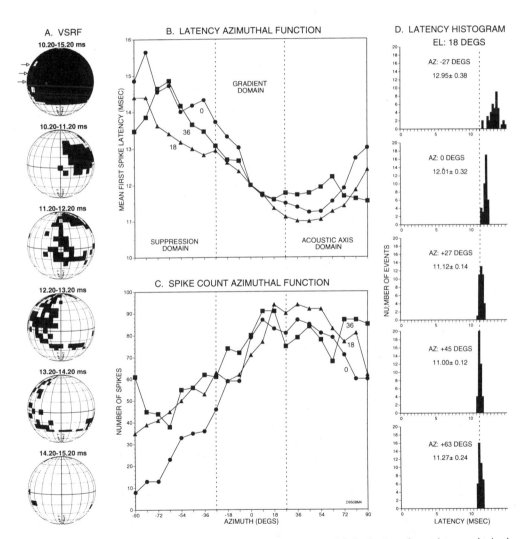

Figure 6. A: Hemifield VSRF from a single AI neuron showing the spatial distribution of onset latency, obtained with a single stimulus at each direction. Plot format same as in Figs. 4 and 5. B-C: Mean latency and total spike count plotted as a function of stimulus azimuth at an elevations of 0, 18 and 36 degrees (arrows on VSRF). D: Histogram of first-spike latency for five directions represented on the azimuthal function at 18 degrees elevation. Forty stimuli were presented at each direction for data shown in B,C &D. D9508M4

acoustic axis. Between +63 and +27 degrees the average latency shifted by no more than 0.27 ms and the temporal jitter (as defined by the std) was minimal. Between +27 deg and the midline the mean latency shifted by nearly one ms, and the temporal jitter increased nearly two-fold. As the signal shifted to -27 degs in the opposite hemifield, the latency continued to lengthen and the histogram continued to widen. The mean and standard deviation of first spike latency within the VSRF are well within the range of values obtained in cat AI using nondirectional tone- or noise-burst stimuli (Brugge et al, 1969; Brugge and Imig, 1978; Phillips, 1985; Phillips and Hall, 1990).

The spike count azimuthal functions derived from this same neuron are nearly mirror images of the latency functions (Fig. 6C). High spike count is associated with short latency, and as spike count falls with displacement of the stimulus across the midline there is a concomitant increase in onset latency. Although such an association between latency and spike count is commonly found for AI neurons, there are exceptions (see e.g. Brugge et al, 1996a). Such exceptions would be expected for neurons for which spike count, but not latency, is a non-monotonic function of stimulus level (Brugge et al, 1969; Phillips, 1985, 1989). Thus, response magnitude (spike rate, spike count or firing probability) and latency may serve jointly as a code for stimulus direction, especially for those cells exhibiting monotonic rate-vs-level functions.

4. DISCUSSION

From the shapes of the azimuthal functions, from the timing of the onset spikes and from the shapes and locations of VSRFs we may infer some ways that direction of a sound in space may be encoded by AI cortical neurons having broad spatial tuning. Referring to Fig. 6, we divide the frontal acoustic hemifield into three spatial domains (dashed lines) that are spanned by the latency and spike-count azimuthal functions. For our present purposes, these domains are defined only in azimuthal terms and only for elevations that pass through the region of the acoustic axis. Thus, we need not deal here with the inevitable distortion in azimuthal distance that comes with comparing data obtained at higher and lower elevations.

The first of these we call the 'acoustic axis domain'. It spans that region in space where the amplitude of a sound is greatest due to the filtering properties of the external ear. Depending on frequency, in the cat it lies between about 30 degs and 90 degs on the azimuth; for all frequencies in the audible range it is found around 18 degs elevation (see Musicant et al, 1990). For frequencies between 9 and 30 kHz, which represents the range of CF of the neurons in our sample, the acoustic axis lies between about 25 degs and 60 degs on the azimuth. The spatial distribution of centers of VSRFs obtained near threshold from our sample population has a similar range (Brugge et al, 1996a). For hemifield neurons, onset latency is uniformly shortest within this domain and, for a monotonic neuron, spike count is uniformly highest as compared with other domains along the azimuth. There is relatively little variation in latency or count within the domain of the acoustic axis. Thus, while relatively high sensitivity (low threshold) to sound may characterize this domain, if the gradient in latency, spike count or both encode signal direction or change in direction then we would expect that directional sensitivity for sounds in this domain would be relatively low.

We refer to the second domain as the 'gradient domain'. It is located within about 25–30 degs of, and often spans, the midline (shown here between plus and minus 27 degrees). Within this domain a small change in stimulus direction along the azimuth results in a relatively large change in onset latency, spike count or both especially for those azimuths that pass near the center of the acoustic axis. Thus, it is within this domain that we would expect that spatial acuity would be relatively high and that auditory sensitivity would fall.

If latency or spike count or both are involved in encoding stimulus direction, then a spatial ambiguity exists for those neurons for which receptive field gradients radiate from the acoustic axis. That is, there are numerous source directions surrounding the acoustic axis for which sounds will evoke the same or very similar latency or spike count. Some of this ambiguity may be reduced or eliminated in those cells exhibiting 'bounded' VSRFs, for with these VSRFs there exists a 'suppression domain' that occupies much of the opposite acoustic hemifield. Sounds that arise from this suppression region inhibit the neuron and thereby work to restrict the receptive field to one hemifield (Brugge et al, 1994). For such neurons sounds are lateralized to one acoustic hemifield.

Of course, the boundaries of these domains are not sharp and will vary somewhat among neurons. Our domains would also have elevational components as well, as can be readily seen in the VSRFs mapped using latency criteria (see e.g. Figs. 4 and 5). Nonetheless, for the great majority of recorded AI cells for which azimuthal functions (Rajan et al, 1990) or VSRFs (Brugge et al, 1994, 1996a,b) have been studied, the acoustic axis domain is located in the contralateral acoustic hemifield, the suppression domain is in the ipsilateral hemifield and the gradient domain tends to straddle the midline.

Our data are in agreement with results of psychophysical studies both with respect to identifying the direction of a sound source and to discriminating between loci of two sound sources in space. In both cat and human the minimum audible angle is relatively small, around 1–3 degrees, for sources straight ahead and substantially larger for sources between 30 degs and 90 degs (Mills, 1958; Martin and Webster, 1987; Heffner and Heffner, 1988). Moreover, human listeners exhibit fewer errors in identifying sound source direction when the source is near the midline than when the source is displaced 40 to 70 degrees laterally (Mackous and Middlebrooks, 1992). It is of further interest to note that cats can execute a saccade with great accuracy a toward a sound source that appears suddenly in the free field provided the sound is within about 30 degs on either side of the midline (Yin and Populin, 1997).

Although the single-neuron results are in qualitative agreement with psychophysical findings, the question remains as to whether the results can also account for the high degree of spatial accuracy and spatial acuity exhibited by listeners in identifying and discriminating sounds in space. Proposing that onset timing by single AI neurons may be involved in the perceptual segregation of transient sounds in space requires that the precision in that timing be of an order that the neural responses to spatially segregated transients are themselves separated in time (see e.g. Phillips and Hall, 1990). As mentioned above, the shape and spread of the latency histogram derived from multiple stimuli at a single source direction in the VSRF is very similar to that obtained from AI neurons under a variety of other listening conditions. Phillips and Hall (1990), using 3 stds as a criterion for the separation of onset latency distributions, estimated the temporal resolution of AI cortical neurons as being between 0.45 and 1.5 ms. This is of the order required of human listeners to discriminate two *temporally* discrete sounds, and probably sufficient to encode the timing of phonetically important acoustic elements of human speech (Phillips, 1993). Discrimination of two *spatially* discrete sound sources near the midline requires a source separation of 1–3 degs, which for the neuron illustrated in Fig. 6 represents a difference in onset latency of around 0.3–0.9 ms. Such a difference in latency falls within the resolving power of AI cells as originally estimated by Phillips and Hall (1990). More laterally the gradients in the VSRF are not as steep, consequently two spatially separated sounds in this region of space may not generate temporally resolveable latency distributions. This may account for the relatively poor spatial acuity experienced by listeners for sounds arriving from these directions. Thus, spatial acuity for transient sounds may depend both on the precision in timing of the onset spike of single AI neurons and the latency gradient within these cells' spatial receptive fields. We are presently testing this notion further in a statistical model that employs latency gradients in broadly-tuned VSRFs of populations of AI neurons (Jenison et al, 1997).

5. REFERENCES

Brugge, J. F., Dubrovsky, N. A., Aitkin, L. M. and Anderson, D. J. (1969) Sensitivity of single neurons in auditory cortex of cat to binaural tonal stimulation; effects of varying interaural time and intensity. J. Neurophysiol. 32, 1005–1024.

Brugge, J. F. and Imig, T. J. (1978) Some relationships of binaural response patterns of single neurons to cortical columns and interhemispheric connections of auditory area AI of cat cerebral cortex. In: Evoked electrical activity in the auditory nervous system. R. Nauton (Ed.) Academic Press, New York, pp. 487–503.

Brugge, J. F. and Reale, R. A. (1985) Auditory Cortex. In: Cerebral Cortex, Association and Auditory Cortices. E. G. Jones and A. Peters (Eds.) Plenum, New York, pp. 229–271.

Brugge, J. F., Reale, R. A. and Hind, J. E. (1996a) The structure of spatial receptive fields of neurons in primary auditory cortex of the cat. J. Neurosci. 16, 4420–4437.

Brugge, J. F., Reale, R. A., and Hind, J. E. (1996b) Auditory cortex and spatial hearing. In: Binaural and spatial hearing in real and virtual environments. R. Gilke and T. Anderson (Eds.) Earlbaum, Mahwah, NJ, pp. 447–473.

Brugge, J. F., Reale, R. A., Hind, J. E., Chan, J. C., Musicant, A. D. and Poon, P. W. (1994) Simulation of free-field sound sources and its application to studies of cortical mechanisms of sound localization in the cat. Hear. Res. 73, 67–84.

Chan, J. C. K., Musicant, A. D. and Hind, J. E. (1993) An insert earphone system for delivery of spectrally shaped signals for physiological studies. J. Acoust. Soc. Am. 93, 1496–1501.

Chen, J., Van Veen, B. D. and Hecox, K. E. (1995) A spatial feature extraction and regularization model for the head-related transfer function. J. Acoust. Soc. Am. 1, 439–452.

Chen, J., Wu, Z. and Reale, R. A. (1994) Applications of least-squares FIR filters to virtual acoustic space. Hear. Res. 80, 153–166.

Heffner, R. S. and Heffner, H. E. (1988) Sound localization acuity in the cat: effect of azimuth, signal duration, and test procedure. Hear. Res. 36, 221–232.

Imig, T. J., Irons, W. A. and Samson, F. R. (1990) Single-unit selectivity to azimuthal direction and sound pressure level of noise bursts in cat high-frequency primary auditory cortex. J. Neurophysiol. 63, 1448–1466.

Jenison, R. L., Reale, R. A. and Brugge, J. F. (1997) Maximum-likelihood estimation error of cortical spatial receptive fields. ARO Midwinter Meeting (Abstract)

Makous, J. C. and Middlebrooks, J. C. (1990) Two-dimensional sound localization by human listeners. J. Acoust. Soc. Am. 87, 2188–2200.

Martin, R. L and Webster, W. R. (1987) The auditory spatial acuity of the domestic cat in the interaural horizontal and median vertical planes. Hear. Res. 30, 239–252.

Middlebrooks, J. C., Clock, A. E., Xu, L. and Green, D. M. (1994) A panoramic code for sound location by cortical neurons. Science

Middlebrooks, J. C. and Pettigrew, J. D. (1981) Functional classes of neurons in primary auditory cortex of the cat distinguished by sensitivity to sound location. J. Neurosci. 1, 107–120.

Mills, A. W. (1958) On the minimum audible angle. J. Acoust. Soc. Am. 30, 237–246.

Musicant, A. D., Chan, J. C. K. and Hind, J. E. (1990) Direction-dependent spectral properties of cat external ear: New data and cross-species comparisons. J. Acoust. Soc. Am. 87, 757–781.

Phillips, D. P. (1985) Temporal response features of cat auditory cortex neurons contributing to sensitivity to tones delivered in the presence of continuous noise. Hear. Res. 19, 253–268.

Phillips, D. P. (1989) Timing of spike discharges in cat auditory cortex neurons: implications for encoding of stimulus periodicity. Hear. Res. 40, 137–146.

Phillips, D. P. (1993) Representation of acoustic events in the primary auditory cortex. J. Exp. Psychol. 19, 203–216.

Phillips, D. P. and Hall, S. E. (1990) Response timing constraints on cortical representation of sound time structure. J. Acoust. Soc. Am. 88, 1403–1411.

Phillips, D. P., Reale, R. A., and Brugge, J. F. (1991) Stimulus processing in the auditory cortex. In: Neurobiology of Hearing: The Central Auditory System. R. A. Altschuler (Ed.) Raven Press, New York, pp. 335–365.

Rajan, R., Aitkin, L. M., Irvine, D. R. F. and McKay, J. (1990) Azimuthal sensitivity of neurons in primary auditory cortex of cats. I. Types of sensitivity and the effects of variations in stimulus parameters. J. Neurophysiol. 64, 872–887.

Reale, R. A., Chen, J., Hind, J. E., and Brugge, J. F. (1996) An implementation of virtual acoustic space for neurophysiological studies of directional hearing. In: Virtual Auditory Space: Generation and Applications. S. Carlile (Ed.) R.G. Landes, Austin, pp. 153–184.

Wightman, F. L. and Kistler, D. J. (1989a) Headphone simulation of free-field listening. I: Stimulus synthesis. J. Acoust. Soc. Am. 85, 858–867.

Wightman, F. L. and Kistler, D. J. (1989b) Headphone simulation of free-field listening. II: Psychophysical validation. J. Acoust. Soc. Am. 85, 868–878.

Yin, T. C. T. and Populin, L. C. (1997) Behavioral and physiological studies of sound localization in the cat. In: Acoustical Signal Processing in the Central Auditory System. J. Syka (Ed.) Plenum, London, pp. 399–406.

35

CORTICAL AND CALLOSAL CONTRIBUTION TO SOUND LOCALIZATION

F. Lepore,[1,2*] P. Poirier,[1] C. Provençal,[1] M. Lassonde,[1,2] S. Miljours,[1] and J. -P. Guillemot[1,3]

[1]Groupe de Recherche en Neuropsychologie Expérimentale
[2]Université de Montréal
C.P. 6128, Succ. Centre-Ville
Montréal, H3C 3J7, Canada
[3]Université du Québec
C.P. 8888, Succ. Centre-Ville
Montréal, H3C 3P8, Canada

ABSTRACT

The corpus callosum, the principal neocortical commissure, allows for the inter-hemispheric transfer of lateralized information between the hemispheres. It is generally accepted that one of the principal functions of the callosum, at least in the visual and somatosensory modalities, is to unite the sensory hemispaces for information projecting to different hemispheres (otherwise known as midline fusion). Two of the principal cues to sound localization in free-field are intensity and time differences for sound arriving to the two ears. Since each ear projects in a preponderant manner to the contralateral hemisphere and since complex sounds are generally analyzed at the cortical level, it is possible that the callosum is required to compare time and intensity differences for information arriving in a biased fashion to separate hemispheres. The aim of the present experiments was to examine this problem at the single cell level using cats and at the behavioral level with human subjects having cortical or callosal pathologies.

Two approaches were used to study how the callosum contributes to this type of binaural interaction: we recorded either callosal fibres directly (callosal efferent neurons) or cells in the callosal zone of A1 (callosal recipient neurons) in normal and callosotomized animals. The animals were anesthetized and recording was carried out in both cases under direct visual control. Stimuli were presented either dichotically through implanted

* Correspondence and reprints requests should be adressed to: Dr. Franco Lepore, Département de Psychologie, Université de Montréal, C.P. 6128, Succ.Centre-Ville, Montréal, Québec, Canada, H3C 3J7. Tél: 514 343 4600; fax: 514 343 5787.

Acoustical Signal Processing in the Central Auditory System
edited by Syka, Plenum Press, New York, 1997

earphones or on a frontally located sound perimeter. Tone bursts or white noise were presented to the two ears and intensity or time differences were varied. Results indicated that callosal efferent neurons or callosal recipient cells appear to prefer sounds coming mainly from the contralateral hemifield or at the midline. This was also confirmed with free-field stimulation. Removing this input through callosal section modifies the distribution of cells in A1 which are tuned to interaural intensity differences and somewhat less those sensitive to interaural time delays.

At the behavioral level, callosal agenesis or hemispherectomized subjects had to identify the apparent location of a sound presented on a frontally positioned perimeter surrounding the head on the horizontal meridian. Either a stationary sound or an apparently moving sound displaced at various velocities, length of trajectory and in the two directions were used. Results indicated, in accordance with the electrophysiological data, that localization performance was poorer in both groups of neurologically deficient subjects than in matched controls.

These results attest to the importance of the corpus callosum to localize sounds in free-field.

1. INTRODUCTION

The ability of an animal or a human subject to localize sound sources in its environment is an ecologically important function. In fact, it has unquestionable survival value since, for the former, it acts as a warning signal for the approach of predators or to detect the presence of otherwise camouflaged prey and, for the latter, to signal impending collisions (as in traffic or a noisy workplace, for example) or to identify the source of an interlocutor (as in a crowd, for example). The auditory system has at its disposal three cues to permit accurate identification of the origin of sound sources. Two depend on the relative position of the ears in the head: intensity cues result from the shadow cast by the head on the partially hidden ear for laterally presented sounds whereas time cues stem from the added time it takes such sounds to travel to the more distal ear. Spectral cues, on the other hand, are created by the diffraction of sound waves by the head and the external ear and vary at each tympanic membrane according to the position of the sound in space.

A number of electrophysiological experiments carried out in various species of animals (some of which are presented in this volume, see Hartung and Sterbing, Kelly, King and Schnupp, Palmer et al., Yin and Populin), has shown that single units at subcortical levels are sensitive to these cues (e.g. Altman, 1968; Boudreau and Tsuchitani, 1968; Brugge et al., 1970; Kuwada, et al., 1979; Semple et al., 1983; Hirsch et al., 1985; Wise and Irvine, 1985; Ivarsson et al., 1988; Rauschecker and Harris, 1989; Irvine and Gago, 1990; Spitzer and Semple, 1991; Poirier et al., 1996; Samson et al., 1996). One may be tempted to conclude that the substrate to localization in free-field is located in these structures and that these are sufficient to assure this function. There are both behavioral and physiological arguments against this oversimplification of the processes and structures involved in localizing sound sources. Thus, at the single-cell level, it has been well established that most units of A1 are concerned with the spatial analysis of contralateral fixed-sound sources of short duration (see Jenkins and Masterton, 1982; Phillips and Brugge, 1985 for reviews). Recordings of A1 units of the cat show that spatially selective neurons are highly sensitive to interaural disparities in time and intensity, responding mainly when these cues mimic sounds presented in the contralateral region of space (e.g. Hall and Goldstein, 1968; Brugge et al., 1969; Kitzes et al., 1980; Phillips and Irvine, 1981; Reale and Kettner, 1986). Similarly, free-field studies suggest that the majority of A1 neurons in cats and monkeys are spatially selective to contralateral stationary sources (Eisenman, 1974; Middlebrooks and Pettigrew, 1981; Imig et al., 1990; Rajan et al., 1990). In addition, it has been shown recently that a subset of spatially selective A1 neurons derive their sensitivity from spectral cues (Samson et al., 1993). Behaviorally, the assumption that the integrity of A1 is essential for the coding of fixed-sound sources of short duration in the horizontal plane has been supported by the fact that its excision affects discrimination for contralaterally presented sound stimuli (Jenkins and Masterton, 1982; Jenkins and Merzenich, 1984; Masterton, this volume). In humans, a number of reports (see Altman, this volume) has confirmed that localization performance is poorer following unilateral hemispheric lesions (Sanchez-Longo, 1957; Sanchez-Longo and Forster, 1958; Klingon and Bontecou, 1966). The cortex, therefore, though possibly not essential, appears nonetheless to contribute to accurate localization performance.

The corpus callosum interconnects most cortical areas in the two hemispheres. In the cat (Imig et al., 1986) and monkey (Pandya and Seltzer, 1986), auditory callosal neurons have been clearly described and, electrophysiologically, appear to produce interhemispheric interactions in the recipient cells which are mainly of the excitatory type (Imig et al., 1986). Although callosal transection has been shown to have little effect on localization behavior in animals, we nevertheless suggest that this structure may contribute significantly to this function for a number of reasons. First, because the interactions seen at the physiological level constitute a sufficient condition for supporting localization behavior. Second, an examination of the lesion experiments suggests that the methodologies may not have been appropriate to bring out more refined deficits in localization since they used tests which were too simple (Jenkins and Masterton, 1982; Jenkins and Merzenich, 1984) and experimental controls which were not completely adequate (see Poirier et al., 1994 for discussion). Third, although there are extensive cross-over points in the auditory system, there is a large bias for neurons to project to the contralateral hemisphere. Since higher order processing of auditory material is carried out by cortical structures in a preponderant fashion either by the contralateral hemisphere or by the functionally specialized one: e.g. language on the left (Geshwind, 1972) and spatial-emotive on the right (Bryden et al., 1985), the operation of bilateral cues allowing for the localization of this material may require the co-activation of single cells with the information present in each hemisphere. This can only be achieved via the principal neocortical commissure, that is, the corpus callosum.

We examined the problem of the cortical and callosal implication in sound localization using two approaches. On the one hand, electrophysiological recordings of single cells were carried out in the primary auditory area A1 using free-field stimuli to determine whether in fact they were sensitive to stimulus location. Moreover, the responses of callosal recipient cells in this area or those of cells projecting through the callosum were recorded to dichotically presented stimuli having various interaural disparities which simulated different sound locations on the horizontal plane. We also analyzed how removing the contribution of the opposite hemisphere by callosal transection affected the response of presumed callosally recipient cells in area A1. On the other hand, using human subjects which were either lacking the callosum (callosal agenesis subjects) or the auditory cortical areas on one side (hemispherectomized subjects), we examined how localization performance was influenced by the absence of these structures.

2. MATERIALS AND PROCEDURES

2.1. Electrophysiological Studies

A detailed description of the procedures is presented elsewhere (Poirier et al., 1995). They were carried out in accordance with the guidelines of the Canadian Council on Animal Care, NIH and the University of Montreal animal care committee. In summary, cats were anesthetized using a mixture of Halothane, nitrous oxide and oxygen ($N_2O:O_2$ in a proportion of 70:30). During recording, they were also paralyzed to prevent movements of the pinnae or contraction of the head musculature and middle ear muscles which might have altered auditory input at the pre-receptor level. They were intubated to allow for artificial respiration and installed in an ear-bars free stereotaxic apparatus placed in a semi-anechoic chamber. In the free-field study, stimuli were presented via a series of 16 loudspeakers fixed to a semi-circular perimeter placed on the horizontal plane crossing the interaural line. The head of the animal was situated in the geometric center of the perimeter at a distance of 50 cm. The centers of the speakers were separated by an interval of approximately 10° and covered 156° of auditory space. Stimuli were broadband noise bursts of 45 to 55 db. For the experiments using dichotic presentations, earphones were sealed in the external meatus through which were delivered two kinds of stimuli: short-duration noise bursts with pre-determined interaural delays (dT: 0, 0.1, 0.4, 1.6, 3.2, 6.4, 10 ms) or pure tones which varied not only with respect to this localization cue but also in terms of interaural intensity differences (dI: 0, 3, 5, 10, 20, 30 db).

The electrophysiological recordings in both A1 and the callosum were carried out using either tungsten micro-electrodes or glass micropipettes filled with NaCl (3M). These were placed on the appropriate structures under visual control. Callosal recipient cells in A1 were identified by stimulating the contralateral hemisphere with an electrical pulse. Only those cells which responded to this stimulation were retained for subsequent analysis. To eliminate the callosal influence on A1 recipient cells, this structure was transected completely at least two months prior to the recording.

2.2. Studies with Human Subjects

All studies were carried out in a quasi-anechoic chamber. Four types of subjects were used: 1-normal control subjects which came from the student population at the university; 2- callosal agenesis subjects, with CT scan identified complete callosal agenesis; 3- hemispherectomized subjects which had undergone complete excision of one hemisphere or its functional disconnection; 4- I.Q., handedness, sex and age matched control subjects to each of the neurologically abnormal individuals. Each subject was seated at the center of the semicircular perimeter described above. Stimuli consisted of broad-band noise bursts of 90 ms duration and 52 db, presented through one of the loudspeakers. Although the head was free to move, the subject was asked to keep the head aligned with the center of the perimeter and fixate a light stimulus situated on top of the central speaker. A warning buzzer informed the subject that a stimulus would be presented. The subject had to point to the perceived location of the sound.

3. RESULTS

3.1. Electrophysiological Studies

Cells recorded in A1 were exquisitely tuned to stimulus location in free-field. Responses varied from units showing extremely fine tuning to a specific location to those sensitive across whole regions of auditory space. Representative examples are shown in figure 1. The cell presented in Fig. 1A appears to be sensitive to a stimulus originating in a fairly restricted region of space, in the present case at 135° in the contralateral field. The more common type of response, however, is illustrated in Fig. 1B, where the cell shows identifiable excitation to a wider though still circumscribed region of space in the contralateral hemifield. These units account for over 60% of the cells recorded. Some cells, which may be subsets of the two given above in terms of the regions of space to which they respond, have the added property that they prefer midline presented stimuli (Fig. 1C). These account for about 10% of all cells recorded. A limited number of cells (14%) preferred ipsilaterally presented stimuli. Finally, about an equal number of cells (13%) could be driven by stimuli which appeared throughout the tested auditory space (Fig. 1D). These cells appear functionally identical to the omnidirectional cells previously described in A1 using free-field stimulations (Middlebrooks et al.,1980).

It is clear from these electrophysiological results that cells in A1 can subserve auditory localization since they are specifically activated by complex sounds presented in limited regions of auditory space. The question remains whether these cells manifest binaural interactions which would suggest that they code spatial locations using the principal

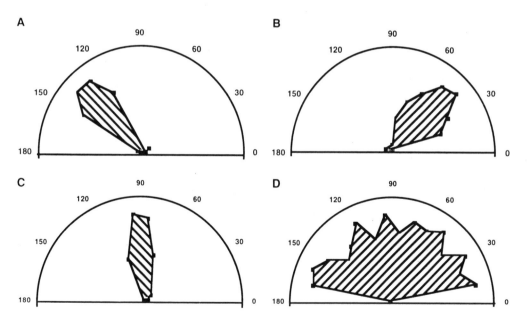

Figure 1. Polar plots of four cortical auditory cells in A1 tested using free-field stimulation. The lenght of each vector in the polar plots represents the percentage of the maximum response (set at 100%) for each of the sixteen positions in the horizontal plane. 12° and 168° represent the two most lateral positions tested. A- Polar plot of a tuned contralateral preferring unit. B- Polar plot of a more broadly tuned ipsilateral preferring unit. C- Midline preferring unit. D- Omnidirectional unit.

physiological cues for this function or whether these binaural integrations are carried out only at subcortical levels, as demonstrated by various researchers. We examined this problem by recording cells in area A1 using dichotically presented stimuli where time and intensity differences between the stimuli presented to each ear were systematically varied. Results indicated that cells in A1 were sensitive to both cues and various disparity profiles could be derived. Thus, some cells showed large interactions mainly of the excitatory type (with the occasional cell manifesting inhibitory interactions) when the two stimuli had zero disparity in either the time or intensity domains. These cells probably signal midline localizations. Others were maximally excited when one ear led the other or when intensity was higher for one ear than for the other. In some cases, the disparity selectivity corresponded to a limited range of interaural differences whereas for others, more intense stimulation of one ear or its precedence with respect to the other always produced the best response. Transposing these results to presumed localization performance, they probably correspond, respectively, to the lateralized and hemifield units described above. Finally, a number of cells showed binaural interactions at all disparities tested, a response profile which would be expected from omnidirectional cells. The relative proportions of these different cell types are presented in table 1. These proportions, however, should not be interpreted in absolute terms but only in relative ones since the units retained for this analysis were also the ones which received callosal activation (see methods above).

Callosal recipient cells in area A1 exhibit physiological properties suggesting that intensity and time differences are used to localize sound in space. Are these characteristics derived from the supplemental inputs to the cell coming from the contralateral hemisphere or are they independent of these inputs? Are callosal cells, that is cells projecting through the callosum, themselves sensitive to these spatial localization cues? The first question was answered by recording in the same region of A1 as above (the callosal recipient zone), but using cats which had had their callosum surgically transected a few months before the recording session. Results showed that although the relative proportions of the different cell types were somewhat different in the normal and callosotomized animals, all were still present in the cortex of the split-brain cats (see table 1). The second question was resolved by recording directly in the callosum the axons of the interhemispheric projecting cells. In this case, for practical reasons (mainly due to the length of the protocol and the difficulty in recording the fibres for extended periods of time), only interaural time differences were examined. Here again, although the relative proportions differed from those recorded in A1 in normal or in callosotomized cats, all cell types were found (see table 1).

Table 1. Relative proportions of cell types recorded directly in the corpus callosum of normal cats or in the callosal recipient zone of auditory area A1 in normal or callosotomized cats. Cell classifications were based on their response profiles to systematic variations of interaural time or intensity differences

Cell type	Interaural time difference			Interaural intensity difference	
	Callosal fibres	A1	A1 split callosal	A1	A1 split callosal
Lateral	28	16	25	76	27
Midline	24	25	25	8	27
Hemifield	22	11	6	1	27
Omnidirectional	12	32	38	10	10
Unclassified	14	16	6	8	9

In summary, therefore, it appears that cells in area A1 of the cat are sensitive to sounds presented in restricted regions of auditory space and that they are sensitive to the principal cues to localization namely binaural intensity and time differences. Although the callosum projects to these cells and may thus be in part responsible for the establishment of interactive effects, and thus localization accuracy, its elimination through surgery does not abolish the characteristic binaural interactions in the recipient cells. Callosal projection cells, that is, cells which send axons through the structure, also show the typical response profiles to variations in interaural delays.

3.2. Studies with Human Subjects

Normal human subjects performed remarkably well using the present experimental paradigm, attesting to its adequacy for carrying out these studies. Two aspects of the results, however, do merit special mention. First, performance in peri-central auditory space was better than at more lateralized positions. This mainly confirms what has been demonstrated by others (Durlach and Colburn, 1978; Oldfield and Parker, 1984; Makous and Middlebrooks, 1990). Second, statistical analyses (ANOVAs) comparing the performance of the subjects in the left and right fields revealed no hemispheric asymmetries for this task. This confirm previous studies but does not confirm one study using similar procedures and which showed right hemisphere superiority (Duhamel et al, 1986).

The performance of the callosal agenesis subjects was compared to that of their appropriately matched controls. These are graphically illustrated in figure 2A. As with normal controls, accuracy was good for both groups and it was better in peri-central than lateral fields. Moreover, the response histograms of the agenesis subjects closely paralleled those of their controls. However, the former were generally more erratic in their ability to point to the apparent origin of the sound in space. It thus appears that the callosum,

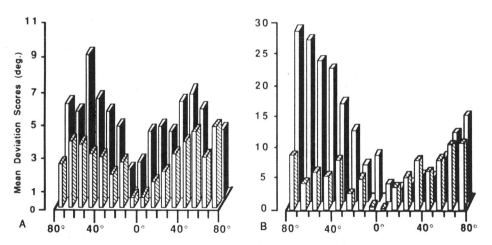

Figure 2. Sound localization performance across the auditory field for the human subjects. A- performance of acallosal subjects (black bars) and their matched controls (dashed bars). B- performance of hemispherectomized subjects (black bars) and their matched controls (dashed bars). Mean deviation scores obtained at each stimulus position are expressed in degrees of arc; in A the right and left hemifield are respectively represented on the right and left side. In B the ipsilateral and contralateral hemifield to the lesioned hemisphere are represented respectively on the right and left side.

though not essential to carry out the task, does contribute to the mechanisms which lead to localization accuracy.

The hemispheric contribution to sound localization appears to be even more crucial. This is illustrated in figure 2B. The bars represent the averaged deviation between the actual and perceived position of the sound (the absolute error). It is clear from this figure that when the sound was presented in the hemifield contralateral to the intact hemisphere, the performance of the hemispherectomized and matched controls is very similar. However, when the sound appeared in the hemifield contralateral to the lesioned hemisphere, the hemispherectomized subjects showed a much more erratic performance then their matched controls.

4. DISCUSSION

Sound localization depends upon disparities in intensity, time and sound spectrum. These cues are integrated at the single-unit level in subcortical structures and thus presumably an animal or a human subject may be able to carry out localization performance without the contribution of cortical areas. Viewed in this manner, the cortex is at best redundant or at worst uninvolved in this function. Our results using both physiological methods in animals and behavioral approaches with humans having different neurological anomalies suggest otherwise.

It is clear that cells in area A1 do respond differentially to sounds presented in various positions in the auditory field. In some cases they are exquisitely tuned to a very narrow region of space. Input from subcortical structures which have already treated the incoming information might account for the results obtained during free-field presentations of the stimuli. However, the use of dichotic stimulation, where disparities in time and intensity can be manipulated, suggests that cells in this area also respond preferentially to limited ranges of interaural differences. Callosal projecting cells, which originate in many cases in A1, are also sensitive to these disparities, reflecting the properties of their area of origin. The cells in the callosal recipient zone, however, are also tuned to interaural time and intensity differences, irrespective of whether the callosum is present or not. It is thus not clear from these physiological results whether the small changes in the relative proportions of cells following a callosotomy would result in altered localization behavior and hence, in the confirmation of the importance of this neocortical commissure to this function.

The behavioral results obtained with the acallosal patients, on the other hand, do attest to the importance of this structure for sound localization. However, the contribution of this interhemispheric link is more supportive than essential: the subjects perform in a fashion which resembles that of their matched controls but they are not as accurate in localizing the sound source. The callosum, or the integrity of the callosal projecting and recepient zones, helps to fine tune this function. These results suggest that cortical areas participate in identifying the origins of a sound in space, a suggestion which is also strongly supported by the results obtained with the hemispherectomized subjects.

ACKNOWLEDGMENTS

These studies were supported by grants from the Natural Sciences and Engeneering Research Council (NSERC) and the Fonds pour la Formation de Chercheurs et l'Aide à la Recherche (FCAR) awarded to F.L., M.L. and J.-P.G.

REFERENCES

Altman, J.A. (1968) Are there neurons detecting direction of sound source motion. Exp. Neurol. 22, 13–25.

Boudreau, J.C. and Tsuchitani, D. (1968) Binaural interaction in the cat superiror olive S segment. J. Neurophysiol. 31, 442–454.

Brugge, J. F., Anderson, D.J. and Aitkin, L.M. (1970) Responses of neurons in the dorsal nucleus of the lateral lemniscus of cat to binaural tonal stimulation. J. Neurophysiol. 33, 441–458.

Brugge, J.F., Dubrovsky, N.A., Aitkin, L.M. and Anderson, D.J. (1969) Sensitivity of single neurons in auditory cortex of cat to binaural tonal stiumlation: effects of varying interaural time and intensity. J. Neurophysiol. 32, 1005–1024.

Bryden, M.P., Ley, R.G. and Sugarman, J.H. (1985) Toward a model of dichotic listening performance. Brain and Cogn. 4, 241–257.

Duhamel J.-R., Pinek, B. and Brouchon, M. (1986) Manual pointing to auditory targets: performances of right versus left handed subjects. Cortex 22, 633–638.

Durlach, N.I. and Colburn, H.S. (1978) Binaural phenomena. In: E.C. Caterrette (Ed.), Handbook of Perception, Hearing, Vol. IV, Academic Press, New York, pp. 365–466.

Eisenman, L.M. (1974) An electrophysiological study in auditory cortex (A1) of the cat using free field stimuli. Brain Res. 75, 203–214.

Geshwind, N. (1972) Language and the brain. Scient. Am., 226, 76–83.

Hall, J.L. and Goldstein, M.H. (1968) Representation of binaural stimuli by single units in primary auditory cortex of unanesthetized cats. J. Acous. Soc. Am. 43, 456–461.

Hirsch, J.A., Chan, J.C. and Yin T.C. (1985) Responses of neurons in the cat's superior colliculus to acoustic stimuli. I. Monaural and binaural response properties. J. Neurophysiol. 53, 726–745.

Imig, T.J., Irons, W.A. and Samson, F.R. (1990) Single-unit selectivity to azimuthal direction and sound pressure level of noise burst in cat high-frequency primary auditory cortex. J. Neurophysiol. 63, 1448–1446.

Imig, T.J., Reale, R.A., Brugge, J.F., Morel, A. and Adrian, H.O. (1986). Topography of cortico-cortical connections related to tonotopic and binaural maps of cat auditory cortex. In: F. Lepore, M. Ptito and H.H. Jasper (Eds.), Two Hemispheres-One Brain: Functions of the Corpus Callosum, Alan R. Liss, New York, pp. 103–115.

Irvarsson, C., De Ribaupierre, Y. and De Ribaupierre, F. (1988) Influence of auditory localization cues on neuronal activity in the auditory thalamus of the cat. J. Neurophysiol. 59, 586–606.

Irvine, D.R. and Gago, G. (1990) Binaural interaction in high-frequency neurons in inferior colliculus of the cat: effects of variations in sound pressure level on sensitivity to interaural intensity differences. J. Neurophysiol. 63, 570–591.

Jenkins, W.M. and Masterton, R.B. (1982) Sound localization: Effects of unilateral lesions in central auditory system. J. Neurophysiol. 47, 987–1016.

Jenkins, W.M. and Merzenich, M.M. (1984) Role of cat primary auditory cortex for sound-localization behavior. J. Neurophysiol. 52, 819–847.

Kitzes, L.M., Wrege, K.S. and Cassady, J.M. (1980) Patterns of response of cortical cells to binaural stimulation. J. Comp. Neurol. 192, 455–472.

Klingon, G.H. and Bontecou, D.C. (1966) Localization in auditory space. Neurology 16, 879–886.

Kuwada, S., Yin, T.C. and Wickesberg, R.E. (1979) Response of cat inferior colliculus neurons to binaural beat stimuli: possible mechanisms for sound localization. Science 206, 586–588.

Makous, J.C. and Middlebrooks, J.C. (1990) Two-dimensional sound localization by human listeners. J. Acous. Soc. Am. 87, 2188–2200.

Middlebrooks, J.C., Dykes, R.W. and Merzenich, M.M. (1980) Binaural response-specific bands in primary auditory cortes (A1) of the cat: topographical organization orthogonal to iso-frequency contours. Brain Res. 181, 31–48.

Middlebrooks, J.C. and Pettigrew, J.D. (1981) Functional classes of neurons in primary auditory cortex of the cat distinguished by the sensibility to sound location. J. Neurosci. 1, 107–120.

Oldfield S.R. and Parker, S.P.A. (1984) Acuity of sound localization: A topography of auditory space. I. Normal hearing conditions. Perception 13, 581–600.

Philipps, D.P. and Brugge, J.F. (1985) Progress of neurophysiology of sound localization. Ann. Rev. Psychol. 36, 245–274.

Philipps, D.P. and Irvine D.R.F. (1981) Response of single neurons in physiologically defined area A1 of cat cerebral cortex: Sensitivity to interaural intensity differences. Hear. Res. 4, 299–307.

Porier, P., Lassonde, M., Villemure, J.-G., Geoffroy, G. and Lepore, F. (1994). Sound localization in hemispherectomized patients. Neuropsychologia 32, 5, 541–553.

Poirier, P., Lepore, F., Provençal, C., Ptito, M. and Guillemot, J.-P. (1995) Binaural noise stimulation of auditory callosal fibers of the cat: response to interaural time delay. Exp. Brain Res., 104, 30–40.

Poirier, P., Samson, F.K. and Imig, T.J. (1996) Directional mechanisms and spatial preferences of single units in the cat's inferior colliculus (IC). Neuroscience abst. 22, 889.

Rajan, R. Aitkin, L.M., Irvine D.R.F. and McKay, J. (1990) Azimuthal sensitivity of neurons in primary auditory cortex of cats I. Types of sensitivity and the effects of variations in stimulus parameters. J. Neurophysiol. 64, 872–887.

Rauschecker, J.P. and Harris, L.R. (1989) Auditory and visual neurons in the cat's superior colliculus selective for the direction of apparent motion stimuli. Brain Res. 490, 56–63.

Reale, R.A. and Kettner, R.E. (1986) Topography of binaural organization in primary auditory cortex of the cat: Effects of changing interaural intensity. J. Neurophysiol. 56, 663–682.

Samson, F.K., Clarey, J.C., Barone, P. and Imig, T.J. (1993) Effects of ear plugging on single-unit azimuth sensitivity in cat primary auditory cortex. I. Evidence for monaural directional cues. J. Neurophysiol. 70, 492–511.

Samson, F.K., Poirier, P., Irons, W.A. and Imig, T.J. (1996) Effects of ear plugging on responses of azimuth-sensitive neurons in medial geniculate body (MGB) and primary auditory cortex (A1) of barbiturate-anesthetized cats. Neuroscience abst. 22, 889.

Sanchez-Longo, L.P. and Forster, F.M. (1958) Clinical significance of impairment of sound localization. Neurology 8, 119–125.

Sanchez-Longo, L.P., Forster, F.M. and Auth, T.L. (1957) A clinical test for sound localization and its applications. Neurology 7, 655–663.

Semple, M.N., Aitkin, L.M. Calford, M.B., Pettigrew, J.D. and Philipps., D.P. (1983). Spatial receptive fields in the cat inferior colliculus. Hear. Res. 10, 203–215.

Spitzer, M.W. and Semple, M.N. (1991) Interaural phase coding in auditory midbrain: Influence of dynamic stimulus features. Science 254, 721–723.

Wise, L.Z. and Irvine, D.R. (1985) Topographic organization of interaural intensity difference sensitivity in deep layers of cat superior colliculus: implications for auditory spatial representation. J. Neurophyiol. 54, 185–211.

36

SOUND LOCALIZATION AND PINNA MOVEMENTS IN THE BEHAVING CAT

Tom C. T. Yin and Luis C. Populin

Department of Neurophysiology and
   Neuroscience Training Program
University of Wisconsin
Madison, Wisconsin 53706

1. INTRODUCTION

The ability to localize the source of a sound is an important function of the auditory system. It is of obvious utility for both prey and predator, as well as for social reasons, to quickly and accurately identify the locus of a sound source. Consequently, the neural mechanisms underlying sound localization have been of much interest to psychophysicists, anatomists and physiologists studying the auditory system. It is probably safe to say that we understand more about the central processing of sound localization cues than of any other auditory function (e.g. pitch perception, vowel discrimination)(see Irvine, 1986; Yin et al. 1997).

Our lab has used the domestic cat for investigations of sound localization in the central auditory system for several reasons: it is a nocturnal predator that must rely heavily on its auditory system for localization, its auditory system is similar to that of humans, and more is known about the cat's central nervous system than any other animal. Although much is known about the physiological and anatomical mechanisms of sound localization in cats, much less work has been done on its behavioral capabilities. Most behavioral studies have involved training on a sound localization task followed by lesions of various brain regions to determine the effect on behavior. Usually, the tasks have either been approach to two or more speakers (Casseday and Neff, 1973; Jenkins and Masterton, 1982) or conditioned avoidance in minimum audible angle tasks (MAA, Martin and Webster, 1987). Heffner and Heffner (1988) measured MAAs using both conditioned avoidance and two-choice testing and found negligible differences between performance on the two tasks. Neither of these tasks measures absolute localization ability, but rather detection of a change or relative positions of two sound sources. Our experimental procedure, in which cats were trained to look at the locus of sound sources, allows a measure of the absolute localization error, which is similar to the head-pointing task used by Thompson and Masterton (1978), Beitel and Kaas (1993), May and Huang (1996).

In recent years a number of laboratories have studied the acoustic filtering properties of the external ears or pinnae in a variety of animals, including humans (Wightman and Kistler, 1989), cats (Musicant et al. 1990, Rice et al. 1992), and ferrets (Carlile 1990). However, there is little information available on the extent to which animals with mobile pinnae actually use them during various acoustic tasks. Heffner and Heffner (1982) described an elephant who held its pinnae out in a stereotyped and unusual position while working on a sound localization task. Heffner and Heffner (1988) also found that the performances of cats on a MAA experiment using both short and long duration stimuli were similar and concluded that movement of the pinnae, which was not possible with the short duration stimulus, had little effect. On the other hand, Jenkins and Masterton (1982) found that monaural localization was improved by longer duration stimuli, which they attributed to the opportunities for scanning movements of the head. It seems likely that pinna movements could have played a part in this improvement. As a first step in uncovering the role of pinna movements, we recorded the movements of the pinnae during active sound localization.

2. METHODS

Domestic cats were trained to look at light and sound sources with the head fixed using operant techniques for a food reward. The cats were positioned in an anechoic room facing an array of speakers (Radio Shack supertweeters) with light emitting diodes (LEDs) positioned in the center of each speaker. The speakers were hidded from view of the cat by a dark screen through which the lit LEDs could be seen and the acoustic stimuli heard. The magnetic search coil technique (Fuchs and Robinson, 1966) was used to measure movements of the eyes and ears. The cats were trained on a variety of oculomotor tasks using visual and/or acoustic stimuli (fixation, standard and delayed saccades, sensory probe). For most of the data presented here, the cats were working on a standard saccade task: after a fixation LED was turned on, the cat was required to fixate it for a set duration (usually 0.5 to 1 sec) at the end of which the LED was turned off and simultaneously a different visual or acoustic target turned on. A saccade to the location of the new target within a temporal window and fixation for another set duration (.5 to 1 sec) were required to receive a reward. The task type and target locations were varied randomly from trial to trial while the fixation durations were varied from task to task. Initially the cats were trained to look at visual stimuli and gradually switched to acoustic targets. In this way we could compare localization behavior using visual and acoustic targets.

All surgical procedures were conducted under pentobarbital or gas (halothane) anesthesia. A device for restraining the head was implanted over the frontal aspect of the skull using dental acrylic and titanium bone screws. Coils of fine, teflon-coated, stainless steel wires were sutured around the eye balls and the ends of the wires led subcutaneously to connectors anchored in the dental acrylic. In addition we also implanted coils on the posterior aspect of each ear to monitor pinna movements.

After recovery from the surgery, the cats started on the behavioral training regime. The eye coils were calibrated behaviorally by turning on red LEDs at known positions in the frontal field. The cats would naturally fixate these LEDs allowing the horizontal and vertical components of the eye movements to be calibrated. A computer monitored eye movements during the task and rewarded the cat only if its eye position remained within a distance from the target for a preset duration. Individual cats exhibited considerable variability, ranging from less than one week to several months, in the speed with which they learned the behavioral task, particularly those involving acoustic targets. Movements of the ear were not behaviorally conditioned. The ear coils were calibrated by rotating identical dummy coils that were placed in the same position as the pinna coils when the cat was alert and fixating straight ahead. Acoustic stimuli were generated using a digital stimulus system and consisted of broad band noise, clicks, or narrow band filtered noises delivered to one of the 15 possible speakers.

Several procedures were necessary to insure that the cat used the acoustic stimulus to localize the sounds rather than the switching transient generated when a speaker was selected: each of the speakers had a separate amplifier so that the background noise of each amplifier was on continuously, the selection of speakers was done early in the behavioral task (before the initial fixation LED was turned on) rather than just before the speaker was turned on, and we selected all speakers rapidly in succession followed by a deselection of all but one speaker, thereby producing two transients from all but one of the 15 speakers. Without these precautions, a cat was able to perform the localization task using the nearly imperceptible artifacts generated in selecting a speaker.

To minimize the possibility that the cat could use differences in the frequency response of individual speakers as a cue to their location, we picked matching impulse responses for the 15 possible speakers from a large original set. In addition in all of the training and experimental runs the levels of the acoustic stimuli delivered were randomly roved over 20 dB.

3. RESULTS

3.1 Auditory Localization

Behavioral data has been collected from 5 cats, trained for 5–6 days/week for periods ranging from one week to 8 months. It is obviously impossible to illustrate much of the large data set collected. Therefore, we will show data that illustrate typical results. Figure 1 shows vertical (upper) and horizontal (lower) eye movement traces made to four visual (left) and corresponding acoustic (right) targets on the vertical and horizontal meridians, respectively. All of these trials and other trials not shown were delivered in random order. Auditory stimuli were long-duration, broad band (0.1–25.0 kHz) noise bursts. In all cases shown here the initial fixation LED was at the center of gaze; it was turned off at time 0 msec, at which time the target LED or speaker was turned on. The arrows on the right indicate the positions of the targets and the brackets show the electronic window within which the eye movement had to land and remain to receive a reward. Figure 1 shows that the cat can make saccadic eye movements to acoustic targets as well as to visual ones, though there are some notable differences. The visual saccades show the expected stereotyped time course and have relatively constant latency. The auditory saccades are more variable both in their latency and in their accuracy. Moreover, there are more corrective saccades, particularly for larger saccade amplitudes.

Close examination of the trajectories of the eye movements made during visual and auditory trials revealed that the kinematics of the eye movements were different. Eye movements to broadband noise stimuli started with a slow ramp component of variable duration, which was followed by a fast saccadic component that kinematically resembled visual saccades. Interestingly, this slow ramp movement was not present when single click stimuli were

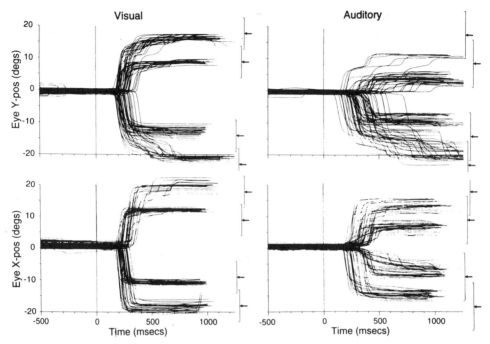

Figure 1. Eye movements to visual (left) and auditory (right) targets using the saccade task. The vertical (upper) and horizontal (lower) components of eye movements to 4 different positions along the vertical and horizontal meridians, respectively, are shown. In all cases the initial fixation LED was at the origin (0°, 0°) and the target came on at 0 msec.

used or when the cat was asked to make delayed saccades, in which there is an additional time, usually about 500 msec, before the cat must execute the eye movement.

3.2 Controls

The data shown in Fig. 1 indicate that the cat is able to make saccadic eye movements to sound sources. We have also done several control experiments to insure that the cats were indeed localizing the sound sources. Several controls are built into the experimental design. During all of the training and experimental runs the cat is presented with a large (ca. 20–40) variety of possible tasks, all of them randomly intermixed. Typically, the cat will have to do visual or auditory fixation trials to a number of different targets and a mixture of visual, auditory, and visual/auditory saccade trials starting from different locations and ending at other locations. In addition the length of time that the cat must fixate the initial target during both fixation and saccade trials is randomly varied. As a result of these randomizations, it is impossible for the cat to predict what the next task will be.

To prevent the cat from memorizing the locations of the speakers, each of the LED/SPKR assemblies could be moved, though this was usually only done at the beginning, not during a training session. No visual cues were available for the auditory tasks since the speakers were hidden from view behind a cloth screen. We were also concerned that the cat might be able to associate and memorize the location of the auditory sources from the locations of the LED. The fact that cats systematically mislocalize the acoustic

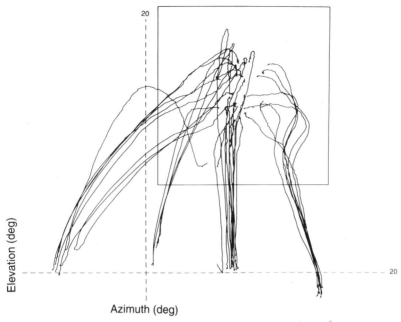

Figure 2. Trajectories on an x-y plane of eye movements made from four different initial positions to an acoustic target at (8°, 13.5°).

targets (Fig. 1) indicates that they do not use the memory of the visual targets as a cue to the auditory ones. In addition we placed acoustic targets at unique positions, off the vertical and horizontal meridians that are the locus for the rest of our targets. Figure 2 shows the trajectories of eye movements made to one such novel target located at (8°, 13.5°) from four different starting positions. In this case, there was never a visual target from this source and yet the cat was able to localize it.

3.3 Movements of the Pinna during Auditory Localization

Recordings of pinna movements using the search coil technique during the sound localization tasks have revealed consistent patterns of movement. Figure 3 shows the vertical and horizontal components of pinna movements during the visual and auditory delayed saccade tasks. In delayed saccades the offset of the initial fixation LED, which is the signal for the cat to execute the saccade, is delayed (by 500 msec in Fig. 3) with respect to the onset of the target. There are several noteworthy features of the pinna movement records. First, during the fixation of the initial LED at the primary position, the pinna usually adopted a standard position even though there were no behavioral contingencies placed on pinna position. Second, there were consistent movements of the pinna even when the tasks were strictly visual in nature, without any auditory stimulus delivered. Third, the movements of the pinna during the visual and auditory saccade were different. During a rightward visual saccade, the right pinna moved to the right at about the same time (250 msec after the signal to move) as the eye movement. During saccades to an acoustic target at the same location, there is a short latency (30–40 msec) rightward pinna movement followed by a smaller rightward movement that occurs at about the same time

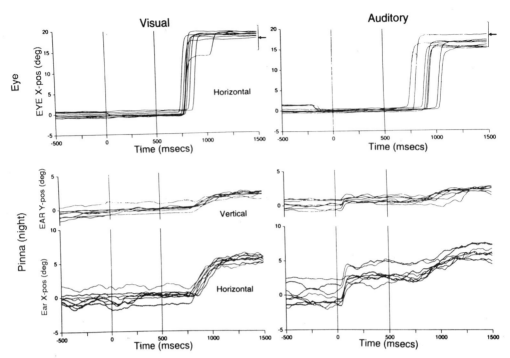

Figure 3. Eye movements (upper trace) and pinna movements (lower two traces) to visual (left) and auditory (right) targets. In all cases the target was located at 18° to the right. The pinna traces show the vertical and horizontal components recorded by a coil implanted in the right ear.

as the eye movement. Fourth, pinna movements are asymmetrical: movements of the ear ipsilateral to the target are larger than movements of the contralateral ear (not shown). Fifth, all of these pinna movements were very consistent and did not habituate, even after thousands of trials over many months of training. Finally, the total movement of the pinna for visual and auditory trials was comparable but for both modalities the magnitude of movement was only a weak function of stimulus eccentricity.

4. DISCUSSION

Since we have paid special attention to eliminating possible artifacts that could provide contaminating cues for the cat to localize the acoustic stimuli, we are confident that our results represent efforts by the cat to look at the sound stimuli. Errors in eye movements, which is the metric that we have used to measure localization performance, could be caused by mislocalizations of the target or by errors in the oculomotor system. Comparison of the accuracy with which visual and auditory localization is performed provides a means of assessing motor errors. Since the errors in auditory localization were generally larger than those in visual saccades, we conclude that these differences largely reflect mislocalizations of the acoustic targets.

The major conclusion of this study is that cats can be trained to look at acoustic targets and that this can provide a convenient preparation for studying neural mechanisms of sound localization. Our results show that localization of auditory targets by eye movements is less accurate and more variable than of comparable visual targets. The results shown in Fig. 2 are important for two reasons: one, it shows that the cats are not localizing auditory targets by associating the remembered location of LEDs with speakers and two, it shows that localization of auditory targets is not affected by initial eye position (Fig. 3) which contradicts the hypothesis set forth by Harris et al. (1980) that cats do not compensate for misalignments of the visual and auditory coordinate systems.

In addition to the deficits in localization, the eye movements to acoustic targets also showed differences in kinematics. To our knowledge, this is the first description of such slow ramp movements preceding saccades. Since the ramp is not present for single clicks or for delayed saccades, we suggest that the slow ramp reflects uncertainty about the target location. We hypothesize that the cat has determined the direction of the target at the time the slow ramp begins, since it almost never ramps in the wrong direction, but needs additional time to compute the position of the target.

Another interesting aspect of our results concern the measurements of pinna movements during these localization tasks. Since the pinna moves with changes in eye position during both visual and auditory tasks, it suggests that the pinna movements are part of a general orienting response rather than a strictly acoustic response. The short latency movement occurs in response to any auditory stimuli even one which is not a target for an eye movement, such as in sensory probe trials, and is therefore likely to be an acoustic reflex.

REFERENCES

Beitel, R.E. and Kaas, J.H. (1993) Effects of bilateral and unilateral ablation of auditory cortex in cats on the unconditioned head orienting response to acoustic stimuli. J. Neurophysiol. 70, 351- 369.

Carlile, S. (1990) The auditory periphery of the ferret. II: The spectral transformations of the external ear and their implications for sound localization. J. Acoust. Soc. Am. 88, 2196–2204.

Casseday, J.H. and Neff, W.D. (1973) Localization of pure tones. J. Acoust. Soc. Am. 54, 365–372.

Fuchs, A.F. and Robinson, D.A. (1966) A method for measuring horizontal and vertical eye movements chronically in the monkey. J. Appl. Physiol. 21, 1068–1070.

Harris, L.A., Blakemore, C. and Donaghy, M. (1980) Integration of visual and auditory space in the mammalian superior. Nature 288, 56–59.

Heffner, R.S. and Heffner, H.E. (1982) Hearing in the elephant (Elephas maximus): Absolute sensitivity, frequency discrimination, and sound localization. J. Comp. Physiol. Psychol. 96, 926–944.

Heffner, R.S. and Heffner, H.E. (1988) Sound localization acuity in the cat: Effect of azimuth, signal duration and test procedure. Hear. Res. 36, 221–232.

Irvine, D.R.F. (1986) The auditory brainstem: processing of spectral and spatial information. Springer-Verlag, Berlin.

Jenkins, W.M. and Masterton, R.B. (1982) Sound localization: effects of unilateral lesions in central auditory system. J. Neurophysiol. 47, 987–1016.

Martin, R.L. and Webster, W.R. (1987) The auditory spatial acuity of the domestic cat in the interaural horizontal and median vertical planes. Hear. Res. 30, 239–252.

May, B.J. and Huang, A.Y. (1996) Sound localization behavior in cats: I. Localization of broadband noise. J. Acoust. Soc. Am. 100, 1059- 1069.

Musicant, A.D., Chan, J.C.K. and Hind, J.E. (1990) Direction-dependent spectral properties of cat external ear: New data and cross-species comparisons. J. Acoust. Soc. Am. 87, 757–781.

Rice, J.J., May, B.J., Spirou, G.A. and Young, E.D. (1992) Pinna-based spectral cues for sound localization in cat. Hear. Res. 58, 132- 152.

Thompson, G.C. and Masterton, R.B. (1978) Brain stem auditory pathways involved in reflexive head orientation to sound. J. Neurophysiol. 41, 1183–1202.

Wightman, F.L. and Kistler, D.J. (1989) Headphone simulation of free- field listening. I: Stimulus synthesis. J. Acoust. Soc. Am. 85, 858–867.

Yin, T.C.T., Joris, P.X., Smith, P.H. and Chan, J.C.K. (1997) Neuronal processing for coding interaural time disparities. In: Binaural and Spatial Hearing in Real and Virtual Environments, pp. 427–445. Editors: R. Gilkey and T. Anderson. Erlbaum Press, New York.

37

GENERATION OF VIRTUAL SOUND SOURCES FOR ELECTROPHYSIOLOGICAL CHARACTERIZATION OF AUDITORY SPATIAL TUNING IN THE GUINEA PIG

K. Hartung[1] and S. J. Sterbing[2]

[1]Lehrstuhl für allgemeine Elektrotechnik und Akustik
[2]Lehrstuhl für allgemeine Zoologie und Neurobiologie
Ruhr-Universität Bochum
Universitätsstr. 150, IC 1-41
44780 Bochum, Germany

1. INTRODUCTION

The sound emitted by a source is distorted by the pinna, head and body. These distortions are caused by diffraction and reflections in the pinna structure and depend on the frequency and the direction of incident. They can be formally described by the transfer function $H(f,\varphi,\upsilon)$ (φ=azimuth, υ=elevation, head related transfer function, HRTF)(Blauert 1983, Wightman 1989). This function is defined as the ratio between the sound pressure at the eardrum and the sound pressure at the position of the head when no animal is present. If the sound is delivered through earphones, the hearing event will be identical to that under free field conditions provided that the sound pressure at the eardrum is the same as in the free field. This can be achieved by filtering the signal of a sound source with the corresponding HRTF of that direction and by compensating the influence of the earphone transfer function and coupling to the ear canal. In the present study we measured the HRTFs of 8 guinea pigs. The individual HRTFs were used to create virtual sound sources (VSS) which simulate free field conditions. The virtual sound sources were used for electrophysiological characterization of spatial tuning in the central nucleus of the guinea pig's inferior colliculus.

2. METHODS

2.1. HRTF Measurement

The guinea pig is placed on a turntable in the center of an anechoic room. The loudspeakers are mounted on an arc ranging from -10 to 90° in 10° steps. After measuring all the elevations for one azimuth the table is turned clockwise. We measure 122 directions of the upper hemisphere. The microphones (Knowles 3046) are positioned a few millimeters within the entrance of the ear canal of the left and the right ear. HRTFs measured at the tympanic membrane can be separated into a term depending on the direction of incidence, and into a second term which contains the non-directional transfer function of the ear canal. The directional part of the HRTF can be measured at a distance of a few millimeters from the entrance of the ear canal. If the sound pressure at this position matches the sound pressure of free field stimulation, the sound pressure at the eardrum will also be the same as in the free field. A random phase noise sequence is generated digitally and repeated 100 times (sampling rate: 50 kHz, sequence duration: 4096 samples, AD/DA-Converter: PowerDac (Tucker Davis)). The response at the microphones is recorded and averaged synchronuously to the output sequence. The transfer function is derived by dividing the discrete fourier transforms of the signal and the response. Reflections and noise are removed by windowing the impulse response (length of window: 128 samples).

2.2. Earphone Measurement and Calibration

The microphones are left in the same position as for the HRTF measurement and the tips of the tubes of the earphones (Beyer DT911) are positioned at the entrance of the ear canal. As the HRTF will be calibrated by this measurement, the position of the earphone has to be the same during the playback of the virtual stimuli. The transfer function from the ports of the headphone to the microphone is measured using the same procedure described above. The HRTFs are calibrated for playback via earphones by convolving the impulse response of the HRTF with the impulse response of the inverse earphone-to-microphone transfer function. The inverse transfer function is calculated using a least-square approximation in the time domain (Proakis and Manolakis 1982). This approach always leads to stable and causal solutions. The error, defined as the difference between the ideal calibration filter and the realized filter, is less than 1 dB.

2.3. Simulations and Experiment

Noise bursts (50 ms duration, 5 ms rise/fall time, 80 dB SPL (38 dB spectrum level)) were filtered using the calibrated HRTF. The anesthetized (ketamine/thiazine mix) guinea pigs (n=7) were fixed with a headholder attached to a stereotaxic bench in an anechoic chamber. Single unit activity of the central nucleus of the inferior colliculus was recorded with high-impedance (4–10 MOhm) glass microelectrodes (filled with 3M KCl). Each virtual direction was presented 5 times via earphones (Beyer DT911) in a pseudo-random order. The median of spike counts is calculated for each direction and tested against the median of the other directions (Fisher exact test, 2x2 contigency table). Additionally, the frequency tuning curves of the single units were measured using pure tone stimulation and their characteristic frequency was determined. Lesions were made at the end of the electrophysiological recordings allowing a 3D-reconstruction of the recording sites after standard histological treatment.

3. RESULTS

We performed an analysis on the monaural and binaural cues of the HRTF. Example monaural cues are presented as the magnitude of the transfer function for the left and the right ear in the median (Fig. 1A) and in the horizontal plane of animal CA6 (Fig. 1B). The transfer function does not change with the direction of incidence for frequencies below

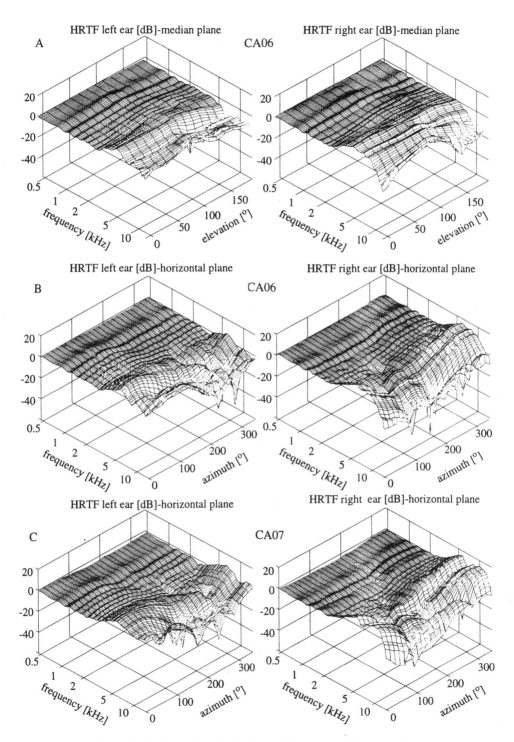

Figure 1. Head related transfer functions of the median and horizontal planes.

1 kHz. In the median plane (Fig.1A) only the right ear shows a notch at approx. 15 kHz for low elevations. For frontal directions the width of the peak around 5 kHz is different for both ears. For the rear directions the HRTF of the left ear contains a notch around 5 kHz which cannot be found for the right ear. Both ears show an attenuation of the high frequencies for the rear directions of incidence. In the horizontal plane (Fig 1B) the shape and the center frequency of the minima and maxima are different for all the directions. The HRTF for the right ear has a broad maximum between 6 and 10 kHz for directions between 45° and 270°. From 90° to 270° a prominent notch around 13 to 15 kHz can be observed. The region of attenuation is much broader at the left ear and covers an area from 195° to 315° azimuth and 7 to 15 kHz. The peak around 5 kHz is much narrower than the one at the right ear. In relation to the body axes this asymmetry looks similar in all the animals that have been tested so far.

Nevertheless the HRTFs of all the animals display significant inter-individual differences. Figure 1C shows the HRTF of a different animal (CA7) for the same directions as presented in Fig. 1B. For the left ear the notches occur at smaller angles (150° to 195°). The peak for frontal directions around 6 kHz is a little broader. For the right ear the frequency band of attenuation is broader and ranges from 6 to 15 kHz, but smaller in space (90° and 210°).

The interaural cues consist of interaural time differences (ITD) and interaural level differences (ILD). Both cues are a function of the direction of incidence and frequency. The ITDs are calculated by filtering the left and right head-related impulse responses with the respective third-octave bandpass filter. The envelopes of the two signals are then cross-correlated. The positions of the maximum of the cross-correlation-function marks the interaural time difference. The ILDs are defined as the energy ratio of the left and right HRIR (head-related impulse response). Frequencies lower than approx. 1 kHz contain no directional information. The shapes of iso-ILD and iso-ITD-contours vary with frequency. The iso-ILD- and iso-ITD-contours are different for the same frequency and have an asymmetric shape relative to the frontal plane (Fig.2). Maximum ILD was 15dB, maximum ITD is 500μs.

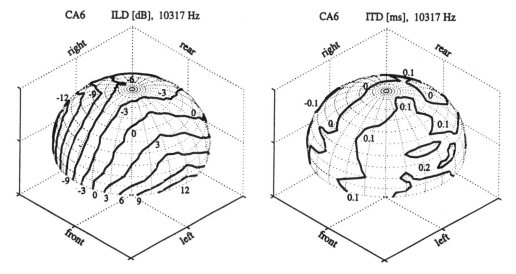

Figure 2. ILD - interaural intensity difference, ITD - interaural time difference.

We recorded n=167 ICc single units, of which n=152 could be activated by the virtual sound source. The other 15 units did not respond to either broadband stimulus, but could be driven by pure tones. From the 152 units, 92.1% (n=140) were significantly spatially tuned. The majority of the units was activated by virtual sound sources (VSS) from the lower contralateral hemifield, but tuning to high elevation VSS and front or back positions with low, intermediate and high elevation occurred frequently (for examples see Fig.3). Only a few neurons (n=23) responded best to ipsilateral VSS. A subset of single units (n=39) was stimulated with both the correct HRTF and the HRTF of a different animal. When these units were stimulated with the VSS of a different individual, they showed either no spatial tuning (n=6) or an altered spatial tuning.

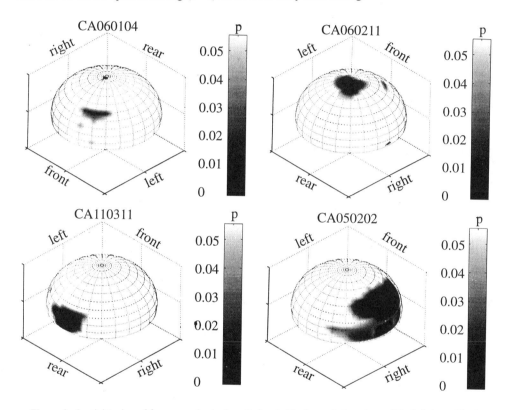

Figure 3. Spatial tuning of four example single units located in the central nucleus of the inferior colliculus.

4. DISCUSSION

Our data strongly supports the psychocaoustical findings that spatial localization is best, when individual head-related transfer functions are used (Wenzel et al. 1993). Experiments with cats have shown that the pinna position affects the HRTF (Young et al. 1996). Domestic guinea pigs do not show pinna movements when they localize external sound sources. At extremely high sound pressure levels leading to acoustic startle response the pinna moves as the animal flinches. The guinea pigs usually turn their heads or entire bodies, when they are localizing a low intensity sound source (personal observations). In the present study rear position tuning and high elevation tuning (e.g. 90°) could be shown for the first time. In contrast to the results of other studies, the spatial receptive fields of most neurons we found in the ICc are smaller than a quadrant (Brugge et al. 1996). In our opinion this is due to the use of individual HRTFs. Using individual HRTFs to generate virtual sound sources is a valuable method for studying the neuronal processing of spatial hearing because it is impossible to create reliable free-field conditions in an experimental electrophysiological set-up. Microdrives, stereotaxic benches, headholders and the surgery, which are necessary for accessing the auditory nuclei or cortical areae influence the HRTF. Furthermore, the domestic guinea pig is a suitable animal model for comparing electrophysiological experiments with human psychoacoustics because of the largely overlapping frequency range and the non-movable pinna.

5. ACKNOWLEDGMENTS

This work was supported by DFG grant GRK 81/2–96 (KOGNET - Cognition, Brain and Neuronal Networks). We wish to thank G.J. Dörrscheidt for providing software for the statistical analysis of the electrophysiological data.

6. REFERENCES

Blauert, J. (1983), Spatial hearing, MIT Press, Cambridge (Mass.).

Brugge, J.F., Reale, R.A., Hind, J.E. (1996), The structure of spatial receptive fields of neurons in primary auditory cortex of the cat. J. Neurosci. 16(14):4420–4437.

Proakis, J.G., Manolakis D.G. (1982), Digital signal processing - principles, algorithms and applications, 2nd edition, MacMillan, New York.

Wenzel, E.M., Arruda, M., Kistler, D.J., Wightman, F.L. (1993), Localization using non-individualized read-related transfer functions, J. Acoust. Soc. Am. 94:111–123.

Wightman, F.L., Kistler, D.J. (1989), Headphone simulation of free-field listening. I. Stimulus synthesis, J.Acoust. Soc. Am. 85:858–865.

Young, E.D., Rice, J.J., Tong, S.C. (1996), Effects of pinna position on read-related transfer functions in the cat, J. Acoust. Soc. Am. 99:3064–3076.

38

SPEECH REPRESENTATION IN THE AUDITORY NERVE AND VENTRAL COCHLEAR NUCLEUS

Quantitative Comparisons

Brad J. May,[1] Glenn S. Le Prell,[2] Robert D. Hienz,[3] and Murray B. Sachs[1,2]

Center for Hearing Sciences and
[1]Department of Otolaryngology-HNS
[2]Department of Biomedical Engineering
[3]Department of Psychiatry and Behavioral Sciences
Johns Hopkins University School of Medicine

1. INTRODUCTION

The representation of speech-like stimuli in the auditory nerve and cochlear nucleus have been the subject of numerous studies over the past 20 years (Blackburn and Sachs, 1990; Geisler, 1988; Kiang and Moxon, 1974; Palmer, et al., 1986; Sachs and Young, 1979; Young and Sachs, 1979). In both the auditory nerve and cochlear nucleus, for example, the spectra of vowels can be represented in terms of average rate (Sachs and Young, 1979) or the temporal patterns of firing (phase-locking; Blackburn and Sachs, 1990; Young and Sachs, 1979) or a combination of the two (Blackburn and Sachs, 1990; Young and Sachs, 1979). In this chapter we focus on average rate representations.

The spectra of complex sounds like vowels are well represented at low sound levels by plots of average discharge rate versus best frequency (BF) across the population of auditory-nerve fibers (rate profiles; (Sachs and Young, 1979). For example Fig. 1A, shows rate-place profiles for a population of high spontaneous rate (SR $\geq$ 20 s/s) auditory-nerve fibers in response to the vowel /ɛ/ (as in "get"; the spectrum of the vowel is shown in Fig. 4). Normalized rate [(rate to the vowel - SR)/(saturation rate to BF tones - SR) = (driven rate)/(driven saturation rate)] is plotted versus BF. At vowel levels of 25 and 35 dB SPL the profiles resemble the spectrum of the vowel with peaks at the formant frequencies (arrows) and troughs in between. However, because of rate saturation (Sachs, et al., 1989) and two-tone suppression (Sokolowski, et al., 1989) the profiles for higher levels lose these spectral features (Sachs and Young, 1979). However, the low SR auditory-nerve fibers (SR $\leq$ 1 s/s) have higher thresholds (Liberman, 1978) and wider dynamic range

Acoustical Signal Processing in the Central Auditory System
edited by Syka, Plenum Press, New York, 1997

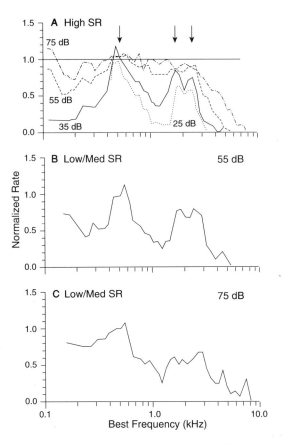

Figure 1. (A) Rate-place profiles for auditory-nerve fiber responses to /ɛ/. Normalized rate averaged over fibers with similar BFs is plotted versus BF. Normalized rate is driven rate to the vowel divided by driven saturation rate and varies form 0 (SR) to 1 (saturation rate). (A) Data from high SR (SR ≥ 20 s/s) fibers at several stimulus levels. (B) and (C) Data for low (SR ≤ 1 s/s) and medium SR (1 < SR < 20 s/s) fibers from the same experiment as in (A). Redrawn from Sachs and Young, 1979.

(Sachs, et al., 1989; Yates, et al., 1990). As shown in Figs. 1B and 1C, these properties allow the low SR fibers to maintain a good rate representation of stimulus spectrum at higher stimulus levels (Sachs and Young, 1979).

It has been suggested that a rate representation that is robust over a broad range of stimulus levels could be formed by differentially weighting the rates of high and low SR fibers (Delgutte, 1982; Winslow, et al., 1987). According to what we have called the "selective listening" hypothesis, a central rate processor should weight its high SR inputs most heavily at low sound levels and its low SR inputs should be given most weight at high sound levels. Indeed, as is shown by the rate profiles in Fig. 2, the population of stellate cells ("chopper" response types; Rhode, et al., 1983; Rouiller and Ryugo, 1984; Smith and Rhode, 1989) in the anteroventral cochlear nucleus (AVCN) is a good candidate to carry out this selective listening process. At the lowest sound level shown, 25 dB SPL, low SR fibers generally do not respond to the second formant; in the region of the second and third formants the rate profile for the chopper units (solid lines) are comparable to that for high SR auditory-nerve fibers (dashed lines). At all higher levels the chopper profiles resemble those of the low and medium SR fibers (dotted lines). Even at 75 dB SPL, where

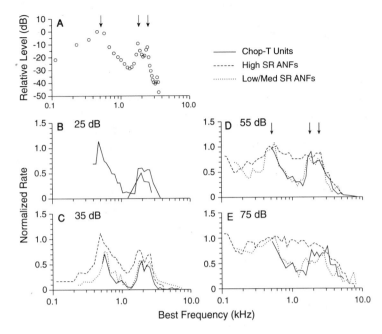

Figure 2. Rate-place profiles for responses to /ɛ/ for auditory-nerve fibers and chop-T units (Blackburn and Sachs, 1989) recorded in the anteroventral-ventral cochlear nucleus. Data shown for four stimulus levels. Solid lines: chop-T; dashed lines: high SR auditory-nerve fibers; dotted lines: low/medium SR auditory-nerve fibers. Redrawn from Blackburn and Sachs, 1990.

rate saturation has obliterated formant peaks in the high SR profiles, the chopper and low/medium SR profiles maintain a clear representations of peaks at the first formant and in the region of the second and third formants (Blackburn and Sachs, 1990).

Conley and Keilson (1995) have plotted, versus BF, differences in rate to two vowels differing only slightly in second formant frequency. Such "rate difference profiles" for high SR auditory-nerve fibers show clear evidence of the formant differences, even at levels where there appears to be no formant information in the rate profiles of Fig.1A (Conley and Keilson, 1995). This improved representation probably reflects the fact that they are making comparisons between the rate responses of a single fiber to different vowels. In doing so they avoid the inter-fiber variability inherent in the true population approach (Conley and Keilson, 1995; Rice, et al., 1995). This result calls into question the need for a selective listening process. In the studies described below we have re-examined the rate representation of vowels in the auditory nerve and in the AVCN in context of modeling the discrimination of formant frequencies. This approach has allowed us to make some more quantitative comparisons of these representations and to reinforce the case for the selective listening hypothesis.

2. PSEUDO-POPULATION METHOD

Because of the difficulty of generating population data like those in Figs. 1 and 2, we developed a pseudo-population approach: A single fiber is used to approximate a population of fibers with different BFs by shifting the stimulus along the frequency axis. The shifting is done by changing the sampling rate of the D/A converter used to generate the stimulus. Increasing (or decreasing) the sampling rate shifts the spectrum up (or down) on a log frequency scale without changing the ratios of the formant frequencies (Oppenheim and Wilsky, 1983). The shift is illustrated in Figs 3A-C, which show the spectrum of a vowel when played back at twice the base synthesis rate (A), at the base synthesis rate (B) and at half the base rate (C). The tuning curve plotted on the same frequency axis shows the changing relationship between a typical fiber's tuning curve and the features of the stimulus spectrum. In interpreting the responses to such frequency-shifted stimuli it is convenient to consider the frequency axis shown in Figs. 3D-F, in which the stimulus spectrum remains fixed and the fiber's BF changes. The tuning curve is shifted to an *effective BF* which is the BF of a fiber that has the same relation to the unshifted spectrum as the actual BF has to the shifted spectrum. Effective BF is given by Eq. 1.

$$\text{Effective BF} = \text{Actual BF} \; \frac{\text{Base synthesis rate}}{\text{Playback sampling rate}} \tag{1}$$

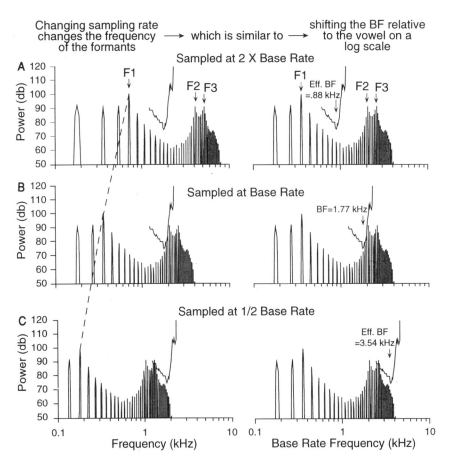

Figure 3. The relationship between a fiber's tuning curve and a vowel spectrum at three playback sampling rates. In the left column plots are shown on the true frequency axis. In the right column plots are shown on the base-rate frequency axis in which the tuning curves are shifted and the spectrum is not. Redrawn from Keilson, et al., 1995.

We will plot results in terms of this effective BF. The interpretation of results with this method assumes that the rate responses of fibers with BFs in the range used (0.5–3.0 kHz) do not depend significantly on BF (see Le Prell, et al., 1996).

3. RATE REPRESENTATION OF VOWELS IN THE AUDITORY NERVE

An example of the results of using this method is shown in Fig. 4 for one high SR fiber (BF = 2.2 kHz). The unshifted spectrum of /ɛ/ is shown (A); the overall level of the vowel is 43 dB SPL. Normalized rate (see above) is plotted versus effective BF for seven different spectrum shifts in (B). Features of the spectrum (formants and troughs) are marked on both plots. For example, the point marked F1 in (B) was obtained with the spectrum shifted so that the first formant was at fiber BF (2.2 kHz); that is, the effective

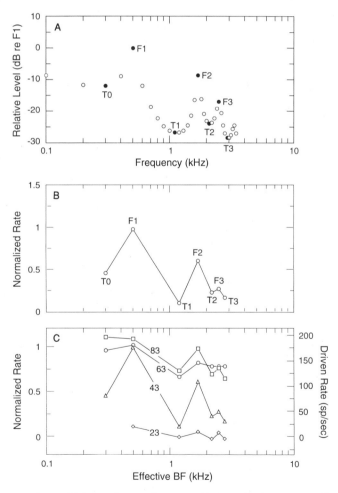

Figure 4. (A) Spectrum of normal /ɛ/ plotted versus frequency. (B) Normalized rate versus effective BF (see text) for a high SR fiber with BF = 2.2 kHz. Spectral features are indicated in both (A) and (B). (C) Same as (B) but for four vowel levels. Redrawn from Le Prell, et al., 1996.

BF was at the first formant (0.5 kHz) as shown. Similarly, for the point marked T1 the spectrum was shifted to put the trough T1 at BF so that the effective BF was at the frequency of T1 (1.10kHz). The correspondence between the plot of normalized rate versus effective BF and the vowel spectrum is clear: there are peaks at effective BFs corresponding to the formants, and a deep trough corresponding to the spectral trough between the first two formants. Figure 4C shows the behavior of rate versus effective BF profiles for this same fiber at four sound levels. It clearly demonstrates the deterioration of the rate representation of vowel spectra at high stimulus levels for this high SR fiber. However, comparison with Fig.1A suggests that the rate representation measured with this pseudo-population approach is better at the highest stimulus levels than that measured with the standard population approach (Fig. 1A). As we suggested above, this improved representation probably reflects the fact that we are avoiding the inter-fiber variability inherent in the true population approach.

The differences in rate representations among the SR groups seen in the population approach (Fig.1) are also evident with the pseudo-population method. Figure 5 shows averaged normalized rate plotted versus effective BF for high, medium and low SR fibers at four stimulus levels. Each plot is the average of the normalized rates of all fibers tested in each SR group. As in Fig. 1, the rate-representation of vowel spectrum is more robust across stimulus levels for the low SR fibers than for the high SR group. Nonetheless, even at the highest level tested there are still spectrally-related peaks and troughs in the high SR pseudo-population profiles which are not seen in the standard population profiles. In order

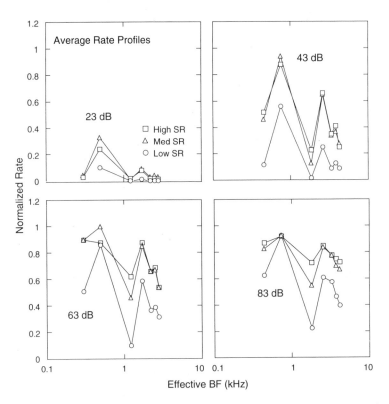

Figure 5. Normalized rate versus effective BF averaged across fibers in three SR groups. Redrawn from Le Prell, et al., 1996.

to compare the spectral information carried by the high SR fibers with those carried by the low SR fibers and to compare representations between auditory-nerve and AVCN, we next consider a quantitative measure of what we will call "sensitivity" of representations to spectral features.

4. MEASURE OF SENSITIVITY TO SPECTRAL FEATURES

A simple relationship between discharge rate and the sound level of individual features (peaks and troughs) of the vowel has allowed us to simulate rate-place profiles for /ɛ/ for the purpose of estimating the discriminability of formant frequencies (Keilson, et al., 1995; May, et al., 1996). This relationship also allows quantitative comparisons of two representations of the same stimulus. Figure 6C shows driven rate plotted versus feature level for the same data as those in Fig. 4. By feature level we mean the level (dB SPL) of the component in the vowel spectrum at the feature frequency. Each point corresponds to the situation in which the vowel spectrum is shifted to put the corresponding feature at fi-

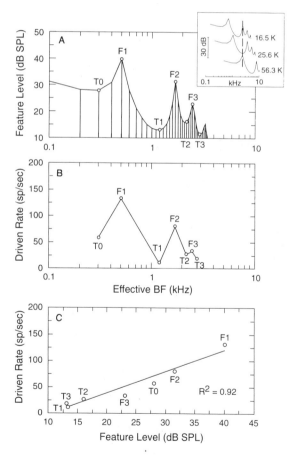

Figure 6. (A) Spectrum of normal /ɛ/ plotted versus frequency. Inset shows vowel spectra shifted to put each of three features at BF (2.2 kHz). (B) Driven rate versus effective BF for a high SR fiber for with BF = 2.2 kHz.; data are from Fig. 4. (C) Driven rate plotted versus feature level (dB SPL). Redrawn from May, et al., 1996.

ber BF (inset); that is, the effective BF is at the feature frequency. The two left most points represent the troughs T1 and T3 and the right most point is the first formant F1. The rate versus feature level function is approximately linear.

Figure 7 shows similar data at three vowel levels. Data are shown for the three SR groups and are averaged across all fibers tested at each level in each group. All of the functions are roughly linear (except for the low SR fibers at 43 dB where five of the features are clearly below fiber thresholds.) In the case of the high SR fibers the slopes decrease with increasing vowel level, reflecting the compression of the rate versus effective BF profiles in Fig. 6. In the case of the low SR fibers there is a roughly parallel shift to the right with increasing vowel level. The shifts seen in each SR group likely reflect the effects of two-tone suppression; for example, notice that at 83 dB SPL the left -most points for the low SR fibers fall well below the 63 dB points at the same feature levels (Sachs and Young, 1979). The functions in Fig. 6 measure the sensitivity of fibers to spectral shape; for convenience we will refer to them as "sensitivity functions".

On the basis of the linear relationships shown in Fig. 7 we may simulate driven rate versus place profiles directly from the vowel spectrum. At any BF we simply convert the spectrum level of the vowel at that BF to driven rate using the appropriate regression line

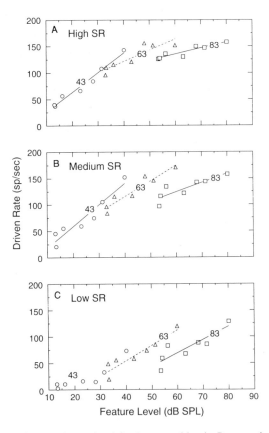

Figure 7. Driven rate plotted versus feature level for three vowel levels. Data are shown for the three SR groups and are averaged across all fibers tested at level in each group. Redrawn from May, et al., 1996.

in Fig. 7. Examples are shown in Fig. 8. Simulated rate-place profiles for high (A) and low (B) SR fibers are shown for vowels at 43 and 83 dB SPL. The compression of the high SR profiles at 83 dB reflects the decrease in the slope of the sensitivity function at high levels. The low SR profile is unaffected by the increase in stimulus level. We now apply this analysis to the representation of vowel spectra in the AVCN.

5. REPRESENTATION OF VOWELS IN THE VENTRAL COCHLEAR NUCLEUS

We have used the pseudo-population method to explore further the representation of vowel spectra in the ventral cochlear nucleus (AVCN). Normalized rate profiles for two classes of choppers, Chop-S and Chop-T are shown in Figure 9. In the experiments represented by these data, only three effective BFs were sampled — at the first and second formants and at the first trough. As was the case in the population data of Blackburn and Sachs (1990; Fig. 2 above), these profiles maintain a good representation of the formant/trough differences, even at sound levels where profiles for the high SR auditory-nerve fibers have seriously deteriorated (Figs. 4 and 5). The maximum driven rates in the chopper profiles are more than twice those in the auditory-nerve profiles.

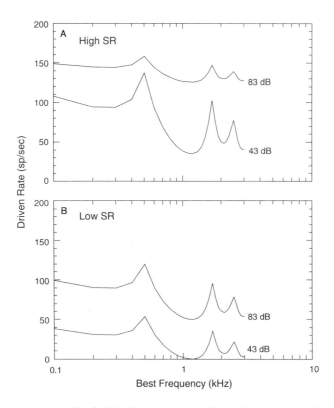

Figure 8. (A) Simulated rate profiles for high SR auditory-nerve fibers. Stimuli are /ɛ/ at 43 and 83 dB SPL. The rate profiles were simulated by applying the linear relationships in Fig. 7 to the vowel spectra. (B) Same as (A) but for low SR fibers.

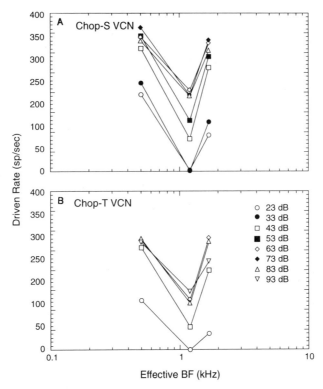

Figure 9. Driven rate plotted versus effective BF for AVCN chop-S (A) and chop-T (B) units in response to /ɛ/. In the experiments represented by these data, only three effective BFs were sampled — at the first and second formants (F1 and F2) and at the first trough (T1; see Fig. 4).

In order to compare the profiles more directly, we plot them in terms of normalized rate in Fig. 10. At levels of 63 dB and higher the profiles for the choppers are similar to those of the low SR fibers (dotted lines). At the lowest sound level the low SR fibers do not respond to the second formant at all, whereas there is a clear spectral representation in the high SR (dashed lines) and chop-T profiles. As we have postulated earlier from our population data (Blackburn and Sachs, 1990), these results suggest that these chopper populations may be performing a selective listening process.

Figure 11 compares the sensitivity functions for the chopper units with those for the high and low SR auditory-nerve fibers. The slopes of the sensitivity functions for the choppers is considerably greater than those of either the low or high SR fibers. This increase in sensitivity is due almost entirely to the increase in absolute driven rates as shown by comparison of the rates to the first formant (right most points). Notice also, that at high vowel levels the sensitivity functions for the choppers are not linear in that the rates to the first and second formants (two right-most points) are equal. That is, fibers with effective BFs at the first and second formants saturate at the same rate and (from Fig. 10) that rate is nearly the same as saturation rates to BF tones. On the other hand, auditory-nerve fibers with effective BF at the second formant saturate at rates less than the saturation rate to BF tones. We have previously attributed this decreased saturation to two-tone suppression (Sachs and Young, 1979).

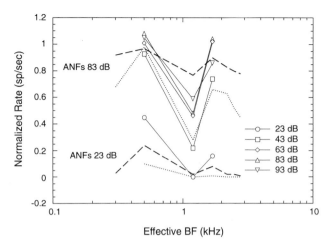

Figure 10. Normalized rate profiles for AVCN chop-T units (solid lines) compared with profiles for high and low SR auditory-nerve fibers (dashed and dotted lines, respectively).

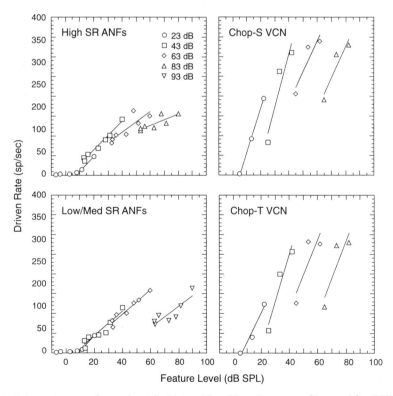

Figure 11. Driven rate versus feature level for high and low SR auditory-nerve fibers and for AVCN chop-S and chop-T units.

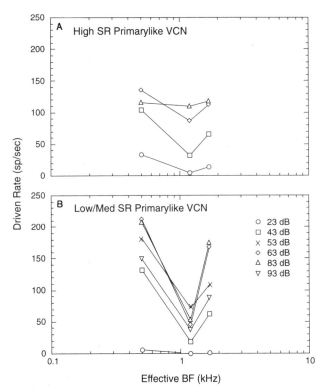

Figure 12. Driven rate plotted versus effective BF for high and low SR AVCN primary-like units in response to /ɛ/. SR classes defined as for the auditory nerve.

Rate profiles for primary-like units in the AVCN are shown in Fig. 12. Units are divided according to SR groups defined as they are for the auditory nerve. Because we studied only units with low (< 3 kHz) BF which give phase-locked responses to BF tones, we were not able to separate primary-like from primary-like with notch units (Blackburn and Sachs, 1989). The behavior of these profiles is similar to those of low and high SR auditory-nerve fibers. The profiles of the high SR primary-like units flatten considerably at high vowel levels while those of the low SR units change little with level. This result is consistent with the observation of Ryugo and his colleagues (Ryugo and Sento, 1991) that suggests that primary-like units receive inputs from either high- or low-SR auditory-nerve fibers, but not both.

Sensitivity functions for primary-like AVCN units are shown in Fig. 13. Those for the high SR AVCN units are similar to those of the high SR auditory-nerve fibers with slopes that decrease with stimulus level. The functions for the low SR AVCN units are considerably steeper than those for the low SR auditory-nerve fibers. It is important to note that these primarylike units do not accomplish selective listening. Selective processing of high and low SR primary-like rate responses is clearly necessary to provide a robust spectral representation over a wide range of stimulus levels at higher levels of auditory processing.

6. DISCUSSION

We have used the pseudo-population paradigm to estimate rate-place profiles for auditory-nerve fibers and AVCN cells in response to the vowel /ɛ/. With this method we

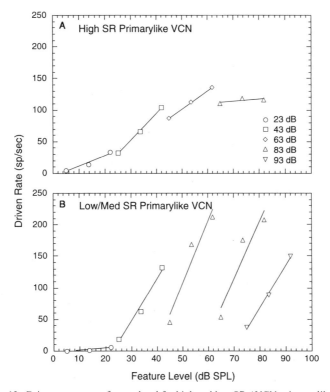

Figure 13. Driven rate versus feature level for high and low SR AVCN primary-like units.

reproduced results of our earlier population studies which showed that rate-place profiles of high SR auditory-nerve fibers deteriorate at high vowel levels and which led us to hypothesize that the CNS performs a selective listening processing of its auditory nerve inputs. However, the pseudo-population data showed that even at high levels there is a clear representation of vowel formants in the high SR profiles. This result leads us to question the need for the selective listening process. A relevant question is whether or not the spectral representation in the rate profiles of high SR auditory-nerve fibers (Fig. 5) is adequate to support behaviorally measured formant frequency discrimination.

In order to address this question we have applied simulations such as those in Fig. 8 to investigate the relative importance of the high and low SR fibers in formant frequency discrimination (May, et al., 1996). A series of vowels was generated with F1 and F3 set at their values for the standard /ɛ/ but with F2 at 1700 Hz (the standard value), 1750 Hz, 1800 Hz, 1900 Hz or 2000 Hz. Rate profiles were generated from the spectra of these vowels as in Fig. 8 and from these simulations we computed rate differences for each vowel pair. We began by asking how discriminable are the rate differences for any one fiber. Motivated by the framework of signal detection theory (Green and Swets, 1974) we computed the measure d' from the simulated rate differences and an empirical estimate of the standard deviations of the associated rates. (d' = rate difference/standard deviation of the rate.) Plots of d' versus BF peak at the places of the second formant frequencies of the two vowels. As expected the maximum d' increases with increasing separation between the formants (standard at 1700 Hz.) For a single fiber the JND is usually defined as |d'| =1. We estimated the JND for second formant frequency difference to be the (interpolated) difference at which d'=1.

We compared the JNDs predicted on the basis of these auditory-nerve simulations with JNDs measured in cats with a positive reinforcement paradigm at comparable sound levels (May, et al., 1996). As shown in Fig. 14, at the lowest sound levels tested (23 and 43 dB) the JND predicted from the rate responses of the single most sensitive high SR fiber is approximately equal to the behavioral JND (about 80 and 50 Hz. respectively). At higher sound levels, however, the predicted JND for high SR fibers increases dramatically while the behavioral JND is roughly constant. At levels of 43 dB and higher, the JND for low SR fibers is nearly equal to the behavioral measure (40–60 Hz.). At 23 dB the low SR fibers generally do not respond to the second formant frequency. These results are consistent with a selective listening model in which JND is determined by high SR fibers at low sound levels and by low SR fibers at high sound levels.

We must be cautious in the interpretation the results of this modeling exercise. The predicted JNDs are computed on the basis of only one (most sensitive) auditory-nerve fiber. Clearly, if the CNS were operating in anything like an optimal fashion it would base discrimination decisions on the responses of more than one fiber. For example, on the basis of the optimum processing of rate responses of all auditory-nerve fibers with BF's between 1 and 5 kHz. Conley and Keilson show that JNDs in second formant frequency of 1 Hz are possible (Barta, 1985; May, et al., 1996). Similarly, May et. al. show that on the basis of optimum processing of the rate responses of only high SR fibers, the predicted JND at 73 dB is less than 10 Hz (May, et al., 1996). Thus, although comparisons based on single fibers suggest that a selective listening scheme may be at play in formant frequency discrimination, simple analysis of discrimination based on populations of fibers shows that the rate responses of high SR fibers alone are sufficient to yield JNDs as low as those measured behaviorally. Furthermore, we have found that the sensitivity to changes in formant frequency in both chopper and primary-like AVCN units is greater than that in the auditory nerve (Figs. 11 and 13). Thus JND estimates based on these units produce estimated JNDs (on the basis of the model outlined above) which are even less than those estimated on the basis of auditory-nerve fibers (Fig.14). Indeed, all of the modeling efforts to explain simple frequency discrimination on the basis of optimum processing of auditory-nerve spike trains produce JNDs that are too small, although a number of trends in the psychophysical data are predicted by the models (Barta, 1985; Hienz, et al., 1993; Siebert, 1970).

There are a number of ways around this problem including drastically limiting the number of fibers whose spike trains are accessible to the central processor. Alternatively, Johnson (Johnson, 1980) has suggested adding noise to the decision process underlying

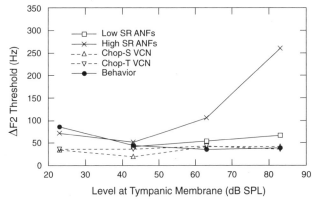

Figure 14. JND for second formant frequency plotted versus vowel level from behavioral measures in cats and from model results like those in Fig.8 for auditory nerve and AVCN. Redrawn from (May, et al., 1996).

the estimates as a way to simulate non-optimum use of auditory-nerve spike trains. In order to account for the large difference between modeled and measured JNDs, the variance of the "processing noise" would have to be much greater than the variance provided by the stochastic pattern of the spike trains. In that case the variance of the decision statistic on which the CNS bases frequency estimates will be approximately the variance of the processing noise. It can be shown that the estimated JND is proportional to this variance (Barta, 1985; Conley and Keilson, 1995). That is, the effect of the processing noise will be to scale the JNDs, so that the trends in the estimates with vowel level will be maintained while the modeled-based functions are shifted vertically in the JND dimension. It is these trends that are suggestive of the need for selective listening.

We have suggested the possibility that the AVCN chopper populations may be performing selective listening. We consider here a simple model on which such selective listening could be based. Chopper responses are recorded from AVCN stellate cells (Bourk, 1976; Rhode, et al., 1983; Rouiller and Ryugo, 1984). These cells typically have four to six long dendrites which exhibit little branching (Cant, 1981; Rhode, et al., 1983; Rouiller and Ryugo, 1984). Details of the somatic and dendritic innervation are not known, but electron microscopic studies suggest two types of stellate cells (Cant, 1981): Type II stellate cells receive a heavier somatic innervation. Both types receive putative inhibitory and excitatory inputs. Liberman traced HRP-filled auditory-nerve fibers to sites of termination in the AVCN (Liberman, 1991). Defining somatic inputs as labeled terminals that were in close proximity to stellate cell bodies and non-somatic inputs as terminals within the neuropil, he found that AVCN stellate cell somata receive more inputs from low and medium SR fibers than from high (see also Fekete, et al., 1982; Rouiller, et al., 1986; Ryugo and Rouiller, 1988). Little is known about the spatial distribution of the SR types on the dendrites of stellate cells.

These anatomical features are consistent with a model of the stellate cell first proposed by (Winslow, et al., 1987) that is capable of performing selective listening (Banks and Sachs, 1991; Lai, et al., 1994; Lai, et al., 1994.). A conceptual form of the model is shown in Fig. 15 (from Lai, et al., 1994.). It is based on excitatory and inhibitory interactions (Koch, et al., 1982). High SR auditory-nerve fibers with BF equal to that of the target stellate cell are assumed to make excitatory synapses on the distal regions of the dendritic tree. Low/medium SR fibers from the same BF region are assumed to form exci-

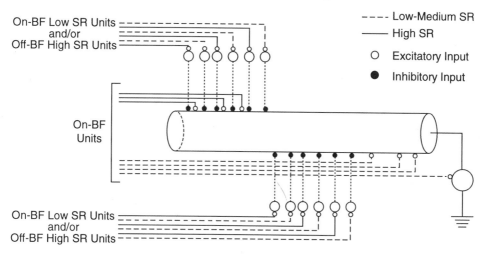

Figure 15. Schematic illustration of a hypothesized neural circuit for performing selective listening processing of auditory-nerve inputs by chopper cells in the AVCN. Redrawn from Lai, et al., 1994.

tatory synapses on the proximal dendrites and/or soma. Off-BF high SR auditory-nerve fibers or on BF low SR fibers are assumed to project to inhibitory interneurons that form inhibitory synapses on the stellate cell. These inhibitory inputs are positioned on the direct path that current must take when flowing from the distal synapses to the soma ("on-path inhibition"). Under these assumptions, at low stimulus levels only the high SR, low threshold auditory-nerve fibers will be activated; thus the distal inputs will drive the cell at low stimulus levels. As stimulus level increases, spread of excitation along the basilar membrane will result in activation of the off-BF fibers and in turn provide on-path inhibitory inputs to the cell. Alternatively, this inhibition could be provided by the on-BF low SR fibers, which are activated at high stimulus levels. On-path inhibition can be very effective in vetoing distal excitation but ineffective in reducing somatic response to more proximal excitation. Thus at high stimulus levels, synaptic current from the distal high SR inputs will be shunted and the low/medium SR auditory-nerve fibers innervating the proximal region will drive the cell.

This model clearly supports the selective listening processing in which the chopper unit output is controlled by high SR auditory-nerve inputs at low stimulus levels and by low SR inputs at high levels. It has been shown, furthermore, that the model can successfully account for post-stimulus-time histograms, rate-level functions and regularity of both chop-S and chop-T units recorded in the AVCN (Banks and Sachs, 1991; Blackburn and Sachs, 1989; Lai, et al., 1994; Lai, et al., 1994.).

In conclusion, the evidence presented here is consistent with the need for selective listening in CNS processing of auditory-nerve inputs. That processing could be occurring in populations of AVCN stellate cells. The sensitivity to spectral features is greater in these cells than in their auditory-nerve fiber inputs.

7. ACKNOWLEDGMENTS

This work was funded by Grant RO1DC00109-22 (Sachs) from the National Institute of Deadness and other Communication Disorders.

8. REFERENCES

Banks, M. I., & Sachs, M. B. (1991). Regularity analysis in a compartmental model of chopper units in the anteroventral cochlear nucleus. J. Neurophysiol., 65, 606–629.
Barta, P. E. (1985) *Testing Stimulus Encoding in the Auditory Nerve*. Ph.D., Johns Hopkins University.
Blackburn, C. C., & Sachs, M. B. (1989). Classification of unit types in the anteroventral cochlear nucleus: post-stimulus time histograms and regularity analysis. J. Neurophysiol., 62, 1303–1329.
Blackburn, C. C., & Sachs, M. B. (1990). The representation of the steady-state vowel /eh/ in the discharge patterns of cat anteroventral cochlear nucleus neurons. J. Neurophysiol., 63, 1191–1212.
Bourk, T. R. (1976) *Electrical Responses of Neural Units in the Anteroventral Cochlear nucleus of the Cat*. Ph.D., Massachusetts Institute of Technology.
Cant, N. B. (1981). The fine structure of two types of stellate cells in the anterior division of the anteroventral cochlear nucleus of the cat. Neuroscience, 6, 2643–2655.
Conley, R. A., & Keilson, S. E. (1995). Rate representation and discriminability of second formant frequencies for /ɛ/-like steady-state vowels in cat auditory nerve. J. Acoust. Soc. Am., 98, 3223–3234.
Delgutte, B. (1982). Some correlates of phonetic distinctions at the level of the auditory nerve. In R. Carlson & B. Granstrom (Eds.), *The representation of speech in the peripheral auditory system* (pp. 131–149). Amsterdam: Elsevier Biomedical Press.
Fekete, D. M., Rouiller, E. M., Liberman, M. C., & Ryugo, D. K. (1982). The central projections of intrecellularly labeled auditory nerve fibers in cats. J. Comp. Neurol., 229, 432–450.

Geisler, C. D. (1988). Representation of speech sounds in the auditory nerve. J. Phonetics, *16*, 20–35.

Green, D. M., & Swets, J. A. (1974). *Signal Detection Theory and Psychophysics*. New York: Krieger.

Hienz, R. D., Sachs, M. B., & Aleszczyk, C. M. (1993). Frequency discrimination in Noise: Comparison of cat cerformance with auditory-nerve models. J. Acoust. Soc. Am., *93*, 462–469.

Johnson, D. H. (1980). The relationship between spike rate and synchrony in responses of auditory-nerve fibers to single tones. J. Acoust. Soc. Am., *68*, 1115–1122.

Keilson, S. E., Richards, V. M., Wyman, B. T., & Young, E. D. (1995). Pitch-tagged spectral representations in the cochlear nucleus. ARO Abstr., *18*, 128.

Kiang, N. Y.-S., & Moxon, E. C. (1974). Tails of tuning curves of auditory nerve fibers. J. Acoust. Soc. Am., *55*, 620–630.

Koch, C., Poggio, T., & Torre, V. (1982). Retinal ganglion: A functional interpretation of dendritic morphology. Phil. Trans. R. Soc. London B, *227*, 227–264.

Lai, Y. C., Winslow, R. L., & Sachs, M. B. (1994). The functional role of excitatory and inhibitory interactions in chopper cells of the anteroventral cochlear nucleus. Neural Computation, *6*, 1127–1140.

Lai, Y. C., Winslow, R. L., & Sachs, M. B. (1994.). A model of selective processing of auditory-nerve inputs by stellate cells of the anteroventral cochlear nucleus. J. Computational Neurosci., *1*, 167–194.

Le Prell, G., Sachs, M. B., & May, B. J. (1996). Representation of vowel-like spectra by discharge rate responses of individual auditory-nerve fibers. Auditory Neurosci., *2*, 275–288.

Liberman, M. C. (1978). Auditory-nerve responses from cats raised in a low-noise chamber. J. Acoust. Soc. Am., *63*, 442–455.

Liberman, M. C. (1991). Central projections of auditory-nerve fibers of differing spontaneous rate. I. Anteroventral cochlear nucleus. J. Comp. Neurol., *313*, 240–258.

May, B. J., Huang, A., LePrell, G., & Hienz, R. D. (1996). Vowel formant frequency discrimination in cats: Comparison of auditory nerve representations and psychophysical thresholds. Aud. Neurosci., *3*, 135–162.

Oppenheim, A. V., & Wilsky, A. S. (1983). *Signals and Systems*. Englewood Cliffs, New Jersey: Prentice-Hall.

Palmer, A. R., Winter, I. M., & Darwin, C. J. (1986). The representation of steady-state vowel sounds in the temporal discharge patterns of the guinea pig cochlear nerve and primary-like cochlear nucleus neurons. J. Acoust. Soc. Am., *79*, 100–113.

Rhode, W. S., Oertel, D., & Smith, P. H. (1983). Physiological response properties of cells labeled intracellularly with horseradish peroxidase in cat ventral cochlear nucleus. J. Comp. Neurol., *213*, 448–463.

Rice, J. J., Young, E. D., & Spirou, G. A. (1995). Auditory-nerve encoding of pinna-based spectral cues: Rate representation of high-frequency stimuli. J. Acoust. Soc. Am., *97*, 1764–1776.

Rouiller, E. M., Cronin-Schreiber, R., Fekete, D. M., & Ryugo, D. K. (1986). The central projections of intracellularly labeled auditory nerve fibers in the cat. J. Comp. Neurol., *249*, 261–278.

Rouiller, E. M., & Ryugo, D. K. (1984). Intracellular marking of physiologically characterized cells in the ventral cochlear nucleus of the cat. J. Comp. Neurol., *255*, 167–186.

Ryugo, D. K., & Rouiller, E. M. (1988). Central projections of intracellularly labeled auditory nerve fibers in the cat: Morphometric correlation with physiological properties. J. Comp. Neurol., *271*, 130–142.

Ryugo, D. K., & Sento, S. (1991). Synaptic connection of the auditory nerve in cats: Relationship between end-bulbs of Held and spherical bushy cells. J. Comp. Neurol., *309*, 35–48.

Sachs, M. B., Winslow, R. L., & Sokolowski, B. A. H. (1989). A computational model for rate-level functions from cat auditory-nerve fibers. Hear. Res., *41*, 61–70.

Sachs, M. B., & Young, E. D. (1979). Encoding of steady-state vowels in the auditory nerve: Representation in terms of discharge rate. J. Acoust. Soc. Am., *66*, 470–479.

Siebert, W. M. (1970). Frequency discrimination in the auditory system: Place or periodicity mechanisms? Proc. IEEE, *58*, 723–730.

Smith, P. H., & Rhode, W. S. (1989). Structural and functional properties distinguish two types of multipolar cells in the ventral cochlear nucleus. J. Comp. Neurol., *282*, 595–616.

Sokolowski, B. A. H., Sachs, M. B., & Goldstein, J. L. (1989). Auditory nerve rate-level functions for two-tone stimuli: Possible relation to basilar membrane nonlinearity. Hear. Res., *41*, 15–124.

Winslow, R. L., Barta, P. E., & Sachs, M. B. (1987). Rate coding in the auditory nerve. In W. A. Yost & C. S. Watson (Eds.), *Auditory Processing of Complex Sounds* (pp. 212–224). Hillsdale, N.J.: Lawrence Erlbaum Assoc.

Yates, G. K., Winter, I. M., & Robertson, D. R. (1990). Basilar membrane nonlinearity determines auditory nerve rate-intensity functions and cochlear dynamic range. Hearing Res., *45*, 203–220.

Young, E. D., & Sachs, M. B. (1979). Representation of steady-state vowels in the temporal aspects of the discharge patterns of populations of auditory-nerve fibers. J. Acoust. Soc. Am., *66*, 1381–1403.

39

PROCESSING OF SPECIES-SPECIFIC VOCALIZATIONS IN THE INFERIOR COLLICULUS AND MEDIAL GENICULATE BODY OF THE GUINEA PIG

J. Syka, J. Popelář, E. Kvašňák, J. Šuta, and M. Jilek

Institute of Experimental Medicine
Academy of Sciences of the Czech Republic
Vídeňská 1083, 142 20 Prague 4, Czech Republic

INTRODUCTION

Speech sounds are complex signals that are characterized by rapid variations in spectral distribution and temporal pattern. In order to understand the neurophysiological basis for the discrimination of speech sounds, several approaches have been used in the past, among them the study of responses of single nerve cells in the auditory system of birds or mammals to natural species-specific vocalizations. Already the first studies performed in the auditory cortex of awake squirrel monkeys suggested that neurons may exist in the cortex which extract specific features of the calls, similar to visual cortex neurons performing a feature extraction (Wollberg and Newman, 1972; Winter and Funkenstein, 1973; Newman and Wollberg, 1973; Manley and Müller-Preuss, 1978; Newman, 1978; Newman and Symmes, 1979). Later these studies were continued in the auditory cortex of the squirrel monkey by Pelleg-Toiba and Wollberg (1991) and most recently in the macaque auditory cortex by Rauschecker et al. (1995). In birds a systematic study of neurons specialized to recognize species-specific calls was performed for example in guinea fowl (Bonke et al., 1979; Scheich et al., 1979). In the auditory neostriatum of this bird neurons were described which respond in a highly selective way to so-called Iambus-like calls and which distinguish these calls from other calls of the guinea fowl. The results of these studies led authors to the conclusion that the band of strongest energy in such wide-band calls plays a key role for neuronal recognition. A specific type of feature extraction exists in bats where the detection of emitted biosonar signals serves for perception of space and detection of prey (as reviewed by Suga ,1996). The auditory periphery of mustached bat shows specializations for detection and frequency analysis of species-specific complex biosonar signals. It contains neurons which are used for fine level-tolerant frequency analysis and others that are used for fine time analysis. The bat's central auditory system processes different types of auditory information in a parallel and hierarchical way and represents these in different areas of the auditory cortex. Specialization of neurons in the anterior auditory cortical fields for detection of self vocalizations was also confirmed in the rufous horseshoe bat (Radtke-Schuller and Schuller, 1995).

Systematic studies of neural coding of such complex sounds as speech started naturally, however, in the first relay of the central aditory system, the auditory nerve, mostly in the laboratory of M.B. Sachs (see his review 1984). These studies have shown that two principles for encoding the acoustical features of speech exist in the auditory periphery: the "rate-place" representations and the "temporal-place" representations. The temporal-place code can represent fine details in the spectra of vowels and stop-consonants (formant frequencies, formant-frequency transitions and pitch). Detailed formant structure of vowels is present in a rate-place code at moderate stimulus levels, but is preserved at high levels only in a small population of low-spontaneous-rate fibers. In later studies from the same laboratory (Wang and Sachs,1994; May et al. this volume) the authors investigated the encoding of speech stimuli in the anteroventral cochlear nucleus in the cat. They found enhanced modulation depth in neurons of the cochlear nucleus in comparison with the responses of auditory nerve fibers.

In contrast to the numerous studies investigating the encoding of vocalizations and speech-like stimuli in the auditory nerve and auditory cortex, only a few papers have reported the responses of neurons in subcortical auditory nuclei such as inferior collliculus (IC) or medial geniculate body (MGB) to these stimuli. Symmes et al. (1980) investigated the responses of MGB neurons in squirrel monkeys to vocalizations and artificial stimuli and found a low selectivity of MGB cells to vocal stimuli. They compared absolute thresholds, response vigor, rate-level functions, binaural interactions, and spectral compositions of responses to both types of stimuli and suggested that MGB neurons process both stimuli in similar ways. Creutzfeldt et al. (1980) observed that in the MGB of unanasthetized guinea pigs, all MGB and cortical neurons responded to a variety of natural calls from the same or other species. The MGB cells responded to more components of a call than did cortical cells. High modulation frequencies within a call, such as those of fundamental frequency, could still be separated in the response of some MGB neurons, but never in those of cortical neurons. Tanaka and Taniguchi (1991) also recorded from MGB neurons in guinea pigs and, in contrast to the findings of Symmes et al.(1980) in squirrel monkeys, they found a low responsiveness of MGB neurons to species-specific vocalizations. In addition they found in neurons responsive to vocalizations discharge patterns which were not predictable from the response to pure tones.

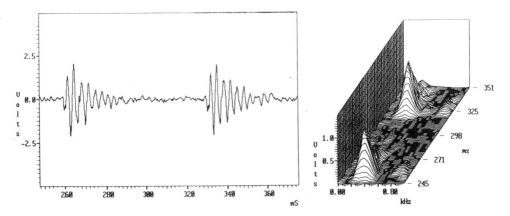

Figure 1. Characteristics of the purr. Left side: voltage versus time envelopes. Right side: sound spectrogram displayed in the waterfall format.

MATERIALS AND METHODS

In our experiments we recorded the extracellular responses of IC and MGB neurons in guinea pigs to pure tones, white noise and four typical species-specific vocalizations (purr, chutter, chirp and whistle). These calls were selected from a large repertoire of guinea pig calls described by Harper (1976). Spontaneous vocalizations were recorded in female pigmented guinea pigs (age 2–24 months) placed in a sound-attenuated room. Recorded calls were analyzed with a high resolution signal analyser B&K 2033 and with a CED 1401plus interface connected to PC 486 (CED program for frequency analysis Waterfall).

Purr is a special call which consists of a bout of regular acoustical impulses with a very small variability of frequency and time parameters (Fig. 1). The acoustical parameters of purr are given in Table 1. Animals express purr in conjunction with mating behavior and when they seek contact. General exploratory activity is accompanied by a frequently occurring chutt or chutter which lasts for 200–300 ms and may appear in a series with variable intercall intervals (Fig. 2., Table 1). The chutter or its parts can sometimes be aperiodic. The calming of an animal in comfortable conditions is accompanied by the occurrence of a brief call chirp. Chirp lasts 40–100 ms and may appear as an isolated call (Fig. 3, Table 1). If the animal expresses a feeling of separation or tries to contact the feeding caretaker, the call typical of exploratory behavior, i.e. chutter, changes to whistle. Whistle usually starts with a rich frequency, chutter-like part, changing to a frequency sweep. At the end the whistle usually finishes as a pure tone (Fig. 4, Table 1).

In our electrophysiological experiments the animals were anaesthetized with ketamine and xylazine (33 mg/kg ketamine and 6.6 mg/kg xylazine) and placed in a sound-proof and anechoic room. Acoustical stimuli were presented in free field conditions through a loudspeaker located in front of the animal. Unit activity was recorded with a glass micropipette; spikes were differentiated on the basis of their amplitude and shape with a CED 1401plus interface. In some cases two or three spikes were recorded from one microelectrode. Responses to four digitally tape recorded guinea pig calls were compared with responses of the same neuron to pure tones at their characteristic frequency (CF) and to white noise bursts. After the experiments the location of the microelectrode tip was determined in histological sections of the brain.

Table 1. Parameters of four typical guinea pig vocalizations

	PURR	CHUTTER	CHIRP	WHISTLE	
				at the begining	at the end
Duration of component [ms]	28.5±2.93	200-300	40-100	300-600	
Number of components in call	10-100	1-10	1	1	
Period between components [ms]	70.1±3.39	300-800			
Fundamental frequency [Hz]	307.6±3.49	800-1500	900-1500	1000-2000	2500-3500
Attenuation of the second harmonics [dB]	-27	-5	-10	0	-40
Attenuation of the third harmonics [dB]	-35	-10	-15	-5	-40

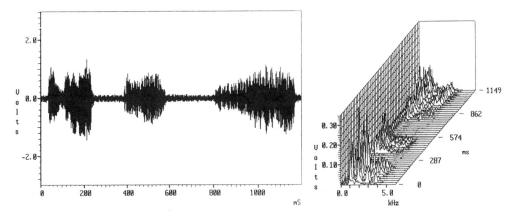

Figure 2. Characteristics of the chutter. Left side: voltage versus time envelopes. Right side: sound spectrogram displayed in the waterfall format.

RESULTS AND DISCUSSION

Altogether, responses were recorded from 120 IC neurons and 220 MGB neurons. In all of these neurons characteristic frequency (CF) was first estimated, and in the majority of them the sharpness of the tuning curve was estimated by computing the Q_{10} value. The CFs ranged in both structures from 0.1 kHz to 30 kHz and Q_{10} values from 0.1 to 8.6. First-peak latencies were measured 10 dB above threshold at the CF and were found to be in the range of 8 to 56 ms (mean 18.85 ms +/- 10.3 ms S.D.) in the IC and in the range 10 to 70 ms (mean 20.44 ms +/- 10.3 ms S.D.) in the MGB. Fig. 5 shows the wide range of thresholds to tones at the CF of investigated neurons in the IC.

Responses of IC and MGB neurons were evaluated with the aid of peri-stimulus time histograms (PSTH). Fig. 6 demonstrates a typical response of an IC neuron to a tone at CF, a white noise burst and four guinea pig calls. The onset response to a tone at CF

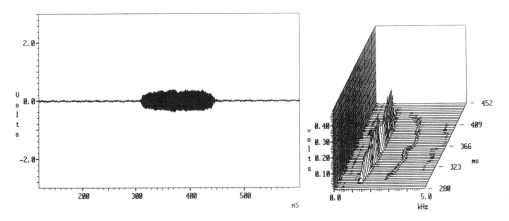

Figure 3. Characteristics of the chirp. Left side: voltage versus time envelopes. Right side: sound spectrogram displayed in the waterfall format.

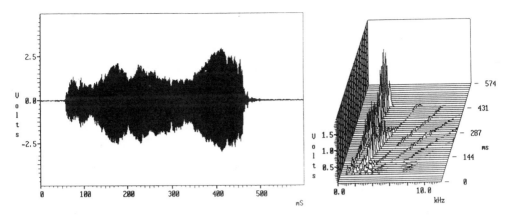

Figure 4. Characteristics of the whistle. Left side: voltage versus time envelopes. Right side: sound spectrogram displayed in the waterfall format.

changes to a sustained response when a white noise burst is used. Responses to all four types of calls are expressed in this neuron. Similar responses were observed in many IC neurons. In general the responsiveness to individual calls was slightly higher in the IC than in the MGB (Fig.7). While 54% of IC neurons responded to all four calls, only 41% of MGB neurons were able to follow all four calls. Similarly higher was the responsiveness to three calls (25% in the IC vs. 20% in the MGB). The lower responsiveness of MGB neurons to species-specific calls suggest that MGB neurons may be more selective for individual calls. In both structures very few neurons were found which did not respond to any of the vocalized sounds (3.5 % in the IC and 5% in the MGB).

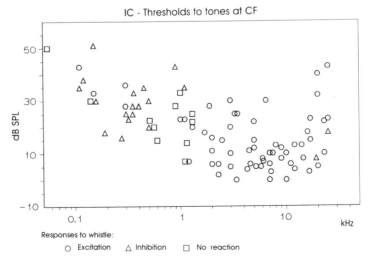

Figure 5. Thresholds of IC neurons in response to tones at the characteristic frequency. Indicated are neurons which do not react to whistle (squares) and those which are inhibited by whistle (triangles).

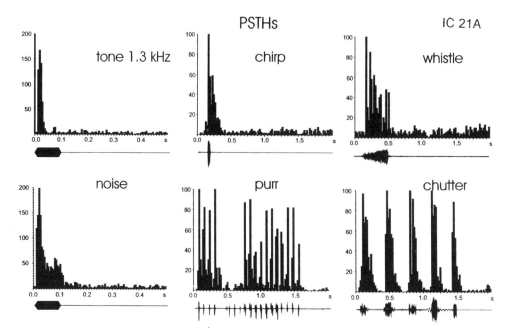

Figure 6. Peri-stimulus time histograms (PSTHs) of an IC neuron in response to a pure tone at the CF, burst of white noise and four typical vocal stimuli. The voltage versus time envelopes of the acoustical signals are shown below the PSTHs.

The selectivity for individual calls in the IC and MGB neurons is demonstrated in Fig. 8. In the IC the average responsiveness is approximately 80%. This means that in the sum of the responses to all four calls, in 20% of the cases the neurons did not react (either by excitatory or by inhibitory reaction). There are, however, some differences between responses to individual calls: least reactive were IC neurons to purr (72% positive reactions), most reactive were IC neurons to chutter (85% positive reactions). The decreased responsiveness seen in the MGB is due to a decreased reactivity to two calls - purr and chutter. Both calls consist of a series of brief sounds with a relatively high repetition rate (approximately 15 Hz for purr). In the case of purr only 50% of neurons produced a response, and in the case of chutter only 57% of neurons reacted positively. In contrast to

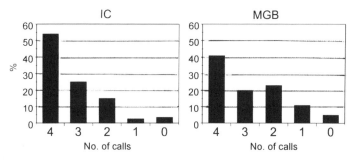

Figure 7. Responsiveness of neurons in the IC and MGB to species-specific calls expressed as the number of calls to which each neuron reacts.

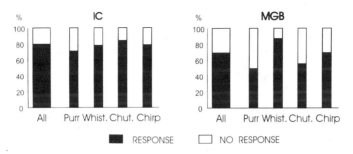

Figure 8. Selectivity of IC and MGB neurons for species-specific vocalizations.

this 87% of MGB neurons were excited by whistle. The average reponsiveness to species-specific calls, therefore, was smaller in the MGB than in the IC (69%). A difference between the IC and MGB responsiveness to calls was also seen in the occurrence of inhibitory reactions. In the IC 9% of the reactions to whistle were inhibitory (Fig. 5) whereas in the MGB inhibitory reactions to vocalizations were observed only in 2% of the neurons and, in this case, to calls other than whistle.

The responses of IC and MGB neurons to individual calls are dependent on their CF. Many neurons with low CF in the IC do not respond to whistle, which is composed of predominantly higher frequencies. Also, inhibitory reactions to whistle are present in the IC almost only in those neurons with a low CF (Fig. 5). Corresponding with this finding is the relatively low reactivity of high-frequency neurons in both IC and MGB to predomi-

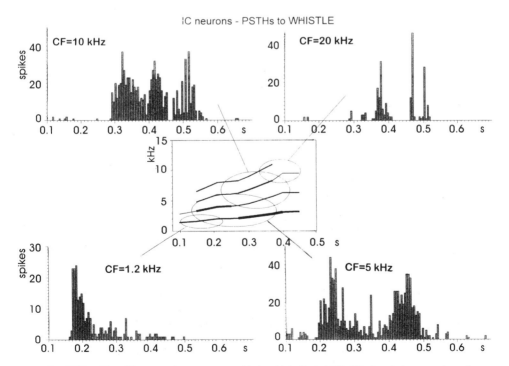

Figure 9. PSTHs of four IC neurons in response to whistle with indicated CFs. The insert shows a schematic spectrogram of a whistle. Indicated are areas of the spectrogram which produce an excitatory reaction in the PSTH.

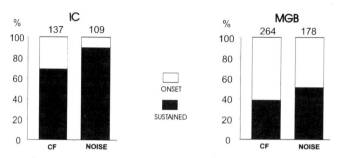

Figure 10. Distribution of onset and sustained types of response to tones and white noise in the IC and MGB

nantly low-frequency calls such as purr and chutter. However, many low-frequency neurons in the MGB did not respond to purr or chutter as well. In principle, the shape of the response to a typical call, the whistle, (which is composed of several harmonics with continuosly increasing frequency) is possible to predict when the CF of the neuron is known. Fig. 9 shows PSTH of four IC neurons with different CFs - 1.2 kHz, 5 kHz, 10 kHz and 20 kHz. The peaks in the individual PSTH correspond with intensity maxima present in the fundamental frequency, second and third harmonics. For example the first peak in the neuron with CF 5 kHz reflects the initial intensity increase in the second harmonic whereas the second peak reflects the stronger part of the fundamental frequency which is present between 300 and 500 ms. The lower reactivity of the neuron with CF 20 kHz corresponds with the small contribution of very high frequencies to the spectrum of the whistle.

In ketamine-anaesthetized animals two basic types of response were found in the IC and MGB on the basis of an evaluation of poststimulus histograms: onset and sustained. Fig. 10 shows the distribution of sustained and onset reactions in the IC and MGB during tonal stimulation at the unit CF and during white noise stimulation (as found 10 dB above threshold). In general, sustained response is more expressed in the IC and during white noise stimulation. The almost 70% occurrence of sustained reaction to tonal stimulation increases to 90% when white noise is used. In the MGB only 38% of neurons react with sustained reaction to tones at the CF, and this increases to 52% with white noise stimuli. The relatively high occurrence of onset reactions in the MGB contrasts with the fact that the responsiveness of MGB neurons is low to a brief call such as purr and high to a long lasting call such as whistle.

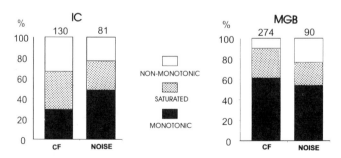

Figure 11. Distribution of monotonic, non-monotonic and saturated rate/level functions in the IC and MGB in response to tones and white noise.

White noise is also more efficient in driving neuronal responses at high sound intensities in the IC (Fig.11). At the CF approximately 30% of neurons display monotonic rate/level (RL) function; in the case of white noise stimulation the occurrence of monotonic RL functions increases to almost 50%. Monotonicity is also more expressed in the MGB than in the IC. Monotonic RL functions were found in the MGB in more than 60% of neurons in response to tones at the CF. A large variability was found in RL functions in the same neuron when different stimuli were compared. Nonmonotonic reaction to a pure tone at the CF can be accompanied in the same neuron by a monotonic or saturated RL function to white noise or vocalization stimuli. Also, thresholds to white noise stimuli and to vocalizations were different from thresholds to tones at the CF and in almost all cases were higher.

Although our data demonstrated in many cases a stronger reactivity to an animal call than to white noise or a pure tone, it is not possible to claim unequivocally on the basis of the RL functions that a selective processing of vocalizations exists in the IC or MGB. One of the generally accepted procedures for distinguishing the specificity of neuronal reaction to vocalization sounds is to compare the reaction to a call with the reaction to the same call, delivered in the reversed order. The left part of Fig. 12 shows poststimulus histo-

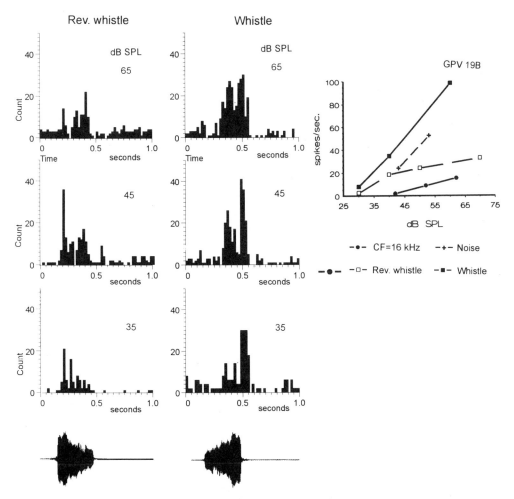

Figure 12. PSTHs of an IC neuron in response to whistle and reversed whistle at three different intensities (left side). Rate/level functions of the same neuron for whistle, reversed whistle, tone at the CF and white noise (right side).

grams of an IC neuron response to whistle and to reversed whistle at different sound intensities while the RL functions are plotted on the right together with the RL functions for the CF and white noise. It is evident that the reaction to whistle is more pronounced than that to the reversed whistle. The specificity of responses to vocalizations was tested in a sample of 40 IC neurons and 208 MGB neurons which were stimulated with the whistle and with the reversed whistle. Different responses were found in 33% of IC neurons and 14% of MGB neurons. This finding suggests that neurons specifically reacting to whistle may exist in the IC and MGB; however, the difference between the reactions to whistle and reversed whistle can be a result of responsiveness to differences in the amplitude and frequency modulation between these two sounds.

Our results demonstrate that species-specific calls elicit in the IC and MGB of the guinea pig a vigorous reaction with some essential differences from the reaction to pure tones; however, the results cannot be interpreted as proof for the high selectivity of these neurons to vocalizations. We are more inclined to believe that like in the cat (Creutzfeldt et al., 1980) or in the squirrel monkey (e.g. Symmes et al., 1980) the processing of such signals in the subcortical nuclei is not highly selective and that the real feature extraction exists at the level of the auditory cortex. This does not correspond with the opinion of Tanaka and Taniguchi (1991) who suggest that neurons in the MGB of guinea pig specifically respond to certain features of the animal calls. Certainly in the IC the selectivity to animal calls is not high, and many discharge patterns of the response to vocalizations can be predicted from the properties of responses to pure tones. However, the experiments of Tanaka and Taniguchi (1991) were performed on awake animals and our guinea pigs were anaesthetized with ketamine.

The low responsiveness of MGB neurons to calls like purr and chutter is not surprising and may be related to a general trend in the progressive reduction in responses to high modulation frequencies in the ascending auditory system (e.g. Rees and Møller, 1983). Similarly, another measure of temporal sensitivity, the ability of neurons to phase lock to pure tones, is also restricted to a lower range of frequencies in the more central parts of the auditory system (Kuwada et al., 1984).

The next step in the analysis of the responsiveness of IC and MGB neurons will be to differentiate among the responses of neurons in individual subdivisions of these structures. In the previous study in the cat (Aitkin et al., 1994) differences in responsiveness to tones, white noise and vocal stimuli were described among three subdivisons of the IC. Also Tanaka and Taniguchi (1991) found some differences in the responses to vocal stimuli among subdivisions of the MGB in the guinea pig

ACKNOWLEDGMENTS

The authors would like to thank Dr. Lindsay Aitkin for his collaboration in the first part of the experiments in the inferior colliculus and Dr. Alexandra Vlkova for her collaboration in recording the guinea pig vocalizations. This study was suported by the grant 309/94/0735 of the Grant Agency of the Czech Republic.

REFERENCES

Aitkin, L., Tran, L. and Syka J. (1994) The responses of neurons in subdivisions of the inferior colliculus of cats to tonal, noise and vocal stimuli. Exp. Brain Res. 98, 53–64.

Bonke, D., Scheich, H. and Langner, G. (1979) Responsiveness of units in the auditory neostriatum of the guinea fowl (Numida meleagris) to species-specific calls and synthetic stimuli. I. Tonotopy and functional zones of field L. J. Comp. Physiol. 132, 243–255.

Creutzfeldt, O., Hellweg, F.C. and Schreiner, C. (1980) Thalamocortical transformation of responses to complex auditory stimuli. Exp. Brain Res. 39, 87–104.

Harper, L.W. (1976) Behavior. In: J. E. Wagner and P.J. Manning (eds.) The Biology of Guinea Pig. Academic Press, New York, pp. 31–51.

Kuwada, S., Yin, T.C.T., Syka, J., Buunen, T.J.F. and Wickesberg, R. (1984) Binaural interaction in low-frequency neurons in the inferior colliculus of the cat. IV. Comparison of monaural and binaural response properties. J. Neurophysiol. 51, 1306–1325.

Manley, J.A. and Müller-Preuss, P. (1978) Response variability of auditory cortex cells in the squirrel monkey. Exp. Brain Res. 32, 171–180.

May, B.J., Le Prell, G.S., Hienz, R.D. and Sachs, M.B. (1997) Speech representation in the auditory nerve and ventral cochlear nucleus: Quantitative comparisons. This volume.

Newman, J.D. (1978) Perception of sounds used in species-specific communication: The auditory cortex and beyond. J. Med. Primatology 7, 98–105.

Newman, J.D. and Symmes, D. (1974) Arousal effects on unit responsiveness to vocalizations in squirrel monkeys. Brain Res. 78, 125–138.

Newman, J.D. and Wollberg, Z. (1973) Multiple coding of species-specific vocalizations in the auditory cortex of squirrel monkeys. Brain Res. 54, 287–304.

Pelleg-Toiba, R. and Wollberg, Z. (1991) Discrimination of communication calls in the squirrel monkey: "call detectors" or "cell ensembles"? J. Bas.&Clin. Physiol.& Pharmacol. 2, 257–272.

Radtke-Schuller, S. and Schuller, G. (1995) Auditory cortex of rufous horseshoe bat: 1. Physiological response properties to acoustic stimuli and vocalizations and the topographical distribution of neurons. Eur. J. Neurosci. 7, 570–591.

Rauschecker, J.P., Tian, B. and Hauser M. (1995) Processing of complex sounds in macaque nonprimary auditory cortex. Science 268, 111–114.

Rees, A. and Møller, A.R. (1983) Responses of neurons in the inferior colliculus of the rat to AM and FM tones. Hear. Res. 10, 301–330.

Sachs, M.B. (1984) Neural coding of complex sounds: speech. Ann. Rev. Physiol. 46, 261–273.

Scheich, H., Langner, G. and Bonke D.(1979) Responsiveness of units in the auditory neostriatum of the guinea fowl (Numida meleagris) to species-specific calls and synthetic stimuli. I. Discrimination of Iambus-like calls. J. Comp. Physiol. 132, 257–276.

Suga, N. (1996) Processing of auditory information carried by species-specific sounds. In: Michael S. Gazzaniga (Ed.) The Cognitive Neurosciences, MIT Press, Cambridge, pp. 295–313.

Symmes, D., Alexander,G.E. and Newman, J.D. (1980) Neural processing of vocalizations and artificial stimuli in the medial geniculate body. Hear. Res. 3, 133–146.

Tanaka, H. and Taniguchi, I. (1991) Responses of medial geniculate neurons to species-specific vocalized sounds in the guinea pig. Jpn. J. Physiol. 41, 817–829.

Wang, X. and Sachs, M.B. (1994) Neural encoding of single-formant stimuli in the cat. II. Responses of anteroventral cochlear nucleus. J. Neurophysiol. 71, 59–78.

Winter, P. and Funkenstein, H.H. (1973) The effect of species-specific vocalization on the discharge of auditory cortical cells in the awake squirrel monkey. Exp. Brain Res. 18, 489–504.

Wollberg, Z. and Newman, J.D. (1972) Auditory cortex of squirrel monkey: response patterns of single cells to species-specific vocalizations. Science 175, 212–214.

40

SPECIES-SPECIFIC VOCALIZATION OF GUINEA PIG AUDITORY CORTEX OBSERVED BY DYE OPTICAL RECORDING

K. Fukunishi, R. Tokioka, T. Miyashita, and N. Murai

Advanced Research Laboratory, Hitachi, Ltd.
Hatoyama, Saitama, 350-03 Japan

1. INTRODUCTION

Neural dynamics, or the neural behavior, of a cortical field has recently become one of the controversial matters related to the coding of sensory information, such as that of auditory, visual and somatosensory information in the mammalian cortex. Temporal neural coding such as dynamical neural assembly coding and dynamical population coding in the cortex have come to be considered much more important than was thought ten years ago (Buzsáki et al. , eds. 1994), but despite extensive physiological study of the auditory cortex, we still do not know how complex sounds like natural sounds are encoded in the auditory cortex nor whether the specialized processing of vocal calls is carried out in the cortex (Pelleg-Toiba and Woll-berg 1990, Wang et al. 1995, Eggermont 1995, McGee et al. 1996).

Spatio-temporal observation of cortical fields by optical imaging with voltage-sensitive dye have shown dynamical movement in the auditory cortical field in response to complex sounds (Fukunishi et al. 1992, 1993, Fukunishi and Murai 1995, Uno et al. 1993). Optical imaging of the responses to species-specific vocalization, for example, has indicated temporal coding of vocal calls (Fukunishi et al. 1996a). That is, it can be said that the idea of call specific neurons or a call specific neural assembly might be question-able. Furthermore, the onset response which is predominant response related with vocal call processing possibly synchronizes with the sound frequency change in the vocal call.

Here we will review recent results obtained by the optical imaging of the responses to complex sounds like vocalizations and will present ideas for the vocal encoding and the vocal processing function in the auditory cortex.

2. METHOD

An optical recording system with a 12x12 photodiode array installed at the eye port of a custom-manufactured microscope was used to detect fluorescence signal changes due to neural responses in the auditory cortex (field A) of guinea pigs. The voltage-sensitive dye RH795 was used to convert the neural membrane potential changes to fluorescent signal changes, and the signals of 128 channels from the photodiode array (pitch: 1.5 mm, gap: 0.2 mm) were amplified and fed into computer after 1–5 kHz parallel sampling. The microscope had bright and transparent lenses on which multilayer metal film had been coated by thermal evaporation (NA: 0.4). Vocal call stimuli as well as tone burst stimuli (rising/ falling time = 10 ms, plateau time = 30 ms) were stored by a computer and delivered by the computer via a 100 kHz A/D and D/A system.

The physiological experiments were carried on male Hartley guinea pigs anaesthetized with ketamine hydrochloride (46 mg/kg, intramuscular) and xylazine hydrochloride (23.3 mg/kg, intramuscular). After the trachea was cannulated, a section of skull over the auditory cortex of the left hemisphere, about 4–5 mm in diameter and about 14 mm lateral and 3 mm posterior to the bregma, was removed and the dura was resected. The voltage-sensitive dye RH795 was applied on the surface of the cortical site .

The timing of acoustic stimuli firing was controlled so as to synchronize them with the electrocardiogram (ECG) and the respirator. The responses to subsequent control trial (no sound stimulus) were subtracted from the responses to trials with a sound stimulus. Successively subtracted data from two to several tens were averaged to improve a signal noise ratio, if necessary. Throughout these experiments, the measured cortical area spread over a square about 3 mm on each side (3 mm square) of the auditory cortex (field A), with each pixel about 0.22 mm by 0.22 mm.

3. RESULTS

3.1 Tone Coding

Although electrophysiological studies of the tonotopical organization in the auditory cortex (field A) of the guinea pig have revealed that there are isofrequency bands, like the low-frequency tones in the rostral part and the high in the caudal, in the dorso-ventral direction across the rostra-caudal axis (Redies 1988). The pattern of the tonotopic organization in the auditory cortex (field A) visualized by optical recording is slightly different from that indicated by these electrophysiology studies. For example, as shown by the optical data of three different animals (Fig. 1), the tonotopical patterns for 1-, 4- and 16-kHz tone bursts in a 3-mm square of the cortical field are not necessarily band-like in dorso-ventral direction. Each tonotopic pattern depicts several combined groups of neural assemblies instead of the isotopic frequency band for the tone. However, the neural arrangement for the low-frequency tones in the rostral part of the field and for a high-frequency tone in the caudal part correspond to the electrophysiological results. Absolute positions of tonotopic positions for frequencies differ from animal to animal.

The temporal characteristics of the observed tonotopical patterns were more stable than the pattern of responses to complex sound stimuli such as clicks. Furthermore, recent high-sensitivity optical recording has revealed that the cortical neurons show diverse spatio-temporal responses to tone burst stimuli in the auditory field. The temporal patterns can be roughly classified into the three types illustrated in Fig. 2, which shows the spatio-temporal responses that a 4-kHz tone burst elicits in the auditory cortex (field A). One of these types represents relatively stable response as previously discussed (Fig. 2 (a)). That is, the tone response pattern, which is usually formed by the several combined groups of neural assembles, expand and contract in the peripheral of its central position as the re-

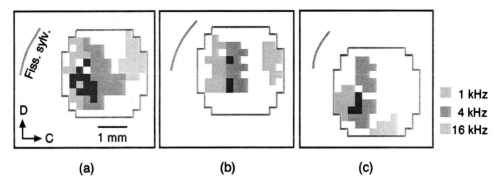

Figure 1. Comparison of the tonotopic organization of the auditory cortex (3 mm by 3 mm, field A) of guinea pig (optically imaged tonotopic areas to 1-, 4-, 16-KHz tone burst by each animal). One pixel area: 0.22 mm by 0.22 mm, D: dorsal, C: caudal.

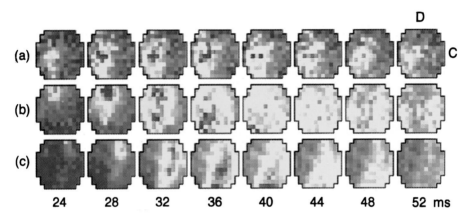

Figure 2. Three typical aspects of temporal pattern for tonotopic response to the same tone (4-KHz) in the auditory cortex: (a), stable; (b), bistable; (c), continuous. (a): averaged 30 times, (b) and (c): averaged 16 times. Red color: excitatory strong amplitude, green color: near zero.

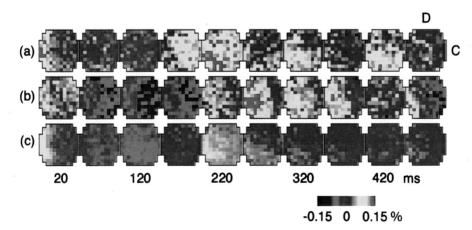

Figure 3. Tonotopic responses and nonstationary responses to the same tone (1-kHz) at the same field A of one animal: (a) and (b), no averaging; (c), averaging (16 times). Color tone shows the normalized response amplitude ($\delta I/I$).

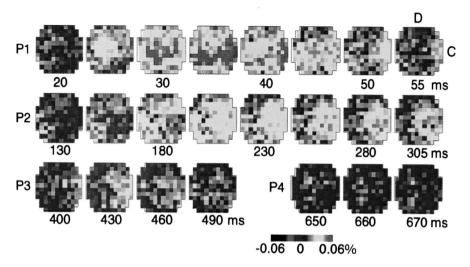

Figure 6. Spatial transient response pattern to the vocal call at P1 (onset) (a), P2 (b), P3 (c) and P4 (offset) (d), phases defined in Fig. 5. Color tone: normalized response amplitude ($\delta I/I$).

sponse time. One of the two other types denotes bistable response, for example, that the two (or more) separated neural assemblies exist predominantly along dorsal-ventral axis and that the response region first expands in the dorsal assembly and then jumps to the ventral center of the neural assembly as shown in Fig. 2 (b). This temporal feature of the tonotopic responses shows the displacement of active neural sites among mainly two or three separate positions in roughly the dorso-ventral axis. The other type of temporal feature of the tonotopic response is a continuous shift of the tonotopic center from dorsal to ventral, as seen in Fig. 2 (c). Thus, it can be said that tonotopical mapping in the auditory cortex has variety of temporal characteristics.

The above optical data were averaged 16 times (Fig. 2a) to 30 times (Fig 2b and c) in order to visualize the response pattern clearly. This data treatment is susceptible of eliminating non-stationary responses from visualizing the tonotopical pattern. Recently, nonaveraging optical imaging has revealed different temporal characteristics associated with the tonotopic behavior of the cortical neural assemblies. Spatio-temporal patterns of the nonaveraging responses to a 1-kHz tone stimulus at the same position in the field A at different several seconds (Fig. 3 (a), (b)) are compared with the pattern by the averaged responses over 16 times (Fig. 3 (c)). Nonaveraged patterns suggest that non-stationary responses which differ with every stimulus are temporally random, but are not necessarily spatially random.

3.2 Vocal Coding

Results of optical experimental performed in order to understand how natural vocalizations are encoded by neurons have suggested significant temporal coding characteristics as discussed already (Fukunishi et al. 1996a). Vocalizations of guinea pigs were classified according to their temporal sound wave patterns and spectra into about six vocal species. The animal's behavior was observed carefully while the animal was hearing these vocal calls, and two of the vocalizations, scream and whistle, were used as natural call stimuli. Here the optical imaging results obtained when using the vocalization scream are elucidated as a representative example.

The temporal sound wave pattern and spectrum sonogram of the vocalization scream are depicted in Fig. 4. The sound wave pattern shows that the call may contain temporal information rather in the later half periods of the vocal calls (total 620 ms duration) with large sound pressure and its change. Spectrum information, on the other hand, may be involved in the first 100 ms of the vocal call with sharp frequency change.

The time course of the responses to the vocal call at the 128 points (pixels) in the auditory cortex (field A) with distinctive entire responses from rostral sites and caudal sites are illustrated in Fig. 5. Onset responses (P1) with a high and acute magnitude were observed over the whole measured region (3 mm square) about 40 ms after emitting the vocal call. The successive excitatory responses (P2) with a relatively long duration but with degraded magnitude were also detected over the whole field after about 200 ms. Furthermore, the slight responses (P3) with a small elevation were recognized predominantly in the dorsal part at about 400s ms. Finally, offset responses (P4) were found in the confined site of the medial rostral part along the ventral-dorsal axis, 30 ~ 60 ms after the call faded out. The onset response P1 and the second excitatory response P2 are steady and convincing in almost the whole cortical field, but other responses are restricted spatially.

Temporally piece-wise responses of spatial neural activity distribution are illustrated at each transient time around the responses, P1, P2, P3, and P4 in Fig. 6. The concentrated responses site of P1, P2 phases are temporally displaced in the cortical field as illustrated in Fig. 6 (a) and (b). The moving occurs in the rostral part from the rostral to the medial at

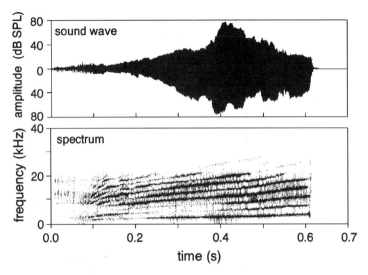

Figure 4. A species-specific vocalization scream of guinea pigs: the sound wave and spectrum.

first, then promptly follows in the opposite part from the caudal to the medial. This means that the neural activities first propagate from the tonotopic site of relatively lower frequencies (~1 kHz) to the tonotopic site of medium frequencies (~ 4 kHz) and successively propagate from the higher frequency (~16 kHz) tonotopic site to the medium frequency tonotopic site (see the tonotopical mapping in this animal (Fig. 1 (a)). On the other hand, as indicated in Figs. 6 (c) and 6 (d), spatial displacements of responses P3 and P4 are not observed. This spatio-temporal property found in significant P1 ~ P4 responses associated with the vocalization scream was similar across animals and similar spatio-temporal responses were observed in response to another vocal call stimulus, whistle.

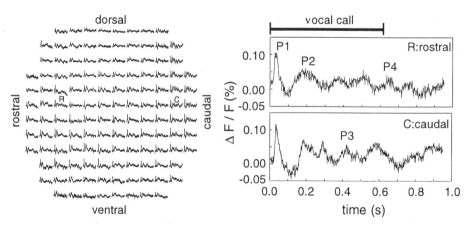

Figure 5. Array display of the optically recorded evoked responses to the vocal call in the auditory cortex (field A), and two typical time courses during ~ 1 sec after onset at a rostral and a ventral positions. P1 (onset), P2, P3, and P4 (offset): significant four response shapes. One pixel area: 0.22 mm by 0.22 mm, 32 times averaged.

4. DISCUSSION

The tonotopical mapping of the mammalian auditory cortex derived from many electrophysiological experiments was static and immovable, but the tone burst responses observed optically have revealed various dynamic signs of tone coding. There are, of course, many aspects of the optical auditory experiments - such as photo toxical effects, anaesthetized conditions, and neural response integration by a photo sensor - that must be clarified. Nevertheless, it can be said that these dynamical optical data correspond unquestionably to the neural responses to the tone stimulus, since the optical response correlate closely with the local field potential at the same measurement position (Tokioka et al. 1996).

The neural responses to vocalization elicited later, when the stimulus was more intense and its frequency was modulated sharply, were less dynamic than those elicited earlier. This suggests that neural encoding for vocalization is closely related to the neural behavior of the intense response phases at the onset (P1) and the 2nd excitatory response (P2). That is, it seems that prediction and feed forward functions in the auditory cortex are involved in the processing of the vocal call.

The spatio-temporal response patterns at phases P1 and P2 displayed similar displacements of neural dynamic behavior in response to the natural call. These dynamic characteristics at the P1 and P2 phases differ from each other, however, when the animal hears a call that has been distorted by eliminating part of the frequency components of the natural call. The P1 neural responses to the distorted call disappeared at the tonotopic site correspond to the frequencies lost. On the contrary, the P2 responses to any distorted call became invisible at the high frequency tonotopic part regardless the frequencies lost, as discussed in (Fukunishi et al. 1996b).

The spatio-temporal response pattern to the vocal call suggested as following: (1) specific neurons are not dedicated to specific vocalizations. (2) vocalization processing depends strongly on the frequency variation in the onset phase, so, temporal coding and feed-forward processing must control vocal processing. (3) specific neurons in response to sound extinguishing exist at the restricted medial site in the auditory field.

REFERENCES

Buzsáki G., Llinás R., Singer W., Berthoz A., Christen Y. (Eds.), 1994, Temporal Coding in the Brain, Springer-verlag Berlin Heidelberg.

Eggermont J. J.,1995, Representation of a voice onset time continuum in primary auditory cortex of the cat, j. Acoust. Soc. Am., 98, 911–920.

Fukunishi K., Murai N., Uno H., 1992, Dynamical characteristics of the auditory cortex of Guinea pig observed with multichannel optical recording, Biol. Cybern., 67, 501–509.

Fukunishi K., Uno H., Murai N., 1993, Spatio-temporal observation of guinea pig auditory cortex with optical recording, Japanese J. Physiol., 43, S61–66.

Fukunishi K., Murai N., 1995, Temporal coding mechanism of guinea pig auditory cortex by optical imaging and its pattern time series analysis, Biol. Cybern., 72, 463–473.

Fukunishi K., Tokioka R., Murai N., 1996a, Neural coding of auditory information in animal brain revealed by optical imaging, Proc. in Neur. Informat. Process.: Amari S. et al. (eds.), Springer-Verlag Singapore, 1281–1284.

Fukunishi K., Tokioka R., Miyashita T., Murai N., 1996b, The coding of species-specific vocalization in guinea pig auditory cortex revealed by optical imaging with dye. (to appear)

McGee T., Kraus N., King C., Nicol T., 1996, Acoustic elements of speech like stimuli are reflected in surface recorded responses over the guinea pig temporal lobe, J. Acoust. Soc. Am., 99, 3606–3614.

Pelleg-Tobia R. and Wollberg Z, 1990, Discrimination of communication calls in the squirrel monkey: "call detectors" or "cell ensembles", J. Basic & Clinic. Physiol. & Pharmacol., 2, 257–272.

Redies H., Sieben U., Creutzfeldt O. D.,1988, Functional subdivisions in the auditory cortex of the guinea pig, J Comparat. Neurol., 282, 473–488.

Tokioka R., Fukunishi K., Murai N., T. Miyashita, 1996, Non-stationary neural response in the auditory cortex of guinea pigs revealed by optical imaging. (to appear)

Uno H., Murai N., Fukunishi K., 1993, The tonotopic representation in the auditory cortex of the guinea pig with optical recording, Neurosci. Lett., 150, 179–182.

Wang X., Merzenich M. M., Beitel R., Schreiner C. H., 1995, Representation of a species-specific vocalization in the primary auditory cortex of the common marmoset: temporal and spectral characteristics, J. Neurophysiol., 74, 2685–2706.

41

CORRELATES OF FORWARD MASKING IN SQUIRREL MONKEYS

P. Müller-Preuss

Department of Neuroendocrinology
Clinical Institute, Max-Planck-Institute for Psychiatry
Munich, Germany

OBJECTIVES

Perception and cognition are, besides intrinsic neural processes, essentially based on information from the organism's peripheral sensory structures. And out of a huge "sensory" environment only those parameters will gain access to perceptive or cognitive brain structures which pass particular windows(filters) shaped by the mechanical, physiological and/or neural properties of the transforming organs(i.e. eyes,ears,skin, nose, etc.). Thereby effects of masking play a considerable part in designing such windows. Mainly due to the temporal nature of acoustic signals but also due to the mechanics of the cochlea as the transforming organ, masking will strongly influence the information parameters available for the auditory system's centripetal structures. Consequently, in the sense of adaptation at an optimal communication system, perceptive or cognitive brain activities will also have influences at the motor side. Thus, in the case of the auditory system, they may be reflected within the production of vocalizations.

Extensive psychoacoustic studies described masking effects(simultaneous as well as forward masking) in man and evaluated the dependence of the detectability of a testpulse from maskerlevel, maskerlength, spectral properties and intervall-times between masker and testpulse(for example see Fastl, 1976). The results clearly indicate at such a window, where the audibility of signals is more or less weakened through masking. With similar approaches numerous neurophysiological experiments were carried out to demonstrate in more detail and for particular structures of the pathway the effects of masking at neuronal level. These studies have been undertaken mainly for peripheral structures like the VIII. nerve or the cochlear nucleus(for example see Delgutte,1988, 1990).

The aim of this study is threefold: firstly to present data about masking effects on the neuronal activity within the colliculus inferior, a central station of the auditory pathway. This presentation is carried out in a more comprehensive and general fashion to allow the elaboration of the second aspect, namely an comparative look for neural correlates of masking in a nonhuman primate with data described for man. This was done in contrasting the reactions of the midbrain neurons with the results of psychoacoustic studies. And a further aspect of the study was to shed light for possible effects of masking within in the mode how vocalizations are produced and used in these nonhuman primates.

METHODS

Recordings

Neuronal recording from the auditory midbrain was carried out with awake monkeys. The animals were restrained in padded contour chair which permitted the head to be fixed and allowed minor body and limb movements. The activity of single cells as well of cell groups(two or more) was recorded extracellularily with tungsten micro-electrodes(1–10 Mohm). Electrode penetrations were stereotactically controlled with a hydraulic microdrive and a x-y micropositioner. For histological verification of some of the electrode tracks electrolytic lesions(5–10 mA, 5–20 ms) were applied at the end of a penetration.

Stimulus Configuration

White noise or tones at units characteristic frequency preceeded a 2msec testpulse. Length of masker was 180msec with 10msec risetime, length of testpulse 2msec, risetime 1msec. Distance from the end of the masker to the beginning of the testpulse was varied between 2msec and 300msec. The relationship of the levels of the two pulses was varied between 40 dB(Lt < = > Lm), absolut level was adjusted along unit properties and was usually between 50 and 70dB.Stimulus configuration resembled those used in psychoacoustic study in man(Fastl 1976) The responses evoked by the testpulse alone served as reference. Stimuli were presented binaurally via a loudspeaker mounted 1.5m in front of the animals head.

Behavioral Studies

The search for possible effects of masking within vocal communictions of Squirrel monkeys is based on a study, where a relationship between the way of the use of non FM-vocalizations and communicative relevance has been demonstrated (Maurus et al. 1989)

RESULTS

The results of the neurophysiological experiments are shown in Fig. 1a. The normalized responses from 52 units out of the central nucleus of the IC are outlined. Effects of masking is shown as a function(dt) of the time between of the end of the masker(L_m) and the beginning of the testtone(L_t) and spike rate. Three different intensity-relationships between testtone and masker have been tested: L_m > = and<L_t. As can be seen, strongest masking effects occur if L_t is about 20 dB smaller than L_m within a dt of 2 - 50msec. After 200 msec only minor masking effects are seen. Lower and upper curves have been extrapolated. Also the maskerlength has influence at the audibility of the testtone, but the effects are smaller than those of the distance: if maskerlength is reduced to only 10 msec, spike rate is about 20% greater than in comparison to a maskerlength of 180msec. Responses shown in the figure have been evoked by noise stimuli. If tones at a units characteristic frequency were the stimulus, masking effects are similar to those of noise. If the testtone was adjusted outside cf, masking effects decreased. This very interesting findings have not systematically evaluated during this study, but they indicate already that effects of masking can be circumvented.by tones modulated in frequency.

Rather this study concentrated at a more comprehensive analysis, to make a comparison with psychoacoustic experiments possible. The results of such a study a shown in fig. 1b and are taken in a modified fashion from Fastl 1976: the detectability of a testtone is outlined as a function of the time between the end of a masker and the beginnig of the testtone. Even if the ordinate parameters are different(spike rate vs. detectability), the abszissa parameters and the stimulus configuration are the same. And as can be seen in the figures the reactions are similar: If the testtone is 5 msec preceeded by a masker with the same sound-intensity, than the detectability or the spike rate is weakened between 20 and 30 dB respectively % rate. The similarity of the curves in the two figures gives evidence that the neuronal activity during this

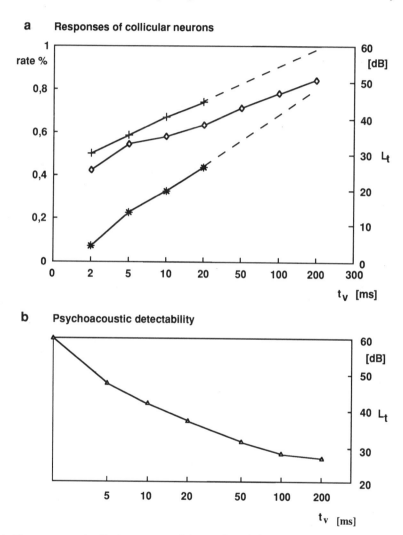

Figure 1. The responses of collicular neurons and the reactions during a psychoacoustic experiment to a forward masking paradigm are shown: in **a** effects of masking are outlined as a function of delay time t_v (time between end of masker L_m and beginning of testtone L_t) and discharge rate. Normalized responses from 52 collicular neurons are shown for three different intensity relationships L_t/L_m: rombs $L_t=L_m$; crosses L_t 20dB> L_m; stars L_t 20dB< L_m. To make data from a and b comparible, a transformation of discharge rate from % in dB values is shown at the right ordinatefor. This transformation has been by experimental testing. **b** shows the reactions of human probands, where the increasing threshold of L_t (detectability) in dependence of decreasing t_v significates masking effects. For example, note that in a for t_v 5 and 10 ms discharge rate is weakened up 40%(depending on L_t/L_m) The corresponding decrease in detectability as estimated from the right ordinate is similar to the values obtained in Fig.b.

masking paradigm may represent a neural correlate for the psychoacoustic phenomenon of forward-masking. With regard to a possible adaptation of the motor side(vocalizations) at these masking phenomena, an review of a behavioral study with Squirrel monkeys have been undertaken, where non frequency-modulated vocalizations were investigated for their communicative meanings(Maurus et al. 1989). According to this study, only particular amplitude changes out of amplitude modulated calls carry the information relevant for species-specific communication. The way how these amplitude changes are placed within the calls gives evidence for a circumvention of inhibiting effects caused by masking: The duration of the amplitude changes has its maximum about 11msec(i.e. a short masker). If such amplitude changes are produced as increments they occur mostly at the beginning of a call (over 90%) and rare at the end. If the change is produced as a decrement, it occurs also often at the beginning or at the end of a call, and if at the end, more often in shorter calls. Furthermore if non AM-calls are considered, frequency of occurence lies between 12 and 75 Hz or 83–13msec respectively (Schott 1975). Also here, epochs of strongest masking(2–10msec) seem to be bypassed.

CONCLUSION

The results show for the colliculus inferior, a central station of the auditory pathway

1. that forward masking causes a windowlike acoustic shadow during which the reception of a "testpulse or a relevant signal" is impossible or strongly weakened.
2. The particular parameter of the window resemble those detected for man; neural correlates of masking thus can be studied modellike in a nonhuman primate even for central auditory stations.
3. Furthermore, certain temporal parameters (length, position) of calls of Squirrel monkeys shown to be of communicative relevance are produced in a manner that effects of forward masking seem to be circumvented. Therefore the findings point at a relationship between structures involved in audition and call production: properties of the transmitting media together with the peripheral sensory organs have a shaping influence at the production of sounds used within species-specific acoustic communication.

ACKNOWLEDGMENTS

This work was supported by the "Deutsche Forschungsgemeinschaft, SFB 204". The author gratefully acknowledges advice and encouragement of H. Fastl and comments of A. Bieser.

REFERENCES

Delgutte, B.(1988) Physiological mechanisms of masking. In H. Duifhuis, W.J. Horst and H.P. Wit(Eds).Basic issues in hearing, Academic Press, London, pp. 204 - 212.
Delgutte, B.(1990) Physiological mechanisms of psychophysical masking:observations from auditory-nerve fibers, JASA, 87, 791 - 809.
Fastl, H. (1976) Temporal masking effects: I. Broadband noise masker. Acoustica 35, 287 - 302.
Maurus, M., Barclay,D., Kühlmorgen, B., Wiesner, E. and Llorach-Forner, V. (1989) Synonomy in Squirrel monkey calls? Language and communication, 9, 69 - 76.
Schott, D. Quantitative analysis of the vocal repertoire of Squirrel monkeys(Saimiri sciureus) (1975) Z. für Tierpsychologie, 38, 225 - 250.

42

CEREBRAL CORRELATES OF HUMAN AUDITORY PROCESSING

Perception of Speech and Musical Sounds

Robert J. Zatorre

Montreal Neurological Institute
McGill University
Montreal, Quebec, Canada

1. INTRODUCTION

Speech and music are perhaps the most interesting way that human cognition makes use of sound. It seems likely that the complex mental operations necessary for the processing of speech and music would demand a correspondingly complex set of neural computations. This paper will review some studies from our laboratory aimed at exploring these issues, utilizing both the traditional behavioral-lesion approach as well as recent brain imaging techniques. Among the latter methods, we have used both functional brain imaging, with positron emission tomography (PET), as well as structural imaging, with magnetic resonance imaging (MRI). These techniques allow us to explore cerebral activation patterns associated with the performance of certain tasks in healthy volunteer subjects, and also allow us to begin to explore structure-function relationships in the brain. We have followed the strategy of adapting or developing behavioral tasks drawn from the fields of psychophysics and cognitive psychology in order to study the neural correlates of a wide range of psychological processes relevant to auditory cognition. The tasks to be described in this paper focus on aspects of phonetic perception, melodic processing, auditory working memory, and auditory imagery. Each of these areas will be discussed in turn, preceded by a brief introduction to functional imaging methods. To conclude, we present some novel findings from structural imaging measures which may have direct relevance to understanding some aspects of functional results.

1.1 Functional Imaging

The development of PET as a technique to measure functional activation represents a major advance for cognitive neuroscience, as it permits for the first time a relatively direct way to investigate changes in cerebral activity patterns as a function of specific task performance in normal subjects. A detailed explanation of the physics of the technique and the image

Acoustical Signal Processing in the Central Auditory System
edited by Syka, Plenum Press, New York, 1997

processing that accompanies it is beyond the scope of this paper (cf. Raichle et al. for a description), but a brief description follows. The basic idea behind the application of PET relevant to our studies is that a short-lived radioactive tracer (oxygen-15) is used to measure cerebral blood flow (CBF) during a 60-second period. A scan is then reconstructed which represents a three-dimensional map of the CBF distribution in the entire brain during that time, with a spatial resolution on the order of 14 to 18 mm. Typically, several such scans are obtained in each individual subject. Data from a group of subjects may then be averaged, after appropriate stereotaxic normalization is applied to correct for differences in brain size, shape, and orientation. Averaged CBF data from a given condition may then be compared to another condition by superimposition of the relevant scans, and application of a pixel-by-pixel subtraction algorithm which detects significantly different areas of CBF in one condition as compared to another (Worsley et al., 1992). Although numerous other approaches to image analysis and paradigm design are also possible, including regression-based techniques (cf. Paus et al., 1996), the studies to be described in the present paper all used the paired-image subtraction method, in which two different conditions are compared directly to one another. The assumption is that the difference image reflects areas of cerebral activity specifically related to the task in question, relative to the baseline condition which typically represents an attempt to control for certain aspects of the task. A final aspect of the functional imaging work presented here is that structural images (MRI) are also obtained for each subject, and are co-registered to the PET images (Evans et al., 1992). These structural images permit much improved anatomical localization, and form the basis for the structural measures of cortical volume to be described at the end of this paper.

2. PHONETIC MECHANISMS IN SPEECH PERCEPTION

The studies of speech processing carried out in our laboratory have focused on understanding the neural mechanisms relevant to the extraction of phonetic units from a speech signal. The psychological literature on speech perception has long maintained that such a process would require a specialized speech module, distinct from other perceptual mechanisms; one specific model of speech processing further emphasizes the possible role of motor-articulatory codes in the perception of speech (e.g., Liberman & Mattingly, 1985). Until recently, it has proven difficult to obtain direct evidence bearing on this question, particularly in terms of specifying the neural bases of these mechanisms. A large literature exists on aphasia, of course, with some relevant findings suggesting the involvement of motor structures in phonetic perceptual disorders (e.g., Blumstein et al., 1977; Gainotti et al., 1982). However, precise anatomical-functional correlations are often difficult to establish, and there is considerable variability in the pattern of deficits observed.

In the first PET study carried out in our laboratory aimed at understanding speech perception (Zatorre et al., 1992), we tested ten normal volunteers using two types of stimuli: pairs of noise bursts, which had been matched acoustically to the syllables to be used subsequently, and pairs of consonant-vowel-consonant real speech syllables. The vowels in any given syllable pair were always different, but the final consonant differed in half of the pairs; in addition, the second syllable had a higher fundamental frequency in half the pairs, and a lower frequency in the other half. The study included five conditions arranged in a subtractive hierarchy (Petersen et al., 1988). The first was a silent baseline; in the noise condition subjects pressed a key to alternate pairs of noise bursts. In the passive speech condition subjects listened to the syllables without making an explicit judgment, and pressed a key to alternate stimulus pairs. In the phonetic condition subjects listened to the same speech stimuli but responded only when the stimuli ended with the same consonant sound (e.g., a positive response would be given to the syllables [bag] and [pig]). In the pitch condition subjects once again listened to the same syllables, but responded only when the second item had a higher pitch than the first.

Our aim was to use the subtractive technique to distinguish between speech-specific neural processes and more general auditory processing mechanisms. In particular, we wished to clarify the role of primary versus secondary auditory regions in speech perception. We hypothesized that simple auditory stimulation (noise bursts) should lead to activation of primary cortex, whereas more complex signals should lead to activity in a broader expanse of cortical areas. Second, we wanted to test the hypothesis that phonological processing depends on left temporo-parietal cortex (as suggested by the earlier PET study of Petersen et al., 1988), by using a phonetic discrimination task. Finally, we attempted to dissociate linguistic from nonlinguistic processing by requiring judgments of pitch changes in the speech syllable, which we hypothesized to involve right-hemisphere mechanisms, in accord with other data from our laboratory (Zatorre, 1988; Zatorre & Samson, 1991).

When the results from the silent baseline condition were subtracted from the noise condition, activation was observed bilaterally approximately within the transverse gyri of Heschl, corresponding to primary auditory cortex (Penhune et al., 1996), as predicted. Subtraction of the noise condition from the passive speech activation pattern yielded several foci along the superior temporal gyrus bilaterally. This region contains several cytoarchitectonically distinct cortical fields responsive to auditory stimulation which receive input both from the medial geniculate nucleus and from corticocortical connections (Brugge & Reale, 1985; FitzPatrick & Imig, 1982). It is therefore likely to be involved in higher-order auditory processing of complex signals. One left-lateralized focus in the posterior superior temporal gyrus was also identified in this comparison. This finding is important in that it implicates this portion of the left temporal lobe, roughly falling within the classical posterior speech region, as being automatically engaged in the processing of speech signals.

Perhaps the most surprising results were obtained in the two active conditions, in which subjects were asked to make specific judgments of either phonetic identity or pitch. In the subtraction of the phonetic condition minus passive speech, activity was largely confined to the left hemisphere: the largest increase was observed in part of Broca's area near the junction with the premotor cortex, and in a superior parietal area. The prediction that pitch processing would involve right-hemispheric mechanisms was confirmed in the pitch condition minus passive speech subtraction, with two foci observed in the right prefrontal cortex. In these latter two subtractions both stimuli and responses were identical; only the nature of the required cognitive processing changed as a function of the instructions. The dissociable patterns of activity observed must therefore reflect the fundamentally different nature of the neural mechanisms involved in analysis of phonetic and pitch information, respectively.

These results were subjected to further scrutiny in a second study (Zatorre et al., 1996a) aimed at replicating and extending the first set of findings. In particular, the finding of increased CBF in a frontal region close to the conventionally defined Broca's area was of special significance, since its involvement in a purely perceptual task has implications for models of speech processing. However, data from any single PET activation comparison must be viewed cautiously, as many different, and perhaps uncontrolled, task dimensions may be responsible for the observed effect (for example, comparing an active discrimination to a passive listening condition entails many differences in cognitive demands, including attention, working memory, response organization, etc.); hence the importance of replicating results under different conditions.

In order to assess these issues we carried out a new task, requiring monitoring of a given target phoneme within a stream of speech syllables (e.g., pressing a key whenever a [b] was perceived). This task is slightly different in its cognitive demands from the original phonetic task (it does not require comparison of pairs of stimuli, for example), but should engage the same phonetic processing system as was recruited by the first task. A comparison of this task to passive listening of syllables yielded very similar activation within the region close to Broca's area previously observed. Furthermore, we carried out a re-analysis of the previous study to compare the phonetic and pitch tasks to one another. Our reasoning was that both tasks require active comparison and decision processes, but differ in terms of the crucial phonetic processing component which we wished to isolate. Once again, this comparison yielded a similar left frontal cortex response.

Figure 1 shows a summary diagram of the various comparisons from our two studies, together with data from two similar experiments conducted by Démonet (1992; 1994). Each symbol represents a stereotaxic point at which maximum CBF activity was reported. It may be seen that the points all cluster fairly closely together, despite coming from different studies with different subjects in different laboratories. It is interesting that all of these foci consistently cluster in the most superior and posterior aspect of cytoarchitectonic area 44, near the border with area 6, rather than to the inferior aspect of the third frontal convolution, or to opercular areas more traditionally associated with Broca's area. Activation in the latter location has been reported with tasks requiring overt (Petrides et al., 1993) or covert (Wise et al., 1991) vocal production. The highly consistent anatomical

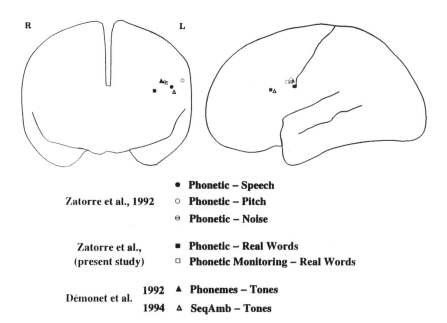

Zatorre et al., 1992	●	**Phonetic – Speech**
	○	**Phonetic – Pitch**
	⊖	**Phonetic – Noise**
Zatorre et al., (present study)	■	**Phonetic – Real Words**
	□	**Phonetic Monitoring – Real Words**
Démonet et al. 1992	▲	**Phonemes – Tones**
1994	△	**SeqAmb – Tones**

Figure 1. Summary diagram demonstrating the location of activation foci near Broca's area identified in PET activation studies by Zatorre et al. (1992; 1996a) placed on coronal (front view) and saggital (lateral view) projections of a left hemisphere. For comparison purposes, the location of foci described by Démonet et al. (1992; 1994) in phonetic discrimination tasks is also shown. The symbols represent the center (most significant voxel) of extended regions of CBF increase identified in each comparison. The brain outlines and major sulci were generated from an averaged MRI data set, transformed into stereotaxic space (Talairach & Tournoux, 1988).

placement of the region shown in Fig. 2 may possibly indicate the existence of a functional subregion within Broca's area related to phonetic operations.

The conclusion that part of Broca's area participates in phonetic processing does not imply that other regions, particularly the left posterior temporal region, do not play a role in the phonetic analysis of speech sounds. Experimental conditions in which subjects listen to speech syllables have consistently yielded asymmetric left posterior temporal CBF increases, as well as bilateral activation in the anterior portion of both superior temporal gyri (Wise et al., 1991; Petersen et al., 1988; Price et al., 1992; Zatorre et al., 1992). Functional MRI data have also corroborated this finding (Binder et al., 1994). Thus, when passive listening is used as the baseline state, any neural activation in these regions would be subtracted away. It seems clear that "passive" listening would include an important phonetic processing component that would be engaged automatically, but that is not observable in subtractions using passive listening as a baseline.

It is reasonable to assume that neural processes in the superior temporal gyri are initially responsible for perceptual analysis of the complex incoming speech stream, since neurophysiological studies of auditory cortices reveal the presence of neuronal populations sensitive to acoustic features that are present in speech sounds, such as frequency modulation (e.g. Whitfield & Evans, 1965), or onset times (e.g. Steinschneider et al., 1995). It is therefore likely that the CBF activation in the left and right anterior superior temporal area observed during "passive" speech reflects the operation of such neural systems. The posterior region of the left superior temporal plane likely plays a special role in speech processing, since this region is not activated by simple tones or noise stimuli (Zatorre et al., 1992, 1994), or by auditory tonal discrimination tasks (Demonet et al., 1992, 1994; Zatorre et al., 1994), but is consistently activated by speech stimuli. The processing carried out within this left posterior temporal area is not fully understood, but probably involves the analysis of speech sounds leading to comprehension, and may operate at the syllabic or whole-word level.

This aspect of speech processing appears to be distinct, however, from processes that engage the network that includes the portion of Broca's area identified above. In the phonetic tasks in question, a relatively abstract pattern-extraction process must take place, since individual phonetic units belonging to the same category may have very different acoustic manifestations. It is therefore apparently insufficient to rely on a whole-syllable representation to perform this type of task; rather, recourse must be made to a specialized mechanism that is able to compute the similarity between phonetic segments that are differently encoded acoustically by virtue of being embedded in syllables with different vowels (Liberman & Mattingly, 1985). We would argue that this type of judgment calls into play the specialized articulatory recoding system whose neural manisfestation is activity in a portion of Broca's area.

3. PROCESSING OF MELODIC PATTERNS

The apparent ease with which most people, including very small children, recognize and reproduce melodies belies the complex nature of music processing. Indeed, perceiving and encoding a melody, or pattern of pitches, entails multiple cognitive operations that include perceptual analysis, abstract pattern-matching, and working memory, among others (Deutsch, 1982; Dowling & Harwood, 1986). Many questions are raised by the ubiquitous human capacity to listen to and perform music. Over the past several years our laboratory has attempted to explore the neural correlates of pitch and melody processes, not only to localize the systems responsible, but also to address whether music may rely on specialized neural operations distinct from those used for speech, and for other auditory processes.

A considerable number of studies have begun to point to the existence of a specialization within the right temporal cortex for processing of certain aspects of pitch. For example, although simple frequency discrimination is affected only slightly or not at all by unilateral cortical lesions in humans (Milner, 1962; Zatorre, 1988) or bilateral lesions in animals (Evarts, 1952; Heffner & Masterton, 1978, Jerison & Neff, 1953), if a pitch judgment requires spectral analysis, then right-hemisphere auditory cortical regions seem to play a special role. Thus, perception of the missing fundamental is affected specifically by right temporal-lobe lesions which invade portions of Heschl's gyri, and not by more restricted anterior temporal-lobe damage or by left temporal excision (Zatorre, 1988). Similar lateralization findings have been reported in tasks requiring processing of complex harmonic structure (Divenyi & Robinson, 1989; Robin et al., 1990; Sidtis & Volpe, 1988). Furthermore, timbre discrimination tasks involving changes in harmonic structure have also yielded consistent evidence favoring right-asymmetric processing, both with temporal-lobe lesioned patients (Milner, 1962; Samson & Zatorre, 1994), as well as with commissurotomized subjects (Tramo & Gazzaniga, 1989).

Short-term retention is another aspect of pitch processing that apparently requires asymmetric mechanisms. Zatorre and Samson (1991) demonstrated that right temporal-lobe excision affected short-term memory for pitch when interfering stimuli were presented between the target and comparison items (Deutsch, 1970). Bilateral ablations in the superior temporal gyrus of the monkey also result in deficits in tonal retention (Colombo et al., 1990; Stepien et al., 1960), and this region is implicated in auditory short-term memory by single-unit data as well (Gottlieb et al., 1989). Zatorre and Samson (1991) also observed that right frontal-lobe damage significantly impaired pitch retention, a finding mirrored in certain animal studies of bilateral frontal ablation (Gross & Weiskrantz, 1962; Iversen & Mishkin, 1973). The latter result may reflect a disruption of functional connectivity between frontal and parietal cortices (Petrides & Pandya, 1988), which may be involved in maintenance of pitch in working memory (Perry et al., 1993; Marin & Perry, in press; see also Chavis & Pandya 1976).

Despite the apparent importance of neural systems within the right cerebral hemisphere for aspects of pitch processing, additional evidence clearly indicates that there are important contributions to musical processes from the left hemisphere as well. The most dramatic evidence for this conclusion comes from studies of patients who suffer from amusia, a specific disorder of recognition of all musical information, including very familiar melodies, which has been observed to occur only after bilateral lesions involving the two superior temporal gyri (Peretz, 1996). Even unilateral left superior temporal lesions can, in some instances, result in mild or moderate melodic processing deficits (Zatorre, 1985; Samson & Zatorre, 1988), particularly for recognition memory tasks (Samson & Zatorre, 1992).

The above studies formed the backdrop for a functional imaging study (Zatorre et al., 1994) whose goal was to better understand the neural basis for perception of melodic patterns and for retention of pitch information in working memory. Our goal was to test two specific hypotheses: (1) that perceiving a novel tonal melody would entail neuronal processing in both left and right superior temporal regions, with a possibly greater contribution from the right; and (2) that right frontal-lobe mechanisms would be engaged when subjects make specific judgments that require retention of pitch over a filled interval.

We tested twelve normal subjects without formal musical training using two classes of stimuli: noise bursts and melodies. The noise bursts were constructed so as to approximate the acoustic characteristics of the melodies in terms of number, duration, inter-stimulus presentation rate, intensity, and onset/offset shape. Sixteen different eight-note tonal melodies were also prepared, all identical in their rhythmic configuration, with the aim of allowing pitch judgments of either the first two notes, or the first and last notes. The last note had a higher pitch than the first in half of the sixteen melodies, and in the other half the last note was higher in pitch.

Four separate conditions were run during each of the four scanning periods. During the first condition, termed the "noise" condition, subjects listened to the series of noise bursts described above, and after each "noise melody" depressed a key to control for motor activity. In the second condition, termed "passive melodies," the subjects were presented with each of the sixteen tonal melodies, and depressed a key after each one, as before. No overt judgments were required, but subjects were instructed to listen carefully to each melody. In the third condition, the "2-note" pitch comparison, subjects listened to the same melodies as before, but this time were instructed to determine whether the pitch of the second note was higher or lower than that of the first note. Finally, in the "first/last" pitch judgment, subjects were asked to compare the pitch of the first and last notes, ignoring the notes in between, and to respond as before, according to whether the pitch rose or fell. Subjects kept their eyes shut throughout the scanning period.

The experiment was set up to permit specific comparisons, accomplished via subtraction of relevant conditions. The first comparison, passive melodies minus noise, permits examination of the cerebral regions specifically active during listening to novel tonal melodies, as opposed to the activation that might be present with any auditory stimulus with similar acoustic characteristics. The principal result indicated a significant CBF increase in the right superior temporal gyrus, anterior to the primary auditory cortex. A much weaker CBF increase was also visible within the left superior temporal gyrus. In addition, and unexpectedly, a significant focus was also identified in the fusiform gyrus of the right hemisphere, within area 19. The finding of an activation within the right superior temporal gyrus while listening to melodies fits in well with our prediction, and likely reflects the specialization of neuronal networks within the right secondary auditory cortices for perceptual analysis of tonal information, consistent with the human lesion evidence reviewed above (Milner, 1962, Zatorre, 1985; 1988). Although subjects were listening "passively," it is evident that they would be extracting perceptual information during this phase, and the CBF changes we observed nostly likely reflect these automatically engaged processes. The weak activity in the left superior temporal area may indicate the additional but perhaps less important or less consistent participation of left temporal cortices in melody processes, which is also suggested by the evidence from lesion studies.

Note that, in this subtraction, no CBF increase was present in the primary auditory cortices beyond that elicited in the control condition. This result is explained by the control condition: by using acoustically matched noise bursts, nonspecific auditory processing can be dissociated from that uniquely elicited by listening to melodies. We previously demonstrated that similar noise bursts result in primary auditory cortical stimulation when contrasted to a silent condition (Zatorre et al., 1992). These findings, together with findings from prior PET studies using speech sounds or tones (Demonet et al., 1992; Petersen et al., 1988; Wise et al., 1991) point to differential activation of primary vs. secondary auditory areas within the superior temporal gyrus, according to the nature of the processing elicited by a given stimulus. Although the noise stimuli proved successful in demonstrating the intended dissociation, caution must be still exercised in interpreting the results, for the noise bursts are clearly not physically identical to the melodic sounds. For example, the noise stimuli contain no periodicity, whereas the tones do; their spectral composition also is quite different. It remains to be established, therefore, which specific features of the melodies may lead to the observed pattern of activation.

Our finding of activation in in the right fusiform gyrus is puzzling. Area 19 is typically described as extrastriate visual cortex (Diamond et al., 1985); there is scant physiological evidence for its direct participation in auditory processing. The possibility that the effect is due to some extraneous visual stimulation is excluded, since scanning was carried out with the subjects' eyes closed. This phenomenon clearly invites further investigation; the possibility that it reflects visual imagery processes elicited indirectly by the melodic patterns in an intriguing one, but this conjecture must await direct evidence before it can be accepted.

The second and third comparisons in this study both used the passive melody condition as the baseline, so that any activation seen represents neural responses beyond those already present during initial listening to the same stimulus materials. Subtraction of the passive conditin from the 2-note condition resulted in significant activation within the right frontal lobe, as predicted. Two separate foci could be distinguished within distinct cytoarchitectonic regions, including Brodmann's areas 47/11, and 6. The first/last-passive melodies subtraction yielded a number of cortical and subcortical activation sites in both hemispheres. Among the more relevant results were CBF increases within the right frontal lobe, consistent with the predictions, including a focus in area 47/11 identical to that observed in the 2-note condition. Of particular interest was an area of significant CBF increase within area 21 of the right temporal lobe, indicating that this condition resulted in greater activity within the right auditory association cortex than already present during passive listening to melodies. As well, we observed an increase in CBF within the right inferior colliculus.

The pattern of results from these conditions implicates frontal-lobe mechanisms in effecting pitch comparisons, as had been predicted, with a particularly important contribution from right-frontal regions. In the first/last minus passive melody comparison, we observed a greater number of separate foci of CBF change over a wider swath of cortical and subcortical territory than was evident in the lower memory-load condition of judging the first two notes; this finding perhaps reflects the complexity and increased cognitive demands of the task, which was also manifested in increased error rate and slower reaction times. Although we are not in a position to interpret all of these foci, one may speculate that the numerous frontal-lobe sites observed might be associated with successful performance of distinct aspects of the task. For example, maintenance of pitch information in working memory might depend on a mechanism separate from that involved with the more "executive" functions required to monitor the presentation of the tones and their temporal order, and to direct the appropriate pitch comparison (see Milner & Petrides, 1984). The right inferior colliculus, known to be an important auditory processing structure (Aitkin, 1986), was also activated in this subtraction, indicating that it too is a component of a specialized distributed network involved in pitch memory.

Putting the results from the various tasks together with the physiological and lesion literature discussed earlier, a preliminary outline of a model to describe the neural substrates associated with pitch processing may be suggested. We may speculate that the primary auditory cortex is chiefly involved in early stages of processing (which might include computation of such signal parameters as pitch, duration, intensity, and spatial location), whereas more complex feature extraction, involving temporally distributed patterns of stimulation, is performed via populations of neurons within the secondary cortices. Neuronal systems located in both temporal lobes likely participate in higher-order perceptual analysis of melodies, but those on the right seem to be particularly important, perhaps because they are specialized to extract the features that are most relevant for melodic stimuli (including, for example, invariant pitch-interval relationships, and spectral characteristics important for pitch and timbre perception). The existence of temporal-lobe neurons with complex response properties (e.g., McKenna et al., 1989) would be in keeping with this idea.

In both pitch judgment conditions we observed significant CBF increases within the right frontal cortex. Only in the first/last comparison, however, did we observe an additional CBF increase in the right temporal lobe, beyond that seen in passive listening. We interpret this result, together with the right frontal activation, as evidence that the high memory load imposed by the first/last task engaged a specialized auditory working memory system, and that this system is instatiated in the brain via interaction of inferior frontal and superior temporal cortices in the right cerebral hemisphere (Marin & Perry, in press). This conclusion would be in accord with our earlier study (Zatorre & Samson, 1991), in which deficits in pitch retention were observed after right frontal and/or temporal-lobe lesions.

4. AUDITORY IMAGERY

Many people, musically trained or not, report a strong subjective experience of being able to imagine music or musical attributes in the absence of real sound input. But subjective reports are of limited use to assess the characteristics of cognitive representations in a scientifically rigorous manner. Therefore, in recent years, psychologists have tried to find more objective means of evaluating the nature of imagery processes. Much of this research has concentrated on the visual domain, and has yielded the conclusion that visual imagery processes operate with similar characteristics to perceptual processes (see Farah, 1988, for a review). This view leads to the hypothesis that perception and imagery may share at least partially, the same neural substrate. Perhaps the nervous system has evolved in such a way that all sensory processing areas, which are normally responsive to environmental input, can also be activated endogenously, i.e., in the absence of external stimulation. If so, then at least a preliminary explanation of the neural basis for imaginal processing would be at hand.

In our laboratory we have examined the aforementioned hypothesis within the context of musical imagery, using both a behavioral lesion approach (Zatorre & Halpern, 1993), and via PET functional imaging (Zatorre et al., 1996b). We adapted a paradigm originally developed by Halpern (1988), in which musically untrained subjects compared the pitch of two lyrics from a familiar, imagined song. (For instance, is the pitch corresponding to "sleigh" higher or lower than that of "snow" in the song "Jingle Bells"?) She varied the distance (number of beats) between the target lyrics chosen, as well as the distance from the beginning of the song of the first lyric of the pair. Response latencies increased systematically as a function of both factors, suggesting that subjects were "mentally scanning" the tune in order to compare the imagined pitches. Thus, she concluded that the temporal pace and ordering of the notes in the real song were preserved in analogous fashion in the image of the song. This result is similar to the conclusion that real-world spatial characteristics are preserved in visual images (Kosslyn, Ball, & Reiser, 1978).

In the first of our neuropsychological imagery studies (Zatorre & Halpern, 1993), we examined whether auditory imagery and perception may share similar neural mechanisms by presenting a modification of the tune scanning task to patients having undergone right or left temporal-lobe excision for the relief of intractable epilepsy. A perceptual version of the task was devised in which the listener made pitch judgments while actually hearing the song. As well, subjects participated in an imagery condition in which judgments of pitch were made to imagined tunes indexed by the lyrics. The results of that study were very clear and striking. While all subjects did better on the perception task compared to imagery, patients with left-temporal excisions showed no deficits whatsoever relative to normal controls, whereas those with damage to the right temporal area were significantly worse than the other groups on both tasks, and by about the same amount on each task. We concluded that structures in the right temporal lobe were crucial for successful performance of both imagery and perception tasks, suggesting the same kind of neuroanatomical parallelism (and by extension functional parallelism) shown by Farah (1988), Kosslyn et al. (1993), and others for visual imagery and perception.

PET methodology allows us to study the neural processes of normal subjects with greater anatomical precision compared to many other physiological techniques, including lesion studies. We therefore designed an experiment to investigate the putative similarity between perceptual and imagery mechanisms. We presented three tasks to a group of 12 normal participants: a visual baseline condition and two active tasks, one termed "perception," the other "imagery." The latter two were similar to those used by Zatorre and Halpern (1993): Two words from a familiar tune were presented on a screen, and the task was to decide if the pitch corresponding to the second word was higher or lower than the pitch corresponding to the first word. In the perceptual task, participants actually heard the song being sung, while in the imagery task they carried out the task with no auditory input. In the basleine task subjects viewed the words and performed a visual length judgment. By subtracting the activation in the visual baseline from both the perception and imagery tasks, we should, in principle, eliminate cerebral activity related to nonspecific processes shared by the two tasks, such as reading words on a screen, making a forced-choice decision, pressing a response key, etc. Thus, any CBF changes still remaining must be due to the unique demands of listening to a tune or imagining it, and making a pitch comparison.

The most striking findings from these subtractions were that for nearly every region demonstrating CBF change in one condition, there was a corresponding CBF peak in the other condition, often within a few millimeters. Most importantly, CBF increases were found bilaterally in the temporal lobes, in both perceptual and imagery conditions, and in the right frontal lobe. In addition, we observed areas of activation in both tasks in the left frontal and parietal lobes, as well as in supplementary motor area (SMA) and midbrain. The similarity in CBF distribution across the two conditions supports the idea that the two processes share a similar neural substrate.

Not surprisingly, highly significant CBF increases were found within the superior temporal gyrus bilaterally when subjects were processing the auditory stimuli for the perceptual task, as compared to the baseline task, in which no auditory stimulation was provided. More interesting is the finding that regions within the superior temporal gyrus were also activated, albeit at a much weaker level, when subjects imagined hearing the stimulus, again as compared to the baseline condition. Note that this latter subtraction entails two entirely silent conditions, so that positive CBF changes in the superior temporal gyri (associative auditory cortices) cannot be due to any external stimulation, but are most likely attributable to endogenous processing.

It is of interest to note that the temporal-lobe activation in the perceptual-visual baseline comparison incorporated primary auditory cortex and extended well into association cortical regions along most of the length of the superior temporal cortices. In contrast, this was not the case for the imagery-baseline comparison: CBF increases in that case occurred exclusively in association cortex (and were of lower relative magnitude). This distinction may be important, and supports the idea that primary sensory regions are responsible for extracting stimulus features from the environment, whereas secondary regions are involved in higher-order processes, which might include the internal representation of complex familiar stimuli.

The activation of the SMA is also of particular interest, given its role in motor processes. This region has consistently shown CBF increases during various types of motor tasks, including speech production tasks (Petersen et al; 1988; 1989). Of greatest relevance to the present study, SMA is also involved when a motor task is only imagined, rather than overtly executed (Rao et al. 1993; Wise et al., 1991). The finding of SMA activation may therefore imply that the SMA is part of a substrate for both overt and covert vocalization, and therefore supports the idea that imagery for songs includes not only an auditory component ("hearing the song in one's head"), probably related to temporal cortical activity, but also a subvocal component ("singing to oneself"), reflected in SMA activity.

Taking the findings of the PET study together with the behavioral lesion study of imagery (Zatorre & Halpern, 1993), we conclude that there is good evidence that perception and imagery share partially overlapping neural mechanisms, and that these include superior temporal auditory regions, as well as motor areas. Clearly, a great deal remains to be learned about the neural correlates of such a complex function as mental imagery, not the least of which is to disentangle verbal from tonal aspects of imagery. The converging evidence provided by the complimentary approaches of lesion and functional imaging appear to be quite powerful, however, and we are thus optimistic about being able to provide a more complete model of this intriguing aspect of cognition.

5. MORPHOMETRY OF AUDITORY CORTEX VIA STRUCTURAL MRI

We conclude the survey of our research into the correlates of human auditory processing by presenting some recent data pertaining to structural measures of auditory cortex (Penhune et al., 1996). Unlike the PET techniques described above, the aim of this research was to characterize the shape, volume, and position of the human primary auditory cortical region in vivo. The findings were somewhat surprising, and have possibly important implications for a better understanding of functional differences, however.

The approach taken was to use three-dimensional MRI scans taken from groups of normal right-handed volunteer subjects, and to label the region of Heschl's gyrus using interactive pixel-marking software that permits simultaneous viewing in all three planes of section, which greatly facilitates accurate anatomical delineation. Heschl's gyrus has long been known to contain the highly granular koniocortex, or primary auditory receiving area (Von Economo & Horn, 1930; Galaburda & Sanides, 1980), but the gross morphology of this region is highly variable across individuals and between hemispheres. Several studies have examined the cytoarchitecture of this region in the human brain, and most agree that the primary cortical region is confined roughly to the medial two-thirds of the most anterior Heschl's gyrus (Galaburda & Sanides, 1980; Rademacher et al., 1993), a conclusion also consistent with measures of evoked responses from depth electrodes within Heschl's gyrus (Liégeois-Chauvel et al., 1991). These data indicate that gyral and sulcal landmarks may serve as consistent boundaries to define the region of interest.

The results from our MRI-based morphometric measures therefore represent an estimate of the position and extent of primary auditory cortex in the human brain, but it is important to note that they also necessarily include non-primary cortical fields (particularly near the lateral edge of the gyrus), since there are no gross morphological features that would permit an exclusion of such areas.

Two sets of data were obtained, each on a different sample of 20 volunteer subjects; the first set underwent MR scanning using a 2-mm slice thickness, the second underwent a higher resolution scanning protocol in which 1-mm thick slices were obtained. T1-weighted images were acquired, then transformed to the standardized stereotaxic space of Talairach and Tournoux (1988) using an automatic three-dimensional cross-correlation algorithm which matches each individual MRI to an average of 305 manually registered images (Collins et al., 1994). Heschl's gyrus was then identified and labelled according to gyral and sulcal features visible in the three-dimensional dataset. This procedure yielded a set of points within the same standardized stereotaxic space for each subject, which may be superimposed to create a probabilistic map of the structure in question, in this case Heschl's gyrus (see Penhune et al., 1996, for further details). In addition, the second, higher-resolution sample of MRI scans also allowed for automatic segmentation of the labelled volumes into gray- and white-matter components. The gray/white boundary was calculated from the histogram of pixel intensity values by taking the midpoint between peak values corresponding to gray and white matter.

The result of greatest relevance to the present review pertains to the estimates of the total volume of Heschl's gyrus, which is corrected for any overall differences in brain size or shape by virtue of the stereotaxic normalization procedure. In the first sample, there was a significant difference between the volume of the left and right Heschl's gyrus (Figure 2, top panel), with 17 of the 20 subjects demonstrating a difference of 10% or more favoring the left side. This finding was replicated in the second sample of subjects, in which the asymmetry was slightly less marked but still strongly significant (Figure 2). Most interesting of all was the outcome of the gray/white segmentation analysis: the differences in volume were found to be confined to the white matter underlying Heschl's gyrus, and not to the volume of cortical tissue within the structure (Fig. 2, bottom panel).

These findings reveal an anatomical asymmetry that arises from a difference in the volume of fibers that carry information to and from the primary auditory cortex, and surrounding regions. However, the asymmetry in white matter may reflect any of a number of underlying neuronal structural differences. With in-vivo MRI we cannot ascertain, for example, whether the asymmetry reflects thalamocortical or corticocortical connections. We also cannot determine if the increased white-matter volume is a consequence of a greater number of axonal elements entering and exiting the primary cortical region, or if it may indicate a higher degree of myelination of these axons. Quantitative cytoarchitectonic studies in the human brain show evidence of differential cellular organization in the left and right auditory cortices. Seldon (1981, 1982), for example, found that cell columns in the left primary auditory cortex are both wider and more widely spaced than those on the right. As well, Hutsler and Gazzaniga (1996) have recently shown that the left primary auditory region has larger layer III pyramidal cells, which would be likely to form larger columns and to send out thicker or more heavily branched axons to other regions of auditory cortex.

The functional significance of such structural asymmetries is not yet clearly established, but it is interesting to speculate that they may be directly related to some of the functional asymmetries observed in many of the studies described in the preceding sections of this paper. In particular, several investigators (e.g. Tallal et al., 1993) have noted that the acoustic parameters necessary to process speech sounds entail rapid acoustic changes, particularly when tracking changes in formant transitions. Conversely, musical stimuli typically involve much slower rates of frequency change. Our data from the MRI study would be consistent with this general explanation if the white-matter volume measures are related to degree of myelination. That is, a greater degree of left-sided myelination could lead to faster transmission of acoustically-relevant information, thereby permitting a specialization of left auditory cortices in the fine-grained analysis of temporal aspects of the signal, which would be highly relevant for decoding of speech sounds.

On the other hand we may speculate that there is a tradeoff between speed of response and spectral selectivity. Neural systems on the left would have a fast rate of re-

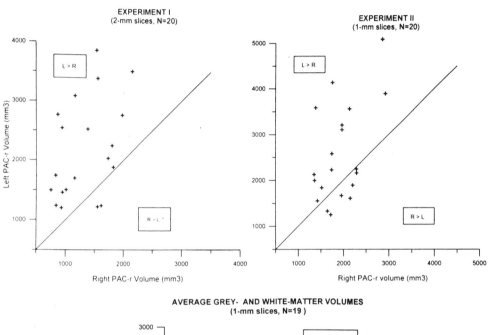

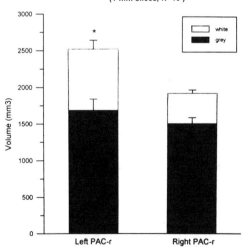

Figure 2. (Top) Scatterplots for volume of Heschl's gyrus, or primary auditory cortex region (PAC-r) in left and right hemispheres for two samples of subjects. The solid line represents values of equal left and right volume; points which fall above the line represent subjects whose left volume was greater than right. (Bottom) Average volumes of gray and white matter for left and right Heschl's gyrus in second sample of twenty subjects scanned with high-resolution MRI protocol; standard errors displayed as vertical bars. Reproduce from Penhune et al. (1996).

sponse, but their spectral tuning function would therefore necessarily be fairly wide-band, as would be appropriate to speech sounds. The right-hemisphere system would have narrower tuning functions, and thus be well-suited to the processing of stimuli containing small frequency differences, but would have a slower rate of integration in the temporal domain. This model could explain why many aspects of pitch processing relevant to music might be predominantly processed by right auditory cortical mechanisms, as reviewed in previous sections, since musical stimuli generally contain relatively slower changes, but small frequency differences are important.

This account of the possible relation between structural and functional asymmetries in the human auditory cortex is at this stage very preliminary and necessarily highly speculative. It is important to put forth such ideas, however, in that they should be testable and verifiable. Perhaps more important, they point to the type of integration of evidence from multiple types of studies (behavioral, functional, structural) that will become more feasible in the near future, thanks to the development of new in-vivo imaging technologies, together with more traditional physiological and anatomical knowledge.

REFERENCES

Aitkin, L.M. (1986) The Auditory Midbrain. Clifton NJ: Humana press.

Binder, J.R., Rao, S.M., Hammeke, T.A. et al. (1994) Functional MRI of human auditory cortex. Annals of Neurology, 35, 662–672.

Blumstein, S.E., Baker, H. and Goodglass, H. (1977) Phonological factors in auditory comprehension in aphasia. Neuropsychologia, 15, 19–30.

Brugge, J.F. and Reale, R.A. (1985). Auditory cortex, in Cerebral Cortex, Vol. 4. Edited by A. Peters and E.G. Jones. New York: Plenum Press, pp. 229–271.

Chavis, D., and Pandya, D.N. (1976) Further observations on corticofrontal pathways in the rhesus monkey. Brain Res. 117: 369–386.

Colombo, M., D'Amato, M.R., Rodman, H.R., and Gross, C.G. (1990) Auditory association cortex lesions impair auditory short-term memory in monkeys. Science 247: 336–338.

Collins, D.L., Neelin, P., Peters, T.M. & Evans, A.C. (1994) Automatic 3D intersubject registration of MR volumetric data in standardized Talairach space. J. Comput. Assist. Tomogr., 18, 192–205.

Démonet, J.-F., Chollet, F., Ramsay, S., Cardebat, D., Nespoulous. J.-L., Wise, R., Rascol, A. and Frackowiak, R. (1992) The anatomy of phonological and semantic processing in normal subjects. Brain, 115, 1753–1768.

Démonet, J.-F., Price, C., Wise, R., and Frackowiack, R.S.J. (1994) A PET study of cognitive strategies in normal subjects during language tasks. Brain, 117, 671–682.

Deutsch, D. (1970) Tones and numbers: Specificity of interference in short-term memory. Science, 168, 1604–1605.

Deutsch, D. (1982) The processing of pitch combinations. In: The Psychology of Music (Deutsch D., ed). New York: Academic Press.

Diamond, I.T., Fitzpatrick, D., and Sprague, J.M. (1985) The extrastraite visual cortex. In: Cerebral Cortex, Vol. 4. (Peters A, Jones EG, eds.), pp. 63–87. New York: Plenum Press.

Divenyi, P.L. and Robinson, A.J. (1989) Nonlinguistic auditory capabilities in aphasia. Brain Lang., 37. 290–326.

Dowling, W.J. and Harwood, D.L. (1986) Music Cognition. Orlando, Fla.: Academic Press.

Evans, A.C., Marrett, S., Neelin, P., Collins, L., Worsley, K., Dai, W., Milot, S., Meyer, E., Bub, D. (1992) Anatomical mapping of functional activation in stereotactic coordinate space. Neuroimage, 1, 43–53.

Evarts, E.V. (1952) Effect of auditory cortex ablation on frequency discrimination in monkey. J. Neurophysiol. 15: 443–448.

Farah, M.J (1988). Is visual imagery really visual? Overlooked evidence from neuropsychology. Psychol. Rev., 95, 307–317.

FitzPatrick, K.A. and Imig, T.J. (1982) Organization of auditory connections: The primate auditory cortex. In: Cortical Sensory Organization, Vol. 3. Edited by C.N. Woolsey. Clifton, N.J.: Humana Press.

Gainotti, G., Miceli, G., Silveri, M.C., and Villa, G. (1982) Some anatomo-clinical aspects of phonemic and semantic comprehension disorders in aphasia. Acta Neurol. Scandinav., 66, 652–665.

Galaburda, A. and Sanides, F. (1980) Cytoarchitectonic organization of the human auditory cortex. J. Comp. Neurol. 190: 597–610.

Gottlieb, Y., Vaadia, E. and Abeles, M. (1989) Single unit activity in the auditory cortex of a monkey performing a short term memory task. Exp. Brain Res. 74: 139–148.

Gross, C.G. and Weiskrantz, L. (1962) Evidence for dissociation of impairment on auditory discrimination and delayed response following lateral frontal lesions in monkeys. Exper. Neurol., 5, 453–476.

Halpern, A.R. (1988). Mental scanning in auditory imagery for tunes. J. Exp. Psychol.: Learn., Mem. Cognit., 14, 434–443.

Heffner, H.E. and Masterton, B. (1978) Contribution of auditory cortex to hearing in the monkey (Macaca mulatta). In: Recent advances in primatology: Vol 1. Behaviour. (Chivers DJ, Herbert J, ed). New York: Academic Press.

Hutsler, J.J. and Gazzaniga, M.S. (1996) Acetylcholinesterase staining in human auditory and language cortices—regional variation of structural features. Cereb. Cortex, 6, 260–270.

Iversen, S.D. and Mishkin, M. (1973) Comparison of superior temporal and inferior prefrontal lesions on auditory and non-auditory tasks in rhesus monkeys. Brain Res., 55, 355–367.

Jerison, H.J. and Neff, W.D .(1953) Effect of cortical ablation in the monkey on discrimination of auditory patterns. Fed. Proc. 12: 237.

Kosslyn, S. M., Ball, T. M. and Reiser, B. J. (1978). Visual images preserve metric spatial information: Evidence from studies of image scanning. J. Exp. Psychol.: Hum. Percep. Perf., 4, 47–60.

Kosslyn, S.M., Alpert, N.M., Thompson, W.L. Maljkovic, V., Weise S.B., Chabris, C.F, Hamilton, S.E. Rauch, S.L., and Buonanno, F.S. (1993). Visual mental imagery activates topographically organized visual cortex: PET investigations. J. Cog. Neurosci., 5, 263–287.

Liberman, A.M. and Mattingly, I.G. (1985) The motor theory of speech perception revised. Cognition, 21, 1–36.

Liegeois-Chauvel C, Musolino A, Chauvel P (1991) Localization of the primary auditory area in man. Brain 114: 139–153.

Marin, O.S.M. and Perry, D.W. (in press) Neurological aspects of music perception and performance. In: The Psychology of Music, 2nd Edition (Deutsch D, ed). New York: Academic Press.

McKenna, T.M., Weinberger, N.M. and Diamond, D.M. (1989) Responses of single auditory cortical neurons to tone sequences. Brain Res. 481: 142–153.

Milner, B. (1962). Laterality effects in audition. In: Interhemispheric Relations and Cerebral Dominance. (Mountcastle VB, ed), pp. 177–195. Baltimore: Johns Hopkins Press.

Milner, B. and Petrides, M. (1984) Behavioural effects of frontal-lobe lesions in man. Trends Neurosci. 7: 403–407.

Paulesu, E., Frith, C.D. and Frackowiak, R.S.J. (1993) The neural correlates of the verbal component of short-term memory. Nature, 362, 342–345.

Paus, T., Perry, D.W., Zatorre, R.J., Worsley, K.J. and Evans, A.C. (1996) Modulation of cerebral blood flow in the human auditory cortex during speech: role of motor-to-sensory discharges. Europ. J. Neurosci., 8, 2236–2246.

Penhune, V.B., Zatorre, R.J., MacDonald, J.D., and Evans, A.C. (1996) Interhemispheric anatomical differences in human primary auditory cortex: Probabilistic mapping and volume measurement from MR scans. Cereb. Cortex, 6, 661–672.

Peretz, I., Kolinsky, R., Tramo, M., Labrecque, R., Hublet, C., Demeurisse, G. and Belleville, S. (1994) Functional dissociations following bilateral lesions of auditory cortex. Brain, 117, 1283–1301.

Perry, D.W., Petrides, M., Alivisatos, B., Zatorre, R.J., Evans, A.C., and Meyer, E. (1993) Functional activation of human frontal cortex during tonal working memory tasks. Soc. Neurosci. Abs., 19, 843.

Petersen, S.E., Fox, P.T., Posner, M.I., Mintun, M., and Raichle, M.E. (1988) Positron emission tomographic studies of the cortical anatomy of single word processing. Nature, 331, 585–589.

Petersen, S.E., Fox, P.T., Posner, M.I., Mintun, M., and Raichle, M.E. (1989) Positron Emission Tomography studies of the processing of single words. J. Cog. Neurosci., 1, 153–170.

Petrides, M. and Pandya, D.N. (1988) Association fiber pathways to the frontal cortex from the superior temporal region in the rhesus monkey. J. Comp. Neurol. 273: 52–66.

Petrides, M., Alivisatos, B., Meyer, M. and Evans, A.C. (1993) Functional activation of the human frontal cortex during the performance of verbal working memory tasks. Proc. Natl. Acad. Sci. USA 90: 878–882.

Price, C., Wise, R., Ramsay, S., Friston, K., Howard, D., Patterson, K., and Frackowiak, R.S.J. (1992) Regional response differences within the human auditory cortex when listening to words. Neurosci. Letters, 146, 179–182.

Rademacher, J., Caviness, V.S., Steinmetz, H. and Galaburda, A.M .(1993). Topographical variation of the human primary cortices: complications for neuroimaging, brain mapping and neurobiology. Cereb. Cortex, 3, 313–329.

Rao, S.M., Binder, J.R., Bandettini, P.A., Hammeke, T.A., Yetkin, F.Z., Jesmanowicz, A., Lisk, L.M., Morris, G.L., Mueller, W.M., Estkowski, L.D., Wong, E.C., Haughton, V.M., and Hyde, J.S. (1993) Functional magnetic resonance imaging of complex human movements. Neurology, 43, 2311–2318.

Robin, D.A., Tranel, D. and Damasio, H. (1990) Auditory perception of temporal and spectral events in patients with focal left and right cerebral lesions. Brain Lang., 39, 539–555.

Samson, S. and Zatorre, R.J. (1988) Discrimination of melodic and harmonic stimuli after unilateral cerebral excisions. Brain Cognit., 7, 348–360.

Samson, S. and Zatorre, R.J. (1992) Learning and retention of melodic and verbal information after unilateral temporal lobectomy. Neuropsychologia, 30, 815–826.

Samson, S. and Zatorre, R.J. (1994) Contribution of the right temporal lobe to musical timbre discrimination. Neuropsychologia, 32, 231–240.

Seldon, H.L. (1981) Structure of human auditory cortex II: axon distributions and morphological correlates of speech perception. Brain Res., 229, 295–310.

Seldon, H.L. (1982) Structure of human auditory cortex III: statistical analysis of dendritic trees. Brain Res., 249, 211–221.

Sidtis, J.J. and Volpe, B.T. (1988) Selective loss of complex-pitch or speech discrimination after unilateral lesion. Brain Lang., 34, 235–245.

Stepien, L.S., Cordeau, J.P. and Rasmussen, T. (1960). The effect of temporal lobe and hippocampal lesions on auditory and visual recent memory in monkeys. Brain, 83, 470–489.

Steinschneider, M., Schroeder, C.E., Arezzo, J.C., and Vaughan, H.G. (1995) Physiologic correlates of the voice-onset time boundary in primary auditory cortex of awake monkey: temporal response patterns. Brain Lang., 48, 326–340.

Talairach, J., and Tournoux, P. (1988) Co-Planar Stereotaxic Atlas of the Human Brain. New York: Thieme.

Tallal, P., Miller, S. and Fitch, R.H. (1992) Neurobiological basis of speech: a case for the preeminence of temporal processing. Ann. N.Y. Acad. Sci., 682, 27–47.

Tramo, M.J. and Gazzaniga, M. (1989) Discrimination and recognition of complex harmonic spectra by the cerebral hemispheres: Differential lateralization of acoustic-discriminative and semantic-associative functions in auditory pattern perception. Soc. Neurosci. Abstr., 15, 1060.

VonEconomo, C. and Horn, L. (1930) Uber Windungsrelief, Masse und Rindenarchitektonik der Supratemporalflache, ihre individuellen und ihre Seitenunterschiede. Z. Neurol. Psychiat., 130, 678–757.

Whitfield, I.C. and Evans, E.F. (1965) Responses of auditory cortical neurones to stimuli of changing frequency. J. Neurophysiol., 28, 655–672.

Wise, R.J., Chollet, F., Hadar, U., Friston, K., Hoffner, E., and Frackowiak, R., (1991) Distribution of cortical neural networks involved in word comprehension and word retrieval. Brain, 114, 1803–1817.

Worsley, K.J., Evans, A.C., Marrett, S., and Neelin, P. (1992) A three-dimensional statistical analysis for CBF activation studies in human brain. J. Cereb. Blood Flow Metab., 12, 900–918.

Zatorre, R.J. (1985) Discrimination and recognition of tonal melodies after unilateral cerebral excisions. Neuropsychologia, 23, 31–41.

Zatorre, R.J. (1988) Pitch perception of complex tones and human temporal-lobe function. J. Acous. Soc. Amer., 84, 566–572.

Zatorre, R.J., Evans, A.C., Meyer, E., and Gjedde. A. (1992) Lateralization of phonetic and pitch processing in speech perception. Science, 256, 846–849.

Zatorre, R.J., Evans, A.C., and Meyer, E. (1994) Neural mechanisms underlying melodic perception and memory for pitch. J. Neurosci., 14, 1908–1919.

Zatorre, R.J. and Halpern, A.R. (1993) Effect of unilateral temporal-lobe excision on auditory perception and imagery. Neuropsychologia, 31, 221–232.

Zatorre, R.J., Halpern, A.R., Perry, D.W., Meyer, E. and Evans, A.C. (1996b) Hearing in the mind's ear: A PET investigation of musical imagery and perception. J. Cognit. Neurosci., 8, 29–46.

Zatorre, R.J., Meyer, E., Gjedde, A., and Evans, A.C. (1996a) PET studies of phonetic processing of speech: Review, replication, and re-analysis. Cereb. Cortex, 6, 21–30.

Zatorre, R.J. and Samson, S. (1991). Role of the right temporal neocortex in retention of pitch in auditory short-term memory. Brain, 114, 2403–2417.

43

SEISMIC COMMUNICATION SIGNALS IN THE BLIND MOLE RAT ARE PROCESSED BY THE AUDITORY SYSTEM

Z. Wollberg, R. Rado, and J. Terkel

Department of Zoology
George S. Wise Faculty of Life Sciences
Tel Aviv University
Tel- Aviv 69978, Israel

INTRODUCTION

The blind mole rat (*Spalax ehrenbergi*) is a solitary, highly aggressive and essentially blind, subterranean rodent that spends its entire life in underground tunnels. Each mole rat excavates its own tunnel system and accidental encounters between two individuals are usually fatal for at least one of them. However, there is a need for females and males to meet during the mating season. To avoid undesired encounters between conspecifics and yet enable contact between males and females for reproductive purposes, a signaling system is necessary to allow communication between individuals inhabiting neighboring tunnels. The question thus arise as to how do mole rats located in different tunnel systems communicate in their underground environment? We and others (Rado et al., 1987; Heth et al., 1987) have previously demonstrated that the blind mole rat produces patterned substrate-borne (seismic) vibrations by tapping its head on the roof of the tunnel. Such tapping behavior has been observed both in nature and under laboratory conditions (Rado et al., 1987). Typically, vibrations produced by a mole rat placed in an experimental Plexiglas tube elicited tapping behavior by another specimen placed in the same tube (yet separated by some barrier) or in a second tube that was in physical contact with the first one (Rado et al., 1987). In such "dialogues", while one mole rat was producing the vibrations the other usually pressed its cheek and lower jaw against the vibrating wall of the tube, and appeared to be "listening". When the two compartments were separated so that no vibrations could cross between them, the dialogue essentially ceased. Based on these behavioral observations and on some unique morphological properties of the mole rat's middle ear and the lower jaw (Rado et al., 1987; Rado et al., 1989; Bruns et al., 1988; Burda et al., 1989) we suggested that blind mole rats use these seismic vibrations for intraspecific communication and that these vibrations are mediated to the auditory system by bone conduction. This notion has been challenged recently by Nevo et al.

(1992) who suggested that the seismic signals are perceived and processed by the somatosensory system. In this study we present electrophysiological evidence corroborating our assumption that the seismic signals used by the blind mole rat for long distance communication are indeed perceived and processed primarily by the auditory system.

METHODS

1. Animals

Adult mole rats of both sexes with no hearing deficiencies, were trapped in the nature and used in this study. They were housed individually in plastic cages under a 14/10 hr light-dark regime and a constant temperature of $22\pm2°C$. Rodent chow was supplied *ad libitum* and twice a week sufficient fresh vegetables and fruit were provided to eliminate the need for drinking water.

2. Recording and Stimulation Procedures

All the experiments were conducted within a double wall and shielded sound attenuation chamber. Event related responses elicited by airborne sounds or vibratory stimuli were recorded differentially from the scalp of anesthetized animals by means of two stainless-steel needle electrodes inserted subcutaneously, one at the vertex and the second over the temporal cortex. A grounding electrode was inserted into the animal's back. Potentials were AC amplified, band-pass filtered (0.02–5.0 kHz), monitored on the face of a CRT, digitized (sampling rate: 10.0 kHz), averaged and stored on a PC for off-line analyses. All the recordings were conducted with the anesthetized animals in a Plexiglas tube (7 cm diameter). Auditory airborne stimuli consisting of 0.2 ms clicks (peak intensity: 120 dB SPL re 20 μPa) were delivered through a speaker facing the aperture of the Plexiglas tube but not touching it, in order to minimize vibration from the sounds. Sound pressure levels were monitored by a calibrated condenser microphone connected to a sound level meter. The acceleration of our standard vibratory stimulus was 7.8 m/s^2 , produced by 0.2 ms square pulses that activated a mechanical mini-shaker placed under the tube at a distance of 60 cm from the animal's head. The tip of the mini shaker was covered with rubber to soften the tap and was adjusted so as to barely maintain contact with the Plexiglas tube. Acceleration was monitored and measured by an accelerometer connected through a charge amplifier to the sound level meter, operating on a vibration measurements mode.

RESULTS

Figure 1 illustrates averaged event related responses elicited by our standard vibratory stimulus and by an airborne click recorded from the same animal. It can be seen that both stimuli evoked an early (within the first 12 ms) multiple peaked response comparable to a typical auditory brain stem evoked response (ABER) and a biphasic middle latency response (MLR). Latencies of the responses elicited by the click and by the vibration were essentially the same. However, the response evoked by the vibration was far greater than the one elicited by the extremely high intensity airborne sound. It is, thus, evident that the vibratory stimulus is much more efficient than the airborne sound in eliciting these responses.

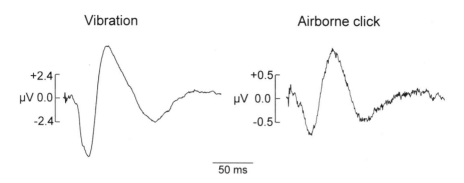

Figure 1. Averaged brainstem response (ABER) and middle latency response (MLR) elicited by repetitive vibratory (7.8 m/sec^2) stimulation (left) and by high intensity (120 dB SPL peak intensity) airborne clicks (right). Stimulation rate in both cases: 0.5 stimuli/sec. Note the similarity in shape and time course of the MLRs in both cases but the marked difference in amplitudes.

To corroborate this finding, and to determine whether the vibrations are indeed picked up primarily by the lower jaw, as we previously suggested, the following experiments were conducted. Two cylindrical Plexiglas tubes, 80 and 30 cm long, were placed on separate tables with their apertures facing each other at a distance of 20 mm so as to prevent physical contact between the two. The mini-shaker was located at the distal end of the longer tube and a calibrated speaker, through which masking noise was delivered, was placed facing the open end of this tube, but not touching it (Fig. 2). The acceleration of the vibrating substrate and the sound pressure produced by the mini-shaker while vibrating, measured near the mole rat's head were 7.8 m/s^2 and 70 dB SPL, respectively. Stimulation rate was 0.5 stimuli/sec.

In a first set of experiments the anesthetized animal was placed inside the short tube with its head positioned in the other tube but not touching it. The animal was thus exposed to the air-borne sounds that the mini-shaker produced while vibrating the tube but not to the vibrations themselves. Under these conditions no visible response was discerned (Fig 2A). This appears to corroborate the low sensitivity of the blind mole rat to airborne sounds as proved electrophysiologically (Bruns et al., 1988) and behaviorally (Bronchti et al., 1989; Heffner and Heffner, 1992). Next, the entire body of the animal, except its head, was located in the vibrating tube while its head was placed in the other tube. Under these conditions the animal was again exposed to the sound produced by the vibrator, but in addition its body was in contact with the vibrating substrate. In this case the vibration elicited an ABER and a prominent MLR (Fig 2B). This response was even greater when the entire body of the animal was located in the non-vibrating tube while its head, especially its lower jaw, was resting on the floor of the vibrating substrate (Fig 2C). While performing these experiments we noticed that the amplitude of the MLR was very much dependent on the precise location of the lower jaw and the pressure it inserted on the vibrating substrate. The firmer was the contact between the jaw and the wall of the tube the greater the response. It was thus evident that the responses were indeed elicited primarily by the vibrations *per se* rather than by the sound that the vibrator was producing, and that these vibrations were picked up predominantly by the lower jaw. Our next question was: are these vibrations perceived and processed by the auditory system as we had previously suggested (Rado et al., 1989) or by the somatosensory system (Nevo et al., 1992)?

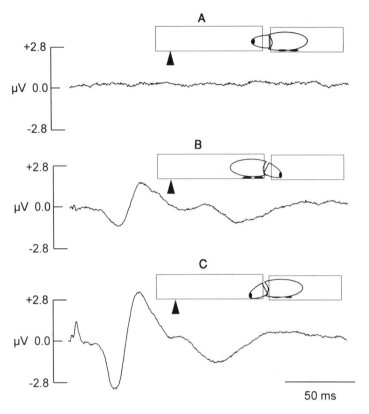

Figure 2. A schematic drawing and actual data demonstrating the difference in the potency of vibratory stimuli vis. a vis. airborne sounds in eliciting an MLR. A. The animal is exposed to the air-borne sounds that the mini-shaker (filled arrow head) produces (70 dB SPL) while vibrating the tube (7.8 m/s^2) but not to the vibrations. B. The animal is exposed to the sound produced by the vibrator and its body is in contact with the vibrating substrate. C. The animal's head, especially its lower jaw, rests on the floor of the vibrating tube. Note that no response was elicited in A whereas in B and C marked ABERs and MLRs were evoked. A maximum response was elicited in C.

In order to answer this question we examined the responses of intact animals and bilaterally deafened animals to vibrations. Intact animals were also examined for their responses while presenting high intensity masking noise. Deafening was achieved by the physical destruction of both middle and inner ears. The experiments with the deafened animals were conducted a few weeks after the surgery, following behavioral recovery from vestibular impairment. Figure 3 depicts the results of one experimental animal before and a few weeks after deafening. In both cases the lower jaw was firmly pressed against the vibrating substrate. It can be seen that the very prominent response that the vibration had elicited in the intact animal disappeared almost completely after deafening. Using high intensity broadband masking noise, in intact animals, yielded similar results (not shown). These results strongly support our suggestion that the vibratory stimuli used by the blind mole rat for intraspecific communication are processed primarily by the auditory system.

Typically, the rate of vibrations within bursts that mole rats produce both in nature and in the laboratory is about 7–12 vibrations/sec. This is a very high rate considering the fact that in animal models the amplitude of auditory MLR, evoked by a train of clicks, is markedly reduced at rates higher than 1/sec (Buchwald et al., 1981). We thus tested the re-

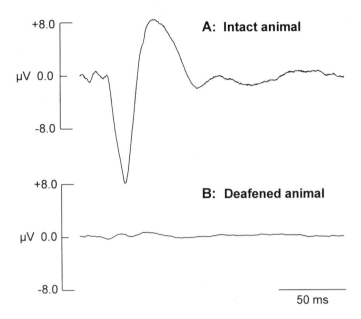

Figure 3. Responses evoked by a vibratory stimulus in an intact (A) and a deafened (B) mole rat. The lower jaw, in both cases is firmly pressed against the vibrating floor of the tube. Note, the residual response in B.

covery cycle of the MLR by using repetitive vibrations. Figure 4 illustrates one case in which three different rates of vibrations were used: 0.5, 1.8 and 4 vibrations/sec. The lower jaw of the tested animal was firmly pressed against the vibrating substrate. It can be seen that at vibration rates higher than 0.5/sec the averaged amplitude of the MLR was extremely reduced. At a rate of 2/sec there was already a marked attenuation of the response and at a rate of 4/sec there was only a residual, hardly visible, response. The ABER in all three cases was essentially not affected.

The fact that MLRs could not follow vibration rates higher than about 4/sec raised the question of how mole rats perceive the natural vibrations that they use for communication? To answer this question we recorded averaged MLRs in mole rats exposed to sequential bursts of vibrations, each consisting of four vibrations at a rate of 7 vibrations/sec and interbursts intervals of 2 seconds. Figure 5 illustrates the results of one such experiment. It can be seen that a maximum response was evoked only by the first vibration in each burst. The amplitudes of all other successive responses declined quite abruptly. The reason for this seeming paradox is simple: averaging many successive trials, at a relatively high rate, also eliminated the responses that were evoked by the first few stimuli (Fig. 4). This phenomenon was avoided in our second experimental paradigm (Fig. 5) in which a longer time interval was given between successive bursts, as in nature.

DISCUSSION

Blind mole rats appear to have two parallel auditory communication systems: one for short distances, based on vocalizations; and the other one for long distances, based on seismic signals. Communication between mother and young and between adults accidentally encountering one another in the same tunnel system is most probably accomplished by vocalization

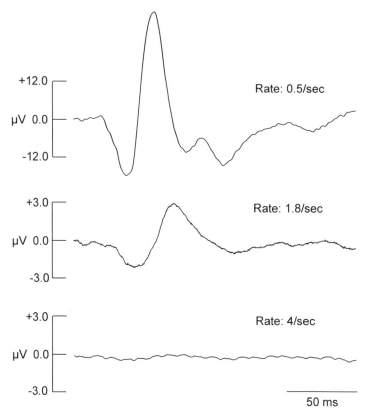

Figure 4. The effect of three different vibratory stimulation rates on the MLR. The lower jaw in all three cases was firmly pressed against the vibrating floor of the tube. Note the almost complete elimination of the response at a rate of 4 vibrations/sec.

(Nevo et al., 1987; Heth et al., 1988). For communication between individuals inhabiting separate and more remote tunnels seismic signalling is used (Rado et al., 1987). Seismic vibrations as means for intraspecific communication has recently been analyzed and described in another subterranean rodent, the Cape mole rat (Narins et al 1992).

Previous studies have shown a limited capacity of the blind mole rat to perceive air borne sounds (Bronchti et al., 1989; Bruns et al., 1988; Heffner and Heffner 1992; Rado et al., 1989). We have suggested that at least for long distance communication this limitation is compensated for by unique behavioral and morphological adaptations for perceiving the vibratory signals. These include the pressing of the lower jaw against the tunnel wall during the vibrational dialogue; a unique physical contact between the lower jaw and the bulla typmanica; and a very peculiar joint between the incus and the periotic bone (Rado et al., 1989). These features form a highly efficient "jaw hearing apparatus" that enables the transmitting of the substrate borne vibrations, with very little energy loss, from the tunnel wall to the cochlea by means of bone conduction.

The notion that the vibratory signals are perceived and processed by the auditory system was further corroborated in the present study by means of electrophysiological procedures. Both brain stem evoked responses and middle latency responses elicited by high intensity airborne clicks and by vibratory stimuli were very much alike in terms of latencies and shape but differed markedly in their amplitudes. The lack of response in the

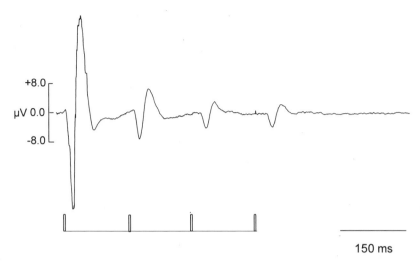

Figure 5. Averaged MLRs evoked by sequential bursts of vibrations, each consisting of four vibrations at a rate of 7 vibrations/sec. Interbursts interval were 2 sec. Note that a maximum response was evoked only by the first vibration in each burst. The amplitudes of all other successive responses declined quite abruptly.

deafened animals and its masking by white noise in intact animals provide additional evidence for the essential role of the auditory system in the processing of the vibratory signals. A maximum response to a vibratory stimulus requires a precise contact between the lower jaw and the vibrating substrate. This is achieved by the mole rat's unique "listening posture", apparently acquired throughout a long evolution of fossorial life. Deaf animals show intensive "jaw hearing" behavior for only a short period after their deafening. In time this behavior becomes more and more scarce (unpublished observation), reflecting possibly the animal's frustration at its inability to perceive the vibrations.

Based on some similar electrophysiological experiments it has been suggested that in the blind mole rat it is the somatosensory system rather than the auditory system that processes the vibratory signals (Nevo et al., 1992). The responses described in Nevo's study were much smaller than those we show here and their latencies were shorter. Unfortunately, in their article they did not mention the rate at which their stimuli were presented. However, if they did use higher rates than 0.5–1.0 stimuli per sec, they may have eliminated the auditory responses due to habituation and the averaging procedure. This is the only explanation we can offer for the contradiction between the results of their study and our own. This possibility is supported by the fact that in our experiments residual MLRs were often observed even when stimulating rates were higher than 2/sec. Low amplitude responses to vibrations also remained in the deafened animals. In both cases these were negligible as compared with the response components that were eliminated by the high rate of stimulation or before deafening. These small amplitude responses may indeed be somatosensory components that are embedded in the overall response but are obscured by the auditory components. When the latter are abolished they become visible.

REFERENCES

Buchwald, J.S. (1990) Animal models of cognitive event-related potentials. In: J.W. Rohrbaugh, R. Parasuraman and R. Johnson, JR. (Eds), Event-related brain potentials. Oxford University Press, pp. 57–75.

Bronchti, G., Heil, P., Scheich, H. and Wollberg, Z. (1989) Auditory pathway and auditory activation of primary visual targets in the blind mole rat (*Spalax ehrenbergi*): I. A 2-deoxyglucose study of subcortical centers. J. Comp. Neurol. 284, 253–274.

Bruns, V., Muller, M., Hofner, W., Heth, G. and Nevo, E. (1988) Inner ear structure and electrophysiological audiograms of the subterranean mole rat *Spalax ehrenbergi*. Hear. Res. 33, 1–10.

Burda, H., Bruns, V. and Nevo, E. (1989) Middle ear and cochlear receptor in the subterranean mole rat *Spalax ehrenbergi*. Hear. Res. 39, 225–230.

Heffner, R.S. and Heffner, H.E. (1992) Hearing and sound localization in blind mole rats (*Spalax ehrenbergi*). Hear. Res. 62, 206–216.

Heth, G., Frankenberg, E., Raz, A. and Nevo, E. (1987) Vibrational communication in subterranean mole rats (*Spalax ehrenbergi*). Behav. Ecol. Sociobiol. 21, 31–33.

Heth, G., Frankenberg, E. and Nevo, E. (1988) Courtship calls of subterranean mole rats (*Spalax ehrenbergi*): physical analysis. J. Mamm. 69, 121–125.

Narins, P.M., Reichman, O.J., Jarvis, J.U.M. and Lewis, E.R. (1992) Seismic signal transmission between borrows of the cape mole-rat, *Georychus capensis*. J. Comp. Physiol. A 170, 13–21.

Nevo, E., Heth, G., Beilis, A. and Frankenberg, E. (1987) Geographic dialects in blind mole rats: Role of vocal communication in active speciation. Proc. Natl. Acad. Sci. USA 84, 3312–3315.

Nevo, E ., Heth, G. and Pratt, H. (1991) Seismic communication in a blind subterranean mammal: A major somatosensory mechanism in adaptive evolution underground. Proc. Natl. Acad. Sci. USA 88, 1256–1260.

Rado, R., Levi, N., Hauser, H., Witcher, J., Adler, N., Intrator, N., Wollberg, Z. and Terkel, J. Seismic signalling as a means of communication in a subterranean mammal. Anim. Behav. 35, 1249–1251.

Rado, R., Himelfarb, M., Arensburg, B., Terkel, J. and Wollberg, Z. (1989) Are seismic communication signals transmitted by bone conduction in the blind mole rat? Hear. Res. 41, 25–30.

44

EVIDENCE FOR RAPID FUNCTIONAL REORGANIZATION IN INFERIOR COLLICULUS AND COCHLEAR NUCLEUS

Richard J. Salvi and Jian Wang

Hearing Research Lab, 215 Parker Hall
SUNY University at Buffalo
Buffalo, New York, 14214

1. INTRODUCTION

During development, the neural circuits in the central auditory pathway are laid down and fine tuned; however, once adulthood is reached, it has generally been assumed that the neural circuits remain stable throughout the life span. Recent studies in the somatosensory and visual system have challenged this hard wired view of the nervous system since extensive remodeling and reorganization occur in both somatosensory and visual pathways when their peripheral inputs are either temporarily or permanently eliminated (Eysel et al., 1980, 1981; Merzenich and Kaas, 1982; Kaas et al., 1990). Because of the relative ease of deafferenting a specific segment of the peripheral receptor surface, the somatosensory system has been one of the most popular systems for studying the reorganization of the central nervous system. A striking example of the rapid reorganization that can occur in the somatosensory cortex is illustrated by studies in which the peripheral inputs from a specific digit or peripheral nerve is temporarily or permanently eliminated by chemical or surgical means (Dostrovsky et al., 1976; Devor and Wall, 1981; Calford and Tweedale, 1988). For example, some neurons in the somatosensory cortex have excitatory receptive fields that are only activated by mechanical stimulation delivered to a specific digit. If the peripheral inputs that activate the neuron are eliminated by amputating or anesthetizing the digit, the neuron does not become "silent" or inactive, but instead, rapidly shifts it excitatory receptive field to an adjacent region of the body surface to which it was originally unresponsive (Calford and Tweedale, 1988). This rapid reorganization is presumably due to the unmasking of the system's intrinsic neural networks which were originally "covered over" by powerful inhibitory circuits. Functional reorganization is not unique to the somatosensory cortex. Indeed, it has been observed at the level of the brainstem and spinal cord following peripheral damage (Dostrovsky et al., 1976; Devor and Wall, 1981).

Cortical somatotopic map are extensively altered by the removal of a segment of the body surface, but the full extent of reorganization may require many weeks or months to

Acoustical Signal Processing in the Central Auditory System
edited by Syka, Plenum Press, New York, 1997

fully express itself (Merzenich et al., 1983; Pons et al., 1991). The maximum extent of re-organization was originally thought extend no more than 1–2 mm, but more recent studies of long-term deafferentation have shown that reorganization can extend up to 10–14 mm (Pons et al., 1991).

Like its counterparts in taction and vision, the tonotopic map of the auditory cortex can be dramatically altered when a segment of the cochlea is damaged or destroyed. While the degree of cortical reorganization appears to be greatest in young animals (Harrison et al., 1991, 1992), extensive restructuring of the tonotopic map can also occur in the adult auditory cortex (Robertson and Irvine, 1989; Willott et al., 1993). The physiological changes observed in auditory cortex have raised a number of important questions regarding possible sources and the mechanisms underlying the functional changes. First, where does the reorganization occur along the auditory pathway? Is it strictly cortical in nature or do more peripheral structures contribute to the remapping? Reorganization of the tonotopic map is likely to involve a dramatic shift in the receptive field of individual auditory neurons. Thus, a second question is to what extent can the receptive fields be modified at different levels of the auditory pathway? Third, how rapidly can these shifts in the receptive fields occur? To begin to address these questions, we have carried out a series of evoked potential and single unit studies in the inferior colliculus (IC) and dorsal cochlear nucleus (DCN) of the chinchilla. The evoked potential technique is a useful method for rapidly identifying regions of altered excitability in the central auditory pathway (Popelar et al., 1987; Syka et al., 1994). Our evoked potential studies have pointed to the inferior colliculus as a region where significant reorganization can take place. Since inhibition plays an important role in deafferentation-induced reorganization in other sensory systems (Hicks and Dykes, 1983), we attempted to selectively alter the inhibitory inputs to neurons in the IC and DCN system by presenting a traumatizing tone above the neuron's excitatory response area.

2. METHODS

The methods for implanting and recording the local evoked potential from the IC of awake chinchillas have been described previously (Salvi et al., 1982; Arehole et al., 1987). During testing, awake animals are restrained in a yoke-like apparatus that maintains the head at a fixed position within the calibrated sound field. Evoked response thresholds and amplitude-level functions were obtained at different frequencies by presenting tone bursts (5 ms rise/fall time) over a wide range of intensities. Evoked potentials were collected from each animal several times before and after exposing the animal to an intense tone that damaged a segment of the cochlea.

Methods for recording from single units in the central nucleus of the inferior colliculus and cochlear nucleus have been described previously (Wang et al., 1996b). Chinchillas, 1 to 2 years of age, were anesthetized with sodium pentobarbital (35 mg/kg i.p.), tracheotomized, and maintained at 37°C using a heating pad. The pinna and bony ear canal were removed and a sound source (Etymotic ER-2) and probe tube microphone (Etymotic ER-7) were placed near the tympanic membrane. Glass microelectrodes (4–15 Mohms) were used to record from single neurons in the ipsilateral DCN or contralateral IC. After a unit was isolated, its response properties were determined. Afterwards, a high level traumatizing tone was presented for 15–20 minutes at a frequency above the unit's characteristic frequency (CF). Immediately after the exposure, the measurement protocol was repeated to determine if any functional changes had occurred.

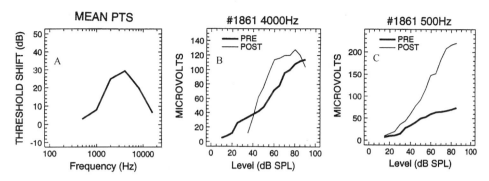

Figure 1. Evoked responses recorded from the IC of the chinchilla following a 5 day exposure to a 2 kHz tone of 105 dB SPL. (A) Mean (n=10) permanent threshold shift of the evoked response. (B) Pre- and post-exposure amplitude-level function at 4000 Hz from a typical chinchilla. (C) Pre- and post-exposure amplitude-level function at 500 Hz.

3. RESULTS AND DISCUSSION

3.1 IC Evoked Response Amplitude Enhancement

To determine if partial cochlear deafferentation resulted in increased excitability in the IC, we measured the amplitude of the evoked response from normal hearing chinchillas and then exposed the animals (n = 10) to a 2 kHz tone at 105 dB SPL for 5 days. The evoked response was remeasured 1–2 months later to determined the degree to which the evoked response thresholds were elevated. The traumatizing exposure caused a mean permanent threshold shift (PTS) of 20–30 dB in the 2–8 kHz range (Figure 1 A) and a loss of hair cells (data not shown) in regions of the cochlear corresponding to the frequencies of hearing loss (Salvi et al., 1990). Although thresholds were elevated in 2–8 kHz region, the amplitude of the response increased rapidly once threshold was exceeded. At suprathreshold levels, the amplitude was either equal to or slightly larger than before the exposure (Figure 1B). The slopes of the input/output functions were generally steeper than normal in the region of hearing loss, but the maximum evoked response amplitude was similar to pre-exposure measurements. Different effects were observed in the low frequency region (0.5–1 kHz) just below the region of hearing loss (Figure 1C). The amplitude of the evoked response increased at a faster than normal rate in this region and at high sound levels, the maximum amplitude was often two to three times larger than pre-exposure values. Although the amplitude of the response was abnormally large, the basic morphology of the evoked response was similar to that seen before the exposure.

3.2 Origins of Enhancement

While the preceding results indicate that cochlear damage results in increased excitability in the IC, it is not clear where the effect is occurring. Insights into the origins of the enhancement phenomenon were obtained from several animals in which recording electrodes were implanted on the round window and in the cochlear nucleus (CN) and IC. After pre-exposure amplitude-level functions were measured, the animals were exposed for 2 hours to a 2.8 kHz tone at 105 dB SPL. Twenty-four hours after the exposure, there was a 60–70 dB threshold elevation at the mid-frequencies (2–8 kHz), but only a slight (≤ 20

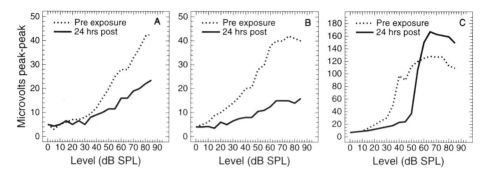

Figure 2. Pre- and post-exposure evoked response amplitude-level functions at 1000 Hz. (A) Compound action potential from the round window (B) evoked response from the CN and (C) evoked response from the IC. Post-exposure measurements obtained 24 h after presenting 2.8 kHz, 105 dB SPL tone for 2 h.

dB) hearing loss at lower frequencies. Figure 2 shows the changes in the 1 kHz amplitude-level functions at 24 hours post-exposure. The compound action potential showed a loss of sensitivity of approximately 15–20 dB and a reduction in amplitude, particularly at high intensities. The amplitude-level function from the cochlear nucleus also showed a 15–20 dB loss in sensitivity and a large reduction in amplitude at suprathreshold levels. A similar loss in sensitivity was observed in recordings from the IC; however, the amplitude of the IC evoked potential increased at an abnormally rapid rate once threshold was exceeded and the response was larger than normal at intensities greater than 55 dB. These results show that response amplitude is reduced at the level of the cochlea and cochlear nucleus; however, response amplitude in the IC was clearly enhanced at suprathreshold intensities.

3.3 Enhancement and Disinhibition

The preceding results imply that central mechanisms are contributing to the enhancement phenomenon seen in the IC (Salvi et al., 1992). The amplitude enhancement is greatest in the low frequency region just below the region of hearing loss suggesting that it may be due to the loss of inhibition between adjacent frequency regions which unmasks the intrinsic excitatory inputs to the IC. We have proposed a highly simplified model to account for the amplitude enhancement in the IC in terms of partially overlapping excitatory and inhibitory inputs that arise from adjacent frequency regions of the cochlea (Figure 3A). The excitatory and inhibitory inputs to the model are shown on the left and the output is shown on the right in Figure 3. The area above solid line (Figure 3A) represents the frequency-intensity combinations that activate the excitatory inputs to a neuron in the central auditory system while the dashed line represents the frequency-intensity combinations that activate the inhibitory inputs to the same neurons. In this example, the characteristic frequency (CF) of the inhibitory input is slightly higher than the CF of the excitatory input. The CF-threshold for inhibition is approximately equal to the CF-threshold for excitation. The shaded regions (Figure 3A) shows the areas where the threshold for inhibition is equal to or lower than the threshold for excitation; we assume that the neuron will not respond to frequency-intensity combinations within the shaded regions because the threshold for inhibition is equal to or lower than the threshold for excitation. The open region represents the area where the excitatory threshold is lower than the inhibitory threshold; the neuron will respond to frequency-intensity combinations in this region be-

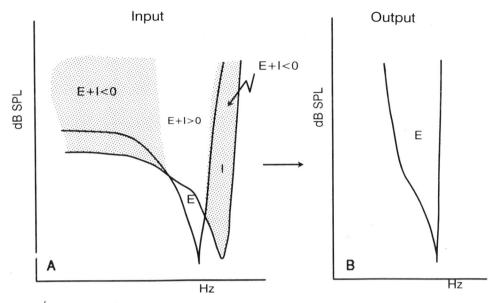

Figure 3. (A) Model illustrating the tuning curve inputs to a neuron in the central auditory pathway that receives overlapping excitatory (E and solid line) and inhibitory inputs (I and solid line). Magnitude of the excitatory input (E) and the inhibitory input (I) is proportional to dB level above the excitatory and inhibitory thresholds. Shaded region show areas where the inhibitory input is equal to or greater than the excitatory inputs. Open area above the excitatory input show the region where excitation is greater than inhibition. (B) Area above solid line shows frequency-intensity combinations that produce an excitatory output. Neuron does not respond to frequency-intensity combinations below the solid line. Note that the response area of the excitatory output is much narrower than the excitatory inputs in (A).

cause excitation is greater than inhibition. In the example shown in Figure 3, we implicitly assume that the excitatory and inhibitory inputs are approximately equal and that the magnitudes of both increase at roughly the same rate as stimulus level increases. In a normal ear, the interaction of the excitatory and inhibitory inputs results in a narrowing of the excitatory output response area, particularly along the low frequency side of the tuning curve. In addition, the excitatory output discharge rate-level function should be nonmonotonic because the inhibitory response becomes activated at suprathreshold levels. (Note that the properties of the model can be adjusted by changing the threshold and/or strength of the excitatory and inhibitory inputs. For example, simply raising the threshold of the inhibitory response area will convert the output from a narrowly-tuned, level-tolerant tuning curve to an open-V tuning curve system with a broad low-frequency tail (Suga and Tsuzuki, 1985; Yang et al., 1992). However, the discharge rate-level functions will be nonmonotonic in regions where the excitatory and inhibitory regions overlap.)

According to the model, if the segment of the cochlea which activates the inhibitory input is damaged or destroyed, the inhibitory input should be reduced or abolished thereby unmasking the full excitatory input to the cell. Two predictions can be made from this model. First, the excitatory output response area should expand mainly along the low frequency side of the tuning curve. Second, the maximum discharge rate within the excitatory response area should increase due to the loss of inhibition which originates from neurons that are tuned to frequencies above the neuron's characteristic frequency.

3.4 Rapid Reorganization and Enhanced Responsiveness in IC Neurons

At least three types of excitatory response areas have been identified in the IC and other regions of the central auditory system (Suga and Tsuzuki, 1985; Yang et al., 1992). Some neurons in the IC have open-V tuning curves with a narrowly tuned, low-threshold tip near CF and a high-threshold, broadly tuned tail. Others have level-tolerant tuning curves which maintain a very narrow tip even at high sound intensities. Finally, some have upper-threshold tuning curves with enclosed response areas, i.e., excitatory responses are elicited at low and moderate intensities, but not at high intensities. Recent neuropharmacologic studies suggest that the narrowly-tuned excitatory response areas of level-tolerant and upper-threshold neurons are shaped by strong inhibitory inputs activated by frequencies above CF (Yang et al., 1992; Wang et al., 1996a). In the experiments described below, we attempted to alter the balance between the excitatory and inhibitory inputs by presenting a traumatizing tone above CF. The traumatizing tone should reduce the inhibitory inputs that are activated by frequencies above CF leading to a loss of inhibition, an expansion of the excitatory response area and an increase in discharge rate.

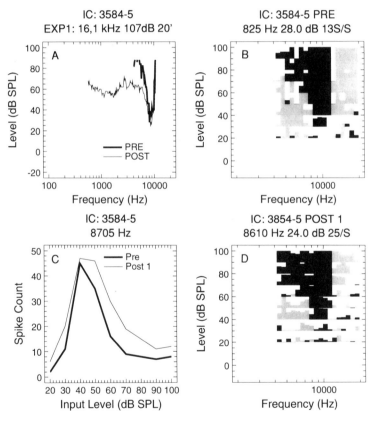

Figure 4. Pre- and post-exposure data from a neuron (3584–5) in the IC. Exposure tone: 16.1 kHz, 107 dB SPL, 20 minutes. (A) Pre-(thick line) and post-exposure (thin line) tuning curves. (B) Pre-exposure response area map. Intensity varied in 10 dB steps. Black areas excitatory, hatched areas inhibitory; height of bars indicates the relative amount of excitation or inhibition. (C) Pre-exposure and post-exposure discharge rate-level functions near CF. Note increase in discharge rate after the exposure. (D) Post-exposure response area map. Parameters as in B. Note decrease in strength and range of inhibitory response area after the exposure.

The thick line in figure 4A shows the tuning curve of a unit in the IC which had a level-tolerant tuning curve consisting of just a narrow tip. The neuron's response area map (Figure 4B) shows an excitatory area that was surrounded by inhibitory regions above and below CF. The threshold for inhibition was similar to the excitatory threshold at CF. The interaction between inhibition and excitation can be seen in the normal discharge rate-level function (Figure 4C) which is strongly non-monotonic at the neuron's CF (8705 Hz).

After measuring the normal discharge patterns, the ear was exposed for 20 minutes to a 107 dB SPL tone at 16.1 kHz; this frequency was located in the inhibitory response area above CF. The spontaneous activity of the neuron was suppressed when the traumatizing tone was initially turned on, but by the end of the exposure, the spontaneous activity had partially recovered. These results suggest that the traumatizing tone damages the region of the cochlea that activates the inhibitory inputs located above CF. Immediately following the exposure, there was a dramatic expansion of the excitatory response area below CF that resulted in a broad, low-frequency tail (Figure 4A). This transformation changed the tuning curve from a narrow, level-tolerant type to a broad, open-V type. In addition, there was a slight upward shift in the unit's CF (8.25 to 8.61 kHz) and about a 4 dB improvement in threshold at CF. The reduction in inhibition and the corresponding increase in excitation can be seen by comparing the pre- and post-exposure response area maps (Figure 4B,D). Note that the traumatizing tone above CF weakened the inhibitory response above and below CF; however, the expansion of the excitatory response was seen mainly below CF. The decrease in inhibition resulted in an increase in the discharge rate near threshold and at suprathreshold levels (Figure 4C). Note that the post-exposure, discharge rate-level function retained its nonmonotonic characteristics implying that inhibition was still present, but at a reduced level.

Approximately 45% of the units in our sample of 40 neurons had level-tolerant or upper-threshold tuning curves. Most (80%) of the units with narrow, level-tolerant tuning curves and all those with upper-threshold (100%) tuning curves developed a broad, low frequency tail after the traumatizing exposure; however, the threshold and width of the tip of the tuning curve was generally unaffected.

Some units in our sample had broad, open-V tuning curves (Figure 5A); however, their discharge rate-level functions were nonmonotonic as illustrated in Figure 5B. Presenting a traumatizing tone above the unit's CF resulted in slight expansion of the post-exposure tuning curve along it high frequency slope as well as a slight improvement in

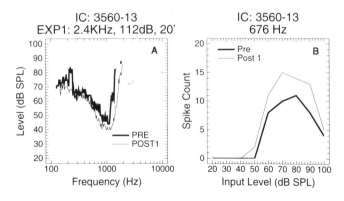

Figure 5. (A) Pre- and post-exposure tuning curves from a neuron with an open-V tuning curves. (B) Pre and post-exposure discharge rate-level functions neuron from neuron shown in A. Exposure: 2.4 kHz, 112 dB SPL, 20 minutes. Note post-exposure expansion of excitatory response area in (A) and increase in maximum discharge rate (B).

threshold near the tip of the tuning curve (Figure 5A). After the exposure, the discharge rate-level function showed a significant increase in discharge rate at suprathreshold levels (Figure 5B). The expansion of the tuning curve and the increase in discharge rate suggest that the traumatizing tone "stripped off" some of the inhibitory inputs that are activated by frequencies above the neuron's CF.

Units with open-V tuning curves and monotonic-saturating discharge rate-level functions showed a different pattern of response to the traumatizing tone above CF as illustrated in Figure 6. Two traumatizing exposures above CF (3 kHz, 105 dB, 20 min.; 4 kHz, 108 dB, 20 min.) did not alter the unit's post-exposure tuning curves (Figure 6A) or its discharge rate-level functions (Figure 6B). Approximately 55% of the units in our sample had open-V tuning curves. Many units with open-V tuning curves and monotonic-saturating rate-level functions were unaffected by the traumatizing exposure. The lack of change in this subpopulation of IC neurons provides an important internal control which eliminates cochlear mechanisms as the source of these changes.

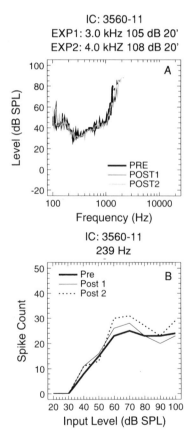

Figure 6. (A) Pre- and post-exposure tuning curves from an IC neuron with an open-V tuning curve (B) Pre- and post-exposure discharge rate-level functions from neuron in A. Exposure 1: 3 kHz, 105 dB SPL, 20 min. Exposure 2: 4 kHz, 108 dB SPL, 20 min..

3.5 Rapid Reorganization and Enhanced Responsiveness in DCN

The functional changes observed in the IC following acoustic overstimulation could be due intrinsic changes in the IC and/or to alternations that occur peripheral to the IC. The IC receives inputs from several nuclei in the ascending auditory pathway, including the contralateral DCN (Osen, 1972; Adams and Warr, 1976; Adams, 1979). Some neurons in the DCN have prominent inhibitory sidebands (Young and Brownell, 1976; Young, 1984). Thus, some of the functional changes seen in the IC may be a reflection of changes occurring in the DCN. We have carried out preliminary experiments to examine this possibility.

Figures 7 illustrates the typical changes seen in DCN neurons with strong inhibitory sidebands above and below CF. The neuron had a level-tolerant tuning curve consisting of a narrowly tuned tip around CF (Figure 7A). The pre-exposure response area consisted of a narrow excitatory response area surrounded by a strong inhibitory area above and below CF (Type III) (Young, 1984). We exposed the neuron twice, once at 7.98 kHz (105 dB, 15 min.) and a second time at 6.57 kHz (111 dB, 15 min.). The first exposure had little effect on the

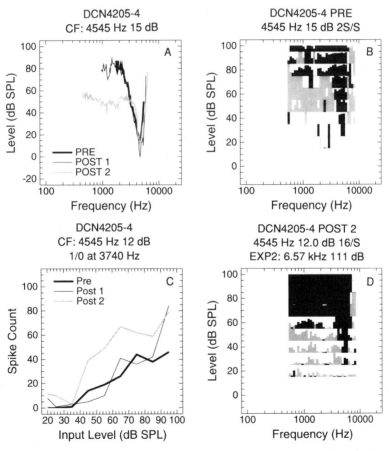

Figure 7. (A) Pre- and post-exposure tuning curves from a DCN neuron with a level-tolerant tuning curve. (B) Pre-exposure response area map from DCN neuron in A. Black areas show excitatory response area; shaded areas show inhibitory response areas. (C) Pre- and post-exposure discharge rate-level functions from neuron shown in A. (C) Post-exposure response area map for neuron shown in A. Note expansion of excitatory response area after exposure (compare to B).

neuron's response properties; however, after the second exposure there was a significant increase in the spontaneous discharge rate, a slight reduction in threshold at CF (Figure 5A), a significant increase in the driven discharge rate (Figure 7C) and an expansion of the excitatory response area along the low-frequency tail of the tuning curve (Figure 7D). All of these changes are consistent with a loss of inhibition. Effects similar to those presented in Figure 7 were typically observed in neurons with narrow, level-tolerant tuning curves and/or strongly nonmonotonic discharge rate-level function.

The enhanced responsiveness seen in the IC and DCN following acoustic overstimulation is similar to the effect seen in audiogenic seizure prone mice that have been primed for seizures using intense acoustic stimulation (Saunders et al., 1972). The traumatizing exposure reduces the amplitude of the cochlear microphonic and the compound action potential from the auditory nerve, but results in a significant increase in the amplitude of the evoked response in IC and the CN (Saunders et al., 1972). It has been suggested that neurons in the DCN of audiogenic seizure prone mice have fewer inhibitory inputs and that the loss of inhibition results in an increase in the amplitude of the evoked response in the CN and IC (Willott et al., 1984). Our single unit data indicate that neurons in the DCN respond more robustly and to a broader range of frequencies when a traumatizing tone reduces the inhibitory inputs above CF. The increased responsiveness that occurs below CF is consistent with the enhanced evoked potential amplitudes seen at frequencies just below the region of hearing loss.

Our single unit data from the IC and DCN indicate that sound-induced cochlear damage above CF can significantly alter the response properties of a subset of neurons that have extremely narrow excitatory response areas, strongly nonmonotonic discharge rate-level functions and inhibitory response area surrounding CF. Neurons in this subgroup typically showed a significant expansion of their excitatory response area below CF and a significant increase in discharge rate. In some cases, particularly in the DCN, the traumatizing exposure actually caused a small improvement in threshold near CF and a slight expansion (< 1/3 octave) of the tuning curve above CF. The increased responsiveness seen on and off CF is consistent with the proposed model outlined in Figure 3.

On the other hand, many neurons in the IC and DCN were largely unaffected by the exposure above CF. Units in this second group generally had open-V tuning curves with an extended low-frequency tail and monotonic discharge rate-level functions. The response properties of these neurons suggest that they are activated predominantly by excitatory inputs (Liberman et al., 1986). Failure to observe a change in responsiveness in this second subgroup is consistent with effects seen at the level of the auditory nerve, i.e., traumatic sound exposures have no effect on the tuning curves of neurons with CFs located below the frequency of the exposure (Liberman et al., 1986). This is an important internal control because it demonstrates that the low-frequency tail that appears on level-tolerant tuning curves following a traumatizing exposure above CF cannot be due to changes in peripheral tuning.

The functional changes seen in some IC and DCN neurons after acoustic overstimulation are most likely due to the unmasking (disinhibition) of intrinsic neural circuits within the auditory pathway. The unmasking effect in the IC could originate in the IC itself and/or at early stages of processing. Neuropharmacologic studies have shown that GABAergic antagonists cause an increase in discharge rate and a broadening of level-tolerant and upper-threshold tuning curves in the IC (Yang et al., 1992). These effects closely resemble those seen in the present study with acoustic overstimulation above CF. Our preliminary results from the DCN suggest that this region may contribute to the rapid functional reorganization seen in the IC. We have not seen any expansion of excitatory response areas in the ventral cochlear nucleus (VCN) after presenting a traumatizing tone above CF. However, units in the VCN with inhibitory response areas show as much as a 25% enhancement in the discharge rate at CF after being exposed to a traumatizing tone above exposure (Boettcher and Salvi, 1993). Thus, the VCN could contribute to the enhancement of on-CF discharge rate. Modifications in on-CF firing rates in the VCN can be mediated by GABAergic inhibition (Palombi and Caspary, 1992).

4. ACKNOWLEDGMENTS

Research supported by NIH grant R01 DC00166-13.

5. REFERENCES

Adams, J. C. (1979) Ascending projections to the inferior colliculus. J. Comp. Neurol. 182, 519–538.

Adams, J. C. and Warr, W. B. (1976) Origins of axons in the cat's acoustic striate determined by injection of horse-radish peroxidase into severed tracts. J. Comp. Neurol. 170, 107–122.

Arehole, S., Salvi, R. J., Saunders, S. S. and Hamernik, R. P. (1987) Evoked response 'forward masking' functions in chinchillas. Hear. Res. 30, 23–32.

Boettcher, F. A. and Salvi, R. J. (1993) Functional changes in the ventral cochlear nucleus following acute acoustic overstimulation. J. Acoust. Soc. Am. 94, 2123–2134.

Calford, M. B. and Tweedale, R. (1988) Immediate and chronic changes in responses of somatosensory cortex in adult flying-fox after digit amputation. Nature 332, 446–448.

Devor, M. and Wall, P. D. (1981) Plasticity in the spinal cord sensory map following peripheral nerve injury in rats. J. Neurophysiol. 55, 679–684.

Dostrovsky, J. O., Millar, J. and Wall, P. D. (1976) The immediate shift of afferent drive of dorsal column nucleus cells following deafferentation: A comparison of acute and chronic deafferentation in gracile nucleus and spinal chord. Exp. Neurol. 52, 480–495.

Eysel, U. T., Gonalez-Aguilar, F. and Mayer, U. (1980) A functional sign of reorganization in the visual system of adult cats: Lateral geniculate neurons with displaced receptive fields after lesions of the nasal retina. Brain Res. 191, 285–300.

Eysel, U. T., Gonzalez-Aguilar, F. and Mayer, U. (1981) Time-dependent decrease in the extent of visual deafferentation in the lateral geniculate nucleus of adult cats with small retinal lesions. Exp. Brain Res. 41, 256–263.

Faingold, C. L., Boersma Anderson, C. A. and Caspary, D. M. (1991) Involvement of GABA in acoustically-evoked inhibition in inferior colliculus. Hear. Res. 52, 201–216.

Harrison, R. V., Nagasawa, A., Smith, D. W., Stanton, S. and Mount, R. J. (1991) Reorganization of auditory cortex after neonatal high frequency cochlear hearing loss. Hear. Res. 54, 11–19.

Harrison, R. V., Smith, D. W., Nagasawa, A. and Mount, R. J. (1992) Developmental plasticity of auditory cortex in cochlear hearing loss: Physiological and psychophysical findings. Adv. Biosci. 83, 625–633.

Hicks, T. P. and Dykes, R. W. (1983) Receptive field size for certain neurons in the primary somatosensory cortex is determined by BAS-mediated intracortical inhibition. Brain Res. 274, 160–164.

Kaas, J. H., Krubitzer, L. A., Chino, Y. M., Langston, A. L., Polley, E. H. and Blair, N. (1990) Reorganization of retinotopic maps in adult mammals after lesions of the retina. Sci. 248, 229–231.

Liberman, M. C., Dodds, L. W. and Learson, D. A. (1986) Structure-function correlation in noise-damaged ears: A light and electron-microscopic study. In: R. J. Salvi, R. P. Hamernik, D. Henderson and V. Colletti (Eds.), Basic and Applied Aspects of Noise Induced Hearing Loss, Plenum Press, pp. 163–177.

Merzenich, M. M. and Kaas, J. H. (1982) Organization of Mammalian somatosensory cortex following peripheral nerve injury. Trends Neurosci. 5, 4428–4436.

Merzenich, M. M., Kaas, J. H., Wall, J. T., Sur, M., Nelson, R. J. and Felleman, D. J. (1983) Progressive change following median nerve section in the cortical representation of the hand in area 3b and 1 in adult owl and squirrel monkeys. Neurosci. 10, 639–665.

Osen, K. K. (1972) Projection of the cochlear nuclei on the inferior colliculus in the cat. J. Comp. Neurol. 144, 355–372.

Palombi, P. S. and Caspary, D. M. (1992) GABAA receptor antagonist bicuculline alters response properties of posteroventral cochlear nucleus neurons. J. Neurophysiol. 67, 738–746.

Pons, T., Garraghty, P. E., Ommaya, A. K., Kaas, J. H., Taub, E. and Mishkin, M. (1991) Massive cortical reorganization after sensory deafferentation in adult macaques. Sci. 252, 1857–1860.

Popelar, J., Syka, J. and Berndt, H. (1987) Effect of noise on auditory evoked responses in awake guinea pigs. Hear. Res. 26, 239–247.

Robertson, D. and Irvine, D. R. F. (1989) Plasticity of frequency organization in auditory cortex of guinea pigs with partial unilateral deafness. J. Comp. Neurol. 282, 456–471.

Salvi, R. J., Ahroon, W. A., Perry, J., Gunnarson, A. and Henderson, D. (1982) Psychophysical and evoked-response tuning curves in the chinchilla. Am. J. Otolaryngology 3, 408–416.

Salvi, R. J., Saunders, S. S., Gratton, M. A., Arehole, S. and Powers, N. (1990) Enhanced evoked response amplitudes in the inferior colliculus of the chinchilla following acoustic trauma. Hear. Res. 50, 245–258.

Salvi, R. J., Powers, N. L., Saunders, S. S., Boettcher, F. A. and Clock, A. E. (1992) Enhancement of evoked response amplitude and single unit activity after noise exposure. In: A. Dancer, D. Henderson, R. J. Salvi and R. Hamernik (Eds.), Noise-Induced Hearing Loss, Mosby Year Book, pp. 156–171.

Saunders, J. C., Bock, G. R., James, R. and Chen, C. S. (1972) Effects of priming for audiogenic seizure on auditory evoked responses in the cochlear nucleus and inferior colliculus of BALB/c mice. Exp. Neurol. 37, 388–394.

Spongr, V. P., Flood, D. G., Frisina, R. D. and Salvi, R. J. (submitted) Quantitative measure of hair cell loss in CBA and C57BL/6 mice throughout their life spans. J. Acoust. Soc. Am., .

Suga, N. and Tsuzuki, K. (1985) Inhibition and level-tolerant frequency tuning in the auditory cortex of the mustached bat. J. Neurophysiol. 53, 1125–1145.

Syka, J., Rybalko, N. and Popelar, J. (1994) Enhancement of the auditory cortex evoked responses in awake guinea pigs after noise exposure. Hear. Res. 78, 158–168.

Wang, J., Salvi, R. J., Caspary, D. M. and Powers, N. (1996a) GABAergic inhibition profoundly alters the tuning and discharge rate of neurons in the primary auditory cortex of the chinchilla. Abstr. Assoc. Res. Otolaryngol. 19, 156.

Wang, J., Salvi, R. J. and Powers, N. (1996b) Rapid functional reorganization in inferior colliculus neurons following acute cochlear damage. J. Neurophysiol. 75, 171–183.

Willott, J. F., Demuth, R. M. and Lu, S.-M. (1984) Excitability of auditory neurons in the dorsal and ventral cochlear nuclei of DBA/2 and C57BL/6 mice. Exp. Neurol. 83, 495–506.

Willott, J. F., Aitkin, L. M. and McFadden, S. L. (1993) Plasticity of auditory cortex associated with sensorineural hearing loss in adult C57BL/6J mice. J. Comp. Neurol. 329, 402–411.

Yang, L., Pollack, G. D. and Resler, C. (1992) GABAergic circuits sharpen tuning curves and modify response properties in the mustache bat inferior colliculus. J. Neurophysiol. 68, 1760–1774.

Young, E. (1984) Response characteristics of neurons of the cochlear nucleus. In: C. Berlin (Ed.), Hearing Science, College-Hill, pp. 423–460.

Young, E. D. and Brownell, W. E. (1976) Responses to tones and noise of single cells in dorsal cochlear nucleus of unanesthetized cats. J. Neurophysiol. 39, 282–300.

45

PLASTICITY OF INFERIOR COLLICULUS AND AUDITORY CORTEX FOLLOWING UNILATERAL DEAFENING IN ADULT FERRETS

David R. Moore, Susan J. France, David McAlpine, Jennifer E. Mossop, and Huib Versnel

University Laboratory of Physiology
Parks Road, Oxford OX1 3PT, United Kingdom

Our studies of the effects of unilateral deafening on the brain have been motivated, in part, by a desire to understand some of the short- and long-term mechanisms underlying clinical hearing loss, and partly by a desire to address basic questions in neurobiology, such as how afferent activity, and its withdrawal and modulation, affect the development and maintenance of neuron form, function, and connectivity. To date, much of our research, and that of others, has focused on the effect of deafening on the immature auditory system. This work has shown (Fig. 1), among other things (see chapter by Rubel, this volume), that surgical removal of one cochlea in neonatal mammals leads to: (a) a loss and shrinkage of neurons in the cochlear nucleus (CN; Hashisaki and Rubel, 1989; Moore, 1990; Moore and O'Leary, 1997; Tierney et al., 1997) and superior olivary complex (SOC; Moore, 1992; Pasic et al., 1994; Moore and Pallas, 1997), (b) the formation of new connections between the CN on the intact side and various target structures in the brainstem (Kitzes et al., 1995; Russell and Moore, 1995) and midbrain (Nordeen et al., 1983; Moore and Kitzes, 1985; Moore, 1994), and (c) an increase in the responsiveness of inferior colliculus (IC; Kitzes, 1984; Kitzes and Semple, 1985; Moore et al., 1993) and primary auditory cortex (AI; Reale et al., 1987) neurons to acoustic stimulation of the ipsilateral, intact ear. In this paper, we focus on the latter, physiological effects of cochlear ablation. In particular, we compare some recent data showing the short- and long-term effects of cochlear ablation in adulthood with data previously obtained following cochlear ablation in infancy.

In 1983, Nordeen, Killackey and Kitzes reported that surgical removal of one cochlea in the newborn gerbil resulted in a dramatic increase, in adulthood, in the proportion of neurons in the IC on the side of the intact ear that could be excited by acoustic stimulation of that ear (see also Kitzes, 1984; Fig. 2). In a subsequent, single-unit study, Kitzes and Semple (1985) showed that the neurons in the normal, adult gerbil IC that were excited by ipsilateral stimulation tended to respond less vigorously, and with narrower dynamic ranges and higher thresholds, than the greater proportion of neurons excited by contralateral stimulation (Fig. 3).

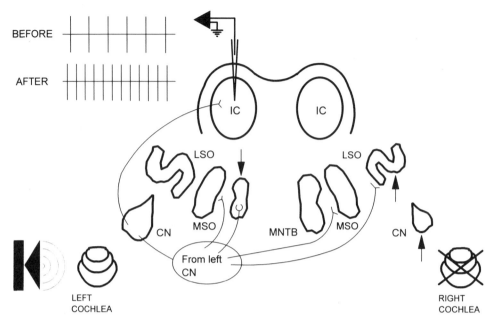

Figure 1. Schematic representation of some effects of neonatal, unilateral cochlear ablation on the auditory brainstem. Note that, for all the experiments reported here, the stimulated, intact ear is referred to the recording site. Thus, in the case illustrated, ipsilateral stimulation is being applied. See text for further details.

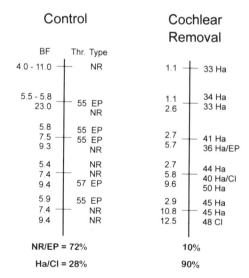

Figure 2. Representative electrode penetrations through the ICC of normally hearing (Control) and neonatally operated gerbils. Each vertical penetration shows, to the left, the best frequency (in kHz) of the neural activity recorded at the site indicated by the horizontal bar in response to tonal stimulation of the (unoperated) ipsilateral ear. To the right is shown the threshold at BF (in dB SPL) and the type of activity recorded at that site. For the Control penetration, non-responsiveness (NR), and volume-conducted, non-spiking evoked potentials (EP) predominated. Spike activity (Ha - 'hash', Cl - 'cluster') was found at many more sites following cochlear ablation. Data from multiple penetrations confirmed that this pattern was general, as indicated by the percentages under each penetration (adapted from Nordeen et al., 1983).

Cochlear ablation in infancy not only increased the number of neurons excited by ipsilateral stimulation, but also led to more vigorous responses, broader dynamic ranges and lower thresholds. In fact, in all these respects, the responses of the neurons to ipsilateral stimulation closely resembled those found in response to contralateral stimulation in the binaurally-intact animal. Kitzes and Semple did not examine the effect of adult cochlear ablation. However, Nordeen et al. (1983) found that cochlear ablation in adulthood did not have any clear effect on the proportion of IC neurons excited by the ipsilateral ear. Moore and Kitzes (1986) found that large CN lesions in adults had no apparent effect on the responsiveness of single neurons in the contralateral IC to stimulation of the intact ear.

Auditory cortical neurons have much the same response bias to contralateral stimulation as that described in the IC, and Reale et al. (1987) showed a qualitatively similar response of primary auditory cortical (AI) neurons to contralateral cochlear ablation in kittens during the first post-natal week. Reale and colleagues also examined the effect of contralateral cochlear ablation on AI excitability in two adult cats. In one of these, no change in the proportion of recording loci excited by ipsilateral stimulation was found 24 hours after the ablation. However, in the other, the proportion of ipsilaterally excited loci increased from 22% to 67% following a similar survival period. Although the authors interpreted the latter result in terms of a possible, long-term conductive hearing loss in that cat, it is also possible that there was a rapid reorganization of the cortex, or some sub-cortical input(s), in direct response to the ablation. These experiments raised two important questions. First, if there is a transition from a developmentally sensitive state to an adult, insensitive state, when and how does this occur? Second, is the adult auditory system really insensitive to unilateral deafening or can it reorganize and, if so, where and how?

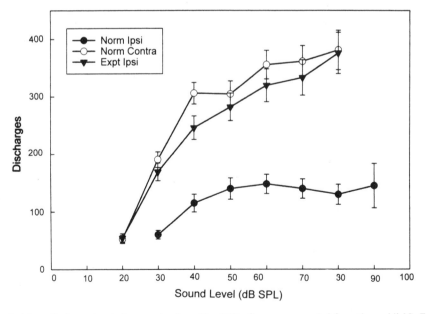

Figure 3. Mean discharge levels for samples (n = 35 - 102) of neurons recorded from the gerbil IC. Each data point shows the mean (±s.e.m.) number of discharges evoked by 50 repetitions of a BF tone delivered to the contra(lateral) or ipsi(lateral) ear of a normally-hearing (Norm) or unilaterally deafened (Expt) adult animal. The Expt animals had one cochlea surgically removed at P2 and recordings and stimulation were performed on the side of the intact ear (see Fig. 1). Neurons in the Expt Ipsi group were more sensitive and responded more strongly than those in the Norm Ipsi group (adapted from Kitzes and Semple, 1985).

We initially sought to address these questions by examining how the animal's age at the time of the cochlear ablation affected the physiological outcome in the mature animal. We performed single and multi-unit recording in the barbiturate anaesthetised adult ferret. We use the ferret primarily because of its immaturity at birth; it doesn't begin to hear until one month post-partum (Moore and Hine, 1992). We can, therefore, more easily examine early developmental events, and manipulate the auditory system earlier in development, than would be possible or convenient in most other mammals. Ferrets receiving cochlear ablation in infancy were born from timed pregnancies and had one cochlea surgically destroyed (see Moore and Kowalchuk, 1988) on postnatal day (P)5, P25 or P40. Other ferrets had a cochlea ablated in adulthood and, like the infant ablated animals, survived for at least 3 months before recording. A further group of adults had the cochlea ablated immediately before recording, and a final group received no inner ear surgery (see Table 1 for details). One aspect of our experimental design that turned out to be crucial was the decision that our control animals (*Acute Adult* groups; Table 1) should receive unilateral cochlear ablation on the day of the recording. In Kitzes' studies, both ears of the control animals were intact, but we argued that, since the experimental animals had a long-standing cochlear ablation, the controls should also be tested with but a single intact ear. For recordings, we used sealed and calibrated delivery systems (Moore et al., 1983), monotic tone stimulation, and standard data acquisition techniques. In some experiments we focused on multi-unit mapping, where qualitative assessments of neural responses were obtained from a large number of loci. In other experiments, we made quantitative measurements on single neurons, including excitatory tuning curves and responses to varying stimulus levels ("rate/level functions").

In the ferret IC, neonatal cochlear ablation led to a statistically higher proportion of excitatory responses than was seen in the *Acute Adult IC* group (Fig. 4), and the proportion of excitatory responses declined as the age at which the cochlea was removed increased. Notice, though, that the proportion of excitatory recording sites in the *Acute Adult IC* group was high - around 70% - in contrast to the binaurally-intact gerbil, where the proportion of excitatory sites was less than 30% (Fig. 2). Initially, we attributed this difference to the previously observed (Moore et al., 1983), higher level of activity in the ferret brain. However, subsequent recordings from AI showed that the adult ferret auditory system responded rapidly and dramatically to cochlear ablation. In those experiments, multiple microelectrode penetrations were made normal to the surface of AI. Although the aim of the experiments was to record from single neurons, rather than to map the cortex extensively, we obtained evidence that the topographic representation of the ipsilateral ear

Table 1. Experimental groups of ferrets used for studies of the effects of unilateral cochlear ablation on IC and AI physiology. *Ipsi* and *Contra* refer to the IC or AI (relative to the intact ear) from which recordings were made (checked in the right columns)

Group:	Age at ablation	Age at recording	Survival	IC?	AI?
Normal Ipsi	—	Adult	—	x	x
Normal Contra	—	Adult	—	x	x
Acute Ipsi	Adult	Adult	Hours	x	x
Chronic Ipsi	Adult	Adult	>3 Months	x	x
P5 Ipsi	P5	Adult	>3 Months	x	
P25 Ipsi	P25	Adult	>3 Months	x	x
P40 Ipsi	P40	Adult	>3 Months	x	

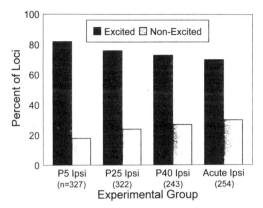

Figure 4. Effect of neonatal and adult cochlear ablation on excitation in the adult ferret IC. Each histogram bar shows the proportion of recording sites, from multiple electrode penetrations, at which tone-evoked spike activity was ("excited") or was not ("non-excited") recorded from the IC on the side of the intact ear to stimulation of that ear (see Fig. 1). Treatment groups are defined in Table 1. Cochlear ablation in neonates led to a higher proportion of excited loci.

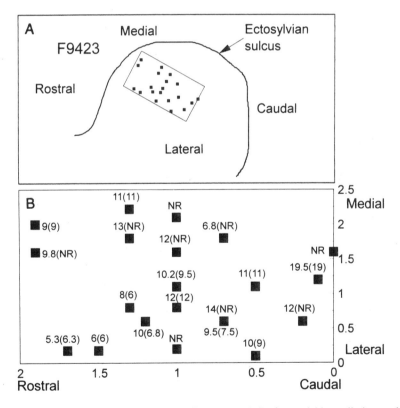

Figure 5. Binaural excitatory, input in the primary auditory cortex (AI) of normal, binaurally-intact adult ferrets. A. Site of electrode penetrations in the left AI (middle ectosylvian gyrus). B. Best frequency (BF) of neurons recorded in individual penetrations, normal to the cortical surface, and distributed across AI as indicated by the inset in A and the axes (scaled in mm). The first number in a pair shows the BF in response to contralateral stimulation and the second number is the ipsilateral BF. NR = no (excitatory) response. In normal ferrets, about half the recorded neurons were unresponsive to ipsilateral stimulation.

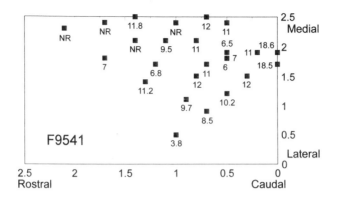

Figure 6. Ipsilateral excitatory, input in AI of an adult ferret that had the contralateral cochlea surgically destroyed 7 hours prior to the commencement of recording. Each data point shows the BF to stimulation of the ipsilateral, intact ear of neurons recorded in electrode penetrations at that point. Details are given in Fig. 5B. Ipsilateral tones produced excitatory responses at almost every recording site.

in AI expanded following ablation of the contralateral cochlea in adult ferrets. In the normal, binaurally-intact ferret, we confirmed the findings of Kelly and Judge (1994). The ipsilateral ear provided excitatory drive to only about 50% of the recording sites that were excited by contralateral stimulation (Fig. 5). In the animals that were unilaterally deafened just prior to recording (*Acute Adult AI*, Table 1), we found a much more widespread distribution of ipsilateral excitation. For example, in the case shown in Fig. 6, excitatory drive was obtained over a large, contiguous region of AI. However, because we focussed on single unit recordings in these experiments, we were also able to show that "acute" cochlear ablation in the adult ferret produced a quantitative reduction in mean unit thresholds (Fig. 7) and an increase in the spontaneous discharge rate (Fig. 8) of cortical neurons on the

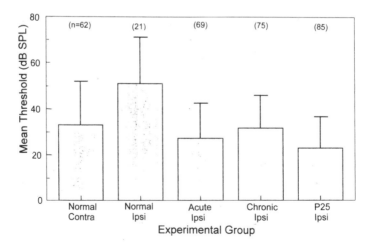

Figure 7. Effect of cochlear ablation in infancy (P25) and adulthood on the mean (± s.d.) threshold of units in the AI of adult ferrets (see Table 1 for definition of groups). Numbers of neurons examined are shown at the top of each histogram bar. In normal, binaurally intact animals, neurons excited by stimulation of the ipsilateral ear had higher thresholds than those excited by stimulation of the contralateral ear. Following cochlear ablation, thresholds to ipsilateral stimulation were significantly reduced in all groups.

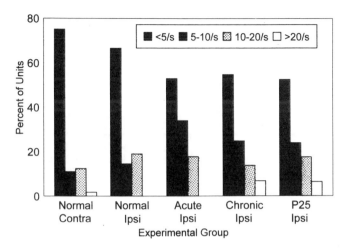

Figure 8. Spontaneous discharge rate of ferret AI neurons. Each bar shows the proportion of neurons recorded for that group (see Table 1) that had the indicated level of discharges in the absence of acoustic stimulation. Spontaneous rates were determined by setting a sampling window late in the interstimulus interval, and averaging over many stimulus conditions. Spontaneous rates were elevated in all lesioned groups, but somewhat more so in the long survival conditions than in the short survival condition.

side of the intact ear, relative to binaurally intact (*Normal*) animals. Note that, in each of these cases, the changes in the *Acute Adult* group were as great as the changes seen following cochlear ablation in infancy (*P25*), or long-term (*Chronic Adult*) cochlear ablation in adults. These results suggest that, in addition to a reorganization of frequency representation following peripheral trauma limited to a part of one cochlea (Robertson and Irvine, 1989; Calford et al., 1993; Willott et al., 1993), the adult auditory system can respond functionally to lesions of the contralateral cochlea.

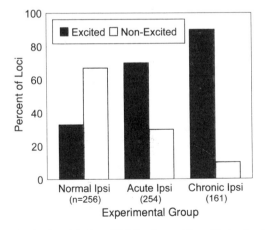

Figure 9. Changes in excitation in the adult ferret IC following cochlear ablation in adulthood and stimulation of the intact, ipsilateral ear (see Fig. 1 and Table 1). For each group, the number of loci sampled is shown below the group name. Cochlear removal lead to a dramatic, short-term increase in ipsilateral responsiveness, and to a further increase in the longer-term (c.f. Fig. 4).

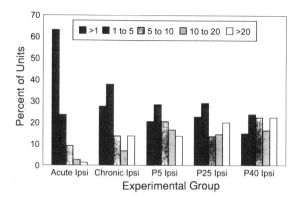

Figure 10. Spontaneous activity of IC neurons following cochlear ablation. In contrast to AI (Fig. 8), IC neurons showed a marked increase in spontaneous level following longer survival.

One of the central tenets of the adult sensory plasticity literature (e.g. Darian-Smith and Gilbert, 1995) is that the organizational changes produced by peripheral lesions mainly occur in the cerebral cortex. However, our recent data has shown that acute co-chlear ablation in adult ferrets also leads to a dramatic reorganization of excitation in the IC ipsilateral to the intact, stimulated ear (Fig. 9). Thus, the main effect of cochlear abla-tion on excitation in the IC is not so much dependent on the age at which the ablation is performed as on the fact that it is performed at all. However, there was no change in IC spontaneous activity following acute, adult ablation (Fig. 10) and, in at least this respect, the cortical effects of acute deafening were more marked than those seen at the midbrain.

The final issue we address is the time course of the acute effect. Most examples of short-term functional reorganization in the nervous system (e.g. Clarey et al., 1996) have been attributed to "unmasking," a rather vague term that implies immediate release from suppression following ablation of an inhibitory input. This unmasking may be contrasted with longer-term changes that include the modification of existing synapses, synaptogene-sis, and the formation of new connections (Darian-Smith and Gilbert, 1994). We have pre-liminary evidence on this issue from our recordings in AI. In one of the ferrets tested following acute ablation (Fig. 11), recordings were made over a long period (50 hours)

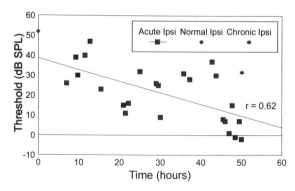

Figure 11. Effect of survival time following cochlear ablation in an adult ferret on the threshold of IC neurons to BF stimulation of the intact, ipsilateral ear (as in Fig. 1). Thresholds declined gradually following the ablation. See text for further details.

from a sample (n = 25) of units having a relatively narrow range of best frequencies (BFs; 8 - 12 kHz). This is the frequency range to which normal ferrets are most sensitive (Kelly et al., 1986a), and neurons with BFs in this range tend to have the most sensitive thresholds (Moore et al., 1983; Kelly et al., 1986b). When the thresholds of these units were plotted against the time since the cochlear ablation they could be reasonably well fit (r = 0.62) by a linear function that showed increasing sensitivity with time. Also shown in Fig. 11 are the mean thresholds of units in the same BF band from the *Normal AI* and *Chronic Adult AI* groups. These data suggest that, instead of the thresholds plummeting immediately after the ablation, as the unmasking hypothesis would seem to predict, changes in sensitivity occur gradually, over a period of many hours. Such a time course for change is consistent with a model that requires cell signalling mechanisms (e.g. for transmitter or receptor expression). In experiments currently underway, we are making detailed maps of the binaurally-intact adult ferret AI using Halothane and nitrous oxide anaesthesia, multi-unit recording, and objective assessment of BF and threshold. The results (Fig. 12) show a clear segregation of large areas of cortex that are responsive or unresponsive to ipsilateral stimulation. Following the initial mapping, we are performing a contralateral cochlear ablation and remapping the same area of cortex. Thus far, we have obtained some evidence of excitatory infilling of the areas previously unresponsive to ipsilateral stimulation prior to the ablation. Those data are consistent with the previous, single unit evidence suggesting that interaurally influenced changes in excitability occur gradually rather than as a result of immediate unmasking.

The data reported in this paper show that unilateral deafening in adult ferrets leads to both a short-term and a persistent expansion of ipsilateral excitation in the auditory midbrain and cortex. Quantitative analysis of single unit recordings shows that, after the initial expansion, there are more subtle, possibly longer-term changes that depend on both the age of the animal at the time that the deafness occurred, and the level of the auditory system under examination. In general, deafness occurring early in life, particularly before the onset of functional hearing, leads to more profound changes in excitation, and auditory

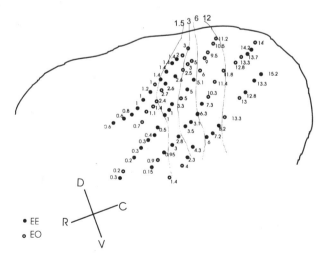

Figure 12. Multi-unit mapping of the normal adult ferret middle ectosylvian gyrus using ipsi- and contralateral stimulation. At each sampling point, the BF was measured, and the response laterality was determined. "EE" sites showed excitatory activity to stimulation of either ear. Neurons at "EO" sites were excited only by the contralateral ear. Note the clustering of EE and EO sites, also reported by other studies (e.g. Kelly and Judge, 1994).

cortical physiology tends to be more affected by cochlear ablation than does the physiology of the auditory midbrain. Finally, we have some evidence that these two factors interact; cochlear ablation later in infancy can produce a level of increased excitation in the AI that is only found in the IC following earlier occurring deafness.

Although these and other physiological data do not provide direct evidence bearing on the mechanisms underlying the responses of the auditory system to deafening, they do allow certain hypotheses to be tested. For example, it has been suggested (Nordeen et al., 1983; D.K. Morest, this meeting) that central auditory changes following cochlear ablation may be due to a degenerative response deriving from the side of the ablation. However, several lines of evidence suggest that this is not the case. The finding reported here that a major change in the excitatory influence of the intact ear can occur following ablation in adulthood shows that the change is not contingent on transneuronal neuron death in the CN on the side of the ablation, which only occurs following ablation very early in life. Moreover, the uptake and retrograde transport of HRP from the IC to that (contralateral) CN is apparently unaffected by the ablation (Moore and Kowalchuk, 1988). Finally, direct lesions of the CN in adult gerbils did not affect the physiological responses of neurons in the contralateral IC to stimulation of the ear on the intact side, either acutely or chronically (Moore and Kitzes, 1986). Another, related hypothesis discussed at the meeting is that, rather than causing a generalised degeneration, unilateral deafening (or other sensorineural hearing loss) may produce a selective disturbance of inhibitory mechanisms in the central auditory system (see papers by Altschuler et al. and Miller et al.; D.M. Caspary, D.K. Morest, personal communications). Most of the evidence for this hypothesis comes from changes in the levels of inhibitory transmitters and their receptors. As mentioned above, the apparently gradual change we observed in ipsilateral excitability is more consistent with this hypothesis than an alternative that posits a rapid ablation of functional inhibition.

REFERENCES

Calford, M.B., Rajan, R. and Irvine, D.R.F. 1993 Rapid changes in the frequency tuning of neurons in cat auditory cortex resulting from pure-tone-induced temporary threshold shift. Neuroscience 55, 953–964.

Clarey, J.C., Tweedale, R. and Calford, M.B. (1996) Interhemispheric modulation of somatosensory receptive fields: Evidence for plasticity in primary somatosensory cortex. Cereb Cortex 6, 196–206.

Darian-Smith, C. and Gilbert, C.D. (1995) Topographic reorganization in the striate cortex of the adult cat and monkey is cortically mediated. J. Neurosci. 15, 1631–1647.

Hashisaki, G.T. and Rubel, E.W (1989) Effects of unilateral cochlea removal on anteroventral cochlear nucleus neurons in developing gerbils. J. Comp. Neurol. 283, 465–473.

Kelly, J.B., Kavanagh, G.L. and Dalton, J.C.H. (1986a) Hearing in the ferret (Mustela putorius): thresholds for pure tone detection. Hear. Res. 24, 269–275.

Kelly, J.B., Judge, P.W. and Phillips, D.P. (1986b) Representation of the cochlea within the primary auditory cortex of the ferret (Mustela putorius). Hear. Res. 24, 111–115.

Kelly, J.B. and Judge, P.W. (1994) Binaural organization of primary auditory cortex in the ferret (Mustela putorius). J. Neurophysiol. 71, 904–913.

Kitzes, L.M. (1984) Some physiological consequences of neonatal cochlear destruction in the inferior colliculus of the gerbil, Meriones unguiculatus. Brain Res. 306, 171–178.

Kitzes, L.M. and Semple, M.N. (1985) Single-unit responses in the inferior colliculus: effects of neonatal unilateral cochlear ablation. J. Neurophysiol. 53, 1483–500.

Kitzes, L.M., Kageyama, G.H., Semple, M.N. and Kil, J. (1995) Development of ectopic projections from the ventral cochlear nucleus to the superior olivary complex induced by neonatal ablation of the contralateral cochlea. J. Comp. Neurol. 353, 341–363.

Moore, D.R. (1990) Auditory brainstem of the ferret: Early cessation of developmental sensitivity to cochlear removal in the cochlear nucleus. J. Comp. Neurol. 302, 810–823.

Moore, D.R. (1992) Trophic influences of excitatory and inhibitory synapses on neurones in the auditory brainstem. NeuroReport 3, 269–272.

Moore, D.R. (1994) Auditory brainstem of the ferret: Long survival following cochlear removal progressively changes projections from the cochlear nucleus to the inferior colliculus. J. Comp. Neurol. 339, 301–310.

Moore, D.R. and Hine, J.E. (1992) Rapid development of auditory brainstem response thresholds in individual ferrets. Dev. Brain Res. 66, 229–235.

Moore, D.R., King, A.J., McAlpine, D., Martin, R.L. and Hutchings, M.E. (1993) Functional consequences of neonatal cochlear removal. Prog. Brain Res. 97, 127–133.

Moore, D. R. and Kitzes, L. M. (1985). Projections from the cochlear nucleus to the inferior colliculus in normal and neonatally, cochlea-ablated gerbils. J. Comp. Neurol. 240, 180–195.

Moore, D. R. and Kitzes, L. M. (1986). Cochlear nucleus lesions in the adult gerbil: Effects on neurone responses in the contralateral inferior colliculus. Brain Res. 373, 268–274.

Moore, D. R. and Kowalchuk, N. E. (1988) Auditory brainstem of the ferret: Effects of unilateral cochlear lesions on cochlear nucleus volume and projections to the inferior colliculus. J. Comp. Neurol. 272, 503–515.

Moore, D.R., Rogers, N.J. and O'Leary, S.J. (1997) Loss of cochlear nucleus neurons following aminoglycoside antibiotics or cochlear removal. Annals Otol. Rhinol. Laryngol., in press.

Moore, D.R. and Pallas, S.L. (1997) Effects of neonatal cochlear removal on neuron size and number in the ferret cochlear nucleus and superior olivary complex. Submitted for publication.

Moore, D. R., Semple, M. N. and Addison, P. D. (1983). Some acoustic properties of neurones in the ferret inferior colliculus. Brain Res. 269, 69–82.

Nordeen, K.W., Killackey, H.P. and Kitzes, L.M. (1983) Ascending projections to the inferior colliculus following unilateral cochlear ablation in the neonatal gerbil, Meriones unguiculatus. J. Comp. Neurol. 214, 144–153.

Pasic, T.R., Moore, D.R. and Rubel, E.W (1994) The effect of altered neuronal activity on cell size in the medial nucleus of the trapezoid body and ventral cochlear nucleus of the gerbil. J. Comp. Neurol. 348, 111–120.

Reale, R.A., Brugge, J.F. and Chan, J.C. (1987) Maps of auditory cortex in cats reared after unilateral cochlear ablation in the neonatal period. Brain Res. 431, 281–290.

Robertson, D. and Irvine, D.R.F. (1989) Plasticity of frequency organization in auditory cortex of guinea pigs with partial unilateral deafness. J. Comp. Neurol. 282, 456–471.

Russell, F.A. and Moore, D.R. (1995) Afferent reorganisation within the superior olivary complex of the gerbil: Development and induction by neonatal, unilateral cochlear removal. J. Comp. Neurol. 352, 607–625.

Tierney, T.S., Russell, F.A. and Moore, D.R. (1997) Abrupt cessation during early postnatal development of afferent dependent survival of gerbil cochlear nucleus neurons. J. Comp. Neurol., 378, 295–306.

Willott, J.F., Aitkin, L.M. and McFadden, S.L. (1993) Plasticity of auditory cortex associated with sensorineural hearing loss in adult C57BL/6J mice. J. Comp. Neurol. 329, 402–411.

46

STATE-DEPENDENT PLASTICITY OF NEURAL RESPONSES TO 70dB CLICKS ALONG CENTRAL, PRIMARY, AUDITORY TRANSMISSION PATHWAYS

Charles D. Woody[*]

Mental Retardation Research Center
Brain Research Institute
UCLA Medical Center
Los Angeles, California 90024

1. SUMMARY

The coding and transmission of acoustic messages is affected profoundly by state-dependent plasticity along central neural transmission pathways. Early studies of spike activity (Engel and Woody, 1972; Olds et al, 1972; Oleson et al 1975; Woody et al, 1976) found evidence of widespread changes in transmissions within the auditory system after conditioning with an acoustic CS. Many forms of conditioning have been shown to produce changes in activity (McCormick and Thompson, 1984; Miller et al, 1982; Recanzone et al, 1993; Scheich et al 1993; Weinberger et al, 1993). Studies of single unit activity have now established that responses to click in the cochlear nucleus, the first central relay of the auditory system, increase in magnitude after eyeblink conditioning with click as a CS, and further changes in activity have been found in the ventral portion of the nucleus after sensitization (Woody et al 1992, 1994). These state-dependent changes in activity will affect transmission at higher levels because all acoustic transmissions must obligatorily pass through the cochlear nucleus.

Short latency responses to 70 dB clicks have recently been found outside the classical, primary (shortest latency), ascending auditory pathways. Finding latencies of response of 4–8 ms in the subcerebellar dentate nucleus and rostral thalamus suggests that these areas might be relays in a novel, primary ascending pathway between cochlear nucleus and cortex. Initiation of the earliest component of the blink CR by a click CS depends on an intact motor cortex. Since CRs are elicited after the removal of classical auditory cortical

---

*Correspondence to: Dr. C.D. Woody, UCLA Medical Center, Room 58-232, NPI, 760 Westwood Plaza, Los Angeles, CA 90024. Tel: (310) 825-0187; fax: (310) 206-5060.

Acoustical Signal Processing in the Central Auditory System
edited by Syka, Plenum Press, New York, 1997

regions, the novel pathway may be as or more important than the classical pathway for rapid transmission of state-dependent acoustic messages to the motor cortex.

2. INTRODUCTION

Spike activity was recorded from single units (n>2500) of conscious cats to determine the pathways over which a click CS was transmitted within 8–12 ms from the ears to the motor (coronal-pericruciate) cortex (Woody et al, 1970; Sakai and Woody, 1980). Layer V pyramidal cells of the motor cortex of cats (Sakai and Woody, 1980) responded to clicks at latencies (8–10 ms) as short as those of the primary auditory cortex (Erulkar et al, 1956; Evans and Whitfield, 1964; Goldstein et al, 1968). Of the many candidate regions for rapid ascending auditory transmission in which responses to click were examined, increases in mean activity within 4–8 ms of presenting clicks were observed in averages from dorsal (DCN) and ventral (VCN) cochlear nucleus, rostral thalamus, and subcerebellar dentate nucleus.

Activity to click was also compared in separate groups of neurons before and after conditioning a blink response to the click as a CS and after sensitization produced by a reversed order of CS-US pairing. Responses to 70 dB hiss stimuli, presented after US delivery during training, were also studied. Significant, state-dependent changes in response were found in DCN, VCN, dentate nucleus, and rostral thalamus that began in the 4–8 post-stimulus period and thus could be transmitted to cortex in time to influence the 8–12 ms state dependent changes found there (Woody et al, 1970, 1991a, 1992, 1994; Wang et al 1991; Aou et al 1992a). Studies of the inferior colliculus and medial geniculate nuclei (Hoang et al 1991; c.f., Astl et al, 1996; Weinberger, 1982) disclosed few responses of comparable rapidity and magnitude; in addition, little change in activity has been thought to occur in the inferior colliculus after conditioning (Kettner and Thompson, 1982).

The findings in dentate and rostral thalamus suggest the existence of a novel, primary (shortest latency) auditory transmission pathway between cochlear nuclei and motor cortex in cats (Fig. 1). Such a pathway could account for the observation that lesions of

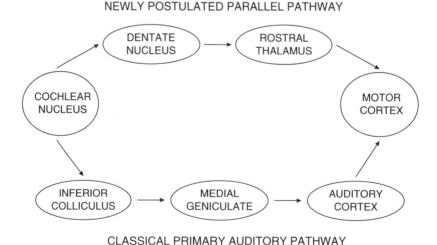

Figure 1. Diagram of postulated primary ascending auditory pathway for primary (shortest latency) transmissions to motor cortex and classical ascending pathway (anatomy simplified) to A_1 cortex.

the posterior cortex involving classical auditory receptive regions produce less impairment of simple auditorimotor transmission than expected and do not prevent development and performance of the blink CR to a click CS (Woody et al, 1974). The blink CR has an onset 20 ms after click (Woody et al, 1970; Aou et al, 1992a). Earlier studies (Woody and Yarowsky, 1972) have shown a 7.5 ms efferent conduction time between cortical electrical stimulation and peripheral muscle contraction (orbicularis oculis).

Evidence has been obtained that coding of enhancement of the response to click after conditioning varies from locus to locus, and can be supported by increases in numbers of CS responsive cells as well as by increases in means of CS-evoked activity. In cortex and rostral thalamus, analysis of subgroups of auditory-responsive cells was required to find the activity reflecting state-dependent enhancement of the earliest component of the responses. In VCN sensitization produced greater enhancement than conditioning. Changes in baseline activity were found in some regions as were increases in intracellularly measured neural excitability. In the motor cortex the latter were replicated by local application of acetylcholine, cGMP, or cGMP-dependent protein kinase, were better related to the potentiated responses to click CS after conditioning than to the changes in baseline activity, and were supported by decreases in a potassium A-current (Aou et al 1992a,b; Woody and Black-Cleworth, 1973; Woody and Gruen, 1988; Woody et al 1991c; Woody, 1996).

3. METHODS

Studies were performed in adult cats weighing between 2.5 and 3.0 kg. Extracellular and intracellular recordings of unit activity were obtained as described in Woody et al (1984). At the conclusion of the studies, the animals received lethal doses of sodium pentobarbital, and serial sections of perfused brain tissue were examined for electrode tracks and biocytin marked cells to confirm the recording loci. Electrodes were pulled from 1.5–2.0 mm (o.d.) theta tubing. When filled with 2% biocytin (Sigma) in 2.5 M KCl and connected on both sides with Ag/AgCl wire, the resistances of the electrodes ranged from 40 to 10 megohms. The animals were surgically prepared under sodium pentobarbital anesthesia (35 mg per kg, intraperitoneally), as described in detail previously (Woody et al, 1970; Woody et al 1991a), to allow later, conscious recording/training sessions using a stereotaxic guide tube and fixation of the head to a stabilizing frame. Penicillin G (150,000 units, i.m.) and benzathine penicillin G (150,000 units, i.m.) were given on the day of surgery, three days later during the recovery period, and at one week intervals thereafter, as needed. During recording/training sessions the bodies of the animals were placed in loose cloth sleeves. The behavior of the animals was continuously observed to evaluate their comfort, and the studies were discontinued if the animals gave any signs of discomfort such as vocalization and hyperactivity. The procedures met APS, USPHS, Society for Neuroscience, and University of California guidelines.

Conditioned blink responses were produced by forward pairing of click as CS with glabella tap and hypothalamic electrical stimulation (interstimulus intervals for tap and onset of four pulse (0.1 ms, 1–5 mA) train of hypothalamic stimulation, 570 and 10 ms respectively; 10 sec intertrial interval) followed 4.4 sec by hiss as DS (CS = conditioned stimulus; DS = discriminative stimulus). Sensitization was produced by a backward conditioning paradigm in which hypothalamic stimulation was presented first, followed 2.5 sec later by click and 570 ms later by glabella tap. The hiss followed the tap by 1.4 sec. The intertrial interval remained 10 sec. Rationale for stimulation of the lateral hypothalamus and parameters of stimulation as well as complete details of the locus stimulated and train-ing-testing protocols have been given earlier (Kim et al, 1983; Hirano et al, 1987).

The clicks were of 70 dB intensity (measured at the ears of the animals at a repeti-tion rate of 100 Hz with General Radio Company dB meter type 1565-A at standard SPL level of 20 μN m^{-2}), and were generated by a rectangular pulse of 1 ms duration, deliv-ered to a loudspeaker placed 1–2 feet in front of the animals. The hiss was of longer (100–200 ms) duration, and was set to the same SPL (single presentations) as click. Power spectral densities of the click and hiss are shown in Xi et al (1994). Amplified earphone recordings of the stimuli were illustrated earlier (Kim et al, 1983). All unit testing was done with CS and DS alone, 4.4 sec apart (click preceding hiss), every 10 sec. as in the baseline recording sessions given prior to conditioning or sensitization. The depth of each studied unit was noted, and training was given after recording each unit if needed to main-tain the behavioral state. The behavioral response was measured electromyographically from the orbicularis oculi muscles as previously described (Aou et al, 1992a). Spike oc-currences were detected with a threshold discriminator (Frederick Haer and Co.; c.f. Woody et al, 1970). Data were analyzed (PDP 11–44) in histograms of ≥ 2 ms bin width for each cell. No differences were found between data obtained from intracellular and ex-tracellular recordings. Thus data from all units were combined when making averages of patterns of activity before and after conditioning and sensitization.

The magnitude of increased activity in response to click and hiss was evaluated us-ing Z scores (Winer, 1971) relative to the standard deviation of the 400 ms prestimulus pe-riod of baseline activity, where 1$\dot{Z}$ was expected to contain 0.841 of the distribution, and 2Z and 3Z were expected to contain 0.977 and 0.9987 of the distribution, respectively.

4. RESULTS

4.1. Latencies of Response

Recordings of activity were made from relays along the ascending pathways by which the click CS might be transmitted within 8–12 ms from cochlear nucleus to motor cortex. Finding too few 4–8 ms mean increases at classical auditory relays such as the in-ferior colliculus and medial geniculate inspired investigation of new areas. (Medial geni-culate had a mean increase at 4–8 ms in one animal, but in the overall group the onset of mean increase was 12–16 ms (Hoang et al, 1991).) Investigation of intralaminar, mid and posterolateral thalamic regions, and caudate nucleus also failed to disclose short (4–8 ms) latency increases in mean activity to click (Woody et al 1991a,b). Histogram averages of activity recorded from candidate relays for ascending transmission that had unequivocal increases in activity in the 4–8 ms post click period are shown in Fig. 2. Those relays were the dentate nucleus and rostral thalamus (Woody et al 1991a. Wang et al 1992). A 4–8 ms mean onset was also found in the superior olivary nucleus (Hoang et al, 1991), but it was not clear whether that reflected ascending transmission. The transmission time between cochlear nucleus and dentate nucleus was not resolved by the 2 ms bin width averages, but a few cells of VCN responded with onsets of 2–4 ms as in others' reports (Kiang et al, 1973), and injection of phaseolus lectin (PHA-L) into the dentate nucleus produced retro-grade marking of somas of fusiform cells in the DCN (Fig. 1 of Wang et al, 1991). Cells such as that in Fig. 3A were readily found in the dentate nucleus with onsets of response preceding those of cells in the motor cortex (Fig. 3B), and the latencies were consistent with expected transmission times through rostral thalamus to motor cortex. The cells of

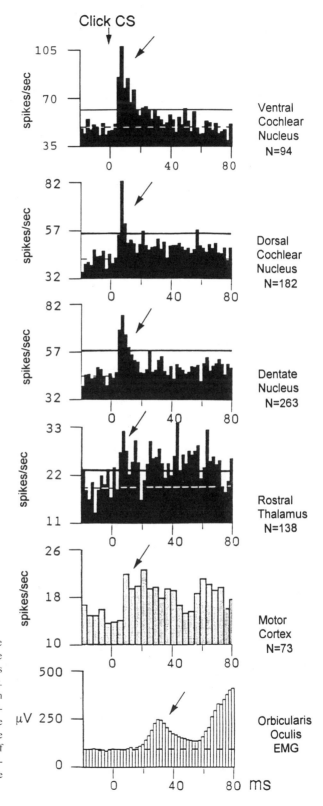

Figure 2. Histogram averages of the spike activity recorded from relays of the rapid transmission pathway and numbers of cells (N) averaged at each relay. Dashed lines show baseline activity in the 380 ms preceding click; solid horizontal lines are 3Z above the baseline mean. EMG activity (μV) is the average of data recorded during performance of the blink CR. It takes 7.5 ms for transmission from cortex to orbicularis muscle (Woody and Brozek, 1969).

the dentate with short latency responses were identified by intracellular injection of bio-cytin as the small (<20 μm soma), relatively aspinous, multipolar projection neurons with beaded dendritic varicosities described by Chan-Palay (1977). Following extracellular in-jection of PHA-L in the dentate nucleus (Fig. 1 of Wang et al, 1991), filling of axonal processes in rostral thalamus as well as VA thalamus and caudate was noted. The two lat-ter regions had increases in mean activity of >12 ms onset to click (Woody et al, 1991a).

Bilateral ablation of the rostral cortex prevented acquisition of short-latency condi-tioned blinking (Fig. 4). The impairment persisted despite extensive training for a period of three months after surgery. Control lesions of comparable size at more caudal regions failed to produce such deficits. These and others' studies with comparable findings (Neff, 1961; Masterton, 1997) predicted the existence of a parallel ascending auditory transmis-sion pathway to motor cortex such as that described in Fig. 1.

4.2. Coding of Changes after Conditioning

A number of variables affecting encoded neural transmissions were noted to change after conditioning. These were baseline firing activity, stimulus evoked activity relative to baseline, and numbers of units responsive to the stimulus. Examples of histograms with

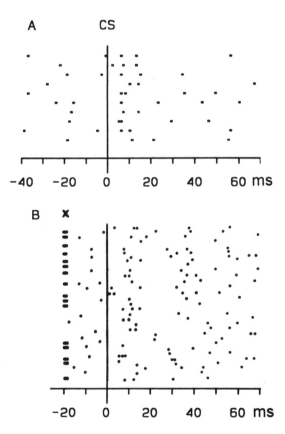

Figure 3. Dot raster displays of spike activity from A: a unit of the dentate nucleus and B: a unit of the motor cor-tex. The click CS was delivered at time 0. The horizontal bars under x in Part B designate the stimulus presenta-tions that elicited a visually observed blink CR. Part B is redrawn from Woody et al (1970).

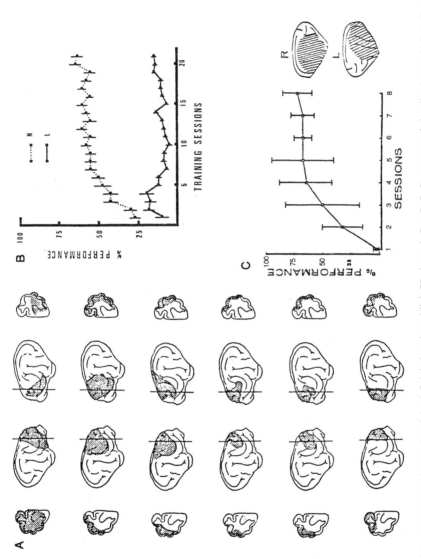

Figure 4. A. Lesions of the motor cortex that produced the deficit in learning a blink CR shown in Part B. B. Percentage of visually observed blink CRs elicited by CSs during training sessions consisting of 150 pairings of click CS and glabella tap US. N - averaged data from normal cats; L - averaged data from the lesioned cats in A. C. Averaged data (as in Part B) from cats with lesions of sensory areas shown to the right. Regions removed from R (right) and L (left) hemispheres are indicated by shading. Redrawn from Woody et al (1974).

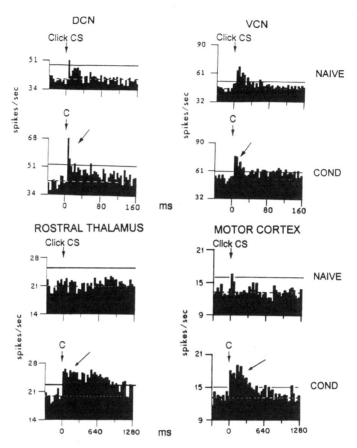

Figure 5. Effects of conditoning on mean neural responses to click. Each histogram is the average of activity from separate groups of units before (NAIVE) and after conditioning (COND). Dorsal cochlear nucleus redrawn from Woody et al, 1991a. Ventral cochlear nucleus redrawn from Woody et al, 1994. Rostral thalamus redrawn from Woody et al 1991b. Motor cortex - redrawn from Aou et al, 1992a.

changes in mean activity to click are shown in Fig. 5. Even though only a few variables could change, each region studied appeared to utilize different combinations of changes to enhance transmission of the click CS versus the hiss DS after conditioning (e.g., Table 2 of Woody et al, 1991b). Also, in some regions early and late components of the unit responses underwent dissimilar changes (c.f. Woody et al 1991a,b).

5. DISCUSSION

The findings suggest that there are two primary ascending pathways by which acoustic information is transmitted from brainstem to cortex and that transmission is characterized by a plasticity that depends on the behavioral state. Responses to click are enhanced relative to that to hiss when the former is used as a CS for conditioning and the latter as a discriminative stimulus (see Aou et al, 1991a, Engel and Woody, 1972, and Woody et al, 1991a,b for examples). A less specific enhancement is observed after sensitization, and the degree of enhancement is small except in the VCN. Potentiation of the transmitted message appears to be coded differently and rather uniquely at each relay by different permutations of the magnitude of peak activity, the magnitude of activity relative to baseline, and the numeric proportion of responsive and unresponsive units. In many regions numerically small subgroups of neurons selectively transmit the acoustic message.

The assumption of stationary processing may be unwarranted in some statistical treatments of spike train data. State-dependent changes can develop rapidly, and can include non-uniform changes in early and later components of response. In addition there will likely be non-uniformities among responses in different groups and types of cells within a single nucleus or area.

Development of short-latency blink CRs is impaired by lesioning the rostral cortex (Fig. 4). Moreover, if 25% KCl is applied to the rostral cortex to produce spreading depression in conditioned animals, the short-latency blink CR is abolished reversibly, returning later after recovery, whereas the unconditioned blink response to glabella tap is maintained throughout (Fig. 7 of Woody and Brozek, 1969). This occurs in the rabbit as well as in the cat (Gutmann et al, 1972). Performance of later components of blink CRs is mediated by other regions of the brain.

Long-latency conditioned eye blink and nictitating membrane movements can develop in decorticate preparations. Buchwald and colleagues (Norman et al, 1974) reported the ability to produce a conditioned blink reflex of 100–300 ms latency in bilaterally hemispherectomized cats using a different associative paradigm with a different auditory CS and a more noxious US. Others (McCormick and Thompson, 1982, 1984; Moore et al, 1982; Yeo et al, 1984) have confirmed this finding in rabbits and extended their investigations to disclose key regions of the brainstem that appear to be involved, including cerebellum and subcerebellar nuclei as well as red nucleus and olivary nucleus circuitry.

Studies of the basis of the startle reflex elicited by loud (100dB) auditory stimuli have disclosed key portions of the subcortical neural circuitry, such as the lateral lemniscal and paralemniscal nuclei, at which startle and conditioning-related acoustic transmissions may interact (Davis et al, 1982).

6. CONCLUSIONS

Conclusions: 1) on the basis of latencies of response to click, the dentate nucleus and rostral thalamus should be considered among the relays that compose the primary (shortest latency), ascending auditory transmission pathway from cochlear nucleus to cerebral cortex; 2) retrograde filling of fusiform cells of the cochlear nucleus and anterograde filling of ascending fiber tracts disclosed by extracellular injection of PHA-L in dentate nucleus (Fig. 1 of Wang et al, 1991) supports this hypothesis; 3) at almost every level of the auditory system, primary as well as later acoustic transmissions are sensitive to changes in behavioral state produced by sensitization or conditioning; 4) this state-dependent plasticity of central auditory transmission is not uniform among regions or even different subgroups of neurons within regions; 5) particularly profound effects of this plasticity are observed in the cochlear nucleus, a relay through which all central acoustic transmissions must pass; 5) sensitization and conditioning may be expected to have different effects on the plasticity of acoustic transmission.

7. ACKNOWLEDGMENT

I thank S. Aou, E. Gruen, M.-C. Xi, X.-F. Wang, and E. Zotova for substantial contributions to the studies. This research was supported in part by HD 05958, NS 25510, the Deafness Research Foundation, and the Exchange Program between St. Petersburg University and the University of California.

8. REFERENCES

Aou, S., Woody, C.D., Birt, D. (1992a) Changes in the activity of units of the cat motor cortex with rapid conditioning and extinction of a compound eye blink movement. *J. Neurosci.* 12:549–559.

Aou, S., Woody, C.D. and Birt, D. (1992b) Increases in excitability of neurons of the motor cortex of cats after rapid acquisition of eye blink conditioning. *J. Neurosci.*, 12:560–569.

Astl, J., Popelář, J., Kvašňák, E. and Syka, J. (1996) Comparison of response properties of neurons in the inferior colliculus of guinea pigs under different anesthetics. *Audiology* 35: 335–345.

Chan-Palay, V. (1977) *Cerebellar Dentate Nucleus* Berlin: Springer, pp. 1–548.

Davis, M., Gendelman, D.S., Tischler, M.D., Gendelman, P.M. (1982) A primary acoustic startle circuit: lesion and stimulation studies. *J. Neurosci.* 2:791–805.

Engel, J. Jr., Woody, C.D. (1972) Effects of character and significance of stimulus on unit activity at coronal - pericruciate cortex of cat during performance of conditioned motor response. *J. Neurophysiol.* 35:220–229.

Erulkar, S.D., Rose, J.E., Davies, P.W. (1956) Single unit activity in auditory cortex of the cat. *Bull. Johns Hopkins Hosp.* 99: 55–86.

Evans, E.F. and Whitfield, I.C. (1964) Classification of unit responses in the auditory cortex of the unanesthetized, unrestrained cat. *J. Physiol.* (London) 171: 476–493.

Goldstein, MH. Jr., Hall, J.L. II, Butterfield, B.O. (1968) Single unit activity in the primary auditory cortex of unanesthetized cats. *J. Acoust. Soc. Amer.* 43: 444–454.

Gutmann, W., Brozek, G., Bures, J. (1972) Cortical representation of conditioned eyeblink in the rabbit studied by a functional ablation technique. *Brain Res.* 40: 203–213.

Hirano, T., Woody, C., Birt, D., Aou, S., Miyake, J., Nenov, V, (1987) Pavlovian conditioning of discriminatively elicited eyeblink responses with short onset latency attributable to lengthened interstimulus intervals. *Brain Res.* 400:171–175.

Hoang, B., Woody, C.D., Wang, X.F., and Gruen, E. A comparison of onset latencies of auditory responses to click in different brainstem areas of the conscious cat. *Abstr. Soc. Neurosci.* 17:303, 1991.

Kettner, R.E. and Thompson, R.F. (1982) Auditory signal detection and and decision processes in the nervous system. *J. Comp. Physiol. Psychol.* 96: 328–321.

Kiang, N.Y.S., Morest, D.K., Godfrey, D.A., Guinan, J.J. Jr., Kane, E.C. (1973) Stimulus coding at caudal levels of the cat's auditory nervous system: I. Response characteristics of single units - In: *Basic Mechanisms in Hearing,* A.R. Moller (Ed.), New York: Academic Press pp. 455–478.

Kim, E.H.-J., Woody, C.D., Berthier, N.E. (1983) Rapid acquisition of conditioned eye blink responses in cats following pairing of an auditory CS with glabella tap US and hypothalamic stimulation. *J. Neurophysiol.* 49:767–779.

Masterton, R.B. (1997) Role of mammalian forebrain in hearing. In: *Acoustical Signal Processing in the Central Auditory System,* J. Syka, Ed. Plenum, New York, pp. 1–17.

McCormick, D.A., Clark, G.A., Lavond, D.G., and Thompson, R.F. (1982) Initial localization of the memory trace for a basic form of learning. *Proc. Natl. Acad. Sci. USA* 79:2731–2735.

McCormick, D.A. and Thompson, R.F. (1984) Neuronal responses of the rabbit cerebellum during acquisition and performance of a classically conditioned nictitating membrane-eyelid response. *J. Neurosci.* 4: 2811–2822.

Moore, J.W., Desmond, J.E. and Berthier, N.E. (1982) The metencephalic basis of the conditioned nictitating membrane response. In: *Conditioning: Representation of Involved Neural Functions.* Ed. C.D. Woody. New York: Plenum Press, pp. 459–482.

Miller, J.M., Pfingst, B.E., Ryan, A.F. (1982) Behavioral modification of response characteristics of cells in the auditory system. In: *Conditioning: Representation of Involved Neural Functions,* C.D. Woody, Ed. Plenum, New York, pp. 345–361.

Neff, W.D. (1961) Neural mechanisms of auditory discrimination. In: *Sensory Communication,* ed Rosenblith, W.A., Cambridge, Mass.: MIT Press, pp.259–278.

Norman, R.J., Villablanca, J.R., Brown, K.A., Schwafel, J.A., Buchwald, J.S. (1974) Classical eyeblink conditioning in the bilateral hemispherectomized cat. *Exp. Neurol.* 44: 363–380.

Olds, J., Disterhoft, J.F., Segal, M., Kornblith, C.L., Hirsh, R. (1972) Learning centers of rat brain mapped by measuring latencies of conditioned unit responses. *J. Neurophysiol.* 35: 202–219.

Oleson, T.D., Ashe, J.H., Weinberger, N.M. (1975) Modification of auditory and somatosensory activity during pupillary conditioning in the paralyzed cat. *J. Neurophysiol.* 38: 1114–1139.

Recanzone, G.H., Schreiner, C.E., Merzenich, M.M. (1993) Plasticity in the frequency representation of primary auditory cortex following discrimination training in adult owl monkey. *J. Neurosci.* 13:87–103.

Sakai, H. and Woody, C.D. (1980) Identification of auditory responsive cells in the coronal-pericruciate cortex of awake cats. *J. Neurophysiol.* 44:223–231.

Scheich, H., Simonis, C., Ohl, F., Tillein, J., Thomas, H. (1993) Functional organization and learning-related plasticity in auditory cortex of the Mongolian gerbil. *Prog. Brain Res.* 97:135–143.

Wang, X.F., Woody, C.D., Chizhevsky, V., Gruen, E., Landeira-Fernandez, J. (1991) The dentate nucleus is a short latency relay of a primary auditory transmission pathway. *NeuroReport* 2:361–364.

Weinberger, N.M. (1982) Sensory plasticity and learning: the magnocellular medial geniculate nucleus of the auditory system. In: *Conditioning: Representation of Involved Neural Functions.* C.D. Woody (Ed.), New York: Plenum, pp. 697–710.

Weinberger, N.M., Javid, R., Lepan, B. (1993) Long-term retention of learning-induced receptive-field plasticity in the auditory cortex. *Proc. Natl. Acad. Sci. (USA)* 90: 2394–2398.

Winer, B.J. (1971) *Statistical Principles in Experimental Design.* New York: MacGraw.

Woody, C.D. (1996) Control of motor behavior acquisition by cortical activity potentiated by decreases in a potassium A-current that increase neural excitability. In: *The Acquisition of Motor Behavior in Vertebrates,* Bloedel, J.R., Ebner, T.J., Wise, S.P. (Eds.), Cambridge (USA): MIT Press, pp. 205–220.

Woody, C.D. and Black-Cleworth, P. (1973) Differences in excitability of cortical neurons as a function of motor projection in conditioned cats. *J. Neurophysiol.* 36: 1104–1116.

Woody, C.D. and Brozek, G. (1969) Changes in evoked responses from facial nucleus of cat with conditioning and extinction of an eye blink. *J. Neurophysiol.* 717–726.

Woody, C.D. and Gruen, E. (1988) Evidence that acetylcholine acts in vivo in layer V pyramidal cells of cats via cyclic GMP and a cyclic GMP-dependent protein kinase to produce a decrease in an outward current. In: *Neurotransmitters and Cortical Function*, M. Avoli, T.A. Reader, R.W. Dykes, and P. Gloor, (Eds.), New York: Plenum Press, pp. 313–319.

Woody, C.D., Gruen, E., Birt, D. (1991c) Changes in membrane currents during Pavlovian conditioning of single cortical neurons. *Brain Res* ., 539:76–84.

Woody, C.D., Gruen, E., McCarley, K. (1984) Intradendritic recording from neurons of motor cortex of cats. *J. Neurophysiol.* 51: 925–938.

Woody, C.D., Gruen, E., Melamed, O., Chizhevsky, V. (1991a) Patterns of unit activity in rostral thalamus of cats related to short latency discrimination between different auditory stimuli. *J. Neurosci* . 11:48–58.

Woody, C.D., Knispel, J.D., Crow, T.J., Black-Cleworth, P. (1976) Activity and excitability to electrical current of cortical auditory receptive neurons of awake cats as affected by stimulus association. *J. Neurophysiol.* 39:1045–1061.

Woody, C.D., Melamed, O., Chizhevsky, V. (1991b) Patterns of activity coding discrimination of auditory stimuli differ between mid- and posterolateral thalamus of cats. *J. Neurosci.* 11:3379–3387.

Woody, C.D., Vassilevsky, N.N., Engel, J. Jr. (1970) Conditioned eye blink: unit activity at coronal-precruciate cortex of the cat. *J. Neurophysiol.* 33:851–864.

Woody, C.D., Wang, X.F., Gruen, E., Landeira-Fernandez J (1992) Unit activity to click CS changes in dorsal cochlear nucleus after conditioning. *NeuroReport* 3:385–388.

Woody, C.D., Wang, X.F., Gruen, E. (1994) Activity to acoustic stimuli increases in ventral cochlear nucleus after stimulus pairing. *NeuroReport* 5:513–515.

Woody, C.D., Yarowsky, P.J. (1972) Conditioned eye blink using electrical stimulation of coronal - precruciate cortex as conditional stimulus. *J. Neurophysiol.* 35:242–252.

Woody, C., Yarowsky, P., Owens, J., Black-Cleworth, P., Crow, T. (1974) Effect of lesions of cortical motor areas on acquisition of conditioned eyeblink in the cat. *J. Neurophysiol.* 37:385–394.

Xi, M-C., Woody, C.D., Gruen, E. (1994) Identification of short latency auditory responsive neurons in the cat dentate nucleus. *NeuroReport* 5:1567–1570.

Yeo, C.H., Hardiman, M.J., Glickstein, M. (1984) Discrete lesions of the cerebellar cortex abolish the classically conditioned nictitating membrane response of the rabbit. *Behav. Brain Res.* 13: 216–266.).

47

CHANGES IN THE CENTRAL AUDITORY SYSTEM WITH DEAFNESS AND RETURN OF ACTIVITY VIA A COCHLEAR PROSTHESIS

Sanford C. Bledsoe, Jr., Shigeyo Nagase, Richard A. Altschuler, and Josef M. Miller

Kresge Hearing Research Institute
Department of Otolaryngology
University of Michigan
Ann Arbor, Michigan 48109-0506

1. INTRODUCTION

It is well established that sensory deprivation during development results in lasting changes in the central nervous system (CNS) (Wiesel and Hubel, 1963). This is certainly the case for the auditory system where in developing animals, sound deprivation and/or cochlear ablation are especially potent means of producing significant neuronal atrophy and reorganization in central structures (Rubel et al., 1984). The loss of afferent input often occurs in the adult auditory system and there is increasing evidence that the mature central auditory system (CAS) also displays substantial changes in structure and function following deafferentation (Gerken, 1979; Rajan et al., 1993, 1996, Willott et al., 1994). However, despite the high incidence of inner ear pathology and a growing utilization of the cochlear prosthesis in deaf adults, little is known about the effects of cochlear damage on the mature CAS, the mechanisms underlying CNS physiological changes produce by deafness, or how deafness induced changes may affect processing with subsequent reactivation of the CAS via a cochlear prosthesis. Elucidating how the CAS responds to such deafferentation is fundamental to our understanding the mechanisms of plasticity and homeostasis of the brain throughout an organisms lifetime.

There is increasing evidence that deafness can induce changes in the mature mammalian central auditory system that alter processing when stimulation is reintroduced through cochlear prostheses (e.g. Snyder et al., 1991; Schwartz et al., 1993; Miller et al., 1992, 1996). Morphological changes include a loss of spiral ganglion cells (Webster and Webster, 1981; Leake and Hradek, 1988; Jyung et al., 1989; Zappia and Altschuler, 1989), changes in cell size in the cochlear nucleus (Webster, 1988; Pasic and Rubel, 1989; Rubel et al., 1990; Hultcrantz et al., 1991; Lustig et al., 1994; Lesperance et al., 1995) and superior olivary complex (Dupont et al., 1996), changes in auditory nerve synapses (Gulley et al., 1978; Rees et al., 1985; Miller et al., 1992), and changes in glia (Canady and Rubel, 1992). Neurochemical changes include changes in GABA (Potashner et al., 1985, Bergman et al., 1989; Helfert et al., 1992; Dupont et al., 1992, 1994; Bledsoe et al., 1995), glycine (Bergman et al., 1989; Dupont et al., 1994) and glycine receptors (Sanes, 1994; Altschuler et al., 1995). These changes, as well as changes in connections, may contribute to functional changes that have been reported with deafness (e.g. Rajan et al., 1996; Salvi et al., 1996), and the changes that occur when chronic electrical stimulation (ES) is re-introduced in the deafened ear (e.g. Miller et al., 1996). Thus deafness results in a decrease in evoked 2-deoxyglucose (2DG) uptake in cells of the inferior colliculus (IC) which can be reversed with chronic stimulation (Schwartz et al., 1993). Snyder et al. (1990) have reported a broadening of the IC representation of restricted cochlear ES following chronic stimulation of the deafened ear. This correlates with observations that when the cochlea is partially destroyed the cochleotopic representation of the remaining portion increases its representation in auditory cortex (Rajan et al., 1993, 1996). At this time, the mechanisms which underlie the changes in responsiveness of CAS cells in deafened animals remain unknown.

In the present study, we have investigated deafness and chronic ES-induced changes in response characteristics of cells of the inferior colliculus (IC) using Fos immunolabeling evoked by ES of the auditory nerve, electrophysiological single-unit recording and chronic in vivo microdialysis of neurotransmitters. Measurements were performed on the central nucleus of the inferior colliculus (CIC) in mature, normal-hearing rats and guinea pigs, and in animals deafened for up to 6 months. Some deafened subjects were studied following a period of reactivation produced by chronic ES.

The expression of the Fos proto-oncogene (c-*fos*) is a useful marker of neural excitatory activity (e.g. Curran and Morgan, 1987; Dragunow and Robertson, 1987; Birder et al., 1991; Sagar and Sharp, 1991; Hoffman et al., 1993) which has been used in the auditory system to label neurons excited by sound (Ehret and Fischer, 1991; Friauf, 1991; Lim et al., 1991; Sato et al., 1992, 1993; Rouiller et al., 1992; Adams, 1995; Brown and Liu, 1995) and electrical stimulation (Vischer et al., 1994, Zhang et al., 1996; Nagase et al., 1997a,b). Moreover, as suggested by these studies (see particularly, Nagase et al., 1997a,b), changes in the number and/or pattern of ES-evoked neurons expressing Fos as a consequence of deafness or of chronic ES suggest changes in central auditory processing. However, these Fos studies did not report data from additional monitors of central processing. The present study, therefore, reports for the first time the influence of 21 days of deafness, with or without chronic ES, on evoked Fos along with electrophysiological recordings of single unit activity and neurochemical measures of neurotransmitters as independent measures of changes in central processing.

2. MATERIALS AND METHODS

2.1. Subjects

A total of 69 Sprague Dawley rats and 46 pigmented guinea pigs (NIH strain, Murphy) were used in this study. The care and use of the animals, consistent with NIH principles, were approved by the University of Michigan Committee on Use and Care of Animals. Animals were divided into deafened and non-deafened groups. In all experiments, anesthetized animals were bilaterally deafened by intracochlear injections of neomycin through the round window as previously described (Bledsoe et al., 1995; Nagase et al., 1997a,b). Both cochleae were slowly perfused over 3 minutes with 30 μl of 10% neomycin sulfate through the round window membrane. Both pre- and post-operative click-evoked auditory brain stem response (ABR) thresholds were recorded to assess the extent of the hearing loss. An immediate threshold shift of 50 dB was defined as the minimum acceptable deficit for inclusion in the study.

2.2. C-fos Studies

Experiments were performed on Sprague Dawley rats (250–350 gm) and Fos immunolabeling in the auditory brain stem was measured in response primarily to monopolar ES via a scala tympani electrode at the base of the cochlea. Additional observations were made to monopolar ES at the apex. Studies were performed in normal hearing subjects, subjects deafened 21 days and 21 day deafened subjects who received 14 days of chronic ES prior to fos study.

For chronic stimulation, electrodes were implanted in deaf animals 7 days after neomycin injections, immediately following which their stimulators were activated. Animals were anesthetized as described above, cochleae exposed via a post-auricular approach and a single Teflon-coated 3T platinum-iridium wire (250 μm diameter ball tip) placed into basal turn scala tympani via a fenestra in the lateral wall made approximately 1.5 mm anterior to the round window, with a reference wire in the middle ear. Chronic ES was provided via charge balanced biphasic 200 μamp square wave pulses at 200 μsec/phase, delivered at 200 Hz, for 4 hr/day. For ABR studies, a stainless steel vertex epidural electrode was placed in the skull midway between bregma and lambda. Electrically-evoked ABRs (EABRs) were recorded in subjects lightly anesthetized with ketamine (75 mg/kg) 5 days after implantation in both normal hearing and deafened animals. Fifty μsec biphasic, charge-balanced square wave stimulation was provided at 100 pulses/sec for each animal. Across the different subject groups stimulation was delivered at 5x the EABR threshold. In selected subjects additional observations were made at 1.5x and 10x EABR threshold. Stimulation was for 90 minutes. Thirty minutes after cessation of ES, animals were heavily anesthetized and perfused through the heart with buffer followed by 4% paraformaldehyde fixative. Brains were processed by conventional immunocytochemical techniques with 50 μm cryostat frontal sections immunoreacted with antibodies to Fos protein (Oncogene Sciences) using the Vectastain ABC immunoperoxidase avidin biotin procedure (Vector Laboratories).

2.3. Electrophysiology Studies

Experiments were performed on both Sprague Dawley rats (250–350 gm) and pigmented guinea pigs (250–400 gm) as previously described (Bledsoe et al., 1982; Bledsoe et al., 1995). Response properties of CIC neurons to ES of the cochlea were recorded using conventional single-unit techniques. Animals were deafened or sham operated and then the responses of CIC neurons to intracochlear ES were obtained after a survival of 1 day, 7 days, 2 weeks, 3 weeks, 1 month, and 3 months. After the appropriate survival time, normal and deafened rats and guinea pigs were studied electrophysiologically under ketamine hydrochloride (100 mg/kg, IM) and xylazine (20 mg/kg, IM) anesthesia. For intracochlear ES, a monopolar stimulating electrode was implanted in the basal turn of the scala tympani. The same type of electrode was used in the c-fos studies. The animals were held in a stereotaxic device, and rectal temperature monitored and maintained at 38 ± 0.5°C with a thermostatically controlled heating pad. Extracellular single-unit activity was recorded in the CIC contralateral to the stimulated cochlea using micropipettes filled with 1.0% Fluoro-Gold in 0.5 M NaCl (10–20 megohm impedance). Data was obtained only from animals in a stable physiological state monitored by heart rate. After each experiment, the brain was processed for histological examination and the resultant Fluoro-Gold spots used to reconstruct single-unit recording sites.

For analysis, the mean number of excited, suppressed and unresponsive cells were calculated for each of the post-deafening survival times. In addition, for driven and suppressed neurons mean values of the various response measures were calculated for each of the post-deafening survival times. Response measures included were: threshold, spontaneous rate, maximum percent change in evoked discharge rate and current level evoking 50% of the maximum discharge rate. For statistical testing, the data were grouped according to the experimental conditions. For all response variables, a nonparametric one-way analysis of variance was performed across experimental conditions with the criterion for statistical significance set at $p<0.05$. If permitted by the analysis of variance, pairwise comparisons were performed using post-hoc Neuman-Kuells multiple range test.

2.4. Microdialysis Studies

In vivo microdialysis techniques were used in normal and deafened guinea pigs to assess changes in γ-aminobutyric acid (GABA), a major inhibitory neurotransmitter in the CIC (Faingold et al., 1991). Studies were conducted using techniques described previously (Bledsoe et al., 1995; Goldsmith et al., 1995) on normal hearing and 30-day deafened guinea pigs (250–400 gm) anesthetized with ketamine hydrochloride (100 mg/kg, IM) and xylazine (20 mg/kg, IM). In addition, chronic microdialysis techniques were used in unanesthetized, alert animals to test the hypothesis that a time-dependent down regulation of GABA levels and release occurs in the CIC after peripheral deafferentation. Beginning one day after implantation of a chronic guide cannula in the CIC, microdialysis samples were obtained at 7 day intervals for 3 weeks to establish baseline levels of amino acids in the dialysate. The animals were then deafened or sham operated and additional microdialysis samples obtained after a survival of 1 day, 7 days, 2 weeks, 3 weeks, 1, 3 and 6 months. An ABR threshold shift of 50 dB again defined the minimum acceptable deficit. The animals were placed in a mild restraining box and a microdialysis probe was inserted into the guide cannula for sampling the extracellular fluid of the CIC. The probe was perfused during and after insertion with Ringer solution of various compositions at a flow rate of 2.0 μl/min. Once the probe was in place, the experimental protocol was the same as we have used successfully in acute preparations (Goldsmith et al., 1995). After a 2-hour stabilization period with standard Ringer solution, 7 consecutive fractions were collected with the perfusate switched to a medium containing a 100 mM concentration of KCl (NaCl reduced to maintain isotonicity) during the collection of the third and fourth fractions. All samples were collected on ice in microcentrifuge vials, rapidly frozen in a dry ice/methanol bath and stored at -20°C until assayed.

Amino acids were measured in standards and 20-μl aliquots of the dialysate by o-phthalaldehyde (OPA)/β-mercaptoethanol derivatization followed by reverse phase, gradient elution high-performance liquid chromatography (HPLC) and fluorometric detection as previously described (Bledsoe et al., 1989). The data were expressed as means ± SEM. The differences between the mean basal release and the KCl-evoked release and between experimental groups and the control animals were tested statistically by two-tailed Student's t-test for paired samples or by analysis of variance with differences between individual means assessed by post-hoc Neuman-Kuells multiple range test, where appropriate. The criterion for statistical significance was $p < 0.05$.

3. RESULTS

3.1. C-fos Findings

In both acute and chronically stimulated 21 day deafened animals, electrically induced Fos-immunoreactive (IR) staining was confined to the nucleus of immunolabeled neurons. No auditory brain stem neurons showed Fos IR in the absence of ES. Cochlear ES-induced Fos IR in nuclei of neurons in the ascending auditory pathways and in neurons of the paralemniscal nuclei, located medially to the dorsal nucleus of the lateral lemniscus (Nagase et al., 1997a,b).

The number of Fos-IR neurons evoked by ES in each auditory brain stem region increased as stimulation intensity increased in both normal hearing and deafened rats. The location of evoked Fos-IR neurons within auditory brain stem nuclei also changed as a function of electrode position, in high frequency tonotopic regions with basal monopolar cochlear ES and in low frequency regions with apical monopolar cochlear ES. While generally, the distribution and patterns of Fos-IR auditory brain stem neurons evoked by ES in 21 day deaf animals were similar to the patterns seen in normal hearing animals, differences were seen in number, density and distribution of IR cells in the contralateral CIC.

No IR cells were seen in the CIC of non-stimulated animals, while 90 minutes of basal monopolar ES evoked Fos-IR neurons in a band in the high-frequency region of the contralateral CIC in both normal and 21-day deafened animals. Following ES at 1.5x threshold, fewer Fos IR neurons were present in sections (3.5/section) of the CIC in 21 day deafened animals than in normal hearing animals (6 cells per section average). On the other hand, with ES at either 5x and 10x EABR, an increased number of evoked Fos IR cells were seen in 21 day deafened animals compared to normal hearing animals (Fig. 1). In the 21-day deafened animals, the band of neurons was much wider, with approximately twice as many IR neurons. The number of Fos-IR cells in an average section of the contralateral CIC totaled 63 in normal hearing rats and 173 after 21-days of deafness. The differences are more easily observed in the camera lucida drawings of these sections in Fig. 1C and 1D. This increased labeling was observed in all animals studied (n=8). No other auditory brain stem areas demonstrated a comparable change in density and distribution of evoked Fos-IR neurons, with deafness after basal monopolar ES.

Interestingly, deafened animals that also received chronic basal monopolar stimulation, demonstrated another significant difference in the pattern of Fos IR neurons in the ICC induced by basal monopolar stimulation at 5x threshold. In addition to an increase in the number of Fos-IR neurons compared to normal hearing rats, there was also a position shift in the location of evoked Fos-IR neurons in the contralateral ICC compared to 21 day deaf animals without chronic stimulation (Fig. 2). Two bands were seen, with the larger (major) band of Fos-IR neurons situated more dorso-laterally, (intermediate frequency area) than the high frequency band induced in 21 day deaf animals which did not receive chronic stimulation. Thus, chronic stimulation induces additional changes than deafness alone.

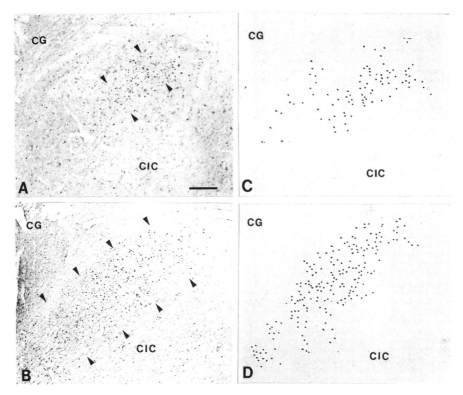

Figure 1. Micrographs and camera lucida drawings of Fos IR cells in cross sections from the CIC in response to contralateral electrical stimulation at 5x EABR threshold in both normal hearing and 21-day deafened rats. A: Micrograph from a normal hearing rat showing a band of Fos IR cells (arrows) in the contralateral CIC. Scale bar represents 100 μm. CG: central gray. B: Micrograph from a 21-day deafened rat showing a broader band (arrows) in the contralateral CIC. C: Camera lucida drawing of micrograph depicted in A (63 IR cells shown). D: Camera lucida drawing of micrograph depicted in B. (173 IR cells shown). Published in NeuroReport 7: 225–229, 1995 and reproduced here with permission.

3.2. Electrophysiological Findings

A total of 224 colliculus cells in 16 normal animals (9 rats, 7 guinea pigs) and 242 cells in 15 animals deafened for 21 days (10 rats, 5 guinea pigs) were studied in the electrophysiological experiments. The data from both normal and deaf animals revealed that IC neurons respond best to 100 Hz sinusoidal ES, exhibit phase-locking, and display varying degrees of non-monotonic behavior to increasing levels of current. Many neurons exhibit varying degrees of adaptive behavior, responding best to initial stimulus presentations and most have a limited dynamic range (Fig. 3). This is consistent with previous electrophysiological studies of auditory neuron responses to ES (van den Honert and Stypulkowski, 1984; Javel et al., 1987; Parkins, 1989) and with models of electrical excitability (Clopton et al., 1983; Colombo and Parkins, 1987). In the sample of 224 units from normal animals, 42% were suppressed by contralateral ES. However, after 21 days of deafness, a remarkable feature was the virtual absence of neurons that were suppressed by contralateral ES. Of 242 units from deaf animals, only 5% showed suppression. This is most likely not due to a sampling bias since 1) we visualized and systematically sampled the CIC; and 2) in normal animals, suppressed units were almost always seen in the same

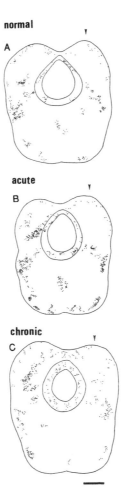

Figure 2. Low-power camera lucida drawings of representative sections through the IC and lateral lemniscus showing Fos immunoreactivity induced by basal monopolar electrical stimulation at 5x EABR threshold in a normal rat (A), in a rat deafened for 21 days (B) and a 21 day deafened rat that received 14 days of chronic stimulation from day 7–21 (C). Arrowheads denote the side ipsilateral to the stimulated cochlea. The data illustrate increased Fos labeling in the contralateral CIC in deafened animals as compared to normal and that the band of Fos-IR neurons evoked in the CIC after chronic stimulation moves dorso-medially towards a lower frequency location. Scale bar = 1 mm.

puncture with driven units (Fig. 3). Figure 3 illustrates results from three units recorded in the same electrode penetration. The discharge rates of two units were suppressed by contralateral stimulation while the third unit displayed a non-monotonic excitatory response.

In Fig. 3, the suppression begins at the same current level (3.4 µA) as the non-monotonic behavior in the driven unit suggesting that the suppressed cells may be simply the responses of non-monotonic cells at high current levels. However, several examples of units recorded in the same electrode penetration in normal rat or guinea pig were obtained in which both suppressed and driven units displayed similar thresholds. In this study, the mean threshold (±S.E.) for driven neurons was 21.6 ± 2.1 µA (n=96) and 20.1 ± 2.3 µA

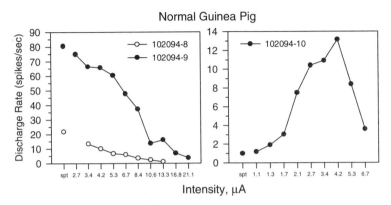

Figure 3. Rate-intensity functions from three single units recorded in the same electrode penetration of the CIC in a normal hearing guinea pig. Results were obtained in response to 100 Hz sinusoidal, monopolar stimulation of the contralateral cochlea. The data illustrate that the discharge rates of many neurons in normal animals are suppressed by electrical stimulation (A) while other neurons are strongly excited and display non-monotonic behavior (B). Spt denotes spontaneous activity. Published in NeuroReport 7: 225–229, 1995 and reproduced here with permission.

(n=43) for suppressed cells. Thus, we conclude that the suppressed cells are not non-monotonic cells at higher current levels.

We have conducted an extensive analysis of the data from normal and 21 day deaf guinea pigs and rats. No difference has been found in the thresholds to 100 Hz sinusoidal current in normal and 21 day deaf animals (Fig. 4). Fast fourier transform analysis on PSTs obtained at 3 and 6 dB above threshold in normal and 21 day deaf animals reveal no difference in the phase-locked response to 100 Hz current. The mean FFT (±S.E.) for normal animals at 3 dB is 0.19 ± 0.01 (n=80) and 0.20 ± 0.02 (n=89) for deaf animals. At 6 dB it is 0.23 ± 0.02 and 0.25 ± 0.03, for normal and deaf subjects, respectively. Figure 5 depicts data on spontaneous activity for normal and deafened guinea pigs and rats. The spontaneous discharge rate for 21-day deaf animals is significantly lower than the rate in

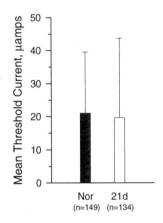

Figure 4. Graphical depiction of the thresholds of units to 100 Hz sinusoidal electrical stimulation in normal and 21 day deaf guinea pigs and rats. Mean values are depicted and the data illustrate that there is no difference in the thresholds for normal and 21 day deaf animals.

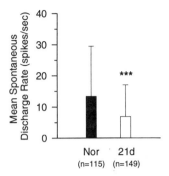

Figure 5. Graphical depiction of the spontaneous activity of units in normal and 21 day deaf guinea pigs and rats. Mean values are depicted and the data illustrate that spontaneous activity in 21 day deaf animals is significantly lower than in normal animals. This is due to a much lower level of spontaneous activity in the deaf rat but not in the deaf guinea pig.

normal animals. This is due to a much lower level of spontaneous activity in the deaf rat. There is no significant difference in spontaneous rate between normal and deaf guinea pigs. Slightly fewer suppressed cells are also observed in the normal rat (37% of responsive cells) than in the normal guinea pig (53% of responsive cells).

The above results have been obtained in a comparison between animals deafened for 21 days and normal animals. To exclude the possibility that the suppression of units in normal animals might result from 'electrophonic' activation of the auditory nerve due to the presence of hair cells (Stevens and Jones, 1939), we carried out experiments in which normal guinea pigs received intracochlear injections of neomycin 2 hours before initiating single-unit recordings. Data was obtained from 34 responsive cells, of which 16 were suppressed (47%) by ES. This result indicates that the suppressive response properties of CIC neurons are not dependent on 'electrophonic' hearing or the presence of a normal functioning inner ear. Substantial numbers of suppressed cells have also been observed in ani-

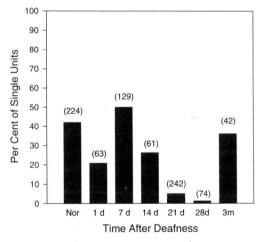

Figure 6. Time-dependent changes in the suppressed population of neurons obtained at 1, 7, 14, 21, 28 days and 3 months of deafness in guinea pigs and rats. The numbers in parentheses provide the numbers of units in each group.

mals deafened for 1 and 7 days (see below.) Importantly, this suggests that the decrease in the number of suppressed units in 21 day deafened animals results from a time-dependent change in the central auditory neuraxis.

Single-unit data on time-dependent changes in the suppressed population of cells has now been obtained at 1, 7 14 and 28 days of deafness in guinea pigs and supports this contention. In addition, we have recorded from 42 responsive units in animals deafened for 3 months. Figure 6 graphically depicts these results along with those from the studies on normal and 21-day deaf animals. The numbers in parentheses provide the numbers of responsive (driven plus suppressed) units in each group. At day 1, 7 and 14 suppressed cells still represent a fairly large percentage of the responsive units present. One day after deafening, the percentage of suppressed cells decreases to 14.3%, which may be due to trauma or deafness-induced hyperexcitability. The percentage recovers by day 7 and then shows a secondary decrease beginning at day 14 which reaches reductions of 5% and 1.3% by days 21 and 28, respectively. Interestingly, 15 of 42 units (36%) studied in animals deafened for 3 months were suppressed by contralateral electrical stimulation indicating that the reduction in the suppressed population of cells may be reversible. This is in keeping with the reversible changes in GABA seen in immunocytochemical and microdialysis studies (see below).

3.3. Microdialysis Studies

In vivo microdialysis data reveal that the potassium stimulated release of GABA from the CIC is markedly suppressed in animals deafened for 30 days (Fig. 7A). In contrast to GABA, the levels of glycine are remarkably similar in normal and 30 day-deafened animals (Fig. 7B). In addition to the acute results depicted in Fig. 7, data from chronic sampling of GABA levels in 6 guinea pigs (3 deafened, 3 normal) indicate that the release of GABA in the deafened animals is markedly decreased 7 and 14 days after deafening, shows signs of recovering at 21 and 28 days and appears fully recovered by 3 and 6 months. These data agree with the transient decrease in GABA immunolabeled cell somata observed with free-floating immunocytochemical techniques (DuPont et al., 1992; 1994).

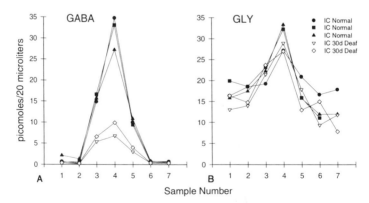

Figure 7. Effect of potassium depolarization on the extracellular levels of GABA and glycine in the CIC from normal and deafened guinea pigs. The data illustrate that 30 days of deafness produces a marked decrease in the basal and stimulated release of GABA (A) but has no effect on the release of glycine (B). Seven consecutive 10-min fractions were analyzed and the efflux of GABA expressed as picomoles/20 µl of dialysate. During the third and fourth fractions standard Ringer solution was switched to medium containing 100 mM KCl. Published in NeuroReport 7: 225–229, 1995 and reproduced here with permission.

4. DISCUSSION

Following loss of inner hair cells and scar formation, degenerative changes take place in the neural elements of the inner ear that may result from the loss of survival factors (see Miller et al, 1996). In the mature mammal, degeneration is not observed in the CAS. Rather, plastic changes occur, that are consistent with a down-regulation of cellular activity, including: a flattening of auditory nerve synapses in the cochlear nucleus, an increase in CAS nucleus neuropil, cell size changes, changes in neurotransmitters and receptors, a decrease in resting and evoked metabolic activity and changes in the spatial pattern of evoked cell responses. These changes may be modified by the reintroduction of activity via a cochlear prosthesis. The findings reported here suggest that at the level of the IC, at least in part, some of these changes may be driven by local changes in the tissue.

Although the CIC receives binaural innervation, most studies of this nucleus have examined the effects of monaural deprivation. Moreover, there are no reports on the effects of deafness on Fos IR. In this study, we found that 21 days after bilateral deafening a marked increase in the number of labeled cells occurs in the CIC in response to basal turn ES of the contralateral cochlea. This suggests a spread of activation or excitability associated with auditory system plasticity. A spread in excitability could be due to a number of factors; these may include neurochemical and cellular changes as well as possible changes in projection and innervation pattern, a change in density of innervation, or synaptic efficiency. In the present study, we have obtained two independent lines of data which suggest the increased Fos expression may be associated with a deafness-induced loss, or downregulation, of inhibition in the tissue of the IC.

One major finding of the present study is that 40% of the collicular neurons are suppressed by monopolar electrical stimulation of the contralateral cochlea. In response to acoustic stimulation, the vast majority of neurons in the IC are excited by contralateral sounds and suppressed by simultaneous ipsilateral stimulation (Kuwada et al., 1984; Semple and Kitzes, 1985). Thus, the suppression by contralateral ES was unexpected and the mechanism(s) underlying its generation not entirely clear. Monopolar electrode configurations are known to produce less discrete activation of the auditory nerve than bipolar configurations (van den Honert and Stypulkowski, 1984, 1987; Ryan et al., 1990). Thus, one explanation is that monopolar stimulation activated more inhibitory centripetal pathways associated with the inhibitory surrounds of neuronal response areas. Alternatively, ES may have activated the ipsilateral inhibitory pathways.

A second major finding of the present electrophysiological studies is the marked decrease in the numbers of collicular neurons which have their discharge rate suppressed by contralateral ES of the cochlea. In normal undeafened animals, 40% of collicular neurons were suppressed by cochlear ES but this decreased to 5% after 21 days of bilateral deafness. The mechanisms underlying this decrease are not known. The high incidence of suppressed cells and the reduction of their numbers in 21 day deaf animals could be explained by an 'electrophonic' excitation of the auditory nerve due to the presence of hair cells in the non-deafened cochlea (Kiang and Moxon, 1972; Stevens and Jones, 1939). We can exclude this possibility, however, on the basis of suppressed cells being present in normal animals injected with intracochlear neomycin immediately prior to obtaining single-unit responses. The loss of suppression, however, is consistent with altered auditory function reported by others in mature animals. For example, with behavioral and electrophysiological techniques hypersensitivity to electrical stimulation has been shown in adult cats following bilateral mechanical lesions of the cochleas (Gerken, 1979) and noise-induced deafness (Gerken et al., 1984). Enhanced neuronal excitability has also been observed in the CN and CIC of mice with genetically determined peripheral damage (Willott et al., 1984) and in C57 mice with noise-induced hearing loss (Urban and Willott, 1979). Other changes have been reported in the CIC including an increase in spontaneous activity, enhanced masking of tones by background noise and altered binaural responses (McFadden and Willott, 1994a,b; Willott, 1994). The basis for these and other CAS plastic changes in mature animals are not known. Willott (1994) has proposed they may reflect, in part, attenuation of neural inhibitory processes.

Several mechanisms could explain the decrease in the number of suppressed units in present study. There could be changes in the cochlea (e.g., loss of cellular elements) that result in altered current spread and activation of the auditory nerve. Another possible mechanism is a spread of excitation in the brain stem due to sprouting and spread of excitatory fibers and terminals within the CIC and/or nuclei sending excitatory inputs to the CIC. Another potential mechanism is a decrease in inhibition in the CIC or nuclei sending projections to the CIC. None of these hypothetical mechanisms are mutually exclusive and all may be occurring to some extent. However, several lines of evidence suggest that reduced inhibition in the IC is likely to be one factor involved. Dupont et al. (1992, 1994) demonstrated that in inhibitory GABAergic innnervation in the guinea pigs IC is significantly reduced 10 to 25 days after unilateral ear destruction. Similarly, eye removal reduces the expression of GABA in the visual cortex (Hendry and Jones, 1988) Using in vivo microdialysis techniques, we have obtained evidence that the basal and potassium stimulated release of GABA from the IC are markedly suppressed in a reversible manner in deafened guinea pigs. A reversible decrease in GABA and its release is consistent with the reversible loss of suppression in the electrophysiological studies. Both lines of evidence are also consistent with the hypothesis that the increased Fos labeling is due to down regulation of inhibition. These results have important implications for cochlear prosthesis research and understanding how the auditory system adapts to sensory deprivation.

Several changes with deafness have been demonstrated in the Fos results of the present study. First, at low basal monopolar stimulation levels (1.5x threshold), less induction of Fos IR in neurons was observed in the CIC after 21 days of deafness, compared to normal hearing animals. This is, in part, consistent with our previous study using 2DG methods, in which we found decreased 2DG uptake evoked by 100 μamps of electrical activity after four or nine weeks of deafness (Schwartz et al., 1993). We hypothesized that this might be due to deafness related changes such as decreased cell size and changes in synapses (Gulley et al., 1978; Rees et al., 1985; Webster, 1988; Rubel et al., 1990; Hultcrantz et al., 1991; Miller et al., 1992; Lustig et al., 1994; Lesperance et al., 1995).

At higher levels of stimulation (5x and 10x EABR threshold) there was an increase in the number of evoked Fos IR cells over that seen in normal hearing animals. With basal cochlear ES this increase was predominantly in the CIC. With apical monopolar and bipolar stimulation the increase was also seen in the lateral superior olive (LSO) and medial nucleus of the trapezoid body, and with bipolar stimulation in the LSO and ventral cochlear nucleus (VCN) (Nagase etal., 1997a,b). While both monopolar apical or basal and bipolar stimulation lead to tonotopically organized increases in Fos IR neurons in the CIC, it is interesting that apical monopolar and bipolar stimulations shows larger deafness-related increases in Fos IR neurons in lower auditory brain stem nuclei. Since bipolar stimulation also provided excitation of the apical cochlea, one could hypothesize that pathways evoked by apical stimulation demonstrate greater plastic changes in superior olivary complex and cochlear nucleus than that shown with basal monopolar stimulation.

Another important result of our studies is that chronic stimulation induces an additional change in the induction of Fos IR neurons from those induced by just a period of time of deafness. In chronically stimulated animals, in addition to a deafness-related increase in the number of Fos IR neurons induced with 5x stimulation, the location of the induced Fos IR cells shifted, compared to both normal hearing animals and acutely stimulated deafened animals. This suggests more complex plastic changes are occurring. Snyder et al. (1990) showed that the average area in the CIC activated by stimulation with electrodes that had been chronically stimulated at 6dB above EABR threshold was approximately twice that observed in non-stimulated deafened animals and prior normal animals. It is also interesting that Irvine and Rajan (1994) report that a restricted cochlear lesion in the basal turn of the adult cat lead to expanded representation in the CIC of the cochleotopic frequencies at the edge of the lesion. They suggested that this change could be a result of reorganization in the IC or of pathways to the IC. A spread in excitability could be due to a number of factors including neurochemical and cellular changes, changes in projection and innervation pattern, or a change in density of innervation or synaptic efficiency. Along these lines, unilateral cochlear ablations in developing animals have been reported to cause degeneration in the ipsilateral colliculus and to enhance neuronal projections to the contralateral IC (Nordeen et al., 1983). In the present study, we have obtained electrophysiological and neurochemical data which indicates that the increased Fos expression in the CIC may be associated with a deafness-induced loss or down regulation of inhibition. The c-fos findings also indicate that chronic electrical stimulation in a deafened auditory system, limited to simple repetitive activation of a restricted population of eighth nerve fibers may lead to expansion of the central representation of the stimulated structures. Both deafness-related changes and changes as a result of chronic stimulation of the deafened auditory pathways appear to occur. In the electrically stimulated deaf subject, the end result likely represents some combination of the two influences.

The current studies indicate that processing of auditory information changes as a consequence of a period of time of deafness and from artificial characteristics of cochlear prosthetic stimulation, which could have important consequences when stimulation is re-introduced either through cochlear prostheses, or perhaps in the future through regenerated hair cells.

5. CONCLUSION

After 21 days of bilateral deafness there is an increase of Fos immunoreactive (IR) cells in the CIC evoked by ES of the auditory nerve. This is complemented by the near elimination of CIC cells that show electrophysiologically suppressed responses to ES of the cochlea and with a decrease in the release of GABA, an inhibitory neurotransmitter in the CIC. These results provide evidence for deafness-induced changes in inhibitory processes occurring in the adult central auditory system and that, at least one important mediating factor, is a downregulation in the local production or release of inhibitory transmitter. These data also indicate that a spread of activation, beyond that induced by deafness, occurs with chronic ES; and chronic ES can induce a shift in the center of excitation of contralateral IC cells. The results have important basic implications for our understanding of plasticity in the CNS, and clinical implications for the reintroduction of hearing in deaf patients with the cochlear prosthesis.

6. ACKNOWLEDGMENTS

These studies were supported by NIDCD Cochlear Prosthesis Program Project Grant DC00274. We thank Jessica Normile, Diane Prieskorn, James Wiler and Dr. John

McLaren for assistance in the experiments. We also thank Margie Conlon for expert secretarial assistance and Jackie Kaufman for helpful comments on the manuscript.

7. REFERENCES

Adams, J.C. (1995) Sound stimulation induces fos related antigen in cells with common morphological properties throughout the auditory brainstem. J. Comp. Neurol. 361, 645–668.

Altschuler, R.A., Raphael, Y., Dupont, J., Sato, K. and Miller, J.M. (1995) Active mechanisms in the response of the auditory system to over or under stimulation. In: A. Flock, D. Ottoson and M. Ulfendahl (Eds.), Active Hearing, Elsevier, pp. 239–256.

Bergman, M., Staatz-Benson, C. and Potashner, S.J. (1989) Amino acid uptake and release in the guinea pig cochlear nucleus after inferior colliculus ablation. Hear. Res. 42, 283–291.

Birder, L.A., Roppolo, J.R., Iadarola, M.J. and deGroat, W.C. (1991) Electrical stimulation of visceral afferent pathways in the pelvic nerve increases c-fos in the rat spinal cord. Neurosci. Lett. 129, 193–196.

Bledsoe, S.C., Jr., McLaren, J.D. and Meyer, J.R. (1989) Potassium-induced release of endogenous glutamate and two as yet unidentified substances from the lateral line of Xenopus laevis. Brain Res. 493, 113–122.

Bledsoe, S.C., Nagase, S., Miller, J.M. and Altschuler, R.A. (1995) Deafness-induced plasticity in the mature central auditory system. NeuroReport 7, 225–229.

Bledsoe, S.C., Rupert, A.L. and Mousehegian, G. (1982) Response characteristics of cochlear nucleus neurons to 500 Hz tones and noise: Findings related to frequency-following potentials. J. Neurophysiol. 47, 113–128.

Brown, M.C. and Liu, T.S. (1995) Fos-like immunoreactivity in central auditory neurons of the mouse. J. Comp. Neurol. 357, 85–97.

Canady, K.S. and Rubel, F.W. (1992) Rapid and reversible astrocytic reaction to afferent activity blockade in chick cochlear nucleus. J. Neurosci. 12, 1001–1009.

Clopton, B.M., Spelman, F.A., Glass, I., Pfingst, B.E., Miller, J.M., Lawrence, P.D. and Dean, D.P. (1983) Neural encoding of electrical signals. Ann. N. Y. Acad. Sci. 405, 146–158.

Colombo, J. and Parkins, C.W. (1987) A model of electrical excitation of the mammalian auditory-nerve neuron. Hear. Res. 31, 287–311.

Curran, T. and Morgan, J.I. (1987) Memories of fos. Bioessays 7, 255–258.

Dragunow, M. and Robertson, H.A. (1987) Kindling stimulation induces c-fos protein(s) in granule cells of the rat dentate gyrus. Nature 391, 441–442.

Dupont, J., Zoli, M., Agnati, L.F. and Aran, J.M. (1992) Morphofunctional changes in guinea pig brainstem auditory nuclei after peripheral deafferentation: A hypothesis about central mechanisms of tinnitus. In: J.M. Aran and R. Dauman (Eds.), Tinnitus, Kluger, Amsterdam, pp. 195–205.

Dupont, J., Bonneau, J.M., Altschuler, R.A. and Aran, J.-M. (1994) GABA and glycine changes in the guinea pig brainstem auditory nuclei after total destruction of the inner ear. Assoc. Res. Otolaryngol. Abstr. 17, 11.

Dupont, J. and Altschuler, R.A. (1996) Glycine and cell size plasticity in the superior olivary complex of the adult guinea pig after partial and total cochlear deafferentation. Assoc. Res. Otolaryngol. Abstr. 19, 154.

Ehret, G. and Fischer, R. (1991) Neuronal activity and tonotopy in the auditory system visualized by c-fos gene expression. Brain Res. 567, 350–354.

Faingold, C.L., Gehlbach, G. and Caspary, D.M. (1991) Functional pharmacology of inferior colliculus neurons. In: R. Altschuler, R. Bobbin, D. Hoffman and B. Clopton (Eds.), Neurobiology of Hearing II: The Central Auditory System. Raven Press, New York, pp. 223–251.

Friauf, E. (1991) C-fos immunohistochemistry reveals no shift of tonotopic order in the central auditory system of developing rats. Soc. Neurosci. Abstr. 17, 123.

Gerken, G.M. (1979) Central denervation hypersensitivity in the auditory system of the cat. J. Acoust. Soc. Am. 66, 721–727.

Gerken, G.M., Saunders, S.S. and Paul, R.E. (1984) Hypersensitivity to electrical stimulation of auditory nuclei follows hearing loss in cats. Hear. Res. 13, 249–259.

Goldsmith, J.D., Kujawa, S.G., McLaren, J.D. and Bledsoe, S.C. Jr. (1995) In vivo release of neuroactive amino acids from the inferior colliculus of the guinea pig using brain microdialysis. Hear. Res. 83, 80–88.

Gulley, R.L., Wenthold, R.J. and Neises, G.R. (1978) Changes in the synapses of spiral ganglion cells in the rostral anteroventral cochlear nucleus of the waltzing guinea pig. Brain Res. 158, 279–294.

Helfert, R.H., Juiz, J.M., Bledsoe, S.C., Bonneau, J.M., Wenthold, R.J. and Altschuler, R.A. (1992) Patterns of glutamate, glycine, and GABA immunolabeling in four synaptic terminal classes in the lateral superior olive of the guinea pig. J. Comp. Neurol. 323, 305–325.

Hendry, S.H. and Jones, E.G. (1988) Activity-dependent regulation of GABA expression in the visual cortex of adult monkeys. Neuron 1, 701–712.

Hoffman, G.E., Smith, M.S. and Verbalis, J.G. (1993) C-fos related immediate early gene products as markers of activity in neuroendocrine systems. Front. Neuroendocrinol 3, 173–213.

Hultcrantz, M., Snyder, R., Robscher, S. and Leake, P. (1991) Effects of neonatal deafening and chronic intracochlear electrical stimulation on the cochlear nucleus in cats. Hear. Res. 54, 272–280.

Irvine, D.R.F. and Rajan, R. (1994) Plasticity of frequency organization in inferior colliculus of adult cats with unilateral restricted cochlear lesions. Assoc. Res. Otolaryngol. Abstr. 17, 21.

Javel, E., Tong, Y.C., Shepherd, R.K. and Clark, G.M. (1987) Responses of cat auditory nerve to biphasic electrical current pulses. Ann. Otol. Rhinol. Laryngol. 96 (Suppl.128), 26–30.

Jyung, R.W., Miller, J.M. and Cannon, S.C. (1989) Evaluation of eighth nerve integrity by the electrically evoked middle latency response. Arch. Otolaryngol. Head Neck Surg. 101, 670–682.

Kiang, N.Y.S. and Moxon, E.C. (1972) Physiological considerations in artificial stimulation of the inner ear. Ann. Otol. Rhinol. Laryngol. (St Louis) 81, 714–730.

Kuwada, S., Yin, T.C.T., Syka, J., Buunen, T.J.F. and Wickesberg, R.E. (1984) Binaural interaction in low-frequency neurons in inferior colliculus of the cat. IV. Comparison of monaural and binaural response properties. J. Neurophysiol. 51, 1306–1325.

Leake, P.A. and Hradek, G.T. (1988) Cochlear pathology of long term neomycin induced deafness in cats. Hear. Res. 33, 11–34.

Lesperance, M.M., Helfert, R.H. and Altschuler, R.A. (1995) Deafness induced cell size changes in rostal AVCN of the guinea pig. Hear. Res. 86, 77–81

Lim, A., Myers, M.W., Miller, J.M. and Altschuler, R.A. (1991) C-fos expression in rat auditory nuclei following high-intensity acoustic stimulation. Soc. Neurosci. Abstr. 17, 163.

Lustig, L.R., Leake, P.A., Snyder, R.L. and Rebscher, S.J. (1994) Changes in the cat cochlear nucleus following neonatal deafening electrical stimulation. Hear. Res. 74, 29–37.

McFadden, S.L. and Willott, J.F. (1994a) Responses of inferior colliculus neurons in C57BL/6J mice with and without sensorineural hearing loss: effects of changing the azimuthal location of an unmasked pure-tone stimulus. Hear. Res. 78, 115–131.

McFadden, S.L. and Willott, J.F. (1994b) Responses of inferior colliculus neurons in C57BL/6J mice with and without sensorineural hearing loss: effects of changing the azimuthal location of a continuous noise masker on responses to contralateral tones. Hear. Res. 78, 132–148.

Miller, J.M., Altschuler, R.A., Dupont, J., Lesperance, M. and Tucci, D. (1996) Consequences of deafness and electrical stimulation on the auditory system. In: R.J. Salvi, D. Henderson, F. Fiorino and V. Colletti (Eds.), Auditory Plasticity and Regeneration, Tieman Med. Publishers, New York, pp. 378–391.

Miller, J.M., Altschuler, R.A., Niparko, J.K., Hartshorn, D.O., Helfert, R.H. and Moore, J.K. (1992) Deafness-induced changes in the central nervous system: reversibility and prevention. In: A.L. Dancer, D. Henderson, R.J. Salvi and R.P. Hamernik (Eds.), Noise-Induced hearing Loss, Mosby Year Book, St. Louis, pp. 130–145.

Nagase, S., Lim, H.H., Miller, J.M., Dupont, J. and Altschuler R.A. (1997a) Expression of fos protein in the auditory brain stem following electrical stimulation of the cochlea. I: Changes with intensity and electrode position. Submitted to Hear. Res.

Nagase, S., Miller, J.M., Prieskorn, D., Dupont, J. and Altschuler R.A. (1997b) Expression of fos protein in the auditory brain stem following electrical stimulation of the cochlea. II: Changes with deafness and stimulation. Submitted to Hear. Res.

Nordeen, K.W., Killackey, H.P. and Kitzes, L.M. (1983) Ascending projections to the inferior colliculus following unilateral cochlear ablation in the neonatal gerbil, Meriones unguiculatus. J. Comp. Neurol. 214, 144–153.

Parkins, C.W. (1989) Temporal response patterns of auditory nerve fibers to electrical stimulation in deafened squirrel monkeys. Hear. Res. 41, 137–168.

Pasic, T.R. and Rubel, E.W. (1989) Rapid changes in cochlear nucleus cell size following blockade of auditory nerve electrical activity in gerbils. J. Comp. Neurol. 283, 474–480.

Potashner, S.J., Lindberg, N. and Morest, D.K. (1985) Uptake and release of gamma-aminobutyric acid in the guinea pig cochlear nucleus after axotomy of cochlear and centrifugal fibers. J. Neurochem. 45, 1558–1566.

Rajan, R. and Irvine, D.R.F. (1996) Features of and boundary conditions for lesion induced reorganization of adult cortical maps. In: R.J. Salvi, D. Henderson, F. Fiorini and V. Colletti (Eds.), Auditory System Plasticity and Regeneration, Thieme Medical Publishers, Inc, NY, pp. 224–237.

Rajan, R., Irvine, D.R.F., Wise, L.Z. and Heil, P. (1993) Effect of partial cochlear lesions in adult cats on the representation of lesioned and unlesioned cochleas in primary auditory cortex. J. Comp. Neurol. 338, 17–49.

Rees, S., Guldner, F.H. and Aitkin, L. (1985) Activity dependent plasticity of postsynaptic density structure in the ventral cochlear nucleus of the rat. Brain Res. 325, 370–374.

Rouiller, E.M., Wan, X.S.T., Moret, V. and Liang, F. (1992) Mapping of c-*fos* expression elicited tones stimulation in the auditory pathways of the rat, with emphasis on the cochlear nucleus. Neurosci. Lett. 144, 19–24.

Rubel, E.W., Born, D.E., Deitch, J.S. and Durham, D. (1984) Recent advances toward understanding auditory system development. In: C. Berlin (Ed.), Hearing Science, College Hill, San Diego, pp. 109–157.

Rubel, E.W., Hyson, R.L. and Durham, D. (1990) Afferent regulation of neurons in brain stem auditory system. J. Neurobiol. 21, 169–196.

Ryan, A.F., Miller, J.M., Wang, Z.X. and Woolf, N.K. (1990) Spatial distribution of neural activity evoked by electrical stimulation of the cochlea. Hear. Res. 50, 57–70.

Sagar, S.M. and Sharp, F.R. (1991) Light induces a Fos-like nuclear antigen in retinal neurons. Molecular Brain Res. 7, 17–21.

Salvi, R.J., Wang, J. and Powers, N. (1996) Rapid functional reorganization in the inferior colliculus and cochlear nucleus after acute cochlear damage. In: R.J. Salvi, D. Henderson, F. Fiorini and V. Colletti (Eds.), Auditory System Plasticity and Regeneration, Thieme Medical Publishers, Inc, New York, pp. 275–296.

Sanes, D.H. (1994) Glycine receptor distribution is dependent on excitatory and inhibition afferents in the gerbil LSO. Assoc. Res. Otolaryngol. Abstr. 17, 10.

Sato, K., Houtani, T., Ueyama, T., Ikeda, M., Yamashita, T., Kumazawa, T. and Sugimoto, T. (1992) Mapping of the cochlear nucleus subregions in the rat with neuronal Fos protein induced by acoustic stimulation with low tones. Neurosci. Lett. 142, 48–52.

Sato, K., Houtani, T., Ueyama, T., Ikeda, M., Yamashita, T., Kumazawa, T. and Sugimoto, T. (1993) Identification of rat brainstem sites with neuronal fos protein induced by acoustic stimulation with pure tones. Acta Otolaryngol. (Stockh) Suppl. 500, 18–22.

Schwartz, D.R., Schacht, J., Miller, J.M., Frey, K. and Altschuler, R.A. (1993) Chronic electrical stimulation reverses deafness-related depression of electrically evoked 2-deoxyglucose activity in the guinea pig inferior colliculus. Hear. Res. 70, 463–477.

Semple, M.N. and Kitzes, L.M. (1985) Single unit responses in the gerbil inferior colliculus: different consequences of contralateral and ipsilateral auditory stimulation. J. Neurophysiol. 53, 1467–1482.

Snyder, R.L., Rebscher, S.J., Cao, K., Leake, P.A. and Kelly, K. (1990) Chronic intracochlear electrical stimulation in the neonatally deafened cat. I: Expansion of central representation. Hear. Res. 50, 7–34.

Snyder, R.L., Rebscher, S.J., Leake, P.A., Kelly, K. and Cao, K. (1991) Chronic intracochlear electrical stimulation in the neonatally deafened cat. II. Temporal properties of neurons in the inferior colliculus. Hearing Res. 56, 246–64.

Stevens, S.S. and Jones, R.C. (1939) The mechanism of hearing by electrical stimulation. J. Acoust. Soc. Am. 10, 261–269.

Urban, G.P. and Willott, J.F. (1979) Response properties of neurons in inferior colliculi of mice made susceptible to audiogenic seizures by acoustic priming. Exp. Neurol. 63, 229–243.

van den Honert, C. and Stypulkowski, P.H. (1984) Physiological properties of the electrically stimulated auditory nerve. II. Single fiber recordings. Hear. Res. 14, 225–243.

van den Honert, C. and Stypulkowski, P.H. (1987) Temporal response patterns of single auditory nerve fibers elicited by periodic electrical stimuli. Hear. Res. 29, 207–222.

Vischer, M.W., Hausler, R. and Rouiller, E.M. (1994) Distribution of Fos-like immunoreactivity in the auditory pathway of the Sprague-Dawley rat elicited by cochlear electrical stimulation. Neurosci. Res. 19, 175–185.

Webster, M. and Webster, D.B. (1981) Spiral ganglion neuron loss following organ of Corti loss. A quantitative study. Brain Res. 212, 17–30.

Webster, D.B. (1988) Conductive hearing loss affects the growth of the cochlear nuclei over an extended period of time. Hear. Res. 32, 185–192.

Wiesel, T.N. and Hubel, D.H. (1963) Single-cell responses in striate cortex of kittens deprived of vision in one eye. J. Neurophysiol. 26, 1003–1017.

Willott, J.F. (1994) Auditory system plasticity in the adult C57BL/6J mouse. (Abstract) Auditory Plasticity and Regeneration: Scientific and Clinical Implications. Terme di Comano, Trento, Italy, May 4–7.

Willott, J.F., Bross, L.S. and McFadden, S.L. (1994) Morphology of the cochlear nucleus in CBA/J mice with chronic severe sensorineural cochlear pathology induced during adulthood. Hear. Res. 74, 1–21.

Willott, J.F., Demuth, R.M. and Lu, S.M. (1984) Excitability of auditory neurons in the dorsal and ventral cochlear nuclei of DBA/2 and C57BL/6 mice. Exp. Neurol. 83, 495–506.

Zappia, J.J. and Altschuler, R.A. (1989) Evaluation of the effect of ototopical neomycin on spiral ganglion cell density in the guinea pig. Hear. Res. 40, 29–38.

Zhang, J.S., Haenggeli, C.A., Tempini, A., Vischer, M.W., Moret, V. and Rouiller, E.M. (1996) Electrically induced Fos-like immunoreactivity in the auditory pathway of the rat : Effects of survival time, duration, and intensity of stimulation. Brain Res. Bull. 39, 75–82.

48

NEURAL PLASTICITY IN PATIENTS WITH TINNITUS AND SENSORINEURAL HEARING LOSS

Alan H. Lockwood, Richard J. Salvi, Mary Lou Coad, David S. Wack, and Brian W. Murphy

The Centers for Positron Emission Tomography and Hearing and Deafness, and the Hearing Research Laboratory
Departments of Neurology, Nuclear Medicine, Communicative Disorders and Sciences, and Department of Veterans Affairs
Western New York Health Care System and State University of New York
University at Buffalo
Buffalo, New York

1. INTRODUCTION

Hearing loss and tinnitus and are two common problems that both increase in prevalence with advancing age. The self-reported incidence of hearing loss reaches 35% by age 75 (Nadol, Jr. 1993). This incidence may be low by a factor of two or more. Experience at the Rochester Institute of Technology suggests that 90% or more of adults in their 70's have hearing loss when standard audiometric tests are used as the test measure (R. Frisina, personal communication). Public health survey data show that the incidence of severe, disabling tinnitus also rises with age and parallels the increase in the incidence of hearing loss (National Center for Health Statistics, 1968). These two problems account for a substantial amount of disability due to the impairments in communication and the attendant psychological impact of isolation, depression, and other symptoms (O'Connor et al. 1987; Hallam et al. 1984).

The origin of tinnitus is unknown. The epidemiological data linking sensorineural hearing loss, cochlear injury, and tinnitus has led some to speculate that tinnitus arises as a result of cochlear injury and is due to abnormal discharges that originate within the cochlea (Kiang et al. 1970). However, since tinnitus may develop as the consequence of damage to the auditory portion of the VIIIth cranial nerve, particularly after surgery for acoustic neuromas, and frequently persists after surgical transection of the nerve, it is likely that central auditory structures are involved in the production of tinnitus.

We have identified a small group of subjects with tinnitus and high frequency sensorineural hearing loss who possess the unique ability to exert substantial volitional control over the self-reported loudness of their tinnitus. Each of these patients indicated that a jaw clench, or similar voluntary contraction of muscles of mastication, affected the loudness of their tinnitus. They were equally divided in terms of whether these oral-facial maneuvers (OFM) increased or decreased tinnitus loudness. One patient reported a long list of maneuvers that affected her tinnitus including sucking back on a tooth with her tongue and the application of pressure to the top of her head with her hand or the hand of another person. This ability to modulate the loudness of tinnitus by motor or somatosensory activity provides additional compelling evidence for significant functional reorganization in the central auditory system.

Each of these patients experienced the sudden onset of tinnitus that had persisted, with little change for several years. At the time of the study, none had found significant relief from these incessant sounds. All of them indicated that tinnitus had become a major disrupting factor in their lives and occupied a central position in their thoughts.

We hypothesized that changes in the loudness of tinnitus would be associated with changes in neural activity in the brain that would be reflected by changes in blood flow to the portions of the brain that mediate tinnitus (Posner et al. 1988). We further hypothesized that, since these patients all had hearing loss, the brain regions that were activated by external tones would activate different brain regions than would be expected in subjects with normal hearing.

2. FUNCTIONAL BRAIN MAPPING

A typical person who weighs 70 kg has a brain that weighs approximately 1.5 kg. Although the brain accounts for only a small portion of the total body weight, it receives 15 - 20% of the cardiac output. This disproportionate distribution of blood flow is due to several factors. Although the brain does not perform "visible work" like the movement of the heart or skeletal muscle, nevertheless it "works" hard all of the time pumping ions across cell membranes. This creates a high energy demand that mandates the existence of a constant supply of oxygen and glucose, the sole metabolic substrates to support normal brain function. The brain is further handicapped by the total absence of significant energy stores. As a consequence, even relatively small changes in the rate of energy expenditure by the brain must be accompanied by corresponding changes in the delivery of metabolic substrates. Thus, by measuring the rate at which blood flows to different brain regions under different experimental conditions, it is possible to map brain regions required to perform the task in question (Posner et al. 1988).

For our studies, we have measured cerebral blood flow (CBF) using positron emission tomography (PET) employing ^{15}O-water as a tracer. After an intravenous bolus injection of ^{15}O-water, the tracer is distributed to various organs in proportion to their blood flow. During a typical experiment, the amount of ^{15}O measured in the brain is a linear function of CBF (Ginsberg et al. 1982; Howard et al. 1983). Thus, images that faithfully depict the tracer content in the brain during an appropriate time interval are a reflection of CBF and brain activity. These images may be used to compare functional states. (Note: the physical principles that underlie the production of PET images are critical elements in this process, but are not considered here.)

2.1 Statistical Parametric Mapping (SPM)

The technique of SPM was developed by Friston, and his associates, in response to the challenge of developing a statistically rigorous process for comparing brain activity on a pixel-by-pixel basis under different experimental conditions (Friston et al. 1995; Friston and Frackowiak, 1991). This complex, multistep process can be summarized as follows. First, the image of brain tracer content (in our case ^{15}O-water) must be edited to remove extra-cerebral activity in the scalp and muscles surrounding the cranial vault (the program considers all activity to be in the brain and does not distinguish brain from non-brain structures). This yields a data set consisting of multiple tomographic images representing the quantitative distribution of the tracer in the brain. In a multi-task study, a separate set of images exists for each task. Tasks may be repeated to increase the statistical power of the process. Next, each individual data set (i.e., a single scan) is examined for evidence of patient movement and realigned to remove effects of movement. The resulting images are then reformatted into standard stereotaxic space without perturbing quantitative relationships among brain structures. The stereotaxic atlas of brain anatomy by Talairach and Tournoux serves as the standard frame of reference (Talairach and Tournoux, 1988). After smoothing these images to minimize the effects of individual differences in brain anatomy and to further improve the signal-to-noise characteristics of the data, a statistical analysis is conducted, based on the general linear model, to identify brain regions where there is a significant variation in tracer content and hence brain activity.

The form of the statistical analysis is determined by weighting each task by a value (usually 0, 1 or -1) to specify the SPM contrast. Thus, if one had three tasks, 1 = resting state, 2 = tone presentation, and 3 = OFM, in a single group, to subtract task 1 activity from task 2 activity the contrast would be: [-1, 1, 0]. An alternative form of notation would be: group(tone - rest) where group describes the study population, e.g., normal. The resulting t statistic is converted to a Z score (mean Z is zero, standard deviation is one) and Z values are mapped into stereotaxic space. A Z threshold is selected so that only those pixels that exceed a predetermined threshold are displayed visually. The resulting SPM {Z} images have x, y, and z coordinates that conform to the Talairach system and display Z scores rather than density or signal strength as used in X-ray CT, or MRI systems. The SPM {Z} images may be superimposed on MRI images or, alternatively, stacked with a subsequent projection of the most significant pixels on to coronal, transaxial, and sagittal planes. An example of an SPM {Z} image, projected on to a sagittal plane is shown in Figure 1. In addition to images, the program returns specific coordinates for pixels with the largest Z scores in regions above the threshold (typically Z = 2.33, equivalent to an omnibus $P = 0.01$) along with the Z scores for these pixels and p values at the threshold chosen by the investigator and after correction for multiple comparisons. The correction for multiple comparisons is complex and determined by the effective number of pixels in the total image set (typically there are 60,000 - 70,000 pixels) and the degree of independence of adjacent pixels (a function of image resolution).

2.2 Experimental Conditions

Written informed consent was obtained from all subjects. Six subjects with normal hearing served as controls. Four patients with high frequency sensorineural hearing loss and the unique ability to exert substantial voluntary control over tinnitus loudness were identified and enrolled in the study. The patients all reported hearing a unilateral, continuous, high-pitched ringing (one in the left, three in the right ear). Each of them was able to modulate the loudness of their tinnitus by performing a voluntary contraction of oral and facial muscles, including the masseter, i.e., an OFM. To characterize their phantom sounds, patients adjusted the frequency and intensity of an external tone to match the pitch and loudness of their tinnitus. Matches occurred at frequencies near the peak of the hearing loss and at sound levels 5–10 dB above threshold. Patients had mild-to-severe, high-frequency (greater than 2000 Hz), cochlear hearing loss (30 - 70 dB Hearing Level), normal middle ear function, no evidence of

Sagittal projection, SPM {Z}

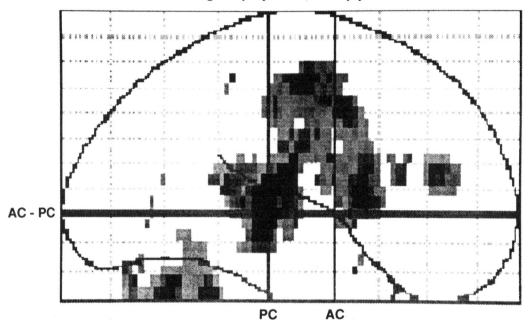

Figure 1. Sites of significant increase in CBF after performing an OFM to cause an increase in tinnitus loudness. Expected sites of increased CBF and neural activity in sensory-motor cortex are shown above the Sylvian fissure and increased CBF and neural activity, corresponding to increases in tinnitus loudness are shown in auditory cortical areas in the superior temporal gyrus, below the Sylvian fissure.

central auditory abnormalities, and no spontaneous otoacoustic emissions. On a separate day, they returned for PET scans according to one of three protocols: 1) for three controls; 500 Hz tones right ear, followed by 2000 Hz tones right ear, followed by 3 resting scans alternating with 3 scans performed during a jaw clench, the common feature in the patient's OFMs (8 scans total); 2) four patients had the same sequence as in (1) with tone stimuli delivered to the ear in which tinnitus was reported (8 scans total); 3) three additional controls had resting and tone sequence scans only (3 scans total). Each task sequence began with the simultaneous intravenous injection of the tracer (70 mCi or less of ^{15}O-water), initiation of the activation procedure (e.g., OFM, tone sequence, etc.) and the collection of the PET data. Multiple frames of PET data were acquired so that the time of arrival of the bolus of activity in the brain could be determined and an image created that consisted of the next 60 seconds of PET data. Tone bursts were presented via Etymotic insert earphones (Etymotic, Elk Grove Village IL) using a Neuroscan Stim system (Neuroscan, Hearndon VA). Scans were edited to remove extra cerebral activity associated with muscle and vascular structures and subjected to SPM analysis.

3. ACTIVATION OF BRAIN BY TONE PULSES

Resting state scans were compared to scans obtained as tone pulses were presented to the right ear with the SPM contrasts: controls(500 Hz - rest) and controls(2000 Hz - rest). The tone pulses resulted in bilateral activation of the transverse temporal gyrus and adjacent portions of the superior temporal gyrus. This bilateral activation of auditory cortex after unilateral sound stimulation provides convincing confirmatory evidence for a complex central auditory system with many decussations. Similar tone activation studies were performed in the three patients who reported right ear tinnitus, i.e. 500 Hz and 2000 Hz stimuli were presented to the right ear only. A difference of differences analysis was performed to identify brain regions activated by the tones in patients, but not in controls by SPM contrasts = [patient(500 Hz - rest) - control(500 Hz - rest)] and [patient(2000 Hz - rest) - control(2000 Hz - rest)]. In the patients, more extensive portions of the auditory cortex were activated by the tone sequences than in controls. This effect was more notable at 2000 Hz than 500 Hz.

4. EFFECTS OF OFMS IN TINNITUS PATIENTS

To identify brain regions where CBF increased in the patients in whom the OFM led to an increase in tinnitus loudness, the following SPM contrast was established: patient(OFM - rest). To identify brain regions where CBF fell in the patients in whom the OFM led to a decrease in tinnitus loudness the following SPM contrast was established: patient(rest - OFM). Both contrasts identified brain regions where CBF rose or fell, respectively, as tinnitus loudness increased or decreased. In both cases, auditory cortical areas were included in regions where the SPM {Z} images showed significant effects. The effect was most pronounced in an analysis of the pooled data from the two subjects who reported decreases in tinnitus loudness during an OFM. The result from the analysis of the data from a single patient are illustrated by Figure 1, which displays a lateral or sagittal projection of an SPM {Z} image from a subject who experienced an increase in tinnitus loudness during the OFM. In the image, two major sites of activity are seen. The first region is confined to the sensory-motor strips (superimposed in this image) and is attributable to the expected increase in CBF due to the continuous voluntary muscle contraction that characterized the OFM. The second is attributable to an increase in CBF in the auditory cortex and is seen as the shaded area that overlies the posterior portion of the superior temporal gyrus.

Anomalous activity in the limbic system, primarily the hippocampus, was identified by two analyses of data from the tinnitus patients. In the (2000 Hz - rest) contrast, activation was seen in the hippocampus of patients that was not evident in data from controls. Similarly, when scans from controls stimulated at 2000 Hz were subtracted scans from patients stimulated at 2000 Hz (SPM contrast: 2000 Hz(patient - control)), a region of significantly increased activity in the hippocampus of the patients was evident. Thus two independent analyses reveal activation in the limbic system of tinnitus patients that is not present in controls.

5. CONCLUSIONS

Our data provide convincing evidence that regional CBF measurements in normal controls and patients can be used to map the functional anatomy of the auditory system. In normal subjects and patients with tinnitus, we were able to demonstrate bilateral activation of the auditory system, particularly auditory cortex, by pure tones played into a single ears. Furthermore, we show that comparisons between normal subjects and patients with abnormalities of the auditory system are feasible. This is shown most clearly by the demonstration that 2000 Hz tones activate more extensive portions of the auditory cortex and adjacent temporal lobe sites in our patients than in controls. These data provide robust evidence for more extensive activity in the brains of our patients. Thus, our PET data from humans with high frequency sensorineural hearing loss and tinnitus confirm data from animals showing invasion of deafferented portions of the cortex by neural impulses that are produced by portions of the cochlea immediately adjacent to the site of cochlear injury(Robertson and Irvine, 1989). We are unable to determine whether these changes are the result of hearing loss, tinnitus, or, as is most likely, a combination of these two factors.

Our data also indicate that PET can be used to objectively identify the spontaneous neural activity in the central auditory system of patients with tinnitus. This is most clearly shown by the results of SPM analysis of the data from the three patients who all reported right ear tinnitus that responds to an OFM. This analysis showed activity in only the left transverse temporal gyrus as the loudness of tinnitus was altered by an OFM. This unilateral, left-sided effect differs from the bilateral activation that was produced by the presentation of tones into the ear in which these subjects reported their tinnitus.

Functional neuroimaging promises to be a powerful tool for the evaluation of patients with tinnitus. It provides a quantitative and objective measure of a phenomenon that has to date defied attempts at rigorous analysis. This approach offers an objective means to assess therapy through the identification of specific anatomical sites, neural pathways, and by inference, the neurotransmitter systems, affected by the mechanisms that produce tinnitus.

ACKNOWLEDGMENTS

We are grateful for the efforts of our many colleagues in the Center for PET. Supported by the American Tinnitus Association and the State University of New York, University at Buffalo.

6. REFERENCES

Friston, K. J. and Frackowiak, R. S. J. (1991) Imaging functional anatomy. In: Lassen NA, Ingvar DH, Raichle ME, and Friberg L.Brain Work and Mental Activity, Munksgaard, Copenhagen.p. 267–277.

Friston, K. J., Holmes, A. P., Worsley, K. J., Poline, J.-P., Frith, C. D., and Frakowiak, R. S. J. Statistical parametric maps in functional imaging: a general linear approach. Hum. Brain Mapp. 2, 189–210.1995.(Abstract)

Ginsberg, M. D., Lockwood, A. H., Busto, R., Finn, R. D., Butler, C. M., Cendan, I. E., and Goddard, J. E. (1982) A simplified *in vivo* autoradiographic strategy for the determination of regional cerebral blood flow by positron-emission tomography: theoretical considerations and validation studies in the rat. J. Cereb. Blood Flow Metabol. 2, 89–98.

Hallam, R., Rachnman, S., and Hinchcliffe, R. (1984) Psychological aspects of tinnitus. In: Rachman S.Contributions to Medical Psychology, Vol. 3, Pergamon Press, Oxford.p. 31–53.

Howard, B. E., Ginsberg, M. D., Hassel, W. R., Lockwood, A. H., and Freed, P. (1983) On the uniqueness of cerebral blood flow measured by the *in vivo* autoradiographic technique and positron-emission tomography. J. Cereb. Blood Flow Metabol. 3, 432–441.

Kiang, N. Y. S., Moxon, E. C., and Levine, R. A. (1970) Auditory nerve activity in cats with normal and abnormal cochleas. In: Wolstenholme GEW and Knight J.Sensorineural hearing loss, J. & A. Churchill, London.p. 241–273.

Nadol, J. B.,Jr. (1993) Hearing loss. New Engl. J. Med. 329, 1092–1102.

National Center for Health Statistics Vital and Health Statistics, Series 11 No. 32, 1968.

O'Connor, S., Hawthorne, M., Britten, S. R., and Webber, P. (1987) Identification of psychiatric morbidity in a population of tinnitus sufferers. J. Laryngol. Otol. 101, 791–794.

Posner, M. I., Petersen, S. E., Fox, P. T., and Raichle, M. E. (1988) Localization of cognitive operations in the human brain. Science. 240, 1627–1631.

Robertson, D. and Irvine, D. R. F. (1989) Plasticity of frequency organization in auditory cortex of guinea pigs with partial unilateral deafness. J. Comp. Neurol. 282, 456–471.

Talairach, J. and Tournoux, P. Co-Planar Stereotaxic Atlas of the Human Brain. Stuttgart and New York: Georg Thieme Verlag, 1988.

49

HAIR CELL LOSS AND SYNAPTIC LOSS IN INFERIOR COLLICULUS OF C57BL/6 MICE

Relationship to Abnormal Temporal Processing

Vlasta P. Spongr,[1] Joseph P. Walton,[2] Robert D. Frisina,[2] Ann Marie Kazee,[3] Dorothy G. Flood,[2] and Richard J. Salvi[1]

[1]Hearing Research Lab
University of Buffalo
Buffalo, New York 14214
[2]University of Rochester
Rochester, New York 14642
[3]SUNY Health Science Center
Syracuse, New York 13210

1. INTRODUCTION

The C57BL/6 mouse is an extremely popular animal model of presbycusis because of its relatively short life span and genetic pattern of high-frequency sensorineural hearing loss (SNHL) (Henry and Chole, 1980; Willott, 1984; Erway et al., 1993) that resembles the age-related hearing loss seen in humans (Nadol, 1993). Presbycusis and SNHL have traditionally been thought of as peripheral disorders that mainly result in the loss of sensitivity and frequency selectivity (Schmiedt and Schulte, 1992). However, recent studies suggest that peripheral pathologies can lead to functional and anatomical changes in the central nervous system (Hall, 1974, 1976; Wightman, 1982; Morest and Bohne, 1983; Willott, 1984; Salvi and Arehole, 1985; Arehole et al., 1987a; Robertson and Irvine, 1989; Salvi et al., 1990) that may contribute to some of the hearing deficits associated with SNHL and presbycusis.

In addition to the loss of sensitivity, there is growing awareness that auditory temporal resolution may be compromised in listeners with SNHL (Fitzgibbons and Wightman, 1982; Salvi et al., 1982; Wightman, 1982). Psychophysical studies indicate that SNHL prolongs the recovery from forward masking (Danahaer et al., 1978; Nelson and Freyman, 1987); however, the neural mechanisms responsible for this are poorly understood. Physiological correlates of forward masking have been studied at several levels of the auditory pathway in normal subjects (Harris and Dallos, 1979; Kramer and Teas, 1982; Arehole et al., 1987a), but relatively little is known about the changes that occur in physiological measures of forward masking in subjects with SNHL. One study found no change in forward masking of the compound action potential of the auditory nerve in animals with noise-induced hearing loss (Gorga and Abbas, 1981). However, the time constant of forward masking of the inferior colliculus evoked response was prolonged in animals with noise-induced threshold shift (Arehole et al., 1987b, 1989). These results suggest that cochlear damage may have a more deleterious effect on the time constant of recovery in the central auditory pathway than in the periphery. The purpose of the present study was twofold. First, to determine if the high-frequency SNHL in the C57BL/6 mouse prolongs the forward masking recovery function of wave P5 of the auditory brainstem response (ABR). Second, to relate the functional deficits in the ABR to morphological changes in the cochlea.

2. METHODS

2.1. ABR

Healthy, C57BL/6 mice, 1 and 8 months of age (n = 10/group), were used as subjects. ABR measurements were obtained from 1-month old and 8-month old mice as described previously (Walton et al., 1995). Animals were tranquilized (taractan), restrained and maintained at 38° C. Subcutaneous electrodes were placed in the vertex (noninverting input), right mastoid (inverting input) and tail (indifferent input). EEG activity was amplified, filtered and averaged (n = 500) for 10 ms. Stimuli were presented through a speaker (Panasonic AA102) at 0° azimuth. The ABR audiogram was measured from 4 to 50 kHz with tone bursts (positive peaks P1-P5). The ABR forward masking function was measured at 12 kHz using a fixed level (20 dB above threshold) probe tone and a forward masker of variable level which preceded the onset of the probe tone by a fixed ΔT. The intensity of the forward masker needed to produce a 50% reduction in the amplitude of wave P5 evoked by the fixed level probe tone was defined as the masked threshold (Lm). Forward masking functions were constructed by plotting the masked threshold as a function of ΔT and a time constant was fit to the data (non-linear least squares iterative procedure, Marquardt-Levenberg algorithm) using an exponential model of the form $Lm = A\ e^{(\Delta T/\tau)}$ (Arehole et al., 1987a; Walton et al., 1995).

2.2. Hair Cell Loss

Mean (n=10) inner hair cell (IHC) and outer hair cell (OHC) loss was determined for each group of 1, 3, 8, 18, and 26-month old C57BL/6 mice using procedures described previously (Spongr et al., 1992). Animals were anesthetized and perfused transcardially (1% glutaraldehyde, 4% paraformaldehyde). Cochleas were removed, post-fixed with osmium tetroxide, decalcified and the organ of Corti dissected out as a surface preparation. Cochleograms were constructed by plotting percent hair cell loss as a function of percent distance from the apex on the basis of the normal hair cell density distribution (Spongr et al., 1994).

2.3. Inferior Colliculus-Light Microscopy

After the transcardial perfusion, the brains (n=15) were removed and cut in the parasagittal and horizontal planes to divide the inferior colliculus into 4 quadrants (Kazee et al., 1995). Blocks were embedded in Spurr epoxy resin and sectioned. Thick sections (1.5 μm) were stained with toluidine blue. Morphometric evaluation of the principal cells of the central nucleus of the inferior colliculus (ICC) was performed on three age groups

(3, 6–18 and 24 months). Measurements were made of neuronal, nuclear and nucleolar area and major and minor axis diameter of principal cells with visible nucleoli.

2.4. Inferior Colliculus-Electron Microscopy

Blocks, trimmed to include only the ICC, were sectioned. Silver thin sections were stained with uranyl acetate and lead citrate and examined on a Hitachi H7100 electron microscope. Entire principal neurons were photographed at low (X9000) and high (X36000) magnification. Measurements were made of somatic membrane perimeter, number and length of synaptic appositions, and synaptic terminal area by tracing the images into a computer using a graphics tablet (Kazee et al., 1995). Only those principal cells which contained visible nuclei were analyzed.

3. RESULTS AND DISCUSSION

3.1. Hair Cell Loss

A cochleogram showing the percentage of IHC and OHC loss as a function of percent distance from the apex was computed for each animal. Data from individual animals were used to compute average (n = 10) cochleograms for 1, 3, 8, 18 and 26-month old C57BL/6 mice. Figures 1 and 2 show the percentage of missing OHCs and IHCs in 20% segments of the cochlea for the five age groups. Hair cell loss progressed along a base-to-apex gradient with age and OHC loss was more severe than IHC loss (Figures 1–2). At 3 months of age approximately 60% and 30% of the OHCs and IHCs respectively were missing in the basal 20% of the cochlea. At 8 months of age, OHC loss ranged from

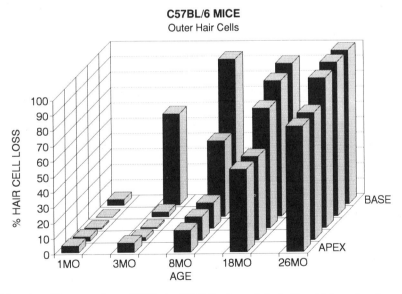

Figure 1. Mean (n = 10) percent loss of OHCs in C57BL/6 mice at 1, 3, 8, 18 and 26 months of age. Each bar shows the percent loss in 20% segments with apical-to-base segments arranged from front-to-back within each age group.

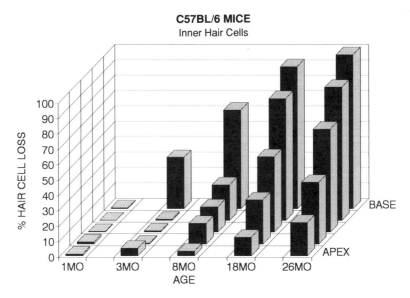

Figure 2. Mean (n = 10) percent loss of IHCs in C57BL/6 mice at 1, 3, 8, 18 and 26 months of age. Each bar shows the percent loss in 20% segments with apical-to-base segments arranged from front-to-back within each age group

40–90% in the basal 40% of the cochlea while IHC loss varied from 20–60%. At 18 months of age, OHC loss was approximately 55% in the apical half of the cochlea and 85–90% in the basal half of the cochlea and IHC loss ranged from 40–90% over the basal 60% of the cochlea. At 26 months of age, more than 80% of the OHCs were missing throughout the cochlea while IHC decreased from almost 100% near the base to approximately 20% near the apex.

3.2. ABR Thresholds

ABR thresholds in 1-month old C57BL/6 mice were similar to those of 1-month old CBA mice which retain normal hearing well into adulthood (Walton et al., 1995). Tone burst thresholds were approximately 25 dB SPL in the 8–12 kHz range and increased to approximately 50 dB at 4 kHz and 50 kHz, respectively. By 8 months of age, the hearing loss at 50 kHz was so great that threshold could not be measured. Hearing losses at 4, 8, 10, 12, 16, 24 and 36 kHz were approximately 18, 15, 12, 28, 28, 37 and 42 dB, respectively (SD: 5–8 dB). The significant, high frequency hearing loss seen in the 8-month old animals parallels the pattern of OHC and IHC loss seen at this age (Figures 1–2).

3.3. ABR Forward Masking

ABR forward masking functions were measured at 12 kHz in 1-month old and 8-month old C57BL/6 mice and a time constant, τ, was fit to the exponential recovery functions for forward masking. The time constants ranged from 66 ms to 98 ms in the 1-month old, normal hearing mice ($r^2 > 0.75$). In contrast, time constants in the 8-month old C57BL/6 mice ranged from 119 ms to 350 ms ($r^2 > 0.74$). Figure 3 shows the forward masking recovery functions from the 1-month old C57BL/6 mouse with the longest ($\tau =$

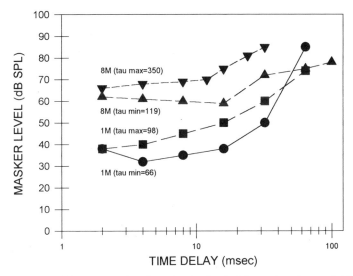

Figure 3. ABR forward masking recovery functions showing level of forward masker needed to produce a 50% reduction in the amplitude of wave P5 of the ABR as a function of the time delay (ΔT) between the masker and probe tone. Forward masking functions shown for the 1-month (circle and square) and the 8-month old (diamonds) C57BL/6 mice with the longest and the shortest time constant. Tau value (τ) is indicated next to each curve.

98 ms) and the shortest time constant ($\tau = 68$ ms). Masked thresholds are relatively constant for ΔTs between 2 and 8 ms, but increase significantly at longer ΔTs. Figure 3 also shows the forward masking recovery function from the 8-month old C57 mouse with the longest time constant ($\tau = 350$ ms) and the shortest time constant ($\tau = 119$ ms). Masked threshold remained nearly constant up to ΔTs of 16 ms and then showed only a moderate increase at longer ΔTs. These results demonstrate that the recovery from forward masking has increased significantly in the 8-month old C57BL/6 mice even though there was only a moderate 25 dB threshold elevation.

3.4. Light Microscopic Analysis

The physiological changes described above were accompanied by significant morphological changes in the ICC. Principal neurons in the ICC of 3-month old mice had mainly oval shaped soma although some rounded and fusiform soma were also seen. Typically two large dendrites at opposite poles of the soma were oriented parallel to the afferent fiber laminae in the inferior colliculus. Multipolar neurons, with numerous, randomly oriented dendrites were also present in the ICC. In young, 3 month old mice, the average neuronal area was 211 μm^2 (Table 1), average nuclear area was 66 μm^2 and average nucleolar area was 9 μm^2. In middle age and old mice, the soma of the neurons were smaller than in young animals. Quantitative morphological analysis of light microscopic data revealed a statistically significant decrease ($p < 0.01$) in neuronal area by middle-age (6–18 months) (Table 1) and a significant decrease ($p < 0.05$) in nuclear area and nucleolar area (data not shown) by old-age (24 months).

Table 1. Mean values and number of neurons evaluated for young, middle age and old C57BL/6 mice. Measures showing a statistically significant change with age indicated by ** (p < 0.01) and * (p < 0.05)

C57BL/6 mouse experimental groups	Young(3 M) (n = 5)	Middle age(6-18 M) (n = 6)	Old(24 M) (n = 4)
Neuronal area (LM)			
n=neurons	284	219	143
microns^2	211	161**	145**
Synapses/neuron			
n=neurons	70	66	72
n=synapses	3.77	1.37**	1.37**
Somatic membrane			
perimeter in microns	42.00	41.63	34.59*
Synaptic terminal			
area in microns^2	0.70	0.50	0.34

** Significant difference from young (p < 0.01)
* Significant difference from young (p < 0.05)

3.5. Electron Microscopic Analysis

Symmetric axosomatic synapses were frequently seen on principal neurons in the ICC. Symmetric synapses contained mainly oval shaped synaptic vesicles, but in some cases, the vesicles were small and round or pleomorphic and irregularly shaped. Symmetric axosomatic synapses often had 2–3 synaptic contacts per terminal. Asymmetric axosomatic synapses, which were less common in the ICC than the symmetric type, usually contained densely packed, round synaptic vesicles and occasionally a few dense core vesicles. Quantitative analysis showed that in 3 month old C57BL/6 mice, the average perimeter of the soma was 42 μm, the average number of synapses per neuron was 3.77, and the mean area of the synaptic terminal was 0.70 μm^2 (Table 1). By middle age, there was a significant decrease (p < 0.01) in the number of synapses per neuron. The loss was evident in both asymmetric and symmetric type synapses. In old animals, there was also significant decrease in somatic membrane perimeter (p < 0.05).

The aged C57BL/6 mouse shows a significant reduction in neuronal area and a dramatic reduction in the number of axosomatic synapses on neurons in the ICC. The massive loss of hair cells in the cochlea may be one factor that contributes to hearing loss and the anatomical changes seen in the ICC. It is interesting to note that the synaptic loss was evident as early as 6–18 months of age. However, the synaptic loss did not vary significantly along the high-to-low frequency tonotopic axis of the ICC. By contrast, hair cell loss showed a distinct tonotopic gradient, being much more severe in the basal, high frequency region than in the apical, low frequency region of the cochlea. This widespread synaptic loss may be one factor that contributes to impaired temporal resolution and the increase in the time constant for forward masking of the mouse ABR.

4. ACKNOWLEDGMENTS

Research supported by NIH Grant: P01 AG 09524 and The International Center for Hearing and Speech Research, RICHS, Rochester, NY.

5. REFERENCES

Arehole, S., Salvi, R. J., Saunders, S. S. and Hamernik, R. P. (1987a) Evoked response 'forward masking' functions in chinchillas. Hear. Res. 30, 23–32.

Arehole, S., Salvi, R. J., Saunders, S. S. and Henderson, D. (1987b) Evoked response 'forward masking' patterns in chinchillas with temporary hearing loss. Hear. Res. 27, 193–205.

Arehole, S., Salvi, R. J., Saunders, S. S. and Gratton, M. A. (1989) Evoked-response forward-masking functions in chinchillas with noise-induced permanent hearing loss. Audiology 28, 92–110.

Danahaer, E., Wilson, M. and Pickett, J. (1978) Backward and forward masking in listeners with severe sensorineural hearing loss. Audiology 17, 324–338.

Erway, L. C., Willott, J. F., Archer, J. R. and Harrison, D. E. (1993) Genetics of age-related hearing loss in mice. I. Inbred and F1 hybrid strains. Hear. Res. 65, 125–132.

Fitzgibbons, P. J. and Wightman, F. L. (1982) Gap detection in normal and hearing-impaired listeners. J. Acoust. Soc. Am. 72, 761–765.

Gorga, M. P. and Abbas, P. J. (1981) AP measurements of short-term adaptation in normal and in acoustically traumatized ears. J. Acoust. Soc. Am. 70, 1310–1321.

Hall, J. G. (1974) Pathological changes in second order auditory neurons after noise exposure and peripheral denervation. Scand. Audiol 4, 31–38.

Hall, J. G. (1976) The cochlear nuclei in monkeys after dehydrostreptomycin or noise exposure. Acta Oto-Laryngol. 81, 344–352.

Harris, D. and Dallos, P. (1979) Forward masking of auditory nerve fibers responses. J. Neurophysiol. 42, 1083–1107.

Henry, K. R. and Chole, R. A. (1980) Genotypic differences in behavioral, physiological and anatomical expressions of age-related hearing loss on the laboratory mouse. Audiology 19, 369–383.

Kazee, A. M., Han, L. Y., Spongr, V. P., Walton, J. P., Salvi, R. J. and Flood, D. G. (1995) Synaptic loss in the central nucleus of the inferior colliculus correlates with sensorineural hearing loss in the C57BL/6 mouse model of presbycusis. Hear. Res. 89, 109–120.

Kramer, S. J. and Teas, D. C. (1982) Forward masking of auditory nerve (N1) and brainstem (wave V) responses in humans. J. Acoust. Soc. Am. 73, 795–803.

Morest, D. K. and Bohne, B. A. (1983) Noise-induced degeneration in the brain and representation of inner and outer hair cells. Hear. Res. 9, 145–151.

Nadol, J. B. (1993) Hearing Loss. New Eng. J. Med 15, 1092–1102.

Nelson, D. A. and Freyman, R. L. (1987) Temporal resolution in sensorineural hearing impaired listeners. J. Acoust. Soc. Am. 81, 709–720.

Robertson, D. and Irvine, D. R. F. (1989) Plasticity of frequency organization in auditory cortex of guinea pigs with partial unilateral deafness. J. Comp. Neurol. 282, 456–471.

Salvi, R. J. and Arehole, S. (1985) Gap detection in chinchillas with temporary high-frequency hearing loss. J. Acoust. Soc. Am. 77, 1173–1177.

Salvi, R. J., Giraudi, D., Henderson, D. and Hamernik, R. P. (1982) Detection of sinusoidally amplitude modulated noise by chinchilla. J. Acoust. Soc. Am. 71, 424–429.

Salvi, R. J., Saunders, S. S., Gratton, M. A., Arehole, S. and Powers, N. (1990) Enhanced evoked response amplitudes in the inferior colliculus of the chinchilla following acoustic trauma. Hear. Res. 50, 245–258.

Schmiedt, R. A. and Schulte, B. A. (1992) Physiologic and histopathologic changes in quiet- and noise-aged gerbil cochleas. In: A. Dancer, D. Henderson, R. J. Salvi and R. P. Hamernik (Eds.), Noise-Induced Hearing Loss, Mosby Year Book, pp. 246–256.

Spongr, V. P., Boettcher, F. A., Saunders, S. S. and Salvi, R. J. (1992) Effects of noise and salicylate on hair cell loss in the chinchilla cochlea. Arch. Otolaryngol. Head Neck Surg. 118, 157–164.

Spongr, V., Powers, N., Flood, D. and Salvi, R. (1994) Difference in hair cell density distribution in the cochlea of young CBA/HSD and C57BL/HSD mice. Abstr. Assoc. Res. Otolaryngol. 17, 91.

Walton, J. P., Frisina, R. D. and Meierhans, L. R. (1995) Sensorineural hearing loss alters recovery from short-term adaptation in the C57BL/6 mouse. Hear. Res. 88, 19–26.

Wightman, F. L. (1982) Psychoacoustic correlates of hearing loss. In: R. P. Hamernik, D. Henderson and R. J. Salvi (Eds.), New Perspectives on Noise-Induced Hearing Loss, Raven Press, pp. 375–394.

Willott, J. F. (1984) Changes in frequency representation in the auditory system of mice with age-related hearing impairment. Brain Res. 309, 159–162.

50

SOUND OVER-EXPOSURE EFFECTS ON 2f$_1$-f$_2$ DISTORTION-PRODUCT OTOACOUSTIC EMISSION ONSET LATENCIES IN RABBIT

Cuneyt O. Kara, Brenda L. Lonsbury-Martin, David Jassir,
Barden B. Stagner, and Glen K. Martin

University of Miami Ear Institute
Miami, Florida

1. INTRODUCTION

It is well-established that acoustic over-stimulation causes hearing loss. Many years of experimental study as well as the examination of temporal-bone specimens from humans have established that loud sounds initially target the outer hair cells (OHCs) by reversibly or permanently damaging them, depending on exposure level (e.g., Bohne & Clark, 1982). The discovery of otoacoustic emissions (OAEs) by Kemp (1978), and the subsequent knowledge that emitted responses were most likely generated by normal-functioning OHCs (Mountain 1980; Siegel & Kim 1982; Horner et al 1985; Schrott et al 1991), suggests that OAEs would make useful detectors of noise-induced cochlear damage. In fact, over the last decade, there have been many experimental studies of noise-induced cochlear effects using the majority of the various subclasses of emissions including spontaneous OAEs (Norton et al 1989), transient OAEs (Anderson & Kemp 1979; Kemp 1982), and distortion-product OAEs or DPOAEs (Zurek et al 1982; Franklin et al 1991; Mensh et al 1993; Sutton et al 1994; Chang & Norton 1996). The primary response parameter measured is typically magnitude in the form of sound pressure level (re 20 µPa), and, together, the previous studies showed the ability of OAEs to qualitatively track the amount of the resulting reduction in emission amplitude along with its recovery time course.

One interest in using objective measures like OAEs to index noise-induced effects is to investigate if they can more sensitively detect the onset of such sequelae to sound over-exposure such as noise-induced hearing loss (NIHL). However, experimental studies which have tracked noise-induced changes in hearing sensitivity simultaneously with changes in the levels of DPOAEs in a closely controlled animal model have failed to identify alterations in OAE levels that precede related changes in behavioral thresholds (e.g., Franklin et al 1991). Nonetheless, other features of DPOAE responses, which might potentially act as early indicators of cochlear pathology, have not, to date, been thoroughly investigated. One such response property is the measure of latency.

Currently, there are two available measures of DPOAE latency. The phase-gradient method first introduced by Kemp and Brown (1983), and further developed by Kimberley et al (1993), defines latency indirectly as the rate of DPOAE-phase change with DPOAE frequency upon sweeping the frequency of one of the primary tones, while keeping the other primary at a constant frequency. With this approach, latency is defined as the slope of the linear-regression line that best fits the data describing DPOAE phase as a function of small changes in one of the primary-tone frequencies. This derived measure, which is also called the group-delay method, can be determined by shifting the frequency of either the f_1, i.e., the swept-f_1 procedure (Kimberley et al 1993), or f_2 (i.e., the swept-f_2 method) primary (O Mahoney & Kemp 1995; Moulin & Kemp 1996a,b).

Another strategy of measuring DPOAE latency, which is called the primary-tone phase-rotation method, was recently introduced by Whitehead et al (1996). With this approach, the influence of the primary tones and associated distortion products including harmonic and intermodulation responses on the DPOAE component of interest are essentially canceled by systematically manipulating the phases of the primary tones. In this manner, the time waveform of the DPOAE is revealed, and the onset latency of the DPOAE can be directly measured from the waveform.

In general, measured group-delay latencies are approximately similar to onset latencies in that they show similar variations with stimulus parameters. That is, both measures are typically a few ms, decrease with increasing frequency and with increasing stimulus level, and are shorter in small laboratory mammals than in humans (Kimberley et al 1993; O Mahoney & Kemp 1995; Moulin & Kemp 1996a,b; Whitehead et al 1996). However, quantitative differences are apparent with onset latencies being closer in value to those measured using the swept-f_1 group-delay method than to the swept-f_2 procedure (Whitehead et al 1996).

To date, the effects of noise over-exposure on DPOAE latency have only been investigated in one study, which used the group-delay technique in a small number of humans. In this analysis, Engdahl and Kemp (1996) determined the effects on DPOAE latency of 10-min, 105-dB SPL narrowband noise exposures, centered at 2 kHz, in five normal-hearing humans. Following exposure, they found that DPOAE latencies evoked by primary tones with f_2/f_1 ratios within the standard range (i.e., 1.2–1.27) showed a tendency to decrease by about 6% (e.g., latencies for 3-kHz DPOAEs decreased on average from about 6.5 to 6.1 ms). Because the amount of data that describe the effects of sound over-exposure on DPOAE latency are limited, and because there are no studies yet measuring these consequences using onset-latency measures, the specific aim of the present study was to determine the effects of moderate sound exposures on DPOAE-onset latency in an animal model.

2. METHODS

2.1. Subjects

Seven young (~3 mos old) adult, female, pigmented rabbits with an average weight of about 2.5 kg served as subjects. Rabbits were obtained commercially from approved vendors, and the experimental protocols described below were reviewed and approved by the School's Institutional Animal Care and Use Committee.

Because ears received more than one over-exposure, and because both ears of each rabbit were studied, a chronic animal preparation was used. One week prior to the initial DPOAE measurements, each animal was surgically implanted with a permanent head-mount device under general anesthesia (ketamine 50 mg/kg, xylazine 10 mg/kg). Once the rabbit was eating and behaving normally, i.e., usually within 5 days of the surgery, DPOAEs from both ears were assessed for their normalcy. Toward this end, routine frequency/level functions (i.e., DPOAE level as a function of the geometric-mean frequency) were determined in the form of DP-grams, i.e., for geometric-mean frequencies ranging from 1.414 to 18.379 kHz (f_2=1.581–20.549 kHz; DPOAE=0.950–12.330 kHz), in 0.1-octave steps (38 frequencies), at primary-tone levels of 45, 55, and 65 dB SPL (L_1=L_2).

The outcome of these first measurements showed that all seven rabbits had normal DPOAEs compared to the laboratory's extensive database. Before each sound-exposure session, ears were checked otoscopically, and any wax removed, and the DP-grams remeasured to ascertain that there were no permanent changes in the normalcy of the DPOAEs. Measurements were performed in a sound-proofed chamber, with the rabbit confined in a standard plastic restrainer, and the animal's head firmly held in position by the secure attachment of the implanted head brace to a rigid bracket fixed to the restrainer. Animals were studied while under the ketamine/xylazine anesthesia noted above in order to minimize the influence of muscular movements on the phase-cancellation method of measuring DPOAE-onset latency.

2.2. Measurement of DPOAEs and Onset Latency

The onset latencies for DPOAEs were measured as described in detail by Whitehead et al (1996). Briefly, primary-tone stimuli were synthesized, and the microphone output sampled, by a 16-bit digital signal processing (DSP) board (Digidesign, Audiomedia) mounted in a personal microcomputer (Macintosh IIci). The stimuli were transduced by two dynamic earspeakers (Etymotic Research, ER-2). The f_2/f_1 ratio was set at 1.25, and the DPOAEs were elicited by four primary-tone levels with L_1=L_2=45, 50, 55, and 60 dB SPL. Ear-canal sound pressure was measured with a microphone assembly (Etymotic Research, ER-10) sealed securely into the outer-ear canal. The speaker-command voltages were ramped on with a raised cosine over 0.1 ms, and the microphone output was sampled at 44.1 kHz for 92.9 ms starting at the end of the stimulus onset ramp using a rectangular window. The samples consisted of 4096 points of 22.7 µs each. The acoustic delay from the speakers to the microphone was approximately 0.9 ms. Thus, the acoustic stimulus onset occurred 0.8 ms after the start of the sample window. Samples containing unusually high levels of noise at frequencies adjacent to the DPOAE frequency were automatically rejected.

To visualize the onset of the $2f_1$-f_2 DPOAEs in the ear-canal sound pressure, the f_1 and f_2 primary tones and other DPOAEs including $2f_2$-f_1, $3f_1$-$2f_2$, $2f_1$, were eliminated by time-domain ensemble averaging. With this approach, the phases of the primary tones were systematically varied between stimulus presentations [see Table 1 in Whitehead et al 1996]. By using a block of n=8 samples and advancing f_1 in 45° steps, and f_2 in 90° steps, i.e., varying the phase of f_2 between samples at twice the rate of the phase of f_1, the phase of the $2f_1$-f_2 DPOAE was left unaffected. Thus, the phases of the primaries and other DPOAEs varied across samples such that they were canceled in the final average, whereas the $2f_1$-f_2 DPOAE averaged normally. For the data presented below, blocks of n=8 samples were acquired eight times, resulting in a final average of n=64 samples.

To reduce noise from frequencies greater than the DPOAE, low-pass filtering using spectral editing was used to improve the signal-to-noise ratio. In this manner, the fast Fourier transform (FFT) of the time waveform was determined, the amplitude of FFT bins more than 0.3 octave above the DPOAE frequency were zeroed, and an inverse FFT was performed to obtain the filtered waveform. In addition, to better measure signals at smaller signal-to-noise ratios, contaminating low-frequency noise was removed by high-pass filtering. Towards this end, the waveform was smoothed using a sliding average over a period equal to one cycle of the DPOAE frequency, and the smoothed waveform subtracted from the raw waveform to remove the low-frequency components. As noted by Whitehead et al (1996), although the low-pass filtering described above had negligible effects on the onset and steady-state portions of the DPOAEs, the high-pass filtering tended to spread out the onset portion of the time waveform over an extra cycle of the DPOAE frequency.

The DPOAE-onset latency was then defined as the time of onset of the DPOAE in the sample window, minus the acoustic stimulus onset time, which occurred at 0.8 ms. The time of DPOAE onset, i.e., the earliest time at which the DPOAE could be detected, was judged objectively using the periodicity of the DPOAE as a guide by comparing the DPOAE waveform to a reference sinusoid at the frequency of the DPOAE. Two objective techniques were used to measure DPOAE onset latencies from the high-pass filtered time waveforms. In one approach called the 'sliding correlation' method, the DPOAE waveform was correlated over a two-cycle period centered around each point in the sample, to the reference DPOAE-frequency sinusoid having the same phase and amplitude as the DPOAE at steady state. The DPOAE onset was arbitrarily defined as the time at the center of the latest two-cycle period (i.e., to avoid any spuriously high correlations that sometimes occurred between the reference sinusoid and noise at the start of the data waveform) for which the correlation coefficient was 0.75.

In the second or 'sliding-amplitude' method, the root-mean-square (rms) amplitude of the time waveform was determined over two-cycle periods centered around each point in the sample. Onset time was arbitrarily defined as the time at the center of the earliest two-cycle period (i.e., spurious noise was not a problem for amplitude assessments) for which the DPOAE level reached 0.625 (i.e., $1-e^{-1}$) of its steady-state value.

2.3. Experimental Protocol

The experimental paradigm consisted of three successive stages that included preexposure, exposure, and postexposure periods. At the beginning of each experiment, DP-grams for the $2f_1-f_2$ DPOAE were measured at the four standard levels (i.e., $L_1=L_2=45, 50, 55, 60$ dB SPL) in 0.1-octave steps in order to confirm that the animal's emissions remained normal. If DPOAEs were normal, the preexposure stage was initiated in which baseline measures of the onset latencies were determined for the four primary-tone levels at one of 12 geometric-mean test frequencies ranging from 1.414–3.249 kHz (i.e., $f_2=1.6-3.7$ kHz). These 'test' DPOAE frequencies encompassed the low-frequency audibility range of the rabbit, and included frequencies from 0.950 to 2.180 kHz. Onset latencies for a specific test frequency were collected for 10 mins postexposure, and average values were computed for each primary-tone level from some 256 determinations.

The ensuing exposure period consisted of the monaural delivery of a pure tone at 95 dB SPL for 3 mins. The level and duration features of the over-exposure tone were based on knowledge from previous sound-exposure experiments in rabbits, which established that these parameters resulted in measurable, but temporary, reductions of about 10–15 dB in DPOAE levels that lasted from 5–10 mins postexposure (Franklin et al 1991; Mensh et al 1993). The exposure frequency was 1/2 octave below the frequency of f_2, and ranged from 1.414 to 3.249 kHz.

Finally, to avoid the initial period of fast recovery, DPOAE-onset latency measures during the postexposure-recovery period were initiated about 30 secs following the end of the exposure at a time when DPOAE levels were expected to stabilize (Mensh et al 1993). However, during the initial recovery period, i.e., from 0–30 sec, DPOAE levels were monitored. To track the recovery process for onset latencies, tests at the four primary-tone levels were systematically interleaved from 30 secs to 10 mins postexposure. In this manner, a complete cycle through all four sets of stimuli was completed in about 3.5 mins. Thus, average onset-latency measures were plotted at center times of about 4, 7, and 10 mins postexposure.

Rabbits were tested bilaterally over several exposure episodes for each ear at different frequencies. Before initiating another exposure, however, it was always necessary that the DPOAEs at all test frequencies be within a standard deviation (typically 2 dB) of the original baseline measures determined at the beginning of the study, i.e., prior to any sound-exposure experience. Overall, 39 exposure episodes were documented, with each ear receiving two to three overstimulations at different frequencies. Across animals, the number of individual exposures at the 12 specific test frequencies varied from two to five.

2.4. Measurement of DPOAE Phase

The phase of DPOAEs with respect to the stimulus tones was easily documented, because the FFT performed on the synchronous DPOAE average produced both amplitude and phase measurements. Thus, an average phase for each set of primary-tone levels was calculated from the preexposure measures, and during the postexposure interval, DPOAE phase was tracked over the entire recovery period. Although it was not possible to monitor both amplitude changes and DPOAE-onset latencies simultaneously given the software limitations at the time of the study, DPOAE phase was an integral part of the FFT analysis of the ear-canal sound pressure, and, thus, it was followed throughout the recovery period from 0 to 10 mins postexposure. Thus, phase responses were measured from the end of the exposure period to about 30 sec, and then at center-time recovery intervals of 4, 7, and 10 mins postexposure.

By applying the following equation to the phase data, which was derived from the formula used to calculate DPOAE group-delay latency in seconds (O Mahoney & Kemp 1995) from the slope of the linear-regression line describing the relation between the phase of the DPOAE as a function of DPOAE frequency, the direction and amount of change in the group-delay latency were computed:

$$\frac{\partial \phi}{\text{frequency (kHz)} \times 360°}$$

That is, a change in phase-gradient latency was equal to the measured change in phase (preexposure average phase minus the postexposure phase) divided by the geometric-mean test frequency in kHz times 360°. These computations were determined for each of the immediate postexposure times up to about 30 secs following the end of sound exposure, and for the recovery-interval times from about 1–10 mins postexposure.

2.5. Control Measures of DPOAE-Onset Latency Determination

One aspect of the measurement procedure that possibly is independent of the effects of over-stimulation on DPOAE latency is the influence of sound-induced reductions in DPOAE level on measures of onset latency. To control for this possibility, a tone at 2-kHz tone was introduced into an artificial cavity that approximated the volume of the rabbit external ear canal at a level that estimated the size of a rabbit DPOAE elicited by 65-dB SPL primaries, i.e., at 30 dB SPL. Following determination of the onset latency using both the sliding-correlation and amplitude methods, the tone was systematically reduced in 5-dB steps to 10 dB SPL, thus, mimicking a 20-dB overall reduction in signal level. The outcome of this control procedure showed that, although onset latency varied from 4.17–4.29 ms for the 0.75 sliding-correlation criterion, the standard deviation at 0.06 ms, or 60 µs, was minimal. For the sliding-amplitude method, onset latency varied similarly from 4.15–4.42 ms, with a standard deviation of 0.1 ms, or 100 µs. Thus, experimental attenuations of the DPOAE-like signal, within a range expected to mimic sound-induced changes in rabbit DPOAEs, did not appear to artifactually decrease or increase response latency by a great amount. At least, changes greater than 100 µs could be treated as nonartifactual findings that were independent of sound-induced changes in DPOAE level. In any case, for the sliding-amplitude computations, the postexposure data were analyzed both in absolute dB units, i.e., as signals that were smaller than their control counterparts, and in normalized dB units, i.e., their levels were increased by external amplification to match control values.

A second control procedure examined the effects of signal rise time on onset latency. To accomplish this goal, a 1.5-kHz tone of 30 dB SPL was introduced into the same artificial cavity with four unique rise times at 1, 2, 4, and 8 ms. Data obtained with the sliding-amplitude method, in particular, was sensitive to increasing signal rise times in that almost a 115% increase in onset latency was measured when rise time was increased by a factor of 4, i.e., from 2 to 8 ms. Such changes in rise time resulted in a corresponding latency change of 5.03–10.79 ms. In comparison, the onset latency determined with the sliding-correlation method increased only by about 26%, i.e., from 4.44–5.62 ms, with the identical extreme change in signal rise time. Thus, the sliding-amplitude method appeared to be very sensitive to changes in signal rise time, whereas the sliding-correlation procedure was less influenced by alterations in DPOAE rise time.

It is important to emphasize that, given the low-frequency range of the experimental measures of sound-damaged latencies, a change from a rise time of 1 to 8 ms would be approximately equivalent to latency changes on the order of 4 ms, which would be unattainable with respect to a latency-shortening outcome, and unlikely in the lengthening direction, with such moderate sound-exposure levels. However, according to the control measures on the effects of signal rise time on DPOAE-onset latency measures, for more realistic changes of about 0.5–1 ms or so, the sliding-correlation method would result in latencies that were increased by about 3–15%, whereas the sliding-amplitude latencies produced latencies that were longer than baseline ones by 17–31%.

To quantify the ability of the sliding-amplitude method to estimate the rise-time of the DPOAE-generation process, latencies were determined for the times when DPOAE level was at 0.5, 0.625, and 0.75 of its full rms amplitude, and the slope of the linear-regression line passing through these points was determined. In this manner, the slope value provided the objective form of DPOAE rise-time.

2.6. Statistical Analyses

Data were transferred electronically from the individual data-acquisition files for each exposure session to a commercially available spreadsheet (Excel, v5.0, Microsoft). In this format, comparisons of the postexposure values of onset latency, group-delay latency, and rise-time slopes with their preexposure control measures were performed using a paired t-test subroutine of a commercially available statistical-software package (StatView, v4.5, Abacus Concepts). However, before these tests were conducted, the processed data were examined empirically to ensure that the onset-latency measures were not confounded by the occurrence of transient noise spikes, which sometimes occurred during the course of an experiment. Thus, for the waveforms determined by the sliding-correlation method, to be accepted into the database, it was necessary that the functions describing the 0.75 correlations remain above this point for, at least, 2 ms. In addition, for the sliding-amplitude method, after reaching the 0.5 normalized rms amplitude, it was necessary for the function describing the growth of the response to continue increasing from the 0.5 to 0.75 rms level, without reversing direction. To satisfy both of these criteria, of the 156 original data points (39 exposures X 4 primary-tone levels), approximately 15% were omitted from the statistical analyses. The majority of these discarded data were associated with the low signal-to-noise related DPOAEs elicited by the least intense stimuli, i.e., the 45-dB SPL DPOAEs. For interpreting the outcome of statistical testing, the level of significance adopted was $p < 0.05$.

3. RESULTS

The immediate reduction in DPOAE level produced by the low-frequency, 3-min, 95-dB SPL tonal exposures was approximately 16 dB, on average, across all four primary-tone levels. There was a tendency for a small level-dependent effect in that the average loss was slightly greater for the 45-dB SPL primaries at 17 dB than the 15-dB reduction exhibited by DPOAEs elicited by the 60-dB SPL stimuli, but this was nonsignificant. By the postexposure time of 30 secs, at the time that the recovery onset-latency measures were initiated, average DPOAE levels had recovered to within about 4 dB of their baseline values, on average, when examined according to either the level of stimulation (i.e., 45 dB SPL: -4.4 dB; 50 dB SPL: -3.9 dB; 55 dB SPL: -3.8 dB; 60 dB SPL: -3.2 dB SPL), or range of exposure frequency (i.e., 1.4–1.7 kHz: -3.1 dB; 1.8–2.2 kHz: -3.8 dB; 2.3–2.8 kHz: -2.8 dB; 2.9–3.3 kHz: -6.2 dB). Within the 10-min postexposure monitoring period, all DPOAEs returned to within about 2–3 dB of their preexposure control levels, except for the DPOAEs within the highest frequency range, which remained about 5 dB, on average, less than their control levels. All the exposure-induced reductions in DPOAE level were significantly different from their corresponding preexposure levels when compared to baseline values at each stage of recovery.

3.1. DPOAE-Onset Latency

The major consequence of brief exposures to moderately intense sound on DPOAE-onset latencies is illustrated in the bargraph plot of Fig 1 comparing for each stimulus level the average DPOAE-onset latencies measured during the three distinct recovery times ending at 4, 7, and 10 mins postexposure. These measures were determined using the sliding-correlation method of determining onset latency. It is clear from this plot and

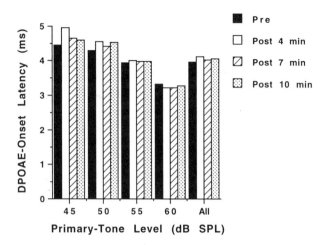

Figure 1. Bargraph plot showing the sound-induced changes in DPOAE-onset latency between the preexposure (solid bar) and the three recovery periods at 4 (open bar), 7 (hatched bar), and 10 (stippled bar) mins postexposure for each of the four levels of test stimulation. The 'All' primary-tone level data represent the overall outcome when all levels of stimulation were combined.

for the mean pre/postexposure latency-difference data displayed in Fig 2 that, for the three lowest primary-tone levels of 45, 50, and 55 dB SPL, the initial effect of sound over-exposure was to increase onset latency by about 0.154 ms, on average. However, the amount of increase in onset latency was level-dependent in that the DPOAEs elicited by the lower-level primaries were increased more than those produced by the higher-level primary tones. Specifically, DPOAEs elicited by the 45-dB primaries were most affected in that their latencies increased by 0.5 ms from an average of 0.4.5 to 5.0 ms, i.e., representing an

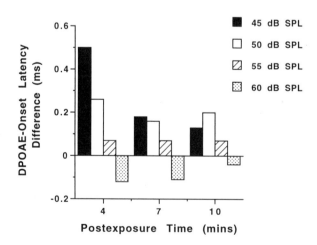

Figure 2. Bargraph plot showing the difference in DPOAE-onset latencies (postexposure minus preexposure value) for each level of stimulation (see legend for defining stimulus level) along with the combined 'All' levels, as a function of postexposure-recovery time. Up direction (+ values) shows an increased latency, whereas down direction (- values) represents a decreased latency.

increase of about 11% over baseline values. The DPOAEs evoked by the mid-level primaries at 50 and 55 dB SPL were less affected, with the associated increases being around
0.2 and 0.06 ms, respectively, which represented an increase of about 2–6%. In contrast to
the outcome determined with low- to mid-level primaries, DPOAE onset-latencies in response to the highest test-stimulation level of 60 dB SPL decreased from their baseline
values by, on average, 0.11 ms. With the exception of the data collected at 10 mins
postexposure for DPOAEs elicited by the 45- and 60-dB SPL primaries, the postexposure
effects observed for each level of stimulation and for the combined stimulus levels, were
statistically significant.

By 7 mins postexposure, the onset latencies for the lowest stimulus levels of 45 and
50 dB SPL were still increased by about 0.160–0.180 ms, on average. However, latencies
in response to the 55-dB SPL primaries were longer by less than 0.07 ms, whereas those
elicited by the 60-dB SPL stimuli remained shorter by about 0.11 ms. At the end of the
10-min postexposure monitoring period, onset latencies elicited by all four primary-tone
levels were within ±0.100 ms of their baseline values.

The plots of Fig 3 illustrate the general effects of sound exposure at the three
postexposure times on the combined level data as a function of frequency. It is clear from
these data that the DPOAEs within the highest frequency range were the least affected in
that they displayed minimal changes from their preexposure latencies. For the lower-fre-
quency DPOAEs, although those within the 1.4–1.7 kHz range showed the most increase
(mean=0.23 ms), proportionally, all DPOAEs between 1.4–2.8 kHz increased similarly by
about 5% over preexposure values.

One factor that could theoretically contribute to the observed longer onset latencies
would be decreased DPOAEs rise times as measured with the sliding-amplitude method.
However, other than for the lowest level of stimulation at 45-dB SPL, none of the slope
measures were significantly different from corresponding baseline values. In fact, rise

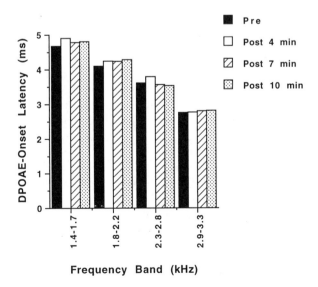

Figure 3. Bargraph plot showing DPOAE-onset latency as a function of DPOAE frequency at the various stages
(see legend) of the exposure paradigm including the preexposure and three recovery periods at 4, 7, and 10 min
postexposure.

times for postexposure DPOAEs tended to decrease from their preexposure values, thus, resulting in increased slopes. Although the direction of change in DPOAE rise times was somewhat dependent on stimulus level, it was independent of frequency.

3.2 DPOAE Group-Delay Latency

The phases for DPOAEs advanced from baseline values following tonal over-exposure, with the amount of phase lead dependent on the levels of the primary tones. Thus, immediately following the exposure period, for the 45-dB SPL tones, phase advanced about 24°, on average. Similarly, phases for DPOAEs elicited by the higher-level primaries advanced from about 15–20° over baseline values. Using the equation noted above (section 2.4) for computing group-delay latency, in contrast to the increase in DPOAE latency observed with the onset-measurement procedure, group latencies decreased by about 0.024 to 0.055 ms (i.e., 24–55 µs), with DPOAEs elicited by 45-dB primaries showing the greatest reduction. However, by 1 min postexposure, DPOAE phase in response to all primary-tone levels was within 15° of baseline, which translated into group latencies that were, on average, only about 0.03 ms (30 µs) shorter than control measures. By 4 mins postexposure, group-delay latencies remained decreased by 0.02 ms, regardless of level or frequency of the primary tones. Overall, for the 4- and 7-min recovery intervals, the observation of sound-induced shorter latencies as measured with the group-delay method was significant. However, for the individual stimulus level and frequency ranges, except for DPOAEs at the highest frequencies, only a few significant differences from control measures were noted (i.e., for the 45- and 60-dB SPL DPOAEs at 4 mins postexposure). By 10 mins postexposure, DPOAE group-delay latencies still were shorter than their preexposure counterparts by 0.010 ms or less. However, the decreased latencies at this later stage of the recovery process were not significantly different from their corresponding control values.

DISCUSSION

The principal finding of the present study was the increase in DPOAE onset-latency following exposure to moderately intense, but relatively brief sounds. In previous experimental studies of the effects of sound overexposure on OAEs, the level parameter has been the most frequently studied response measure, with the major outcome being a reduction in amplitude. Only recently have other OAE-response features such as latency been closely examined (Kimberley et al 1993; O Mahoney & Kemp 1995; Moulin & Kemp 1996a,b; Whitehead et al 1996), and, to date, there has only been one systematic study of the effects of over-exposure on the latency-response measure. In particular, the previous group-delay findings of Engdahl and Kemp (1996) in humans noted a slight decrease in latencies for mid-frequency DPOAEs following a brief exposure to a moderately intense narrowband noise centered at 2 kHz. Thus, it was surprising that the present measures of DPOAE-onset latency showed increased latencies following moderate overstimulation with pure tones. There are, however, clearly many notable differences between the two studies including subject species (rabbits vs humans), type of exposure (tonal vs noise band), and, most importantly, measurement procedure (onset vs group-delay latency). It is notable that when the comparable measure of DPOAE latency was derived from the present data, group-delay latencies also decreased as they did in the earlier Engdahl and Kemp (1996) study. However, in the present work, the decreased latencies computed from the DPOAE-phase measures were not very different from control measures. In addition, they were about an order of magnitude less than either the observed changes in onset latencies, or the comparable decreases noted in the Engdahl and Kemp (1996) study, i.e., they were in tens rather than in hundreds of microseconds.

Increased onset latencies are also in contrast to the expected outcome that damaged cochlea, in the absence of normal sensitivity and reduced frequency-filtering capabilities, should process acoustic stimuli more rapidly due to the absence of a time-consuming healthy cochlear-amplifier mechanism. This expectation is based on the assumption for linear-filter systems that reduced frequency selectivity is predictably coupled with decreased delays or, equivalently, phase leads. Because DPOAE-rise time measured within a few minutes of the exposure period was relatively unchanged by the exposure episodes, it is unlikely that the dynamics of the DPOAE-generation process itself, once initiated, contributed to the sound-induced increase in onset latency. However, although the increased onset latencies were an unexpected consequence of the tonal over-stimulations, there is some precedent for other measures of cochlear biologic activity to exhibit similar findings. For example, Ruggero et al (1993) reported noise-induced phase lags, i.e., in the direction of increased response times, near the best-frequency site for the basilar-membrane response measured with laser techniques. Moreover, Cooper and Rhode (1992) had previously observed similar results in the cat. Ruggero et al (1996) attributed such unanticipated findings to a sound-induced increased elasticity of the cochlear partition, which would ordinarily lead to longer latencies.

Because it is unknown if the immediate effects of sound over-exposure on onset latency were similar to the present findings determined by a few minutes postexposure, a primary goal of future studies will be to initiate the measurement of onset-latencies immediately following the exposure to ensure that a short-lasting initial decrease in latency was not present. In this manner, the findings using a tonal-overstimulation paradigm can be more directly compared to those of Engdahl and Kemp (1996). Moreover, given that the observed increase in onset latency did not appear to completely recover during the short postexposure period, other follow-up work in rabbits will be aimed at tracking the sound-exposure effects over longer postexposure intervals in order to more completely describe the recovery-time course. Finally, basic studies aimed at discerning the stages of DPOAE generation associated with each type of latency measure will continue in order to better understand the contrasting effects of sound overstimulation on group-delay versus onset latency.

ACKNOWLEDGMENTS

Portions of this work were supported by grants from the Public Health Service (DC00613, DC/ES03114). Dr. Kara was supported by a research fellowship provided by The Scientific and Technical Research Council of Turkey (TUBITAK) under the auspices of the NATO Science Fellowships Programme. The authors thank Mayte T Ruiz and Geoffrey M Waxman, for assisting in data processing, and in the preparation of the manuscript and related illustrations.

REFERENCES

Anderson, S.D., and Kemp, D.T. (1979): The evoked cochlear mechanical response in laboratory primates. Arch Otorhinolaryngol 224:47–54.

Bohne, B.A., and Clark, W.W. (1982): Growth of hearing loss and cochlear lesion with increasing duration of noise exposure. In: New Perspectives on Noise-Induced Hearing Loss, R.P. Hamernik, D. Henderson. R. Salvi (eds.). New York: Raven, pp283–303.

Chang K.W., and Norton, S.J. (1996): The effects of continuous versus interrupted noise exposures on distortion product otoacoustic emissions in guinea pigs. Hear Res 96:1–12.

Cooper, N.P., and Rhode, W.S. (1992): Basilar membrane mechanics in the hook region of cat and guinea-pig cochleae: Sharp tuning and nonlinearity in the absence of baseline position shifts. Hear Res 63:163–190.

Engdahl, B., and Kemp, D.T. (1996): The effect of noise exposure on the details of distortion product otoacoustic emissions in humans. J Acoust Soc Am 99:1573–1587.

Franklin, D.J., Lonsbury-Martin, B.L., Stagner, B.B., and Martin, G.K. (1991): Altered susceptibility of $2f_1$-f_2 acoustic-distortion products to the effects of repeated noise exposure in rabbits. Hear Res 53:185–208.

Horner, K.C., Lenoir, M. and Bock, G.R. (1985): Distortion product otoacoustic emissions in hearing-impaired mutant mice. J Acoust Soc Am 78:1603–1611.

Kemp, D.T. (1978): Stimulated acoustic emissions from within the human auditory system. J Acoust Soc Am 64:1386–1391.

Kemp, D.T. (1982): Cochlear echoes: Implications for noise-induced hearing loss. In: New Perspectives on Noise-Induced Hearing Loss, R.P. Hamernik, D. Henderson, and R. Salvi (eds.). New York: Raven Pr, pp189–207.

Kemp, D.T., and Brown, A.M. (1983): A comparison of mechanical nonlinearities in the cochleae of man and gerbil from ear canal measurements. In: Hearing—Physiological Bases and Psychophysics, R. Klinke, and R. Hartmann (eds.). Berlin: Springer-Verlag, pp82–88.

Kimberley, B.P., Brown, D.K., and Eggermont, J.J. (1993): Measuring human cochlear travelling wave delay using distortion product emission phase responses. J Acoust Soc Am 94:1343–1350.

Mensh, B.D., Patterson, M.C., Whitehead, M.L., Lonsbury-Martin, B.L., and Martin, G.K. (1993): Distortion-product otoacoustic emissions in rabbit: I. Altered susceptibility to repeated pure-tone exposures. Hear Res 70:50–64.

Moulin, A., and Kemp, D.T. (1995): Interpreting phase of distortion product otoacoustic emissions in humans. Br J Audiol Abstr 29:65.

Moulin, A., and Kemp, D.T. (1996a): Multicomponent acoustic distortion product otoacoustic emission phase in humans. I. General characteristics. J Acoust Soc Am 100:1617–1639.

Moulin, A., and Kemp, D.T. (1996b): Multicomponent acoustic distortion product otoacoustic emission phase in humans. II. Implications for distortion product otoacoustic emissions generation. J Acoust Soc Am 100:1640–1662.

Mountain, D.C. (1980): Changes in endolymphatic potential and crossed olivocochlear bundle stimulation alter cochlear mechanics. Science 210:71–72.

Norton, S.J., Mott, J.B., and Champlin, C.A. (1989): Behavior of spontaneous otoacoustic emissions following intense ipsilateral acoustic stimulation. Hear Res 38:243–258.

O Mahoney, C.F., and Kemp, D.T. (1995): Distortion product otoacoustic emission delay measurement in human ears. J Acoust Soc Am 97:3721–3735.

Ruggero, M.A., Rich, N.C., and Recio, A. (1993): Alteration of basilar membrane responses to sound by acoustic overstimulation. In: Biophysics of Hair Cell Sensory Systems, H. Duifhuis, J.W. Horst, P. van Dijk, and S.M. van Netten (eds.). Singapore: World Scientific Publishing, pp258–264.

Ruggero, M.A., Rich, N.C., Robles, L., and Recio, A. (1996): The effects of acoustic trauma, other cochlear injury, and death on basilar-membrane responses to sound. In: Scientific Basis of Noise-Induced Hearing Loss, A. Axelsson, P-A Hellstrom, H. Borchgrevink, D. Henderson, R.P. Hamernik, and R.J. Salvi (eds.). New York: Thieme, pp23–35.

Schrott, A., Puel, J-L, and Rebillard, G. (1991): Cochlear origin of $2f_1$-f_2 distortion products assessed by using two types of mutant mice. Hear Res 52:245–253.

Siegel, J.H., and Kim, D.O. (1982): Efferent neural control of cochlear mechanics? Olivocochlear bundle stimulation affects cochlear biomechanical nonlinearity. Hear Res 6:171–182.

Sutton, L.A., Lonsbury-Martin, B.L., Martin, G.K., and Whitehead, M.L. (1994): Sensitivity of distortion-product otoacoustic emissions in humans to tonal over-exposure: Time course of recovery and effects of lowering L_2. Hear Res 75:161–174.

Whitehead, M.L., Stagner, B.B., Martin, G.K., and Lonsbury-Martin, B.L. (1996): Visualization of the onset of distortion-product otoacoustic emissions, and measurement of their latency. J Acoust Soc Am 100:1663–1679.

Zurek, P.M., Clark, W.W., and Kim, D.O. (1982): The behavior of acoustic-distortion products in the ear canals of chinchillas with normal or damaged ears. J Acoust Soc Am 72:774–780.

51

DISTORTION-PRODUCT OTOACOUSTIC EMISSIONS RELATED TO PHYSIOLOGICAL AND PSYCHOPHYSICAL ASPECTS OF HEARING

Michael Ganz and Hellmut von Specht

Department of Experimental Audiology
Otto-von-Guericke University of Magdeburg
39120 Magdeburg, Germany

1. INTRODUCTION

Acoustic signal processing in the central auditory system is determined by what the cochlea's output to the auditory nerve contains. Once we understand signal preprocessing in the auditory periphery better evaluation of central processing is possible. In this respect, distortion-product otoacoustic emissions (DPOAE) are a beneficial tool of audiological investigation. Use of acoustically simple signals like pure tones enables nonlinear signal processing in the cochlea to be characterized. Measurements of DPOAE can demonstrate a great many interesting properties which are caused by both physiological and morphological behaviour of the cochlea.

DPOAE are generated within the mammalian cochlea upon stimulation with two primary tones of appropriate frequencies. These emissions consist of new frequencies which are not present at the eliciting stimuli. They arise from nonlinear micromechanical characteristics of the cochlea at a specific place bearing a precise mathematical relation to the frequencies of the two primary tones. The largest DPOAE in the human cochlea occurs at the frequency of $2f_1$-f_2.

Supplementary to these objective measurements, psychophysical investigations include the perception of sound. The properties of cubic difference tones are highly similar to those of DPOAE. Experimental results in combination with theoretical considerations can elucidate the link between psychophysical data and DPOAE.

The questions we are interested in are:

- How can we use DPOAE to obtain information on the behaviour of the cochlea?
- How is sound processing in the cochlea related to psychophysical perception of sound?

Acoustical Signal Processing in the Central Auditory System
edited by Syka, Plenum Press, New York, 1997

The first question is aimed at signal preprocessing within the auditory periphery while the second question refers to link to central signal processing.

In this context, it is suggested that pitch perception be determined by cochlear micromechanics. The properties of DPOAE and their dependence on primary frequencies and levels reflect the discrepancy between melodic and harmonic pitch.

The correlation between the width of critical bands and the dependences of DPOAE on primary levels and frequencies is discussed in detail. An assumption is made to the extent that the motion of the basilar membrane may be estimated from DPOAE properties.

2. METHODS

All measurements of DPOAE were conducted with an ER 10-C probe system (ETYMOTIC RESEARCH) in a group of 15 normal hearing adults. The data presented are median values. We only used DPOAE that were more than 6 dB above the noise floor. All measurements were performend in a sound booth. The primary tones were characterized by their frequencies (f_1 and f_2, where $f_1 < f_2$) and their sound pressure levels L_1 and L_2. In general, the level of DPOAE can be described be as a function $L_{DP} (L_1, L_2, f_1, f_2)$. To simplify these dependences it is common practice to employ the frequency ratio $r = f_2/f_1$ so as to eliminate the dependence on f_1. Using the parameter r is helpful when comparing the levels of DPOAE at different frequencies f_2. The level L_{DP} is approximately in the range from 40 to 60 dB below the primary levels.

In the first session the levels of DPOAE were measured by varying the levels L_1 and L_2. The range of L_1 was from 28 to 79 dB SPL, in increments of 3 dB. The range of L_2 was from 5 to 60 dB SPL, in increments of 5 dB. These measurements were performed at $f_2 = $ 1, 2, 4, and 8 kHz and fixed $r = 1.19$. In the second session the levels of DPOAE were measured by varying the frequency ratio r within the range from 1.04 to 1.27. The combinations of primary levels used are specified in Fig. 3. For evaluation of results it is necessary that the localization of the source of DPOAE be known. The site of generating DPOAE on the basilar membrane is located at the place where the higher frequency f_2 will cause the maximum amplitude of basilar membrane motion. The localization of the source was obtained from suppressing experiments (Kummer et al., 1995) and our model calculations.

3. RESULTS

Measurement of level dependence with fixed f_2 and r is the first step in an approach aimed at obtaining information about DPOAE. Figure 1 shows the $L_{DP} (L_1, L_2)$ at different frequencies of $f_2 = 1$ and 8 kHz, and fixed $r = 1.19$.

At fixed L_2 the level of DPOAE increased with L_1 and reached a maximum. With further increase in L_1 the level L_{DP} decreased. We cannot account for the decrease in L_{DP} with higher levels of L_1. At fixed L_1 and increasing L_2 the behaviour of L_{DP} was similar but the slope was not so steep. Comparison of the plots for different values of f_2 revealed that for increasing f_2 the line of maximum L_{DP} shifted to higher levels L_1. The most important result is a rule as to how L_1 and L_2 should be chosen to get a maximum L_{DP}. From the present results this function can be approximated as $L_1 = 0.58 \, L_2 + L_0$. The values of L_0 at different frequencies f_2 are given in Table 1. This linear approximation is consistent with the experimental findings shown in Fig. 2.

When the frequency ratio r is varied at constant levels L_1 and L_2 the maximum L_{DP} strongly depends on the frequency ratio $r = f_2/f_1$ of applied tones. The maximum occurred at different values of r dependent on f_2. At high frequencies f_2 the value chosen for r should be smaller than at low frequencies. These results are shown in Fig. 3. When the level function described above was used at each frequency f_2 the value of r was nearly constant. Other investigators (see, for example, Harris et al. 1989) have collected data at equal primary levels $L_1 = L_2$. Their data showed a shift of r towards smaller values with decreasing primary levels.

In clinical use this choice of primary levels is helpful when generating moderately high levels of DPOAE even if primary levels are very low. Adopting this approach provides meaningful information about the behaviour of the cochlea at the place of f_2 at levels $L_2 < 5$ dB SPL. Qualitatively, it is possible to estimate the hearing threshold by measuring L_{DP}.

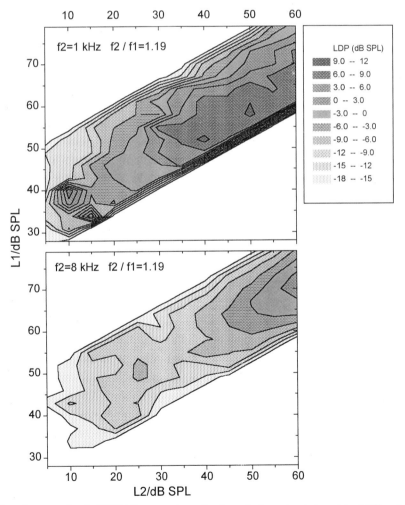

Figure 1. Sound pressure level of DPOAE measured as a function of primary levels L_1 and L_2 at different frequencies $f_2 = 1$ and 8 kHz, and fixed $r = 1.19$. All levels are sound pressure levels.

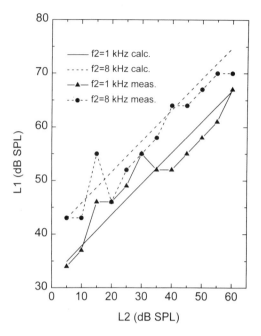

Figure 2. Primary level L_1 as a function of L_2 described above (solid and dashed line) and the measured values of L_1 at maximum L_{DP} (lines with symbols).

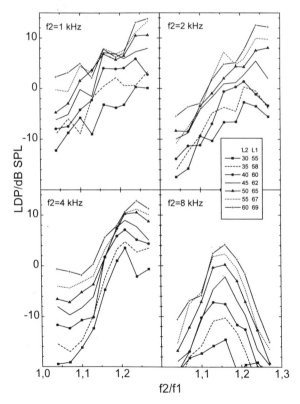

Figure 3. Level of DPOAE measured as a function of frequency ratio $r = f_2/f_1$ at different frequencies $f_2 =1, 2, 4$, and 8 kHz and different primary level combinations.

Table 1. Values of L_0 at different
frequencies f_2

f_2	L_0
1 kHz	32 dB SPL
2 kHz	35 dB SPL
4 kHz	37 dB SPL
8 kHz	40 dB SPL

4. DISCUSSION

The micromechanical structure of the basilar membrane and nonlinear properties of sound processing in human cochlea cause cubic distortions which are measurable objectively as DPOAE, and psychophysically as cubic difference tones. Comparison of DPOAE with psychophysically measured cubic difference tones (Zwicker and Feldtkeller, 1967; Zwicker, 1982) has revealed similarities as to the pattern of curves and the dependence on parameters. Figure 4 shows the levels of DPOAE as a function of L_1.

In the light of the similarities of DPOAE to cubic difference tones, it may be concluded that sound perception is governed by preprocessing of sound in the inner ear.

The critical band width at frequencies above 500 Hz is 0.2 f, i. e. if $r < 1.2$ the primaries are within the critical band. The spatial distance of maximum amplitudes is constant for r = const. For r = 1.19 the distance is about 1 mm. At high frequencies the maximum L_{DP} occurred if $r < 1.2$. The level L_1 has to be increased when a maximum L_{DP} with increasing frequencies is to be generated. Assuming that the same energy input is necessary at each frequency to generate an identical L_{DP} it may be concluded that the slope of basilar membrane oscillations is steeper and the area of overlap where high oscillations occur is narrower at high frequencies. These findings are not easy to explain. Generally, the assumption is accepted that the mechanical properties of the cochlea change with frequency in a linear relation. On the grounds of this assumption one should not expect any differences of DPOAE vs. frequency to occur.

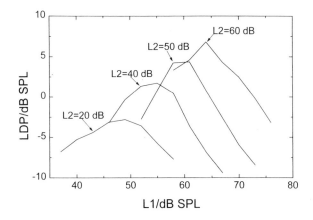

Figure 4. Level of DPOAE measured as a function of primary level L_1, and L_2 as the parameter. These results are very similar to psychoacoustically measured cubic difference tones.

All natural scales of the basilar membrane (i.e. spatial distribution of hair cells, critical band rate, pitch sensation, harmonic pitch) are linear functions of length of the basilar membrane or frequency-place transformation. The only exception is the melodic pitch. Melodic intervals become narrower with increasing frequency. This effect is probably attributable to L_{DP} being a function of r. Cubic difference tones are relevant to pitch sensation, and a shift of maximum L_{DP} with increasing f changes the perception of intervals.

REFERENCES

Harris, F.B., Lonsbury-Martin, B.L. Stagner, B.B., Coats, A.C. and Martin, G.K.: Acoustic distortion products in human: Systematic changes in amplitude as a function of f_2/f_1 ratio. J. Acoust. Soc. Am. 85 (1) January 1989, 220-229

Kummer, P., Janssen, T. and Arnold, W.: Suppression tuning characteristics of the $2 f_1$-f_2 distortion-product otoacoustic emissions in humans. J. Acoust. Soc. Am. 98 (1), July 1995, 197-210

Zwicker, E. and Feldtkeller, R.: Das Ohr als Nachrichtenempfänger. S. Hirzel Verlag, Stuttgart, 1967

Zwicker, E.: Psychoakustik. Springer-Verlag, Berlin Heidelberg NewYork, 1982

52

FUNCTIONAL ORGANIZATION OF THE AUDITORY CORTEX IN THE CONGENITALLY DEAF WHITE CAT

Rainer Hartmann, Rainer Klinke, and Silvia Heid

Physiologisches Institut III
J.W. Goethe-Universität Frankfurt
D-60590 Frankfurt/M, Germany

Cochlear-implants have developed to the most successful neuroprosthesis. In most cases patients benefit substantially from such a device. However, there is one important exception, the implantation of prelingually deaf adults. In these patients the results are disappointing. The reason for this can be easily hypothesised: During early childhood the central auditory system has to acquire strategies for the evaluation of sound stimuli in general and for the processing in language in particular. It is assumed that these processes of central auditory maturation have to take place within certain time-windows, so called critical periods and that later on a deficit is beyond recovery. Consequently the results of cochlear implantation in prelingually deaf children are substantially more successful (Lehnhardt and Bertram, 1991; Lenarz et al., 1994; Blamey, 1995; Dowell et al., 1995) and at first sight one would have to recommend an early implantation in connatally deaf infants. Obviously such an implantation is not generally possible. There are problems with an early and safe diagnoses of deafness, there are surgical problems with an implantation of a rigid device into a growing head and there will always be cases where a correct early diagnosis was missed by one or the other reason. Thus one would like to know when and how rigid are the critical periods in humans. Is there an age beyond of which an implantation is useless and what is this age?

Secondly there is little information about the functional status of the central auditory system, if deprived of its peripheral input from birth or even—so to speak—from conception. As the discouraging results of an implantation of prelingually deaf adults show, there must be functional deficits which prevent a successful use of implants in these cases. Studying the functional status of the central auditory system in cases of peripheral auditory deprivation in an animal model would thus also render results relevant for the understanding of central nervous development and maturation.

In our view an adequate model for studying the above questions is the deaf white cat. This animal suffers from a congenitally deafness (Mair, 1973; Schwarz and Higa, 1982; Heid et al., 1995) and the animals have most likely never had any useful auditory percept. This view is based on the observation that in these deaf cats the organ of Corti is

Acoustical Signal Processing in the Central Auditory System
edited by Syka, Plenum Press, New York, 1997

absent by the age of three weeks (Bosher and Hallpike, 1965; Heid et al., 1995; Heid et al., to be published). At this time, normal cats start to produce microphonics and compound action potentials at intensities above 80 dB SPL (Moore, 1981; Romand, 1983). At birth the kitten cochlea is immature. During the late first and during the second week the cochlea undergoes maturation and by the end of third week the above mentioned compound action potentials are found in normal kittens. In the deaf white cat, however, the cochlea degeneration takes place between two and three weeks of age, so that an episodal hearing, if at all existing, would require very loud sound stimuli which are not normally present in an animal house. One longitudinal study in our group attempting to record acoustically evoked brain stem responses (ABR) has shown that ABRs could only be found on postnatal day 24 and with click stimuli above 114 dB SPL peak equivalent.

For the time being there was a total of 85 cats in our stock. Their hearing thresholds were first tested by the aid of click-evoked brain stem responses in the age of three weeks and followed up later. 70% of these animals showed no ABRs at click levels exceeding 115 dB SPL peek equivalent and were classified as connatally deaf. Their was never a later acquisition of some hearing function observed in these animals. The remaining 30% of animals showed different hearing levels from severe hearing impairment to nearly normal hearing with a tendency of further impairment of hearing with age.

The results reported here were exclusively taken from animals classified as connatally deaf.

As already mentioned the organ of Corti is absent over the entire length of the cochlea in deaf animals. This is not the case for the ganglion cells. Fig. 1 gives an example of the preservation of the spiral ganglion. Six deaf cats were compared with six normal cats for this purpose. At an age of about two years the preservation of ganglion cells is as follows: 90% in the hook portion and the first half turn, 55% in the second half turn, 27% in the third half turn, 45% in the fourth half turn and nearly 100% in the fifth apical half turn. Thus for the purpose of electrical stimulation of the basal cochlea there are 90% of the normal population of afferent fibres available, a number which is much higher than with animal models in which the cochlea was damaged by pharmacological means (Leake and Hradek, 1988).

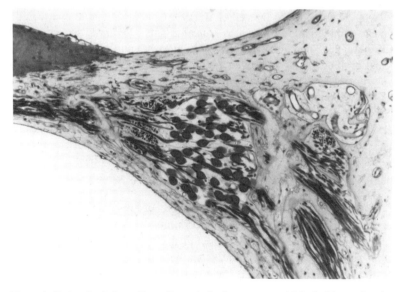

Figure 1. Reduced spiral ganglion cell population in a two year old deaf white cat, basal turn.

For the assessment of the functional status of the auditory cortex in DWCs, the cochlear afferents were electrically stimulated and different types of recordings were made from the auditory cortex.

For this purpose a NUCLEUS-22 electrode was inserted into the scala tympani under barbiturate anaesthesia (pentobarbital as necessary for stable long-term anaesthesia after initial induction with ketamin and xylacine, artificial ventilation of the animal). This human type electrode can be introduced at least up to 6 mm from the round window, so that electrode pair 1/2 reaches the 7 kHz region of normal cochlea and pair 7/8 is close to the 16–18 kHz region according to the data of Liberman (1982). Electrical stimulation was performed by a constant current source generally in a bipolar mode. Rectangular biphasic pulses 200 μs/phase were used, repetition rate between 1 and 13 per second.

For control purposes single fibre recordings from primary afferent fibres were made. The auditory nerve was located by a posterior fossa approach. Conventional glass micro-electrodes were used.

The auditory cortex was exposed by a wide opening of the skull over the contralateral auditory field so that there was free access to the regions between and around the suprasylvian and the posterior and anterior ectosylvian sulci. Thus the primary auditory cortex (AI) and the anterior auditory field (AAF) were exposed. The gross morphology of the auditory cortex in DWC does not reveal any differences to normal cats. The upper end of the posterior ectosylvian sulcus was used as the reference position 0 for all recordings. A computer controlled X-Y-Z motordrive was mounted on the stereotaxic frame, the recording plane X-Y was tilted by 35° as shown in Fig. 2.

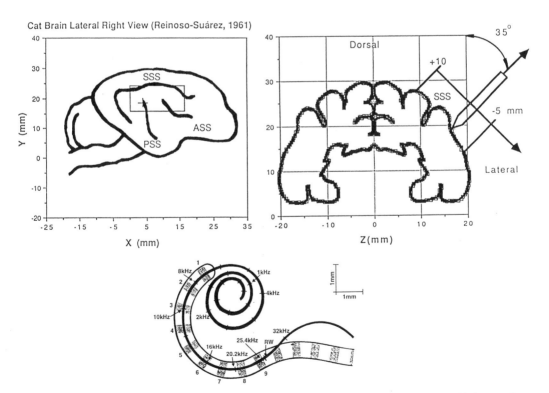

Figure 2. Top left: Lateral view of cat brain with the auditory cortex after Reinoso-Suarez (1961) with the recording area and reference point at the dorsal end of the posterior supra-sylvian sulcus. Top right: Frontal slice of the brain with tilted recording plane. Bottom: Cochlear spiral of a cat with frequency representation after Liberman (1982) and the inserted NUCLEUS-22 electrode array.

The X-axis is caudal-rostral, the Y-axis lateral-medial. The motordrive positioned the recording electrodes. These were multi-electrode arrays consisting of 2x5 silverchloride balls for recording electrically evoked cortical surface activities. Field potentials within penetration tracks were recorded by glass microelectrodes with an open tip filled with Ringers' and an impedance of <5 MΩ. Multi- and single unit activity was recorded by glass insulated (>10 MΩ) tungsten microelectrodes.

One of the interests of the present work is the functional status of the central auditory pathways in absence of peripheral inputs and, if functional to some extent, the possible existence of a cochleotopic representation in AI and AAF. For this purpose it has first to be shown that intracochlear electrical stimulation in the DWC by bipolar electrodes renders a sufficient channel separation in the activity of the primary auditory afferents. Fig. 3 illustrates the results. The threshold currents necessary for activation of single fibres are shown with monopolar upper panel) and bipolar (lower panel) stimulation. It can be seen that with monopolar stimulation there is hardly any spatial separation in efficacy of the

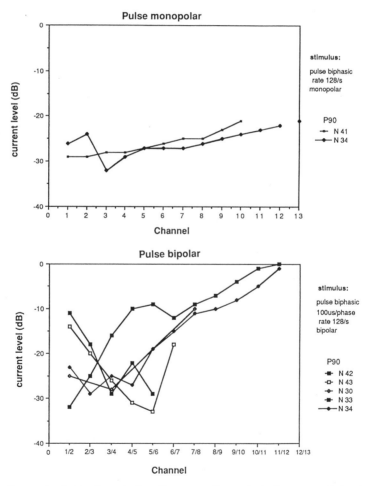

Figure 3. Top: Current thresholds of two single auditory nerve fibers of a deaf cat monopolarly stimulated at different electrodes of the NUCLEUS-22 electrode. Bottom: Current threshold functions of 5 single auditory nerve fibres with bipolar stimulation of electrode pairs indicated.

electrodes. This is different with bipolar stimulation. The thresholds do depend on the site of the electrical stimulus, that is, there is spatial tuning. In fortunate cases, the slopes of the tuning can be up to 13 dB/mm. Thus threshold differences of an apical electrode pair (1/2) versus a basal one (7/8), which have a distance of 4.5 mm can be quite considerable. For this reason in the following consideration always comparisons are made between stimuli applied to electrodes 1/2 and 7/8. Note that in Fig. 3 one of the neurones illustrated (N34) was stimulated monopolarly and bipolarly as well.

Electrical stimulation of primary afferents does infact activate the auditory cortex. This is shown in Fig. 4. This figure sketches the surface of the auditory cortex contralateral to the stimulated cochlea. The sulci are indicated, the dots represents points where recordings were made with surface electrodes. The evoked cortical responses recorded in some of these spots are also displayed in Fig. 4. Stimulation in this case was performed with electrodes (1/8) that is in a wide spacing in order to activate as many afferent fibres as possible.

One can see from the figure that the auditory cortex of a DWC can infact be activated through stimulation of primary auditory afferents. Outside of the auditory areas (not shown in the figure) the potentials are very small or can hardly be recorded at all. The figure also shows that the shapes of the potentials evoked change over the auditory areas al-

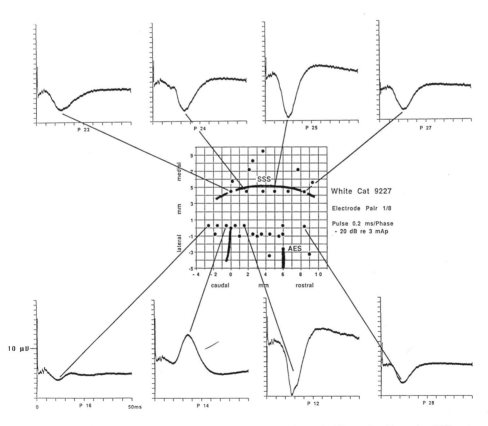

Figure 4. Electrically evoked cortical potentials using bipolar electrode pair 1/8 at a fixed intensity. Different recording positions on the contralateral auditory cortex of DWC 9227. Electrical stimulation with 300 μA biphasic pulses, 0.2 ms /phase, presented at a rate of 2/s.

though in this case the stimulation in the cochlea was not spatially selective as the stimulation electrodes were widely spaced. The potentials to be recorded are bi- or triphasic, the maximum amplitude is found between the anterior and posterior ectosylvian sulci.

A more quantitative evaluation of these findings is given in Fig. 5. Here the amplitudes and latencies for the electrically evoked potentials recorded from caudal, central and rostral sites are shown (P14, P12 and P28, see Fig. 4). It can be seen that the thresholds for stimulation electrodes 1/2 are lower than for electrode 7/8 at caudal and central recording sites but that in the rostral regions the thresholds are similar. The amplitude functions have initially steep slopes and finally reach saturation. In caudal and central regions the intensity functions can have a non-monotonic shape and form a saddle. Saturation may not be reached in these cases with the stimulus currents available. Amplitude are maximal in the centre and reach values of 300 µV.

The latencies of the first dominant peak of the cortical responses are in the order of 20 ms with low intensity stimuli and decline to 13 ms, a value, which is reached when the amplitude have come to their maximum value. In the central and rostral fields the 13 ms latencies remain constant even with high stimulus levels. In the caudal part, however, the latency functions are more complex and latencies less than 13 ms can be found indeed.

In case there is some cochleotopic organization of the auditory cortex in DWC different intracochlear stimulation sites should be more effective in evoking cortical responses than others. This would most easily show up in graphs that depict the threshold differences for different intracochlear electrode pairs. A 50 µV cortical response was used as a threshold criterion, thresholds were calculated from the amplitude-level functions mentioned. A typical result from one DWC is presented in Fig. 6. The upper two graphs depict the thresholds for cortical responses recorded at different caudal-rostral (X) and lateral-medial (Y) directions. The necessary current level for the evocation of a 50 µV response is plotted in the Z-axis. It can be seen in the upper graphs that generally the more

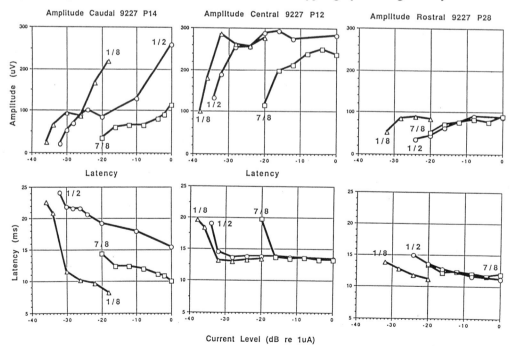

Figure 5. Amplitude and latency input-output functions of electrically evoked surface potentials recorded from three cortical sites of DWC9227 using bipolar scala tympani pairs 1/8, 1/2 and 7/8.

caudal recording sites have lower thresholds, electrode pair 1/2 is more effective by 5–10 dB. However, the slopes of the threshold-planes differ and a sufficiently detailed description of relative efficacy of the stimulus electrodes can only be gained by a subtraction of the threshold values 7/8–1/2. The results of such a computation is given in the lower panel. This graph clearly shows that the different is neither zero nor a constant value. Rather the difference changes from caudal to rostral. This means that electrode pair 1/2 is relatively more effective in caudal recording sites whereas pair 7/8 is more effective rostrally. This finding therefore is an indication for a cochleotopic input of the auditory cortex also in the connatally deaf cat!

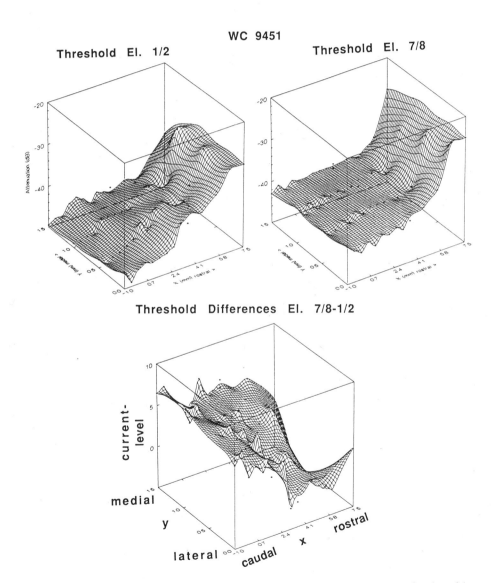

Figure 6. Current thresholds of electrically evoked cortical responses in a deaf white cat as a function of the surface of the auditory cortex from caudal to rostral and lateral to medial. Top left: Electrical stimulation pair 1/2. Top right: Electrical stimulation pair 7/8. Bottom: Calculated threshold difference (threshold level 7/8 minus level 1/2) Threshold criterion 50 μV.

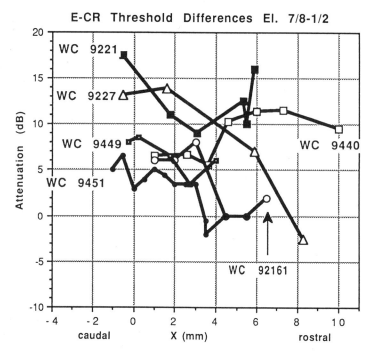

Figure 7. Threshold difference functions (current threshold level 7 / 8 minus 1 / 2) of cortical responses in caudal rostral direction from 6 DWCs at the contralateral auditory cortex.

The above mentioned differences are shown for a total of a 6 DWCs in Fig. 7. Only the values along one single caudal-rostral dimension are shown. Although there is some scatter in the data, the tendency is clear: The differences decrease from caudal to rostral, undergo a local minimum and raise again towards the rostral edge of the recording field. We will come back for a possible explanation later.

In order to gain more insights into the generation of the cortical potentials the electrically evoked activity was recorded in different cortical layers. For this reasons penetrations were made perpendicular to the surface of the cortex up to a depth of 4 mm. Glass electrodes were used. Field potentials were recorded every 300 μm evoked by medium level electrical stimulation through electrode pair 1/2 and 7/8. Field potentials manly represent synaptic activity characterized by EPSPs and IPSPs in the different layers. From the potentials recorded, the one-dimensional Current Source Density (CSD) was calculated as performed for the visual cortex by e.g. Mitzdorff and Singer (1978). Under the assumption that the main current flow is perpendicular to the cortical layers a positive CSD represents an outward flow a negative one an inward flow of current in neurones. An illustration is provided in Fig. 8. In the left hand panels the filed potentials recorded can be seen with stimulation through electrode pairs 1/2 and 7/8. The field potentials change strongly while penetrating from the cortical surface to a depth of 1 mm, whereas in greater depths changes can visually be hardly detected. Consequently the CSD signals (right hand panels) are strong down to about 1.2 mm and then became weaker. This indicates strong synaptic activity in cortical layers 2 and 3 (corresponding of a depth of 1–1.6 mm) and some weak activity in layers 5 and 6, as shown by the weaker CSD signal.

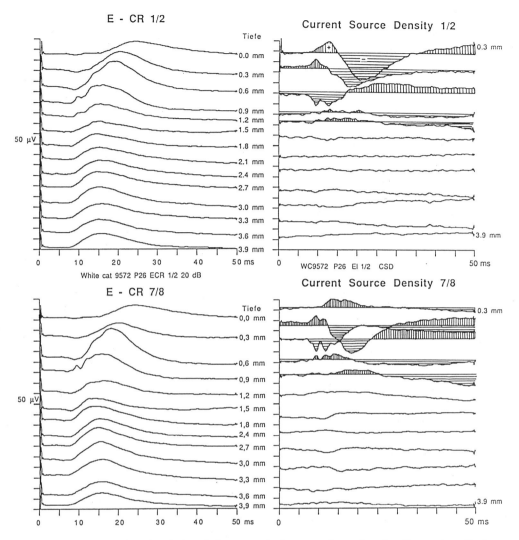

Figure 8. Left: Depth profile of evoked cortical responses in the centre of contralateral auditory cortex with electrical cochlear stimulation of electrode pairs 1 / 2 or 7 / 8. Right: Calculated current source density profiles of the corresponding cortical responses.

The synaptic activity represented by the strong current densities goes of course along with the release of action potentials. These could be recorded as multi- or single-unit activity with the aid of tungsten electrodes. Fig. 9 shows a simultaneous recording of field potentials and multi-unit activity. It can be seen that neurones are mainly active during the leading slope of the field potentials.

A corresponding evaluation of two electrically driven cortical single units are presented in Fig. 10. The unites are stimulated through intracochlear electrodes 1/2 or 7/8 respectively and recorded at depths of about 1.2 mm. Neurone N6 frequently responds with two action potentials per electric stimulus whereas unit N7 generally only releases one spike. The post stimulus time histograms reveal that there is hardly any spontaneous activ-

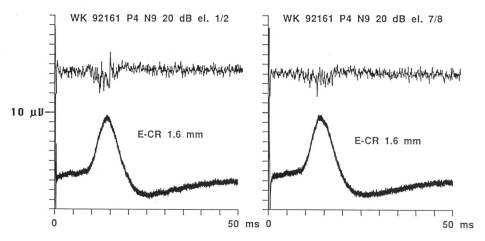

Figure 9. Comparison of multi-unit and field potential recordings. Left: Electrical stimualtion pair 1/2. Right: Electrical stimulation pair 7/8. Pulse stimulation -20 dB re 3 mA, recording depth 1.6 mm, y-scale 10μV/div.

ity and that the action potentials are strictly locked to the electrical stimulus with a latency between 7 and 13 ms depending on stimulus electrodes and current level. If a unit possessed some spontaneous activity, the electric stimulus led to a post-stimulus suppression of activity leading to a silent interval of 30 and more ms after the evoked action potential. The intensity functions and the latencies of these units' activities are further illustrated in Fig. 11. The entrainment increases over a very narrow dynamic range of 3–5 dB after which saturation is reached at least under the barbiturate anaesthesia used for these experiments. The latencies decline to a minimum of 7 ms, this is shorter than the latencies of the field potentials. It seems that an initial action potential causes a volley of activity in a subsequent neuronal circuit, indicated by the field potential.

A question to be discussed further is the cochleotopic organization of the auditory cortex in the DWCs and its relation to the tonotopic organization in normal hearing cats. For this comparison Fig. 12 was prepared. The frequency representation in a caudal-rostral axis in a cat auditory cortex AI is plotted as taken from the papers by Harrison et al. (1991), Merzenich and Reid (1974), Merzenich et al. (1975) and Rajan et al. (1991). The zero-point chosen is our reference point zero at the upper edge of the posterior ectosylvian sulcus. Secondly the second cochlear electrode array used in our experiments plotted against Liberman's (1982) frequency map of the cat cochlea (lower panel). It can be seen that electrode pair 1/2 in a normal cochlea would lie close to the 7 kHz place, whereas pair 7/8 would be close to the 18 kHz. On the other hand the slope of the cortical frequency map is about 1 oct/mm taken as an average of the data of the above authors. Thus, if there is a cochleotopic map in DWC corresponding to the frequency representation in normal cat cortex, then stimulus electrodes 1/2 would be represented about 2 mm rostral of our reference point and pair 7/8 correspondingly 3.5 mm rostral of the reference. In principle the thresholds for cortical potentials evoked by electrical cochlear stimulation in our DWCs are as in the left part of Fig. 13, upper panel. Electrode 1/2 is more effective caudally; the threshold curve is V-shaped. Similarly electrode pair 7/8 produces a V-shaped threshold curve except for a rostral shift and higher absolute thresholds. The reverse is postulated for the AAF. There the tonotopic map in normal cats

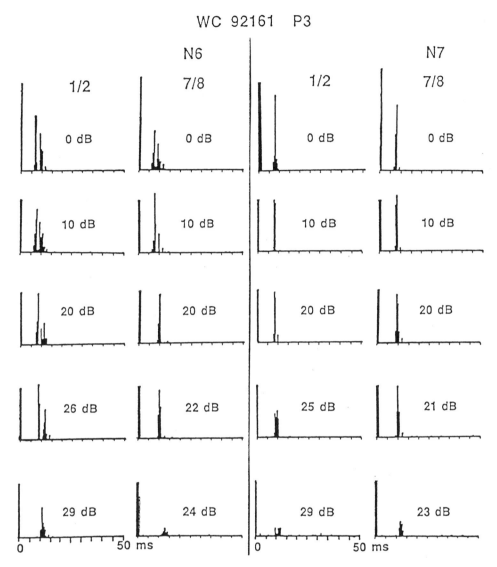

Figure 10. Electrically evoked single unit activity for two AI neurones recorded from the same DWC in the auditory cortex contralateral to the stimulated cochlea. Post-stimulus-time histograms from threshold to saturation current levels at two cochlear stimulation sites (left 1/2, right 7/8).

is reversed indeed (Merzenich et al., 1975). The same is assumed to be the case for the DWC. Thus two similar threshold curves have to be constructed for electrically evoked cortical potentials in the AAF (Fig. 13, right upper panel). From these threshold curves a common threshold curve was constructed displayed in the lower panel of Fig. 13. Accepting inter-animal differences in cortical maps of DWCs which would justify some caudal-rostral shift of the threshold difference curves of Fig. 7 (i.e. of the cochleotopic maps of these animals) one can easily fit the data found with the threshold curve constructed in the lower panel of Fig. 13. Going from caudal cortical places to rostral ones,

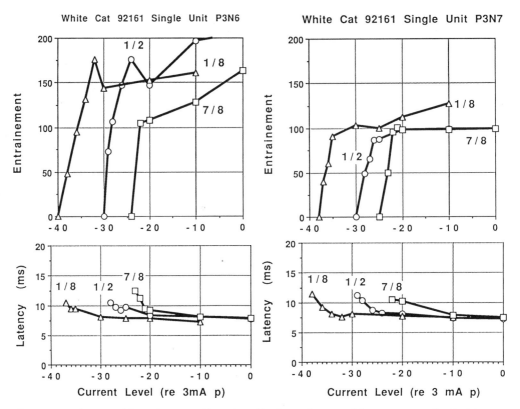

Figure 11. Action potential entrainment and latencies of leading action potentials versus current level of 4 neurones from one DWC recorded in the same track. Narrow bipolar pulsatile electrical stimulation (100 repetitions) using electrode pair 1/2 or 7/8 is compared with the wider bipolar electrode pair 1/8.

an apical to basal cochleotopic map is therefore assumed for AI in DWCs and a basal to apical one in AAF.

In conclusion, a sufficient number of afferent nerve fibres survive in adult DWCs to permit electrical stimulation in a place dependent manner. The spatial resolution is sufficient for safe separation of intracochlear stimulation channels on the level of the auditory nerve. This electrical stimulation evokes cortical responses over AI and AAF which decline rapidly outside the auditory cortex. Similarly multi-unit and single-unit activity can be evoked by electrical stimulation of the cochlea.

There are threshold differences in caudal-rostral direction for stimulation at different cochlear places indicative of a cochleotopic organization of the auditory cortex in DWC. It is concluded that the development of the gross wiring of afferent auditory pathways does not depend on an intact acoustic input, rather spontaneous activity seems to do for this purpose. It will have to be shown whether and how this gross wiring is influenced and eventually changed with chronic electrical stimulation of the auditory nerve in these animals.

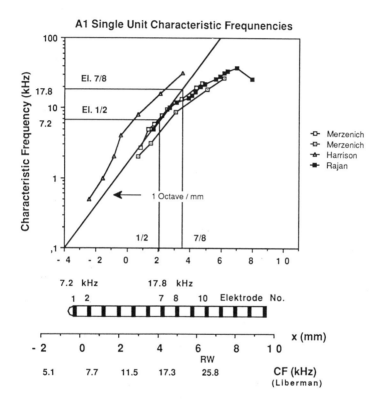

Figure 12. Top: Representation of characteristic frequencies of single unit activity of the primary auditory cortex of cats in caudal-rostral direction (redrawn from Merzenich et al. 1975; Harrison et al. 1991; Rajan et al. 1993). Straight lines indicate the corresponding frequencies of the electrical stimulation sites 1 /2 and 7/8. Bottom: NU-CLEUS-22 electrode array with the CF positions according to Liberman (1982)

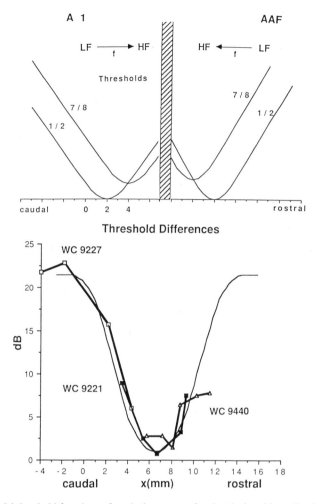

Figure 13. Top: Model threshold functions of cortical responses for electrical cochlear stimulation with electrode pairs 1/ 2 and 7/8. Note the inverse representation from high frequencies (HF) to low frequencies (LF) in the frontal field AAF. Bottom: Combined difference - threshold model function (thin line) with three measured threshold - difference functions from DWCs. The X-position offset and the threshold offset was adjusted to fit one model function because of inter-animal variations.

ACKNOWLEDGMENT

This work was supported by the Deutsche Forschungsgemeinschaft (SFB 269). The participation of Dr. R. K. Shepherd, Melbourne in a part of these experiments is gratefully acknowledged.

REFERENCES

Blamey, P.J. (1995) Factors affecting auditory performance of postlinguistically deaf adults using cochlear implants: Etiology, age, and duration of deafness. NIH Consensus Development Conference Cochlear Implants in Adults and Children. pp. 15–20.

Bosher, S.K. and Hallpike, C.S. (1965) Observations on the histological features, development and pathogenesis of the inner ear degeneration of the deaf white cat. Proc. Royal Soc. Lond. 162: 147–170.

Dowell, R.C., Blamey, P.J. and Clark, G.M. (1995) Potential and limitations of cochlear implants in children. Ann. Otol. Rhinol. Laryngol. 104 (Suppl. 166) 324–327.

Harrison, R.V., Nagasawa, A., Smith, D.W., Stanton, S. and Mount, R.J. (1991) Reorganization of auditory cortex after neonatal high frequency cochlear hearing loss. Hear. Res. 54, 11–19.

Hartmann, R., Topp, G. and Klinke, R. (1984) Discharge patterns of cat primary auditory fibers with electrical stimulation of the cochlea. Hear. Res. 13: 47–62.

Heid, S., Jaehn-Siebert, T.K., Klinke, R., Hartmann, R., Langner, G. (1996) Afferent projection patterns in the auditory brain stem in normal cats and congenitally deaf white cats. Hear. Res. in press.

Heid, S., Jaehn-Siebert, T.K., Klinke, R. Hartmann, R. and Langner, G. (1996) Quantitative analysis of projections from cochlear nucleus to inferior colliculus in connatally deaf white cat by means of retrograde transport of a fluorescent tracer. Proc. 24th Goettingen Neurobiol. Conf. Vol. II, 515.

Heid, S., Hartmann, R. and Klinke, R. (1995) Degeneration of the spiral ganglion in the deaf white cat. Proc. 23rd Goettingen Neurobiol. Conf. Vol. II, 593.

Heil, P., Rajan, R. and Irvine, D.R.F. (1992) Sensitivity of neurons in cat primary auditory cortex to tones and frequency-modulated stimuli. II: Organization of response properties along the isofrequency dimension. Hear. Res. 63: 135–156.

Lehnhardt, E. and Bertram, B. (Eds.): Rehabilitation von Cochlea-Implant-Kindern. Springer Berlin 1991.

Leake, P.A. Hradek, G.T. (1988) Cochlear pathology of long term neomycin induced deafness in cats. Hear. Res. 33: 11–34.

Lenarz, T., Lehnhardt, E. and Bertram, B.: 2nd CI-Workshop, Hannover. Thieme Verlag Stuttgart, 1994.

Liberman, M.C. (1982) The cochlear frequency map for the cat: Labeling auditory-nerve fibers of known characteristic frequency. J. Acoust. Soc. Am. 72: 1441–1449.

Mair, I.W.S. (1973) Hereditary deafness in the white cat. Acta Otolaryngol. Suppl. 314: 1–53.

Merzenich, M.M. and Reid, M.D. (1974) Representation of the cochlea within the Inferior Colliculus of the cat. Brain Res. 77: 397–415.

Merzenich, M.M., Knight, P.L. and Roth, G.L. (1975) Representation of cochlea within primary auditory cortex in the cat. J. Neurophysiol. 38: 231–249.

Mitzdorf, U. and Singer, W. (1978) Prominent excitatory pathways in the cat visual cortex (A17 and A18): A current source density analysis of electrically evoked potentials. Expl. Brain Res. 33, 371–394.

Moore D. R. (1981) Development of the cat peripheral auditory system: Input-output functions of cochlear potentials. Brain Res. 219: 29–44.

Rajan, R., Irvine, D.R.F., Wise, L.Z. and Heil, P. (1991) Effect of unilateral partial cochlear lesions in adult cats on the representation of lesioned and unlesioned cochleas in primary auditory cortex. J. Comp. Neurol. 338, 17–49.

Reinoso-Suarez, F. (1961) Topographischer Hirnatlas der Katze, Merck, Darmstadt.

Romand, R. (1983) Development in the frequency selectivity of auditory nerve fibers in the kitten. Neurosci. Letters 35, 271–276.

Schwartz, I.R. and Higa, J.F. (1982) Correlated studies of the ear and brainstem in the deaf white cat: Changes in the spiral ganglion and the medial superior olivary nucleus. Acta Otolaryngol. 93, 9–18.

53

RESPONSES OF THE AUDITORY NERVE TO HIGH RATE PULSATILE ELECTRICAL STIMULATION

Comparison between Normal and Deafened Rats

J. S. Zhang,[1] M. W. Vischer,[2] C. A. Haenggeli[1*] and E. M. Rouiller[1†]

[1]Institute of Physiology
University of Fribourg
Rue du Musée 5, CH-1700 Fribourg, Switzerland
[2]ENT Clinic
University Hospital, Inselspital
CH-3010 Bern, Switzerland

INTRODUCTION

The new speech-processing strategy, the so called "Continuous Interleaved Sampling" (CIS) strategy, used in cochlear implant subjects is based on a paradigm of stimulation at high rates where 1000–1500 pulses/s (pps) are used on each channel (Wilson et al., 1991; McDermott et al., 1992). Recent psychophysical studies on cochlear implant subjects have demonstrated, with the CIS as compared to previous paradigms of stimulation, an improvement in speech perception (Wilson et al., 1991; McDermott et al., 1992; Pelizzone et al., 1995). However, electrical stimulation of the auditory nerve (AN) at such high rates will affect its responses because of the neurons' refractory period. Furthermore, prolonged electrical stimulation at high rate might influence neurons' metabolism, possibly leading to fatigue. Therefore, study on the physiological behavior of the AN when stimulated by cochlear implant at higher rates is important to better understand the speech coding procedure, as well as to provide insight into the safety of such new speech processing strategy. A cochlear implant animal model has been established in the rat (Hall, 1990; Vischer et al., 1994; Zhang et al., 1996), in order to address such issues that might not be conducted routinely in patients.

* Present Address: ENT Clinic, Cantonal University Hospital, Rue Micheli-du-Crest 24, CH-1211 Genève 4, Switzerland.
† Correspondence to: E.M. Rouiller at the address above. Tel: (+41) 26 300 86 09; fax: (+41) 26 300 97 34; E-mail: Eric.Rouiller@unifr.ch.

Acoustical Signal Processing in the Central Auditory System
edited by Syka, Plenum Press, New York, 1997

The goal of the present study was to establish the response properties of the AN to high rate pulsatile stimulation in normal rats, as assessed by its electrically evoked compound action potentials (ECAP), defined as the first vertex positive peak and negative trough (P1, N1 waves) of the electrically evoked auditory brainstem responses (EABR). These properties will then be compared to those derived from rats deafened prior to implantation, which are likely to represent a better model to the deaf ear of cochlear implant patients.

METHODS

1. Surgical Preparation and Neomycin Treatment

Adult Long-Evans or Sprague-Dawley rats (n = 16) were deeply anesthetized (sodium pentobarbital, 6.5 mg/100 g, i.m.), and two stimulating electrodes (Teflon-coated platinum iridium wires) were implanted unilaterally in the apical turn and basal turn of the cochlea, respectively (Vischer et al., 1994; Zhang et al., 1996). Four of the rats received a unilateral local treatment with neomycin before implantation. After drilling the holes for electrode insertion, the cochlea was perfused with 30 µl of a solution of 1% neomycin (2 rats) or 4% neomycin (2 rats) in saline. The neomycin administration was repeated 6 times at intervals of 5 min. Following the perfusion with neomycin, the implantation was continued as described above.

2. Stimulation Paradigms and Electrophysiological Recordings

The stimuli, consisting of 20 µs pulses, were delivered in an alternating bipolar "apex to base" configuration at a rate of 50 pps. The EABR, averaged for 600 presentations (Fig.1 A), were recorded from a "vertex" screw serving as an active electrode and a reference electrode inserted behind the periotic capsule contralateral to the implanted ear (Vischer et al., 1994). The growth function of the ECAP was established on the basis of recordings made at 10 increasing intensity levels (100 to 1000 µA by steps of 100 µA) for both implanted untreated and implanted neomycin treated rats (Fig.1 B, C and D).

3. Pulsatile Stimulation Paradigm

The effects of high rate stimulation were studied by delivering electrical stimuli in the form of pulse trains which lasted 200 ms, followed by a pause of 200 ms before the next train (50% duty cycle). This duty cycle was chosen because it better approximates the stimulus waveforms delivered to cochlear implant patients listening to speech (Tykocinski et al., 1995) than continuous stimulation. The train consisted of consecutive monophasic square waves of 20 µs separated from each other by interpulse intervals ranging from 10–0.5 ms, corresponding to intra-train pulsatile rates ranging from 100 pps up to 2000 pps, by steps of 100 pps. The intensity of stimulation was usually 500 µA above the ECAP threshold. An average ECAP was derived from 300 train presentations in one polarity followed by another 300 train presentations in the opposite polarity.

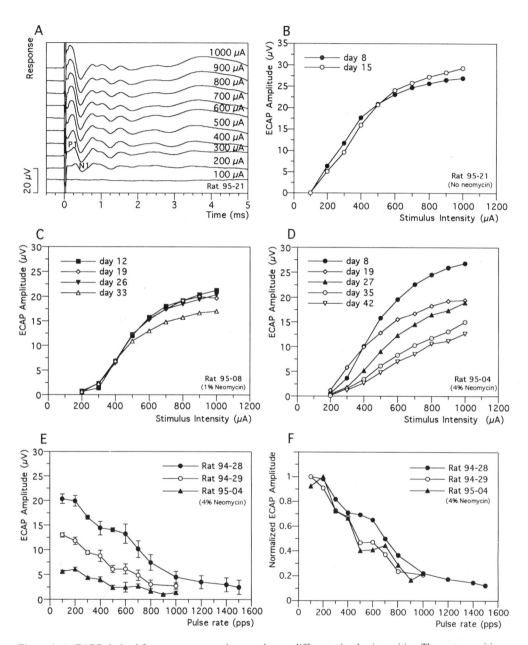

Figure 1. A. EABR derived from a non-neomycin treated rat at different stimulus intensities. The vertex positive wave and the following negative trough of ECAP are labeled with P1 and N1, respectively. B. Growth function curves of the ECAP obtained from two different recording experiments on the same animal as shown in panel A. C & D. Growth function curves of ECAP derived from different recording experiments in deafened rats. Note that the rat in panel C was treated with 1% neomycin while the rat in panel D was treated with 4% neomycin. E. Average ECAP amplitude as a function of pulse rate, in normal and neomycin treated (95–04) rats. F. Same data as in panel E, but the amplitude of ECAP was normalized (value of 1 given to the maximal ECAP amplitude observed).

4. Cochlear Histology

The animals were sacrificed, at the end of the experiments, by intraperitoneal administration of a lethal dose of sodium pentobarbital. The rats were perfused transcardially with 100 ml saline followed by 1000 ml of 4% paraformaldehyde in 0.1M phosphate buffer, pH 7.4 (PB). The brain and both cochleae were dissected and saved for histological processing. Following decalcification (10 days in 22.3% EDTA), infiltration (basic resin/liquid + activator) and plastic embedding (120 min polymerization), the cochleae were cut at 0.5 μm thickness in a plane parallel to the midmodiolar plane. Every tenth section was mounted, Nissl stained and coverslipped.

12 cochleae were categorized in three groups of animals: intact, implanted, implanted and neomycin treated (Table I). The area of the Rosenthal's canal was measured in sections taken from apical, middle and basal turns, using the software NIH Image 1.58. Blood vessels or isolated bony structures within Rosenthal's canal larger than 100 μm^2 were not included in the area computation. All spiral ganglion neurons (SGN) containing a nucleus were counted, excluding those cells showing cytoplasmic vacuoles or other degenerative changes. Type I and type II SGN were counted separately, on the basis of standard criteria (Keithley and Feldman, 1979). The total volume densities of SGN were estimated in implanted untreated and implanted neomycin treated cochleae and compared to the density of intact cochleae (Table I).

Table I. SGN counts in three types of cochleae: intact (right side), implanted, implanted and neomycin treated; SGN density comparisons of implanted, implanted and neomycin treated cochleae with that of intact cochleae

| Cochlea ID | Nb of electrodes | Neomycin | Spiral ganglion neurons | | | Density (Nb/mm^3) |
			Type I	Type II	Total	
94-45 R	—	—	14 458	1 076	15 534	91 951
95-08 R	—	—	11 020	500	11 520	64 836
95-09 R	—	—	11 756	1 054	12 810	80 769
95-11 R	—	—	12 454	815	13 269	70 594
95-22 R	—	—	12 307	793	13 100	86 717
95-26 R	—	—	10 979	685	11 664	82 289
93-22 L	2	—	11 664	717	12 381	97 788
93-24 L	2	—	12 454	1 000	13 454	83 800
95-26 L	3	—	5 873	315	6 188	46 151
95-08 L	3	1%	9 256	685	9 940	62 974
95-09 L	3	1%	9 963	946	10 909	65 312
95-11 L	3	4%	5 211	185	5 396	35 391

Density comparison	Intact	Implanted	1% Neomycin	4% Neomycin
Average	79 526	75 913	64 143	35 391
Stdev.	10 108	26 707	1 653	—
T-test	—	0.77	0.09	—
% Difference	—	-5%	-19%	-55%

RESULTS

Typical EABR traces averaged for 600 single pulses delivered at different current intensities are displayed in Fig.1 A, as representative of a normal rat (95–21). The threshold of the EABR was defined as the lowest intensity at which the ECAP was visually detected. The threshold ranged from 150 to 200 μA across individual rats. The ECAP amplitude and latency were the main parameters to characterize the response of the AN to electrical stimulation of the cochlea. The growth functions in Fig.1 B demonstrate in the same rat the progressive increase of the ECAP amplitude for increasing stimulus intensities. The two growth functions derived from two different experiments a week apart demonstrate the stability of the AN responses in this rat following cochlear implantation. In contrast, the growth functions established in neomycin treated rats exhibited a progressive decrease of the ECAP amplitude from one experiment to the next (Fig.1 C and D), in particular in the case of treatment with neomycin at high concentration (Fig.1 D). This decrease of ECAP amplitude reflected a progressive loss of SGN due to the neomycin treatment.

The SGN counts, shown in Table I, confirm the idea that the SGN loss resulting from neomycin treatment is the main reason for the ECAP amplitude decrease (Fig.1 C and D). As seen in Table I, there was only a slight loss of SGN (5%) in the left implanted cochlea of normal rat and the inner ear structures remained relatively unchanged as compared to the right intact cochlea. It can be concluded that the implantation itself does not affect significantly the cochlea except in the close vicinity of the implanted electrodes where a disruption of the organ of Corti occurred, followed by local SGN degeneration (e.g. 95–26 L in Table I). In sharp contrast, the total number of SGN dropped by a proportion of 19% and 55% after treatment with 1% and 4% neomycin of left cochlea, respectively, as compared to the right intact cochlea. Accordingly, the organ of Corti and other inner ear structures were severely disrupted and the number of SGN was significantly reduced (compare Fig. 2 A and B).

In the train paradigm experiments, the ECAP amplitudes were measured for individual pulses in the train and then averaged giving a mean ECAP amplitude for each pulse rate. The average amplitude of ECAP decreased with increasing pulse rate (Fig.1 E), both in normal and neomycin treated rats. Furthermore, the ECAP, irrespective of the neomycin treatment, showed a comparable relative decrease for increasing pulse rates: the ECAP amplitude was reduced by more than 50% at rates around 600–700 pps and then reached to a plateau (80–85% decrease) as the pulse rate reached 1000 pps (see the normalized data in Fig.1 F).

DISCUSSION AND CONCLUSIONS

In intact untreated cochleae, the number of SGN obtained here (Table I) is consistent with a previous report (Keithley and Feldman, 1979). The local administration of 1% (not shown) and 4% neomycin (Fig.2 B) led to nearly complete loss of hair cells. Furthermore, neomycin treatment resulted in a shrinkage of cochlear structure and a slight degeneration of SGN at low concentration of neomycin (1%), while at high concentration (4%) the loss of SGN was severe. These results are consistent with previous studies (Hall, 1990; Leake et al., 1992) except that the concentration of neomycin used in the present study was relatively low. These data furthermore verified the notion, as Hall (1990) proposed, that the degree of survival of SGN in the rat is correlated with the magnitude of ECAP.

For increasing pulse rates, the average ECAP amplitude decreased (Fig.1 E and F) whereas the average ECAP latency increased (not shown). This indicates that the reduction of recruitment of the AN fibers began at low pulse rate (200 pps), and then progressively became more prominent at higher pulse rate to reach a plateau at pulse rate of 1000 pps. A certain proportion of AN fibers are in their refractory period of the response to the preceding pulse and therefore can not be recruited by the next pulse presented at a short interval. In neomycin deafened rats, although the number of SGN is lower than in normal rats, the decrease of the ECAP amplitude in percentage as a function of pulse rates is comparable to that observed in normal rats. These data indirectly provide evidence that the re-

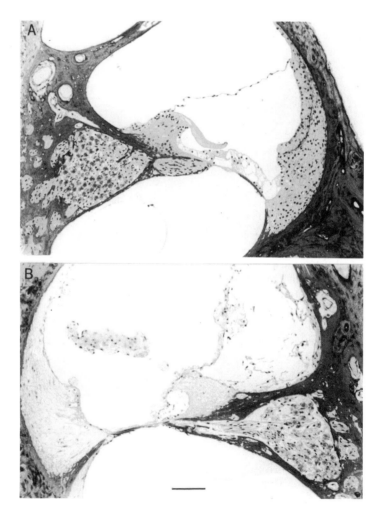

Figure 2. Photomicrographs showing histology of the right intact cochlea (A) and that of the corresponding left cochlea which was implanted and perfused with 4% neomycin (B). A. Portion of a midmodiolar section of the right intact cochlea showing the normal organ of Corti and high density of SGN. B. Portion of a midmodiolar section of the left implanted cochlea after treatment with 4% neomycin. Note the total disruption of the organ of Corti and the significant loss of SGN. Bar = 100 μm (for panels A and B).

duced population of SGN in cochlear implant patients behave as the normal population of SGN, in response to pulsatile stimulation at high rates. Although the refractory period induces a decrease of the number of AN fibers recruited at high rates, the number of SGN activated at each pulse is sufficient to code with precision the rapid variations in amplitude of speech. Although stimulation at low rate might activate more SGN, it provides a worse speech recognition than the CIS strategy (Wilson et al., 1991; Pelizzone et al., 1995). This suggests that the ability to represent with precision the fine temporal properties of speech (at high stimulation rates as in the CIS) is more important than having a large number of SGN recruited (as obtained by stimulation at low rates), even if the population of functional SGN has already been reduced by the deafness.

ACKNOWLEDGMENTS

The authors thank A. Tempini, C. Roulin and V. Moret for excellent technical assistance, J. Corpataux and B. Morandi for taking care of the rats in the animal room, R. Staub for software, B. Aebischer for hardware, A. Pisani and A. Gaillard for mechanics. This research project was supported by the Swiss National Science Foundation (Grant No 31-28572.90; 32-36482.92; 31-43422.95; 31-45731.95), the Roche Research Foundation (Basle) and the CIBA-GEIGY Jubiläums Stiftung (Basle, Switzerland). C.A. Haenggeli was financially supported during one year by the Sandoz-Stiftung zur Förderung der Medizinisch-Biologischen Wissenschaften (Basle, Switzerland).

REFERENCES

Hall, R.D. (1990) Estimation of survival spiral ganglion cells in the deaf rat using the electrically evoked auditory brainstem response. Hear. Res. 45, 123–136.

Keithley, E.M. and Feldman, M.L. (1979) Spiral ganglion cell counts in an age-graded series of rat cochleas. J. Comp. Neurol. 188, 429–442.

Leake, P.A., Snyder, R.L., Hradek, G.T. and Rebscher, S.J. (1992) Chronic intracochlear electrical stimulation in neonatally deafened cats: Effects of intensity and stimulating electrode location. Hear. Res. 64, 99–117.

McDermott, H.J., McKay, C.M. and Vandali, A.E. (1992) A new portable sound processor for the University of Melbourne/Nucleus Limited multielectrode cochlear implant. J. Acoust. Soc. Am. 91, 3367–3391.

Pelizzone, M., Boex-Spano, C., Sigrist, A., Francois, J., Tinembart, J., Degive, C. and Montandon, P. (1995) First field trials with a portable CIS processor for the ineraid multichannel cochlear implant. Acta Otolaryngol. (Stockh) 115, 622–628.

Tykocinski, M., Shepherd, R.K. and Clark, G.M. (1995) Reduction in excitability of the auditory nerve following electrical stimulation at high stimulus rates. Hear. Res. 88, 124–142.

Vischer, M.W., Häusler, R., Rouiller, E.M. (1994) Distribution of Fos-like immunoreactivity in the auditory pathway of the Sprague-Dawley rat elicited by cochlear electrical stimulation. Neurosci. Res. 19, 175–185.

Wilson, B.S., Finley, C.C., Lawson, D.T., Wolford, R.D., Eddington, D.K. and Rabinowitz, W.M. (1991) Better speech recognition with cochlear implants. Nature 352, 236–238.

Zhang, J.S., Haenggeli, C.A., Tempini, A., Vischer, M.W., Moret, V. and Rouiller, E.M. (1996) Electrically induced Fos-like immunoreactivity in the auditory pathway of the rat: effects of survival time, duration and intensity of stimulation. Brain Res. Bull. 39, 75–82.

54

ACOUSTICALLY AND ELECTRICALLY EVOKED AUDITORY BRAINSTEM RESPONSES

Similarities and Differences

H. von Specht,[1] Z. Kevanishvili,[2] M. Hey,[1] R. Muehler,[1] and K. Begall[1]

[1]Department of Experimental Audiology
Otto-von-Guericke University of Magdeburg
39120 Magdeburg, Germany
[2]Centre of Audiology
380079 Tbilisi, Georgia

1. INTRODUCTION

In early models of the generation of the human auditory brainstem response (ABR), the successive components were considered to arise in consecutive structures of the auditory pathway (e.g. Lev and Sohmer, 1972). Waves I, II, III, IV, and V were in particular believed to originate from the 8th nerve, cochlear nuclei, superior olivary complex, lateral lemniscus and inferior colliculus, respectively. The sources of the latest ABR constituents, i.e. of Waves VI and VII, were connected to still further relays: the medial geniculate body and the acoustic radiation, respectively. Later, these early models underwent alterations (Kevanishvili, 1980; Caird and Klinke, 1984). Both neurophysiological and neurosurgical evidence was presented substantiating subcollicular or even sublemniscal origin of the main ABR complex waves including Wave V. The loci of generation of the successive components were shifted downward in proportion. Wave VI, for instance, has been stated to be related to the inferior colliculus (Hall, 1992). Recording of electrically evoked ABR in cochlear implant patients offers a unique opportunity to revise the models of ABR generation.

2. METHODS

The considerations of this paper are based on recordings of acoustically evoked ABR in normal hearing subjects and of electrically evoked ABR in patients with a Nucleus 22-channel cochlear implant. Standard techniques were used in both ABR recordings. The electrodes were placed on the vertex and on both mastoids. The band-pass filter settings of the amplifiers were 100–3000 Hz. A speech processor interface was employed

Acoustical Signal Processing in the Central Auditory System
edited by Syka, Plenum Press, New York, 1997

for programming purposes and used to connect the speech processor of the cochlear implant to the evoked potential equipment.

3. RESULTS AND DISCUSSION

The ABR elicited to higher-intensity acoustical stimuli comprises seven components: Waves I - VII. As documented recently (Abbas and Brown, 1991; Hall, 1992; Kasper et al., 1992), the electrically evoked ABR registered in patients with cochlear prosthesis shows pronounced Waves II, III and V. Moreover, the electrically evoked ABR comprises dubious, if any, Wave IV and lacks Waves VI and VII (Fig. 1). The different pattern of the electrically evoked ABR also revealed in the present study cannot be explained by difficulties in identifying the waves. This problem has only to be taken into account with regard to Wave VII, the most labile ABR component of the highest threshold in wave identification. This, however, cannot be adopted for Waves IV and VI for two reasons: (1) The amplitude of the electrically evoked ABR is generally high, reaching or exceeding 1 μV. The acoustically evoked ABR of similar magnitudes, as a rule, exhibits all the deflections, including Waves IV and VI. (2) The threshold of Wave IV is usually lower than that of Wave II, while the threshold of Wave VI lags behind that of Wave II systematically and matches that of Wave III closely. If the dubious appearance or absence of Waves IV and VI were related to detection problems, then Waves II and III too should be indistinct or missing.

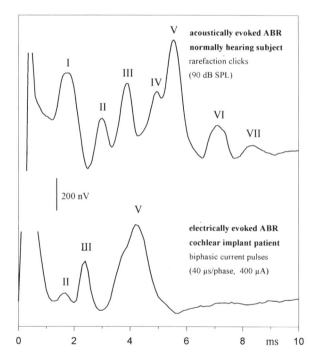

Figure 1. Acoustically and electrically evoked ABR. In the electrically evoked ABR, Wave I is obscured due to overlapping of stimulus artifact; peak latencies are shorter due to direct stimulation of auditory nerve fibres avoiding middle- and inner-ear mechanisms; Wave IV is dubious and Wave VI and VII are absent, probably for reasons outlined in the paper (ipsilateral recordings, averaging number: N=2000, stimulus repetition rate 17/s).

To understand the reason of the peculiar shape of the ABR in implant patients we need to consider the mechanism of its generation. In a revision of the simplified early models of successive ABR components linked with the consecutive brain structures, the more accurate assumption of the subcollicular origin of the whole ABR complex calls for downward shifting of sequential origins of ABR components (Kevanishvili, 1980; Moore, 1987; Hall, 1992; Møller, 1994). Thus, Wave III being connected with the trapezoid body, Wave IV has been reported to be the outcome of the next structure, the lateral lemniscus. Wave V being linked with the lateral lemniscus, Wave VI has been attributed to the following formation, the brachium of the inferior colliculus, and Wave VII to a still more proximal structure.

Both the former and contemporary theories fail to account for the dubious appearance of Wave IV and absence of Waves VI and VII in the electrically evoked ABR. Nor do the dubious appearance of Wave IV and absence of Waves VI and VII fit the idea of their initiation in the structures, following the sources of Waves III and V, respectively. In both electrical and acoustical stimulation, it would appear that activation of the generators of Waves III and V (the trapezoid body and the lateral lemniscus, respectively) be followed by activation of the consecutive structures, the lateral lemniscus and both the brachium of the inferior colliculus and still more proximal structures, respectively. This, in turn, should result in initiation of Waves IV, VI, and VII if the sequential principle was correct. When pondering the dubious appearance or absence of Wave IV, one should recall its particular sensitivity to the phase of the onset of the acoustic stimulus: an increase with stimuli of a rarefaction phase and a decrease or elimination with those of a condensation phase (Kevanishvili and Aphonchenko, 1981).

Rarefaction and condensation induce opposite displacements of the basilar membrane. In implant patients the basilar membrane displacements and the whole traveling wave mechanism are lacking. The phase information is not directed towards the phase-sensitive neural units (presumably in the superior olivary complex, e.g. in the medial olivary nucleus; Moore, 1987). Hence, they are either not properly activated or not activated at all. Accordingly, the respective potential, Wave IV, either is not properly initiated or not initiated at all.

When judging the accustomed theories to be inappropriate for explanation of the absence of Waves VI and VII, an alternative concept should be equally considered. It was advanced in the eighties (Thümmler et al., 1981; Pantev and Pantev, 1982), but did not find followers and was actually forgotten. The concept denies origination of Waves V, VI, and VII at sequential brainstem levels. Instead, Waves V thru VII are assumed to be the outcome of the same structures, discharging due to repeated activation from basal and more apical parts of the cochlea, respectively. Repeated activation governing the repeated dischargings and, hence, the gaps between Waves V, VI, and VII, are related to the time required for the traveling wave to pass from the basal part of the cochlea to the consecutive regions of the apical part. The concept, while simplifying intricate cochlear and retrocochlear events, well explains the results of investigations with selective masking and the derived response techniques (Don and Eggermont, 1978; Thümmler and Tietze, 1984). As the cutoff frequency of the high-pass masking noise is successively lowered in these investigations, together with a prolongation of the wave latencies the earlier peaks begin to disappear. These findings suggest that the whole cochlear partition contributes to the ABR, however the supplies of basal and apical parts are different (Fig. 2).

The absence of Waves VI and VII in the electrically evoked ABR substantiates the non-traditional concept of their origin. Proceeding from this concept, the absence of Waves VI and VII could satisfactorily be explained. In patients with cochlear prosthesis the traveling wave mechanism is missing. The applied current pulses trigger auditory nerve fibres of both basal and apical parts of the cochlea simultaneously, but not succes-

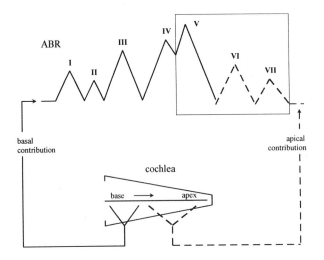

Figure 2. Schematic representation of contributions to the ABR from basal and apical parts of the cochlea. The basal part more contributes to Waves I-V, the apical part to Waves VI and VII. The common frame over Waves V, VI, and VII signifies their origination in the same brainstem structure.

sively. Hence, the respective generator structure in the brainstem is activated synchronously, but not repeatedly. This promotes initiation of high-amplitude Wave V while the absence of the repeated dischargings determines the lack of later components: Waves VI and VII.

ACKNOWLEDGMENTS

This study was supported by Bundesministerium für Bildung, Wissenschaft, Forschung und Technologie (FKZ: 07 NBL 04/01 ZZ9505).

REFERENCES

Abbas, P.J. and Brown, C.J. (1991) Electrically evoked auditory brainstem response: Growth of response with current level. Hear. Res. 51, 123–138.

Don, M. and Eggermont, J.J. (1978) Analysis of the click-evoked brainstem potentials in man using high-pass noise masking. J. Acoust. Soc. Am. 63, 1084–1092.

Hall, J.W. (1992) Handbook of Auditory Evoked Responses. Allyn and Bacon, Boston, pp. 575–579.

Kasper, A., Pelizzone, M. and Montandon, P. (1992) Electrically evoked auditory brainstem responses in cochlear implant patients. ORL 54, 285–294.

Kevanishvili, Z. (1980) Sources of the human brainstem auditory evoked potential. Scand. Audiol. 10, 75–82.

Kevanishvili, Z. and Aphonchenko, V. (1981) Click polarity inversion effects upon the human brainstem auditory evoked potential. Scand. Audiol. 10, 141–147.

Lev, A. and Sohmer, H. (1972) Sources of averaged neural responses recorded in animal and human subjects during cochlear audiometry (Electrocochleogram). Arch. Klin. Exp. Ohr.- Nas.- u. Kehlk. Heilk. 201, 79–90.

Møller, A.R. (1994) Neural generators of auditory evoked potentials. In: Jacobson, J. (ed.), Principles and Applications in Auditory Evoked Potentials. Allyn and Bacon, Boston, pp. 23–46.

Moore, J.K. (1987) The human auditory brainstem as a generator of auditory evoked potentials. Hear. Res. 29, 33–43.

Pantev, Ch. and Pantev, M. (1982) Derived brain stem responses by means of pure-tone masking. Scand. Audiol. 11, 15–22.

Thümmler, I., Tietze, G. and Matkei, P. (1981) Brain-stem responses when masking with wide-band and high-pass filtered noise. Scand. Audiol. 10, 255–259.

Thümmler, I. and Tietze, G. (1984) Derived acoustically evoked brainstem responses by means of narrow-band and notched-noise masking in normal hearing subjects. Scand. Audiol. 13, 129–137.

55

DISORDERS OF SOUND SOURCE LOCALIZATION AND AUDITORY EVOKED POTENTIALS IN PATIENTS WITH TEMPORAL EPILEPSY

J. A. Altman, L. M. Kotelenko, and S. F. Vaitulevich

I.P.Pavlov Institute of Physiology
Russian Academy of Sciences
St.Petersburg, Russia

INTRODUCTION

Many data in literature give evidence that damages of the higher auditory centers (temporal areas of the cortex in humans) result in changes of auditory localization (Baru and Karaseva, 1972; Altman, 1978; Durlach et al., 1981; Hari, 1995). In particular, our earlier work (Altman et al., 1987) showed that a local damage of the cortical temporal area was followed by significant disorders in ability of localization of moving auditory images in patients suffering from temporal epilepsy.

As is known, in many cases of epilepsy the pathological process involves not only cortical areas of the temporal lobe but also its medio-basal structures, especially the hippocampus.

Since hippocampus is one of the brain centers taking important part in perception of the outer space parameters (O'Keefe and Nadel, 1978; O'Keefe, 1983) it seems of interest to evaluate a possibility of its influence on localization function performance.

On the other hand as is well known, the temporal cortical area of the brain hemispheres is one of the main generators of the long-latency auditory (N_1-P_2) evoked potentials (AEPs). It was established that in the human intact brain these electrical responses to sound signals reflect some processes related to sound source localization (see review Altman and Vaitulevich, 1990). As to the auditory evoked potentials in cases of brain pathology, the up-to-date data in literature describe mainly distortion of the short-latency auditory evoked potentials. There are only single observations concerning the long-latency AEPs in cases of brain pathology which show the possibility of augmentation or diminution of some of the AEP-components (Drake et al., 1986).

With all this in mind, the task of the present investigation was formulated while studying patients with temporal epilepsy: i/ to estimate the role of the temporal area of the auditory cortex and medio-basal structures in disorders of localization of the moving fused auditory images /FIs/; ii/ to compare possible disorders of the localization function with corresponding AEP-changes.

LATERALIZATION OF MOVING FIs (EXPERIMENT 1)

Subjects And Methods

31 patients, aged 18–39, with diagnosed temporal epilepsy (polymorphic paroxysms up to several times per day) were investigated. Epileptic focus locations were determined on the basis of complex investigations of the patients, including clinical, psychological, electrophysiological and X-ray examinations. The investigations were performed with the background anticonvulsive therapy. The patients were differentiated in two groups.

In 14 patients of the first group the initial focus of epileptic activity was localized in temporal cortex area (on the left in 5 patients, on the right - 6, bilateral - in 3 others). Two of the patients were treated surgically, which showed the pole of the temporal lobe.

17 patients of the second group suffered from the damage of both temporal area of the cortex and the medio-basal structures (on the left in 8 patients, on the right in 5, bitemporal - in 4 others). In 10 of the 17 patients the damage localization was supported during surgery treatment. The pole of the temporal lobe, anterior part of the hippocampus and amygdala were ectomized under the control of electrocortico- and subcorticograms, during the open surgery as just these regions had shown pronounced convulsive activity. In 7 other patients localization of convulsive focuses was determined on the basis of the above complex investigation.

Dichotic stimulation by clicks through the earphones, causing a sensation a fused auditory image (FI) was used in this test. The clicks were produced by a square-wave pulse generator (each pulse of 0.4 ms duration) with 2 independent channels. The click rate was 60 clicks per sec. Binaurally presented click trains with an interaural time delay (ΔT_{max}= 0.05, 0.2, 0.4, 0.6 ms) linearly changing from these maximal values to zero were employed. In case of healthy subjects FI gradually shifted from the leading ear to the midline (Altman et al., 1987). The click trains of 1 sec duration were used. Intensity of the sound signals was controlled by attenuators.

At first monaural hearing thresholds were determined for the right and left ears, separately, in response to the click train.

The intensity of sound signals was 40 dB above the hearing threshold (40 dB SL i.e. 40 dB above sensation level) as measured at each ear.

According to the previous instruction the subjects had to watch the movement of the fused auditory image (FI) and, after the signal cessation, to show at the head surface the points of the movement beginning and of its end. Positions of these points were fixed and thus the length of the FI movement trajectory was evaluated in degrees. The results were averaged over 5 measurements.

Control data were obtained on 9 healthy subjects aged from 21 to 35 years, with normal hearing.

Statistical processing was performed with t-criterion by Student-Fisher (with the significance level p<0.05) to compare difference significance level between the means.

The quantitative evaluation of the length of the trajectory of FI movement in detail, was considered in the paper (Altman et al., 1987).

Results

In healthy subjects at the initial ΔT_{max}=0.05 ms the FI movement trajectory was very short and located near to the head midline. With augmentation of the T_{max} the FI movement trajectory increased and achieved about 90 degrees at ΔT_{max}= 0.6 ms. These results fully coincided with those described earlier (Altman et al., 1987).

Investigation of the first group patients with damages of the cortical temporal areas shows that with increase of the initial interaural time delay the value of FI movement trajectory rises. However, though in all the patients hearing thresholds were normal, in 13 of them revealed were changes of FI movement perception. At low values of interaural time delay (ΔT_{max}=0.05 or 0.2 ms) trajectory estimation was difficult since it was not stable. Therefore Table 1 presents only results comparable with the data obtained on healthy subjects (ΔT_{max}=0.4 and 0.6 ms).

As it can be seen from Table 1, in most cases perception changes manifest themselves as shortening of trajectory of perceived FI movement. The trajectory shortening results from the fact that the patients do not perceive the signal at the initial area of FI movement. Change of the movement direction (from the head midline to the ear) does not change trajectory length of the perceived signal. As before, the patients do not hear the sound at the area near the leading ear.

Special feature of moving FI perception in patients with right-sided damages is a shift of the movement end point beyond the head midline for 5–20 degrees. In a number of cases this results in a lengthening of the movement trajectory as compared to the healthy subjects.

In patients with left-sided damages the trajectory length at ΔT_{max}=0.6 ms does not differ from that of healthy subjects (Table 1). However at lower values of ΔT_{max} the trajectory is shortened, estimation of its length is less stable, though differences with healthy subjects are not significant (Table 1).

In three subjects with bitemporal damages trajectory length (on the right and on the left) was near to that of healthy subjects (Fig. 1, C).

In the second group of patients, with damages of both the temporal cortex area and the medio-basal structures significant shortenings of the FI movement trajectory, as com-

Table 1. Trajectories of the FI movement (in degrees) in healthy subjects and patients with different localization of the damage foci

	Group	n	Interaural time delay, ΔT, in ms			
			0.4		0.6	
			Leading ear			
			Left	Right	Left	Right
Control	Healthy subjects	9	68 ± 10	69 ± 9	85 ± 3	85 ± 3
I	Right temporal cortex area damage	6	$100 \pm 6^*$	$84 \pm 5^*$	$97 \pm 7^*$	$95 \pm 1^*$
	Left temporal cortex area damage	5	63 ± 7	69 ± 7	73 ± 14	85 ± 3
II	Right temporal cortex and mediobasal structures damage	7	17 ± 5	39 ± 13	41 ± 13	32 ± 9
	Right temporal cortex and mediobasal structures damage	9	24 ± 6	29 ± 8	27 ± 4	23 ± 12

* There were observed cases when the FI movement was finished on the opposite side of the head.
Repetition rate of 25 Hz within the train.

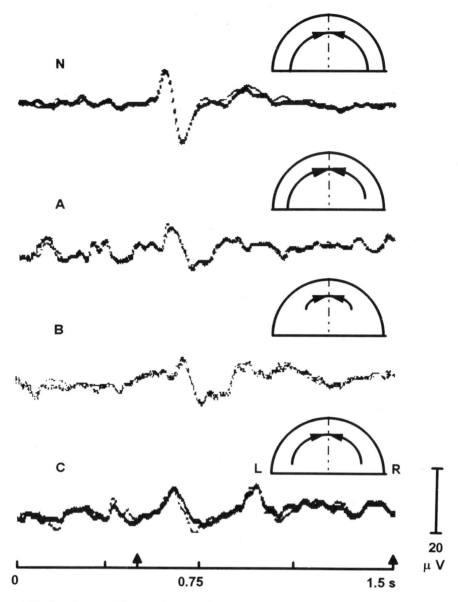

Figure 1. The long-latency auditory evoked potentials to moving auditory images in healthy subject (N) and in patients with monofocal unilateral (A), multifocal unilateral (B) and bilateral (C) damages of the temporal cortical areas. On the abscissa - time (s). Vertical arrows on the abscissa - beginning and cessation of the stimulus. The insert - direction and position of the perceived trajectory of the FIs movement (arrows). L - left side, R - right side of the head.

pared with healthy subjects, were found at ΔT_{max} of 0.02, 0.4 and 0.6 ms (p<0.01). In patients of this group the greatest changes of the ability to perceive FI movement could be observed (Table I).Drastic narrowing of the subjective acoustic space could be observed as a full absence of perception of the FI movement. In such cases the FI images could be localized by patients either at the head midline or somewhere between it and the ear i.e. as unmoved FI. If the FI movement was perceived, a shift of the trajectory beyond the head midline could be observed, as well as mistakes in estimation of the movement direction.

Most often the shortening of the FI movement was due to a shift of the point of the movement beginning to the point of its cessation. In patients with right-sided epileptic focuses the greatest shortening was observed at the contralateral (left) side. In patients with left-sided damages, localization function disorders, were very significant, so that it was difficult to show definitely the side of the greatest disorders.

Repeated investigations of 7 patients in one and two years after the surgical treatment were performed. A certain ability of improvement in estimating localization parameters of the FIs was found. More precise localization of moving FIs could be observed. These improvements were found in cases when normalization of both the EEG and results of clinical investigations took place. On the other hand, estimation of the FI localization parameters became even more difficult for patients if in the course of time their clinical appearances and EEG characteristics became worse (Fig. 2, C).

AUDITORY LONG-LATENCY EVOKED POTENTIALS (EXPERIMENT II)

Subjects and Methods

24 patients, aged 14–37, with diagnosed temporal epilepsy lasting for 7–25 years were investigated under conditions of background anticonvulsive drug therapy. For control 14 healthy subjects with normal hearing were investigated. Complex clinical investigation, as described for the Experiment I, allowed to differentiate two groups of patients.

The first, group included 14 patients with cortical damages. In 4 of these patients the epileptic foci were localized in the temporal cortical areas (in 2 cases in the left and in 2 cases in the right). In 6 of these patients bitemporal epileptic focuses were established, and in 4 others (in 3 cases on the left, and in one case on the right) unilateral multifocal epilepsy was diagnosed.

The second group included 10 patients with initial foci of convulsive activity both in the temporal cortex and medio-basal structures of the brain. Eight patients of the group were treated surgically, the pole of the temporal lobe and amygdalo-hippocampal complex were ectomized on the left in four cases and on the right in four others; besides other cortical areas with pronounced convulsive activity were aspirated. Two patients of this group declined the surgery treatment.

AEPs were recorded under conditions of presentation of the sound signals modelling moving sound sources, as it was described in Experiments I. Click trains were presented once/8 sec.

The subject sat comfortably in a special chair throughout the experiment with the head fixed on head-holder and the eyes shut. All the experiments were performed in a sound-attenuated and electrically shielded dark chamber. The only instruction to the subjects was to sit quietly with eyes shut. About every 8 min the experimental chamber was opened and the subject was given a rest in his chair for 5–10 min; the head helmet with electrodes was not touched and their positions were unchanged. The active electrodes, at C3 and C4, were used for recording AEPs. The corresponding reference electrodes were placed on earlobes ipsilateral to the active electrodes, the earthing electrodes was fixed on the forehead. An amplifier of 0.3–30 Hz band-width and 3.5×10^3–2×10^5 gain was used. AEPs was averaged and recorded according to a standard procedure using a PC. Time

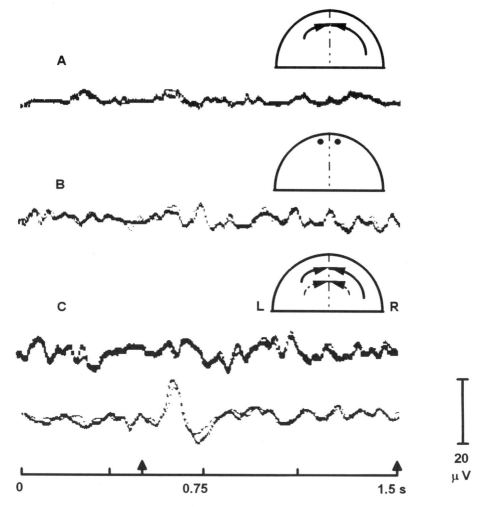

Figure 2. The long-latency auditory evoked potentials to moving auditory images in patients. A - with unilateral, B - bilateral damages of the temporal cortical areas and hippocampus; C - unilateral damages of the temporal cortical areas and hippocampus before and after surgical treatment. The inserts on C - trajectories of the FI movement before (solid lines) and after surgical treatment (dashed lines). Other designations as in Fig.1.

analysis was equal to 1500 ms. Sixty responses were averaged with bin width was equal to 5 ms. During the first 500 ms prestimulus activity was recorded. The peak-to-peak amplitude of the N_1-P_2 complex of the AEP was measured as well as the peak latencies of both components.

The data obtained on patients were compared with those for healthy subjects.

Results

Auditory evoked potentials in patients corresponded in general to AEPs of healthy subjects as far as their waveform was concerned (Fig. 1).

A more detailed description of the results obtained in different groups of patients is following.

In the first group of 14 patients with cortical epileptic foci two subgroups could be also differentiated. The first subgroup consisted of 8 patients in whom one-sided epilepsy was diagnosed. In 4 of them only the temporal lobe was damaged. The AEPs did not differ significantly from the AEPs recorded in healthy subjects, whose the mean amplitude of N_1-P_2 complex was 21.8 ± 4.1 μV. The perception of the moving FIs were near to normal (Fig.1, A). In other 4 patients of this subgroup the EEG-activity was changed not only in the temporal region but also in the parietal, frontal and central ones. All these patients showed AEPs of 9–10 μV mean amplitude, their waveform distorted (Fig.1, B).

In 6 patients with bitemporal pathological foci AEPs could be recorded readily (Fig.1, C) though in 5 of them N_1-P_2 amplitude was lower than its mean value in healthy subjects for "moving" sound signals.

Of special interest were the results obtained for the sixth patient of this subgroup. In this single case AEP-amplitude showed great augmentation (up to 40–45 μV) as compared with that for healthy subjects, both on the left and on the right.

In the second group were studied 10 patients with simultaneous involvement of the temporal lobe cortex and amygdalo-hippocampal complex in the pathological process.

In 6 patients with one-sided pathological foci low-amplitude AEPs of 5–6 μV could be recorded following stimulation with "moving" sound signal. These AEPs could be hardly distinguished from the background activity and their wave-form was distorted (Fig.2, A).

In three of four patients with supposed bilateral damages involving both the cortical and medio-basal structures AEPs could not be recorded (Fig.2, B). In the fourth patient a distorted low-amplitude AEP was evoked by sound signals, it could be hardly distinguished from the high-amplitude background activity. This patient declined the surgery treatment.

Two patients of second group showed nearly normal AEPs (both in their configuration and amplitude) after surgery treatment (Fig.2, C). Their clinical appearance improved to a great extent with no paroxysms and with significant normalization of their EEGs where paroxysmal activity was diminished significantly.

The peak latency values of the N_1 and P_2 components in patients of all three groups as a rule usually showed no significant differences from those in healthy subjects.

DISCUSSION

The results obtained showed that epileptic damages of the temporal cortex and medio-basal structures of the brain hemispheres resulted in disorders of localization of moving fused auditory images (FIs), as well as in pronounced changes of the auditory evoked potentials (AEPs). At first a deficit of the FI-localization in epileptic patients will be considered.

It is obvious that in patients suffering from temporal epilepsy there are disorders of moving FI perception. The character and degree of the revealed disorders are obviously determined by localization and amount of brain lesions. Directional hearing disorders are least pronounced in cases of relatively isolated damage of temporal areas of the brain cortex. Different character and degree of disorders connected with right- and left-sided damages support the data about special role of the right hemisphere in estimation of spatial position of the sound signal (Altman et al., 1987; 1995).

Meanwhile, additional involvement of the medio-basal structures of the brain hemispheres in pathological process strengthened the deficit of the localization function. As was shown, with simultaneous damages of the temporal cortex and medio-basal structures, pronounced narrowing of the subjective space within which FI movement could be perceived was found. These results agree with the idea that hippocampus takes a significant part in perception of the outer space parameters (O'Keefe and Nadel, 1978; Smith and Milner, 1981; O'Keefe, 1983; Olton et al., 1989) and allow to suggest a certain role of hippocampus in changing localization function.

It should be mentioned that there are electrophysiological data which showed that 92% of the cats' hippocampal neurones specifically responded to different positions of the "unmoved" and "moving" sound signals with pronounced afterdischarges (Altman et al., 1990). As is known afterdischarges specific for FI different positions are also characteristic of the auditory cortex neurones (Altman, 1978). Consequently, in impulse activity of neurones both of the auditory cortex and hippocampus observed is selective maintenance of traces of sound source position. Thus the both structures can participate in processes of memory formation.

However it is known from numerous clinical data that origin of temporal epilepsy and development of the disease are often determined by functional and morphological changes which develop mainly in hippocampus. Existence of convulsive activity focus in hippocampus as a rule leads to change of electrical activity recorded in different cortical areas, with which hippocampus has numerous connections (O'Keefe and Nadel, 1985).

With prolonged development of the disease, which is especially characteristic of the second group patients, in different cortical areas influenced by hippocampal focus of convulsive activity a series of interconnected and independent foci of stable pathological excitation is formed. These foci are not suppressed by anticonvulsive drugs and are powerful generators of afferent impulsion; the latter greatly influences functioning majority of areas of the new cortex and interferes with activity of different cerebral systems.

Thus it can be suggested that significant disorders of the directional hearing in patients with formed epileptic system (cortical temporal area - hippocampus) can be not only the result of changing specific involvement of hippocampus in processing information concerning afferent stimuli (Wicklegren, 1968; O'Keefe and Nadel, 1978; Stern, 1981; Smith and Milner, 1981; O'Keefe, 1983). Changes of directional hearing, being not the result of auditory sensitivity disorders in patients, can be probably determined to a greater extent by disorders of processes of binaural interaction. These disorders develop as a result of influence of the hippocampal convulsive activity focus on different areas of the brain cortex. Meanwhile it is known that binaural hearing in human demands normal integrative activity of a number of brain areas which are phylogenetically younger than the area of primary projections.

The second point of the discussion concerns AEPs in epileptic patients with cortical and medio-basal damages.

However some special features of AEPs were characteristic for all the groups of patients.

1. A certain interrelation was traced between the level of the prestimulus synchronized background activity and the N_1-P_2 peak-to-peak amplitude: the higher was the synchronized activity level the lower was the AEP-amplitude.
2. The degree of AEP-changes corresponded to a certain extent to the degree of disorders in perception of sound signal spatial characteristics.

Such successive increase of disorders following increase in quantity of structures involved in epileptic system was also observed with AEP recording in the same patients. It is obvious that damage of temporal areas of the brain cortex does not exclude the possibility of AEP formation though results in amplitude decrease of N_1-P_2 components. Simultaneous damage of cortical temporal areas and hippocampus as well as development of multifocal epilepsy results in destruction of the responses or in their absence (especially with bitemporal damages).

All this allows to conclude that there is a pronounced parallelism and probably interconnection between the extent of disorders of perception of spatial position moving auditory images and AEP changes: the more grave are the clinical manifestations and the more is the pathological process spread, the greater are the deficit of the localization function and destruction of the AEPs, down to their disappearance.

One more observation seems of importance. Even in cases when no AEPs could be recorded following severe pathological process, the patients showed normal hearing thresholds. Whatever neurophysiological mechanisms could underlie this phenomenon, it evidences that AEPs do not connected directly with processes of forming normal hearing sensitivity, but sooner they are connected with some more complex processes of the auditory function.

Though there is no single point of view concerning AEP origin, many authors are inclined to the opinion that one of the important generators of these responses are associative structures of the human brain hemispheres (Näätänen and Picton, 1987). In case this is true, then our earlier point of view is supported that final stages of recognition of sound signal spatial parameters are going beyond the centres of the classic auditory pathway (Altman, 1988).

In conclusion it should be mentioned that these results could serve as a diagnostic and prognostic criterion in cases of temporal epilepsy.

Supported by RFFI 96-04-48030[*]

REFERENCES

Altman, J.A. (1978) Sound Localization (Neurophysiological Mechanisms). The Belton Institute for Hearing Research, Chicago, 188 pp.

Altman, J.A. (1988) Information processing Concerning moving Sound Sources in the Auditory Centres and Its Utilisation by brain integrative an motor Structures. In: J.Syka, B.Masterton (Eds.). Auditory Pathway Structure and Function, Plenum Press, New York-London, 349–354.

Altman, J.A., Rosenblum, A.S. and Lvova, V.G. (1987) Lateralization of a moving auditory image in patients with focal damage of the brain hemispheres. Neuropsychologia, 25, 2, 435–442.

Altman, J.A. and Vaitulevich, S.F. (1990) Auditory image movement in evoked potentials. EEG and Clin. Neurophysiol. 75, 323–333.

Altman, J.A., Vasilieva, M.Yu. and Kotelenko, L.M. (1990) Responses of hippocampal neurones to signals modelling spatial positions of a sound source. Sensory Systems. 4, 4, 407–414. (In Russian, English translation by Plenum Press).

Altman, J.A., Vaitulevich, S.F., Kotelenko, L.M., Fedko, L.I. and Sanotzkaya, N.N. (1995) Estimation of the space characteristics of the sound stimuli in patients with temporal epilepsy. Human Physiology. 21, 1, 54–61. (in Russian, translated in English by Plenum Press).

Baru, A.V., Karaseva, T.A. (1972) The Brain and Hearing (Hearing Disturbances Associated with Local Brain Lesion). In: A.R.Luria (Ed.), Neurophysiology Series, New York-London, 116 pp.

Drake, M.E., Burgess, R.J., Jelety, J.J. and Ford, E.E. (1986) Long-latency auditory event-related potentials in epilepsy. Clin. Electroencephalogr. 17, 1, 10–13.

Durlach, N.I., Thompson, C.L. and Colburn, H.S. (1981) Binaural interaction in impaired listeners. A review of past research. Audiology, 20, 3, 181–211.

Hari, R. (1995) Illusory directional hearing in humans. Neuroscience Letters. 189, 29–30.

O'Keefe, J. (1983) Spatial memory within and without the hippocampal system. In: W.Seifert (Ed.), Neurobiology of the Hippocampus. Academic Press, New York, pp. 375–403.

O'Keefe J. and Nadel, L. (1978) The hippocampus as a cognitive map. Oxford University Press, 543 pp.

* The authors are greatly appreciative to the staff of the Bekhterev Psychoneurological Institute (St.Petersburg): Prof. V.A.Shustin, Dr. L.I.Fedko, Dr. I.S.Tets for clinical examination and diagnostics.

Näätänen, R. and Picton, T. (1987) The N_1 wave of the human electric and magnetic response to sound: A review and an analysis of the component structure. Psychophysiology. 24, 375–425.

Olton, D.S., Wible, C.G. and Pang, K. (1989) Hippocampal cells have mnemonic correlates as well as spatial ones. Psychobiology. 17, 3, 228–229.

Stern, L.D. (1981) A review of theories of human amnesia. Memory and Cognition. 9, 3, 247–262.

Wicklegren, W.A. (1968) Sparing of short-term memory in an amnesic patient: implications for strength theory of memory. Neuropsychologia. 6, 3, 235.

56

REDUCTION OF BACKGROUND NOISE IN HUMAN AUDITORY BRAINSTEM RESPONSE BY MEANS OF CLASSIFIED AVERAGING

Roland Mühler and Hellmut von Specht

Department of Experimental Audiology
Otto-von-Guericke-University of Magdeburg
Magdeburg, Germany

INTRODUCTION

The most widely used method for the recording of evoked potentials (EP) is the classical technique of ensemble averaging in the time domain. The use of averaging calls for a number of requirements to be met (random noise, independence of noise and signal, deterministic signal in all sweeps). If these requirements are satisfied and the noise is stationary throughout the period of examination, the well-known improvement of the signal-to-noise ratio (SNR) with the square root of the number of sweeps is obtained. The stationarity is frequently violated by either muscular activity (artifacts) or slow changes of the subjects state of relaxation. Two methods are used to minimize the undesirable effects of noise instability on the SNR: The first and most common method used is artifact rejection: Elimination of realizations with signal amplitudes exceeding a certain level from the averaging process. The second method used is weighted averaging: Weighting of realizations or blocks of realizations inversely to the estimated power of background noise (Hoke et al. 1984). Optimal artifact rejection requires a prior knowledge of the noise distribution. Unavoidable underestimation of the signal amplitude is the main disadvantage of weighted averaging (Lütkenhöner et al. 1985). In this paper classified averaging will be presented as an alternative approach to optimal reduction of residual background noise.

METHODS

Sets of four thousand single responses of normal hearing subjects to monaural stimulation of the left and the right ear at 90 dB and 70 dB (nHL) were recorded from

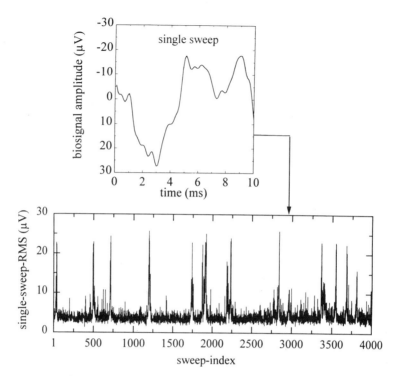

Figure 1. Plotting the fine-structure of the spontaneous brain activity by calculating single-sweep RMS-plots: The RMS of each sweep is plotted as a function of the sweep-index

electrodes at vertex (active) and mastoid (reference). Stimuli were 100µs clicks presented by means of headphones. All single responses were stored on hard disk for off-line data processing. To show the influence of background noise instability on the SNR the RMS-amplitude of each sweep was calculated and plotted against the sweep-index as shown in Figure 1 and 2.

The signal-to-noise ratio of the averaged ensemble is estimated by the inverse single point variance of the ensemble (Don et al. 1984). Figure 2 shows the linear increase of SNR, if the power of background noise is stationary. High-noise-sweeps reduce the SNR in some cases considerably. The proposed method of Classified averaging is based on the interchangeability of realizations within the ensemble, and is accomplished by (1) the classification of all realizations according to their estimated power of background noise, and (2) successive averaging of this classified ensemble. Starting with low-noise realizations the SNR will increase, and after reaching a maximum, SNR starts to decrease because of including more noisy realizations in the ensemble average. This maximum is used to interrupt the process of averaging at an optimal number of realizations as shown in Fig. 3.

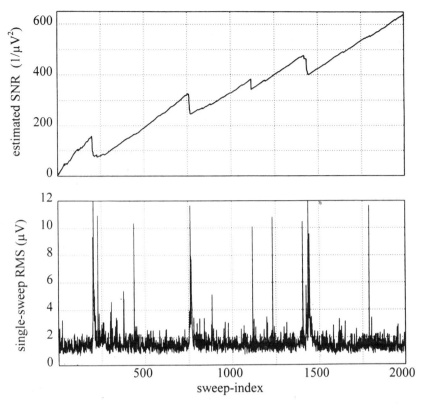

Figure 2. Influence of high-noise sweeps on the estimated signal to noise ratio (SNR) of the averaging result: The SNR (estimated by the inverse single point variance of the ensemble) increases linear with N while high-noise sweeps can reduce the SNR considerably.

RESULTS

An off-line implementation of the method of classified averaging is presented, illustrating an application to human auditory brainstem responses (ABR) in 20 subjects. Single-sweep RMS-functions (Fig. 4) and single-sweep SNR-functions were calculated off-line by application of a conventional averaging procedure (using artifact rejection as well as weighted averaging) and by application of classified averaging. The SNR's achieved by these three methods were estimated by calculating the F_{SP}-Value as proposed by Don et al. (1984).The SNR's were compared to the SNR's achieved with conventional averaging without any suppression of high-noise sweeps. The mean values and standard deviations of SNR-improvement shown in Figure 5 demonstrate the benefit of the proposed method.

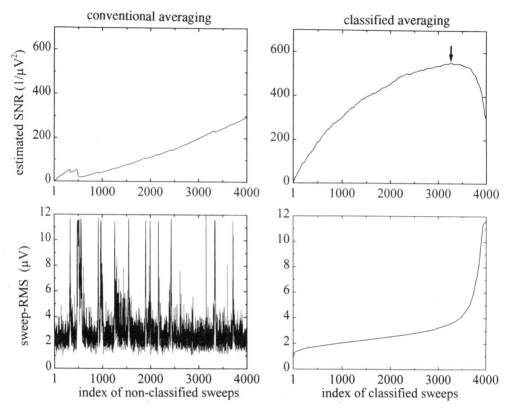

Figure 3. The principle of classified averaging (right column) compared to a classical averaging procedure (left column): Single-sweep RMS values (lower row) and estimated SNR (upper row) are plotted as a function of sweep-index. The stop of averaging in the classified method is marked by an arrow.

Monte Carlo Simulation

A simple model of inhomogenity with two noise populations was used (Lütkenhöner et al. 1985): A number of N epochs of zero mean normal distributed noise is considered to have a low variance VAR^L in $(1-\alpha) \cdot N$ cases and a high variance $VAR^H = c^2 \cdot VAR^L$ in $\alpha \cdot N$ cases (α: artifact-ratio, c: amplitude-ratio). Using this assumption the SNR-ratios for classified averaging, weighted averaging and artifact rejection are plotted as a function of artifact-ratio α. The plots in Figure 6 show the advantage of classified averaging especially in cases of high artifact ratios.

CONCLUSIONS

The presented investigations have shown that classified averaging can be used to increase the signal-to-noise ratio of evoked potential registrations particularly in ensem-

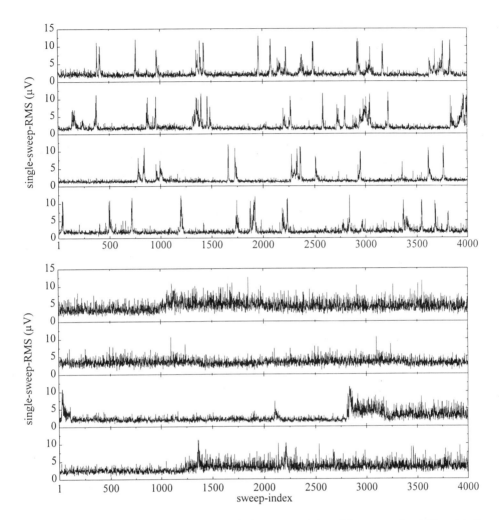

Figure 4. Single-sweep RMS-functions for two typical cases, consisting of four successive ABR-registrations with 4000 sweeps. (Top rows: Low-level spontaneous activity interrupted by short artifacts. Bottom rows: Slow changes in spontaneous activity.)

bles with inhomogeneous noise distribution. Therefore it can be a useful tool for the recording of evoked potentials in the fields of clinic and research.Using a personal computer the proposed method can today only be applied off-line. With improved sorting procedures and increased computer power, on-line implementations will be possible in the near future.

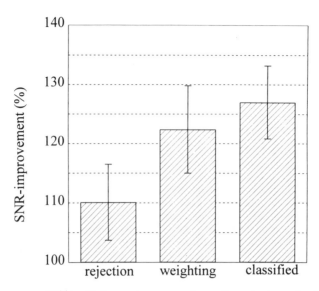

Figure 5. SNR-improvement by classified averaging compared to artifact rejection and weighted averaging. The SNR-values of 160 ABR registrations in 20 subjects are normalized at the SNR-values achieved with conventional averaging without any suppression of high-noise sweeps

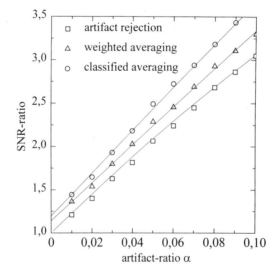

Figure 6. SNR-ratio for classified averaging, weighted averaging and artifact rejection as a function of artifact ratio. Results of a Monte Carlo simulation with a model using two noise populations (amplitude ratio c=5).

REFERENCES

Don, M.; Elberling, C.; Waring, M. (1984) Objective Detection of Averaged Auditory Brainstem Responses. Scand. Audiol.13, 219–228

Hoke, M., Ross, B., Wickesberg, R.E., Lütkenhöner, B. (1984) Weighted Averaging - Theory and Application to Electric Response Audiometry. Electroenceph. clin. Neurophys. 57, 484–489

Lütkenhöner, B., Hoke, M., Pantev, Ch. (1985) Possibilities and Limitations of Weighted Averaging. Biological Cybernetics 52, 409–416

PARTICIPANTS

Altman, Jacob A.
I.P. Pavlov Inst. of Physiology, Russian
  Academy of Sciences,
Nab. Makarova 6, St. Petersburg 199034,
  RUSSIA

Altschuler, Richard A.
University of Michigan, Kresge Hearing
  Research Inst., 1301 E. Ann St., Ann
  Arbor,
MI 48109–0506, USA

Aran, Jean-Marie
Lab. d'Audiologie Experimentale et
  Clinique, Hopital Pellegrin, Place Amelie
  Raba Leon, Bordeaux Cedex, FRA-33076,
  FRANCE

Backus, Kurt Harald
Department of General Zoology, University
  of Kaiserslautern
P.O.B. 3049, D-67653 Kaiserslautern,
  GERMANY

Bajo Lorenzana, Victoria Maria
University of Salamanca, Faculty of
  Medicine, Department of Celullar Biology
  & Pathology, Av. Campo Charro, s./n.,
  37007 Salamanca, SPAIN

Bibikov, Nikolay
Acoustics Institute, Shvernik st. 4, Moscow,
  117036, RUSSIA

Biebel, Ulrich W.
Zool. Inst. of TH - Darmstadt,
  Schnittspahnstr. 3, 64287 Darmstadt,
  GERMANY

Bleeck, Stefan
Zool. Inst. of TH - Darmstadt,
  Schnittspahnstr. 3, 64287 Darmstadt,
  GERMANY

Brugge, John F.
University of Wisconsin, 275 Medical
  Sciences Bldg., Madison, WI 53711, USA

Brückner, Susanne
Florida State University, Department of
  Psychology, Program in Neuroscience,
  Tallahassee, FL 32306 -1051, USA

Carretta, Donatella
Universite de Lausanne, Institut de
  Physiologie, Laboratoire de
  Neuro-Heuristique,
7 Rue du Bugnon, Lausanne, CH-1005,
  SWITZERLAND

Caspary, Donald M.
Southern Illinois University School of
  Medicine, Department of
  Pharmacology-1222,
801 N. Rutledge Street, Springfield, IL
  62702, USA

Cransac, Herve
URA CNRS 1447, Laboratoire de
    Physiologie, Faculte de Medecine,
8 Avenue Rockefeller, F- 69373 Lyon Cedex
    08, FRANCE

Dlouhá, Olga
Phoniatric Clinic, 1st Medical Faculty,
    Charles University, Zitná 24, 120 00
    Prague 2,
CZECH REPUBLIC

Druga, Rastislav
Department of Anatomy, 2nd Medical
    Faculty, Charles University, José Martího
    31, 162 52 Prague 6
CZECH REPUBLIC

Eggermont, Jos J.
University of Calgary, Dept. of Psychology,
    2500 University Drive N.W.,
Calgary, T2N 1N4, CANADA

Evans, Edward F.
Dept. of Communication and Neuroscience,
    University of Keele,
Staffs ST5 5BG, Keele, UNITED
    KINGDOM

Festen, Joost
Ac. Hospital, Vrije Universiteit, Dept.
    Audiology,
P.O. Box 7057, 1007 MB, Amsterdam
THE NETHERLANDS

Fischer, Catherine
Hopital Neurologique, Laboratoire
    Neurologie Fonctionnelle, 59 Boulevard
    Pinel,
69003 Lyon, FRANCE

Friauf, Eckhard
University of Frankfurt, Zentrum der
    Physiologie,
Theodor-Stern-Kai 7, D-60590 Frankfurt a.
    M., GERMANY

Frisina, Robert D., Jr.
University of Rochester School of
    Medicine, Dept. of Otolaryngology,
601 Elmwood Avenue, Box 629, Rochester,
    NY 14642–8629, USA

Fukunishi, Kohyu
Advanced Res. Laboratory, Hitachi, Ltd.,
    Hatoyama, Saitama 350 - 03, JAPAN

Gaese, Bernhard
Dept. of Animal Physiology, University of
    Tübingen,
Auf der Morgenstelle 28, D-72076
    Tübingen, GERMANY

Ganz, Michael
Exp. Audiology , Med. Faculty, Otto - von -
    Guericke - Univ., Leipziger Str. 44,
39120 Magdeburg, GERMANY

Gleich, Otto
HNO - Klinik Univ. Regensburg, FJS -
    Allee 11, 93042 Regensburg, GERMANY

Glendenning, Karen K.
Department of Psychology, Florida State
    University, Tallahassee, Florida
    32306–1051, USA

Godfrey, Donald A.
Medical College of Ohio, Department of
    Otolaryngology, P.O. Box 10008
Toledo, OH 43699–0008, USA

Hancock, Kenneth
Boston University, Department of
    Biomedical Engineering, 44 Cummington
    St., Rm 417, Boston, MA 02215–2407,
    USA

Hari, Riitta
Brain Research Unit, Low Temperature
    Laboratory, Helsinki University of
    Technology,
02150 Espoo, FINLAND

Hartmann, Rainer
Zentrum der Physiologie,
Theodor-Stern-Kai 7, D-60590, Frankfurt
a. M., GERMANY

Hartung, Klaus
Lehrstuhl für allg. Elektrotechnik und
Akustik, Ruhr-Universität Bochum,
Universitätsstr. 150, 44780 Bochum,
GERMANY

Häusler, Udo
Deutsches Primatenzentrum, Abt.
Neurobiologie, Kellnerweg. 4, 37077
Göttingen,
GERMANY

Herve-Minvielle, Anne
University of Lausanne, Institute of
Physiology, Rue du Musee 5, 1700
Fribourg, SWITZERLAND

Hill, Kenneth G.
Developmental Neurobiology Gr., Research
School of Biological Sciences,
Australian National University , Canberra
2601, AUSTRALIA

Hyson, Richard L.
Florida State University, R-54, Department
of Psychology, Program in Neuroscience,
Tallahassee, FL 32308, USA

Jilek, Milan
Institute of Experimental Medicine,
Academy of Sciences, Vídeňská 1083,
142 20 Prague 4,
CZECH REPUBLIC

Kaiser, Alexander
Institut für Zoologie, Technische Universität
München, Lichtenbergstr. 4, 85747
Garching,
GERMANY

Kapteyn, Theo S.
Ac. Hospital Vrije Universiteit, Dept.
Audiology, Post Box 7057, 1007 MB
Amsterdam,
THE NETHERLANDS

Kelly, Jack B.
Carleton University, Laboratory of Sensory
Neuroscience, Psychology Department,
Ottawa, Ontario K1S 5B6, CANADA

King, Andrew J.
University of Oxford, Laboratory of
Physiology, Parks Road, Oxford OX1
3PT, UNITED KINGDOM

Klinke, Rainer H.
J.W. Goethe-Universität, Zentrum der
Physiologie, Theodor-Stern-Kai 7,
D-60590 Frankfurt a. M., GERMANY

Kopp, Cornelia
University of Leipzig, Zoological
Department, Neurobiology Lab.,
Talstrasse 33,
D-04103 Leipzig, GERMANY

Kosmal, Anna
Department of Neurophysiology, Nencki
Institute of Experimental Biology, 3
Pasteur St.,
02–093 Warsaw, POLAND

Kretzschmar, Roswitha
University of Leipzig, Institute of Zoology,
Neurobiology Lab., Talstrasse 33, 04103
Leipzig, GERMANY

Kučerová, Jitka
Institute of Pathological Physiology,
Medical Faculty, Charles University,
Barrandova 31,
301 43 Plzeň, CZECH REPUBLIC

Kvašňák, Eugen
Institute of Experimental Medicine,
Academy of Sciences, Vídeňská 1083,
142 20 Prague 4,
CZECH REPUBLIC

Langner, Gerald
Zoological Inst. of THD, Schnittspahnstr. 3,
  64287 Darmstadt, GERMANY

Lepore, Franco
University of Montreal, C.P. 6128, Station
  Centreville, Montreal, Quebec, H3C 3J7
CANADA

Lockwood, Alan H.
VA Medical Center, SUNY at Buffalo,
  Center for PET(115P), VA Hospital,
3495 Bailey Ave, Buffalo, NY 14215, USA

Lonsbury-Martin, Brenda L.
University of Miami Ear Institute, Dept.
  Otolaryngology (M805), P.O. Box
  016960, Miami,
FL 33101, USA

Malmierca, Manuel S.
Department of Cellular Biology &
  Pathology, Faculty of Medicine,
  University of Salamanca, Avda. Campo
  Charro, s/n, 37007 Salamanca, SPAIN

Masterton, R. Bruce
Florida State University, Department of
  Psychology, Tallahassee, FL 32306–1051,
  USA

McAlpine, David
MRC Institute of Hearing Research,
  University Park, Science Road,
  Nottingham NG7 2RD,
UNITED KINGDOM

Miller, Josef M.
University of Michigan, Kresge Hearing
  Research Inst., 1301 E. Ann Street, Ann
  Arbor,
MI 48109–0506, USA

Moore, David R.
University of Oxford, University Lab. of
  Physiology, Parks Road, Oxford OX1 3PT,
UNITED KINGDOM

Morest, D. Kent
University of Connecticut, Health Center,
  Center for Neurological Sciences,
  263 Farmington Ave, Farmington, CT
  06030, USA

Muehler, Roland
Exp. Audiology, Med. Faculty, Otto - von -
  Guericke - Univ., Leipziger Str. 44
  39120 Magdeburg, GERMANY

Müller-Preuss, Peter
Dept. of Neuroendocrinology, Max Planck
  Institute for Psychiatry, Kraepelin Str. 2,
  80804 München, GERMANY

Oertel, Donata
University of Wisconsin, Dept. of
  Neurophysiology, 1300 University Ave,
  Madison, WI 53706, USA

Oliver, Douglas
Univ. of Connecticut, Health Center,
  Department of Anatomy, 263 Farmington
  Avenue,
Farmington, CT 06030–3405, USA

Osen, Kirsten
Univ. of Oslo, Dept. of Anatomy, Inst. of
  Basic Medical Sciences, P. 0. Box 1105
  Blindern,
0317 Oslo, NORWAY

Palmer, Alan R.
University of Nottingham, MRC Inst. of
  Hearing Research, University Park,
  Nottingham, NG7 2RD, UNITED
  KINGDOM

Pollak, George D.
University of Texas at Austin, Dept. of
  Zoology, Austin, TX 78712, USA

Poon, Paul W. F.
Department of Physiology, National Cheng
  Kung University, Medical School,
  1 University Road, Tainan 70101, TAIWAN

Popelář, Jiøí
Institute of Experimental Medicine,
  Academy of Sciences, Vídeňská 1083,
  142 20 Prague 4,
CZECH REPUBLIC

Pozzoli, Uberto
Universite de Lausanne, Institut de
  Physiologie, Laboratoire de
  Neuro-Heuristique,
7 Rue du Bugnon, Lausanne, CH-1005,
  SWITZERLAND

Puil, Ernest
Univ. of British Columbia, Dept. of
  Pharmacology & Therapeutics U.B.C.,
2176 Health Sciences Mall, Vancouver,
  B.C., V6T IZ3, CANADA

Rauschecker, Josef P.
Georgetown Institute for Cognitive and
  Computational Sciences,
The Research Building EPO4, 3970
  Reservoir Road NW, Washington, DC
  20007–2197, USA

Rees, Adrian
University of Newcastle Upon Tyne, Dept.
  of Physiology, The Medical School,
Framlington Place, Newcastle-Upon-Tyne,
  NE2 4HH, UNITED KINGDOM

Rouiller, Eric M.
Institute of Physiology, University of
  Fribourg, Rue du Musee 5, CH-1700
  Fribourg, SWITZERLAND

Robertson, Donald
Department of Physiology, The University
  of Western Australia, Nedlands,
Western Australia 6709, AUSTRALIA

Rubel, Edwin W.
University of Washington, VMB Hearing
  Research Center, Box 357923, Seattle, WA
  98195, USA

Rübsamen, Rudolf
University of Leipzig , Zoological Institute,
  Neurobiology Laboratory,
Talstrasse 3, D-04103 Leipzig, GERMANY

Rybalko, Natalia
Institute of Experimental Medicine,
  Academy of Sciences, Vídeňská 1083,
  142 20 Prague 4,
CZECH REPUBLIC

Sachs, Murray B.
Biomed Eng. Ctr. for Hearing Sci, Johns
  Hopkins School of Med., Traylor 710,
720 Rutland Avenue, Baltimore, MD 21205,
  USA

Saldaña, Enrique
Dept. of Cell Biology & Pathology, Faculty
  of Medicine, University of Salamanca,
37007 Salamanca, SPAIN

Salvi, Richard
SUNY University of Buffalo, Hearing
  Research Lab., 215 Parker Hall, Buffalo,
  NY 14214, USA

Scheich, Henning
Federal Institute for Neurobiology,
  Brenneckestr. 6, P.O. Box 1860,
D-39008 Magdeburg, GERMANY

Schwarz, Dietrich W.F.
Univ. of British Columbia, 2211 Westbrook
  Mall, Acute Care Unit Room F-153,
Vancouver, B.C., V6T 2B5, CANADA

Šejna, Ivan
ENT Clinic, Bulovka, Budínova 2, 180 81
  Prague 8, CZECH REPUBLIC

Skřivan, Jiří
1st Medical Faculty, Charles University,
  ENT Clinic, U nemocnice 2, 128 08
  Prague 2,
CZECH REPUBLIC

von Specht, Hellmut
Otto - von -Guericke - Univ., Exp.
    Audiology, Med. Faculty,
    Leipziger Str. 44,
D-39120 Magdeburg, GERMANY

Spongr, Vlasta P.
University of Buffalo, Hearing Research
    Lab., 215 Parker Hall, Buffalo, NY 14214,
    USA

Štěpánek, Jan
Akustika Praha Ltd., U Okrouhlíku 5
150 00 Prague 5, CZECH REPUBLIC

Sterbing , Susanne J.
Dept. Zoology and Neurobiology,
    Ruhr-Univ. Bochum, Universitätsstr. 150,
    44780 Bochum,
GERMANY

Šuta, Daniel
Institute of Experimental Medicine,
    Academy of Sciences, Vídeňská 1083,
    142 20 Prague 4,
CZECH REPUBLIC

Syka, Josef
Institute of Experimental Medicine,
    Academy of Sciences, Vídeňská 1083,
    142 20 Prague 4,
CZECH REPUBLIC

Vaitulevich, Svetlana F.
I.P. Pavlov Institute of Physiology, Russian
    Academy of Sciences,
Nab. Makarova 6, 199034, St. Petersburg,
    RUSSIA

Valvoda, Jaroslav
1st Medical Faculty, Charles University,
    ENT Clinic, U nemocnice 2, 128 08
    Prague 2,
CZECH REPUBLIC

Vischer, Mattheus
ENT Clinic, University of Bern, Inselspital,
    CH 3010 Bern, SWITZERLAND

Wenthold, Robert J.
Lab. of Neurochemistry, National Institutes
    of Health, 36 Covent Drive, Building 36,
Room 5DOB, Bethesda, MD 20892, USA

Wollberg, Zvi
Dept. of Zoology, Tel Aviv University, Tel
    Aviv 69978, ISRAEL

Woody, Charles D.
Anatomy, Psychiatry and Biobehavioral
    Sciences, UCLA Medical Center, Rm 58 -
    258, NPI,
760 Westwood Plaza, Los Angeles, CA
    90024, USA

Yin,Tom
Dept. Neurophysiology, Univ. of Wisconsin,
    273 Medical Science Bldg., Madison,
WI 53706, USA

Young, Eric D.
Johns Hopkins University, School of Med.,
    Dept. of Biomedical Engineering,
720 Rutland Avenue, Baltimore, MD 21205,
    USA

Zatorre, Robert J.
Montreal Neurological Institute, McGill
    University, 3801 University St., Montreal,
Quebec H3A 2B4, CANADA

Zhang, Jinsheng
Institute of Physiology, Dept. of
    Neurobiology, University of Fribourg,
    Rue du Musee 5,
CH-1700 Fribourg, SWITZERLAND

International Symposium on Acoustical Signaling Processing in the Central Auditory System
held September 4–7, 1996
Prague, Czech Republic

INDEX